# Nanocarriers for Nucleic Acids and Proteins

At the frontier of modern medicine lies a revolution in drug delivery systems that operates at the scale of billionths of a meter. *Nanocarriers for Nucleic Acids and Proteins* presents a comprehensive exploration of these groundbreaking technologies that are reshaping therapeutic approaches across medical science.

From fundamental concepts to cutting-edge applications, this comprehensive volume brings together world-class researchers to provide an in-depth examination of:

- Design principles and characteristics of various nanocarrier systems
- Advanced delivery mechanisms for nucleic acids and proteins
- Breakthrough applications in gene therapy and cancer immunotherapy
- Critical aspects of translating laboratory success to clinical implementation

Featuring detailed coverage of lipid-based, polymer-based, inorganic, and bio-inspired nanocarriers, this essential resource bridges the gap between theoretical understanding and practical application. Whether you are developing new therapeutic approaches, optimizing delivery systems, or exploring the frontiers of nanomedicine, this comprehensive guide provides the insights and frameworks needed to advance your work.

# Nanocarriers for Nucleic Acids and Proteins

Edited by
Luis H. Reyes, Juan C. Cruz and Yashwant V. Pathak

CRC Press
Taylor & Francis Group
Boca Raton London New York

CRC Press is an imprint of the
Taylor & Francis Group, an **informa** business

First edition published 2026
by CRC Press
2385 NW Executive Center Drive, Suite 320, Boca Raton FL 33431

and by CRC Press
4 Park Square, Milton Park, Abingdon, Oxon, OX14 4RN

*CRC Press is an imprint of Taylor & Francis Group, LLC*



ISBN: 9781032750583 (hbk)
ISBN: 9781032752792 (pbk)
ISBN: 9781003473183 (ebk)

DOI: 10.1201/9781003473183

Typeset in Times
by Newgen Publishing UK

**Juan:**
*To Carolina Agudelo,*
*whose love, wisdom, and encouragement*
*have sustained me through every draft,*
*every doubt, and every dawn.*
*Your brilliant mind challenges me to think deeper,*
*your gentle heart teaches me to write with compassion,*
*and your unwavering faith carries me forward*
*when words feel impossible to find.*
*This book, like so much else in my life,*
*would not exist without you—*
*my muse, my anchor, my brightest star.*
*Every page bears the invisible imprint*
*of your boundless support and inspiration*
*through all these beautiful years together.*

**Luis:**
*To my precious daughter Emma—you make the world brighter,*
*and my beautiful wife Priscila—you make everything possible.*

**Yashwant V. Pathak:**
*Dedicated to the memory of Dr Keshav Baliram Hedgewar and my parents Vishnupant and Shalini Pathak who gave meaning to my life, my wife Seema gave a positive meaning and my son Dr Sarvadaman who gave a golden lining to my life. I owe a great deal to my wonderful friends from Indonesia Dean Dewi Melani Hariyadi, Airlangga University, Dean Umi, Den and Vice rector Dr Christina Avanthi, Pak Fran of Ubaya University and many more hard to mention all their names.*

## Contents

## Editors' Bio

**Luis H. Reyes**, PhD, is an associate professor at Universidad de Los Andes (Bogotá, Colombia), where he leads research at the intersection of biological engineering and therapeutic delivery systems. He holds a BS in Chemical Engineering from Universidad Industrial de Santander (Colombia) and a PhD in Chemical Engineering from Texas A&M University.

Dr Reyes brings extensive postdoctoral experience from prestigious institutions, including the National Renewable Energy Laboratory (Golden, Colorado), where he contributed to sustainable energy research. His postdoctoral work at Texas A&M University focused on optimizing carotenoids production through metabolic engineering and process development. At the Pontificia Universidad Javeriana's Institute for the Study of Inborn Errors of Metabolism, he advanced research in metabolic disorders and therapeutic approaches.

His current research program spans multiple domains of biotechnology and biomedical engineering. He leads an interdisciplinary team developing bionanoconjugated vehicles for therapeutic molecule delivery, with applications in gene and enzyme replacement therapies. His work also encompasses the discovery and production of peptides for medical, food, and petrochemical applications, alongside innovative research in food engineering, including sensory design of beer and dairy protein alternatives.

His interdisciplinary approach combines tools from molecular and synthetic biology, reverse engineering of microorganisms, laboratory adaptive evolution, and microbiology to design and develop innovative bioprocesses and bioproducts.

**Juan C. Cruz**, PhD, is a distinguished researcher and engineer at the Department of Biomedical Engineering, Universidad de Los Andes (Bogotá, Colombia), where he has been advancing the frontiers of nanobiotechnology since 2016. He holds a BS in Chemical Engineering from the National University of Colombia and earned his PhD in Chemical Engineering from Kansas State University, where his innovative work established a new platform for enzyme immobilization for applications in non-aqueous media.

Dr Cruz's career spans both academic excellence and industrial innovation. As a Postdoctoral Research Fellow at The Johns Hopkins University, he conducted pioneering research on peptide–peptide interactions in model lipid bilayers, employing advanced techniques such as Förster Resonance Energy Transfer and Neutron Diffraction. His work included the synthesis and modification of peptide sequences with various chemistries, including fluorophores and polymers.

In the industrial sector, Dr Cruz served as Head of New Product Development at CarboQuímica, where he led transformative initiatives in nanomaterial applications for the plastics industry. His achievements include developing multifunctional nanomaterials for plastic processing additives and successfully implementing a new family of green stabilizers. His leadership in R&D+i encompassed strategic planning, team management, and the application of TRIZ and Design Thinking methodologies to solve complex technical challenges.

His academic career includes positions as Senior Lecturer at the University of Adelaide and Assistant Professor at Universidad de Antioquia, where he contributed to both research and education in chemical and biomedical engineering. Currently, he brings more than 15 years of expertise in industrial and medical biotechnology to his research programs. His work focuses on the development of nanobiocatalysts, fundamental studies of protein interactions (protein–protein, protein–surface, and protein–lipid), design of nanoparticulate vehicles for drug delivery, and investigation of membrane-active peptides.

His multidisciplinary approach combines expertise in plastics engineering, innovation management, advanced microscopy techniques, and biotechnology, making him a leading voice in the development of novel biomedical technologies and nanomaterial applications.

**Yashwant V. Pathak** received his PhD from Nagpur University, India, Executive MBA and Masters in Conflict Management, Sullivan University, Louisville, KY, USA. He has over 45 years' experience in Academia and Industry. Presently, he is Professor and Associate Dean for Faculty Affairs at Taneja College of Pharmacy, University of South Florida, Tampa, FL, USA. Yashwant has published over 400 research papers, reviews and chapters with close to 5000 citations in Google Scholar. He edited over 70 books in nanotechnology, nutraceuticals, drug delivery systems, artificial intelligence, and artificial neural network. He has keen interest in conflict management and edited several books in the areas of organizational conflict management, handbook of organizational conflict management, Eastern and Indigenous perspectives in conflict management, RSS way of resolving conflicts and these two volumes on nonprofit organizations' conflict management.

He is recipient of several international and national awards including Fulbright Senior Scholar 2016–2017 at Indonesia for 6 months, Fulbright Specialist Scholar 2019 at Witwatersrand University, South Africa, Adjunct Professor, Witwatersrand University, South Africa, Adjunct Professor, Faculty of Pharmacy,

Airlangga University, Surabaya, Indonesia, Fellow of American Association for advancement of Sciences (FAAAS), Fulbright Specialist Universidad Distrital Francisco José de Caldas, Bogotá, Colombia, 2023, Fulbright Senior Scholar, Masaryk University, Czech Republic 2024. He is actively involved in many nonprofit organizations including HSS.org, ICHRRF.org, IACDSC.org, Uberoi Foundation for Religious Studies, and so on.

## Contributors

**Seth Kwabena Amponsah**
Department of Medical Pharmacology
University of Ghana Medical School
Accra, Greater Accra Region, Ghana

**Dipanjana Ash**
Department of Pharmaceutics
BCDA College of Pharmacy and Technology
Kolkata, West Bengal, India

**Biswajit Basu**
Department of Pharmaceutical Technology, School of Health & Medical Sciences
Adamas University
Barasat, Kolkata, West Bengal, India

**Stiven Castellanos**
Department of Biomedical Engineering
Universidad de Los Andes,
Bogotá, Colombia

**Shilpa Chatterjee**
Department of Pharmaceutical Technology, School of Health & Medical Sciences
Adamas University
Barasat, Kolkata, West Bengal, India

**Nishant B. Chopade**
Department of Pharmaceutics
Dr. Rajendra Gode College of Pharmacy
Malkapur, Buldhana, Maharashtra, India

**Ruby Christian**
Faculty of Pharmacy
Dharmsinh Desai University
Nadiad, Gujarat, India

**Juan C. Cruz**
GIB Research Group, Department of Biomedical Engineering and Grupo de Diseño de Productos y Procesos (GDPP), Department of Chemical and Food Engineering
Universidad de los Andes
Bogotá, Colombia

**Aparoop Das**
Department of Pharmaceutical Sciences
Dibrugarh University
Dibrugarh, Assam, India

**Dibyajyoti Das**
Pratiksha Institute of Pharmaceutical Science
Panikhaiti, Guwahati, Assam, India

**Ayon Dutta**
Department of Pharmaceutical Technology
Bengal School of Technology
Sugandha, Delhi Road, Hooghly, West Bengal, India

**Prashant K. Deshmukh**
Department of Pharmaceutics
Dr. Rajendra Gode College of Pharmacy
Malkapur, Buldhana, Maharashtra, India

**Saptasikha Gogoi**
Centre for Biotechnology and Bioinformatics
Dibrugarh University
Dibrugarh, Assam, India

**Urvashee Gogoi**
Department of Pharmaceutical Sciences
Dibrugarh University
Dibrugarh, Assam, India

**Paula Guzmán-Sastoque**
Department of Biomedical Engineering
Universidad de Los Andes
Bogotá, Colombia

**Vikas Jain**
Research & Development – Formulation Division,
Alembic Research Center
Alembic Pharmaceuticals Ltd.
Alembic Road, Vadodara, Gujarat, India

**Dipanjan Karati**
Department of Pharmaceutical Technology, School of Pharmacy
Techno India University
Kolkata, West Bengal, India

**Seema Kohli**
Department of Pharmacy
Government Kalaniketan Polytechnic College
Jabalpur, Madhya Pradesh, India

**Nidhi Kotadiya**
Independent Researcher
NJ, USA

**Lalit Kumar**
Department of Pharmaceutics
National Institute of Pharmaceutical Education and Research
Hajipur, Vaishali, Bihar, India

**Vinit Mahajan**
Department of Experimental Pathology, Pittsburgh Liver Research Center
University of Pittsburgh
Pittsburgh, PA, USA

**María Camila Monsalve**
Department of Biomedical Engineering
Universidad de Los Andes
Bogotá, Colombia

**Mahesh P. More**
Novel Formulation Laboratory
Sekkei Bio Pvt Ltd.
Bangalore, Karnataka, India

**Swarupananda Mukherjee**
Department of Pharmaceutical Technology
NSHM Knowledge Campus
Kolkata, West Bengal, India

**Tosha Pandya**
Department of Pharmaceutical Technology, Institute of Pharmacy
L J University
Sarkhej, Ahmedabad, Gujarat, India

**Kinjal Parikh**
Faculty of Pharmacy
The Maharaja Sayajirao University of Baroda
Kalabhavan, Vadodara, Gujarat, India

**Komal Parmar**
Associate Professor
ROFEL Shri G. M. Bilakhia College of Pharmacy
Vapi, Gujarat, India

**Jayvadan K. Patel**
Vice-President
Viesain Pharma LLC
GA, USA

**Kalyani Pathak**
Department of Pharmaceutical
Sciences Dibrugarh University
Dibrugarh, Assam, India

**Manash Pratim Pathak**
Faculty of Pharmaceutical Sciences
Assam Down Town University
Panikhaiti, Guwahati, Assam, India

**Yashwant V. Pathak**
Taneja College of Pharmacy
University of South Florida
Tampa, FL, USA
Faculty of Pharmacy
Airlangga University
Surabaya, Indonesia

**Bhupendra Prajapati**
Department of Pharmaceutics
Parul Institute of Pharmacy
Faculty of Pharmacy
Parul University
Waghodia, Vadodara, Gujarat, India

**Cristian F. Rodríguez**
Department of Biomedical
Engineering
Universidad de Los Andes
Bogotá, Colombia

**Coryna Rodriguez-Bazurto**
Department of Biomedical Engineering
Universidad de Los Andes
Bogotá, Colombia

**Juan C. Rojas-Hernández**
Department of Biomedical
Engineering
Universidad de Los Andes
Bogotá, Colombia

**Satish Rojekar**
Mount Sinai Center of Translational Medicine and Pharmacology
Icahn School of Medicine at Mount Sinai
New York, NY, USA

**Jon Jyoti Sahariah**
Department of Pharmaceutical Sciences
Dibrugarh University
Dibrugarh, Assam, India

**Riya Saikia**
Department of Pharmaceutical Sciences
Dibrugarh University
Dibrugarh, Assam, India

**Satish Shinde Sampada**
Department of Pharmaceutics
National Institute of Pharmaceutical Education and Research
Hajipur, Vaishali, Bihar, India

**Frederick Sarfo-Antwi**
Department of Biochemistry and Molecular Medicine
University for Development Studies
Tamale, Ghana

**Sinjini Sarkar**
Shobhaben Pratapbhai Patel School of Pharmacy and Technology Management
SVKM's Narsee Monjee Institute of Management Studies
Vile Parle (W), Mumbai, Maharashtra, India

**Ankita Alice Singh**
Department of Pharmacy
Government Kalaniketan Polytechnic College
Jabalpur, Madhya Pradesh, India

**Sukanya Sonowal**
Department of Life Sciences
Faculty of Biological Sciences
Dibrugarh University
Dibrugarh, Assam, India

**Bhavna Sonule**
Department of Nuclear Medicine
Manipal Academy of Higher Education
Nagpur, Maharashtra, India

**Disha Suthar**
Sharda School of Pharmacy
Pethapur, Gandhinagar, Gujarat

**Jennifer Suurbaar**
Department of Biochemistry and Molecular Medicine
School of Medicine and Health Sciences
University for Development Studies
Tamale, Ghana

**Mitul Talala**
Independent Researcher
NJ, USA

**Nishant Talati**
Pharmacist at CVS Pharmacy
Suffolk County, NY, USA

**Shahin Vhora**
Navinta LLC
Ewing, NJ, USA

## Preface

The field of nanomedicine has witnessed remarkable advancements in recent years, particularly in the development of nanocarriers for drug delivery. These nanoscale vehicles, typically ranging from 1 to 100 nm in size, offer innovative solutions to many challenges in delivering nucleic acids and proteins to their target sites within the body. The potential of these minute carriers to revolutionize therapeutic approaches is both exciting and daunting, as we stand on the precipice of a new era in medicine.

This book, *Nanocarriers for Nucleic Acids and Proteins*, provides a comprehensive overview of the current state of nanocarrier technology. It brings together contributions from leading researchers, offering readers insight into the latest developments in this rapidly evolving field. From fundamental concepts to cutting-edge applications, the volume spans the breadth of this interdisciplinary domain.

The book begins with an introduction to the fundamental concepts of nanocarriers for biomolecule delivery (Chapter 1). Chapter 2 then explores the critical aspects of nanocarrier design, showing how size, shape, and surface properties influence their behavior in biological systems. This chapter also introduces stimuli-responsive nanocarriers, which can release their payload in response to specific environmental triggers.

Subsequent chapters focus on specific types of nanocarriers, each with its unique properties and potential applications. Chapter 3 explores lipid-based nanocarriers, including liposomes and solid lipid nanoparticles, discussing their preparation methods, characterization techniques, and applications. The versatility of these systems is remarkable, mimicking cellular structures to deliver their precious cargo. Chapter 4 shifts to polymer-based nanocarriers, such as dendrimers and hydrogels, highlighting their flexibility in design and functionality. These synthetic marvels can be tailored with precision, offering a level of control that was unimaginable just a few decades ago.

Inorganic nanocarriers, including gold nanoparticles and quantum dots, are the subject of Chapter 5, with a focus on their unique properties for theranostic applications. These particles, often smaller than a virus, can be used to diagnose and treat diseases simultaneously—a true paradigm shift in medical intervention. Chapter 6 introduces carbon-based nanocarriers, discussing their preparation, functionalization, and potential in drug delivery and gene therapy. The ability of these carriers to traverse biological barriers is nothing short of revolutionary.

The book then progresses to more specialized systems. Chapter 7 explores virus-inspired nanocarriers, nature's own design repurposed for therapeutic use. Chapter 8 delves into nucleic acid nanocarriers, where the building blocks of life become the vehicles of healing. Protein-based nanocarriers are the focus of Chapter 9, exploring how protein cages and self-assembling peptides can be engineered for drug delivery. The elegance of these bio-inspired systems is a testament to the ingenuity of researchers in the field.

Chapter 10 provides an overview of nanocarrier functionalization strategies, a critical aspect in developing effective delivery systems. This chapter bridges the gap between design and application, showcasing the myriad ways nanocarriers can be optimized for their intended purpose. Chapters 11 through 14 explore the application of nanocarriers for delivering specific types of biomolecules, including nucleic acids and proteins, and their potential in gene therapy and cancer immunotherapy. These chapters paint a vivid picture of the future of medicine, where treatments are precisely targeted and personalized.

The final chapter (Chapter 15) addresses the challenges of translating nanocarrier technologies from the laboratory to the clinic, discussing regulatory considerations, scale-up issues, and ethical implications. It serves as a sobering reminder of the complexity involved in bringing these groundbreaking technologies to patients.

This book aims to serve as a comprehensive resource for researchers, students, and professionals working in nanomedicine, drug delivery, and related fields. By providing a thorough examination of current research and future directions, we hope to stimulate further advancements in this promising area of biomedical science. The journey from concept to clinic is long and arduous, but the potential rewards—in terms of human health and scientific understanding—are immeasurable.

# 1 Introduction to Nanocarriers for Nucleic Acids and Proteins

*Juan C. Cruz, Yashwant V. Pathak, and Luis H. Reyes*

## 1.1 INTRODUCTION TO NANOBIOTECHNOLOGY

Nanobiotechnology merges nanotechnology and biology to address complex biomedical challenges. At the nanoscale (1–100 nm), materials exhibit unique properties that enable precise manipulation of biological systems. In drug delivery, nanoparticles are engineered to target specific cells or tissues, enhancing therapeutic efficacy while minimizing side effects. Notably, lipid nanoparticles (LNPs) have been instrumental in delivering mRNA vaccines against COVID-19, revolutionizing vaccine development (Hou et al., 2021). These nanoparticles improve drug solubility and bioavailability, opening avenues for treating previously intractable diseases.

Nanobiotechnology also advances diagnostic techniques. Nanosensors and quantum dots enable high-sensitivity detection of biomarkers at early disease stages, crucial for timely intervention in conditions like cancer and infectious diseases (Khazaei et al., 2023). In regenerative medicine, nanomaterials mimic the extracellular matrix, promoting cell growth and tissue regeneration. These bio-nano scaffolds are essential in developing artificial tissues and organs, potentially alleviating the shortage of donor organs (Fathi-Achachelouei et al., 2019). Nanostructured biomaterials integrate seamlessly with human tissues, advancing organ repair and regeneration.

Recent developments include the exploration of nanorobots for precise drug delivery and surgical procedures, pushing the boundaries of medical innovation (Ye & Sun, 2023). These advancements underscore nanobiotechnology's profound impact on medicine—from precise drug delivery to innovative diagnostics and regenerative solutions. As research progresses, the field continues to bridge nanotechnology and biology, addressing some of today's most pressing healthcare challenges.

These tiny scales confer on materials a sufficiently higher surface area and quantum effects that enable them to engage biological systems with great specificity. Such specificity can become of much consequence when one considers therapeutic agents like nucleic acids—siRNA, mRNA, and plasmid DNA—that are constitutively labile and subject to enzymatic digestion within the body. Enter nanocarriers tasked with safeguarding these fragile molecules just so they reach their targets intact. For instance, LNPs have transformed nucleic acid delivery by encapsulating genetic materials from degradation and thus offering an easier way of entering the cells. That approach has been very successful in the mRNA vaccines against COVID-19, whereby safe and efficient delivery of mRNA was delivered by LNPs (Hou et al., 2021).

Another pathway for nucleic acid delivery involves polymeric nanoparticles made from biodegradable substances, including poly(lactic-co-glycolic acid) (PLGA). By decorating their surfaces with targeting ligands, such nanoparticles are capable of targeting specific cells, hence resulting in enhanced uptake and effectiveness of the therapeutic agents carried therein (Rehman et al., 2022). In the tree-like branched structures of dendrimers lies a flexible platform for binding nucleic acids. Their design thus allows for control over size and functionality with extreme care, hence enhancing delivery precision. This technology also benefits proteins, which are very important but at the same time sensitive therapeutic agents (Mintzer & Grinstaff, 2010).

Nanogels protect the proteins with a hydrophilic network to prevent degradation, thus allowing for controlled release. Among all nanoparticles, gold nanoparticles have emerged as the most biocompatible and easily functionalized ones. They can deliver proteins inside cells directly for increased therapeutic efficiency with minimum side effects (Zhang et al., 2012). Nanoparticles of porous silica are capable of taking up substantial quantities of proteins and releasing them in a stimulus-specific fashion, demonstrating the potential for release upon changes in pH (Mohan et al., 2024). Nanocarrier design is much more than protection and delivery; these are highly specialized mechanisms for overcoming biological barriers. With attachments of molecules like antibodies or peptides on the surface, nanocarriers achieve targeted delivery—secret attack—only the diseased cells are under attack while leaving others healthy. Targeted drug delivery does not only maximize therapeutic effectiveness but also minimizes side effects (Yang et al., 2024).

Besides, nanocarriers are engineered to allow endosomal escape. Upon cellular uptake, they can disrupt the endosomal membranes and release into the cytoplasm before being degraded. Coating usually occurs with substances such as polyethylene glycol (PEG), which extends their circulation time in the body by giving them stealth properties because of reduced recognition and clearance through the immune system (Gisbert-Garzaran et al., 2021). Emerging platforms in nanobiotechnology are opening up new frontiers. Exosomes are natural vesicles secreted from cells that, nowadays, are being developed into biocompatible nanocarriers with inherent targeting properties. They can carry nucleic acids and proteins across all biological barriers and have therefore emerged as a promising path toward various future therapies (Kalluri & LeBleu, 2020).

DOI: 10.1201/9781003473183-1

Similarly, DNA origami nanostructures are an advanced approach in which the DNA molecules are folded into well-defined shapes instead of their normal random structure and can encapsulate therapeutic agents, releasing them in response to specific triggers. These deliveries point to the huge leap in concepts concerning therapeutic delivery. Through the principles of nanobiotechnology, we can design nanocarriers that work in harmonious relay with the biological entity of the body, thus opening new vistas for the treatment of diseases that, until now, were considered beyond the realm of treatment. This stream of continuous research and innovations has brilliant future prospects in medicine and might change the whole treatment and patient care policies.

The development of nanobiotechnology represents a chronicle with landmark, pathbreaking discoveries and inventions forming the foundation that molded the topological view of biomedical applications into their current forms. It all began with the prophetic lecture of Richard Feynman in 1959 entitled "Plenty of Room at the Bottom," where he mentioned the concept of manipulating matter with regard to atoms (Feynman, n.d.). Thus, the idea sowed the seeds for future excursions into the realm of nanoscale dimensions. Until the discovery of Scanning Tunneling Microscopy (STM) by Gerd Binnig and Heinrich Rohrer in the 1980s, a breakthrough that finally made it possible for scientists to see single atoms, there was no progress toward this dream (Binnig & Rohrer, 1984). This was an important invention in the field of microscopy that finally gave way to the tools that had to be used to manipulate and understand material at the nanoscale level.

As the field matured, the development of Atomic Force Microscopy (AFM) further expanded the capability to study biological molecules in unprecedented detail, thus extending our capabilities to study in great detail a wide range of biological molecules. Advances in materials science and microscopy opened up doors for novel approaches in drug delivery and diagnostics. With the beginning of the 1990s, liposomal drug delivery systems started to be marketed, and in 1995 the Food and Drug Administration (FDA) approved Doxil, a liposomal doxorubicin, as the first nanomedicine on the market (Barenholz, 2012). The success of Doxil showed how nanocarriers can improve the therapeutic index of existing drugs by enhancing delivery to target tissues while minimizing systemic toxicity.

With this momentum, different nanocarrier platforms were developed to include polymeric nanoparticles, dendrimers, and quantum dots that enabled controlled release and improved targeting of therapeutic agents. Farokhzad and Langer (2009) further developed gold nanoparticles and carbon nanotubes that added to the toolkit of biomedical applications and provided unique optical and electronic properties that could be harnessed for imaging and sensing (Farokhzad & Langer, 2009). In recent years, this area began to reach new heights with the appearance of mRNA vaccines using LNPs as vectors (Dreaden et al., 2011). The rapid development and deployment of COVID-19 vaccines by Pfizer-BioNTech and Moderna underlined the transformative potential of nanocarriers in therapeutics (Horejs, 2021). These LNPs protect fragile mRNA molecules and facilitate their uptake by host cells to show their nanobiotechnology contribution to critical participation in modern medicine.

Each of these historical milestones has helped shape the modern landscape in nanocarrier technology for biomedical applications. From Feynman's early musings to today's sophisticated nanoengineered systems, the journey has been one of relentless innovation-seeking at the foot of nanotechnology–biology hybrids.

Biological systems fundamentally function on a nanoscale architecture, as proteins, nucleic acids, and cellular organelles possess nanometric dimensions. The intrinsic operation at the nanoscale implies that the nanoparticulation of carriers aligns their dimensions with those of biological molecules, facilitating seamless interactions. By operating at the nanoscale, we can create carriers that can be designed to mimic natural biomolecules, thereby allowing them to interact not only more effectively but also with great specificity.

One of the major advantages is brought about by size compatibility. Nanocarriers, owing to their minuscule dimensions, can cross through complex biological barriers within the body, such as cellular membranes, to intracellular targets where nucleic acids and proteins carry out their functions (Salvati et al., 2011). Their size enables them to take advantage of existing cellular uptake mechanisms, such as endocytosis, thus increasing the rate of cellular internalization and delivery efficiency.

The larger nanocarriers possess a significantly greater surface area relative to their volume. With an increase in the superficial area, there is a higher functionalization with targeting ligands, polymers, or other therapeutic molecules. Thus, the nanocarriers can be engineered so that they identify given cell types and receptors, and bind with or within them, hence enhancing specific targeting and reducing off-target effects (Duncan & Gaspar, 2011).

In other words, with nanoscale carriers, unique physicochemical properties not observable in bulk materials may be expressed. This includes enhanced permeability—through the crossing of biological barriers like the blood–brain barrier (BBB), which is of vital importance for the treatment of central nervous system disorders (Saraiva et al., 2016). More importantly, nanocarriers can protect degradable biological molecules, including nucleic acids, from nucleases and proteases in the bloodstream to ensure that more of the therapeutic agent reaches its target (Peer et al., 2007).

These nanodimensions in size let the carriers take advantage of the enhanced permeability and retention (EPR) effect observed in tumor tissues; tumors usually have leaky vasculature that permits nanocarriers to accumulate with much more ease at tumor sites. This increases the local concentration of the therapeutic agent and enhances its efficacy (Maeda et al., 2000). By taking advantage of nanoscale phenomena, we are capable of designing carriers that improve the delivery of therapeutic molecules, modulate their release profiles, and respond to specific intra-body stimuli, such as pH or enzymatic activity changes. It is this level of control that leads to greatly improved therapeutic outcomes with minimized side effects and improved patient compliance (Blanco et al., 2015).

## 1.2 FUNDAMENTALS OF NANOCARRIERS

### 1.2.1 Definition and Characteristics of Nanocarriers

Nanocarriers can be defined as nanoparticles engineered to deliver therapeutic agents, including nucleic acids and proteins, to target cells or tissues. Nanoparticles possess a diminutive size and extensive surface area, facilitating unique interactions with biological systems, thereby offering an optimal approach to address shortcomings in traditional delivery methods. The performance of nanocarriers in delivering therapeutic nucleic acids and proteins depends on the understanding of their defining characteristics. Certain key attributes that have a bearing on their functionality and delivery include the following:

- *Size:* Nanocarriers are typically 10–100 nm in size. A size of this magnitude allows them to cross various biological barriers and to be internalized by cells through paths that are endocytic in nature, as reported by Blanco et al. (2015). This size influences biodistribution, cellular uptake, and clearance from systemic circulation.
- *Surface Charge:* This is important for interaction with cellular membranes and biological molecules. Positively charged cationic nanocarriers can enhance the delivery of nucleic acids due to enhanced electrostatic interactions with negatively charged cell membranes and nucleic acids (Guo et al., 2024). High positive charge may, however, cause toxicity and non-specific interaction.
- *Surface Chemistry:* The coating of the surface with the polymer, such as PEG, enhances stability and prolongs circulation time due to a lower degree of protein adsorption and lesser recognition by the immune system (Immordino et al., 2006). Targeting ligand(s) can also be attached to the surface for specific delivery to certain cells or tissues.
- *Shape:* While the majority of nanocarriers are spherical, other shapes may include rods or filaments that could influence cellular uptake and eventually biodistribution. Non-spherical particles tend to exhibit prolonged circulation due to their hydrodynamic properties and enhanced accumulation within the target tissue. Functionalization targets molecules like antibodies, peptides, or aptamers that attach with a vector to largely enhance its specificity toward cells or tissues. This improves the therapeutic efficacy of such vectors. An example is given by (Zhang et al., 2015). Functionalization can be done to enable a response against pH or enzyme stimuli to enable controlled release at the target site (Mura et al., 2013).

All of these characteristics have a cumulative effect on how nanocarriers will interact with biological environments, thus determining whether or not they can accomplish their task of delivering therapeutic agents both effectively and safely.

### 1.2.2 Size-Dependent Properties of Nanoparticles

Size dependence implies that nanoscale materials have different properties from the bulk. Such examples include quantum effects, higher reactivity, and mechanical properties that greatly influence the function of nanoparticle nanocarriers in biomedical applications. One of the most important factors here is the quantum effects emerging in nanosized systems. Because of the dimensions' very small size, electrons become considerably confined, and quantized energy levels are a result, along with changes in photophysical properties, electrical conductivity, and magnetic behavior. For example, quantum dots can exhibit fluorescence whose wavelength is tunable by size, thus becoming excellent tools for bioimaging and diagnostics (Alivisatos, 2003).

The other critical property is the higher surface area-to-volume ratio: as the size of the particles decreases, there is a remarkable increase in surface area relative to volume. Because of this higher surface area, greater interaction with biological molecules is possible, which enhances the loading capacity for therapeutic agents such as nucleic acids and proteins (Cho et al., 2008). This also allows better functionalization of nanoparticles with targeting ligands—for example, polymers or other molecules—which can enhance specificity and stability. Indeed, due to the changed mechanical properties at the nanoscale, nanoparticles have different capabilities in their ability to penetrate physiological barriers intracorporeally. For example, specific flexibility and deformability enable some nanoparticles to pass through tissues at tight junctions, enabling higher penetration and distribution within the target sites (Discher & Eisenberg, 2002).

Size-dependent cellular uptake is another important factor. Cells take up nanoparticles through mechanisms such as endocytosis. The efficiency of such mechanisms is strongly dependent on the size of the particles. Nanoparticles are generally best for cellular uptake within the range of 10–100 nm. It is a balancing act, because while smaller sizes ensure better internalization, it might also lead to lesser clearance by the immune system (Jiang et al., 2008). Size and shape influence *in vivo* biodistribution and pharmacokinetics: while small nanoparticles deeply penetrate tissues, larger ones may stay longer in circulation.

The size-related properties are also influencing the nanocarrier's ability to bypass the immune system. Nanoparticles can be engineered to avoid recognition by immune cells, thereby prolonging their circulation time and increasing the possibility of reaching the target, as stated by (Zamboni et al., 2012). Size, together with surface properties, determines the way nanoparticles interact with plasma proteins, forming a sort of "protein corona" able to change their biological identity and functionality. Such size-dependent properties can be exploited to design nanocarriers that enable the release of therapeutic agents in controlled ways, responding to specific stimuli, thereby generally improving therapeutic efficacies in drug delivery. It is possible to further develop highly specialized delivery systems for use in targeted medical applications by refining such nanoparticle size and surface properties.

### 1.2.3 Types of Nanocarriers

#### *1.2.3.1 Organic Nanocarriers*

Organic nanovehicles comprise liposomes and polymeric nanoparticles, which are critical in enabling the therapeutic delivery of nucleic acids and proteins. These are typically prepared from biocompatible and biodegradable materials such as lipids and respective polymers. The rationale for engineering design is to improve such agents' stability, bioavailability, and/or targeting ability (Figure 1.1).

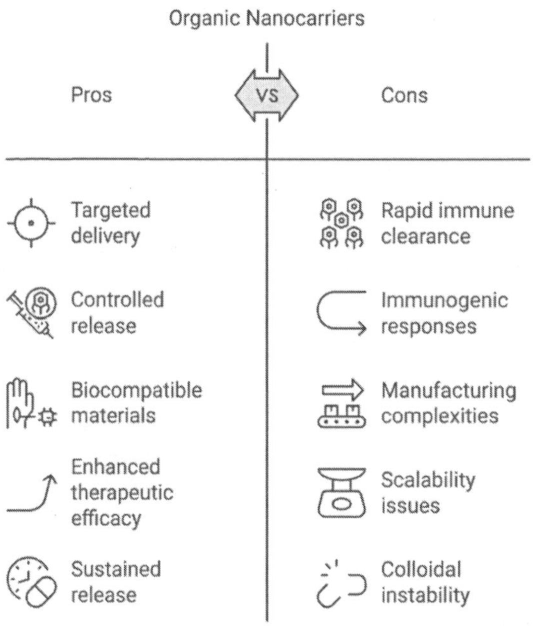


**Figure 1.1** Pros and cons of organic nanocarriers in biomedical applications. This figure presents a comparative analysis of the advantages and disadvantages of organic nanocarriers used in drug delivery and therapeutic interventions. Organic nanocarriers, such as liposomes and polymeric nanoparticles, offer targeted delivery, controlled and sustained release, biocompatibility, and enhanced therapeutic efficacy, making them ideal for delivering sensitive molecules like nucleic acids and proteins. However, challenges such as rapid immune clearance, immunogenic responses, manufacturing complexities, scalability issues, and colloidal instability pose significant obstacles to their widespread clinical use.

Liposomes are spherical vesicles with a lipid bilayer, just like cell membranes. Their amphiphilic nature gives them the capacity to encapsulate both hydrophilic and hydrophobic molecules, protecting the sensitive nucleic acids and proteins from degradation through interaction with the biological environment (Allen & Cullis, 2013). Targeting the liposome surface with ligands or polymers ensures specific delivery to particular cells or tissues, allowing for increased therapeutic efficacy with minimal side effects.

Polymeric nanoparticles, made from materials such as PLGA, provide controlled and sustained release of therapeutic agents. Nanoparticles would therefore be able to release cargo at a predetermined rate because they are degradable at a fixed rate over time, which can be critical in therapies using precisely correct dosing over very long periods of time (Danhier et al., 2012). Surface functionalization of polymeric nanoparticles increases cellular uptake and targeting specificity, thereby enhancing the delivery of nucleic acids and proteins to desired sites within the body.

Though organic nanocarriers have their advantages, there are certain limitations. One major drawback is its potential rapid clearance by a part of the immune system called the mononuclear phagocyte system, or MPS, leading to reduced circulation time and less therapeutic effect (Moghimi et al., 2005). Other characteristics that can affect stromal transport impair the colloidal stability of these carriers in the circulation, which, in turn, may lead to premature leakage from the carrier before reaching the target site.

Another drawback involves possible immunogenic responses. Components of liposomes or polymers may be immunogenic, hence limiting the treatment effects by adverse reactions (Immordino et al., 2006). Also, there are manufacturing complexities and scalability issues because, for the production of consistent and reproducible nanocarriers, one has to have precise control of size, composition, and surface characteristics.

#### *1.2.3.2 Inorganic Nanocarriers*

Inorganic nanocarriers, with gold nanoparticles and quantum dots as their most important representatives, have unique optical, magnetic, and electrical properties that can be irreplaceable in imaging and therapeutic applications. There are a number of strategies for optimizing these carriers to enhance the delivery and targeting of active therapeutic agents without causing toxicity (Figure 1.2).

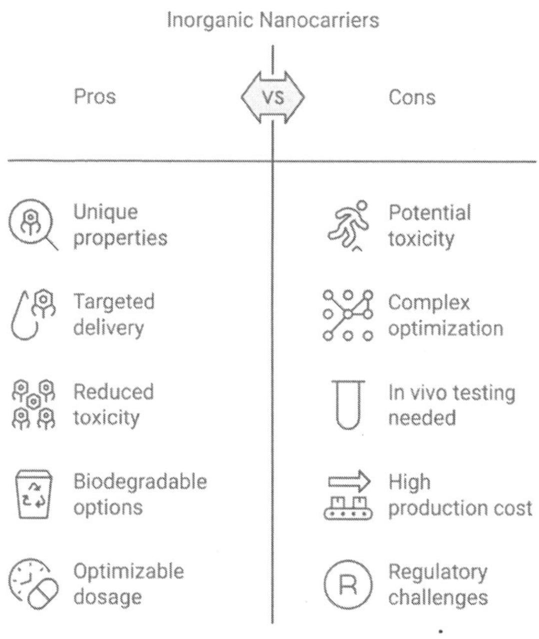


**Figure 1.2** Pros and cons of inorganic nanocarriers in drug delivery systems. This figure highlights the advantages and challenges associated with inorganic nanocarriers in therapeutic applications. Inorganic nanocarriers, such as gold nanoparticles and quantum dots, offer unique properties including targeted delivery, reduced toxicity, biodegradable options, and customizable dosage, making them valuable in precision medicine. However, they present challenges such as potential toxicity, complex optimization, the need for extensive *in vivo* testing, high production costs, and significant regulatory hurdles that must be addressed for their successful translation into clinical practice.

First, surface modification is critically important. Coating the surface of inorganic nanoparticles with biocompatible materials like PEG reduces their immunogenicity and toxicity, extending the circulation time within the blood (Jokerst et al., 2011). Functionalization of the surface by targeting ligands, such as antibodies or peptides, allows nanocarriers to home to the diseased cells, thereby increasing the therapy's selectivity (Bardhan et al., 2011). Second, the size and shape of the nanoparticles control their biodistribution and cellular uptake. Nanoparticles from 10 to 100 nm can avoid clearance by the MPS effectively and are readily internalized by cells (Jain et al., 2012). Moreover, spherical nanoparticles typically exhibit less toxicity than rod-shaped nanoparticles because of different cellular uptake mechanisms. Third, biodegradable inorganic materials could ensure minimal long-term toxicity. For example, silica nanoparticles can be surface engineered to degrade into harmless silicic acid under physiological conditions; this can reduce the accumulation of nanoparticles in organs and tissues to low levels that may result in potentially harmful side effects (He et al., 2011).

##### *1.2.3.2.1 Optimization of Dosage*

The therapeutic efficacy and toxicity can be stabilized by optimizing the quantity of therapeutic agent loaded onto the nanocarriers and managing the release kinetics. Thus, controlling the release kinetics can achieve therapeutic efficacy with minimal toxicity by fine-tuning the amount of therapeutic agent loaded onto the nanocarriers (Khan et al., 2019). This safety profile is further improved by the incorporation of stimuli-responsive features, which trigger the drug release by specific physiological conditions, such as pH or temperature. Finally, in-depth *in vivo* testing and toxicological studies are needed for complete elucidation of interactions between inorganic nanocarriers and the biological systems. Extensive studies will enable the detection of possible side effects and provide insight into how to redesign nanocarriers in response to safety concerns (Nel et al., 2006).

#### *1.2.3.3 Hybrid Nanocarriers*

Hybrid nanocarriers are innovative systems made up of organic–inorganic hybrid materials that combine the best of both worlds with the goal of providing advantages in a variety of biomedical applications. Hybrid nanocarriers combine the biocompatibility of the organic component with the functional properties of the inorganic substance, resulting in synergistic delivery of nucleic acids and proteins (Yanar et al., 2023) (Figure 1.3).

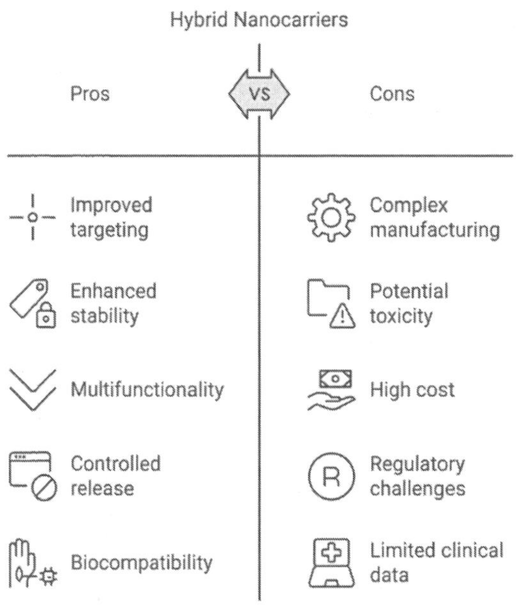


**Figure 1.3** Pros and cons of hybrid nanocarriers in biomedical applications. This figure outlines the advantages and challenges associated with hybrid nanocarriers, which combine the best features of organic and inorganic materials. Hybrid nanocarriers offer improved targeting, enhanced stability, multifunctionality, controlled release, and biocompatibility, making them a versatile platform for drug delivery and therapeutic applications. However, the complexity of their manufacturing process, potential toxicity, high cost, regulatory challenges, and limited clinical data present significant barriers to their widespread adoption in clinical practice.

Organic components, such as lipids or polymers, provide a non-toxic and biodegradable matrix that reduces toxicity and enhances the biocompatibility of the system. They enclose active therapeutic agents and reduce the degree of degradation, ensuring also controlled release. Inorganic constituents of the hybrid structure provide added properties, such as gold nanoparticles, quantum dots, or magnetic nanoparticles, which provide optical responsiveness, magnetic resonance, and electrical conductivity (Yetisgin et al., 2020).

The hybrid nanocarriers are capable of the following upon the combination of both:

- *Improve Targeting:* Surface modification by organic ligands allows for specific targeting of particular cells or tissues. Contribute the inorganic component to imaging and tracking, ensuring that the carriers reach their proper destinations.
- *Improve Stability and Efficacy:* By providing structural integrity, the inorganic materials protect nucleic acids and proteins during circulation. This would provide stability, allowing higher concentrations of the therapeutic agent to arrive at the target site.
- *Enable Multifunctionality:* Hybrid carriers can frequently perform both therapeutic and diagnostic functions simultaneously, such as drug delivery, allowing for real-time imaging of their distribution and release.
- *Stimuli-Responsive Release:* Because of the presence of inorganic components, such systems could be made sensitive to external stimuli—light, heat, or magnetic field—ensuring on-demand and site-specific release of therapeutic agents.

These synergies ensure that hybrid nanocarriers avoid problems associated with purely organic or inorganic systems and provide a flexible platform for enabling innovative drug delivery and personalized medicine (Chehelgerdi et al., 2023).

### 1.2.4 Physicochemical Properties Affecting Nanocarrier Behavior

Among all the physicochemical properties affecting nanocarrier behavior in biological systems, size and shape are the most influential. These properties also determine the nanocarrier's interaction with cells or tissues in general, which in turn determines its biodistribution, cellular uptake, and removal from the body (Figure 1.4).

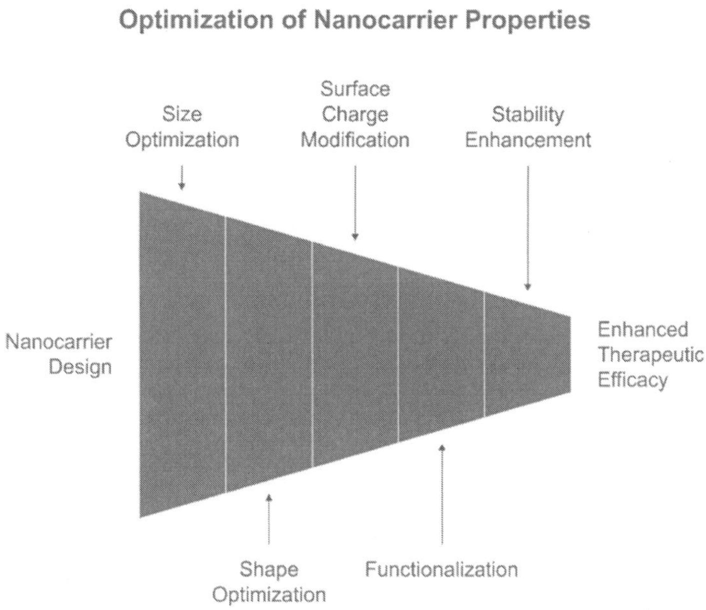


**Figure 1.4** Optimization of nanocarrier properties for enhanced therapeutic efficacy. This figure illustrates the critical design factors that contribute to optimizing nanocarriers for drug delivery systems. The design and performance of nanocarriers can be enhanced through size and shape optimization, surface charge modification, functionalization, and stability enhancement. These optimizations lead to improved therapeutic outcomes by enabling better targeting, prolonged circulation, and controlled release of the therapeutic agents, ultimately increasing the overall efficacy of nanocarriers in biomedical applications.

#### *1.2.4.1 Size and Shape*

For many aspects described below, the size and shape of nanocarriers are the most critical factors that influence delivery efficiency and therapeutic outcome.

##### *1.2.4.1.1 Biodistribution and Tissue Penetration*

- *Size Impact:* From 10 to 100 nm, nanocarriers exhibit the dimensions ideally needed for systemic delivery. Particles below 5 nm in dimension pass out quickly through the kidneys, while particles above 200 nm are cleared through sequestration via the spleen and liver (Blanco et al., 2015). Reduced size can access deep behind tissues and penetrate biological barriers more effectively, thereby reaching target sites more easily.
- *Shape Effect:* The shape of nanocarriers is determined by their ability to navigate through the extracellular matrix and the vasculature. For example, rod-shaped or filamentous particles, compared with their spherical counterparts, have demonstrated enhanced margination and adhesion to vascular walls, thereby enhancing accumulation at target sites (Toy et al., 2014).

##### *1.2.4.1.2 Cellular Uptake Mechanisms*

- *Size Dependence:* The nanocarriers are mainly taken up by the cells through a process called endocytosis, which is size-dependent. Particles of about 50 nm are usually internalized more effectively by the cells. Larger particles may not readily be internalized, hence reducing delivery efficiency (Zhang et al., 2015).
- *Shape Dependence:* Generally symmetric, spherical nanoparticles are internalized easily. On the other hand, there are particles of other shapes that have longer circulation times, but at the cost of slower cellular uptake. By design, the shape-engineered nanoparticle takes advantage of specific uptake pathways (Gratton et al., 2008).

##### *1.2.4.1.3 Circulation Time and Clearance*

- *Size Effect:* There are intermediate-sized nanocarriers that can evade both rapid renal clearance and uptake by the MPS, allowing them to remain in circulation longer. The extended time in the bloodstream allows more time for the carriers to reach the targeted tissues (Longmire et al., 2008).
- *Shape Factor:* Filamentous or elongated particles were observed to exhibit longer circulation times compared to spherical ones because their geometry impairs recognition and elimination by immune cells (Geng et al., 2007).

##### *1.2.4.1.4 Therapeutic Efficiency*

- *EPR Effect:* The tumor tissue's vasculature is typically leaky, allowing nanocarriers of a certain size to accumulate, preferably in tumor tissues. Particles in the nanorange of 30–200 nm have been seen to exploit this EPR effect and improve the drug delivery to tumors (Maeda et al., 2013).
- *Improved Targeting Efficiency:* Optimizing carrier size and shape might help to engineer nanocarriers for enhanced binding affinity toward target cells and tissues, resulting in improved therapeutic efficacy while causing minimal off-target effects (Skowicki et al., 2024).

#### ***1.2.4.2 Surface Charge and Functionalization***

Therefore, surface modification of nanocarriers is one of the important aspects of improving targeting accuracy and reducing off-target effects in therapeutic delivery. The addition of specific functional groups and surface charge adjustments improves the precise interaction between nanocarriers and target cells/tissues.

##### *1.2.4.2.1 Influence of Surface Charge*

The surface charge of nanocarriers has a high impact on their stability, biodistribution, and cell uptake.

- *Cationic Nanocarriers:* Cationic nanocarriers interact well with the negatively charged cell membranes and nucleic acids. Positive charge promotes cellular internalization and facilitates agents like siRNA and DNA. On the other hand, very high positive charge might cause non-specific interaction with healthy cells and proteins, leading to toxicity and immunogenicity (Yang & Luo, 2023).

- *Negative or Neutral Charge:* Though neutral or slightly negative charge on the surface diminishes the non-specific binding to proteins in the blood, it increases the circulation time as well as the stability of nanocarriers (Immordino et al., 2006). It can be done by coating the surface with hydrophilic polymers like PEG, also called PEGylation.

##### *1.2.4.2.2 Functionalization with Targeting Ligands*

Functionalization refers to the attachment of specific targeting ligands on the surface of nanocarriers to achieve active targeting.

- *Antibodies and Antibody Fragments:* Because nanocarriers link antibodies that recognize tumor-specific antigens, nanocarrier attachment to cancer cells will occur more selectively, thereby improving targeting accuracy (Nobs et al., 2004).
- *Peptides and Aptamers:* It is possible to design small peptides and nucleic acid aptamers to attach to special receptors overexpressed on diseased cells to enhance cellular uptake via receptor-mediated endocytosis (Zhao et al., 2020).
- *Small Molecules:* Through attachment, small molecules such as folic acid target the folate receptor, which is often overexpressed in many cancers, thereby increasing delivery to those cells carrying such receptors (Lu & Low, 2012).

##### *1.2.4.2.3 Benefits of Surface Modification*

- *Improved Specificity:* Functionalized nanocarriers target specific cells. Increasing the local concentration of the therapeutic agent, thereby improving efficacy, is also enhanced (Elumalai et al., 2024).
- *Reduced Off-Target Effects:* There are fewer side effects because of the minimum off-target effects; it spares the healthy cells and hence is normally a better therapeutic index (Niculescu & Grumezescu, 2022).
- *Improved Cellular Uptake:* Targeting ligands provides a means of entry into cells through specific pathways, thereby increasing nucleic acid and protein delivery efficiency (Yan et al., 2024).
- *Stealth Properties:* Surface modifications, such as PEGylation, reduce recognition by the MPS, which can prolong the circulation time and increase the possibility of reaching target tissues (Raza et al., 2017).

##### *1.2.4.2.4 Issues and Considerations*

There is supposed to be an optimal balance between charge and toxicity. While cationic surfaces enhance the uptake, they also appear to disrupt the membrane, leading to toxicity. A proper balance should be achieved to maximize delivery with minimum adverse effects (Sanità et al., 2020).

- *Stability of Ligand Attachments:* The linkage between targeting ligands and nanocarriers should be stable in circulation but allow for effective interaction at the target site.
- *Manufacturing Complexity:* Functionalization increases production time due to the need for careful control to ensure regularity and efficacy.

#### ***1.2.4.3 Stability and Degradation***

The balance between stability and controlled degradation remains extremely critical toward the effectiveness and safety of nanocarriers in therapeutic delivery. While it is important for nanocarriers to be stable in circulation to avoid the release of their payload before time attains destinations, they ought to degrade appropriately to release the therapeutic agents upon reaching their destination.

##### *1.2.4.3.1 Strategies to Optimize Stability and Controlled Degradation*

- *Surface Treatment: Application of Protective Coatings:* Coating with hydrophilic polymers, such as PEG, reduces protein adsorption and recognition by the immune system, hence increasing the stability of nanocarriers. This process is described as PEGylation: this prolongs circulation time and minimizes clearance through MPS (Owens & Peppas, 2006).
- *Application of Biodegradable Materials:* The use of a biodegradable polymer such as PLGA in forming nanocarriers provides for controlled degradation. Materials made up of such polymers degrade through

hydrolysis into biocompatible monomers, which pass through metabolism before being excreted by the body. According to Danhier et al. (2012), further fine-tuning of the degradation rate can be achieved through alterations in the composition and molecular weight of the polymer.

##### *1.2.4.3.2 Incorporation of Stimuli-Responsive Elements*

The design of nanocarriers that respond to specific physiological stimuli ensures degradation/payload release only under specified conditions:

- *pH-Sensitive Materials:* These are materials that degrade or change configuration due to a difference between pH values of the normal and pathologic tissues, for example, the tumor environment. This allows for active targeting by releasing a payload (Binauld & Stenzel, 2013).
- *Enzyme-Responsive Linkages:* Cleavage of incorporated linkages by enzymes, which are overexpressed in target tissues, results in controlled degradation (Hu et al., 2014).
- *Redox-Sensitive Systems:* The utilization of disulfide bonds, which cleave under reductive conditions of the cytoplasm and enable intracellular release of the payload, has been employed by redox-sensitive systems. This approach has been utilized by Duan et al. (2013).

While in circulation, cross-linking of nanocarrier components improves stability and prevents premature degradation. Cleavable cross-linkers may allow the nanocarrier to remain intact during circulation but degrade at the site of action upon arrival (Kataoka et al., 2012). Nanocarriers with the ability to control composition and thickness prepared by layer-by-layer (LbL) assembly have been constructed for enhanced stability. Such outer layers are engineered to degrade under specific conditions; this facilitates the release of the inner payload (De Koker et al., 2012).

The modulation of size, shape, and surface charge in nanocarriers influences their interaction with biological hosts and stability; for instance, particle size optimization reduces opsonization and clearance, thereby enhancing circulation stability (Blanco et al., 2015). Nanocarriers can be protected by shells or nanocapsules to avoid degradation. Such shells may be made to disassemble under some stimulus conditions, with the ensuing release of their content in a controlled fashion (Torchilin, 2014).

## 1.3 NANOCARRIERS FOR NUCLEIC ACIDS

Delivering nucleic acids as a therapeutic agent poses considerable problems due to several major *in vivo* barriers. These obstacles include enzymatic degradation by nucleases, large size, and negative charge, all of which lead to a general inability to cross cellular membranes.

### 1.3.1 Challenges in Nucleic Acid Delivery

Naked DNA and RNA rapidly undergo degradation due to the presence of nuclease enzymes in systemic blood circulation and extracellular fluids. These enzymes cleave the nucleic acids at a rapid rate, rendering them ineffective even before they could penetrate the target cells. According to Mehta et al. (2023), the severe instability of nucleic acids in biological environments greatly restricts their therapeutic potential. Also, the cell membrane is a phospholipid bilayer that acts as a crucial permeation barrier to large molecules with negative charge, including nucleic acids (Kotzé et al., 1999). Due to their phosphate backbone, which imparts a negative charge, they are repelled by the cell surface, which is equally negatively charged, and passive diffusion across the membrane is highly ineffective. As a result, when nucleic acids are administered without a delivery system, they have a low cellular intake. Naked nucleic acids could also trigger an immune response because the body identifies them as noxious agents. Immune activation leads to inflammation that dislodges therapeutic agents from the body, reducing treatment efficacy. In severe cases, such an immune response could have toxic side effects (De Jong et al., 2007).

The problem can be resolved by nanocarriers, which encapsulate nucleic acids and protect them from enzymatic degradation. LNPs physically protect nucleic acids from nuclease activity, increasing their half-life in circulation and augmenting their probabilities of reaching the target cells (Gilleron et al., 2013). Changing their superficial properties promotes cellular internalization, such as endocytosis. For example, cationic nanocarriers, through electrostatic interactions, interact with the negatively charged cell membrane and promote internalization. This has been developed one step further to where the attachment of ligands such as antibodies or peptides on the nanocarriers' surface can target a certain site and induce receptor-mediated endocytosis, which enhances specificity in cellular uptake (Shi et al., 2023; Won et al., 2011). Nanocarriers can neutralize the negative charge of nucleic acids through complexation with positively charged polymers or lipids. It reduces electrostatic repulsion from the cell membrane, allowing for close interaction and intracellular uptake (Zhang et al., 2007). Coating nanocarriers with hydrophilic polymers, such as PEG, can make them "invisible" to the immune system by avoiding opsonization and eventual phagocytic clearance. This stealth property prolongs circulation time and improves biodistribution (Shi et al., 2021).

Moreover, nanocarriers can be engineered for the delivery of nucleic acids to target cells or tissues with minimal off-target effects. The addition of targeting ligands on nanocarriers imparts the recognition capability and specific binding to receptors overexpressed on diseased cells (Liu et al., 2022). Targeted delivery enhances therapeutic efficacy with reduced systemic toxicity. Intracellularly, nucleic acids have to escape the endosome into either the cytoplasm or nucleus—wherever they exert a therapeutic effect. Nanocarriers can be engineered to contain endosomolytic agents or acid-labile materials that disrupt the endosomal membrane, allowing nucleic acids to be released into the cytoplasm (Ahmad et al., 2019).

### 1.3.2 Types of Nucleic Acids Used in Therapy

In general, therapeutic nucleic acids include a range of molecules that express different features in structure and function; this is important for defining the better design for their nanocarriers. Currently, the most commonly used therapeutic nucleic acids include siRNA, miRNA, mRNA, and pDNA.

#### *1.3.2.1 Small Interfering RNA*

siRNA molecules are small, usually 21–23 nucleotides long, double-stranded RNA. They initiate RNA interference, RNAi, which results in the knockdown of specific genes by degrading complementary mRNA sequences (Elbashir et al., 2001). Because of its small and double-stranded nature, siRNA necessitates nanocarriers to protect it against enzymatic degradation while enabling efficient cellular uptake and escape from endosomes.

##### *1.3.2.1.1 Design Implications for siRNA Nanocarriers*

- *Protection against Degradation:* Nanocarriers must protect siRNA from the nucleases present in circulation.
- *Improved Cellular Uptake:* Complexes of cationic lipids or polymers with siRNA can also serve as a delivery vehicle for improved cellular entry.
- *pH-Responsive Materials for Endosomal Escape:* These materials, upon cellular internalization, help in releasing siRNA into the cytosol.

#### *1.3.2.2 MicroRNA*

miRNAs are single-stranded, endogenous RNA molecules usually 18–25 nucleotides in length. They regulate gene expression by binding to target mRNAs, leading to translational repression or degradation (Bartel, 2004). miRNAs have therapeutic applications involving the use of miRNA mimics or antagomiRs that modulate the activity of miRNAs in diseases.

##### *1.3.2.2.1 Design Implications for miRNA Nanocarriers*

- *Targeted Delivery:* Precision is key to avoiding off-target effects; nanocarriers often incorporate targeting ligands that guarantee miRNA delivery into target cells.
- *Sustained Stability:* miRNA must be shielded from degradation by stable encapsulation or complexation.
- *Controlled Release:* Over time, sustained release profiles help to maintain therapeutic levels.

#### *1.3.2.3 Messenger RNA*

mRNA therapies involve the delivery of mRNA sequences that encode therapeutic proteins. Unlike DNA-based therapies, mRNA operates in the cytoplasm and does not integrate into the genome; this reduces potential safety risks associated with its use (Sahin et al., 2014). However, mRNA is relatively large and unstable, making its delivery a challenge.

##### *1.3.2.3.1 Design Implications for mRNA Nanocarriers*

- *Size Accommodation:* The nanocarriers must be able to encapsulate large mRNA molecules.
- *Stabilization:* This is done by ionizable lipids or polymers, which protect mRNA from degradation by RNases.

- *Efficient Translation:* Delivery systems should ensure the release of mRNA into the cytoplasm to allow translation to take place.

#### *1.3.2.4 Plasmid DNA*

pDNA is double-stranded, circular DNA that is used to deliver genes either for replacement or addition therapies. For this sort of therapy to be effective, pDNA must reach the cell nucleus (Kay et al., 2001).

##### *1.3.2.4.1 Design Implications for pDNA Nanocarriers*

- *Nuclear Targeting:* Nuclear localization signals can be implanted on nanocarriers to enhance nucleus transportation.
- *Protection from DNases:* Encapsulation helps protect pDNA against enzymatic degradation.
- *Large Payload Capacity:* Nanocarriers must accommodate the larger pDNA size compared to RNA molecules.

##### *1.3.2.4.2 Influence of Structural and Functional Differences*

- *Size and Complexity:* Larger-sized nucleic acids, such as mRNA and pDNA, require nanocarriers with higher loading capacity and robust protection mechanisms.
- *Chemical Stability:* RNA molecules are susceptible to degradation, so carriers that provide better stabilization are required.
- *Cellular Targeting:* siRNA and miRNA act in the cytoplasm, while pDNA must be delivered to the nucleus, necessitating the use of targeting signals.
- *Immunogenicity:* Unmodified mRNA and pDNA are immunogenic, so nanocarriers may need to include components that modulate these effects.

### 1.3.3 Design Considerations for Nucleic Acid Nanocarriers

In the process of delivering nucleic acids to target cells, several critical design parameters are concerned with the design of appropriate nanocarriers. These parameters ensure that nanocarriers can protect nucleic acids from degradation, achieve efficient cellular uptake, and provide targeted delivery to specific cells or tissues (Figure 1.5).

Nucleic acids are highly susceptible to the degrading action of nucleases within the bloodstream and extracellular environment. Nanocarriers must protect these therapeutic molecules by creating a protective environment. This is achieved by encapsulating nucleic acids into LNPs or polymeric nanoparticles to protect them from enzymatic degradation (Mao et al., 2010). Materials that strongly interact with nucleic acids, such as cationic polymers, stabilize the complexes formed (Hadianamrei & Zhao, 2022).

It is necessary for nucleic acids to be taken up efficiently in cells to express their therapeutic effects. Surface modification of nanocarriers with positive charges increases their interactions with negatively charged cell membranes, thus enhancing cellular uptake (Oh & Park, 2009). The size optimization into a range of 10–200 nm optimizes endocytosis and penetration while minimizing the chances of clearance by the immune system (Blanco et al., 2015). Additionally, modifying nanocarriers with cell-penetrating peptides (CPPs) enhances their intracellular delivery by facilitating translocation across cell membranes (Presente & F. Dowdy, 2013).

Targeted delivery is of the utmost importance for enhancing therapeutic efficacy while reducing side effects. Nanocarriers can be functionalized with targeting molecules—such as antibodies, aptamers, or small molecules—that direct them to receptors on target cells (Kwon et al., 2008). Engineering nanocarriers to release their therapeutic payload in response to specific stimuli, such as pH or enzymes present in diseased tissues, further enhances targeting accuracy (Lee et al., 2003). PEG modification of nanocarrier surfaces can reduce non-specific uptake by the immune system, further extending circulation time and increasing the chances of reaching target tissues (Moghimi et al., 2001).

Within the cell, many nucleic acids are sequestered in endosomes prior to their release into the cytoplasm by the nanocarriers. The use of materials sensitive to the acidic environment of the endosome can cause membrane disruption and nucleic acid release (Varkouhi et al., 2010). For example, the proton sponge effect caused by polymers like polyethylenimine (PEI) provokes endosomal swelling and rupture, allowing successful delivery to the cytoplasm (Benjaminsen et al., 2012).

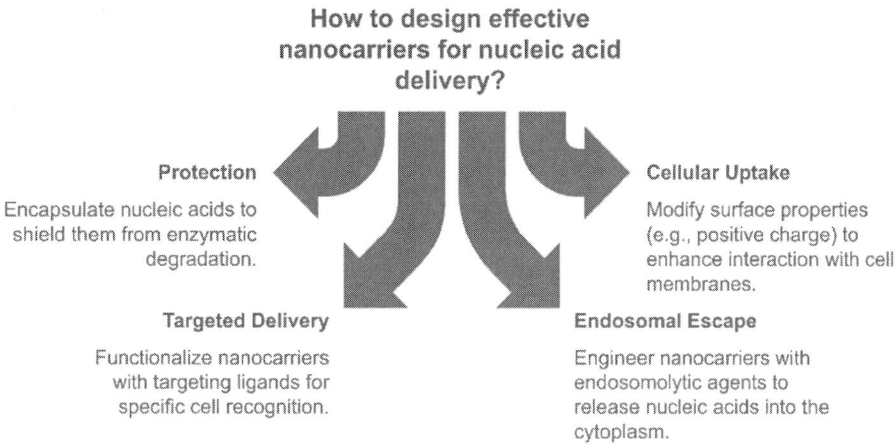


**Figure 1.5** Key design strategies for effective nanocarriers in nucleic acid delivery. This figure illustrates four essential strategies for designing nanocarriers that optimize the delivery of nucleic acids, which include (1) protection, by encapsulating nucleic acids to shield them from enzymatic degradation; (2) cellular uptake, achieved through surface modifications (e.g., positive charge) to enhance interactions with cell membranes; (3) targeted delivery, where nanocarriers are functionalized with targeting ligands for specific cell recognition; and (4) endosomal escape, by engineering nanocarriers with endosomolytic agents that facilitate the release of nucleic acids into the cytoplasm after internalization. These strategies collectively improve the efficacy of nucleic acid therapies by overcoming biological barriers.

To achieve safety and tolerability, nanocarriers should be designed with biodegradable materials that degrade into non-toxic byproducts, minimizing long-term toxicity (Danhier et al., 2012). Additionally, avoiding components that may be recognized by the immune system further enhances the biocompatibility of nanocarriers, reducing unwanted adverse reactions (Fan et al., 2023).

Nanocarriers developed for clinical applications must be manufacturable at scale with guaranteed quality (Patel et al., 2021). Robust synthesis methods ensure reproducibility in batches, while strict quality control protocols help maintain the integrity of nanocarrier formulations throughout production (Buya et al., 2024).

### 1.3.4 Examples of Successful Nucleic Acid Nanocarriers

The enormous success of LNPs with mRNA vaccines, and in particular, those from Pfizer-BioNTech and Moderna against COVID-19, has shown the transforming power of nanocarriers in the clinic. These vaccines have demonstrated one way through which LNPs can successfully deliver nucleic acids to achieve high immune responses and very high efficacy rates (Baden et al., 2020; Polack et al., 2020). There were a number of factors that contributed to the clinical success of LNPs. One such factor was efficient delivery and uptake. LNPs are designed so that mRNA molecules are located within the core, quite well protected against nuclease-mediated degradation in the bloodstream. Such a design allows for the effective uptake of LNPs by host cells through endocytosis and the concomitant access of mRNA to the cytoplasm, where translation into the desired protein occurs (Hassett et al., 2019).

Improvement in stability and protection is also an important factor. The lipid components of LNPs, especially the ionizable lipids, form stable complexes with negatively charged mRNA, protecting its integrity during storage, handling, and administration of the concocted vaccine preparation (Hou et al., 2021). While these vaccine types are administered intramuscularly, LNPs are specifically tailored to target the antigen-presenting cells in the local tissue to enhance the immune response. PEG is incorporated on the surface of the nanoparticles, serving to prolong the circulation time and reduce non-specific interactions (Samaridou et al., 2020).

Scalability and manufacturability have formed another important attribute to their success. The manufacturing processes of the LNPs were optimized for large-scale productions, which has been a crucial determinant in ensuring that supplies meet the global demand for the vaccines. For example, methods such as microfluidic mixing allow nanoparticles to be generated consistently and in a reproducible manner (Cullis & Hope, 2017). Preclinical and clinical trials have also established safety and tolerability. LNP-formulated mRNA vaccines have demonstrated an acceptable safety profile, with the majority of the side effects being ranked as mild to moderate. The use of biodegradable and biocompatible lipids minimizes long-term toxicity problems (Jackson et al., 2020). This success underlines several perspectives that could be useful in the design of future nanocarriers. For instance, the chemical tailoring of the lipid components allows perspectives in enhancing delivery to target tissues or cell types.

Design improvements of ionizable lipids have enabled improved endosomal escape and reduced the required dose (Sun & Lu, 2023). Besides, the placing of targeting ligands on the surface of LNPs would redirect them to target cells, like cancerous cells or cells within tissues that are hardly reached, improving their therapeutic outcome (Safari & Zarnegar, 2013). Other areas of interest involve improvements in mRNA stability. Modifications to the mRNA itself, such as the incorporation of nucleotide analogs, can improve stability and promote translation at lower doses, with less immune activation versus the mRNA itself, while modifications to the delivery vehicle can be used to limit off-target effects (Pardi et al., 2018). Improvements in manufacturing techniques include the development of scalable and cost-effective methods, including continuous production, which will be important for the broader adoption of nanocarrier-based therapies in general (Su & Kang, 2020). In addition, there is still one challenge to be overcome: immunogenicity. Research on the immunogenicity of nanocarriers and their components will make possible the design of systems in ways to reduce unwanted immune responses, thus improving the safety profile of nanosized drug delivery systems (Bahl et al., 2017).

### 1.3.5 Mechanisms of Cellular Uptake and Endosomal Escape

The entrance of nanocarriers into cells and the release of their cargo is pivotal in effective nucleic acid delivery. Two major steps—enhancement of cellular uptake and facilitating endosomal escape—play an important role in the success of these delivery systems.

Different strategies can be developed to enhance cellular uptake for nanocarriers. Surface modification with targeting ligands is considered one of them. Receptor-mediated endocytosis could be facilitated by the attachment of ligands such as antibodies, peptides, or aptamers onto the nanocarrier surface. This method enables nanocarriers to recognize and bind receptors that are overexpressed on target cells, hence increasing specificity (Kwon et al., 2008). Other factors that are crucial in enhancing uptake include particle size and shape optimization. Nanocarriers designed within the optimal size range, usually between 50 and 200 nm, are internalized easily by cellular elements. Moreover, the shape of the nanocarrier could affect internalization pathways—for example, rod-shaped particles may be internalized differently compared to spherical ones (Gratton et al., 2008). Another approach is the modification of the surface charge; slightly positively charged nanocarriers interact more effectively with the negatively charged cell membrane, thus improving the uptake of the particle. However, excess positive charge can induce toxicity; thus, a proper balance has to be maintained (Oh & Park, 2009). The decoration of nanocarriers with CPPs is another strategy that promotes nucleic acid intracellular delivery by allowing nanocomplexes to translocate through cellular membranes via energy-independent pathways, which further promotes the intracellular delivery of nucleic acids (Nakase et al., 2012).

Intracellularly, nanocarriers are often trapped in endosomes and need strategies for efficient release of nucleic acids into the cytoplasm. One method includes pH-sensitive materials. These materials could be collectively designed to respond in the acidic environment of the endosomes at pH ~ 5–6 that might result in structural changes of the nanocarrier, disruption of the endosomal membrane, and finally release of the cargo (Du et al., 2015). Another potent mechanism is the proton sponge effect, where one could incorporate appropriate amounts of polymers such as PEI (Benjaminsen et al., 2012). These are the polymers that buffer the acidified environment of the endosome, leading to an influx of ions and water into the polymer particles. This influx causes the polymer particles to swell and rupture the endosome. Membrane-disruptive peptides, which become active under acidic conditions, are also employed to disrupt the endosomal membrane and thereby easily release nucleic acids into the cytoplasm (Bechinger & Lohner, 2006). Another sophisticated strategy involves photochemical internalization, which employs light-sensitive compounds to trigger the activation of nanocarriers. These compounds force endosomal escape upon irradiation with light and allow for precise control of the release process (Soe et al., 2020). These approaches might be combined in such a way that it would be possible to engineer nanocarriers to be optimal for both cellular uptake and endosomal escape.

Targeting ligand, pH-responsive material, and endosomal escape agent combined into one multifunctional nanocarrier—an approach that improves the overall nucleic acid delivery efficiency (Pandey et al., 2024). Nanocontainers prepared by an LbL assembly approach offer excellent capabilities for the precise control of surface properties and functional components of these assemblies toward the optimization of targeting and release profiles (De Koker et al., 2012). For instance, a stimuli-responsive system could be designed to be sensitive to specific intracellular conditions, such as reductive environments or enzymatic activities, under which the nucleic acids are released only within target cells. This concept has been demonstrated by Du et al. (2015). These carefully engineered strategies allow nanocarriers to overcome biological barriers, thus ensuring that nucleic acids are effectively delivered into the desired cells while maximizing the therapeutic potential with minimal unwanted side effects.

## 1.4 NANOCARRIERS FOR PROTEINS

In terms of delivery challenges, proteins are unique therapeutic agents due to their complexity and extreme sensitivity. Maintaining biological activity and ensuring stability during the delivery process are essential for their efficacy.

### 1.4.1 Challenges in Protein Delivery

There are numerous significant challenges associated with protein delivery. Of those, the most important is denaturation and degradation. Due to changes in temperature, pH, and other environmental problems, proteins can lose their three-dimensional (3D) structure—a process called denaturation. Moreover, therapeutic proteins have a very short-lived half-life in the body due to enzymatic digestion by proteases, which often makes the loss of their functional level possible (Wang et al., 2006). The other challenge is immunogenicity as several therapeutic proteins have a tendency to activate immune responses. This could lead to the generation of neutralizing antibodies, which interfere either with the efficacy of proteins or lead to adverse immune responses (Hermeling et al., 2004). Another challenge arises due to poor cellular uptake. Protection from accessing intracellular targets occurs by virtue of poor permeability through cellular membranes due to their large size and hydrophilic nature (Haddadzadegan et al., 2022). The circulating half-life is rather short for many proteins because of rapid clearance by the reticuloendothelial system (RES), thus further narrowing their therapeutic window (Shirian et al., 2024).

Nanocarriers have come to offer innovative solutions to these challenges. Another one of the key advantages nanocarriers possess is protecting proteins against denaturation and degradation. This means that once a protein is encapsulated within the nanocarriers, it becomes shielded from harsh conditions in the environment as well as enzymatic assault. In this regard, the lipid nature of nanocarriers, represented by liposomes and solid LNPs, provides shielding for proteins—per se, a coating composed of a double layer of lipids—that may contribute to the stabilization of proteins. Massignan et al. (2023) have presented the role of nanocarriers in enhancing cellular uptake. The nanoparticles, which are polymeric in nature, and the nanoemulsions are designed to be endocytosis-supportive with distinct shapes. This enables proteins to more effectively reach intracellular targets (Elsabahy & Wooley, 2012).

Nanocarriers can also reduce immunogenicity through surface modification with PEG, a process termed PEGylation. This process protects the protein from the immune system, reducing its immunogenicity and allowing it to circulate longer in the bloodstream (Veronese & Pasut, 2005). Another key feature of the nanocarriers is the controlled release. Nanocarriers can be engineered for the controlled and sustained delivery of proteins, maintaining therapeutic levels over extended periods. Some nanocarriers may be stimuli-sensitive; these are designed to release their protein cargo in response to specific triggers, such as changes in pH or enzymes present in target tissues (Ortiz et al., 2024). Nanocarriers can also be functionalized with different targeting ligands, such as antibodies or peptides, to be directed toward a specific needy cell or tissue. This increases therapeutic efficacy while off-target effects are minimized (Todaro et al., 2023). Applying nanocarriers provides solutions to major challenges in protein delivery, such as stability, bioavailability, and efficacy, considerably improving protein therapeutics. This indeed opens ways for the development of more effective treatments in a wide circle of medical fields.

### 1.4.2 Types of Therapeutic Proteins

The concept of therapeutic proteins enlists a wide variety of molecules with different functions that shape the design requirements for nanocarriers. Among the major classes are enzymes, antibodies, and growth factors; all perform specific functions in treatments.

Enzymes are the biological catalysts for the chemical reactions that take place in the human body. They are administered as medical therapy to take the place of missing or malfunctioning enzymes, for example, in enzyme replacement therapies in diseases such as Gaucher's disease (Desnick & Schuchman, 2002). In the development of nanocarriers for enzymes, several critical factors have to be taken into account. Firstly, the activity of the enzyme is paramount as enzymes are generally sensitive to environmental conditions, which may lead to their denaturation, thereby losing their activity. Nanocarriers need to provide a stable environment that maintains the 3D structure of the enzyme and hence its catalytic activity (Yu et al., 2015). A very important feature is controlled release, as enzymes often need to be delivered at specific sites within the body. This means nanocarriers should be able to release their enzyme payload upon receiving a stimulus like pH changes or enzymatic triggers in target tissues (Mizukami et al., 2017). In addition, most of the therapeutic enzymes are of non-human origin, which may evoke immune responses. Surface modification of nanocarriers with biocompatible polymers may act to reduce the immunogenicity of nanocarriers (Zhang et al., 2016).

Another major class of therapeutic proteins includes antibodies, which are integral components of the immune system designed to recognize and bind to specific antigens. They have wide applications in treatments against cancer, autoimmune diseases, and infections (Ecker et al., 2014).

Consequently, the design of nanocarriers for antibody delivery becomes rather challenging. It is necessary to keep the correct orientation of the antibodies and their binding potential because their antigen-binding sites need to be free and available. Some examples of immobilizing antibodies by using nanocarriers were given without blocking their binding capability with their respective target antigens. Another critical issue related to the stability of antibodies is that they tend to aggregate and degrade (Chames et al., 2009). In this regard, nanocarriers should provide a microenvironment that protects the antibodies' stability. As stated by Shire et al. (2004), this will serve as an additional benefit. Also, the use of antibodies may enhance targeted delivery (Shire et al., 2004). Conjugated to the surface of nanocarriers, they direct the carrier toward specific cells or tissues, hence increasing specificity and improving cellular uptake (Zhao et al., 2020).

Growth factors are signaling proteins that, through binding with specific membrane-bound receptors, regulate cell proliferation, differentiation, and survival; hence, growth factors are widely used in regenerative medicine and tissue repair (Anitua et al., 2011). Nanocarriers designed for growth factor delivery should protect such sensitive proteins from degradation mediated by proteolytic enzymes. Entrapment of growth factors into nanocarriers protects them from enzymatic degradation. This becomes crucial in the case of growth factors because they are part of a complex signaling process (Yu et al., 2015). Engineering of nanocarriers with the capability to provide sustained or sequential release to be able to mimic natural release patterns of these proteins. The materials used for nanocarriers also have to be biocompatible. In tissue engineering, for instance, they need to support cell adhesion and cell proliferation (Briquez et al., 2015).

Designing nanocarriers for therapeutic proteins needs modification based on the specific role played by those proteins, including stability and release profile. For enzymes and antibodies, especially, stability during transport is critical. As a matter of fact, recent studies have focused on self-assembling protein nanocarriers that could enhance the bioavailability while preserving the structure of therapeutic proteins (Li & Champion, 2022). Stimuli-responsive systems, for instance, release their cargo according to environmental signals such as pH or enzymes. This represents a great leap forward in optimizing therapeutic effects for growth factors, whose activities are finely tuned by both concentration and timing (Qin et al., 2019). Antibodies, on the other hand, often play dual roles both as targeting agents and as therapeutics and thus impose additional multifunctional requirements on nanocarriers at no compromise to efficacy during the delivery process (Knauer et al., 2019). Finally, proteins that might induce some immune effects also require nanocarriers with the capability to camouflage or regulate the immune responses to reduce side effects, as was previously seen in drug delivery with protein-based nanostructures (Krishnan, 2020).

### 1.4.3 Design Considerations for Protein Nanocarriers

Nanocarriers for protein delivery should be developed by strategies that ensure that protein structure is maintained, rates of release are effective, and targeting of tissues occurs. Proteins are fragile molecules; they may lose their functions by denaturation, degradation, or aggregation. Thus, maintaining stability and bioactivity throughout the process of delivery is a challenge (Muslykhah et al., 2024) (Figure 1.6).

Among these encapsulation techniques, proteins are protected, and their stability and bioactivity are maintained. Proteins encapsulated within the nanocarriers, such as within liposomes, polymeric nanoparticles, or nanogels, are protected against severe physiological conditions and enzymatic degradation that may cause denaturation through changes in their native conformation. Knauer et al. (2019) illustrated that mild preparation methods include self-assembly and electrostatic complexation, which avoid extreme conditions of pH, temperature, or organic solvents, which in turn normally destabilize protein structure (Knauer et al., 2019).

Surface modification and functionalization also play a critical role. Coating nanocarriers with PEG—a process known as PEGylation—improves their stability by preventing protein adsorption and reducing immune recognition. In general, this prolongs the circulation time and diminishes immunogenicity (Veronese & Pasut, 2005). The attachment of targeting ligands to the surface of nanocarriers enables one to achieve specific delivery into target tissues, increasing therapeutic efficacy while further minimizing off-target effects (Zhao et al., 2020).

This precision in protein delivery is further imparted with controlled release mechanisms. The stimuli-responsive nanocarriers undergo environmental pH changes, temperature fluctuations, or enzymatic action to release at the right site and keep the proteins in an inactive state until reaching their intended site. Biodegradable polymers, such as PLA, allow for sustained release—often maintaining therapeutic levels of proteins for weeks—and hence decreasing dosing frequency (Liu et al., 2024).

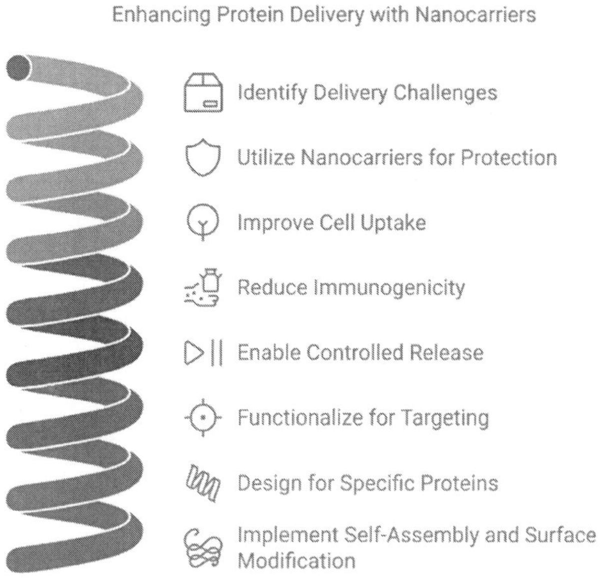


**Figure 1.6** Strategies to enhance protein delivery using nanocarriers. This figure summarizes key approaches to improve the delivery of therapeutic proteins through nanocarriers. These strategies include (1) identifying delivery challenges, (2) utilizing nanocarriers to protect proteins from degradation, (3) enhancing cellular uptake, (4) reducing immunogenicity, (5) enabling controlled and sustained release, (6) functionalizing nanocarriers for targeted delivery, (7) designing nanocarriers for specific protein structures, and (8) implementing self-assembly and surface modifications to ensure stability and bioactivity. These approaches collectively enhance the efficacy and precision of protein-based therapies.

Generally, to stabilize protein structure during formulation and delivery, the addition of stabilizing agents, such as trehalose or other sugars, amino acids, or polymers, to the formulation helps in maintaining the protein tertiary structure (Olgenblum et al., 2024). Proteins freeze-dried within nanocarriers can have improved stability during storage and transportation while retaining bioactivity up to the time of administration (Tchessalov et al., 2023).

Further, immunogenicity is minimized. Encapsulation and surface modification can mask the immunogenic regions of the protein against recognition, thereby causing minimal immune response (Harris & Cohen, 2024). Besides, construction with biocompatible materials minimizes unwanted side effects and intends to keep the functionality of a protein upon delivery (Mitchell et al., 2020).

Another important determinant of successful protein delivery is the ability to optimize nanocarrier properties, such as size and surface charge. While smaller particles generally may penetrate better through tissues, the surface charge can confer different interactions with cell membranes (Blanco et al., 2015). It is also important that the design of soft nanocarriers avoids protein aggregation. This may be achieved by controlling protein loading density and incorporating anti-aggregation agents to decrease protein–protein interactions (Mahler et al., 2008).

Another important aspect is the improvement of cellular uptake. As an example, CPPs attached to nanocarriers enable the effective transport across cellular membranes due to their properties for enhancing intracellular delivery, as observed by Nakase et al. (2012). Another way by which nanocarriers could be modified is by targeting ligands that attach to certain receptors on the cell surface. In this instance, receptor-mediated endocytosis is favored, and many more cells are taken up, increasing delivery efficiency, as stated by Rojekar et al. (2024).

### 1.4.4 Examples of Successful Protein Nanocarriers

Nanocarriers, including polymeric micelles and nanogels, have exhibited remarkable efficacy in the delivery of therapeutic proteins in both preclinical and clinical investigations. These developments provide essential insights for the design of future protein nanocarrier systems, hence improving their efficacy in many therapeutic applications.

A primary problem in protein distribution is maintaining the stability and bioactivity of proteins throughout the procedure. Nanocarriers mitigate this issue by establishing a protective milieu that safeguards proteins against degradation and denaturation. An exemplary instance is the application of polymeric micelles for enzyme administration. Micelles composed of amphiphilic block copolymers have effectively facilitated the delivery of enzymes. This method involves encapsulating the enzyme within the hydrophobic core of the micelle, safeguarding it from proteolytic degradation, while the hydrophilic shell enhances solubility and extends circulation duration. Research, like that conducted by Kim et al. (2001), demonstrates that this approach improves enzyme stability and increases treatment efficacy in oxidative stress models. This example demonstrates how nanocarriers can maintain protein bioactivity by protecting them from detrimental circumstances during transport.

An additional crucial element is the attainment of controlled and sustained release of therapeutic proteins, which aids in maintaining effective concentrations throughout time and diminishes the necessity for frequent dosage. Nanogels—cross-linked hydrophilic polymer networks—have been utilized for the delivery of growth hormones such as insulin-like growth factor-1 (IGF-1). These nanogels facilitate the prolonged release of IGF-1, enhancing tissue regeneration and repair in myocardial infarction models (Xu et al., 2024). This example illustrates the capability of nanocarriers to improve therapeutic efficacy by sustaining stable protein concentrations at the target location over prolonged durations.

Moreover, precise administration and minimizing immunogenicity are essential for enhancing the safety and efficacy of protein-based treatments. PEGylated liposomes have been utilized to provide interferon-alpha, a therapeutic agent for viral infections and certain malignancies. The use of PEG in these liposomal formulations diminishes immunogenicity and prolongs circulation duration by reducing immune system identification and clearance (Haghighat et al., 2024). This approach enhances therapeutic effectiveness while reducing adverse effects. Surface modification of biocompatible polymers such as PEG enhances targeting accuracy and diminishes adverse immune reactions, resulting in improved patient outcomes.

For proteins that must operate within cells, efficient cellular uptake and endosomal escape are crucial. Nanocarriers enhanced with CPPs have been employed to transport therapeutic proteins, including caspase-3, into cancer cells. CPPs facilitate cellular absorption and enable the release of proteins from endosomes, hence permitting therapeutic proteins to trigger apoptosis in cancer cells (Farkhani et al., 2014). This method illustrates how the functionalization of nanocarriers with CPPs or analogous compounds might enhance intracellular transport, hence broadening the possibilities of protein-based therapeutics.

Moreover, responsiveness to environmental stimuli might augment the accuracy and efficacy of protein delivery. For instance, pH-responsive nanoparticles have been engineered to release anticancer therapeutics exclusively within the acidic microenvironment of malignancies. This tailored release enhances the concentration of the therapeutics at the tumor location while decreasing systemic exposure and mitigating negative effects (Xu et al., 2024). This study demonstrates how stimuli-responsive nanocarriers provide targeted and controlled release in reaction to particular physiological situations, enhancing specificity and therapeutic effectiveness, which could be eventually implemented for therapeutic proteins.

In anticipation of the future, many ways can be employed to enhance protein nanocarrier systems. Primarily, maintaining protein stability will be a concern, with encapsulating methods and material selection being crucial in preventing degradation and sustaining bioactivity. Furthermore, integrating targeting methods, such as functionalizing nanocarriers with antibodies or peptides, would facilitate precise distribution to targeted tissues or cells, hence enhancing therapeutic efficacy. Developing nanocarriers for sustained or stimuli-responsive release, contingent upon the therapeutic protein and disease setting, is crucial for optimizing release patterns.

Enhancing intracellular delivery can be achieved by altering nanocarriers with CPPs or analogous compounds to boost cellular absorption and facilitate endosomal escape. Employing biocompatible and biodegradable materials will reduce toxicity and unfavorable immunological responses, hence enhancing overall patient safety. Ultimately, establishing scalable manufacturing techniques and guaranteeing reproducibility will be essential for transferring these technologies into clinical applications and complying with regulatory norms.

Integrating these insights into forthcoming protein nanocarrier research may yield more efficacious, safer medicines and improved patient outcomes across various medical domains.

## 1.5 APPLICATIONS IN BIOMEDICAL RESEARCH

Nanocarriers have revolutionized *in vitro* diagnostics by markedly improving the sensitivity and specificity of diagnostic assays. Their capacity to enhance signals, adhere to targets with exceptional precision, and provide swift detection has resulted in the creation of more precise and effective diagnostic assays (Figure 1.7).

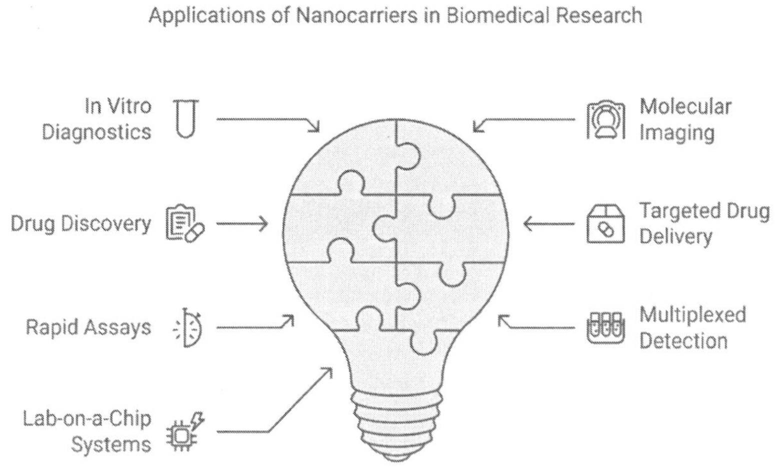


**Figure 1.7** Applications of nanocarriers in biomedical research. This figure highlights the wide-ranging applications of nanocarriers in the field of biomedical research. Nanocarriers are being utilized in various areas, including *in vitro* diagnostics, molecular imaging, drug discovery, targeted drug delivery, rapid assays, multiplexed detection, and lab-on-a-chip systems. These applications are pivotal in enhancing the sensitivity, specificity, and efficacy of biomedical interventions, helping to advance precision medicine and improve therapeutic outcomes.

### 1.5.1 Molecular Imaging and Diagnostics

Nanocarriers enhance the sensitivity of diagnostic tests owing to their elevated surface area-to-volume ratio, adjustable surface chemistries, and capacity for biomolecular functionalization. These attributes enable the amplification of detection signals, enhancement of target selectivity, and acceleration of assay kinetics, rendering them optimal for swift and accurate diagnostics (Mendes et al., 2022).

Nanocarriers primarily boost diagnostic sensitivity by means of signal amplification. Nanoparticles, including gold, quantum dots, and magnetic particles, can transport several signal-generating molecules or have inherent optical characteristics that amplify detection signals. Quantum dots, semiconductor nanoparticles with adjustable fluorescence, are recognized for their exceptional brightness and photostability. These characteristics render them very proficient at identifying low-abundance biomarkers. Quantum dots, when coupled to antibodies or nucleic acids, have markedly enhanced sensitivity relative to conventional fluorophores (Alivisatos, 2003).

Nanocarriers can be modified with ligands such as antibodies, aptamers, or peptides that selectively bind to target analytes. This focused strategy minimizes non-specific interactions, thereby diminishing background noise and enhancing the specificity of diagnostic assessments. Magnetic nanoparticles can be coated with antibodies that selectively bind and concentrate target molecules from intricate biological samples, thereby greatly improving detection limits and enabling the identification of uncommon biomarkers (Khizar et al., 2022).

Due to their nanoscale dimensions, nanocarriers demonstrate enhanced diffusivity, hence expediting their interactions with target analytes. This enhanced efficiency reduces incubation durations, accelerating the total diagnostic procedure. Rapid assays are crucial in point-of-care environments, where prompt results are vital for timely clinical decision-making. Nanoparticle-based lateral flow tests can rapidly identify disease markers, facilitating real-time decision-making and enhancing patient care (Lou et al., 2022).

Nanocarriers facilitate multiplexed detection, permitting the concurrent investigation of numerous biomarkers within a single test. Diverse nanocarriers can be labeled with unique optical or magnetic signatures, enabling the simultaneous detection of several targets. The multiplexing capability is essential for thorough diagnostics, including the profiling of numerous disease indicators to enhance diagnostic precision (Geiss et al., 2008).

Encapsulating sensitive biomolecules in nanocarriers safeguards them against environmental degradation, hence improving the stability and shelf life of diagnostic reagents. This safeguard guarantees uniform assay performance throughout time, rendering nanocarriers optimal for prolonged diagnostic testing. Gold nanoparticles exhibit colorimetric changes following aggregation, facilitating rapid visual identification of target analytes without the necessity for intricate instrumentation (Saha et al., 2012).

Magnetic nanoparticles can be controlled with magnetic fields to effectively isolate and concentrate target molecules, enhancing the sensitivity and rapidity of immunoassays (Khizar et al., 2022). Quantum dots and dye-doped nanoparticles provide intense, sustained fluorescence signals, rendering them exceptionally efficient for identifying low-abundance targets in assays like fluorescence in situ hybridization (FISH) (Alivisatos, 2003).

The integration of nanocarriers with microfluidic devices has facilitated the development of lab-on-a-chip systems that perform intricate diagnostic procedures on small platforms. This integration facilitates automated, high-throughput analysis utilizing minimal sample amounts, hence enhancing the speed and ease of *in vitro* diagnostics (Behera et al., 2023).

Nanocarriers are essential for enhancing *in vivo* imaging by acting as contrast agents in several imaging modalities, including magnetic resonance imaging (MRI), fluorescence imaging, positron emission tomography (PET), and computed tomography (CT). These carriers improve disease detection by augmenting the contrast between healthy and diseased tissues, resulting in more precise and accurate imaging.

Nanocarriers typically measure 10–100 nm in diameter, a dimension that facilitates their effective circulation in the bloodstream, evasion of swift clearance by the RES, and targeting to tissues via the increased permeability and retention (EPR) effect. The size is crucial for optimizing circulation duration and enhancing imaging precision (Blanco et al., 2015). The form also influences performance. Spherical nanocarriers are prevalent due to their straightforward production and advantageous distribution; however, rod-shaped carriers may exhibit extended circulation durations and enhanced targeting capabilities (Geng et al., 2007).

Nanocarriers are frequently coated with hydrophilic polymers such as PEG to diminish immune recognition and prolong their circulation duration (Owens & Peppas, 2006). Surface functionalization with targeting ligands, such as antibodies or peptides, facilitates accurate delivery to targeted tissues, enhancing imaging specificity by ensuring contrast chemicals accumulate in sick regions (Niculescu & Grumezescu, 2022). Efficient nanocarriers must contain or bind imaging chemicals, such as fluorescent dyes, radioactive isotopes, or contrast-enhancing metals such as gadolinium. These medicines must maintain stability within the nanocarrier to prevent premature release and provide dependable delivery to the target region, which is essential for consistent imaging outcomes (Liu et al., 2007).

Nanocarriers composed of biocompatible, biodegradable materials reduce toxicity and immunological responses. Upon completion of imaging, the carriers are expected to decompose into innocuous byproducts, thus mitigating the possibility of long-term buildup and adverse effects (Lei et al., 2022). Nanocarriers that transport numerous imaging agents can markedly enhance signal sensitivity, facilitating disease detection. Certain nanocarriers are engineered to activate their agents only at the target site, hence improving contrast and reducing background noise, resulting in crisper imaging outcomes (Miao et al., 2017).

Nanocarriers can be designed to integrate various imaging agents, facilitating multimodal imaging. A solitary nanocarrier could amalgamate both MRI and PET agents, providing a more holistic diagnostic viewpoint (Lee et al., 2011). Theranostic nanocarriers provide concurrent diagnosis and treatment (Janib et al., 2010). To preserve image quality, nanocarriers must exhibit stability within the body. This entails inhibiting aggregation in the circulation and resisting enzymatic breakdown until the target site is reached (Mura et al., 2013).

Superparamagnetic iron oxide nanoparticles (SPIONs) improve MRI contrast by modifying the relaxation periods of adjacent protons. Coating these particles with polymers and targeting ligands increases their stability and specificity, hence enhancing the detection of malignancies and inflammatory regions (Shubayev et al., 2009). Quantum dots are optimal for fluorescence imaging because of their luminosity and photostability. When modified with targeting molecules, they can identify cancer cells for high-resolution tumor imaging (Gao et al., 2004).

Gold nanoparticles, due to their elevated atomic number, function as efficient contrast agents for CT imaging. Their alteration for biocompatibility improves the imaging of vascular structures and malignancies, providing sharper diagnostic images (Hainfeld et al., 2004). Hybrid nanocarriers merge organic and inorganic materials to facilitate multimodal imaging, such as the combination of optical and PET imaging within a single platform, hence providing a more cohesive diagnostic methodology (Lee et al., 2011).

### 1.5.2 Drug Discovery and Development

Nanocarriers greatly enhance drug discovery and development by boosting high-throughput screening (HTS), enabling targeted drug administration, and increasing the efficiency of preclinical and clinical investigations. Their distinctive characteristics expedite the identification of promising therapies while enhancing the assessment and safety of potential pharmaceuticals.

Nanocarriers have transformed HTS, an essential phase in drug discovery. Researchers may now simultaneously test hundreds of chemicals by facilitating the creation of nanoarray-based systems. These technologies immobilize nanocarriers containing potential medications onto microarrays, facilitating quick, parallel evaluation of biological activity (MacBeath & Schreiber, 2000). This approach significantly minimizes the time and resources needed in the preliminary screening phase. Furthermore, nanocarriers augment the sensitivity and specificity of assays by elevating the local concentration of candidate drugs, thereby enhancing their interaction with target biomolecules (Rissin et al., 2010). Their nanoscale dimensions facilitate assay downsizing, conserve chemicals, and enable automation, thus expediting the screening procedure.

Nanocarriers markedly enhance the targeted delivery of drug candidates, hence improving the assessment of therapeutic efficacy and safety profiles during development. Encapsulating poorly soluble pharmaceuticals into nanocarriers enhances solubility and stability, thereby boosting absorption and bioavailability (Kumar et al., 2020). Nanocarriers can be functionalized with targeting ligands to direct medicine delivery to specific disease locations. This accuracy diminishes off-target effects and mitigates toxicity, facilitating more precise evaluations of a drug's therapeutic potential (Petros & DeSimone, 2010).

Nanocarriers can be designed for regulated or stimuli-responsive drug release, aiding researchers in the investigation of pharmacokinetics and pharmacodynamics under various situations. This control is crucial for enhancing medication release profiles and assessing therapeutic efficacy (Kamaly et al., 2016). Nanocarriers facilitate the simultaneous administration of various therapeutic drugs, hence streamlining the investigation of combination therapies. This feature is especially significant in oncology, where combination therapies are frequently required to improve therapeutic effectiveness (Alghamdi et al., 2022).

The integration of nanocarriers into pharmaceutical research enhances progress in personalized medicine. By customizing nanocarriers to target specific genetic or molecular markers, medicines can be designed that are more efficacious for individual patients. This method provides a more individualized treatment, enhancing both results and patient satisfaction (Wilhelm et al., 2016). Nanocarriers enhance the precision of drug behavior models in biological systems, hence diminishing attrition rates. They facilitate the early identification of ineffectiveness or toxicity, thereby averting late-stage failures and conserving essential time and resources (Moghimi et al., 2005).

Nanocarriers augment the therapeutic index of pharmaceuticals by increasing efficacy at the target location while reducing systemic exposure and adverse effects. This accuracy is essential for converting novel chemicals into effective clinical treatments (Peer et al., 2007). Nanocarriers can traverse biological barriers, including the BBB, which poses a considerable obstacle in administering medications to the central nervous system. Nanocarriers enable distribution to previously inaccessible locations, creating new opportunities for treating disorders impacting the brain and other regions shielded by barriers (Saraiva et al., 2016).

Nanocarriers' capacity to enhance therapeutic specificity and predictability may also streamline the regulatory approval process. Improved targeting accuracy and diminished toxicity profiles facilitate the demonstration of safety and efficacy to regulatory authorities, expediting the approval of novel medicines (Ventola, 2017).

### 1.5.3 Tissue Engineering and Regenerative Medicine

Nanocarriers have emerged as potent instruments in tissue engineering and regenerative medicine, facilitating the controlled and targeted delivery of growth factors, genes, and other bioactive compounds. Nanoscale systems improve tissue engineering effectiveness by safeguarding therapeutic compounds from degradation, facilitating controlled release, and influencing cellular activity crucial for tissue regeneration and repair. Growth factors are essential signaling molecules that govern activities including cell proliferation, differentiation, and migration, all of which are critical for tissue regeneration. Nanocarriers facilitate the regulated and prolonged release of growth factors at the designated spot, sustaining therapeutic concentrations for extended durations and minimizing the necessity for frequent administration (Briquez et al., 2015).

By encapsulating growth factors, nanocarriers safeguard these delicate molecules from enzymatic breakdown, preserving their biological activity. Furthermore, nanocarriers can be included in scaffolds or hydrogels, facilitating localized administration of growth factors. This focused strategy diminishes systemic exposure, hence decreasing side effects and enhancing therapeutic results (Langer & Peppas, 2003). Gene therapy has the capacity to induce cells to synthesize restorative proteins directly. Nanocarriers can transport genes that encode these proteins or growth factors to cells within tissue engineering constructions, facilitating prolonged synthesis of therapeutic proteins at the target location.

In contrast to viral vectors, which present safety concerns, nanocarriers offer a safer, non-viral alternative characterized by reduced immunogenicity and toxicity (Liu et al., 2011). Nanocarriers enhance cellular absorption and safeguard nucleic acids from degradation, therefore improving transfection efficiency and gene expression, which increases the efficacy of gene therapy in tissue engineering (Pack et al., 2005). Nanocarriers can be modified with bioactive ligands to enhance cell adhesion, proliferation, and differentiation. Modifying nanocarriers with RGD peptides (arginine–glycine–aspartic acid) improves cell adhesion and integration into scaffolds (Hersel et al., 2003). Nanocarriers can transport chemicals that prompt stem cells to differentiate into specific cell types, thus aiding in the creation of functional tissue constructions (Silva et al., 2004).

Therapeutic proteins and nucleic acids are susceptible to enzymatic breakdown within the body. Nanocarriers provide a protective shield, guaranteeing that these bioactive chemicals arrive at their target cells unscathed. Polymeric nanocarriers, such as those composed of PLGA, offer a biodegradable framework that maintains the stability of medicinal medicines (Danhier et al., 2012). Lipid-based nanocarriers, including liposomes and solid LNPs, provide biocompatible environments that preserve the integrity of encapsulated molecules during transport (Plaza-Oliver et al., 2021).

Nanocarriers can be engineered to release therapeutic drugs in reaction to particular physiological cues, including pH, temperature, or enzymatic activity, commonly found in injured tissues. These intelligent delivery devices guarantee accurate release at the designated spot, enhancing efficacy while reducing side effects (Cheng et al., 2013). Moreover, stimuli-responsive systems can adjust to the healing process by releasing bioactive molecules corresponding to various stages of tissue repair (Uskoković & Ghosh, 2016).

The integration of nanocarriers into tissue engineering scaffolds promotes both the delivery of medicinal medicines and the characteristics of the scaffolds. Nanocarriers can transport angiogenic substances like vascular endothelial growth factors (VEGFs), facilitating the development of blood vessels essential for tissue viability and integration (Richardson et al., 2001). Nanocarriers can enhance the mechanical properties of scaffolds, hence reinforcing the structural integrity essential for efficient tissue regeneration (Shi et al., 2010).

Nanocarriers delivering bone morphogenetic proteins (BMPs) have demonstrated the ability to induce osteogenesis in bone defects, thus facilitating bone repair and mineralization (Zhao et al., 2010). This method has been effective in enhancing bone tissue engineering results. Chitosan-based nanocarriers containing transforming growth factor beta (TGF-β) have been employed to promote chondrogenic differentiation of stem cells, essential for cartilage tissue regeneration. This application demonstrates potential in cartilage restoration methodologies (Kim et al., 2023). Nanocarriers administering epidermal growth factor (EGF) augment keratinocyte proliferation and migration, expediting wound closure and facilitating skin regeneration. These carriers offer substantial advantages for wound healing and skin tissue engineering (Boateng & Catanzano, 2015).

### 1.5.4 Biosensors and Bioelectronics

Nanocarriers are essential for improving the sensitivity and efficacy of biosensors, enabling the detection of biomolecules at extremely low concentrations. Their unique characteristics, including elevated surface area, adjustable surface chemistries, and quantum phenomena, markedly enhance biosensor efficacy, rendering them exceptionally successful in diverse biomedical applications.

Gold and silver nanoparticles are frequently employed for their plasmonic properties, which increase optical signals in methods such as surface-enhanced Raman scattering (SERS) and surface plasmon resonance (SPR). These effects facilitate the identification of biomolecules at femtomolar concentrations, significantly enhancing sensitivity (Anker et al., 2008). Quantum dots, semiconductor nanoparticles with adjustable fluorescence, improve fluorescence-based detection owing to their luminosity and photostability. Conjugation of quantum dots to biomolecules enhances test sensitivity, enabling the detection of low-abundance targets (Medintz et al., 2005).

Nanocarriers provide an extensive surface area conducive to the immobilization of biorecognition components, including antibodies and aptamers. The extensive surface area enhances the probability of interaction with target molecules, thereby augmenting the sensitivity of biosensors (Li et al., 2019). Nanostructured materials, such as carbon nanotubes and graphene, are used in electrodes to augment electron transfer rates and enhance the sensitivity of electrochemical biosensors (Wang & Musameh, 2004).

In field-effect transistor (FET) biosensors, nanocarriers such as silicon nanowires or carbon nanotubes function as the conductive channel. Biomolecule binding to the surface alters the electrical characteristics of nanowires or nanotubes, facilitating label-free detection of biomolecules at extremely low concentrations (Patolsky et al., 2006).

Magnetic nanoparticles can be modified with capture ligands to extract and concentrate target proteins from intricate materials. This magnetic separation improves detection limits by concentrating the target molecules (Gijs, 2004). Magneto-resistive sensors can subsequently identify these magnetic nanoparticles, hence enhancing sensitivity (Rarokar et al., 2024).

Nanozymes—nanoscale carriers exhibiting enzyme-like activity—catalyze processes that produce observable signals, enhancing responses and facilitating ultra-sensitive detection (Wu et al., 2018). Nanocarriers can induce cascade reactions, resulting in many phases of signal amplification that enhance biosensor performance (Shi et al., 2016).

Nanocarriers are utilized in Förster resonance energy transfer (FRET) devices. Nanocarriers function as donors or acceptors in these systems, influencing energy transfer efficiency in the presence of target biomolecules, resulting in quantifiable fluorescence alterations and facilitating sensitive detection (Clapp et al., 2005). Nanocarriers can be integrated into lab-on-a-chip devices, improving the interaction between targets and detecting elements. These microfluidic devices minimize the necessary sample amount and provide swift, sensitive detection of biomolecules, proving particularly advantageous in point-of-care diagnostics (Sachdeva et al., 2021).

Gold nanoparticles facilitate electron transfer in glucose oxidase-based sensors, allowing for the detection of glucose at minimal concentrations. These sensors are very beneficial in diabetes care, offering accurate and timely glucose level measurement (Wu et al., 2007). DNA detection quantum dot-labeled probes, in conjunction with nanostructured electrodes, facilitate the identification of specific DNA sequences at picomolar concentrations. This methodology is essential for genetic diagnostics and the identification of gene mutations (Driskell et al., 2011).

Magnetic nanoparticles are utilized to isolate and concentrate low-abundance proteins from intricate biological materials, including blood. This improves the sensitivity of immunoassays for the early identification of illness biomarkers, facilitating earlier diagnosis of several disorders (Rarokar et al., 2024).

## 1.6 ADVANCES AND LIMITATIONS OF NANOCARRIERS

Understanding the benefits and constraints of nanocarriers is crucial for enhancing their design and utilization in therapeutic delivery, diagnostics, and biomedical research.

### 1.6.1 Improved Bioavailability and Targeting

Nanocarriers markedly enhance the solubility and stability of therapeutic medicines, particularly hydrophobic pharmaceuticals that face challenges due to inadequate water solubility. Encapsulating these medications within the hydrophobic core of nanocarriers, such as LNPs or polymeric micelles, enhances their dispersibility in biological fluids (Kalepu & Nekkanti, 2015). This encapsulation safeguards pharmaceuticals against enzymatic destruction and extends their circulation duration, enhancing their therapeutic efficacy. Nanocarriers augment the pharmacokinetic characteristics of pharmaceuticals by stabilizing them in circulation and enhancing their distribution. Surface changes, including PEGylation (the incorporation of PEG), diminish immune system recognition and extend circulation duration (Blanco et al., 2015). The prolonged circulation allows nanocarriers additional time to reach target tissues, hence improving therapeutic efficacy.

- *Passive Targeting:* This exploits the EPR effect, wherein tumors, characterized by their compromised vasculature and inadequate lymphatic drainage, facilitate the accumulation of nanocarriers (Maeda et al., 2008).
- *Active Targeting:* Nanocarriers can be functionalized with ligands, such as antibodies or peptides, that specifically bind to receptors overexpressed on target cells. This approach enhances cellular absorption by target cells and minimizes off-target effects, thereby augmenting therapy efficacy (Prajapati et al., 2024).

The dimensions and surface characteristics of nanocarriers facilitate cellular absorption, frequently via endocytosis. This is essential for administering medications that must function within cells, such as those aimed at cancer or viral infections (Oh & Park, 2009). Nanocarriers can surmount biological obstacles that restrict medicine delivery, including the BBB. Nanocarriers can traverse the BBB to administer medicines to the central nervous system by employing receptor-mediated transcytosis or by temporarily disrupting tight junctions between endothelial cells (Saraiva et al., 2016). Nanocarriers can be designed for controlled release, either progressively or in reaction to specific stimuli (e.g., pH, temperature, or enzyme presence). This guarantees that therapeutic concentrations are sustained at optimal levels for extended periods, hence decreasing dosage frequency and mitigating side effects (Kamaly et al., 2016).

Some nanocarriers, especially those composed of non-biodegradable materials, may induce chronic toxicity. Accumulation in organs, including the liver and kidneys, may result in inflammation or injury (Xuan et al., 2023). Comprehensive safety assessments are crucial to alleviating these hazards during the development phase. While PEGylation can diminish immune recognition, the recurrent application of nanocarriers may still elicit immunological responses, resulting in the generation of anti-drug antibodies (ADAs). These ADAs can neutralize the therapeutic drug, thus diminishing its efficacy, and may potentially induce hypersensitivity reactions (Talamini et al., 2021).

Attaining elevated drug loading and encapsulation efficiency continues to be a challenge, particularly with big or hydrophobic compounds. Insufficient drug-loading capacity may restrict treatment efficacy, necessitating more modification of nanocarrier designs to enhance delivery (Chehelgerdi et al., 2023). The fabrication of nanocarriers can be costly and intricate, especially when employing sophisticated materials or surface alterations. Enhancing production capacity while preserving quality poses a hurdle for clinical translation. Furthermore, adhering to rigorous regulatory standards can impede the development and approval timeline (Al-Thani et al., 2024). Certain nanocarriers may exhibit instability under specific conditions, resulting in aggregation or premature payload release. This diminishes their shelf life and necessitates specific storage conditions, complicating general utilization (Su & Kang, 2020).

### 1.6.2 Controlled Release and Sustained Effect

Nanocarriers have emerged as fundamental components in contemporary treatments owing to their capacity to regulate drug release, sustain therapeutic concentrations over time, diminish the necessity for frequent dosing, and mitigate adverse effects. Enhancing treatment outcomes depends on the meticulous regulation of drug release kinetics, an essential element in various medicinal applications.

The design procedure of nanocarriers with tailored release kinetics commences with the selection of materials. Biodegradable polymers such as PLGA, polycaprolactone (PCL), and chitosan enable precise modulation of drug release through the alteration of degradation rates. The degradation rate of the polymer is influenced by factors including molecular weight, monomer makeup, and crystallinity, which subsequently impact the drug release profile, whether rapid or prolonged (Danhier et al., 2012). Lipid-derived materials, including solid LNPs and lipid nanocapsules, present an alternative solution. Modifying the lipid content and crystallinity of these carriers enables a regulated and steady release, especially advantageous for hydrophobic pharmaceuticals (Mehnert & Mäder, 2012).

Surface modification is equally significant in controlling drug release. Nanocarriers coated with hydrophilic polymers such as PEG attenuate drug transport, leading to prolonged therapeutic effects (Gref et al., 1994). Furthermore, the integration of stimuli-responsive polymers provides enhanced control. These polymers respond to environmental alterations, including pH, temperature, or enzymatic activity. pH-sensitive polymers release medications in acidic environments, such as tumors, facilitating targeted and regulated administration (Koo et al., 2011). The nanocarrier structure is a crucial element. Creating carriers with a core–shell structure, wherein the outer shell modulates drug diffusion and safeguards the inner core, facilitates meticulous control over the timing and location of drug release (Xu et al., 2015). Adjusting the porosity of nanocarriers, such as mesoporous silica nanoparticles, enables precise modulation of drug-loading capacity and release rates through alterations in pore size, volume, and surface area (Slowing et al., 2008).

The interplay between pharmaceuticals and polymers is essential for attaining the intended release profile. Pharmaceuticals can be covalently attached to the polymer matrix via degradable linkages, such as ester or amide bonds, facilitating a controlled release as these bonds progressively degrade (Cheng et al., 2023). Similarly, ionic interactions between the pharmaceutical agent and the carrier material, exemplified by polyelectrolyte complexes, facilitate the precise modulation of drug release rates through the alteration of contact strength (Li et al., 2001). Diverse encapsulation methods augment control over medication release. LbL construction, in which nanocarriers are fabricated by successively depositing oppositely charged polymers, facilitates meticulous regulation of the carrier's thickness and the timing of drug release (De Geest et al., 2008). Altering factors in emulsion-based synthesis techniques, such as solvent evaporation rates or emulsifier concentrations, can impact the size and drug distribution within nanocarriers, therefore affecting drug release (Bouchemal et al., 2004).

For intricate treatment approaches, nanocarriers can be designed to administer several pharmaceuticals. Sequential release systems are engineered to deliver medications at variable rates, typically by including layers within the carrier or employing materials that degrade at distinct velocities (Wang & Kohane, 2017). Co-delivery systems enhance this concept by facilitating the simultaneous or regulated release of many therapeutic agents, including medicines and genes, which is particularly advantageous for combination cancer therapy (Aborode et al., 2024). The release of drugs driven by external stimuli presents a more novel method of medication delivery. Nanocarriers containing magnetic nanoparticles can be remotely triggered to release their contents upon exposure to magnetic fields, allowing for precise control over the timing and location of medication delivery (Liu et al., 2019). Nanocarriers that react to ultrasound or light provide non-invasive and precisely regulated drug delivery (Kaushik et al., 2024).

The potential uses of nanocarriers are extensive. In cancer therapy, controlled release carriers facilitate the prolonged delivery of chemotherapeutic drugs directly to tumors, therefore reducing systemic toxicity while preserving effective therapeutic concentrations at the target location (Shi et al., 2016). Sustained drug release systems for chronic diseases, such as diabetes or hypertension, diminish the necessity for frequent dosage, hence enhancing patient compliance and treatment outcomes (Singh et al., 2019). Nanocarriers that release antigens in a regulated fashion can replicate the normal course of diseases, resulting in augmented immune responses and increased vaccine efficacy (Lozano et al., 2023).

Nonetheless, numerous obstacles persist. Increasing the production of nanocarriers while ensuring uniform release profiles is challenging due to inconsistencies in synthesis methods. Ensuring consistency across production batches is crucial for clinical application. Stability and storage provide considerable issues. Preserving the structural integrity of nanocarriers during storage while ensuring their release qualities necessitates meticulous regulation of environmental parameters such as temperature and humidity. Ultimately, compliance with regulatory criteria for nanocarrier-based therapeutics necessitates comprehensive safety and efficacy assessments, emphasizing constant quality control and reproducibility.

### 1.6.3 Scalability and Manufacturing Challenges

Increasing the production of nanocarriers while ensuring consistent quality is a significant problem as these technologies transition from laboratory research to clinical applications. With the rising demand for nanocarrier-based therapeutics, it is crucial to tackle the intricacies of large-scale manufacturing to guarantee their effective incorporation into clinical applications.

A primary problem in this transformation is attaining uniformity between batches. Fluctuations in essential factors, including size, shape, surface charge, and drug loading, can profoundly affect biological performance and therapeutic efficacy. Maintaining stringent control over these features during extensive production is challenging, frequently resulting in discrepancies that undermine the safety and efficacy of the final product (Ventola, 2017). A significant challenge is the scalability of manufacturing processes. Methods refined for small-scale laboratory manufacturing frequently do not directly apply to industrial sizes. Alterations in parameters such as mixing efficiency, heat transfer, and reaction kinetics during scale-up might modify the characteristics of nanocarriers, necessitating process redesign to maintain consistent product quality (Fernandes et al., 2023a).

Furthermore, the synthesis of nanocarriers entails intricate procedures such as nanoprecipitation, emulsification, and self-assembly, each necessitating meticulous regulation of various parameters. Consistently replicating these conditions on a wider scale poses considerable technological difficulties, frequently requiring specialized apparatus to sustain the precision attained in smaller batches (Mahajan & Bhattacharya, 2024). Quality control and characterization introduce an additional layer of complications. Although comprehensive characterization is essential for guaranteeing the safety and efficacy of nanocarriers, numerous analytical techniques employed in research environments are too time-consuming or expensive for industrial-scale manufacturing. Establishing scalable, swift, and dependable analytical procedures are crucial for preserving quality while ensuring efficiency (Gabor, 2011).

Regulatory compliance poses an additional barrier. Nanocarriers must adhere to rigorous regulatory standards that necessitate comprehensive documentation of production procedures, quality control protocols, and product specifications. Increasing output while complying with these criteria requires substantial expenditure in regulatory affairs and quality assurance (Anselmo & Mitragotri, 2019). Stability challenges, such as aggregation, degradation, and alterations in physicochemical properties with time, further hinder the large-scale manufacture of nanocarriers. Maintaining long-term stability during storage and distribution is essential for therapeutic applications, although it poses a considerable challenge that must be overcome to preserve product quality over time (Danaei et al., 2018).

The production expense becomes a significant challenge. The materials and techniques required for nanocarrier manufacture can be costly, and it is essential to balance cost with quality and efficacy to enhance the accessibility of nanocarrier-based therapeutics (Mitchell et al., 2020). Various solutions can be implemented to address these difficulties. Process optimization and standardization, via Quality by Design (QbD) and Good Manufacturing Practice (GMP), can delineate important parameters and institute consistent methods. Automation and sophisticated control systems can improve reproducibility and diminish human error (Ali et al., 2022).

Moreover, continuous manufacturing technologies, including microfluidics, present a viable option. Continuous flow techniques enhance control over the synthesis of nanocarriers, promoting scalable manufacturing and assuring uniformity. These approaches may also lower expenses relative to conventional batch production (Raval et al., 2019). Advanced analytical methods such as dynamic light scattering (DLS), nanoparticle tracking analysis (NTA), and high-performance liquid chromatography (HPLC) can be modified for high-throughput quality control, guaranteeing that goods comply with requisite requirements at industrial scales (Kaushik et al., 2023).

Enhancing the stability of nanocarriers via surface changes, lyophilization, and the incorporation of stabilizers is essential for maintaining long-term product quality. Excipients that inhibit degradation can further improve stability during storage (Kim et al., 2022). Engaging with regulatory agencies early in the development phase helps facilitate the pathway to market approval. Interacting with authorities guarantees that production procedures conform to regulatory standards, hence minimizing delays or impediments in the approval process (Anselmo & Mitragotri, 2019).

Strategies for cost reduction are of similar significance. Acquiring more economical raw materials, enhancing production methods, and increasing yields can reduce production costs, hence rendering nanocarrier-based therapies more cost-effective. As production increases, economies of scale are expected to decrease per-unit costs (Mitchell et al., 2020). Ultimately, investments in infrastructure and personnel training are essential for upholding superior manufacturing standards. Specialized facilities and proficient workers are crucial for maintaining consistent quality in large-scale nanocarrier production (Buya et al., 2024).

In summary, although the production of large-scale nanocarriers has certain problems, these can be mitigated through process optimization, new technologies, enhancements in quality control, regulatory coordination, and cost-effective solutions. With enough infrastructure and experience, it is feasible to scale production to satisfy clinical requirements.

## 1.7 REGULATORY CONSIDERATIONS

Understanding the regulatory framework is crucial for advancing nanocarriers from research to clinical use. Regulatory agencies are increasingly concentrating on the distinct characteristics of nanoparticles to guarantee the safety and efficacy of nanocarrier-based medicines. As nanomedicine progresses, regulatory frameworks pertaining to nanomaterials are adapting to confront the issues presented by these novel therapeutic systems.

### 1.7.1 Present Regulatory Framework for Nanomedicines

The regulatory framework for nanomedicines is influenced by the necessity to assess the unique properties of nanocarriers, encompassing their nanoscale dimensions, surface chemistry, and interactions with biological systems. These aspects affect the classification, assessment, and eventual approval of nanocarrier-based medicines.

Regulatory mandates profoundly influence the development and approval process for nanocarrier-based medicines, guaranteeing their safety, efficacy, and quality. These regulations affect multiple critical facets of growth. Regulatory agencies, such as the US FDA and the European Medicines Agency (EMA), have defined nanoparticles, influencing regulatory processes. The FDA characterizes nanomaterials as compounds with at least one dimension ranging from 1 to 100 nm (Food and Drug Administration, 2014). This classification imposes additional scrutiny and special regulatory guidelines on nanomedicines, affecting the evaluation of therapies for safety and efficacy.

Regulatory agencies necessitate comprehensive safety and efficacy assessments due to the distinctive characteristics of nanocarriers, including their augmented surface area, improved permeability, and potential for innovative biological interactions. Thorough preclinical investigations are essential to evaluate pharmacokinetics, biodistribution, toxicity, and interactions with biological systems (European Medicines Agency, 2011). Nanocarriers may pose unrecognized dangers, rendering comprehensive clinical testing essential. Regulatory bodies evaluate these risks using comprehensive clinical data to guarantee that nanocarriers do not present unexpected safety concerns.

Maintaining uniform quality in the production of nanocarriers is a significant regulatory issue. Agencies mandate compliance with GMP, ensuring stringent control over essential parameters influencing nanocarrier characteristics—such as size, surface charge, and drug encapsulation (Food and Drug Administration, 2017). Variability in these parameters may result in inconsistencies in efficacy and safety, rendering stringent quality control and manufacturing standards essential for clinical application.

Regulatory authorities, such as the FDA, have published guidance materials to elucidate regulatory expectations for developers. The FDA's advice on "Drug Products, Including Biological Products, that Contain Nanomaterials" delineates suggestions for the development, characterization, and assessment of nanomedicines (Food and Drug Administration, 2017). This guidance establishes a framework for evidencing safety, efficacy, and quality, influencing the entire development process from preclinical testing to clinical trials and quality assurance.

Nanocarrier-based therapies frequently exhibit attributes of both pharmaceuticals and devices, resulting in intricate regulatory processes. The dual nature of these goods can complicate the determination of their regulatory classification—whether as a medicine, biologic, or device. This complexity may prolong the evaluation process and necessitate further discussions (Bawa, 2011). Effectively traversing these routes necessitates a comprehensive comprehension of the various regulatory frameworks pertinent to the multifunctional characteristics of nanocarriers.

Developers aiming to launch nanocarrier-based medicines in worldwide marketplaces must adhere to diverse regulatory restrictions. Variations in regulatory mandates across areas, including the USA, Europe, and Asia, provide problems for achieving cross-border compliance (Tinkle et al., 2014). These variances might hinder the development process as companies must comply with the varied safety, efficacy, and quality criteria of each location. Attaining worldwide regulatory harmonization is a protracted goal; however, in the interim, developers must customize their tactics to conform to the regulatory frameworks of each region to prevent delays in the approval process.

### 1.7.2 Safety Assessment of Nanocarriers

Evaluating the safety of nanocarriers necessitates customized techniques that account for their distinct physicochemical characteristics and possible long-term impacts on human health and the environment. Conventional toxicological approaches may inadequately address the intricacies of nanocarriers, requiring a thorough strategy to guarantee their safety and obtain regulatory approval. A thorough safety evaluation of nanocarriers must assess their interactions with biological systems, long-term effects, and environmental consequences. These evaluations must be customized to the unique characteristics of nanocarriers and encompass numerous essential elements.

The initial phase of safety assessment involves the comprehensive physicochemical characterization of nanocarriers. Essential aspects like size, shape, surface charge, aggregation state, and chemical composition must be meticulously analyzed, as these attributes affect the behavior of nanocarriers in biological systems (Guidance on Sample Preparation and Dosimetry for the Safety Testing of Manufactured Nanomaterials, 2012). For instance, size influences biodistribution and cellular absorption, whereas surface charge affects interactions with cell membranes and proteins. Comprehending these features is essential for forecasting the safety and effectiveness of nanocarriers.

*In vitro* toxicity studies serve as preliminary assessments to detect potential cytotoxicity, genotoxicity, and immunotoxicity. These investigations evaluate the interactions of nanocarriers with cells, proteins, and other biological molecules, providing preliminary markers of toxicity. Moreover, *in vitro* testing clarifies processes of cellular damage, including oxidative stress and membrane disruption, which can guide subsequent *in vivo* studies (Cevaal et al., 2022). These tests are crucial for the early identification of risk prior to advancing to animal models.

Animal studies are essential for evaluating the biodistribution, pharmacokinetics, and toxicity of nanocarriers *in vivo*. Both acute and chronic toxicity should be assessed, emphasizing primary organs including the liver, kidneys, lungs, and brain (Fadeel et al., 2018). These investigations offer insights into the metabolism, clearance, and possible accumulation of nanocarriers, which are essential for assessing their long-term safety in humans. The diminutive size of nanocarriers may result in tissue buildup over time, potentially causing long-term detrimental effects. Long-term exposure studies are essential to evaluate the persistence of nanocarriers in the body and the potential for delayed harm resulting from chronic accumulation (Oberdörster et al., 2005). These investigations are particularly significant for nanocarriers employed in chronic therapy or those capable of traversing biological barriers, such as the BBB.

Assessing the environmental impact of nanocarriers is essential, as they may infiltrate ecosystems during production, utilization, or disposal. Environmental assessments evaluate the degradability and ecological dangers of nanocarriers to confirm that they do not endanger the environment (Kah et al., 2019). These studies investigate the behavior of nanocarriers in aquatic, terrestrial, and atmospheric environments, along with their potential for bioaccumulation in species, while guaranteeing adherence to environmental safety regulations.

Nanocarriers can provoke immunological responses, rendering immunotoxicity studies a critical component of safety reviews. These evaluations encompass tests for complement activation, cytokine release, and hypersensitive reactions, aiding in the identification of unwanted immunological responses (Dobrovolskaia & McNeil, 2013). This testing is especially pertinent for nanocarriers that utilize materials or surface modifications potentially identified as foreign by the immune system.

Studies on reproductive and developmental toxicity are necessary to ascertain if nanocarriers adversely affect fertility or induce developmental anomalies. These studies are essential for nanocarriers employed in therapies for pregnant women or in prolonged treatments (Hardy et al., 2018). Assessing the potential for teratogenic effects and ensuring that nanocarriers do not disrupt reproductive processes are essential factors for regulatory approval.

To guarantee reproducibility and streamline regulatory approval, it is imperative to implement standardized and verified testing methodologies. International entities, such as the Organization for Economic Co-Operation and Development (OECD), have established protocols expressly for the assessment of nanomaterials (Guidance on Sample Preparation and Dosimetry for the Safety Testing of Manufactured Nanomaterials, 2012). These methodologies provide uniform safety assessments across diverse nanocarrier formulations, guaranteeing dependable and similar outcomes.

A thorough risk assessment methodology is essential for the safety evaluations of nanocarriers. This methodology integrates hazard identification, dose-response assessment, exposure evaluation, and risk characterization to deliver a comprehensive understanding of the possible dangers associated with nanocarriers. Risk management techniques are formulated to mitigate recognized concerns, hence enabling the safe application of nanocarriers in clinical settings (Shatkin & Kim, 2015).

### 1.7.3 Ethical Considerations in Nanomedicine

Ethical considerations in nanomedicine, particularly regarding the development and use of nanocarrier technologies, are crucial for ensuring responsible and equitable progress. Critical concerns, including informed consent, equitable access, long-term societal ramifications, and regulatory monitoring, must be addressed to foster trust in nanomedicine and ensure its advantages are broadly disseminated across society. Comprehensive frameworks are essential to address the ethical concerns related to nanocarrier technologies, encompassing concepts such as patient autonomy, equity, beneficence, and justice. These frameworks ought to promote responsible innovation and societal involvement.

An ethical framework for nanomedicine must emphasize informed consent and maintain patient autonomy. Clear communication is crucial, as patients must be thoroughly educated about the attributes of nanocarrier-based therapy, encompassing the associated risks, advantages, and uncertainties (Lin & Allhoff, 2008). Due to the intricacies of nanotechnology, healthcare practitioners must guarantee that patients fully comprehend their therapy to make educated and independent decisions (Wasti et al., 2023).

Fair access to nanocarrier medicines is an essential element of any ethical framework. Access to nanomedicine may disproportionately advantage affluent individuals or nations, thereby intensifying healthcare disparities (Resnik & Tinkle, 2007). Ethical frameworks must facilitate equitable resource distribution, ensuring that marginalized populations are not overlooked. Internationally, these frameworks must also tackle access inequalities between rich and poor nations, promoting initiatives to guarantee global accessibility (Petersen et al., 2007).

Beneficence and nonmaleficence are essential ethical concepts in the utilization of nanocarrier technologies. Ethical frameworks must mandate thorough risk–benefit analyses to weigh potential advantages against hazards, ensuring that patient welfare is prioritized (Wasti et al., 2023). Furthermore, prolonged safety surveillance is crucial, as the impacts of nanocarriers may not be immediately apparent. Continuous supervision is required to safeguard patient health over time (Resnik & Tinkle, 2007).

Justice regarding nanocarrier technology entails guaranteeing equitable access and instituting stringent regulatory frameworks. Ethical frameworks must advocate for rigorous regulatory monitoring to protect public health (Mnyusiwalla et al., 2003). Moreover, societal engagement is vital. Engaging the public in debates about the consequences of nanomedicine cultivates trust and aligns scientific progress with societal values (Pidgeon & Rogers-Hayden, 2007).

The implementation of nanocarrier technologies may need the collection and analysis of sensitive patient information. Ethical frameworks must safeguard privacy and confidentiality, complying with data protection rules such as General Data Protection Regulation (GDPR) and Health Insurance Portability and Accountability Act (HIPAA) (Miller & Tucker, 2014). Patients should be certain that their health information is safeguarded and that nanocarriers do not infringe upon their privacy.

Ethical considerations pertain to the research and development of nanocarrier technology. Responsible innovation should direct research, ensuring transparency in reporting and a dedication to preventing conflicts of interest (Poel van de & Royakkers, 2011). Animal utilization in preclinical investigations must adhere to the 3Rs principles—Replacement, Reduction, and Refinement—to mitigate injury and guarantee ethical treatment (Ferdowsian & Beck, 2011). Ethical norms in research are fundamental to fostering trust and honesty in the development of nanomedicine.

Global cooperation is essential to align ethical norms and regulatory frameworks for nanomedicine. As nanocarrier technologies advance globally, ethical frameworks must encourage international collaboration to establish uniform regulatory processes, thereby facilitating the development of safe and effective nanomedicine worldwide (Hunt & Mehta, 2013).

## 1.8 FUTURE PERSPECTIVES

### 1.8.1 Emerging Trends in Nanocarrier Design

The domain of nanocarriers is advancing swiftly, with novel technologies poised to surmount existing constraints and transform therapeutic applications. Advancements in material science, biotechnology, and interdisciplinary collaboration are facilitating the creation of more advanced, efficient, and precise delivery systems. These developments are expected to enhance the efficacy, safety, and adaptability of nanocarriers in medicine, hence generating new opportunities for the treatment of complicated diseases.

Numerous advanced design methodologies demonstrate considerable potential in improving the accuracy, flexibility, and efficacy of nanocarriers for drug administration. Stimuli-responsive nanocarriers, referred to as "smart" nanocarriers, are designed to release therapeutic chemicals in reaction to particular internal or external stimuli. These stimuli may encompass alterations in pH, redox gradients, enzymatic activity, temperature, light, or magnetic fields (Mura et al., 2013). In oncological therapy, pH-responsive systems have garnered interest owing to the acidic milieu characteristic of neoplastic tissues. These nanocarriers can selectively discharge their therapeutic payload at the tumor location, improving therapy efficacy while minimizing damage to healthy tissues (Sun et al., 2017). Enzyme-responsive nanocarriers utilize the overexpression of certain enzymes in pathological tissues to trigger medication release, thus enhancing targeted therapy (Hu et al., 2014).

Nanocarriers activated by external stimuli, such as light or magnetic fields, provide enhanced precision. These devices provide geographic and temporal manipulation of medication delivery through external controls, permitting non-invasive therapy of localized disorders. Light-activated carriers utilize photosensitive polymers to release pharmaceuticals upon exposure to light, rendering them appropriate for targeted therapy (Dai et al., 2013). Magnetic nanocarriers, containing magnetic nanoparticles, can be maneuvered and heated by magnetic fields to accurately deliver medications (Busquets et al., 2015).

Multifunctional nanocarriers are engineered to execute various functions—such as targeted distribution, diagnostics, and therapy—within a singular platform. These carriers are pivotal to theranostics, a discipline that integrates therapeutic and diagnostic activities (Kashyap et al., 2023). The capacity to incorporate real-time surveillance of drug administration and therapeutic outcomes is a significant benefit of multifunctional carriers, facilitating tailored therapies (Lim et al., 2014). Moreover, co-delivery systems that carry various therapeutic agents, including pharmaceuticals and gene treatments, can improve the treatment of complicated diseases such as cancer by leveraging synergistic effects (Hopkins et al., 2022).

Gold nanoparticles serve as multifunctional nanocarriers, functioning simultaneously as imaging agents and drug carriers, enhancing imaging quality and facilitating therapeutic delivery (Amaral et al., 2024). Furthermore, polymeric nanocarriers modified with targeting ligands can improve delivery specificity by attaching to cell receptors and transporting imaging chemicals for diagnostic applications (Bao et al., 2013). Biomimetic nanocarriers are engineered to replicate natural biological structures or processes, frequently by enveloping nanoparticles in cell membranes to evade immune recognition. These carriers utilize biological features to enhance therapeutic efficacy and safety (Fang et al., 2018).

Nanocarriers enveloped in red blood cell membranes can circumvent the immune system, prolonging their circulation duration and diminishing the probability of immunological reactions (Lotfabadi et al., 2024). Moreover, biomimetic carriers can acquire the targeting characteristics of the source cells, hence improving the specificity of medication delivery. Cancer cell membrane-coated nanoparticles can utilize homologous targeting to selectively return to and deliver medications to the original cancer cells (Chen et al., 2016). Instances of this biomimetic strategy encompass nanoparticles coated with red blood cell membranes, which merge the biocompatibility of red blood cells with the efficacy of synthetic nanocarriers, thus enhancing medication administration (Hu et al., 2011). Cancer cell membrane-coated nanoparticles utilize biological targeting mechanisms to improve therapy accuracy (Zhang et al., 2018).

Nanorobots exemplify an advanced methodology in the design of nanocarriers. These nanoscale devices possess the ability for autonomous movement and can traverse intricate biological environments to deliver treatments with unparalleled precision (Soto et al., 2020). Nanorobots possess the capability to transform targeted medicine delivery by accurately identifying sick cells and reducing off-target effects. These devices can be designed to react to certain biomarkers or environmental cues, administering medications alone at the designated locations. Externally guided active targeting technologies, such as magnetically driven nanorobots, enhance therapy precision by directing them to sick regions (Li et al., 2017). Enzyme-driven nanomotors are an innovation that uses biological enzymes for propulsion, enabling nanocarriers to traverse more efficiently toward designated targets. This method has potential for enhancing medication delivery in difficult settings, such as solid tumors (Xu et al., 2020).

### 1.8.2 Integration with Other Technologies

The amalgamation of nanocarriers with advanced technologies like artificial intelligence (AI) and 3D printing is set to transform their development and utilization in healthcare. Utilizing the analytical prowess of AI alongside the customization possibilities of 3D printing, researchers can markedly improve the design, optimization, and implementation of nanocarrier-based therapies, facilitating highly tailored and successful treatments.

The integration of AI and 3D printing with nanocarriers presents multiple benefits, fostering advancements in personalized medicine, precision manufacturing, and on-demand therapeutic solutions. AI enhances data-driven optimization in the design and development of nanocarriers. AI greatly improves the efficiency and accuracy of these systems using machine learning algorithms and predictive models. AI systems analyze extensive datasets concerning the characteristics and behaviors of nanocarriers. Employing machine learning models, AI can forecast interactions between nanocarriers and biological systems, facilitating the advancement of more efficient and targeted medication delivery systems (Hamilton & Kingston, 2023). AI-driven predictive modeling enhances targeted efficiency by anticipating interactions with particular cell types, reducing side effects, and improving therapeutic precision (Tan et al., 2023).

AI can enhance HTS of nanocarrier formulations, swiftly finding the most suitable candidates for subsequent development (Serov & Vinogradov, 2022). This methodology decreases the time and expenses linked to conventional experimental techniques. Furthermore, AI can evaluate patient-specific data, customizing nanocarrier-based therapeutics for particular patients, hence promoting customized medicine (Kim et al., 2020). AI-driven methodologies enhance therapeutic outcomes by aligning the optimal nanocarrier system with a patient's genetic, physiological, and pathological characteristics.

3D printing facilitates the development of highly tailored nanocarrier-based drug delivery systems, providing enhanced control over design, composition, and drug release characteristics. This approach could transform the manufacture and utilization of nanocarriers in therapeutic environments. 3D printing offers unmatched accuracy in the fabrication of nanocarrier-loaded devices, enabling the creation of drug delivery systems with meticulously regulated geometry and compositions. This modification facilitates the creation of individualized dosage regimens and drug release profiles adapted to the specific requirements of each patient (Goole & Amighi, 2016). Furthermore, 3D printing facilitates the fabrication of intricate structures that integrate nanocarriers, applicable in tissue engineering and regenerative medicine (Murphy & Atala, 2014).

A primary advantage of 3D printing is its capacity to swiftly prototype nanocarrier systems, hence expediting research and development processes. This capacity enables researchers to rapidly iterate designs, optimizing formulations within a reduced timeframe (Lim et al., 2018). Furthermore, 3D printing facilitates point-of-care manufacturing, allowing for the on-site production of nanocarrier-based medicines, therefore mitigating storage and transit challenges and enhancing patient accessibility (Awad et al., 2019).

The amalgamation of AI and 3D printing cultivates a synergistic framework for enhancing the design, production, and utilization of nanocarriers. This integration facilitates adaptive, data-informed production processes that improve both accuracy and customization. AI-driven 3D printing facilitates real-time optimization of printing parameters and material choices, hence improving the efficacy of nanocarrier systems. Integrating AI algorithms into the 3D printing process enables researchers to forecast and modify variables like material qualities, layer thickness, and drug loading, so ensuring that the final product adheres to precise design parameters (Trenfield et al., 2018). This adaptive manufacturing method enhances consistency and quality, enabling the creation of high-performance nanocarriers for therapeutic applications. The integration of AI's data analysis capabilities with the customization potential of 3D printing enables the creation of patient-specific therapeutic solutions. Patient imaging data can be utilized to create 3D-printed implants or devices embedded with nanocarriers, customized to a patient's specific anatomical and clinical circumstances (Ozbolat & Yu, 2013). Moreover, the incorporation of sensors into these 3D-printed devices facilitates the development of intelligent, adaptive drug delivery systems. These devices can modulate drug release in real-time according to physiological feedback, providing a highly individualized and adaptive therapeutic strategy (Mandrycky et al., 2015).

### 1.8.3 Challenges and Opportunities in Clinical Translation

Connecting laboratory research with clinical practice poses hurdles and substantial prospects for nanocarrier technologies. The intricacy of these systems, regulatory challenges, biocompatibility issues, and financial impediments are significant constraints. Nonetheless, numerous ways can enhance their effective implementation in clinical environments. The therapeutic application of nanocarriers necessitates surmounting technical, regulatory, and economic obstacles. These challenges can be mitigated by interdisciplinary collaboration, regulatory involvement, sophisticated manufacturing techniques, and efficient stakeholder communication.

Nanocarriers exhibit considerable complexity; increasing their production while ensuring consistency presents substantial problems. The synthesis methods employed—namely, nanoprecipitation, emulsification, and self-assembly—render the attainment of batch-to-batch consistency challenging on an industrial scale (Fernandes et al., 2023b). Moreover, the characterization of their physicochemical features, such as size, shape, surface charge, and drug loading, is crucial for regulatory approval; however, it is a labor-intensive endeavor (Faria et al., 2018).

A significant impediment in clinical translation is the absence of defined regulatory frameworks for nanomedicine. The regulatory frameworks for nanocarriers are still developing, resulting in uncertainty in the approval process (Tinkle et al., 2014). Establishing long-term safety and efficacy is particularly difficult, as nanocarriers display distinct behaviors in biological systems that remain inadequately comprehended (Souto et al., 2024). This necessitates extensive preclinical and clinical investigations that adhere to rigorous safety and efficacy criteria.

Nanocarriers pose concerns including toxicity, immunogenicity, and off-target effects. These systems can elicit immunological responses, accumulate in tissues, or interact unpredictably with biological systems (Fadeel et al., 2018). Managing biodistribution and clearance presents a significant problem, as it is crucial to ensure that nanocarriers effectively reach their targets and are properly cleared from the body to ensure patient safety (Blanco et al., 2015).

Elevated development expenses represent a substantial impediment to the commercialization of nanocarrier technologies. The financial commitment necessary for research, development, and clinical trials may dissuade corporations from exploring these medicines (Etheridge et al., 2012). Moreover, it is essential to exhibit distinct advantages over current medicines to secure acceptability from healthcare professionals and patients, highlighting the necessity for cost-effectiveness and enhanced efficacy.

Interdisciplinary collaboration is fundamental for tackling the complexities of nanocarrier systems. Chemists, biologists, engineers, doctors, and regulatory specialists must collaborate to address technical, biological, and regulatory difficulties (Souto et al., 2024). Public–private collaborations can offer resources and enable the conversion of research into commercially viable goods. Establishing standardized techniques for the characterization and testing of nanocarriers can facilitate regulatory approval processes. This entails the formulation of explicit standards for safety, efficacy, and quality evaluations (Rauscher et al., 2019). Early and continuous engagement with regulatory bodies during the development phase ensures that nanocarrier development conforms to regulatory requirements, hence minimizing the possibility of approval delays (Anselmo & Mitragotri, 2019).

Utilizing sophisticated manufacturing technologies, including continuous manufacturing and microfluidic systems, can enhance the scalability and uniformity of nanoparticle carrier production (Shen et al., 2023). Continuous manufacturing enhances control over production processes, guaranteeing uniform product quality. Stringent quality control methods are essential to comply with regulatory criteria. Extensive safety evaluations, encompassing *in vitro* and *in vivo* research, are essential for the effective translation of nanocarriers. These investigations must assess toxicity, biodistribution, and immunological responses (Fadeel et al., 2018). Meticulously planned clinical trials are vital to establish the safety and efficacy of nanocarriers, which is requisite for regulatory approval and patient safety.

Engaging and instructing stakeholders—healthcare professionals, patients, and policymakers—is essential for promoting the acceptability of nanocarrier technology. Transparent communication about the advantages and disadvantages of nanocarriers fosters trust and facilitates informed decision-making (Etheridge et al., 2012). It is imperative to minimize the expenses related to nanocarrier development. Enhancing production processes to reduce waste and increase cost efficiency might decrease total development costs. Acquiring money via grants, investments, and governmental incentives can facilitate the advancement of nanocarrier-based therapeutics.

Notwithstanding the hurdles, substantial prospects exist for the clinical translation of nanocarrier technologies. One of the most potential uses is personalized medicine, wherein nanocarriers can be customized to administer medications according to unique patient profiles, hence enhancing therapeutic outcomes. Nanocarriers facilitate enhanced targeting and regulated drug release, hence minimizing negative effects and augmenting efficacy. Moreover, nanocarriers present novel therapeutic approaches, including gene therapy and immunotherapy, offering alternative solutions for conditions that are challenging to manage with traditional treatments.

## 1.9 CONCLUSIONS

Nanocarriers represent a significant advancement in the field of therapeutic nucleic acid and protein delivery. These systems offer highly controlled and targeted delivery mechanisms, overcoming many of the inherent challenges associated with traditional therapeutic approaches. Their ability to encapsulate and protect unstable molecules, such as mRNA, siRNA, and therapeutic proteins, while facilitating targeted delivery, underscores their transformative potential in contemporary medicine.

Nanocarriers enhance therapeutic efficacy through the optimization of several factors, including size, shape, surface charge, and functionalization. Such design parameters enable nanocarriers to penetrate biological barriers, evade immune system detection, and deliver therapeutic agents directly to target tissues. This minimizes off-target effects and improves bioavailability. Furthermore, advancements in nanotechnology have led to the development of stimuli-responsive nanocarrier systems, which release therapeutic agents in response to specific physiological conditions, thereby further enhancing treatment outcomes.

In the context of personalized medicine, nanocarriers hold the potential to revolutionize patient-specific therapies. By incorporating targeting ligands and customizable surface chemistries, these systems enable precise delivery of therapeutics based on an individual's molecular profile. This specificity has the potential to reduce systemic toxicity and improve patient outcomes by tailoring treatments to the unique needs of each patient.

The clinical success of LNPs in mRNA vaccines, particularly during the COVID-19 pandemic, has demonstrated the viability of nanocarriers in achieving rapid global deployment and high efficacy. As drug delivery systems continue to be optimized and efforts to minimize side effects intensify, nanocarriers are poised to play a pivotal role in the future of biomedicine. Their scalability, safety, and therapeutic efficacy underline their importance in the advancement of gene and protein therapies, as well as their broader applications in areas such as tissue engineering and molecular diagnostics.

Continued research and innovation in the design of nanocarriers, particularly with regard to hybrid systems and biocompatible materials, are opening new opportunities for addressing previously untreatable diseases. As technological progress continues, the integration of nanocarriers with other advanced biomedical platforms may accelerate the development of therapies that are not only more effective but also less invasive, reshaping the future of healthcare.

The next frontier in nanocarrier technology will likely focus on improving biocompatibility, enhancing targeting specificity, and reducing potential immunogenicity. Advances in bioengineering and materials science will be essential to optimizing these systems for widespread clinical use. Future research will need to explore how nanocarrier technologies can be further refined to improve patient-specific therapies, lower treatment costs, and enhance the overall efficacy of medical interventions across various fields.

## REFERENCES

Aborode, A. T., Oluwajoba, A. S., Ibrahim, A. M., Ahmad, S., Mehta, A., Osayawe, O. J. et al. (2024). Nanomedicine in cancer therapy: advancing precision treatments. *Advances in Biomarker Sciences and Technology, 6*, 105–119. https://doi.org/10.1016/j.abst.2024.06.003

Ahmad, A., Khan, J. M., & Haque, S. (2019). Strategies in the design of endosomolytic agents for facilitating endosomal escape in nanoparticles. *Biochimie, 160*, 61–75. https://doi.org/10.1016/J.BIOCHI.2019.02.012

Alghamdi, M. A., Fallica, A. N., Virzì, N., Kesharwani, P., Pittalà, V., & Greish, K. (2022). The promise of nanotechnology in personalized medicine. *Journal of Personalized Medicine, 12*(5), 673. https://doi.org/10.3390/jpm12050673

Ali, F., Neha, K., & Parveen, S. (2022). Current regulatory landscape of nanomaterials and nanomedicines: a global perspective. *Journal of Drug Delivery Science and Technology, 80*, 104118. https://doi.org/10.1016/j.jddst.2022.104118

Alivisatos, P. (2003). The use of nanocrystals in biological detection. *Nature Biotechnology 2004 22:1, 22*(1), 47–52. https://doi.org/10.1038/nbt927

Allen, T. M., & Cullis, P. R. (2013). Liposomal drug delivery systems: from concept to clinical applications. *Advanced Drug Delivery Reviews, 65*(1), 36–48. https://doi.org/10.1016/J.ADDR.2012.09.037

Al-Thani, A. N., Jan, A. G., Abbas, M., Geetha, M., & Sadasivuni, K. K. (2024). Nanoparticles in cancer theragnostic and drug delivery: a comprehensive review. *Life Sciences, 352*, 122899. https://doi.org/10.1016/j.lfs.2024.122899

Amaral, M. N., Nunes, D., Fortunato, E., Martins, R., Rodrigues, C., Faísca, P., et al. (2024). Gold nanoparticles for photothermal therapy – influence of experimental conditions on the properties of resulting AuNPs. *Journal of Drug Delivery Science and Technology, 101*, 106215. https://doi.org/10.1016/j.jddst.2024.106215

Anitua, E., Alkhraisat, M. H., & Orive, G. (2011). Perspectives and challenges in regenerative medicine using plasma rich in growth factors. *Journal of Controlled Release, 157*(1), 29–38. https://doi.org/10.1016/j.jconrel.2011.07.004

Anker, J. N., Hall, W. P., Lyandres, O., Shah, N. C., Zhao, J., & Van Duyne, R. P. (2008). Biosensing with plasmonic nanosensors. *Nature Materials, 7*(6), 442–453. https://doi.org/10.1038/nmat2162

Anselmo, A. C., & Mitragotri, S. (2019). Nanoparticles in the clinic: an update. *Bioengineering & Translational Medicine*, *4*(3), e10143. https://doi.org/10.1002/btm2.10143

Awad, A., Fina, F., Trenfield, S., Patel, P., Goyanes, A., Gaisford, S., & Basit, A. (2019). 3D printed pellets (MiniPrintlets): A novel, multi-drug, controlled release platform technology. *Pharmaceutics*, *11*(4), 148. https://doi.org/10.3390/pharmaceutics11040148

Baden, L. R., El Sahly, H.M., Essink, B., Kotloff, K., Frey, S., Novak, R., et al. (2020). Efficacy and safety of the mRNA-1273 SARS-COV-2 vaccine. *New England Journal of Medicine, 384*(5), 403–416. https://doi.org/10.1056/nejmoa2035389

Bahl, K., Senn, J. J., Yuzhakov, O., Bulychev, A., Brito, L. A., Hassett, K. J. et al. (2017). Preclinical and clinical demonstration of immunogenicity by mRNA vaccines against H10N8 and H7N9 influenza viruses. *Molecular Therapy, 25*(6), 1316–1327. https://doi.org/10.1016/j.ymthe.2017.03.035

Bao, G., Mitragotri, S., & Tong, S. (2013). Multifunctional nanoparticles for drug delivery and molecular imaging. *Annual Review of Biomedical Engineering*, *15*(1), 253–282. https://doi.org/10.1146/annurev-bioeng-071812-152409

Bardhan, R., Lal, S., Joshi, A., & Halas, N. J. (2011). Theranostic nanoshells: from probe design to imaging and treatment of cancer. *Accounts of Chemical Research*, *44*(10), 936–946. https://doi.org/10.1021/AR200023X

Barenholz, Y. (2012). Doxil®--the first FDA-approved nano-drug: Lessons learned. *Journal of Controlled Release: Official Journal of the Controlled Release Society*, *160*(2), 117–134. https://doi.org/10.1016/J.JCONREL.2012.03.020

Bartel, D. P. (2004). MicroRNAs: genomics, biogenesis, mechanism, and function. *Cell*, *116*(2), 281–297. https://doi.org/10.1016/S0092-8674(04)00045-5

Bawa, R. (2011). Regulating nanomedicine – can the FDA handle it? *Current Drug Delivery*, *8*(3), 227–234. https://doi.org/10.2174/156720111795256156

Bechinger, B., & Lohner, K. (2006). Detergent-like actions of linear amphipathic cationic antimicrobial peptides. *Biochimica Et Biophysica Acta (BBA) – Biomembranes*, *1758*(9), 1529–1539. https://doi.org/10.1016/j.bbamem.2006.07.001

Behera, P. P., Kumar, N., Kumari, M., Kumar, S., Mondal, P. K., & Arun, R. K. (2023). Integrated microfluidic devices for point-of-care detection of bio-analytes and disease. *Sensors & Diagnostics*, *2*(6), 1437–1459. https://doi.org/10.1039/d3sd00170a

Benjaminsen, R. V, Mattebjerg, M. A., Henriksen, J. R., Moghimi, S. M., & Andresen, T. L. (2012). The possible "Proton sponge" effect of polyethylenimine (PEI) does not include change in lysosomal pH. *Molecular Therapy*, *21*(1), 149–157. https://doi.org/10.1038/mt.2012.185

Binauld, S., & Stenzel, M. H. (2013). Acid-degradable polymers for drug delivery: a decade of innovation. *Chemical Communications*, *49*(21), 2082–2102. https://doi.org/10.1039/C2CC36589H

Binnig, G., & Rohrer, H. (1984). Scanning tunneling microscopy. *Physica B+C*, *127*(1–3), 37–45. https://doi.org/10.1016/S0378-4363(84)80008-X

Blanco, E., Shen, H., & Ferrari, M. (2015). Principles of nanoparticle design for overcoming biological barriers to drug delivery. *Nature Biotechnology, 33*(9), 941–951. https://doi.org/10.1038/nbt.3330

Boateng, J., & Catanzano, O. (2015). Advanced therapeutic dressings for effective wound healing—a review. *Journal of Pharmaceutical Sciences*, *104*(11), 3653–3680. https://doi.org/10.1002/jps.24610

Bouchemal, K., Briançon, S., Perrier, E., & Fessi, H. (2004). Nano-emulsion formulation using spontaneous emulsification: Solvent, oil and surfactant optimisation. *International Journal of Pharmaceutics*, *280*(1–2), 241–251. https://doi.org/10.1016/j.ijpharm.2004.05.016

Briquez, P. S., Hubbell, J. A., & Martino, M. M. (2015). Extracellular matrix-inspired growth factor delivery systems for skin wound healing. *Advances in Wound Care, 4*(8), 479–489. https://doi.org/10.1089/wound.2014.0603

Busquets, M. A., Estelrich, J., & Sánchez-Martín, M. J. (2015). Nanoparticles in magnetic resonance imaging: from simple to dual contrast agents. *International Journal of Nanomedicine, 10,* 1727–1741. https://doi.org/10.2147/ijn.s76501

Buya, A. B., Mahlangu, P., & Witika, B. A. (2024). From lab to industrial development of lipid nanocarriers using quality by design approach. *International Journal of Pharmaceutics X, 8*, 100266. https://doi.org/10.1016/j.ijpx.2024.100266

Cevaal, P. M., Roche, M., Lewin, S. R., Caruso, F., & Faria, M. (2022). Experimental quantification of interactions between drug delivery systems and cells *in vitro*: a guide for preclinical nanomedicine evaluation. *Journal of Visualized Experiments, 187*. https://doi.org/10.3791/64259

Chames, P., Van Regenmortel, M., Weiss, E., & Baty, D. (2009). Therapeutic antibodies: successes, limitations and hopes for the future. *British Journal of Pharmacology, 157*(2), 220–233. https://doi.org/10.1111/j.1476-5381.2009.00190.x

Chehelgerdi, M., Chehelgerdi, M., Allela, O. Q. B., Pecho, R. D. C., Jayasankar, N., Rao, D. P. et al. (2023). Progressing nanotechnology to improve targeted cancer treatment: overcoming hurdles in its clinical implementation. *Molecular Cancer*, *22*(1), 169. https://doi.org/10.1186/S12943-023-01865-0

Chen, Z., Zhao, P., Luo, Z., Zheng, M., Tian, H., Gong, P., et al. (2016). Cancer cell membrane–biomimetic nanoparticles for homologous-targeting dual-modal imaging and photothermal therapy. *ACS Nano*, *10*(11), 10049–10057. https://doi.org/10.1021/acsnano.6b04695

Cheng, R., Meng, F., Deng, C., Klok, H., & Zhong, Z. (2013). Dual and multi-stimuli responsive polymeric nanoparticles for programmed site-specific drug delivery. *Biomaterials*, *34*(14), 3647–3657. https://doi.org/10.1016/j.biomaterials.2013.01.084

Cheng, R., Wang, S., & Santos, H. A. (2023). Acid-labile chemical bonds-based nanoparticles for endosome escape and intracellular delivery. *Biomedical Technology, 3*, 52–58. https://doi.org/10.1016/j.bmt.2023.01.001

Cho, K., Wang, X., Nie, S., Chen, Z., & Shin, D. M. (2008). Therapeutic nanoparticles for drug delivery in cancer. *Clinical Cancer Research: An Official Journal of the American Association for Cancer Research*, *14*(5), 1310–1316. https://doi.org/10.1158/1078-0432.CCR-07-1441

Clapp, A. R., Medintz, I. L., & Mattoussi, H. (2005). Förster resonance energy transfer investigations using quantum-dot fluorophores. *ChemPhysChem*, *7*(1), 47–57. https://doi.org/10.1002/cphc.200500217

Cullis, P. R., & Hope, M. J. (2017). Lipid nanoparticle systems for enabling gene therapies. *Molecular Therapy*, *25*(7), 1467–1475. https://doi.org/10.1016/j.ymthe.2017.03.013

Dai, Y., Xiao, H., Liu, J., Yuan, Q., Ma, P., Yang, D. et al. (2013). In vivo multimodality imaging and cancer therapy by near-infrared light-triggered trans-platinum pro-drug-conjugated upconverison nanoparticles. *Journal of the American Chemical Society*, *135*(50), 18920–18929. https://doi.org/10.1021/ja410028q

Danaei, M., Dehghankhold, M., Ataei, S., Davarani, F. H., Javanmard, R., Dokhani, A. et al. (2018). Impact of particle size and polydispersity index on the clinical applications of lipidic nanocarrier systems. *Pharmaceutics*, *10*(2), 57. https://doi.org/10.3390/pharmaceutics10020057

Danhier, F., Ansorena, E., Silva, J. M., Coco, R., Breton, A. L., & Préat, V. (2012). PLGA-based nanoparticles: an overview of biomedical applications. *Journal of Controlled Release*, *161*(2), 505–522. https://doi.org/10.1016/j.jconrel.2012.01.043

De Geest, B. G., De Koker, S., Sukhorukov, G. B., Kreft, O., Parak, W. J., Skirtach, A. G. et al. (2008). Polyelectrolyte microcapsules for biomedical applications. *Soft Matter, 5*(2), 282–291. https://doi.org/10.1039/b808262f

De Jong, S., Chikh, G., Sekirov, L., Raney, S., Semple, S., Klimuk, S. et al. (2007). Encapsulation in liposomal nanoparticles enhances the immunostimulatory, adjuvant and anti-tumor activity of subcutaneously administered CpG ODN. *Cancer Immunology, Immunotherapy: CII, 56*(8), 1251–1264. https://doi.org/10.1007/S00262-006-0276-X

De Koker, S., Hoogenboom, R., & De Geest, B. G. (2012). Polymeric multilayer capsules for drug delivery. *Chemical Society Reviews, 41*(7), 2867. https://doi.org/10.1039/c2cs15296g

Desnick, R. J., & Schuchman, E. H. (2002). Enzyme replacement and enhancement therapies: lessons from lysosomal disorders. *Nature Reviews Genetics*, *3*(12), 954–966. https://doi.org/10.1038/nrg963

Discher, D. E., & Eisenberg, A. (2002). Polymer vesicles. *Science*, *297*(5583), 967–973. https://doi.org/10.1126/SCIENCE.1074972

Dobrovolskaia, M. A., & McNeil, S. E. (2013). Understanding the correlation between in vitro and in vivo immunotoxicity tests for nanomedicines. *Journal of Controlled Release*, *172*(2), 456–466. https://doi.org/10.1016/j.jconrel.2013.05.025

Dreaden, E. C., Mac key, M. A., Huang, X., Kang, B., & El-Sayed, M. A. (2011). Beating cancer in multiple ways using nanogold. *Chemical Society Reviews*, *40*(7), 3391–3404. https://doi.org/10.1039/C0CS00180E

Driskell, J. D., Jones, C. A., Tompkins, S. M., & Tripp, R. A. (2011). One-step assay for detecting influenza virus using dynamic light scattering and gold nanoparticles. *The Analyst*, *136*(15), 3083. https://doi.org/10.1039/c1an15303j

Du, J., Lane, L. A., & Nie, S. (2015). Stimuli-responsive nanoparticles for targeting the tumor microenvironment. *Journal of Controlled Release, 219*, 205–214. https://doi.org/10.1016/j.jconrel.2015.08.050

Duan, X., Xiao, J., Yin, Q., Zhang, Z., Yu, H., Mao, S., & Li, Y. (2013). Smart pH-sensitive and temporal-controlled polymeric micelles for effective combination therapy of doxorubicin and disulfiram. *ACS Nano*, *7*(7), 5858–5869. https://doi.org/10.1021/NN4010796

Duncan, R., & Gaspar, R. (2011). Nanomedicine(s) under the microscope. *Molecular Pharmaceutics*, *8*(6), 2101–2141. https://doi.org/10.1021/MP200394T

Ecker, D. M., Jones, S. D., & Levine, H. L. (2014). The therapeutic monoclonal antibody market. *MAbs*, *7*(1), 9–14. https://doi.org/10.4161/19420862.2015.989042

Elbashir, S. M., Harborth, J., Lendeckel, W., Yalcin, A., Weber, K., & Tuschl, T. (2001). Duplexes of 21-nucleotide RNAs mediate RNA interference in cultured mammalian cells. *Nature, 411*(6836), 494–498. https://doi.org/10.1038/35078107

Elsabahy, M., & Wooley, K. L. (2012). Design of polymeric nanoparticles for biomedical delivery applications. *Chemical Society Reviews*, *41*(7), 2545. https://doi.org/10.1039/c2cs15327k

Elumalai, K., Srinivasan, S., & Shanmugam, A. (2024). Review of the efficacy of nanoparticle-based drug delivery systems for cancer treatment. *Biomedical Technology, 5*, 109–122. https://doi.org/10.1016/J.BMT.2023.09.001

Etheridge, M. L., Campbell, S. A., Erdman, A. G., Haynes, C. L., Wolf, S. M., & McCullough, J. (2012). The big picture on nanomedicine: the state of investigational and approved nanomedicine products. *Nanomedicine Nanotechnology Biology and Medicine*, *9*(1), 1–14. https://doi.org/10.1016/j.nano.2012.05.013

European Medicines Agency. (2011). *Reflection paper on nanotechnology-based medicinal products for human use* (Issue EMA/CHMP/79769/2011). https://etp-nanomedicine.eu/wp-content/uploads/2018/10/reflection-paper-nanotechnology-based-medicinal-products-human-use_en-1.pdf

Fadeel, B., Farcal, L., Hardy, B., Vázquez-Campos, S., Hristozov, D., Marcomini, A., et al. (2018). Advanced tools for the safety assessment of nanomaterials. *Nature Nanotechnology*, *13*(7), 537–543. https://doi.org/10.1038/s41565-018-0185-0

Fan, D., Cao, Y., Cao, M., Wang, Y., Cao, Y., & Gong, T. (2023). Nanomedicine in cancer therapy. *Signal Transduction and Targeted Therapy*, *8*(1), 1–12. https://doi.org/10.1038/s41392-023-01536-y

Fang, R. H., Kroll, A. V, Gao, W., & Zhang, L. (2018). Cell membrane coating nanotechnology. *Advanced Materials*, *30*(23). https://doi.org/10.1002/adma.201706759

Faria, M., Björnmalm, M., Thurecht, K. J., Kent, S. J., Parton, R. G., Kavallaris, M., et al. (2018). Minimum information reporting in bio–nano experimental literature. *Nature Nanotechnology*, *13*(9), 777–785. https://doi.org/10.1038/s41565-018-0246-4

Farkhani, S. M., Valizadeh, A., Karami, H., Mohammadi, S., Sohrabi, N., & Badrzadeh, F. (2014). Cell penetrating peptides: efficient vectors for delivery of nanoparticles, nanocarriers, therapeutic and diagnostic molecules. *Peptides, 57*, 78–94. https://doi.org/10.1016/j.peptides.2014.04.015

Farokhzad, O. C., & Langer, R. (2009). Impact of nanotechnology on drug delivery. *ACS Nano*, *3*(1), 16–20. https://doi.org/10.1021/NN900002M

Fathi-Achachelouei, M., Knopf-Marques, H., Ribeiro da Silva, C. E., Barthès, J., Bat, E., Tezcaner, A. et al. (2019). Use of Nanoparticles in Tissue Engineering and Regenerative Medicine. *Frontiers in Bioengineering and Biotechnology, 7*, 453043. https://doi.org/10.3389/FBIOE.2019.00113

Ferdowsian, H. R., & Beck, N. (2011). Ethical and scientific considerations regarding animal testing and research. *PLOS One*, *6*(9), e24059. https://doi.org/10.1371/journal.pone.0024059

Fernandes, C., Jathar, M., Sawant, B. K. S., & Warde, T. (2023a). Scale-up of nanoparticle manufacturing process. In *Pharmaceutical Process Engineering and Scale-up Principles* (pp. 173–203). Springer.

Fernandes, C., Jathar, M., Sawant, B. K. S., & Warde, T. (2023b). Scale-up of nanoparticle manufacturing process. In *AAPS Introductions in the Pharmaceutical Sciences* (pp. 173–203). https://doi.org/10.1007/978-3-031-31380-6_12

Feynman, R. P. (n.d.). *Plenty of Room at the Bottom*. Presentation at American Physical Society, Pasadena, December 1959.

Food and Drug Administration. (2014). *Guidance for industry considering whether an FDA-regulated product involves the application of nanotechnology contains nonbinding recommendations*. www.fda.gov/RegulatoryInformation/Guidances/ucm257698.htm

Food and Drug Administration. (2017). *Drug products, including biological products, that contain nanomaterials—guidance for industry*. www.fda.gov/regulatory-information/search-fda-guidance-documents/drug-products-including-biological-products-contain-nanomaterials-guidance-industry

Gabor, F. (2011). Characterization of nanoparticles intended for drug delivery. *Scientia Pharmaceutica*, *79*(3), 701–702. https://doi.org/10.3797/scipharm.br-11-01

Gao, X., Cui, Y., Levenson, R. M., Chung, L. W. K., & Nie, S. (2004). In vivo cancer targeting and imaging with semiconductor quantum dots. *Nature Biotechnology, 22*(8), 969–976. https://doi.org/10.1038/nbt994

Geiss, G. K., Bumgarner, R. E., Birditt, B., Dahl, T., Dowidar, N., Dunaway, D. L. et al. (2008). Direct multiplexed measurement of gene expression with color-coded probe pairs. *Nature Biotechnology*, *26*(3), 317–325. https://doi.org/10.1038/nbt1385

Geng, Y., Dalhaimer, P., Cai, S., Tsai, R., Tewari, M., Minko, T., & Discher, D. E. (2007). Shape effects of filaments versus spherical particles in flow and drug delivery. *Nature Nanotechnology 2007 2:4*, *2*(4), 249–255. https://doi.org/10.1038/nnano.2007.70

Gijs, M. A. M. (2004). Magnetic bead handling on-chip: new opportunities for analytical applications. *Microfluid Nanofluid*, *1*, 22–40. https://doi.org/10.1007/s10404-004-0010-y

Gilleron, J., Querbes, W., Zeigerer, A., Borodovsky, A., Marsico, G., Schubert, U. et al. (2013). Image-based analysis of lipid nanoparticle-mediated siRNA delivery, intracellular trafficking and endosomal escape. *Nature Biotechnology*, *31*(7), 638–646. https://doi.org/10.1038/NBT.2612

Gisbert-Garzaran, M., Lozano, D., Matsumoto, K., Komatsu, A., Manzano, M., Tamanoi, F., & Vallet-Regi, M. (2021). Designing mesoporous silica nanoparticles to overcome biological barriers by incorporating targeting and endosomal escape. *ACS Applied Materials and Interfaces*, *13*(8), 9656–9666. Please replace with: https://doi.org/10.1021/acsami.0c21507

Goole, J., & Amighi, K. (2016). 3D printing in pharmaceutics: a new tool for designing customized drug delivery systems. *International Journal of Pharmaceutics*, *499*(1–2), 376–394. https://doi.org/10.1016/j.ijpharm.2015.12.071

Gratton, S. E. A., Ropp, P. A., Pohlhaus, P. D., Luft, J. C., Madden, V. J., Napier, M. E. et al. (2008). The effect of particle design on cellular internalization pathways. *Proceedings of the National Academy of Sciences of the United States of America*, *105*(33), 11613–11618. https://doi.org/10.1073/pnas.0801763105

Gref, R., Minamitake, Y., Peracchia, M. T., Trubetskoy, V., Torchilin, V., & Langer, R. (1994). Biodegradable Long-Circulating polymeric nanospheres. *Science*, *263*(5153), 1600–1603. https://doi.org/10.1126/science.8128245

Guo, S., Wang, J., Wang, Q., Wang, J., Qin, S., & Li, W. (2024). Advances in peptide-based drug delivery systems. *Heliyon*, *10*(4), e26009. https://doi.org/10.1016/J.HELIYON.2024.E26009

Haddadzadegan, S., Dorkoosh, F., & Bernkop-Schnürch, A. (2022). Oral delivery of therapeutic peptides and proteins: technology landscape of lipid-based nanocarriers. *Advanced Drug Delivery Reviews*, *182*, 114097. https://doi.org/10.1016/j.addr.2021.114097

Hadianamrei, R., & Zhao, X. (2022). Current state of the art in peptide-based gene delivery. *Journal of Controlled Release: Official Journal of the Controlled Release Society*, *343*, 600–619. https://doi.org/10.1016/J.JCONREL.2022.02.010

Haghighat, Z. A., Safekordi, A., Ardjmand, M., & Akbarzadeh, A. (2024). Exploring the antitumor efficacy of PEGylated liposomes loaded with licorice extract for cancer therapy. *Current Cancer Drug Targets*, *24*. https://doi.org/10.2174/0115680096292153240416115744

Hainfeld, J. F., Slatkin, D. N., & Smilowitz, H. M. (2004). The use of gold nanoparticles to enhance radiotherapy in mice. *Physics in Medicine and Biology*, *49*(18), N309–N315. https://doi.org/10.1088/0031-9155/49/18/n03

Hamilton, S., & Kingston, B. R. (2023). Applying artificial intelligence and computational modeling to nanomedicine. *Current Opinion in Biotechnology*, *85*, 103043. https://doi.org/10.1016/j.copbio.2023.103043

Hardy, A., Benford, D., Halldorsson, T., Jeger, M. J., Knutsen, H. K., More, S. et al. (2018). Guidance on risk assessment of the application of nanoscience and nanotechnologies in the food and feed chain: part 1, human and animal health. *EFSA Journal*, *16*(7). https://doi.org/10.2903/j.efsa.2018.5327

Harris, C. T., & Cohen, S. (2024). Reducing immunogenicity by design: approaches to minimize immunogenicity of monoclonal antibodies. *BioDrugs*, *38*(2), 205–226. https://doi.org/10.1007/s40259-023-00641-2

Hassett, K. J., Benenato, K. E., Jacquinet, E., Lee, A., Woods, A., Yuzhakov, O. et al. (2019). Optimization of lipid nanoparticles for intramuscular administration of mRNA vaccines. *Molecular Therapy—Nucleic Acids, 15*, 1–11. https://doi.org/10.1016/j.omtn.2019.01.013

He, Q., Zhang, Z., Gao, F., Li, Y., & Shi, J. (2011). In vivo biodistribution and urinary excretion of mesoporous silica nanoparticles: effects of particle size and PEGylation. *Small (Weinheim an Der Bergstrasse, Germany)*, *7*(2), 271–280. https://doi.org/10.1002/SMLL.201001459

Hermeling, S., Crommelin, D. J. A., Schellekens, H., & Jiskoot, W. (2004). Structure-Immunogenicity relationships of therapeutic proteins. *Pharmaceutical Research*, *21*(6), 897–903. https://doi.org/10.1023/b:pham.0000029275.41323.a6

Hersel, U., Dahmen, C., & Kessler, H. (2003). RGD modified polymers: biomaterials for stimulated cell adhesion and beyond. *Biomaterials*, *24*(24), 4385–4415. https://doi.org/10.1016/s0142-9612(03)00343-0

Hopkins, C., Javius-Jones, K., Wang, Y., Hong, H., Hu, Q., & Hong, S. (2022). Combinations of chemo-, immuno-, and gene therapies using nanocarriers as a multifunctional drug platform. *Expert Opinion on Drug Delivery*, *19*(10), 1337–1349. https://doi.org/10.1080/17425247.2022.2112569

Horejs, C. (2021). From lipids to lipid nanoparticles to mRNA vaccines. *Nature Reviews. Materials*, *6*(12), 1075–1076. https://doi.org/10.1038/S41578-021-00379-9

Hou, X., Zaks, T., Langer, R., & Dong, Y. (2021). Lipid nanoparticles for mRNA delivery. *Nature Reviews Materials, 6*(12), 1078–1094. https://doi.org/10.1038/s41578-021-00358-0

Hu, C. J., Zhang, L., Aryal, S., Cheung, C., Fang, R. H., & Zhang, L. (2011). Erythrocyte membrane-camouflaged polymeric nanoparticles as a biomimetic delivery platform. *Proceedings of the National Academy of Sciences*, *108*(27), 10980–10985. https://doi.org/10.1073/pnas.1106634108

Hu, Q., Katti, P. S., & Gu, Z. (2014). Enzyme-responsive nanomaterials for controlled drug delivery. *Nanoscale*, *6*(21), 12273–12286. https://doi.org/10.1039/C4NR04249B

Hunt, G., & Mehta, M. D. (Eds.). (2013). *Nanotechnology: Risk, Ethics and Law*. Routledge.

Immordino, M. L., Dosio, F., & Cattel, L. (2006). Stealth liposomes: review of the basic science, rationale, and clinical applications, existing and potential. *International Journal of Nanomedicine*, *1*(3), 297.

Jackson, L. A., Anderson, E. J., Rouphael, N. G., Roberts, P. C., Makhene, M., Coler, R. N. et al. (2020). An mRNA vaccine against SARS-CoV-2—Preliminary report. *New England Journal of Medicine, 383*(20), 1920–1931. https://doi.org/10.1056/nejmoa2022483

Jain, S., Hirst, D. G., & O'Sullivan, J. M. (2012). Gold nanoparticles as novel agents for cancer therapy. *The British Journal of Radiology*, *85*(1010), 101. https://doi.org/10.1259/BJR/59448833

Janib, S. M., Moses, A. S., & MacKay, J. A. (2010). Imaging and drug delivery using theranostic nanoparticles. *Advanced Drug Delivery Reviews*, *62*(11), 1052–1063. https://doi.org/10.1016/j.addr.2010.08.004

Jiang, W., Kim, B. Y. S., Rutka, J. T., & Chan, W. C. W. (2008). Nanoparticle-mediated cellular response is size-dependent. *Nature Nanotechnology*, *3*(3), 145–150. https://doi.org/10.1038/nnano.2008.30

Jokerst, J. V., Lobovkina, T., Zare, R. N., & Gambhir, S. S. (2011). Nanoparticle PEGylation for imaging and therapy. *Nanomedicine (London, England)*, *6*(4), 715–728. https://doi.org/10.2217/NNM.11.19

Kah, M., Tufenkji, N., & White, J. C. (2019). Nano-enabled strategies to enhance crop nutrition and protection. *Nature Nanotechnology*, *14*(6), 532–540. https://doi.org/10.1038/s41565-019-0439-5

Kalepu, S., & Nekkanti, V. (2015). Insoluble drug delivery strategies: review of recent advances and business prospects. *Acta Pharmaceutica Sinica B*, *5*(5), 442–453. https://doi.org/10.1016/j.apsb.2015.07.003

Kalluri, R., & LeBleu, V. S. (2020). The biology, function, and biomedical applications of exosomes. *Science (New York, N.Y.)*, *367*(6478). https://doi.org/10.1126/SCIENCE.AAU6977

Kamaly, N., Yameen, B., Wu, J., & Farokhzad, O. C. (2016). Degradable controlled-release polymers and polymeric nanoparticles: mechanisms of controlling drug release. *Chemical Reviews*, *116*(4), 2602–2663. https://doi.org/10.1021/acs.chemrev.5b00346

Kashyap, B. K., Singh, V. V, Solanki, M. K., Kumar, A., Ruokolainen, J., & Kesari, K. K. (2023). Smart nanomaterials in cancer theranostics: challenges and opportunities. *ACS Omega*, *8*(16), 14290–14320. https://doi.org/10.1021/acsomega.2c07840

Kataoka, K., Harada, A., & Nagasaki, Y. (2012). Block copolymer micelles for drug delivery: design, characterization and biological significance. *Advanced Drug Delivery Reviews*, *64*(SUPPL.), 37–48. https://doi.org/10.1016/J.ADDR.2012.09.013

Kaushik, A., Khan, S., Pharasi, N., & Mani, S. (2024). Dual pH and ultrasound responsive nanocarriers: a smart approach in cancer theranostics. *Journal of Drug Delivery Science and Technology, 95*, 105560. https://doi.org/10.1016/j.jddst.2024.105560

Kaushik, A., Srinivasan, S. S., & Mishra, Y. K. (2023). *Analytical Techniques for Biomedical Nanotechnology*. IOP Publishing. https://doi.org/10.1088/978-0-7503-3379-5

Kay, M. A., Glorioso, J. C., & Naldini, L. (2001). Viral vectors for gene therapy: the art of turning infectious agents into vehicles of therapeutics. *Nature Medicine*, *7*(1), 33–40. https://doi.org/10.1038/83324

Khan, I., Saeed, K., & Khan, I. (2019). Nanoparticles: properties, applications and toxicities. *Arabian Journal of Chemistry*, *12*(7), 908–931. https://doi.org/10.1016/J.ARABJC.2017.05.011

Khazaei, M., Hosseini, M. S., Haghighi, A. M., & Misaghi, M. (2023). Nanosensors and their applications in early diagnosis of cancer. *Sensing and Bio-Sensing Research, 41*, 100569. https://doi.org/10.1016/J.SBSR.2023.100569

Khizar, S., Al-Dossary, A. A., Zine, N., Jaffrezic-Renault, N., Errachid, A., & Elaissari, A. (2022). Contribution of magnetic particles in molecular diagnosis of human viruses. *Talanta, 241*, 123243. https://doi.org/10.1016/j.talanta.2022.123243

Kim, B., Hosn, R. R., Remba, T., Yun, D., Li, N., Abraham, W., Melo, M. B. et al. (2022). Optimization of storage conditions for lipid nanoparticle-formulated self-replicating RNA vaccines. *Journal of Controlled Release, 353*, 241–253. https://doi.org/10.1016/j.jconrel.2022.11.022

Kim, H., Kim, E., Lee, I., Bae, B., Park, M., & Nam, H. (2020). Artificial intelligence in drug discovery: a comprehensive review of data-driven and machine learning approaches. *Biotechnology and Bioprocess Engineering*, *25*(6), 895–930. https://doi.org/10.1007/s12257-020-0049-y

Kim, S. C., Kim, D. W., Shim, Y. H., Bang, J. S., Oh, H. S., Kim, S. W. et al. (2001). In vivo evaluation of polymeric micellar paclitaxel formulation: toxicity and efficacy. *Journal of Controlled Release*, *72*(1–3), 191–202. https://doi.org/10.1016/s0168-3659(01)00275-9

Kim, Y., Zharkinbekov, Z., Raziyeva, K., Tabyldiyeva, L., Berikova, K., Zhumagul, D. et al. (2023). Chitosan-based biomaterials for tissue regeneration. *Pharmaceutics*, *15*(3), 807. https://doi.org/10.3390/pharmaceutics15030807

Knauer, N., Pashkina, E., & Apartsin, E. (2019). Topological aspects of the design of nanocarriers for therapeutic peptides and proteins. *Pharmaceutics, 11*(2), 91. https://doi.org/10.3390/pharmaceutics11020091

Koo, H., Huh, M. S., Sun, I., Yuk, S. H., Choi, K., Kim, K., et al. (2011). In vivo targeted delivery of nanoparticles for theranosis. *Accounts of Chemical Research, 44*(10), 1018–1028. https://doi.org/10.1021/ar2000138

Kotzé, A. F., Lueßen, H. L., De Boer, A. G., Verhoef, J. C., & Junginger, H. E. (1999). Chitosan for enhanced intestinal permeability: prospects for derivatives soluble in neutral and basic environments. *European Journal of Pharmaceutical Sciences: Official Journal of the European Federation for Pharmaceutical Sciences, 7*(2), 145–151. https://doi.org/10.1016/S0928-0987(98)00016-5

Krishnan, U. M. (2020). *Protein and Peptide Nanostructures for Drug and Gene Delivery*. Elsevier. https://doi.org/10.1016/b978-0-08-102850-6.00013-9

Kumar, R., Dalvi, S. V, & Siril, P. F. (2020). Nanoparticle-based drugs and formulations: current status and emerging applications. *ACS Applied Nano Materials, 3*(6), 4944–4961. https://doi.org/10.1021/acsanm.0c00606

Kwon, E. J., Bergen, J. M., & Pun, S. H. (2008). Application of an HIV GP41-derived peptide for enhanced intracellular trafficking of synthetic gene and sIRNA delivery vehicles. *Bioconjugate Chemistry, 19*(4), 920–927. https://doi.org/10.1021/bc700448h

Langer, R., & Peppas, N. A. (2003). Advances in biomaterials, drug delivery, and bionanotechnology. *AIChE Journal, 49*(12), 2990–3006. https://doi.org/10.1002/aic.690491202

Lee, D., Koo, H., Sun, I., Ryu, J. H., Kim, K., & Kwon, I. C. (2011). Multifunctional nanoparticles for multimodal imaging and theragnosis. *Chemical Society Reviews, 41*(7), 2656–2672. https://doi.org/10.1039/c2cs15261d

Lee, E. S., Shin, H. J., Na, K., & Bae, Y. H. (2003). Poly(L-histidine)-PEG block copolymer micelles and pH-induced destabilization. *Journal of Controlled Release, 90*(3), 363–374. https://doi.org/10.1016/S0168-3659(03)00205-0

Lei, W., Yang, C., Wu, Y., Ru, G., He, X., Tong, X., et al. (2022). Nanocarriers surface engineered with cell membranes for cancer targeted chemotherapy. *Journal of Nanobiotechnology, 20*(1). https://doi.org/10.1186/s12951-022-01251-w

Li, J., De Ávila, B. E., Gao, W., Zhang, L., & Wang, J. (2017). Micro/nanorobots for biomedicine: delivery, surgery, sensing, and detoxification. *Science Robotics, 2*(4). https://doi.org/10.1126/scirobotics.aam6431

Li, S., Ma, L., Zhou, M., Li, Y., Xia, Y., Fan, X. et al. (2019). New opportunities for emerging 2D materials in bioelectronics and biosensors. *Current Opinion in Biomedical Engineering, 13*, 32–41. https://doi.org/10.1016/j.cobme.2019.08.016

Li, Y., & Champion, J. A. (2022). Self-assembling nanocarriers from engineered proteins: design, functionalization, and application for drug delivery. *Advanced Drug Delivery Reviews, 189*, 114462. https://doi.org/10.1016/j.addr.2022.114462

Li, Y., Pei, Y., Zhang, X., Gu, Z., Zhou, Z., Yuan, W., et al. (2001). PEGylated PLGA nanoparticles as protein carriers: synthesis, preparation and biodistribution in rats. *Journal of Controlled Release, 71*(2), 203–211. https://doi.org/10.1016/s0168-3659(01)00218-8

Lim, E., Kim, T., Paik, S., Haam, S., Huh, Y., & Lee, K. (2014). Nanomaterials for theranostics: recent advances and future challenges. *Chemical Reviews, 115*(1), 327–394. https://doi.org/10.1021/cr300213b

Lim, S. H., Kathuria, H., Tan, J. J. Y., & Kang, L. (2018). 3D printed drug delivery and testing systems — a passing fad or the future? *Advanced Drug Delivery Reviews, 132*, 139–168. https://doi.org/10.1016/j.addr.2018.05.006

Lin, P., & Allhoff, F. (2008). *Nanotechnology & Society*. Springer. https://doi.org/10.1007/978-1-4020-6209-4

Liu, B., Shi, T., Yang, H., Zhang, J., & Li, Y. (2024). Size-dependent effect on controllable release and field insecticidal efficacy of diamide insecticide polylactic acid microspheres delivery systems against Ostrinia nubilalis. *International Journal of Biological Macromolecules, 278*, 134659. https://doi.org/10.1016/j.ijbiomac.2024.134659

Liu, J. F., Jang, B., Issadore, D., & Tsourkas, A. (2019). Use of magnetic fields and nanoparticles to trigger drug release and improve tumor targeting. *Wiley Interdisciplinary Reviews Nanomedicine and Nanobiotechnology*, *11*(6). https://doi.org/10.1002/wnan.1571

Liu, M., Wang, L., Lo, Y., Shiu, S. C. C., Kinghorn, A. B., & Tanner, J. A. (2022). Aptamer-enabled nanomaterials for therapeutics, drug targeting and imaging. *Cells*, *11*(1). https://doi.org/10.3390/CELLS11010159

Liu, X., Jin, X., & Ma, P. X. (2011). Nanofibrous hollow microspheres self-assembled from star-shaped polymers as injectable cell carriers for knee repair. *Nature Materials*, *10*(5), 398–406. https://doi.org/10.1038/nmat2999

Liu, Y., Miyoshi, H., & Nakamura, M. (2007). Nanomedicine for drug delivery and imaging: a promising avenue for cancer therapy and diagnosis using targeted functional nanoparticles. *International Journal of Cancer*, *120*(12), 2527–2537. https://doi.org/10.1002/ijc.22709

Longmire, M., Choyke, P. L., & Kobayashi, H. (2008). Clearance properties of nano-sized particles and molecules as imaging agents: considerations and caveats. *Nanomedicine, 3*(5), 703–717. https://doi.org/10.2217/17435889.3.5.703

Lotfabadi, A. S., Abadi, B., & Rezaei, N. (2024). Biomimetic nanotechnology for cancer immunotherapy: state of the art and future perspective. *International Journal of Pharmaceutics*, 123923. https://doi.org/10.1016/j.ijpharm.2024.123923

Lou, D., Fan, L., Jiang, T., & Zhang, Y. (2022). Advances in nanoparticle-based lateral flow immunoassay for point-of-care testing. *View*, *3*(1). https://doi.org/10.1002/viw.20200125

Lozano, D., Larraga, V., Vallet-Regí, M., & Manzano, M. (2023). An overview of the use of nanoparticles in vaccine development. *Nanomaterials, 13*(12), 1828. https://doi.org/10.3390/nano13121828

Lu, Y., & Low, P. S. (2012). Folate-mediated delivery of macromolecular anticancer therapeutic agents. *Advanced Drug Delivery Reviews*, *64*(SUPPL.), 342–352. https://doi.org/10.1016/J.ADDR.2012.09.020

MacBeath, G., & Schreiber, S. L. (2000). Printing proteins as microarrays for High-Throughput Function determination. *Science*, *289*(5485), 1760–1763. https://doi.org/10.1126/science.289.5485.1760

Maeda, H., Bharate, G., & Daruwalla, J. (2008). Polymeric drugs for efficient tumor-targeted drug delivery based on EPR-effect. *European Journal of Pharmaceutics and Biopharmaceutics*, *71*(3), 409–419. https://doi.org/10.1016/j.ejpb.2008.11.010

Maeda, H., Nakamura, H., & Fang, J. (2013). The EPR effect for macromolecular drug delivery to solid tumors: improvement of tumor uptake, lowering of systemic toxicity, and distinct tumor imaging in vivo. *Advanced Drug Delivery Reviews*, *65*(1), 71–79. https://doi.org/10.1016/J.ADDR.2012.10.002

Maeda, H., Wu, J., Sawa, T., Matsumura, Y., & Hori, K. (2000). Tumor vascular permeability and the EPR effect in macromolecular therapeutics: a review. *Journal of Controlled Release*, *65*(1–2), 271–284. https://doi.org/10.1016/S0168-3659(99)00248-5

Mahajan, K., & Bhattacharya, S. (2024). The advancement and obstacles in improving the stability of nanocarriers for precision drug delivery in the field of nanomedicine. *Current Topics in Medicinal Chemistry*, *24*(8), 686–721. https://doi.org/10.2174/0115680266287101240214071718

Mahler, H., Friess, W., Grauschopf, U., & Kiese, S. (2008). Protein aggregation: pathways, induction factors and analysis. *Journal of Pharmaceutical Sciences*, *98*(9), 2909–2934. https://doi.org/10.1002/jps.21566

Mandrycky, C., Wang, Z., Kim, K., & Kim, D. (2015). 3D bioprinting for engineering complex tissues. *Biotechnology Advances*, *34*(4), 422–434. https://doi.org/10.1016/j.biotechadv.2015.12.011

Mao, S., Sun, W., & Kissel, T. (2010). Chitosan-based formulations for delivery of DNA and siRNA. *Advanced Drug Delivery Reviews*, *62*(1), 12–27. https://doi.org/10.1016/J.ADDR.2009.08.004

Massignan, F., Gyulai, G., Horváti, K., Bősze, S., & Kiss, É. (2023). Effective nanoparticulate-type encapsulation delivery system for hydrophilic proteins and peptides. *EXPRESS Polymer Letters*, *18*(1), 72–87. https://doi.org/10.3144/expresspolymlett.2024.6

Medintz, I. L., Uyeda, H. T., Goldman, E. R., & Mattoussi, H. (2005). Quantum dot bioconjugates for imaging, labelling and sensing. *Nature Materials, 4*(6), 435–446. https://doi.org/10.1038/nmat1390

Mehnert, W., & Mäder, K. (2012). Solid lipid nanoparticles. *Advanced Drug Delivery Reviews, 64*, 83–101. https://doi.org/10.1016/j.addr.2012.09.021

Mehta, M., Bui, T. A., Yang, X., Aksoy, Y., Goldys, E. M., & Deng, W. (2023). Lipid-based nanoparticles for drug/gene delivery: an overview of the production techniques and difficulties encountered in their industrial development. *ACS Materials Au, 3*(6), 600–619. https://doi.org/10.1021/acsmaterialsau.3c00032

Mendes, B. B., Conniot, J., Avital, A., Yao, D., Jiang, X., Zhou, X. et al. (2022). Nanodelivery of nucleic acids. *Nature Reviews Methods Primers*, *2*(1). https://doi.org/10.1038/s43586-022-00104-y

Miao, Q., Xie, C., Zhen, X., Lyu, Y., Duan, H., Liu, X. et al. (2017). Molecular afterglow imaging with bright, biodegradable polymer nanoparticles. *Nature Biotechnology*, *35*(11), 1102–1110. https://doi.org/10.1038/nbt.3987

Miller, A. R., & Tucker, C. E. (2014). Privacy protection, personalized medicine and genetic testing. *SSRN Electronic Journal*. https://doi.org/10.2139/ssrn.2411230

Mintzer, M. A., & Grinstaff, M. W. (2010). Biomedical applications of dendrimers: a tutorial. *Chemical Society Reviews*, *40*(1), 173–190. https://doi.org/10.1039/B901839P

Mitchell, M. J., Billingsley, M. M., Haley, R. M., Wechsler, M. E., Peppas, N. A., & Langer, R. (2020). Engineering precision nanoparticles for drug delivery. *Nature Reviews Drug Discovery, 20*(2), 101–124. https://doi.org/10.1038/s41573-020-0090-8

Mizukami, S., Kashibe, M., Matsumoto, K., Hori, Y., & Kikuchi, K. (2017). Enzyme-triggered compound release using functionalized antimicrobial peptide derivatives. *Chemical Science*, *8*(4), 3047–3053. https://doi.org/10.1039/c6sc04435b

Mnyusiwalla, A., Daar, A. S., & Singer, P. A. (2003). Mind the gap: science and ethics in nanotechnology. *Nanotechnology*, *14*(3), R9–R13. https://doi.org/10.1088/0957-4484/14/3/201

Moghimi, S. M., Hunter, A. C., & Murray, J. C. (2001). Long-circulating and target-specific nanoparticles: theory to practice. *Pharmacological Reviews*, *53*(2), 283. http://pharmrev.aspetjournals.org/content/53/2/283.abstract

Moghimi, S. M., Hunter, A. C., & Murray, J. C. (2005). Nanomedicine: current status and future prospects. *FASEB Journal: Official Publication of the Federation of American Societies for Experimental Biology*, *19*(3), 311–330. https://doi.org/10.1096/FJ.04-2747REV

Mohan, A., Santhamoorthy, M., & Lee, Y. C. (2024). Recent advances in the pH-responsive organic–inorganic mesoporous hybrid silica for targeted drug delivery. *European Polymer Journal, 206*, 112783. https://doi.org/10.1016/J.EURPOLYMJ.2024.112783

Mura, S., Nicolas, J., & Couvreur, P. (2013). Stimuli-responsive nanocarriers for drug delivery. *Nature Materials 2013 12:11*, *12*(11), 991–1003. https://doi.org/10.1038/nmat3776

Murphy, S. V, & Atala, A. (2014). 3D bioprinting of tissues and organs. *Nature Biotechnology*, *32*(8), 773–785. https://doi.org/10.1038/nbt.2958

Muslykhah, U., Phupaboon, S., Suriyapha, C., Matra, M., & Wanapat, M. (2024). Encapsulation of protein-based bioactive from black soldier fly for ruminant feeding. *Journal of Agriculture and Food Research, 18*, 101325. https://doi.org/10.1016/j.jafr.2024.101325

Nakase, I., Tanaka, G., & Futaki, S. (2012). Cell-penetrating peptides (CPPs) as a vector for the delivery of siRNAs into cells. *Molecular BioSystems*, *9*(5), 855. https://doi.org/10.1039/c2mb25467k

Nel, A., Xia, T., Mädler, L., & Li, N. (2006). Toxic potential of materials at the nanolevel. *Science (New York, NY), 311*(5761), 622–627. https://doi.org/10.1126/SCIENCE.1114397

Niculescu, A. G., & Grumezescu, A. M. (2022). Novel tumor-targeting nanoparticles for cancer treatment-a review. *International Journal of Molecular Sciences*, *23*(9). https://doi.org/10.3390/IJMS23095253

Nobs, L., Buchegger, F., Gurny, R., & Allémann, E. (2004). Current methods for attaching targeting ligands to liposomes and nanoparticles. *Journal of Pharmaceutical Sciences*, *93*(8), 1980–1992. https://doi.org/10.1002/JPS.20098

Oberdörster, G., Oberdörster, E., & Oberdörster, J. (2005). Nanotoxicology: an emerging discipline evolving from studies of ultrafine particles. *Environmental Health Perspectives*, *113*(7), 823–839. https://doi.org/10.1289/ehp.7339

OECD. (2012). *Guidance on sample preparation and dosimetry for the safety testing of manufactured nanomaterials*. https://doi.org/10.1787/ed430e1d-en

Oh, Y. K., & Park, T. G. (2009). siRNA delivery systems for cancer treatment. *Advanced Drug Delivery Reviews*, *61*(10), 850–862. https://doi.org/10.1016/J.ADDR.2009.04.018

Olgenblum, G. I., Hutcheson, B. O., Pielak, G. J., & Harries, D. (2024). Protecting proteins from desiccation stress using molecular glasses and gels. *Chemical Reviews*, *124*(9), 5668–5694. https://doi.org/10.1021/acs.chemrev.3c00752

Ortiz, A. C., Jackson, N., Buelvas, N., Jerez, A., López-Muñoz, R. A., Morales, J. et al. (2024). Release kinetics approach of stimuli-responsive mesoporous silica nanocarriers: pH-sensitive linker versus pH-sensitive framework. *Journal of Drug Delivery Science and Technology, 91*, 105212. https://doi.org/10.1016/j.jddst.2023.105212

Owens, D. E., & Peppas, N. A. (2006). Opsonization, biodistribution, and pharmacokinetics of polymeric nanoparticles. *International Journal of Pharmaceutics*, *307*(1), 93–102. https://doi.org/10.1016/J.IJPHARM.2005.10.010

Ozbolat, I. T., & Yu, N. Y. (2013). Bioprinting toward organ fabrication: challenges and future trends. *IEEE Transactions on Biomedical Engineering*, *60*(3), 691–699. https://doi.org/10.1109/tbme.2013.2243912

Pack, D. W., Hoffman, A. S., Pun, S., & Stayton, P. S. (2005). Design and development of polymers for gene delivery. *Nature Reviews Drug Discovery*, *4*(7), 581–593. https://doi.org/10.1038/nrd1775

Pandey, P., Arya, D. K., Deepak, P., Ali, D., Alarifi, S., Srivastava, S., et al. (2024). αvβ3 integrin and folate-targeted pH-sensitive liposomes with dual ligand modification for metastatic breast cancer treatment. *Bioengineering*, *11*(8), 800. https://doi.org/10.3390/bioengineering11080800

Pardi, N., Hogan, M. J., Porter, F. W., & Weissman, D. (2018). mRNA vaccines—a new era in vaccinology. *Nature Reviews Drug Discovery*, *17*(4), 261–279. https://doi.org/10.1038/nrd.2017.243

Patel, D. M., Patel, N. N., & Patel, J. K. (2021). Nanomedicine scale-up technologies: feasibilities and challenges. In *Emerging Technologies for Nanoparticle Manufacturing* (pp. 511–539). Springer.

Patolsky, F., Zheng, G., & Lieber, C. M. (2006). Nanowire-based biosensors. *Analytical Chemistry*, *78*(13), 4260–4269. https://doi.org/10.1021/ac069419j

Peer, D., Karp, J. M., Hong, S., Farokhzad, O. C., Margalit, R., & Langer, R. (2007). Nanocarriers as an emerging platform for cancer therapy. *Nature Nanotechnology 2007 2:12*, *2*(12), 751–760. https://doi.org/10.1038/nnano.2007.387

Petersen, A., Anderson, A., Wilkinson, C., & Allan, S. (2007). Nanotechnologies, risk and society. *Health Risk & Society*, *9*(2), 117–124. https://doi.org/10.1080/13698570701306765

Petros, R. A., & DeSimone, J. M. (2010). Strategies in the design of nanoparticles for therapeutic applications. *Nature Reviews Drug Discovery*, *9*(8), 615–627. https://doi.org/10.1038/nrd2591

Pidgeon, N., & Rogers-Hayden, T. (2007). Opening up nanotechnology dialogue with the publics: risk communication or "upstream engagement"? *Health Risk & Society*, *9*(2), 191–210. https://doi.org/10.1080/13698570701306906

Plaza-Oliver, M., Santander-Ortega, M. J., & Lozano, M. V. (2021). Current approaches in lipid-based nanocarriers for oral drug delivery. *Drug Delivery and Translational Research*, *11*(2), 471–497. https://doi.org/10.1007/s13346-021-00908-7

Poel van de, I. R., & Royakkers, L. M. M. (2011). *Ethics, Technology, and Engineering: An Introduction.* Wiley-Blackwell. http://search.ebscohost.com/login.aspx?direct=true&scope=site&db=nlebk&db=nlabk&AN=510116

Polack, F. P., Thomas, S. J., Kitchin, N., Absalon, J., & others. (2020). Safety and efficacy of the BNT162B2 mRNA COVID-19 vaccine. *New England Journal of Medicine, 383*(27), 2603–2615. https://doi.org/10.1056/nejmoa2034577

Prajapati, A., Rangra, S., Patil, R., Desai, N., Jyothi, V. G. S. S., Salave, S. et al. (2024). Receptor-Targeted nanomedicine for cancer therapy. *Receptors, 3*(3), 323–361. https://doi.org/10.3390/receptors3030016

Presente, A., & Dowdy, S. F. (2013). PTD/CPP peptide-mediated delivery of siRNAs. *Current Pharmaceutical Design*, *19*(16), 2943–2947. https://doi.org/10.2174/1381612811319160008

Qin, X., Yu, C., Wei, J., Li, L., Zhang, C., Wu, Q., et al. (2019). Rational design of nanocarriers for intracellular protein delivery. *Advanced Materials*, *31*(46). https://doi.org/10.1002/adma.201902791

Rarokar, N., Yadav, S., Saoji, S., Bramhe, P., Agade, R., Gurav, S. et al. (2024). Magnetic nanosystem a tool for targeted delivery and diagnostic application: current challenges and recent advancement. *International Journal of Pharmaceutics X, 7*, 100231. https://doi.org/10.1016/j.ijpx.2024.100231

Rauscher, H., Roebben, G., Mech, A., Gibson, P., Kestens, V., Linsinger, T. et al (2019). *An overview of concepts and terms used in the European Commission's definition of nanomaterial* (Issue EUR 29647 EN). https://doi.org/10.2760/459136

Raval, N., Maheshwari, R., Kalyane, D., Youngren-Ortiz, S. R., Chougule, M. B., & Tekade, R. K. (2019). Importance of physicochemical characterization of nanoparticles in pharmaceutical product development. In *Elsevier eBooks* (pp. 369–400). https://doi.org/10.1016/b978-0-12-817909-3.00010-8

Raza, K., Kumar, P., Kumar, N., & Malik, R. (2017). Pharmacokinetics and biodistribution of the nanoparticles. In Surendra Nimesh, Ramesh Chandra, and Nidhi Gupta (Eds.), *Advances in Nanomedicine for the Delivery of Therapeutic Nucleic Acids*, 165–186. Woodhead Publishing. https://doi.org/10.1016/B978-0-08-100557-6.00009-2

Rehman, U., Parveen, N., Sheikh, A., Abourehab, M. A. S., Sahebkar, A., & Kesharwani, P. (2022). Polymeric nanoparticles-siRNA as an emerging nano-polyplexes against ovarian cancer. *Colloids and Surfaces. B, Biointerfaces, 218*. https://doi.org/10.1016/J.COLSURFB.2022.112766

Resnik, D. B., & Tinkle, S. S. (2007). Ethics in nanomedicine. *Nanomedicine, 2*(3), 345–350. https://doi.org/10.2217/17435889.2.3.345

Richardson, T. P., Peters, M. C., Ennett, A. B., & Mooney, D. J. (2001). Polymeric system for dual growth factor delivery. *Nature Biotechnology, 19*(11), 1029–1034. https://doi.org/10.1038/nbt1101-1029

Rissin, D. M., Kan, C. W., Campbell, T. G., Howes, S. C., Fournier, D. R., Song, L. et al. (2010). Single-molecule enzyme-linked immunosorbent assay detects serum proteins at subfemtomolar concentrations. *Nature Biotechnology, 28*(6), 595–599. https://doi.org/10.1038/nbt.1641

Rojekar, S., Gholap, A. D., Togre, N., Bhoj, P., Haeck, C., Hatvate, N. et al. (2024). Current status of mannose receptor-targeted drug delivery for improved anti-HIV therapy. *Journal of Controlled Release, 372*, 494–521. https://doi.org/10.1016/j.jconrel.2024.06.002

Sachdeva, S., Davis, R. W., & Saha, A. K. (2021). Microfluidic Point-of-Care Testing: commercial landscape and future directions. *Frontiers in Bioengineering and Biotechnology, 8*. https://doi.org/10.3389/fbioe.2020.602659

Safari, J., & Zarnegar, Z. (2013). Advanced drug delivery systems: nanotechnology of health design A review. *Journal of Saudi Chemical Society, 18*(2), 85–99. https://doi.org/10.1016/j.jscs.2012.12.009

Saha, K., Agasti, S. S., Kim, C., Li, X., & Rotello, V. M. (2012). Gold nanoparticles in chemical and biological sensing. *Chemical Reviews, 112*(5), 2739–2779. https://doi.org/10.1021/cr2001178

Sahin, U., Karikó, K., & Türeci, Ö. (2014). mRNA-based therapeutics — developing a new class of drugs. *Nature Reviews Drug Discovery, 13*(10), 759–780. https://doi.org/10.1038/nrd4278

Salvati, A., Åberg, C., dos Santos, T., Varela, J., Pinto, P., Lynch, I. et al. (2011). Experimental and theoretical comparison of intracellular import of polymeric nanoparticles and small molecules: toward models of uptake kinetics. *Nanomedicine: Nanotechnology, Biology, and Medicine, 7*(6), 818–826. https://doi.org/10.1016/J.NANO.2011.03.005

Samaridou, E., Heyes, J., & Lutwyche, P. (2020). Lipid nanoparticles for nucleic acid delivery: current perspectives. *Advanced Drug Delivery Reviews, 154–155*, 37–63. https://doi.org/10.1016/j.addr.2020.06.002

Sanità, G., Carrese, B., & Lamberti, A. (2020). Nanoparticle surface functionalization: how to improve biocompatibility and cellular internalization. *Frontiers in Molecular Biosciences, 7*, 587012. https://doi.org/10.3389/fmolb.2020.587012

Saraiva, C., Praça, C., Ferreira, R., Santos, T., Ferreira, L., & Bernardino, L. (2016). Nanoparticle-mediated brain drug delivery: overcoming blood-brain barrier to treat neurodegenerative diseases. *Journal of Controlled Release: Official Journal of the Controlled Release Society, 235*, 34–47. https://doi.org/10.1016/J.JCONREL.2016.05.044

Serov, N., & Vinogradov, V. (2022). Artificial intelligence to bring nanomedicine to life. *Advanced Drug Delivery Reviews, 184*, 114194. https://doi.org/10.1016/j.addr.2022.114194

Shatkin, J. A., & Kim, B. (2015). Cellulose nanomaterials: life cycle risk assessment, and environmental health and safety roadmap. *Environmental Science Nano, 2*(5), 477–499. https://doi.org/10.1039/c5en00059a

Shen, Y., Gwak, H., & Han, B. (2023). Advanced manufacturing of nanoparticle formulations of drugs and biologics using microfluidics. *The Analyst, 149*(3), 614–637. https://doi.org/10.1039/d3an01739g

Shi, J., Kantoff, P. W., Wooster, R., & Farokhzad, O. C. (2016). Cancer nanomedicine: progress, challenges and opportunities. *Nature Reviews. Cancer*, *17*(1), 20–37. https://doi.org/10.1038/nrc.2016.108

Shi, K., Dou, B., Yang, J., Yuan, R., & Xiang, Y. (2016). Cascaded strand displacement for non-enzymatic target recycling amplification and label-free electronic detection of microRNA from tumor cells. *Analytica Chimica Acta, 916*, 1–7. https://doi.org/10.1016/j.aca.2016.02.034

Shi, L., Zhang, J., Zhao, M., Tang, S., Cheng, X., Zhang, W. et al. (2021). Effects of polyethylene glycol on the surface of nanoparticles for targeted drug delivery. *Nanoscale*, *13*(24), 10748–10764. https://doi.org/10.1039/D1NR02065J

Shi, M., Kretlow, J. D., Nguyen, A., Young, S., Baggett, L. S., Wong, M. E., et al. (2010). Antibiotic-releasing porous polymethylmethacrylate constructs for osseous space maintenance and infection control. *Biomaterials*, *31*(14), 4146–4156. https://doi.org/10.1016/j.biomaterials.2010.01.112

Shi, P., Cheng, Z., Zhao, K., Chen, Y., Zhang, A., Gan, W., & Zhang, Y. (2023). Active targeting schemes for nano-drug delivery systems in osteosarcoma therapeutics. *Journal of Nanobiotechnology*, *21*(1), 1–27. https://doi.org/10.1186/S12951-023-01826-1

Shire, S. J., Shahrokh, Z., & Liu, J. (2004). Challenges in the development of high protein concentration formulations. *Journal of Pharmaceutical Sciences*, *93*(6), 1390–1402. https://doi.org/10.1002/jps.20079

Shirian, J., Hockla, A., Gleba, J. J., Coban, M., Rotenberg, N., Strik, L. M., et al. (2024). Improving circulation Half-Life of therapeutic candidate N-TIMP2 by unfolded peptide extension. *Biomolecules*, *14*(9), 1187. https://doi.org/10.3390/biom14091187

Shubayev, V. I., Pisanic, T. R., & Jin, S. (2009). Magnetic nanoparticles for theragnostics. *Advanced Drug Delivery Reviews*, *61*(6), 467–477. https://doi.org/10.1016/j.addr.2009.03.007

Silva, G. A., Czeisler, C., Niece, K. L., Beniash, E., Harrington, D. A., Kessler, J. A. et al. (2004). Selective differentiation of neural progenitor cells by High-Epitope Density nanofibers. *Science*, *303*(5662), 1352–1355. https://doi.org/10.1126/science.1093783

Singh, A. P., Biswas, A., Shukla, A., & Maiti, P. (2019). Targeted therapy in chronic diseases using nanomaterial-based drug delivery vehicles. *Signal Transduction and Targeted Therapy*, *4*(1). https://doi.org/10.1038/s41392-019-0068-3

Skowicki, M., Tarvirdipour, S., Kraus, M., Schoenenberger, C. A., & Palivan, C. G. (2024). Nanoassemblies designed for efficient nuclear targeting. *Advanced Drug Delivery Reviews, 211*, 115354. https://doi.org/10.1016/J.ADDR.2024.115354

Slowing, I. I., Vivero-Escoto, J. L., Wu, C., & Lin, V. S. (2008). Mesoporous silica nanoparticles as controlled release drug delivery and gene transfection carriers. *Advanced Drug Delivery Reviews*, *60*(11), 1278–1288. https://doi.org/10.1016/j.addr.2008.03.012

Soe, T. H., Watanabe, K., & Ohtsuki, T. (2020). Photoinduced endosomal escape mechanism: a view from photochemical internalization mediated by CPP-photosensitizer conjugates. *Molecules*, *26*(1), 36. https://doi.org/10.3390/molecules26010036

Soto, F., Wang, J., Ahmed, R., & Demirci, U. (2020). Medical micro/nanorobots in precision medicine. *Advanced Science*, *7*(21). https://doi.org/10.1002/advs.202002203

Souto, E. B., Blanco-Llamero, C., Krambeck, K., Kiran, N. S., Yashaswini, C., Postwala, H. et al. (2024). Regulatory insights into nanomedicine and gene vaccine innovation: safety assessment, challenges, and regulatory perspectives. *Acta Biomaterialia, 180*, 1–17. https://doi.org/10.1016/j.actbio.2024.04.010

Su, S., & Kang, P. M. (2020). Recent advances in nanocarrier-assisted therapeutics delivery systems. *Pharmaceutics*, *12*(9), 837. https://doi.org/10.3390/pharmaceutics12090837

Sun, D., & Lu, Z. (2023). Structure and function of cationic and ionizable lipids for nucleic acid delivery. *Pharmaceutical Research*, *40*(1), 27–46. https://doi.org/10.1007/s11095-022-03460-2

Sun, Q., Ojha, T., Kiessling, F., Lammers, T., & Shi, Y. (2017). Enhancing tumor penetration of nanomedicines. *Biomacromolecules*, *18*(5), 1449–1459. https://doi.org/10.1021/acs.biomac.7b00068

Talamini, L., Matsuura, E., De Cola, L., & Muller, S. (2021). Immunologically inert nanostructures as selective therapeutic tools in inflammatory diseases. *Cells, 10*(3), 707. https://doi.org/10.3390/cells10030707

Tan, P., Chen, X., Zhang, H., Wei, Q., & Luo, K. (2023). Artificial intelligence aids in development of nanomedicines for cancer management. *Seminars in Cancer Biology, 89*, 61–75. https://doi.org/10.1016/j.semcancer.2023.01.005

Tchessalov, S., Maglio, V., Kazarin, P., Alexeenko, A., Bhatnagar, B., Sahni, E., et al. (2023). Practical advice on scientific design of freeze-drying process: 2023 update. *Pharmaceutical Research*, *40*(10), 2433–2455. https://doi.org/10.1007/s11095-023-03607-9

Tinkle, S., McNeil, S. E., Mühlebach, S., Bawa, R., Borchard, G., Barenholz, Y. et al. (2014). Nanomedicines: addressing the scientific and regulatory gap. *Annals of the New York Academy of Sciences*, *1313*(1), 35–56. https://doi.org/10.1111/nyas.12403

Todaro, B., Ottalagana, E., Luin, S., & Santi, M. (2023). Targeting peptides: the new generation of targeted drug delivery systems. *Pharmaceutics*, *15*(6), 1648. https://doi.org/10.3390/pharmaceutics15061648

Torchilin, V. P. (2014). Multifunctional, stimuli-sensitive nanoparticulate systems for drug delivery. *Nature Reviews Drug Discovery*, *13*(11), 813–827. https://doi.org/10.1038/nrd4333

Toy, R., Peiris, P. M., Ghaghada, K. B., & Karathanasis, E. (2014). Shaping cancer nanomedicine: the effect of particle shape on the in vivo journey of nanoparticles. *Nanomedicine (London, England), 9*(1), 121. https://doi.org/10.2217/NNM.13.191

Trenfield, S. J., Awad, A., Goyanes, A., Gaisford, S., & Basit, A. W. (2018). 3D printing pharmaceuticals: drug development to frontline care. *Trends in Pharmacological Sciences*, *39*(5), 440–451. https://doi.org/10.1016/j.tips.2018.02.006

Uskoković, V., & Ghosh, S. (2016). Carriers for the tunable release of therapeutics: etymological classification and examples. *Expert Opinion on Drug Delivery*, *13*(12), 1729–1741. https://doi.org/10.1080/17425247.2016.1200558

Varkouhi, A. K., Scholte, M., Storm, G., & Haisma, H. J. (2010). Endosomal escape pathways for delivery of biologicals. *Journal of Controlled Release*, *151*(3), 220–228. https://doi.org/10.1016/j.jconrel.2010.11.004

Ventola, C. L. (2017). Progress in nanomedicine: approved and investigational nanodrugs. *P&T: A Peer-Reviewed Journal for Formulary Management*, *42*(12), 742–755. https://pubmed.ncbi.nlm.nih.gov/29234213/

Veronese, F. M., & Pasut, G. (2005). PEGylation, successful approach to drug delivery. *Drug Discovery Today*, *10*(21), 1451–1458. https://doi.org/10.1016/s1359-6446(05)03575-0

Wang, J., & Musameh, M. (2004). Carbon nanotube screen-printed electrochemical sensors. *The Analyst*, *129*(1), 1. https://doi.org/10.1039/b313431h

Wang, W., Singh, S., Zeng, D. L., King, K., & Nema, S. (2006). Antibody structure, instability, and formulation. *Journal of Pharmaceutical Sciences*, *96*(1), 1–26. https://doi.org/10.1002/jps.20727

Wang, Y., & Kohane, D. S. (2017). External triggering and triggered targeting strategies for drug delivery. *Nature Reviews Materials, 2*(6). https://doi.org/10.1038/natrevmats.2017.20

Wasti, S., Lee, I. H., Kim, S., Lee, J., & Kim, H. (2023). Ethical and legal challenges in nanomedical innovations: a scoping review. *Frontiers in Genetics, 14*. https://doi.org/10.3389/fgene.2023.1163392

Wilhelm, S., Tavares, A. J., Dai, Q., Ohta, S., Audet, J., Dvorak, H. F. et al (2016). Analysis of nanoparticle delivery to tumours. *Nature Reviews Materials*, *1*(5). https://doi.org/10.1038/natrevmats.2016.14

Won, Y. W., Yoon, S. M., Lee, K. M., & Kim, Y. H. (2011). Poly(oligo-D-arginine) with internal disulfide linkages as a cytoplasm-sensitive carrier for siRNA delivery. *Molecular Therapy: The Journal of the American Society of Gene Therapy*, *19*(2), 372–380. https://doi.org/10.1038/MT.2010.242

Wu, B., Hou, S., Yin, F., Zhao, Z., Wang, Y., Wang, X., et al. (2007). Amperometric glucose biosensor based on multilayer films via layer-by-layer self-assembly of multi-wall carbon nanotubes, gold nanoparticles and glucose oxidase on the Pt electrode. *Biosensors and Bioelectronics*, *22*(12), 2854–2860. https://doi.org/10.1016/j.bios.2006.11.028

Wu, J., Wang, X., Wang, Q., Lou, Z., Li, S., Zhu, Y. et al. (2018). Nanomaterials with enzyme-like characteristics (nanozymes): next-generation artificial enzymes (II). *Chemical Society Reviews*, *48*(4), 1004–1076. https://doi.org/10.1039/c8cs00457a

Xu, H., Medina-Sánchez, M., Maitz, M. F., Werner, C., & Schmidt, O. G. (2020). Sperm micromotors for cargo delivery through flowing blood. *ACS Nano*, *14*(3), 2982–2993. https://doi.org/10.1021/acsnano.9b07851

Xu, Q., Xiao, Z., Yang, Q., Yu, T., Deng, X., Chen, N. et al (2024). Hydrogel-based cardiac repair and regeneration function in the treatment of myocardial infarction. *Materials Today Bio*, 100978. https://doi.org/10.1016/j.mtbio.2024.100978

Xu, S., Zhang, H., Qian, Z., & Yuan, W. (2024). pH-Responsive injectable self-healing hydrogels loading Au nanoparticles-decorated bimetallic organic frameworks for synergistic sonodynamic-chemodynamic-starvation-chemo therapy of cancer. *Journal of Colloid and Interface Science, 675*, 746–760. https://doi.org/10.1016/j.jcis.2024.07.039

Xu, X., Ho, W., Zhang, X., Bertrand, N., & Farokhzad, O. (2015). Cancer nanomedicine: from targeted delivery to combination therapy. *Trends in Molecular Medicine*, *21*(4), 223–232. https://doi.org/10.1016/j.molmed.2015.01.001

Xuan, L., Ju, Z., Skonieczna, M., Zhou, P., & Huang, R. (2023). Nanoparticles-induced potential toxicity on human health: applications, toxicity mechanisms, and evaluation models. *MedComm*, *4*(4). https://doi.org/10.1002/mco2.327

Yan, S., Na, J., Liu, X., & Wu, P. (2024). Different targeting ligands-mediated drug delivery systems for tumor therapy. *Pharmaceutics*, *16*(2). https://doi.org/10.3390/PHARMACEUTICS16020248

Yanar, F., Carugo, D., & Zhang, X. (2023). Hybrid nanoplatforms comprising organic nanocompartments encapsulating inorganic nanoparticles for enhanced drug delivery and bioimaging applications. *Molecules*, *28*(15), 5694. https://doi.org/10.3390/MOLECULES28155694

Yang, E. C., Divine, R., Miranda, M. C., Borst, A. J., Sheffler, W., Zhang, J. Z. et al. (2024). Computational design of non-porous pH-responsive antibody nanoparticles. *Nature Structural & Molecular Biology*, *31*(9), 1404–1412. https://doi.org/10.1038/s41594-024-01288-5

Yang, J., & Luo, G. F. (2023). Peptide-based vectors for gene delivery. *Chemistry*, *5*(3), 1696–1718. https://doi.org/10.3390/CHEMISTRY5030116

Ye, Q., & Sun, J. (2023). Nanorobots for drug delivery, surgery, and biosensing. In Ki-Taek Lim and Kamel A. Abd-Elsalam *(Eds.), Nanorobotics and Nanodiagnostics in Integrative Biology and Biomedicine*, 15–34. Springer. https://doi.org/10.1007/978-3-031-16084-4_2

Yetisgin, A. A., Cetinel, S., Zuvin, M., Kosar, A., & Kutlu, O. (2020). Therapeutic nanoparticles and their targeted delivery applications. *Molecules (Basel, Switzerland)*, *25*(9). https://doi.org/10.3390/MOLECULES25092193

Yu, M., Wu, J., Shi, J., & Farokhzad, O. C. (2015). Nanotechnology for protein delivery: overview and perspectives. *Journal of Controlled Release, 240*, 24–37. https://doi.org/10.1016/j.jconrel.2015.10.012

Zamboni, W. C., Torchilin, V., Patri, A. K., Hrkach, J., Stern, S., Lee, R. et al. (2012). Best practices in cancer nanotechnology: perspective from NCI nanotechnology alliance. *Clinical Cancer Research: An Official Journal of the American Association for Cancer Research, 18*(12), 3229–3241. https://doi.org/10.1158/1078-0432.CCR-11-2938

Zhang, P., Sun, F., Liu, S., & Jiang, S. (2016). Anti-PEG antibodies in the clinic: current issues and beyond PEGylation. *Journal of Controlled Release, 244*, 184–193. https://doi.org/10.1016/j.jconrel.2016.06.040

Zhang, Q., Dehaini, D., Zhang, Y., Zhou, J., Chen, X., Zhang, L. et al. (2018). Neutrophil membrane-coated nanoparticles inhibit synovial inflammation and alleviate joint damage in inflammatory arthritis. *Nature Nanotechnology*, *13*(12), 1182–1190. https://doi.org/10.1038/s41565-018-0254-4

Zhang, S., Gao, H., & Bao, G. (2015). Physical principles of nanoparticle cellular endocytosis. *ACS Nano*, *9*(9), 8655–8671. https://doi.org/10.1021/acsnano.5b03184

Zhang, S., Zhao, B., Jiang, H., Wang, B., & Ma, B. (2007). Cationic lipids and polymers mediated vectors for delivery of siRNA. *Journal of Controlled Release: Official Journal of the Controlled Release Society*, *123*(1), 1–10. https://doi.org/10.1016/J.JCONREL.2007.07.016

Zhang, X. D., Wu, D., Shen, X., Liu, P. X., Fan, F. Y., & Fan, S. J. (2012). In vivo renal clearance, biodistribution, toxicity of gold nanoclusters. *Biomaterials*, *33*(18), 4628–4638. https://doi.org/10.1016/J.BIOMATERIALS.2012.03.020

Zhao, L., Weir, M. D., & Xu, H. H. (2010). An injectable calcium phosphate-alginate hydrogel-umbilical cord mesenchymal stem cell paste for bone tissue engineering. *Biomaterials*, *31*(25), 6502–6510. https://doi.org/10.1016/j.biomaterials.2010.05.017

Zhao, Z., Ukidve, A., Kim, J., & Mitragotri, S. (2020). Targeting strategies for tissue-specific drug delivery. *Cell*, *181*(1), 151–167. https://doi.org/10.1016/J.CELL.2020.02.001

# 2 Design Principles of Nanocarriers

## *Size, Shape, and Surface Properties, Stimuli-Responsive Systems*

*Juan C. Cruz and Luis H. Reyes*

### 2.1 INTRODUCTION

Nanocarriers are one of the most important developments in therapeutics' delivery systems of nucleic acids and proteins. Targeted drug-delivery systems should always consider the size and shape of the nanocarriers to have optimum size, shape, surface properties, and stimuli responsiveness. This will include fundamental aspects of the design of nanocarriers, enlightening their paramount role in increasing the therapeutic efficacy and specificity of drug delivery for other complex diseases, but primarily cancer.

The size of nanocarriers plays a pivotal role in their biodistribution, cellular uptake, and clearance from the body. Nanocarriers in the 10–200 nm size range can use the enhanced permeability and retention (EPR) effect to target the tumor efficiently. In this effect, the vasculature of the tumor tissues is enough to be considered leaky, allowing nanocarriers to accumulate passively [1]. In addition, the circulation half-life of the nanocarriers, which this size would strongly influence, might be subject to being taken up by the mononuclear phagocyte system (MPS) and undergoing renal clearance [2].

The shape of the nanocarrier highly determines how it will interact with the biological environment: cellular internalization and biodistribution. Namely, nanocarriers that are non-spherical in shape (rods or discs) preferably accumulate in the tumor tissue and thus may be more effective than spherical ones in treating the outcomes [3]. The aspect ratio and the surface curvature of the nanocarrier may further enhance their permeation through the biological barriers and help them evade immune recognition [4].

In addition, surface modification of the nanocarriers comprises targeting ligands or stealth coatings, such as polyethylene glycol (PEG), to increase the specificity of the nanocarrier and avert host defenses. This way, those nanocarriers with targeting ligands are fixed to specific cell types or tissues; hence, the delivery efficiency to the site of action is increased. Besides, the use of nanocarriers protects against opsonization by surface modification, which allows them to have a long time of exposure in circulation in the blood [5]. Furthermore, the use of nanocarriers allows protection against opsonization by surface modifications.

Thus, the stimuli-responsive nanocarriers become an excellent choice for delivering the therapeutic payload in response to a specific cue, such as pH, temperature, or enzymatic activity, that allows controlled drug release at the site of action. For example, this can be employed using the acidic tumor microenvironment or endosomal pH for drug release, thereby enhancing the therapeutic index while concomitantly diminishing systemic toxicity [6].

The design principles of nanocarriers are integral to their function as drug-delivery systems (Figure 2.1). In terms of size, shape, surface properties, and incorporation of stimuli-responsive elements, nanocarriers are becoming suitable for reaching the target delivery of nucleic acids and proteins in varying kinds of diseases. The remaining problems in the area include better specificity in targeting, reduced off-target effects, and increased stability of nanocarriers. Further development in the design of nanocarriers would probably bring the greatest revolution in therapeutic delivery, helping to realize effective and personalized medical treatments.

#### 2.1.1 Size

The size of nanocarriers is a crucial parameter in their design; it dramatically influences the biodistribution, cellular uptake, and clearance of nanocarriers and, eventually, will affect the therapeutic efficacy of drug-delivery systems. Understanding the optimal size ranges of nanocarriers is thus important in rational design because it will enable these systems to pass effectively through the complex biological environment to the targeted sites.

The biodistribution of nanocarriers is heavily dependent on their size. For example, most tumors have been shown to be highly leaky [2]; therefore, nanoparticles (NPs) in the range of 10–100 nm preferentially accumulate in tumor tissues through the effect of EPR. This size-dependent accumulation can significantly impact the therapeutic efficacy of anticancer drugs. Otherwise, when the size of NPs exceeds 200 nm, the MPS is more likely to recognize and remove them. This reduces their time of circulation in the body, and they may thus not reach the site of action [1].

In addition to the above, the nanocarrier's size also influences the cells' uptake. Internalization into the cells of particles below 50 nm in size might be much better than that of particles exceeding 50 nm through several endocytosis routes. For example, it has been reported that particles of 50 nm have shown higher uptake rates by cancer cells, thus increasing efficiency in drug delivery to target sites [7]. However, NPs of a


DOI: 10.1201/9781003473183-2

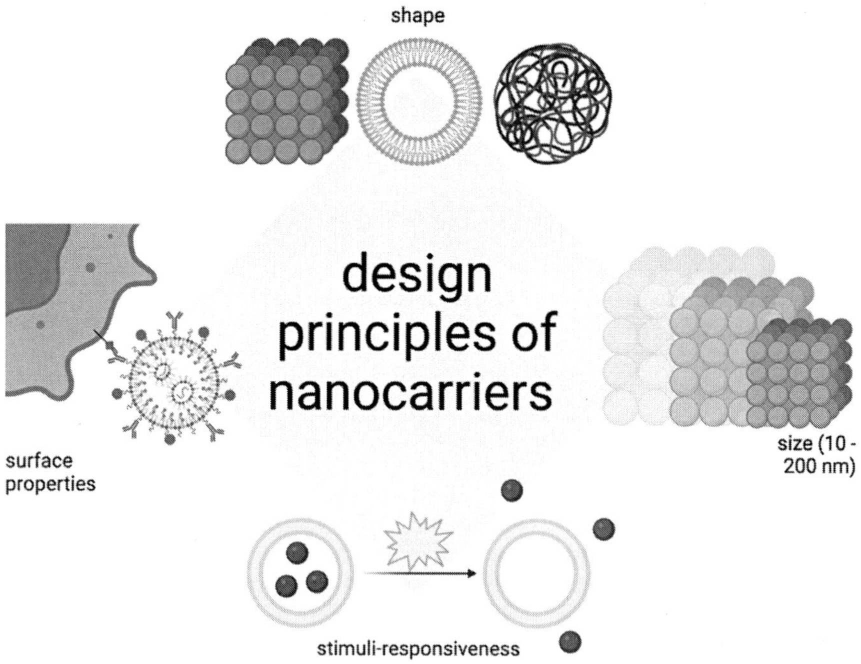


**Figure 2.1** Design principles of nanocarriers. This schematic illustrates the key factors influencing nanocarrier design and function: shape (represented by spherical, liposomal, and polymeric structures), size (ranging from 10 to 200 nm), surface properties (depicted by a functionalized nanoparticle with various ligands), and stimuli responsiveness (shown as a nanocarrier releasing its payload upon stimulation). These design principles are crucial for optimizing nanocarrier performance in drug delivery, affecting biodistribution, cellular uptake, and therapeutic efficacy. The interplay between these factors determines the nanocarrier's ability to overcome biological barriers and achieve targeted drug delivery.

size lower than 10 nm can rapidly escape from the target tissue due to diffusion and subsequently be excreted via the kidneys, thus compromising the delivery of the therapeutic payload into the target cells [8].

The clearance of nanocarriers from the body is another crucial aspect influenced by their size. Particles smaller than 10 nm are quickly cleared from circulation by renal filtration, whereas those larger than 200 nm are predominantly cleared by the liver and spleen [9]. A general optimal size between 10 and 100 nm is always preferable to effectively penetrate tissues at a trade-off for the most effective circulation time without being cleared too early, thereby maximizing therapeutic efficacy.

Ideal size ranges for these are from 10 to 100 nm and are preferred for application and targeting tissue. In the case of tumor targeting through the EPR effect, sizes ranging around 10–100 nm of the nanocarriers are preferably used, allowing deep penetration into tumor tissue and efficient cell uptake. Those of a size toward the upper limits of this range might find the best balance between tissue accumulation and clearance in applications where passive targeting for long circulation times is required, for example, toward cardiovascular diseases or inflammatory sites [10].

Size is one of the most important criteria influencing nanocarriers' biodistribution, cellular uptake, and clearance, thereby affecting therapeutic success. Therefore, size tuning of the nanocarriers is crucial for optimizing their delivery efficiency with reduced off-target effects, which increases therapeutic efficacy. Future developments in the design of the nanocarrier will further delve into the complex interplay of NP size in biological interactions, essential for drug-delivery systems to be optimized against the broadest range of therapeutic applications.

### 2.1.2 Shape

The shape factor in nanocarrier design remarkably influences their biological interaction, including circulation time, cellular internalization, and targeting ability. Later research on other geometrical forms, such as spherical, rod-shaped, and many other geometries, demonstrated that the exhibited behaviors were geometrical form-dependent, which can be harnessed to optimize drug delivery and therapeutic outcomes.

The size, shape, and flexibility of nanocarriers significantly influence their blood circulation time. Historically, spherical NPs were preferred for their ease of synthesis and uniformity. Recent studies, however, show that elongated shapes like rods or worms demonstrate increased circulation times due to reduced uptake by the MPS compared to spherical NPs. This extended circulation enhances the nanocarrier's ability to reach and accumulate at the target site, improving therapeutic efficacy [11, 12].

Additionally, the nanocarrier's shape also affects the cell's internalization. Because of endocytosis, in a spherical-shaped NP, penetration into a cell would be evident without any doubt. For instance, rod-shaped or other elongated NPs were better internalized and followed several routes, including macropinocytosis or receptor-mediated endocytosis, depending on the aspect ratio and surface modifications [13]. This enhanced and selective uptake can be leveraged to improve drug delivery to specific cell types or tissues.

Geometry has been recognized as an important determinant of the targeting capability of nanocarriers. Spherical NPs target tumor sites passively through EPR, but rod-shaped and other nonspherical NPs have evidenced better-targeting abilities. Geometries of this kind may help better interact with some cell receptors and increase the binding affinity of target cells. For example, rod-shaped NPs can easily penetrate deeper into tumors and thus are more effective in producing improved therapeutics for cancer [14].

Further studies of several geometrical shapes in the case of NPs bring out quite interesting features and possible applications. NPs with a spherical shape are especially remarkable for facile syntheses and effective cellular uptake [15]. However, these shapes have shorter circulation times and a lower precision of targeting than other shapes [16]. On the other hand, rod-shaped NPs have the advantage of distribution over time and have been attributed to their elongated form. This property supports their presence not only in the bloodstream but also increases the penetration depth of the tumors, which allows better localization of the drug and may, in turn, improve therapeutic efficiency [17].

The exploration of NP designs extends beyond rod-shaped and spherical forms, incorporating discoidal, worm-like, or cubic shapes. Each shape offers unique opportunities for optimizing nanocarrier design, allowing for the precise tuning of formulations in combination with specific compounds. This variety in shape enables the development of distinct biodistribution profiles, targeting capabilities, and cellular uptake mechanisms, catering to specific therapeutic needs. Such diversity in NP design is crucial for enhancing the effectiveness of targeted drug-delivery systems, ultimately improving therapeutic outcomes [18, 19].

### 2.1.3 Surface Properties

All the surface features of nanocarriers, like their charge and changes that make them more soluble, stable, and specific to targets in the immune system and clearance, are due to their strong interaction with living systems. Therefore, these properties could be refined for better biocompatibility, targeting efficiency, and therapeutic efficacy toward effective drug-delivery applications.

It has been reported that the surface charge of nanocarriers considerably affects their interaction with biological membranes, leading to consequences for their biodistribution, cellular uptake, and clearance. Dey et al. [20] point out that the surface charge is the most crucial factor in modulating the physiology of the macrophage, and it has to be adequately optimized for gene delivery and mRNA vaccine delivery. Surface changes like PEGylation and ligand attachment were used to do this. These changes are meant to make the nanocarriers more stable and soluble while also making them more specific to their target.

On the other hand, Gajbhiye et al. [21] explained that PEGylation greatly extends the circulation time of the nanocarriers because it dramatically decreases their uptake by the Reticuloendothelial System (RES). They regarded it as a rather predominant and critical mechanism that results in upregulation of nanocarrier uptake by the brain. Furthermore, Milani et al. [22] go further to outline how the cellular uptake of targeted nanocarriers, along with their nanotoxicity and site specificity, has been upwardly modulated toward lower toxicity and higher site specificity as one strategy of general therapeutic enhancement.

The interaction of nanocarriers with the immune system is profoundly affected by their surface chemistry. Vincent et al. [23] further illustrate that surface chemistry-mediated modulation of adsorbed albumin folding state may be one of the mechanisms to guide the clearance of nanocarriers by the distinct *in vivo* macrophage subset, hence promoting evasion or nonspecific clearance. This finding is pivotal for designing nanocarriers to avoid premature clearance and achieve targeted delivery. Additionally, attaching antibodies to the nanocarrier surface first ensures they work and can target cells when the biomolecular corona is added later. This also suggests that the properties of the surface are conserved, and the functionality toward targeting the nanocarriers can even be provided in the presence of the biomolecular corona [24].

Nanocarriers' surface properties, such as charge, solubility, and stability, are fundamental in interacting with the immune system. These properties directly affect how well they target and treat diseases. Nanocarriers can function better for specific therapeutic uses by engineering and functionalizing their surfaces to fine-tune specific properties. This improves drug delivery and treatment effectiveness. This approach to surface engineering improves the delivery and function of nanocarriers and opens new avenues for overcoming the challenges of biological systems for efficient, targeted drug delivery. As a result, there is a growing focus on surface modification techniques to develop nanocarriers with improved performance in clinical applications [23, 25].

## 2.2 INTRODUCTION TO STIMULI-RESPONSIVE NANOCARRIERS

Stimuli-responsive nanocarriers represent a new shift in targeted drug-delivery technology and stand at the frontier of such development. They offer the most sophisticated system using environmental cues to release therapeutic agents where they are most required. These are well-tuned systems to a vast number of internal and external stimuli, going from biochemical ones, such as pH and redox potential, to physical ones, like temperature, magnetic fields, and light (Figure 2.2). At this line, the response is the basis for high efficiency in treatments that will offer, at the same time, dramatically improved therapeutic outcomes and minimize the risk of unwanted systemic side effects.

The hallmark of these nanocarriers is their fine-tuned ability to deliver adroitly in a controlled mode in response to different physiological statuses or outside environments. This precise targeting mechanism has not made an enormous difference in ensuring that therapeutic agents stay where they are needed most and that only a small amount of drugs reach healthy tissues to avoid harmful side effects and get the most out of the therapy. For example, pH-sensitive nanocars can be designed in such a way that they can make use of the acidity difference between the normal tissue and the acidic microenvironment of the tumors. It thereby facilitates selective antitumor drug release at the tumor site. Efficiency is very high in this particular method of drug release, according to Sethuraman et al. [26].

Among the intrinsic bodily conditions, the nanocarriers are responsible for detecting are changes in pH, redox potential, and the concentration of specific enzymes, among others. These are the conditions at the pathological site, such as the acidic pH found in tumor environments and the high glutathione concentration in cancer cells, which may be used as a trigger point for the release of drugs. Handa et al. [27] thus have given compelling evidence on the effectiveness of utilizing these internal stimuli for targeted therapy. External stimuli include temperature, magnetic fields, light, and ultrasound, which underlie non-invasive therapeutic interventions, allowing conduction and possessing more precision of time. Guo and You [28] created a new type of light-responsive nanocarriers that are mostly excited by near-infrared (NIR) light. These nanocarriers have much potential for controlled drug release while penetrating deep into tissues with the least damage to healthy tissue around them. This is an auspicious direction for controlled drug delivery.

Targeted and controlled drug delivery by stimuli-responsive nanocarriers points to the utmost necessity in this area of several specialties, including cancer therapy. This means the combination of redox, light, and pH stimuli with cyclodextrin-based nanocarriers offers a versatile ground for the controlled release mechanism of the chemotherapy agent [29]. Meanwhile, dual stimuli of pH and redox conditions have attracted more attention, such as being highly demanded to a greater extent in advancing technology for

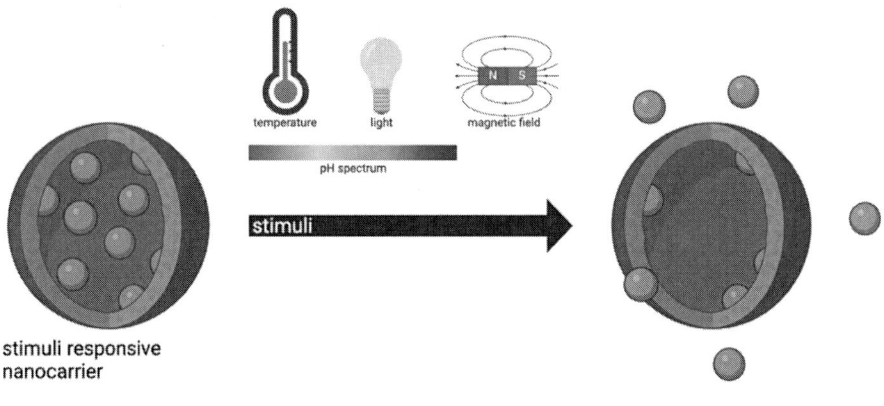


**Figure 2.2** Stimuli-responsive nanocarriers and their activation mechanisms. This figure demonstrates the concept of stimuli-responsive drug release from nanocarriers. On the left, a loaded nanocarrier is shown before stimulation. Upon exposure to specific stimuli (represented by icons for temperature, light, magnetic field, and pH), the nanocarrier undergoes structural changes, leading to the release of its payload (orange spheres). The pH spectrum indicates the range of acidic to basic environments that can trigger response. This design allows for precise control over drug release, enhancing therapeutic efficiency and reducing off-target effects in applications such as cancer treatment.

nanocarrier development [30, 31]. This will ensure the delivery of the drugs to tumor sites more effectively and specifically, marking the advent of new, improved strategies for treating cancer [32, 33].

The continuous evolution and refinement of stimuli-responsive nanocarrier designs bear immense promise for the future of medical treatment, especially toward more exact, targeted therapeutic interventions. Technologies such as multi-stimuli-responsive biocompatible magnetic nanocarriers, responsive nanostructures [34], and stimulus-responsive antibacterial therapy [35] underscore the potential of such designs to revolutionize how diseases are treated. The development of stimuli-responsive image-guided nanocarriers [36] and nanoassemblies targeting tumors and their microenvironment [37] further illustrates the importance of these innovations in healthcare, highlighting the pivotal role of surface engineering and functionalization toward efficient targeted drug delivery.

### 2.2.1 Design and Application of pH-Responsive Systems

The pH-responsive nanocarriers target all the unique acidic conditions prevailing in the complex and heterogeneous tumor microenvironment, as well as those present across cellular compartments, including the endosome and lysosome, to optimize the site-specific therapeutic agent delivery. Thus, these advanced systems become crucial for increasing drug-delivery selectivity and efficiency, and these applications turn out to be very promising, mainly in the scope of cancer treatment.

Thus, pH-responsive nanocarriers may act through a different mechanism under acidic conditions, including protonation, charge inversion, or breaking chemical bonds. Those mechanisms could result in the conformational changes of nanocarriers, and because of that, they are able to deliver the drug with precision to the places of their action. For example, the acid microenvironment, typical in tumor tissues, may trigger functional groups' protonation at the nanocarrier's surface. As a consequence of protonation, these groups can reverse the charge and thereby increase the uptake of nanocarriers by cells or directly release the encapsulated drug in the targeted tissue or cellular compartment [38]. The strategic design of such pH-responsive systems ensures therapeutic agent delivery in a controllable way right at the target site, minimizing side effects while maintaining maximal therapeutic efficacy.

For instance, pH-responsive magnetic nanocarriers designed with charge reversal capability can guide subcellular doxorubicin delivery for chemotherapeutic drug delivery to the target, which is the acidic environment of the tumor [38], and also to multi-stimuli-responsive behaviors. This kind of micellar nanoassembly, therefore, is responsive to pH and acid-labile group incorporation into the hydrophobic cores, which presents a very promising possibility for the target release of drugs in a sequence triggered by acidic pH found within a tumor site or intracellular compartments [39]. The principle of pH responsiveness is of the essence when developing nano self-assemblies of anticancer drugs forming structured carriers, which can disassemble under an acidic environment for therapeutic payload release [40]. The second major advantage is the more generalized application of the pH-responsive micro- and nanocarrier system strategy because it offers a versatile platform for delivering a more significant number of therapeutic agents [41]. Lastly, modern design concepts for advanced micelles include charge-conversional and pH-sensitive polymeric micelles, ensuring more effective intracellular drug delivery to facilitate better drug therapeutic efficacy [42].

A recent study by Kanamala et al. [43] has shown that nanocarriers are an effective pH-sensitive drug vehicle for targeting tumors. They conclude that such nanocarriers overcome the drawbacks of conventional chemotherapy, making it possible for improved therapeutic outcomes, as they are formulated with pH-sensitive biomaterials to allow drug release from the carrier system at the tumor site, both extracellularly and intracellularly. Similarly, Nogueira-Librelotto et al. [44] furthered the study of pH-responsive chitosan–tripolyphosphate NPs functionalized with a pH-sensitive amphiphile for increased intracellular uptake and the resultant antitumor activity of doxorubicin. This underlines the critical, innovative function of pH-responsive systems in cancer therapy.

This is a very active area of research, with recent progress targeting drug-delivery systems that are acid-sensitive when exposed to the pH conditions of the tumor microenvironment or even when exposed to the usual pH values in endosomes and lysosomes. The pH-responsive nanocarriers developed by Wu et al. [45] can change their programmed size, and these carriers are supposed to enhance the safety and therapeutic efficacy of antitumor drugs by showing adaptability to pH changes. In further research, Prajapati et al. [40] designed pH-responsive nano self-assemblies from 2-hydroxyoleic acid and explored them as a novel platform for the targeted delivery of chemotherapeutics to cancer cells to allow localization in cancer cells and minimize toxic side effects.

All these examples have shown that leading strategies are being used in relation to pH-responsive nanocarriers in the field of targeted drug delivery for cancer therapy. One of the main advantages of pH-responsive nanocarriers is the possibility of achieving either selectivity toward targets or reactivity in the acidic environment characteristic of the tumor or toward the pH values existing in the cell. This allows controlled drug release, together with localized drug targeting. This way, systemic toxicity is significantly reduced, and at the same time, improved patient care outcomes are observed [46]. The versatility of cyclodextrin-based nanocarriers may find applicability in the encapsulation of redox, light, and pH stimuli, thus depicting the same. Controlled release mechanisms of the respective chemotherapeutic agents and bringing new inventive strategies in this area of cancer therapy [47]. Moreover, dual stimuli, such as pH and redox conditions, have gained huge research attention and are forthcoming due to the demands of nanocarrier technology. Such developments further assure more effective and specific delivery of drugs into neoplasmic sites, heralding the first glimmers of novel, improved strategies in the treatment of cancer [48].

pH-responsive nanocarriers have allowed new scopes in the stimuli-responsive drug-delivery landscape. Such nanocarriers are designed in such a way that they can release drugs at a controlled rate, exactly at the site of the tumor, using the specific acidic microenvironment of the tumor. This specificity becomes possible due to the unique composition of tumor tissues, unlike healthy cells. This means the respective pH-sensitive nanocarriers could target cancer cells with much-improved, enhanced accuracy, significantly improving the ability of anticancer treatments carried out inside cancer cells and, at the same time, apparently reducing risks associated with systemic toxicity [46]. These pH-responsive nanocarriers create three main processes: protonation, charge inversion, and the cleavage of chemical bounds. These processes cause changes in conformation that make it easier for therapeutic agents to get to the right place where they need to go. For instance, the tumor's acidic environment protonates the nanocarrier's functional groups, inducing charge reversal, which is characteristic of promoting cellular uptake, or drugs are released directly in the respective tissue [48].

Constantly, very active research and development in this area are expanding the horizons of personalized medicine and precision medicine and opening new pathways for the treatment of many diseases. In leading innovation, pH-responsive systems promise to improve the healthcare outcome of a product that offers safety and effective treatment tailored to the unique needs of an individual patient [49].

### 2.2.2 Temperature and Light-Responsive Systems

Within just a couple of years, advanced smart nanocarriers have become increasingly of central importance in the research of targeted drug delivery with the development of nanotechnology. With that in place, smart nanocarriers are tuned in such a way that they are sensitive to certain stimuli that include temperature and light, therefore guiding and governing the site-specific release of therapeutic agents within the tumor site to increase precision and effect of treatment, with collateral damage to healthy tissues minimized [50]. For example, the slight temperature change we discussed earlier may cause the temperature-responsive nanocarriers to change how permeable they are. This will ensure that drugs are released at temperatures between the healthy tissue and the tumor site. The constituents of this unique characteristic are those materials that can physically or chemically change at given temperatures, which will permit the drug to be released in a targeted and controlled way [51].

Light-responsive nanocarriers work in a similar way to spatial drug release. In this case, light-responsive nanocarrier activation might occur only within a tumor; thus, therapeutic agents will be released only at the place of interest. It not only increases drug concentration at the tumor site for increased therapeutic effect but also, in a way, significantly reduces the systemic circulation of highly potent drugs, which thereby reduces toxicity risk [52]. A temperature- and light-responsive unit was thus implemented into a single nanocarrier platform for synergistic effects toward efficient anticancer phototherapy. The nanocarriers with two functions can use light and high temperatures to break down drugs more efficiently in photothermal molecules, improving cancer cell treatment [52].

This progressive improvement and application of these stimuli-responsive nanocarriers underline these technologies' potential usefulness in revolutionizing cancer treatment approaches. Greater development in this area of research holds promise toward expanding the tenets of personalized and precision medicine, thereby opening new opportunities to improve health outcomes within the healthcare setting via targeted treatment of cancer and other diseases [53].

In the field of drug-delivery systems, there has been a significant advancement in the form of temperature-sensitive liposomes and NPs. These systems are designed to launch the release of the drug at specific temperatures, which is a significant development. This is a targeted approach and can, therefore, be of great value in cancer therapeutics, where it is indispensable to deliver therapeutic agents specifically into the tumor sites while leaving the rest of the system with minimized side effects. Choi et al. [54] first set off the development of radially porous structured silica NPs to demonstrate their potential to load large amounts of drugs for temperature-controlled release. On such an issue, it was found that the synthesized NPs have immense potential for the controlled release of ibuprofen, one of the most common anti-inflammatory and analgesic drugs. This innovation presented an application of practical value with regard to temperature-sensitive nanocarriers and improved drug-delivery efficiency.

Pandey et al. [55] also looked into the behavior of temperature-responsive fluorescent polymer nanoparticles (TRFNPs) to learn more about the temperature-responsive system. The designed NPs not only have the ability to act as effective cellular imaging agents but also indicate the controlled release of drugs into the living cells of an anticancer drug. Thus, they point out another vast potential for applying TRFNPs in the domains of diagnostics and therapeutics in nanomedicine. Bearing these bases in mind, recent research has followed up on different applications and novelties brought by temperature-sensitive nanocarriers. Lohe and Umekar [56] have provided thorough and elaborate review articles on the thermosensitive liposomal drug-delivery system, focusing on the system's crucial role in improving the precision of therapeutic interventions, particularly cancer treatment. This review underlines the importance of ongoing research in optimizing these systems for clinical applications.

Sam et al. [57] reviewed the versatile applications of temperature-responsive nanogels in drug delivery. These nanogels react to temperature through physical changes and outline the new trend in drug release mechanisms. Such potentialities make these candidates very potential for stimuli-responsive drug delivery that may specifically target a very nuanced field, such as cancer therapy, where very controlled drug release would finally yield significantly enhanced therapeutic outcomes. This mini-review emphasizes, on the one hand, some current advances in the niche but also opens a discussion regarding the possible integration of temperature-responsive nanogels into myriads of drug-delivery systems, pointing out their significance within modern pharmacotherapy.

The temperature-responsive NPs developed by Prawatborisut et al. [58] are also the first of their kind and can contribute to the selective binding of apolipoproteins from human plasma. The results showed how specific the target was and how temperature-responsive NPs might be able to target and bind a particular protein. This could open the door to targeted delivery and therapy. Lastly, Yuba's [59] review paper gave an in-depth look at the planning and creation of functional liposomes for new biomedical uses involving materials that respond to stimuli. It highlighted several areas where further research could be done in this area. The study will further highlight that temperature-sensitive liposomes, or NPs, are considered very flexible drug carriers as they can be modified according to the drug's therapeutic needs. In conclusion, these studies show that temperature-sensitive nanocarriers for drug delivery are a dynamic and ever-changing landscape that gives us new and exciting ways to think about this technology's role in the future of personalized medicine and targeted cancer therapies.

Light-responsive materials designed for precision medicine offer a new revolution, allowing for the non-invasive, on-demand, site-specific delivery of therapeutic agents with unprecedented spatial and temporal control. Smart materials activate the system from the outside, allowing precise location of drug-delivery nanocarriers where the medicine is needed and, in this case, sparing side effects associated with general, systemic drug administration. Introduced first by Liu et al. [60], singlet-oxygen-responsive linkers equipped with nanoscale coordination polymers are pioneering. This design lets light-controlled drug release happen for the first time when light is used with cancer therapy and achieves precise control of the therapeutic output.

Building on this earlier work, Xiang et al. [61] developed UV-responsive diblock copolymer-coated upconversion NPs. Thus, the upconversion NPs showed potential for photo-controlled drug delivery and gave a perfect example of the adaptation and efficiency of light-responsive materials in drug release. This opens the door to using the Hao Ma group's red-light-sensitive Donor–Acceptor Stenhouse Adduct (DASA) polymer with high water stability to control drug release in aqueous environments [62]. This horizon of light-responsive polymers has great potential for more effective and safer drug-delivery systems, whether they are established or not.

Rivero-Buceta et al. [63], on the other hand, utilized the "ship in a bottle" principle in one of their inventions and manufactured camptothecin porous materials to achieve a controlled and photoinduced release of light. Using light as a stimulus to release chemotherapeutic agents, they describe a work that emphasizes current strategies in innovative drug-delivery systems. The purpose of this work was to improve the precision of the application of these agents in the treatment of cancer. The first of these pioneering works underlined the great potential of light-responsive materials for developing controlled drug-delivery systems. Such materials will open new avenues in the progress of personalized medicine, wherein treatments could be applied only in the place of interest with extremely high precision for better efficacy and safety.

Smart, temperature, and light-stimuli-responsive nanocarriers represent considerable breakthroughs in research for drug delivery. Using temperature-sensitive liposomes, NPs, and light-responsive materials together could help fine-tune drug release rates for a therapeutic intervention that is both safe and effective. In this regard, further research and development are, therefore, required to be carried out in the field to actualize the optimum potential of such highly innovative, sophisticated drug-delivery systems in clinical practice.

### 2.2.3 Magnetic and Ultrasound-Responsive Systems

Integrating magnetic and ultrasound-responsive mechanisms into nanocarriers signifies a pivotal advancement in targeted therapy. These sophisticated systems introduce innovative solutions for drug delivery, utilizing non-invasive control methods to ensure precise delivery of therapeutic agents to the intended site, thereby minimizing side effects. The ability to finely tune drug release through external magnetic fields or acoustic waves significantly enhances therapeutic efficiency while prioritizing patient safety.

Wang et al. [64] highlighted the potential of combining magnetic targeting with focused ultrasound for intracerebral hemorrhage therapy. This dual approach enables concentrated therapeutic efforts in the affected area, improving treatment outcomes. Pawar et al. [34] emphasized integrating magnetic and ultrasound-responsive elements in nanocarrier designs for controlled drug release, showcasing their ability to be manipulated precisely in response to specific external stimuli.

In breast cancer treatment, Edwards et al. [65] explored ultrasound-responsive nanocarriers, suggesting a paradigm shift in treatment approaches. These carriers offer a way to reduce the side effects associated with traditional treatments while increasing the concentration of therapeutic agents directly at the tumor site. Zhu et al. [66] reviewed recent developments in magnetic nanocarriers for tumor therapy, focusing on the applications and specificity of magnetic iron oxide NPs in targeted drug delivery. This review underlines the significant role of magnetic nanoparticles (MNPs) in enhancing the specificity and efficiency of cancer treatments.

Additionally, Chapla et al. [67] discussed the potential of microbubble-NP complexes for ultrasound-enhanced drug delivery. This combination leverages the strengths of both microbubbles and NPs, offering a novel strategy for enhancing drug delivery across biological barriers and showcasing the capabilities of ultrasound-responsive systems in medical applications.

These studies highlight the advancements and promise of magnetic and ultrasound-responsive nanocarriers in nanomedicine. By offering precise, controlled drug-delivery mechanisms, these systems pave the way for more effective and less invasive therapeutic interventions, marking a new era in targeted therapy and personalized medicine.

MNPs stand as a cornerstone in advancing nanomedicine, especially for targeted drug delivery and controlled release. Their unique responsiveness to external magnetic fields allows for precise localization of drug delivery, significantly enhancing therapeutic efficacy and reducing systemic toxicity. This strategic drug-delivery approach overcomes challenges like nonspecific distribution and its resultant side effects. Liu et al. [68] introduced Cu9S5@mSiO2@Fe3O4-PEG nanocarriers capable of facilitating magnetic resonance imaging alongside chemo/photothermal therapy. With the ability to use near-infrared photothermal therapy, this multifunctional platform shows how MNPs can be used for both the diagnosis and treatment of cancer.

Ghorbani et al. [69] designed ternary-responsive nanocarriers for targeted delivery of thiol-containing anticancer drugs, illustrating the potential of MNPs to improve therapeutic efficiency and patient-friendly cancer therapies. As emphasized by da Silva Oliveira and de Oliveira Conceicão [70], magnetic drug-carrying NPs play a key role in treating cancer. They allow therapeutic agents to be magnetically targeted directly to tumor sites, which improves treatment outcomes.

Burmistrov et al. [71] explored the magneto-mechanical actuation capabilities of single-domain MNPs for controlled drug release, showcasing the flexibility of MNPs in creating responsive drug-delivery systems. Stanicki et al. [72] provided an update on the applications and characteristics of magnetic iron oxide NPs in drug-delivery systems, highlighting the evolving landscape of MNP-based therapeutics. Haghighattalab et al. [73] created a smart system for sustained drug release using a silk fibroin hydrogel reinforced with MNPs. This system shows how biocompatibility and responsiveness can be combined to create new drug-delivery solutions. Patel et al. [74] discussed the broad applications of MNPs in addressing therapeutic challenges, emphasizing their essential role in advancing targeted therapies.

These studies collectively underline the transformative impact of MNPs in drug delivery, offering new solutions that promise to revolutionize targeted therapy and controlled release mechanisms. Continuous research and development in this area are opening new doors, advancing the frontier of personalized and precision medicine.

Ultrasound-responsive nanocarriers represent a cutting-edge technology in drug delivery, using acoustic waves to control drug release. This method enhances tissue penetration and allows for minimally invasive therapeutic applications. Nanocarriers responsive to ultrasound are remarkably versatile, marking a significant advancement in nanomedicine by providing new solutions to improve therapeutic efficacy and patient compliance. Chapla et al. [67] highlighted the utility of microbubble-NP complexes in ultrasound-enhanced cargo delivery. Their research outlines a promising strategy for targeted tissue penetration and therapeutic enhancement. It utilizes the unique interactions between ultrasound waves and microbubbles for precise drug release and enhanced drug delivery to specific tissues.

Mukhopadhyay et al. [75] focused on the potential of ultrasound-responsive liposomes and polymeric micelles in cancer therapy. Their findings emphasized the selective nature of these carriers, significantly reducing toxicity and enabling controllable drug release patterns. This selectivity positions ultrasound-responsive nanocarriers as preferred vehicles for delivering anticancer drugs, offering a hopeful route to minimize side effects and improve therapeutic efficiency in cancer treatment. Tagai et al. [76] investigated drug-delivery systems with high responsiveness to ultrasound using metal NP-encapsulated liposomes. This approach aimed to enhance the response of liposomal carriers to ultrasound waves, thereby improving the efficacy of targeted drug delivery and controlled release mechanisms.

Additionally, the fabrication and evaluation of ultrasound-responsive emulsions loaded with paclitaxel for targeted chemotherapy were examined. This innovative strategy seeks to utilize ultrasound to trigger the release of chemotherapeutic agents directly at the tumor site, presenting a promising solution for enhancing cancer treatment efficacy while minimizing adverse effects. Ayana et al. [77] explored the development of ultrasound-responsive nanocarriers specifically for breast cancer chemotherapy. Their work demonstrates the potential of ultrasound-responsive nanocarriers to significantly improve the delivery and efficacy of chemotherapy agents in treating breast cancer, showcasing a compelling example of the therapeutic applications of these advanced nanocarriers.

These studies collectively highlight the transformative potential of ultrasound-responsive nanocarriers for targeted drug delivery and controlled release. As research in this field continues to evolve, the integration of ultrasound-responsive mechanisms in nanocarrier design is poised to revolutionize therapeutic interventions, offering more efficient, targeted, and patient-friendly treatment options.

## 2.3 CHALLENGES AND FUTURE DIRECTIONS

The development and use of nanocarriers for drug delivery and diagnostics have recently advanced thanks to advancements in nanotechnology and a better understanding of biological processes. However, transitioning these innovations from the laboratory to clinical application is a complex journey fraught with challenges that must be diligently addressed to fully realize their transformative potential. Li et al. [78] discussed the challenges and opportunities presented by nanomaterials in mRNA-based therapeutics, highlighting the need to solve issues related to stability, delivery efficiency, and immune response modulation. This underscores the importance of overcoming these obstacles to maximize the therapeutic potential of nanocarriers in gene therapy.

Cai et al. [79] focused on the physicochemical properties, pharmacokinetics, toxicology, and applications of nanocarriers, emphasizing the need for a perfect balance to utilize them clinically. This study points to the significance of understanding nanocarriers' physicochemical and biological interactions to enhance therapeutic effectiveness and reduce toxicity. Ahmad et al. [80] extended the analysis into precision nanotoxicology in drug development, addressing customized multifunctional nanocarriers' safety and toxicity implications. This highlights the critical need for safety assessments in developing nanocarriers for clinical use.

Dri et al. [81] provided insights into the state of clinical trials with nanomedicines and nanocarriers, including regulatory requirements and physicochemical critical quality attributes. Their analysis of the regulatory landscape offers perspectives on the conditions needed for the successful transition of nanocarriers from the bench to the bedside, emphasizing the necessity of adhering to regulatory standards.

These contributions underline the complexity of translating nanocarrier innovations into clinical applications. A multidisciplinary approach involving nanotechnology, biology, pharmacology, toxicology, and regulatory science is essential to overcome the identified difficulties. The ongoing contribution of these advancements promises significant patient benefits across various therapeutic areas as the field of nanomedicine continues to proliferate.

The field of nanocarrier design faces several critical challenges that need to be addressed for a successful transition from laboratory innovation to clinical applications. These challenges encompass various aspects of nanocarrier development, including reproducibility, scalability, regulatory issues, and the safety and biocompatibility of these systems.

Fernandes et al. [82] tackled the scale-up challenges associated with NP manufacturing processes. The difficulty of consistently producing nanocarriers with uniform size, shape, and functional properties presents a significant hurdle. This inconsistency impacts the scalability of nanocarrier manufacturing processes, which are essential for clinical applications. Patel et al. [83] discussed the complexities involved in the preformulation of nanoformulations and the intricacies in their selection, design, and preparation. The complexity of nanocarriers introduces substantial regulatory challenges, necessitating thorough characterization and standardization to ensure their safety, efficacy, and quality. The regulatory frameworks need to evolve to adequately address the unique challenges posed by nanomedicine.

Patel et al. [84] explored the challenges and feasibilities of scaling up nanomedicine technologies. Despite the promising potential of nanocarriers for targeted delivery and minimizing side effects, concerns regarding their long-term safety and biocompatibility remain. Understanding the interactions of nanocarriers with biological systems is crucial to mitigating their potential immunogenicity and toxicity. Cury et al. [85] highlighted the importance of designing nanocarriers to enhance drug absorption through the gastrointestinal tract, addressing one of the many challenges requiring innovative nanocarrier applications.

These works collectively emphasize the complex landscape of nanocarrier design and development. Addressing these challenges necessitates a multidisciplinary approach that includes insights from material science, pharmacology, toxicology, and regulatory science. As the field progresses, ongoing research and collaboration among scientists, clinicians, and regulatory bodies will be crucial in overcoming these obstacles, paving the way for the successful clinical translation of nanocarriers and realizing their full potential in enhancing patient outcomes.

The future of nanocarrier design is bright, with several promising directions poised to overcome current limitations and significantly enhance therapeutic outcomes. Among the notable advancements are the development of multi-stimuli-responsive systems, the integration of artificial intelligence (AI) and machine learning (ML) into nanocarrier design and optimization, and the move toward personalized nanomedicine. Modern nanogels are being engineered to respond to multiple stimuli, such as pH, temperature, or magnetic fields, offering enhanced drug release and targeting control. For example, recent research has demonstrated the effectiveness of multi-stimuli-responsive and cell membrane-camouflaged Aggregation-Induced Emission (AIE) nanogels for chemo-photothermal synergistic therapy of tumors, highlighting their adaptability and responsiveness in complex biological environments [86].

Applying AI and ML technologies in nanocarrier design opens new frontiers, enabling predictions about nanocarrier behavior, optimizing formulations, and personalizing therapies based on patient-specific data. This approach has already shown promise, particularly in magnetic hyperthermia research, where AI applications have improved our understanding and use of nanocarriers for targeted therapies [87]. Tailoring nanocarrier systems to individual patient profiles and disease states is an emerging trend that could transform treatment strategies. By maximizing efficacy and minimizing adverse effects, personalized nanomedicine focuses on patient-centered approaches, offering solutions based on individual health data [88].

Stimuli-responsive multifunctional nanomedicine presents a novel strategy for enhancing chemotherapy, particularly for complex conditions like glioblastoma. By improving blood-to-brain trafficking and tumor targeting, these approaches highlight the importance of multifunctional and responsive nanocarriers in addressing treatment complications [89].

These advancements in nanomedicine, including AI and ML, multi-stimuli-responsive systems, and personalized approaches, are expected to play pivotal roles in overcoming the current challenges of nanocarrier design and application. These future directions signify a shift towards more sophisticated, efficient, and patient-specific nanotherapeutic strategies, potentially revolutionizing drug delivery and diagnostics.

Advancing in nanomedicine necessitates a multifaceted approach to surmount the inherent challenges of transitioning nanocarrier technologies from lab to clinical settings. This journey is a collaborative effort that melds diverse expertise, state-of-the-art manufacturing, and sophisticated computational models to unlock nanocarriers' full therapeutic potential, making them fundamental to advanced healthcare solutions. Nanomedicine's complexity demands a multidisciplinary team incorporating materials science, biology, pharmacology, and regulatory science. Each discipline contributes unique insights crucial for tackling nanocarrier design, safety assessment, and therapeutic application challenges. Fostered by open communication and shared objectives, interdisciplinary teams can spur innovation, from novel nanocarrier designs to strategies for mitigating risks and boosting clinical efficacy.

Realizing nanomedicine's promise requires advancements in manufacturing technologies for scalable, reproducible, and high-quality nanocarrier production. Emerging nanofabrication techniques, along with continuous process optimization, are essential for meeting clinical demands. Developing standardized protocols and quality controls will aid regulatory approval, transitioning nanomedicine from experimental therapies to global treatments. Integrating AI and ML opens unparalleled opportunities for enhancing nanocarrier design and optimization. These computational tools can predict nanocarrier behavior within biological systems, optimize formulations across numerous parameters, and tailor therapy regimens to individual patient data, accelerating nanomedicine development and introducing a new paradigm in precision medicine.

The future of nanomedicine lies in personalized therapeutic solutions. Tailoring nanocarrier systems to individual patient profiles and disease states promises to revolutionize treatment strategies. This personalization demands a deep understanding of disease molecular mechanisms and integrating patient-specific data into the nanocarrier design process, resulting in more effective, safer, and patient-friendly therapies. While the path forward for nanomedicine is challenging, it is filled with opportunities for breakthrough advancements. We can unlock nanocarriers' potential by fostering interdisciplinary collaboration, pushing manufacturing technologies forward, harnessing computational insights, and moving toward personalized medicine. This concerted effort will advance nanomedicine and herald a new healthcare era where nanotechnology significantly improves patient outcomes and life quality.

## 2.4 CONCLUSIONS

The exploration and utilization of nanocarriers for delivering nucleic acids and proteins represent a significant leap in medical therapeutics, unveiling novel strategies to circumvent traditional drug-delivery challenges. Through meticulous design and innovative breakthroughs, nanocarriers have demonstrated exceptional potential for enhancing nucleic acid and protein delivery systems' effectiveness. Their capability to precisely target disease sites, minimize off-target effects, and systemic toxicity marks a considerable advancement in therapeutic modalities.

Nanocarriers' ability to be engineered to target specific cells or tissues using biological pathways is a major breakthrough that improves the therapeutic index of the drugs they carry and ensures direct delivery to the right cells or tissues. Additionally, integrating stimuli-responsive materials into nanocarrier designs enables controlled therapeutic agent release in response to specific environmental triggers at precisely needed locations. Moreover, encapsulating therapeutic agents within nanocarriers significantly boosts their stability and solubility, facilitating the delivery of otherwise challenging compounds.

Despite these advancements, challenges such as reproducibility, scalability, and regulatory compliance remain. Addressing these challenges necessitates ongoing research and innovation in various areas. Developing multi-stimuli-responsive nanocarriers promises finer control over drug release and targeting, adapting dynamically within complex biological environments. Integrating AI and ML into nanocarrier design could revolutionize this field, enabling behavior prediction, formulation optimization, and personalized therapies based on individual patient data. Furthermore, personalizing nanomedicine to align with individual patient profiles aims to maximize efficacy and minimize adverse effects, heralding a shift toward patient-centric therapeutic approaches.

Overcoming current obstacles and releasing nanocarriers' full therapeutic potential are crucial as the field develops as a result of multidisciplinary research, technological advancements, and clinical insights. Although the path from lab to clinical application is fraught with challenges, the transformative potential for medical therapeutics is unparalleled. It heralds a new era of precision medicine and targeted therapies, potentially revolutionizing disease treatment methodologies.

## REFERENCES

1. M. Danaei, M. Dehghankhold, S. Ataei, F. Hasanzadeh Davarani, R. Javanmard, A. Dokhani, et al., "Impact of particle size and polydispersity index on the clinical applications of lipidic nanocarrier systems," *Pharmaceutics*, vol. 10, no. 2, p. 57, 2018

2. N. Hoshyar, S. Gray, H. Han, and G. Bao, "The effect of nanoparticle size on in vivo pharmacokinetics and cellular interaction," *Nanomedicine*, vol. 11, no. 6, pp. 673–692, 2016.

3. N. P. Truong, M. R. Whittaker, C. W. Mak, and T. P. Davis, "The importance of nanoparticle shape in cancer drug delivery," *Expert Opin Drug Deliv*, vol. 12, no. 1, pp. 129–142, 2015.

4. M. Frey, S. Bobbala, N. Karabin, and E. Scott, "Influences of nanocarrier morphology on therapeutic immunomodulation," *Nanomedicine*, vol. 13, no. 14, pp. 1795–1811, 2018.

5. D. Milani, U. Athiyah, D. M. Hariyadi, and Y. V. Pathak, "Surface modification of nanocarriers for specific cell targeting for better therapeutic effect," in *Surface Modification of Nanoparticles for Targeted Drug Delivery*, Springer, Cham, pp. 355–368, 2019.

6. Y. He, X. Fan, X. Wu, T. Hu, F. Zhou, S. Tan, et al., "pH-Responsive size-shrinkable mesoporous silica-based nanocarriers for improving tumor penetration and therapeutic efficacy," *Nanoscale*, vol. 14, no. 4, pp. 1271–1284, 2022.

7. W. Zhang, R. Taheri-Ledari, F. Ganjali, S. Shadi Mirmohammadi, F. Sadat Qazi, M. Saeidirad, et al., "Effects of morphology and size of nanoscale drug carriers on cellular uptake and internalization process: a review," *RSC Adv*, vol. 13, no. 1, pp. 80–114, 2023.

8. R. G. R. Pinheiro, A. J. Coutinho, M. Pinheiro, and A. R. Neves, "Nanoparticles for targeted brain drug delivery: what do we know?," *Int J Mol Sci*, vol. 22, no. 21, p. 11654, 2021.

9. D. Peer, J. M. Karp, S. Hong, O. C. Farokhzad, R. Margalit, and R. Langer, "Nanocarriers as an emerging platform for cancer therapy," *Nano-enabled medical applications,* pp. 61–91, 2020.

10. B. Ouyang, W. Poon, Y. Zhang, Z. P. Lin, B. R. Kingston, A. J. Tavares, et al., "The dose threshold for nanoparticle tumour delivery," *Nat Mater*, vol. 19, no. 12, pp. 1362–1371, 2020.

11. A. C. Anselmo. M. Zhang, S. Kumar, D. R. Vogus, S. Menegatti, M. E. Helgeson, et al., "Elasticity of nanoparticles influences their blood circulation, phagocytosis, endocytosis, and targeting," *ACS Nano*, vol. 9, no. 3, pp. 3169–3177, 2015.

12. J. Lu, X. Gao, S. Wang, Y. He, X. Ma, T. Zhang, et al., "Advanced strategies to evade the mononuclear phagocyte system clearance of nanomaterials," in *Exploration*, Wiley Online Library, 2023, p. 20220045.

13. B. D. Chithrani, A. A. Ghazani, and W. C. W. Chan, "Determining the size and shape dependence of gold nanoparticle uptake into mammalian cells," *Nano Lett*, vol. 6, no. 4, pp. 662–668, 2006.

14. Y. Li, M. Kröger, and W. K. Liu, "Shape effect in cellular uptake of PEGylated nanoparticles: comparison between sphere, rod, cube and disk," *Nanoscale*, vol. 7, no. 40, pp. 16631–16646, 2015.

15. W. Borzęcka, P. M. R. Pereira, R. Fernandes, T. Trindade, T. Torres, and J. P. C. Tome, "Spherical and rod shaped mesoporous silica nanoparticles for cancer-targeted and photosensitizer delivery in photodynamic therapy," *J Mater Chem B*, vol. 10, no. 17, pp. 3248–3259, 2022.

16. Z. Chu, S. Zhang, B. Zhang, C. Zhang, C.Fang, I. Rehor, et al., "Unambiguous observation of shape effects on cellular fate of nanoparticles," *Sci Rep*, vol. 4, no. 1, p. 4495, 2014.

17. G. F. Chin, M. Misran, P. Y. Tang, and P. F. Lee, "Synthesizing gold nanoparticles in spherical and rod shapes for targeting cancer treatment," in *The 15th International Conference on Biomedical Engineering: ICBME 2013, 4th to 7th December 2013*, Singapore, Springer, 2014, pp. 259–262.

18. P. Lagarrigue, F. Moncalvo, and F. Cellesi, "Non-spherical polymeric nanocarriers for therapeutics: the effect of shape on biological systems and drug delivery properties," *Pharmaceutics*, vol. 15, no. 1, p. 32, 2022.

19. H. Hu et al., "Reduction-responsive worm-like nanoparticles for synergistic cancer chemo-photodynamic therapy," *Mater Today Bio*, vol. 18, p. 100542, 2023.

20. A. K. Dey, A. Nougarède, F. Clément, C. Fournier, E. J. -Marche, M. Escude, et al., "Tuning the immunostimulation properties of cationic lipid nanocarriers for nucleic acid delivery," *Front Immunol*, vol. 12, p. 722411, 2021.

21. K. R. Gajbhiye, A. Pawar, K. R. Mahadik, and V. Gajbhiye, "PEGylated nanocarriers: a promising tool for targeted delivery to the brain," *Colloids Surf B Biointerfaces*, vol. 187, p. 110770, 2020.

22. D. Milani, U. Athiyah, D. M. Hariyadi, and Y. V. Pathak, "Surface modification of nanocarriers for specific cell targeting for better therapeutic effect," in *Surface Modification of Nanoparticles for Targeted Drug Delivery*, pp. 355–368, 2019.

23. M. P. Vincent, S. Bobbala, N. B. Karabin, M. Frey, Y. Liu, J. O. Navidzadeh, et al., "Surface chemistry-mediated modulation of adsorbed albumin folding state specifies nanocarrier clearance by distinct macrophage subsets," *Nat Commun*, vol. 12, no. 1, p. 648, 2021.

24. M. Tonigold, J. Simon, D. Estupiñán, M. Kokkinopoulou, J. Reinholz, U. Kintzel, et al., "Pre-adsorption of antibodies enables targeting of nanocarriers despite a biomolecular corona," *Nat Nanotechnol*, vol. 13, no. 9, pp. 862–869, 2018.

25. S. Priya, V. M. Desai, and G. Singhvi, "Surface modification of lipid-based nanocarriers: a potential approach to enhance targeted drug delivery," *ACS Omega*, vol. 8, no. 1, pp. 74–86, 2022.

26. V. Sethuraman, K. Janakiraman, V. Krishnaswami, and R. Kandasamy, "Recent progress in stimuli-responsive intelligent nano scale drug delivery systems: a special focus towards pH-sensitive systems," *Curr Drug Targets*, vol. 22, no. 8, pp. 947–966, 2021.

27. M. Handa, A. Singh, S. J. S. Flora, and R. Shukla, "Stimuli-responsive polymeric nanosystems for therapeutic applications," *Curr Pharm Des*, vol. 28, no. 11, pp. 910–921, 2022.

28. X. Guo and J. You, "Near infrared light-controlled therapeutic molecules release of nanocarriers in cancer therapy," *J Pharm Investig*, vol. 47, pp. 297–316, 2017.

29. X. Li, J. Liu, and N. Qiu, "Cyclodextrin-based polymeric drug delivery systems for cancer therapy," *Polymers (Basel)*, vol. 15, no. 6, p. 1400, 2023.

30. Q. Lu, "Bioresponsive and multifunctional cyclodextrin-based non-viral nanocomplexes in cancer therapy: building foundations for gene and drug delivery, immunotherapy and bioimaging," *Environ Res*, p. 116507, 2023.

31. O. A. A. Alabrahim, S. A. Fahmy, and H. M. E.-S. Azzazy, "Stimuli-responsive cucurbit[n]uril-based supramolecular nanocarriers for delivery of chemotherapeutics," *ACS Appl Nano Mater*, vol. 6, no. 5, pp. 3139–3158, 2023.

32. N. Kang, S. Son, S. Min, H. Hong, C. Kim, J. An, et al., "Stimuli-responsive ferroptosis for cancer therapy," *Chem Soc Rev*, vol. 52, no. 12, pp. 3955–3972, 2023.

33. E. Lee and E. S. Lee, "Development of pH-Responsive Hyaluronic Acid-Conjugated Cyclodextrin Nanoparticles for Chemo-/CO-Gas Dual Therapy," *Pharmaceutics*, vol. 15, no. 7, p. 1818, 2023.

34. V. Pawar, P. Maske, A. Khan, A. Ghosh, R. Keshari, M. Bhatt, et al., "Responsive nanostructure for targeted drug delivery," *J Nanotheranost*, vol. 4, no. 1, pp. 55–85, 2023.

35. N. Bag, S. Bardhan, S. Roy, J. Roy, D. Monda, B. Guo, et al., "Nanoparticle-mediated stimulus-responsive antibacterial therapy," *Biomater Sci*, vol. 11, no. 6, pp. 1994–2019, 2023.

36. X. Chen, S. Lei, J. Lin, and P. Huang, "Stimuli-responsive image-guided nanocarriers as smart drug delivery platforms," *Expert Opin Drug Deliv*, vol. 19, no. 11, pp. 1487–1504, 2022.

37. C. S. Sia, H. P. Lim, B. T. Tey, B.-H. Goh, and L. E. Low, "Stimuli-responsive nanoassemblies for targeted delivery against tumor and its microenvironment," *Biochim Biophys Acta (BBA)—Rev Cancer*, vol. 1877, no. 5, p. 188779, 2022.

38. H. Zhang, Y. Xue, J. Huang, X. Xia, M. Song, K. Wen, et al., "Tailor-made magnetic nanocarriers with pH-induced charge reversal and pH-responsiveness to guide subcellular release of doxorubicin," *J Mater Sci*, vol. 50, pp. 2429–2442, 2015.

39. A. Moreno, J. C. Ronda, V. Cádiz, M. Galià, G. Lligadas, and V. Percec, "pH-responsive micellar nanoassemblies from water-soluble telechelic homopolymers endcoding acid-labile middle-chain groups in their hydrophobic sequence-defined initiator residue," *ACS Macro Lett*, vol. 8, no. 9, pp. 1200–1208, 2019.

40. R. Prajapati, M. Gontsarik, A. Yaghmur, and S. Salentinig, "pH-responsive nano-self-assemblies of the anticancer drug 2-hydroxyoleic acid," *Langmuir*, vol. 35, no. 24, pp. 7954–7961, 2019.

41. S. Nowag and R. Haag, "pH-responsive micro-and nanocarrier systems," *Angew Chem Int Ed*, vol. 53, no. 1, 2014.

42. G. Liu, M. Li, C. Zhu, Q. Jin, Z. Zhang, and J. Ji, "Charge-conversional and pH-sensitive PEGylated polymeric micelles as efficient nanocarriers for drug delivery," *Macromol Biosci*, vol. 14, no. 9, pp. 1280–1290, 2014.

43. M. Kanamala, W. R. Wilson, M. Yang, B. D. Palmer, and Z. Wu, "Mechanisms and biomaterials in pH-responsive tumour targeted drug delivery: a review," *Biomaterials*, vol. 85, pp. 152–167, 2016.

44. D. R. Nogueira-Librelotto, L. E. Scheeren, M. P. Vinardell, M. Mitjans, and C. M. B. Rolim, "Chitosan-tripolyphosphate nanoparticles functionalized with a pH-responsive amphiphile improved the in vitro antineoplastic effects of doxorubicin," *Colloids Surf B Biointerfaces*, vol. 147, pp. 326–335, 2016.

45. W. Wu, L. Luo, Y. Wang, Q. Wu, H.Dai, J. Li, et al., "Endogenous pH-responsive nanoparticles with programmable size changes for targeted tumor therapy and imaging applications," *Theranostics*, vol. 8, no. 11, p. 3038, 2018.

46. S. Bhattacharya, B. G. Prajapati, and S. Singh, "A critical review on the dissemination of PH and stimuli-responsive polymeric nanoparticular systems to improve drug delivery in cancer therapy," *Crit Rev Oncol Hematol*, vol. 185, p. 103961, 2023.

47. J. Pilch, A. Potęga, A. Kowalczyk, A. Kasprzak, P. Kowalik, P. Bujak, et al., "pH-responsive drug delivery nanoplatforms as smart carriers of unsymmetrical bisacridines for targeted cancer therapy," *Pharmaceutics*, vol. 15, no. 1, p. 201, 2023.

48. K. Brewer and A. Blencowe, "Composition-property relationships of pH-responsive poly[(2-vinylpyridine)-co-(butyl methacrylate)] copolymers for reverse enteric coatings," *Pharmaceutics*, vol. 15, no. 2, p. 454, 2023.

49. S. Chu, X. Shi, Y. Tian, and F. Gao, "pH-responsive polymer nanomaterials for tumor therapy," *Front Oncol*, vol. 12, p. 855019, 2022.

50. S. Luo, Z. Lv, Q. Yang, R. Chang, and J. Wu, "Research progress on stimulus-responsive polymer nanocarriers for cancer treatment," *Pharmaceutics*, vol. 15, no. 7, p. 1928, 2023.

51. H. Hemmatpour, V. Haddadi-Asl, T. C. Q. Burgers, F. Yan, M. C. A. Stuart, C. Reker-Smit, et al., "Temperature-responsive and biocompatible nanocarriers based on clay nanotubes for controlled anti-cancer drug release," *Nanoscale*, vol. 15, no. 5, pp. 2402–2416, 2023.

52. X. Zhao, S. Wang, Y. Ma, W. Liu, H. Zhao, J. Di, et al., "Synergistic release of photothermal molecules from nanocarriers induced by light and hyperthermia benefits efficient anticancer phototherapy," *Anal Chem*, vol. 94, no. 49, pp. 17160–17168, 2022.

53. M. Ashrafizadeh, A. Zarrabi, A. Bigham, A. Taheriazam, Y. Saghari, S. Mirzaei, et al., "(Nano) platforms in breast cancer therapy: drug/gene delivery, advanced nanocarriers and immunotherapy," *Med Res Rev*, vol. 43, no. 6, pp. 2115–2176, 2023.

54. Y. Choi, J. Kim, S. Yu, and S. Hong, "pH- and temperature-responsive radially porous silica nanoparticles with high-capacity drug loading for controlled drug delivery," *Nanotechnology*, vol. 31, no. 33, p. 335103, 2020.

55. N. Pandey, J. U. Menon, M. Takahashi, J. Hsieh, J. Yang, K. T. Nguyen, et al., "Thermo-responsive fluorescent nanoparticles for multimodal imaging and treatment of cancers," *Nanotheranostics*, vol. 4, no. 1, p. 1, 2020.

56. S. Lohe and M. Umekar, "Thermosensitive liposomal drug delivery systems: a review," *J Pharm Res Int*, vol. 34, no. 63, pp. 40–49, 2022.

57. R. Sam, M. Divanbeigi Kermani, M. Ohadi, S. Salarpour, and G. Dehghannoudeh, "Different applications of temperature responsive nanogels as a new drug delivery system mini review," *Pharm Dev Technol*, vol. 28, no. 5, pp. 492–500, 2023.

58. M. Prawatborisut, J. Oberländer, S. Jiang, R. Graf, Y. Avlasevich, S. Morsbach, et al., "Temperature-responsive nanoparticles enable specific binding of apolipoproteins from human plasma," *Small*, vol. 18, no. 3, p. 2103138, 2022.

59. E. Yuba, "Development of functional liposomes by modification of stimuli-responsive materials and their biomedical applications," *J Mater Chem B*, vol. 8, no. 6, pp. 1093–1107, 2020.

60. Y. Liu, P. Bhattarai, Z. Dai, and X. Chen, "Photothermal therapy and photoacoustic imaging via nanotheranostics in fighting cancer," *Chem Soc Rev*, vol. 48, no. 7, pp. 2053–2108, 2019.

61. J. Xiang, X. Tong, F. Shi, Q. Yan, B. Yu, and Y. Zhao, "Near-infrared light-triggered drug release from UV-responsive diblock copolymer-coated upconversion nanoparticles with high monodispersity," *J Mater Chem B*, vol. 6, no. 21, pp. 3531–3540, 2018.

62. H. Ma, W. Li, H. Fan, and J. Xiang, "A red-light-responsive DASA–polymer with high water stability for controlled release," *Polymers (Basel)*, vol. 15, no. 11, p. 2489, 2023.

63. E. Rivero-Buceta, M. E. Encheva, B. Cech, E. Fernandez, G. Sastre, C. C. Landry, et al., "Light-activated controlled release of camptothecin by engineering porous materials: the ship in a bottle concept in drug delivery," *Nanoscale*, vol. 15, no. 30, pp. 12506–12517, 2023.

64. T. Wang, H. Lei, X. Li, N. Yang, C. Ma, G. Li, et al., "Magnetic targeting nanocarriers combined with focusing ultrasound for enhanced intracerebral hemorrhage therapy," *Small*, vol. 19, no. 17, p. 2206982, 2023.

65. I. A. Edwards, F. De Carlo, J. Sitta, W. Varner, C. M. Howard, and P. P. Claudio, "Enhancing targeted therapy in breast cancer by ultrasound-responsive nanocarriers," *Int J Mol Sci*, vol. 24, no. 6, p. 5474, 2023.

66. J. Zhu, J. Wang, and Y. Li, "Recent advances in magnetic nanocarriers for tumor treatment," *Biomed Pharmacother*, vol. 159, p. 114227, 2023.

67. R. Chapla, K. T. Huynh, and C. E. Schutt, "Microbubble–nanoparticle complexes for ultrasound-enhanced cargo delivery," *Pharmaceutics*, vol. 14, no. 11, p. 2396, 2022.

68. B. Liu, X. Zhang, C. Li, F. He, Y. Chen, S. Huang, et al., "Magnetically targeted delivery of DOX loaded Cu9S5@mSiO2@Fe3O4-PEG nanocomposites for combined MR imaging and chemo/photothermal synergistic therapy," *Nanoscale*, vol. 8, no. 25, pp. 12560–12569, 2016.

69. M. Ghorbani, H. Hamishehkar, H. Hajipour, N. Arsalani, and A. A. Entezami, "Ternary-responsive magnetic nanocarriers for targeted delivery of thiol-containing anticancer drugs," *New J Chem*, vol. 40, no. 4, pp. 3561–3570, 2016.

70. E. F. da Silva Oliveira and D. C. de Oliveira Conceição, "Magnetic drug-carrying nanoparticles in cancer treatments," *Caderno de ANAIS HOME*, vol. 1, pp. 76–83, 2023.

71. I. Burmistrov, M. Veselov, A. Mikheev, T. Pallaeva, T. Bukreeva, N. Klyachko, et al., "Single domain magnetic nanoparticles as a magneto-mechanical actuators for remote drug release from polyelectrolyte microcapsules," in *Medical Sciences Forum*, MDPI, 2022, p. 90.

72. D. Stanicki, T. Vangijzegem, I. Ternad, and S. Laurent, "An update on the applications and characteristics of magnetic iron oxide nanoparticles for drug delivery," *Expert Opin Drug Deliv*, vol. 19, no. 3, pp. 321–335, 2022.

73. M. Haghighattalab, A. Kajbafzadeh, M. Baghani, Z. Gharehnazifam, B. M. Jobani, and M. Baniassadi, "Silk fibroin hydrogel reinforced with magnetic nanoparticles as an intelligent drug delivery system for sustained drug release," *Front Bioeng Biotechnol*, vol. 10, p. 891166, 2022.

74. A. G. Patel, R. B. Jotania, and M. F. Desimone, "Magnetic nanoparticles for drug delivery applications," *Mater Res Foundat*, vol. 143, pp. 233–252, 2023.

75. D. Mukhopadhyay, C. Sano, N. AlSawaftah, R. El-Awady, G. A. Husseini, and V. Paul, "Ultrasound-mediated cancer therapeutics delivery using micelles and liposomes: a review," *Recent Pat Anticancer Drug Discov*, vol. 16, no. 4, pp. 498–520, 2021.

76. Y. Tagai, K. Kajie, Z. Peng, K. Shimba, Y. Miyamoto, T. Shibata, et al., "A study on DDS with high response to ultrasound using metal nanoparticle-encapsulated liposomes," *IEEJ Trans Electron Inform Sys*, vol. 143, no. 7, pp. 712–713, 2023.

77. G. Ayana, J. Ryu, and S. Choe, "Ultrasound-responsive nanocarriers for breast cancer chemotherapy," *Micromachines (Basel)*, vol. 13, no. 9, p. 1508, 2022.

78 D. Li, Q. Liu, M. Yang, H. Xu, M. Zhu, Y. Zhang, et al., "Nanomaterials for mRNA-based therapeutics: challenges and opportunities," *Bioeng Transl Med*, vol. 8, no. 3, p. e10492, 2023.

79. X. Cai, M. Jin, L. Yao, B. He, S. Ahmed, W. Safdar, et al., "Physicochemical properties, pharmacokinetics, toxicology and application of nanocarriers," *J Mater Chem B*, vol. 11, no. 4, pp. 716–733, 2023.

80. A. Ahmad, M. Imran, and N. Sharma, "Precision nanotoxicology in drug development: current trends and challenges in safety and toxicity implications of customized multifunctional nanocarriers for drug-delivery applications," *Pharmaceutics*, vol. 14, no. 11, p. 2463, 2022.

81. D. A. Dri, F. Rinaldi, M. Carafa, and C. Marianecci, "Nanomedicines and nanocarriers in clinical trials: surfing through regulatory requirements and physico-chemical critical quality attributes," *Drug Deliv Transl Res*, vol. 13, no. 3, pp. 757–769, 2023.

82. C. Fernandes, M. Jathar, B. K. S. Sawant, and T. Warde, "Scale-up of nanoparticle manufacturing process," in *Pharmaceutical Process Engineering and Scale-up Principles*, Springer, 2023, pp. 173–203.

83. K. K. Patel, A. K. Agrawal, and S. Singh, "Preformulation challenges: the concept behind the selection, design and preparation of nanoformulations," in *Nanoformulations in Human Health: Challenges and Approaches*, Springer, 2020, pp. 43–71.

84. D. M. Patel, N. N. Patel, and J. K. Patel, "Nanomedicine scale-up technologies: feasibilities and challenges," in *Emerging Technologies for Nanoparticle Manufacturing*, Springer, 2021, pp. 511–539.

85. B. S. F. Cury, A. B. Meneguin, F. G. Prezotti, F. I. Boni, and V. M. de Oliveira Cardoso, "Nanocarriers for oral drug delivery," in *Nanocarriers for Drug Delivery: Concepts and Applications*, Elsevier, 2021, pp. 127–151.

86. L. Zhang, Z. Wang, R. Zhang, H. Yang, W. Wang, Y. Zhao, et al., "Multi-stimuli-responsive and cell membrane camouflaged aggregation-induced emission nanogels for precise chemo-photothermal synergistic therapy of tumors," *ACS Nano*, vol. 17, no. 24, pp. 25205–25221, 2023.

87. M. Osial and A. Pregowska, "The application of artificial intelligence in magnetic hyperthermia based research," *Future Internet*, vol. 14, no. 12, p. 356, 2022.

88. M. Tian, X. Xin, R. Wu, W. Guan, and W. Zhou, "Advances in intelligent-responsive nanocarriers for cancer therapy," *Pharmacol Res*, vol. 178, p. 106184, 2022.

89. C. Martins, M. Araújo, A. Malfanti, C. Pacheco, S. Smith, B. Ucakar, et al., "Stimuli-responsive multifunctional nanomedicine for enhanced glioblastoma chemotherapy augments multistage blood-to-brain trafficking and tumor targeting," *Small*, vol. 19, no. 22, p. 2300029, 2023.

# 3 Lipid-Based Nanocarriers for Nucleic Acids and Proteins

## *Liposomes, Solid Lipid Nanoparticles, Nanostructured Lipid Carriers*

*Biswajit Basu, Ayon Dutta, Dipanjana Ash, Bhupendra Prajapati, and Swarupananda Mukherjee*

### 3.1 Introduction

Over the past few decades, a number of drug-delivery technologies have attracted scientific interest. Nanomedicine is one of the most promising approaches which fascinate the drug-delivery system. Nanomedicine comprises the delivery of pharmaceutical compounds, such as therapeutic ingredients, to the targeted site of the human body with improved efficacy and safety. This approach also involves improving the therapeutic efficacy by improving pharmacokinetics and localized delivery with the proper release of the drug. This chapter mainly focuses on the importance of nanocarriers or nanoparticle-based systems for the delivery of nucleic acid and protein in human body. Nowadays, the applications of protein-based drug delivery have increased the management of gene-related therapy, including cancer, autoimmune diseases, neuropathological conditions, etc. In those cases, lipid-based nanocarriers improved the delivery of the protein or nucleic acid molecules in the particular site of the body [1]. Protein or lipid-based nanoparticles are a more promising drug-delivery approach due to their biodegradability, stability, ability to modify the surface of the particles, simplicity of controlling particle size, and lack of toxicity-related concerns like immunogenicity. This study incorporates the different classifications of lipid nanoparticles for the delivery of protein and nucleic acids (RNA/mRNA) in several gene therapy-based treatments in medical fields. Liposome is one of the most unique nanoparticles for the delivery of protein substances. The entrapment of the protein molecules with the liposome nanoparticle totally depends on the structure and compositions of the different components used for the development of the liposome, such as oil, surfactant, co-surfactant, etc. The different encapsulation efficiency of protein in liposomes and strategies of delivery makes it a promising delivery approach. Solid lipid nanoparticles (SLNs), as well as nanostructured lipid carriers (NLC), also play an important role in the delivery of the protein-based substance. The most important parameters of the SLN and NLC are the unique loading and release capacity of the drug or protein substance with proper maintenance of stability and biocompatibility, which is broadly mentioned in this chapter [2].

#### 3.1.1 Importance of Lipid-Based Nanocarriers in Nucleic Acid and Protein Delivery

In recent days, numerous advanced nanotechnology platforms have been developed for the purpose of diagnosis and treatment of several diseases. One of them, lipid-based nanocarriers, is the most promising approach for the transportation of nucleic acid and protein. Recent studies have found more applications of nucleic acids (RNA/mRNA) in several gene therapy-based treatments in medical fields. It is a very crucial biological process to regulate gene expression. However, recent clinical studies showed some major issues in the results of the expression of the different classes of nucleic acids as a mediator in the case of the management of gene-related medical conditions, including cancer, autoimmune diseases, neuropathological conditions, etc. One of the most crucial treatment approaches is RNA-based treatment [1]. The reason that freestanding RNA molecules are short-lived when given systemically is that they break down quickly, and RNA medications have less target specificity, which raises the possibility of inducing immunogenic reactions. So, in this case, the lipid-based nanocarriers facilitate the delivery of nucleic acids in case of the different management of the diseases [3]. Thus, lipid-based nanocarriers naturally interact favorably with the cell membrane to promote RNA uptake by the cell. Lipid-based nanocarrier systems consist of various biocompatible and biodegradable lipids and phospholipids, which do not need any further chemical treatment and are easily available in commercial markets [2]. For the delivery of mRNA, numerous materials have been established, such as polymers, protein derivatives, lipids, and lipid-like compounds. Specifically, lipid nanoparticles have been well studied and have had a successful clinical debut for the administration of small compounds, small interfering RNA (siRNA) medications, and messenger RNA (mRNA) [4].

#### 3.1.2 Lipid-Based Formulations Classification Figure

Lipid-based formulations are a very effective drug delivery system nowadays due to their effective size and site-specific targeted system. This lipid-based system is used to formulate different pharmaceutical formulations like topical, oral, pulmonary, or parenteral delivery. These types of formulations can be modified in various approaches on the basis of the disease conditions and product cost, stability, toxicity, and efficacy. This lipid formulation classification was added in the year 2006. The primary goal of the Lipid Formulation

DOI: 10.1201/9781003473183-3

Classification (LFCs) is to make it easier to interpret *in vivo* research, which will, therefore, make it easier to determine which formulations are best for particular medications [5]. The LFCs and their characteristic features with their merits and demerits are presented in Table 3.1, and different types of lipid-based drug delivery systems are mentioned in Figure 3.1.

## Table 3.1 Types of lipid-based formulation with merits and demerits

| Sl. no | Type of formulation | Ingredients | Characteristics | Merits | Demerits | References |
|---|---|---|---|---|---|---|
| 1 | Type 1 | Oils without surfactants | Digestion is required for nondispersing. | Generally recommended as safe | The drug's formulation has a low solvent capacity unless it is very lipophilic. | [5] |
| 2 | Type 2 | Oils and water-insoluble surfactants | SNEDDS were developed without water-soluble components. | Unlikely to lose its ability to dissolve solvents | Show turbidity during o/w dispersion. | [5] |
| 3 | Type 3 | Oil, surfactants and co-solvents | SNEDDS/ SMEDDS were developed with the help of water-soluble components. | Clearly dispersed in the solvent with drug or without drug | Potential loss of solvent capacity during dispersion and reduced digestibility | [6] |
| 4 | Type 4 | Co-solvents and water-soluble surfactants | Usually, formulation disperses to create a micellar solution. | Formulations have sufficient solvent capacity for many drugs | Possible loss of solvent capacity during dispersion; perhaps indigestible | [7] |

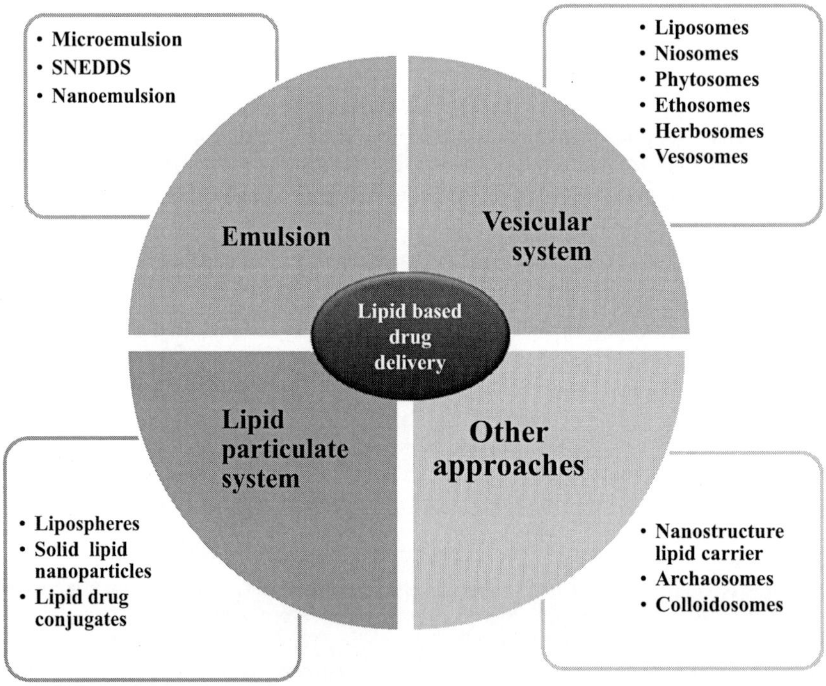


**Figure 3.1** Different types of lipid-based drug delivery system.

## 3.2 LIPOSOMES: STRUCTURE AND FUNCTION

### 3.2.1 Liposome Composition and Architecture

#### *3.2.1.1 Liposomes with Lipids and Phospholipids*

Liposomes are multilayer spherical vesicles constructed from lipid bilayers, or diacyl-chain phospholipids, self-assembling in aqueous solutions. Phospholipids are amphiphilic molecules—each has a hydrophilic "head" and a hydrophobic "tail"—which drives their self-assembly into bilayer membranes. Both natural and synthetic phospholipids can therefore be used to form liposomes [6]. The properties of liposomes, such as particle size, stiffness, fluidity, stability, and electrical charge, are all significantly affected by their lipid compositions. For example, liposomes made from naturally occurring unsaturated phosphatidylcholine, such as that found in eggs or soybeans, offer extremely permeable and low stable characteristics [7]. However, liposomes based on saturated phospholipids, like dipalmitoyl phosphatidylcholine, produced stiff, nearly impermeable bilayer structures. The hydrophilic group found in lipids can have one of three charges: positive, negative, or zwitterionic (containing both charges in one molecule). Stability is provided by the hydrophilic group's charge through electrostatic repellents. Lipids' hydrophobic groups differ in terms of saturation, acyl chain length, and symmetry [8].

#### *3.2.1.2 Liposomes with Natural Lipids*

Natural lipids can be obtained from various natural sources like soya bean and egg yolk. In the presence of the hydrocarbon chain is unsaturated, natural phospholipids are less stable than synthetic phospholipids when liposomes are being prepared [9]. Among various types of fatty acids in natural phospholipids, most are saturated fatty acids such as palmitic acid. Palmitic oleic acids constitute 92% of total fatty acid composition with a presence of polyunsaturated fatty acids and egg phospholipids consist of 40% of 1-palmitoyl-2-oleoylphosphatidylcholine. In phosphatidylethanolamine (PE) and phosphatidylserine (PS), the main saturated acid was stearic, but in the other lipids, it was palmitic. Additionally, approximately 95% of the fatty acid pattern obtained from soybeans is composed of palmitic acid (C16:0), stearic acid (C18:0), oleic acid (C18:1), linoleic acid (C18:2), and linolenic acid (C18:3) [10]. Phosphatidylcholine, phosphatidylethanolamine, and phosphatidylcholine lipids exist at both the $\alpha$- and $\beta$-positions of the glycerol moiety of these phospholipids because they make up more than 50% of the total acids [11].

#### *3.2.1.3 Liposomes with Synthetic Lipids*

Synthetic phospholipids are developed by some specific chemical modifications in the nonpolar and polar regions of the natural phospholipids. This modification helps to provide more amount of phospholipids for the commercial production of liposomes. Stearic and/or palmitic fatty acids are the main sources of saturated synthetic phospholipids. Additionally, unsaturated fatty acids in either one or both hydrocarbon chains, as well as mixed fatty acids, can be used to create synthetic phospholipids [12].

#### *3.2.1.4 Liposomes with Steroids*

Steroids basically consisted of four ring-based hydrophobic lipids in the structure. For example, cholesterol is integrated into the lipid bilayer of liposomes, and it is typically used as the main steroid in liposome synthesis at a ratio of less than 30% of total lipids to optimize liposome rigidity and stability [13]. The comparison study found that the combination of cholesterol and beta-sitosterol reduces the liposome membrane fluidity and induces the zeta potential, causing frequent changes in the particle size [12].

#### *3.2.1.5 Liposomes with Surfactants*

The role of surfactants in liposomal formulations is to modify the encapsulation efficacy and also help to improve the release profile by reducing the surface tension. Surfactants are single acyl-chain amphiphiles that increase the deformability of nano-vessels by disrupting the lipid bilayer of liposomal nanoparticles [14]. Sodium cholate, span 60, span 80, and tween 60, are commonly used surfactants in the liposomal formulation. Some surfactants also induce the skin permeation ability of the therapeutic agent in liposomal formulation. Liposomes are flexible mostly due to the edge activator (surfactant) [15]. Vesicles' lipid bilayers can be changed by the edge activator, making them more malleable. Compared to traditional liposomes, these nanovesicles are able to rapidly change their shape in response to osmotic pressure using minimal energy. Additionally, transepidermal drug flow was observed to be enhanced in ultra-deformable liposomes, making them a more appropriate nanovehicle for topical antihypertensive delivery [16].

### 3.2.2 Encapsulation Benefit of Nucleic Acids and Proteins in Liposomes

#### *3.2.2.1 siRNA*

Spliceosome-associated factor 1 (*Sart1*) siRNA was recently employed by Pan et al. to cure pulmonary fibrosis by repressing M2 macrophage polarization. The delivery of *Sart1* siRNA-loaded liposomes by intratracheal injection to an animal model of bleomycin (BLM)-induced pulmonary fibrosis produced results that indicated the liposomes accumulated in macrophages and prevented fibrosis in the mice via polarizing M2 macrophages [17]. In a different study, Wang and associates treated glomerulonephritis by decreasing the inflammatory response by focusing on the p38 *MAPK* and p65 pathways in the glomerulus. P38 *MAPK* and p65 siRNAs were complexed with polye thyleneimine (PEI) to promote siRNA condensation and facilitate endosomal escape. The complex is put itnto DSPE-PEG2000, DOPE, and CHEM liposomes, and then octa-arginine (R8) peptides are added to the liposomes to make them cationic [18].

#### *3.2.2.2 miRNA*

Liposomal-based system helps to promote the transport of miRNA in the specific targets of the cell. To deliver miRNA, a variety of nanocarriers have been studied, including liposomes, inorganic nanoparticles, solid lipid nanocarriers, lipopolyplexes, polyplexes, and dendrimers. Liposomes have been used in several studies as a miRNA delivery mechanism. Vakhshiteh et al., for instance, developed liposomes using DSPC: DSPE-PEG2000 for *miR-34a* delivery into breast cancer cells via DOTAP: CHOL. The findings indicated that the proliferation, invasion, and migration of breast cancer cells were significantly inhibited by *miR-34a*-loaded liposomes functionalized with cyclic arginine–glycine–aspartic acid (RGD) peptide. The miRNA-34a was used as a tumor-suppressing agent which is a component of the p53 transcriptional network. These suppressing and inhibition effects were very superior to liposomes-loaded non-targeted *miR34a*. [19]. Cui and colleagues synthesized cationic liposomes from 1,2-dioleoyl-3-trimethylammonium-propane (chloride) (DOTAP), 1,2-dioleoyl-sn-glycero-3-phosphoethanolamine (DOPE), and cholesterol (CHOL) to introduce *miR-7* into ovarian cancer cells cultured *in vitro*. The capacity of cationic liposomes to transfer *miR-7* into ovarian cancer cells that reside in the cell proliferation, invasion, and migration of ovarian cancer was demonstrated by the results [20].

#### *3.2.2.3 shRNA*

shRNA is synthesized and manufactured in the nuclei of cells and moved to the cytoplasm where it makes silencing complex with RNA for further activity. Both shRNA and miRNA are processed by the same RNAi mechanism. Internal mismatches in miRNAs usually result in secondary structural bulges. On the other hand, shRNAs lack bulges and feature wide internal base pairing [21]. Recently iRGD-modified liposomes have been developed to deliver eukaryotic translation initiation factor 3i (*eIF3i*) shRNA. The targeted liposomes prevented lung metastasis from melanoma and demonstrated great selectivity and effectiveness of elF3i shRNA delivery [22].

#### *3.2.2.4 CRISPR/Cas9*

CRISPR/Cas9 cannot be directly used as a treatment approach due to the poor pharmacokinetics and pharmacodynamics properties. It is interesting to note that liposomes show promise as CRISPR/Cas9 delivery vehicles [23]. A CRISPR/Cas9 delivery method based on PEGylated liposomes was recently published by McMillan et al. These scientists looked at the possibility of using PEGylated liposomes to transfer and maintain CRISPR/Cas9 plasmids outside of the bloodstream. They found that for up to six hours, PEGylated liposomes can shield plasmids from hydrolysis in serum [24]. Moreover, after four hours of injection into the blood, no DNA was found. So, using the freeze-dried matrix technique for cancer therapy, CRISPR/Cas9 with the sgRNA specific to *HPV E6* and *E7* oncogenes was entrapped inside the particle, resulting in effective tumor growth reduction and treatment *in vivo* [25].

### 3.2.3 Targeting Strategies and Surface Modifications

Liposomes have the potential to act as medication delivery vehicles for a variety of diseases, including cancer treatment. Tumor therapies are thought to be one of the greatest uses for liposomes. Many human cancer cell types, including those from the bladder, lung, ovarian, and breast, express epidermal growth factor receptor (EGFR), which is well-known to be an important therapeutic target. It is also known that this receptor is associated with a number of serious illnesses with poor prognosis. Different approaches have been found that some immunoliposomes consist of antibodies directly linked with the phospholipids by surface adsorption and it can also be linked with covalent bonding. Some specific ingredients like PEG are used as a linker

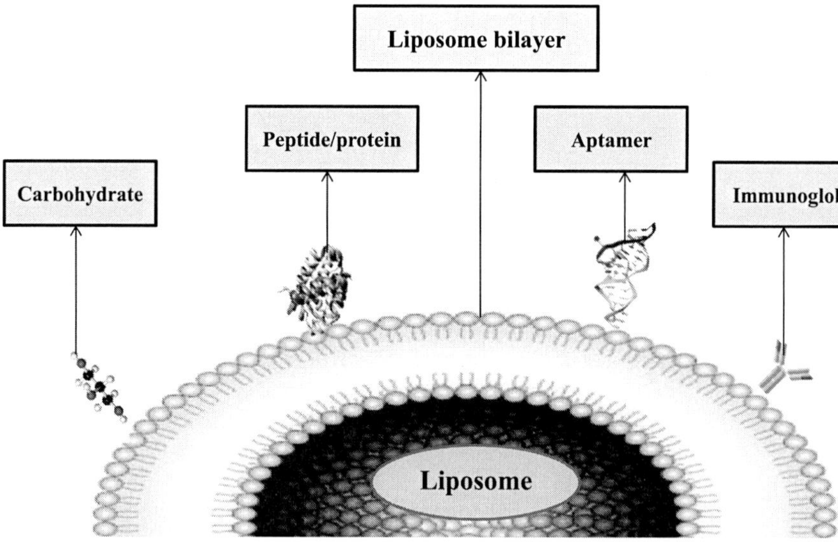


**Figure 3.2** Surface modification of liposomes with different modifications.

which helps to bind the antibody in the surface of liposomes. Other innovative strategies use engineered proteins or antibody fragments (much smaller than full-length antibodies) that are specifically conjugated to the surface of liposomes to enable targeted delivery [26]. These techniques have been used to create various nanoparticle antibody fragments and smaller liposomes, which serve as crucial biological instruments in the fields of therapy and diagnosis. Those are very promising techniques that target drug delivery with the surface modification of protein and liposomes. Important targets using such liganded liposomes have been proposed for the treatment of various diseases, including infectious and autoimmune diseases, neurodegenerative diseases, inflammatory and EGFR-positive cancers, and cardiovascular and neurological disorders. Some specific drugs like doxorubicin linked with the immunoliposomes targeted against EGFR showed promising results on animal xenografts model. In addition to entire antibodies or antibody fragments, peptide ligands have also been shown to interact specifically with receptors found on the neo-vasculature of tumors. This connection was used to directly intervene with genes, other proteins, or chemotherapy agents. Those are the modern techniques and target strategies for proteins, nucleic acids, and antibodies with liposomes [27]. Figure 3.2 represents the overview of surface modification of different protein components with the liposomes.

### 3.2.4 Applications in Nucleic Acid and Protein Delivery

The number of nucleic acid-loaded liposome treatments that have advanced to the clinical phases of treatment has demonstrated that liposome is a promising treatment approach for a wide range of disorders over the past 20 years. Precision RNA-based therapy has reached its potential with the FDA's approval of ONPATTRO® for the treatment of hereditary transthyretin-mediated amyloidosis (*hATTR* amyloidosis). ONPATTRO® is a siRNA (double-stranded) loaded liposome that basically targets the sequence of three untranslated sections of the *TTR* mRNA, as a result silencing the expression of transthyretin [28]. A genetic mutation in the transthyretin (*TTR*) gene causes HATTR amyloidosis. The mutation results in a mutant *TTR* protein that quickly misfolds, forming clumps that accumulate as amyloid in a variety of tissues and organs, primarily the liver. The patisiran siRNA composed of DLin-MC3-DMA, PEG2000-C-DMG, CHOL, and DSPC is loaded into liposomes. Initially, apolipoprotein E opsonizes the liposome, allowing it to enter liver cells via the apolipoprotein receptor. Conversely, the ionizable lipid DLin-MC3-DMA facilitates the release of patisiran siRNA into the cytoplasm [29]. The mRNA is destroyed through an RNAase H-dependent route after the RISC machinery cleaves the double-stranded siRNA. An antisense strand directs the RISC complex to engage specifically into the preserved sequence in the 3′untranslated part of the *TTR* mRNA. As a result, whether it is the wild type or the mutant variant, the transcription of the *TTR* protein decreases. Studies have indicated that it is generally well tolerated, yet it may cause mild to moderate infusion-related responses [30].

## 3.3 SOLID LIPID NANOPARTICLES (SLNS)

### 3.3.1 Formation and Characteristics of SLNs

Solid lipid nanoparticles are the most promising colloidal carriers in the pharmaceutical field. This is a new-generation drug-delivery approach where lipid or oil has been transformed into solid lipid to form submicron-sized lipid emulsions. SLN is very attractive and effective in the pharmaceuticals and cosmetics industry due to its distinctive properties, such as small size, large surface area, and high drug loading capacity [31]. Nowadays, SLNs are more popular as novel drug carriers for intravenous routes. The sub-micron colloidal carriers, or SLNs, are made of physiological lipids with the help of water or an aqueous surfactant solution as continuous media. Basically, the SLNs are composed of a combination of solid, lipid, emulsifier, and water or other solvent. Triglycerides (tri-stearin), partial glycerides (Imwitor), fatty acids (stearic acid, palmitic acid), steroids (cholesterol), and waxes (cetyl palmitate) are examples of the lipids that can be utilized as nanoparticle preparations. To stabilize lipid dispersion, a variety of emulsifiers have been utilized alone and in combination (Pluronic F 68, F 127). In this case, particle agglomeration may be more effectively prevented by the addition of emulsifiers. The fact that the lipid matrix in SLN is composed of physiological lipids lowers the risk of both acute and long-term toxicity, which is definitely an advantage. Different methods of preparation for the formation of SLNs include hot homogenization, cold homogenization, ultrasonication, solvent evaporation, microemulsion, spray drying method, double emulsion method, and using supercritical fluid [32]. During the formation of SLNs, the adhesive forces between two liquids serve as the primary driving factor behind the production of solid lipid nanoparticles. Due to the lower adhesive forces than those involving gases, the interfacial tension between two liquids is often smaller than their surface tension. The surface free energy of interfacial tension is made up of molecules near the interface, which agitate and create a spherical system to minimize the surface free energy. SLNs bear a striking resemblance to emulsions, in which the aqueous phase is combined with melted solid lipids that have been employed in place of the oil phase. Agitation at high speed results in the formation of droplets of the dispersed phase to develop in the dispersion medium. Finally, the surfactant is added as a third component to form stable SLNs [33].

### 3.3.2 Loading and Release of Nucleic Acids and Proteins in SLNs

Microchannels are the most promising methods for the incorporation of nucleic acid or proteins in the SLNs by a narrow-size distribution method. Furthermore, it has been demonstrated that this approach facilitates continuous complexation between cationic lipids and DNA during the formation of SLNs, making it comparatively labor-insensitive due to the reduction of typically needed stages. Lobovkina et al. took an alternative route, employing a hydrophobic ion pairing strategy to create a complex between siRNA and DOTAP before the SLNs were assembled as a whole. It is interesting to note that this method makes the siRNA more lipophilic, which facilitates a more effective nucleic acid incorporation into the lipid core of the SLNs.

### 3.3.3 Stability and Biocompatibility of SLNs

This has already been reported that SLNs are very much stable for more than three years. The stability of the SLNs depends on the different excipients and materials like lipid materials, surfactant concentration, and temperature optimization during the preparation. Those parameters reflect stability and storage. During preparation and storage, triglycerides undergo polymorphic transitions between the α, β′, and β crystal forms [34]. The chain length affects the kinetics of the polymorphic transitions, where slower the crystallization process for the longer chain than shorter chain triglycerides. Gel formation can occur in SLNs occasionally, and the gelling tendency of these materials is highly dependent on b0 modification caused by exposure to light, temperature, and shear stress. In various studies, solid lipid nanoparticles showed gel formation in various destabilizing factors, which causes decreased zeta potential. Nevertheless, there are a number of stability problems with SLNs, and the drug may hydrolyze in an aqueous solution. Thus, drying is the most promising option, which can increase the stability of SLNs during storage [35]. Lyophilization and spray drying are the most suitable drying techniques for the SLNs. Recently, SLNs were prepared using the electrospray technique, which produced a dry SLN powder directly from this technique. SLNs are formulated as a powder that may be placed into tablets, capsules, or pellets, making these materials very beneficial for medication delivery. However, because SLN formulations provide relatively sparse "burst release" of the medication by uncontrolled particle development via coagulation or agglomeration, their applicability may be limited. Ideal crystal lipid matrices, found in SLNs, transport the loaded medication in its molecular form between fatty acid chains. During storage of SLNs, enhancement of crystal structure affects the release of loaded drug solution, which is a huge drawback of solid lipid nanoparticles [33].

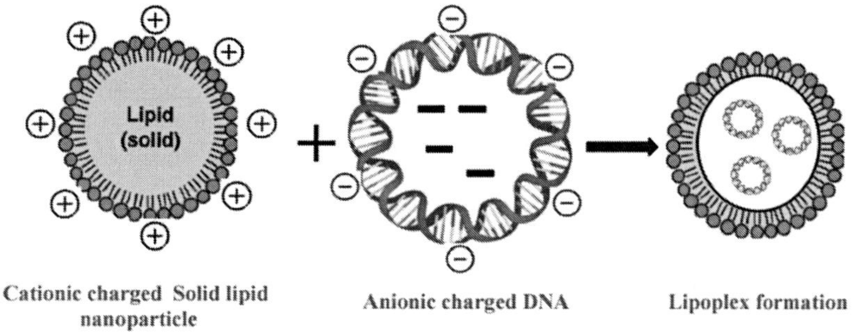


**Figure 3.3** Formation of lipoplex complex with the interaction of SLN and DNA.

### 3.3.4 Applications in Nucleic Acid and Protein Delivery

Recently, the use of SLNs as gene delivery vehicles has grown in popularity. The potential for electrostatic interaction between the positive and negative charges of the lipid and DNA results in the creation of a complex known as a lipoplex, shown in Figure 3.3. The ability of cationic SLNs to form a compound with DNA plasmids was demonstrated by Carrillo et al. The siRNA-PEG/SLN, developed by Jin et al., has no systemic toxicity and can pass through the blood–brain barrier to reach the tumor location. Montana et al. used cationically modified SLN as RNA carriers for *Paracentrotus lividus* bep3 in a different investigation, and they assessed their viability as a nonviral gene delivery vector [36].

## 3.4 NANOSTRUCTURED LIPID CARRIERS (NLCS)

### 3.4.1 Advantages of NLCs Over Traditional Lipid Nanoparticles

NLCs have the potential to overcome the general limitation compared with the traditional and conventional lipid carriers. It is well known that NLCs are a novel class of lipid nanoparticles that provide stable drugs with enhanced drug loading and release characteristics. The possible advantages of the NLC are discussed in the subsections below [37].

#### *3.4.1.1 Higher Drug Loading and Entrapment Efficiency*

In the case of SLNs, the amount of drug soluble in the melted lipid before the particle formation is higher than in the final SLN, and the large concentration in the melt could result in drug expulsion in case of the cooling process [38]. On the other hand, in NLCs, the lipid nanoparticle's solid matrix has a nano-oil portion where the drug solubility is greater, enhancing the capacity for total drug loading. Consequently, the amount of liquid lipid in NLCs has a significant impact on the entrapment efficiency because it creates many crystal defects in solid lipids and issues in the highly ordered crystal matrix, which creates enough space for a significant amount of drug to lodge effectively [39].

#### *3.4.1.2 Optimization of Drug Release Pattern*

NLCs show a biphasic drug release pattern, which consists of an initial rapid drug release followed by a steady, persistent release of the drug. The drug is first released in a burst form due to the formation of a drug-enriched casing by the liquid lipid found in the outer layers of the nanoparticles [40]. Oil-enriched outer layers, in contrast to SLN, exhibit much greater drug solubility for lipophilic medicines. Consequently, a greater quantity of drug may be readily loaded and released by drug diffusion or matrix erosion. The solid lipid core releases the medication slowly after the first, quicker release phase. It is interesting to note that in NLCs, release profiles may be adjusted based on the nature of the lipid matrix, for example, by changing the ratio of liquid to solid lipid content [41].

#### *3.4.1.3 More Stability During Storage*

The concepts of NLC, or non-crystalline lipid matrices, which are solid in nature but not crystalline, are derived from the fact that crystallization causes the expulsion of the drug. Utilizing unique blends of liquid and solid lipids, the particles cool down and solidify but do not crystallize [42]. Apart from affecting the particle size, entrapment, and *in vitro* drug release characteristics, the liquid lipids integrated with the solid lipids in NLC, also solve the issue of possible long-term storage modifications, particularly those related to crystallinity and polymorphism. As crystallization forms the cause of supersaturation, the presence of liquid lipids is expected to maintain the solid lipid's subsaturated state, hence reducing crystallization [43].

#### *3.4.1.4 Maximum Drug Loading Potential at the Minimum Surfactant Level*

NLCs have optimal stability, trapping, and release characteristics when stabilized at the lowest feasible surfactant concentration. In many cases, 0.5–1% of the surfactant is enough to produce stable NLC of lipophilic drugs. Additionally, compared to liquid emulsion, where there is very little scope for experimentation with excipients, a full variety of accepted state excipients is available. This feature makes NLC a considerably better formulation strategy than LE in situations when there is a worry about the high and sparing usage of surfactants [41].

### 3.4.2 NLC Formulation and Design

Basically, NLCs are formulated with a combination of lipids, surfactants, organic solvents, and other agents like surface modifiers. Lipid is the main ingredient of NLCs, which controls drug loading capacity, duration of action, and formulation stability. The formulation of NLC requires lipids such as waxes, steroids, triglycerides, diglycerides, monoglycerides, etc. For the preparation of NLC, biodegradable, nontoxic and GRAS-recommended lipids are required. The selection of proper lipids is crucial before using them in the formation of NLC. Suitable lipids improve the drug entrapment and loading efficacy in nanocarriers [44]. The addition of surfactant with proper concentration is also important during the formation of NLC. It has been found that the choice of surfactant has a significant impact on the toxicity, physical stability, and crystallinity of NLCs. The choice of surfactant depends on the route of administration, hydrophilic–lipophilic balance (HLB), or hydrophilic–lipophilic balance. Because of their amphipathic character, surface active agents or emulsifiers are adsorbed on the interface, where they reduce the tension between the lipid and aqueous phases. During the formulation of NLC, the most common aspect is the crystallization of colloids, causing the solidification of particles and the increase in the surface area of the particles and crystallization occurs so that the whole system becomes unstable. In that case, the addition of surfactant improves the quality and stability of the product [45]. The use of the required HLB or rHLB system is very crucial during the selection of surfactant for NLC formulation. Basically, rHLB defines the value or amount of emulsifiers or surfactants required for the formation of NLC. The HLB value of the emulsifier, or the rHLB value for lipid, is required for proper emulsification, or the lowering of interfacial tension between the oil and water phases which provides stable nanocarriers. Some other excipients such as organic salt ionic polymers that are required may be added as counter-ions in the formulations of NLC to overcome the problems of encasing of water-soluble drug molecules. Another type of excipients used in NLC formulations is called surface modifiers, which reduce the excipients' phagocytic absorption by the macrophages in the reticuloendothelial system (RES) [46]. To increase the duration of therapeutic molecules' residence time in the systemic circulation, hydrophilic polymers like PEG, poloxamines, or poloxamers are applied in the formulation of NLC. For the production of NLCs, some specific methods, such as higher-pressure homogenization, hot homogenization and cold homogenization, solvent emulsification evaporation and diffusion method, microemulsion, phase inversion, etc., are employed. Those production techniques help to formation of NLC from the different formulations and designs of nanocarriers [47].

### 3.4.3 NLC Applications in Gene Delivery

NLCs, which are typically classified into two groups, viral vectors and non-viral vectors, can be employed as a transfer gene carrier. The restricted amount of transport DNA, tumorigenicity, immunogenicity of the carrier, and greater expenses of the viral vector limit its use even if it can transfer genes effectively. Some applications of NLC as gene delivery systems are presented in Table 3.2.

## 3.5 FUTURE PERSPECTIVES AND CONCLUSION

Currently, a number of nanotechnology platforms are being developed with the goal of enhancing medicine delivery, particularly in the fight against cancer. Lipid-based nanoparticles are among the most extensively researched nanocarriers and are among the most promising options for medication delivery. Specifically, this approach is employed for the design and development of anticancer nanocarriers. As per our opinion, efforts should be focused on the development of the nanoparticles depending on the production methods, the creation of repeatable nanocarrier procedures, large-scale manufacturing, and the attachment of targeting molecules such as peptides to the surface of nanoparticles and proper selection of lipids for the unique drug release pattern. We summarize that in the future, the demand for phospholipids will increase for the drug release of nanoparticles. Bolalipids have a few special characteristics above glycerol-based phospholipids, which are utilized to make liposomes, and their uses as nanocarriers are still in their infancy. Bolalipid drug-delivery assemblies' future is thus still unknown. Similarly, because of their comparatively stable components and likely simplicity of drug encapsulation, SLN and NLC are particularly appealing possibilities for drug delivery. However, a lot of study is also being done on their potential in the clinical arena [55].

## Table 3.2 Nanolipid career-based gene delivery

| Sl. no. | Loaded gene | Drug name | Route of administration | Utility | Result | Reference |
|---|---|---|---|---|---|---|
| 1. | Specific siRNA duplexes for EGFR | Paclitaxel | Intravenous and inhalation | Lung cancer treated with chemotherapy and gene therapy. | Cancer cells demonstrate the efficient transmission of siRNA and medication. *In vivo* research indicates that inhalation leads to increased lung distribution, but intravenous injection was unable to reach the tumor-affected lung. | [48] |
| 2. | pDNA with enhanced green fluorescence protein | Temozolomide | Intravenous | Therapy for gliomatosis cerebri combination | Greater gene transfection efficiency was associated with a fourfold reduction in the IC50 value. Co-loaded NLC exhibits a threefold increase in tumor inhibition *in vivo* compared to standard medication. | [49] |
| 3. | Fluorescent siRNA AF488–231 | | Topically | Topical delivery for wound healing | Collagen scaffold coupled with gene-encapsulated NLC exhibits prolonged siRNA administration and ERK-1 protein downregulation. | [50] |
| 4. | Antigens of the Zika virus that encode rvRNA | | Intramuscular | A single, low-dose Zika vaccination | A stable vaccine in two vials is made and fine-tuned to deliver maximal immunogenicity at minimum dosage. In mice, NLC loaded rvRNA offers defense against the Zika virus. | [51] |
| 5. | TNF-α siRNA | Tacrolimus | Topically | Therapy for psoriasis | Studies on cellular transfection reveal the existence of RNA in the cytoplasm. *In vivo* investigations demonstrate a sevenfold reduction in the expression of the cytokine TNF-α in mice. | [52] |

(*continued*)

## Table 3.2 (Continued)

| Sl. no. | Loaded gene | Drug name | Route of administration | Utility | Result | Reference |
|---|---|---|---|---|---|---|
| **6.** | MMP3, CCL12, and HIF1A siRNAs | Prostaglandin E | Inhalation | Idiopathic pulmonary fibrosis treatment | According to *in vivo* research, breathing in a mouse model causes a reduction in body mass, which is a sign of a decrease in pulmonary fibrosis. Low tissue damage is attained during lung delivery, and death is seen. | [53] |
| **7.** | pDNA enriched with a plasmid for the green fluorescent protein. | Paclitaxel | Intravenous | For the lung cancer treatment | Transferrin-modified cationic surface NLC was made. *In vitro* and *in vivo* investigations confirm high gene transfection and increased antitumor efficacy with little cytotoxicity. | [54] |

## REFERENCES


1. Yi Xue, H., P. Guo, W.-C. Wen, and H. L. Wong, *Lipid-based nanocarriers for RNA delivery. Current pharmaceutical design*, 2015. **21**(22): p. 3140–3147.
2. Pathak, K., L. Keshri, and M. Shah, *Lipid nanocarriers: influence of lipids on product development and pharmacokinetics. Critical Reviews™ in Therapeutic Drug Carrier Systems*, 2011. **28**(4).
3. Kumar, L.D. and A.R. Clarke, *Gene manipulation through the use of small interfering RNA (siRNA): from in vitro to in vivo applications. Advanced drug delivery reviews*, 2007. **59**(2-3): p. 87–100.
4. Hou, X., T. Zaks, R. Langer, and Y. Dong, *Lipid nanoparticles for mRNA delivery. Nature Reviews Materials*, 2021. **6**(12): p. 1078–1094.
5. Shrestha, H., R. Bala, and S. Arora, *Lipid-based drug delivery systems. Journal of pharmaceutics,* 2014. 2014.
6. Wu, X., X. Dai, Y. Liao, M. Sheng, and X. Shi, *Investigation on drug entrapment location in liposomes and transfersomes based on molecular dynamics simulation. Journal of molecular modeling*, 2021. **27**: p. 1–10.
7. Nakhaei, P., R. Margiana, D. O. Bokov, W. K. Abdelbasset, M. A. Jadidi Kouhbanani, R. S., Varma, et al., *Liposomes: structure, biomedical applications, and stability parameters with emphasis on cholesterol.* Frontiers in bioengineering and biotechnology, 2021. **9**.
8. Large, D.E., R. G. Abdelmessih, E. A. Fink, and D. T. Auguste, *Liposome composition in drug delivery design, synthesis, characterization, and clinical application. Advanced drug delivery reviews*, 2021. **176**: p. 113851.

9. Daraee, H., A. Etemadi, M. Kouhi, S. Alimirzalu, and A. Akbarzadeh, *Application of liposomes in medicine and drug delivery. Artificial cells, nanomedicine, and biotechnology*, 2016. **44**(1): p. 381–391.

10. Li, J., T. Zhang, C. Wang, Z. Huang, X. Luo, and Y. Deng, *A review on phospholipids and their main applications in drug delivery systems. Asian journal of pharmaceutical sciences*, 2015. **10**(2): p. 81–98.

11. Lordan, R., A. Tsoupras, and I. Zabetakis, *Phospholipids of animal and marine origin: Structure, function, and anti-inflammatory properties. Molecules,* 2017. **22**(11): p. 1964.

12. Nsairat, H., D. Khater, U. Sayed, F. Odeh, A. Al Bawab, and W. Alshaer, *Liposomes: Structure, composition, types, and clinical applications. Heliyon*, 2022. **8**(5).

13. Briuglia, M.-L., C. Rotella, A. McFarlane, and D. Cipolla, *Influence of cholesterol on liposome stability and on in vitro drug release. Drug delivery and translational research*, 2015. **5**: p. 231–242.

14. Cipolla, D., H. Wu, I. Gonda, S. Eastman, T. Redelmeier, and H.-K. Chan, *Modifying the release properties of liposomes toward personalized medicine. Journal of pharmaceutical sciences*, 2014. **103**(6): p. 1851–1862.

15. Lee, E.H., A. Kim, Y. K. Oh, and C.-K. Kim, *Effect of edge activators on the formation and transfection efficiency of ultradeformable liposomes. Biomaterials*, 2005. **26**(2): p. 205–210.

16. Souto, E.B., A. S. Macedo, J. Dias-Ferreira, A. Cano, A. Zielińska, and C. M. Matos, *Elastic and ultradeformable liposomes for transdermal delivery of active pharmaceutical ingredients (APIs). International Journal of Molecular Sciences*, 2021. **22**(18): p. 9743.

17. Pan, T., Q. Zhou, K. Miao, L. Zhang, G. Wu, J. Yu, et al., *Suppressing Sart1 to modulate macrophage polarization by siRNA-loaded liposomes: a promising therapeutic strategy for pulmonary fibrosis. Theranostics*, 2021. **11**(3): p. 1192.

18. Wang, Y., Q. Wu, J. Wang, L. Li, X. Sun, Z. Zhang, et al., *Co-delivery of p38α MAPK and p65 siRNA by novel liposomal glomerulus-targeting nano carriers for effective immunoglobulin a nephropathy treatment. Journal of controlled release*, 2020. **320**: p. 457–468.

19. Vakhshiteh, F., E. Khabazian, F. Atyabi, S. N. Ostad, Z. Madjd, and R. Dinarvand, *Peptide-conjugated liposomes for targeted miR-34a delivery to suppress breast cancer and cancer stem-like population. Journal of Drug Delivery Science and Technology*, 2020. **57**: p. 101687.

20. Cui, X., K. Song, X. Lu, W. Feng, and W. Di, *Liposomal delivery of MicroRNA-7 targeting EGFR to inhibit the growth, invasion, and migration of ovarian cancer. ACS omega*, 2021. **6**(17): p. 11669–11678.

21. Lambeth, L.S. and C.A. Smith, *Short hairpin RNA-mediated gene silencing.* siRNA Design: Methods and Protocols, 2013: p. 205–232.

22. Xiao, W., W. Zhang, H. Huang, Y. Xie, Y. Zhang, X. Guo, et al., *Cancer targeted gene therapy for inhibition of melanoma lung metastasis with eiF3i shRNA loaded liposomes. Molecular Pharmaceutics*, 2019. **17**(1): p. 229–238.

23. Zhen, S. and X. Li, *Liposomal delivery of CRISPR/Cas9. Cancer gene therapy*, 2020. **27**(7): p. 515–527.

24. Chaverra-Rodriguez, D., D. Chaverra-Rodriguez, V. M. Macias, G. L. Hughes, S. Pujhari, Y. Suzuki, et al., *Targeted delivery of CRISPR-Cas9 ribonucleoprotein into arthropod ovaries for heritable germline gene editing. Nature communications*, 2018. **9**(1): p. 3008.

25. Jubair, L., S. Fallaha, and N.A. McMillan, *Systemic delivery of CRISPR/Cas9 targeting HPV oncogenes is effective at eliminating established tumors. Molecular Therapy*, 2019. **27**(12): p. 2091–2099.

26. Sapra, P. and T. Allen, *Ligand-targeted liposomal anticancer drugs. Progress in lipid research*, 2003. **42**(5): p. 439–462.

27. Khan, A.A., K. Allemailem, S. A. Almatroodi, A. Almatroudi, and A. H. Rahmani, *Recent strategies towards the surface modification of liposomes: an innovative approach for different clinical applications. 3 Biotech*, 2020. **10**: p. 1–15.

28. Al Shaer, D., A. Al Musaimi, F. Albericio, and M. de la Torre, *2021 FDA TIDES (peptides and oligonucleotides) harvest. Pharmaceuticals,* 2022. **15**(2): p. 222.

29. Zhang, X., V. Goel, and G.J. Robbie, *Pharmacokinetics of Patisiran, the first approved RNA interference therapy in patients with hereditary transthyretin-mediated amyloidosis. The Journal of Clinical Pharmacology*, 2020. **60**(5): p. 573–585.

30. Nsairat, H., W. Alshaer, F. Odeh, E. Esawi, D. Khater, A. A. Bawab, et al., *Recent advances in using liposomes for delivery of nucleic acid-based therapeutics. OpenNano*, 2023. **11**: p. 100132.

31. Cavalli, R., O. Caputo, and M.R. Gasco, *Solid lipospheres of doxorubicin and idarubicin. International journal of pharmaceutics*, 1993. **89**(1): p. R9–R12.

32. Mukherjee, S., S. Ray, and R. Thakur, *Solid lipid nanoparticles: a modern formulation approach in drug delivery system. Indian journal of pharmaceutical sciences*, 2009. **71**(4): p. 349.

33. Duan, Y., A. Dhar, C. Patel, M. Khimani, S. Neogi, P. Sharma, et al., *A brief review on solid lipid nanoparticles: Part and parcel of contemporary drug delivery systems. RSC advances*, 2020. **10**(45): p. 26777–26791.

34. Luzzati, V., A. Tardieu, and T. Gulik-Krzywicki, *Polymorphism of lipids. Nature*, 1968. **217**(5133): p. 1028–1030.

35. Bunjes, H., K. Westesen, and M.H. Koch, *Crystallization tendency and polymorphic transitions in triglyceride nanoparticles. International journal of pharmaceutics*, 1996. **129**(1-2): p. 159–173.

36. Dolatabadi, J.E.N., H. Valizadeh, and H. Hamishehkar, *Solid lipid nanoparticles as efficient drug and gene delivery systems: recent breakthroughs. Advanced pharmaceutical bulletin*, 2015. **5**(2): p. 151.

37. Müller, R., M. Radtke, and S. Wissing, *Nanostructured lipid matrices for improved microencapsulation of drugs. International journal of pharmaceutics*, 2002. **242**(1-2): p. 121–128.

38. Üner, M., *Preparation, characterization and physico-chemical properties of solid lipid nanoparticles (SLN) and nanostructured lipid carriers (NLC): their benefits as colloidal drug carrier systems. Die pharmazie-an international journal of pharmaceutical sciences*, 2006. **61**(5): p. 375–386.

39. Zhuang, C.-Y., N. Li, M. Wang, X.-N. Zhang, W.-S. Pan, J.-J. Pe, et al., *Preparation and characterization of vinpocetine loaded nanostructured lipid carriers (NLC) for improved oral bioavailability. International journal of pharmaceutics*, 2010. **394**(1-2): p. 179–185.

40. Mühlen, A.z., E. z. Mühlen, H. Niehus and W. Mehnert, *Atomic force microscopy studies of solid lipid nanoparticles. Pharmaceutical research*, 1996. **13**: p. 1411–1416.

41. Khan, S., M. Baboota, J. Ali, S. Khan, R. S. Narang, and J. K. Narang, *Nanostructured lipid carriers: An emerging platform for improving oral bioavailability of lipophilic drugs. International journal of pharmaceutical investigation*, 2015. **5**(4): p. 182.

42. Muchow, M., P. Maincent, and R.H. Müller, *Lipid nanoparticles with a solid matrix (SLN®, NLC®, LDC®) for oral drug delivery. Drug development and industrial pharmacy*, 2008. **34**(12): p. 1394–1405.

43. Tiwari, R. and K. Pathak, *Nanostructured lipid carrier versus solid lipid nanoparticles of simvastatin: comparative analysis of characteristics, pharmacokinetics and tissue uptake. International journal of pharmaceutics*, 2011. **415**(1-2): p. 232–243.

44. Shah, R., D. Eldridge, E. Palombo, and I. Harding, *Lipid nanoparticles: Production, characterization and stability*. Vol. 1. 2015: Springer.

45. Noor, N.M., N. Sheikh, S. Somavarapu, and R. Taylor, *Preparation and characterization of dutasteride-loaded nanostructured lipid carriers coated with stearic acid-chitosan oligomer for topical delivery. European Journal of Pharmaceutics and Biopharmaceutics*, 2017. **117**: p. 372–384.

46. Shah, M.R., M. Imran, and S. Ullah, *Lipid-based nanocarriers for drug delivery and diagnosis*. 2017: William Andrew.

47. Chauhan, I., M. Yasir, M. Verma, and A. P. Singh, *Nanostructured lipid carriers: A groundbreaking approach for transdermal drug delivery. Advanced pharmaceutical bulletin*, 2020. **10**(2): p. 150.

48. Garbuzenko, O.B., A. Kuzmov, O. Taratula, S. R. Pine, and T. Minko, *Strategy to enhance lung cancer treatment by five essential elements: inhalation delivery, nanotechnology, tumor-receptor targeting, chemo-and gene therapy. Theranostics*, 2019. **9**(26): p. 8362.

49. Wang, H., S. Liu, L. Jia, F. Chu, Y. Zhou, Z. He, et al., *Nanostructured lipid carriers for MicroRNA delivery in tumor gene therapy. Cancer Cell International*, 2018. **18**: p. 1–6.

50. Tezgel, Ö., N. DiStasio, V. Laghezza-Masci, A.-R. Taddei, A. Szarpak-Jankowska, R. Auzély-Velty, et al., *Collagen scaffold-mediated delivery of NLC/siRNA as wound healing materials. Journal of Drug Delivery Science and Technology*, 2020. **55**: p. 101421.

51. Erasmus, J.H., A.P. Khandhar, J. Guderian, B. Granger, J. Archer, M. Archer, et al., *A nanostructured lipid carrier for delivery of a replicating viral RNA provides single, low-dose protection against Zika. Molecular Therapy*, 2018. **26**(10): p. 2507–2522.

52. Viegas, J.S.R., F. Garcia Praça, A. L. Caron, I. Suzuki, A. V. Pupo Silvestrini, W. S. Garcia Medina, et al., *Nanostructured lipid carrier co-delivering tacrolimus and TNF-α siRNA as an innovate approach to psoriasis. Drug Delivery and Translational Research*, 2020. **10**: p. 646–660.

53. Garbuzenko, O.B., V. Ivanova, V. Kholodovych, D. C. Reimer, K. R. Reuhl, E. Yurkow, et al., *Combinatorial treatment of idiopathic pulmonary fibrosis using nanoparticles with prostaglandin E and siRNA (s). Nanomedicine: Nanotechnology, Biology and Medicine*, 2017. **13**(6): p. 1983–1992.

54. Khan, S., A. Sharma, and V. Jain, *An overview of nanostructured lipid carriers and its application in drug delivery through different routes. Advanced Pharmaceutical Bulletin*, 2023. **13**(3): p. 446.

55. Puri, A., K. Loomis, B. Smith, J.-H. Lee, A. Yavlovich, E. Heldman, et al., *Lipid-based nanoparticles as pharmaceutical drug carriers: from concepts to clinic. Critical Reviews™ in Therapeutic Drug Carrier Systems*, 2009. **26**(6).

# 4 Polymer-Based Nanocarriers

## *Dendrimers, Polymeric Micelles, Hydrogels*

*Juan C. Cruz and Luis H. Reyes*

### 4.1 INTRODUCTION

Polymer-based nanocarriers are revolutionizing drug-delivery systems in the fast-paced field of nanobiotechnology. Dendrimers, polymeric micelles, and hydrogels stand out as up-and-coming innovations. These nanocarriers offer a range of advantages: they enhance drug solubility and bioavailability, enable site-specific targeting, and provide controlled release mechanisms while reducing toxicity. By overcoming the limitations of traditional drug-delivery methods, nanocarrier-equipped pharmaceuticals pave the way for more targeted therapies with fewer side effects for patients. This drug administration paradigm shift exemplifies nanobiotechnology's transformative potential in modern medicine.

In the pharmacological context, there are two major obstacles: solubility and bioavailability. Low solubility in water is a common problem for many drugs and thus presents limited absorption and low therapeutic value (1). Notably, nanotechnology drug-delivery systems using polymer-based nanocarriers are among the most prominent areas. In such nanocarriers, the solubility and bioavailability of the drugs may therefore be increased. For instance, dendrimers are versatile systems for carrying drugs formed using drug encapsulation, increasing poorly water-soluble drugs' solubility.

Another critical feature is targeted drug delivery, where the therapeutic is delivered at the disease site without affecting the rest, lowering the overall side effects. Polymeric micelles formed by the self-assembly of block copolymers have offered great promise for this application. Core–shell structures allow the encapsulation of hydrophobic drugs in the core, which could be delivered to sites, for example, tumor cells, in a targeted manner with the help of receptor–ligand interactions at the shell (2).

Regulation of the drug release from the formulation is needed to maintain an appropriate drug level in the body over an extended period to improve treatment efficacy and patient compliance. Hydrogels are exceptionally hydrophilic polymer networks that are capable of imbibing large amounts of water or biological fluids and hence finding wide use in systems for the controlled release of encapsulated drugs in response to outer stimuli, such as pH, temperature, or enzymes present in the diseased environment (3).

Designing drug-delivery systems with minimal toxicity is crucial for their clinical success. Nanocarriers must be engineered to be biocompatible and biodegradable, ensuring safe metabolism after delivering their therapeutic payload and facilitating proper bodily excretion. This aspect is vital in preventing the accumulation of potentially harmful materials and mitigating adverse reactions. Researchers can develop more effective and safer drug-delivery systems by focusing on these key factors, ultimately improving patient outcomes and expanding treatment options for various diseases (4).

Recent advances in nanobiotechnology have built upon the foundational work in the field, highlighting the potential of polymer-based nanocarriers in drug-delivery systems. Current research and development efforts are focused on optimizing these nanocarriers for a broad spectrum of therapeutic applications. Scientists are working diligently to overcome existing challenges to unlock new medical possibilities. By refining the design and functionality of nanocarriers, researchers are paving the way for more effective treatment interventions across various medical conditions. This ongoing innovation promises to revolutionize drug delivery, potentially improving patient outcomes and expanding the reach of targeted therapies.

Nanocarriers have revolutionized drug delivery by improving therapeutics' bioavailability, targeting efficiency, and pharmacokinetics. Various nanocarriers, including liposomes, dendrimers, polymeric nanoparticles, solid lipid nanoparticles, and nanostructured lipid carriers, interact differently with biological systems and exhibit varied delivery efficiencies.

Liposomes, spherical vesicles with a lipid bilayer, mimic cellular membranes, facilitating effective drug delivery and reducing toxicity. Their biocompatibility and ability to encapsulate hydrophilic and hydrophobic drugs make them highly versatile (5). With their branched structure, dendrimers offer precise control over size and surface functionality, allowing for targeted delivery and reduced off-target effects (6). Polymeric nanoparticles are designed for controlled release, thereby providing sustained drug release and improving patient compliance (7). Solid lipid nanoparticles and nanostructured lipid carriers improve the solubility of hydrophobic drugs and offer an alternative to polymeric systems, with advantages in biodegradability and lower cytotoxicity (8).


DOI: 10.1201/9781003473183-4

The interaction of these nanocarriers with biological systems—ranging from cellular uptake mechanisms to biodistribution and clearance—is critical in determining their delivery efficiency. Factors such as size, surface charge, and hydrophobicity/hydrophilicity balance play important roles in these interactions. Optimizing these characteristics can enhance the delivery efficiency of nanocarriers, making them more effective in targeting specific tissues or cells, crossing biological barriers, and releasing their payload at the desired site of action. Polymer-based nanocarriers have emerged as excellent alternate systems for the conventional drug-delivery system and represent revolutionary, transformative advantages leading to improved therapeutic outcomes. Hallmark features of the most celebrated nanocarriers developed with the help of polymers include controlled drug release, improved drug stability, and target specificity.

Controlled release of drugs is the primary application of polymer-based nanocarriers, allowing the control of pharmacokinetics. This allows the drug to be released sustainably, maintaining a therapeutic level in the bloodstream over a long period with minimal requirement for frequent dosing. Such controlled release mechanisms are convenient and important for improving patient adherence to treatment regimens. These nanocarriers also play an important role in enhancing the stability of the drug. Polymer-based nanocarriers coat the drug with active substances to protect it from enzymatic and hydrolytic degradation. Such protection is especially important for sensitive molecules that begin to degrade before reaching their target sites in the body (9).

Adopting polymer-based nanocarriers offers significant advantages beyond those previously discussed, particularly in enhancing drug bioavailability. These nanocarriers excel at improving the solubility of hydrophobic drugs, facilitating their optimal transport across biological barriers. This enhancement is crucial, as many drugs face challenges in their administration due to low solubility, a point emphasized in recent research by Gouthami et al. (10). Perhaps even more promising is the potential for targeted delivery to specific tissues, cells, or pathological sites. Through surface engineering with various ligands or antibodies, these nanocarriers can be designed to bind with specific cellular receptors, enabling precise payload delivery to the intended site of action. This targeted approach contrasts traditional transmembrane methods, offering a more focused therapeutic impact while minimizing systemic side effects. Such precision is especially valuable in treating localized diseases like cancer, as noted by Zhu et al. (11). Combining improved bioavailability with targeted delivery, polymer-based nanocarriers are poised to revolutionize drug administration, potentially increasing treatment efficacy while reducing unwanted effects.

The features of these polymer-based nanocarriers seem much more advanced and better when compared to traditional ones that lack high accuracy and efficiency. Thus, the benefits that came with some specific nanocarriers are, in turn, improving drug therapies' efficacy and, in this way, changing the quality of life for patients.

Polymer-based nanocarriers are revolutionizing biomedical research, particularly in delivering nucleic acids and proteins. These biomolecules, known for their instability in physiological environments, are finding new life through sophisticated carrier systems. Nanocarriers preserve their integrity and functionality by shielding these delicate molecules from enzymatic degradation until they reach their cellular targets. This protective mechanism is crucial for maintaining biological activity during transit through the body, a significant advancement highlighted in recent work by Mashal et al. (12). Moreover, these nanocarriers are adeptly modified to increase cellular uptake and promote the escape of molecules from endosomes, markedly elevating the delivery success of nucleic acids and proteins. Through the strategic addition of specific ligands or alterations to their surface properties, these nanocarriers substantially boost intracellular delivery and transfection success, thereby proving indispensable in gene therapy advancements (13).

The dynamic nature of polymer-based nanocarriers is further demonstrated by their ability to react to physiological stimuli, including pH shifts, temperature changes, or enzymatic actions. Such reactive capabilities facilitate the targeted and controlled release of the enclosed nucleic acids or proteins, effectively reducing side effects and optimizing therapeutic results. This is crucial in advancing precision medicine, as emphasized in an earlier study (14). Additionally, these nanocarriers enhance the pharmacokinetic characteristics of nucleic acids and proteins, extending their circulation within the bloodstream. This prolongation improves the likelihood of the therapeutic agents reaching their intended destinations and enhances their availability, which is essential for successful treatment outcomes (15).

Lastly, polymer-based nanocarriers offer essential protection against the thermal and mechanical challenges that nucleic acids and proteins might face during storage and transport. This protective measure is crucial for preserving the therapeutic potency of these biomolecules until they are ready for clinical deployment, a critical factor highlighted by Yuan and Li in 2017 (16). The comprehensive functionality of polymer-based nanocarriers underscores their transformative impact in biomedicine, especially in gene therapy and protein-based treatments. Their unparalleled ability to protect, deliver, and stabilize therapeutic agents makes them a cornerstone in the evolution of disease treatment methodologies (Figure 4.1).

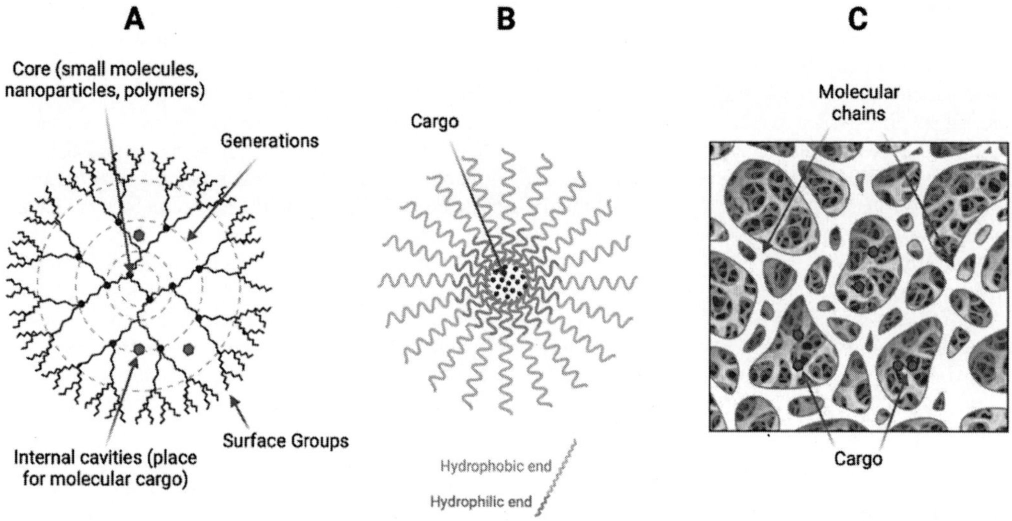


**Figure 4.1** Structural comparison of polymer-based nanocarriers. (A) Dendrimer: Highly branched structure with a core, multiple generations of branching units, internal cavities for cargo, and surface groups amenable for functionalization. (B) Polymeric micelle: Amphiphilic structure with a hydrophobic core (pink) for drug encapsulation and a hydrophilic shell (gray) for stability in aqueous environments. (C) Hydrogel: Cross-linked polymer network forming a porous, water-retaining structure capable of dispersing drug molecules throughout its matrix.

## 4.2 FUNDAMENTALS OF POLYMER-BASED NANOCARRIERS

### 4.2.1 Basic Concepts and Definitions

Nanobiotechnology demands a specialized approach to polymer design, tailoring materials to meet the exacting standards of medical applications. These medical-grade polymers distinguish themselves from their industrial counterparts through fundamental properties: biocompatibility, biodegradability, and adaptability for functionalization. Such characteristics ensure optimal performance within biological systems. Biocompatibility is a cornerstone requirement, guaranteeing that these materials can coexist harmlessly within the body without triggering adverse immune responses. This quality sets medical polymers apart from many industrial variants, which often contain substances unsuitable for biological integration. Researchers create materials that seamlessly blend into biological environments by engineering polymers to be inherently compatible with living systems. This fundamental compatibility enables medical polymers to perform their intended functions without disrupting the delicate balance of the body's internal systems. The stark contrast between these specialized polymers and their industrial counterparts underscore the precision and care invested in their development, reflecting the unique demands of medical applications in nanobiotechnology (17).

Polymers used in nanobiotechnology are often designed to be biodegradable, decomposing into non-toxic substances that the body can quickly expel. This characteristic is particularly important for applications such as drug delivery, where removing the polymer after its job is as crucial as its initial function. This contrasts with industrial polymers, which often prioritize durability and cost-effectiveness over environmental and biological safety (18). One of the unique features of nanobiotechnology polymers is their ability to be functionalized. This procedure entails chemically altering the polymers to incorporate precise functional groups, such as targeting agents, imaging markers, or drugs, thereby augmenting their interaction and efficacy within biological systems. This tailor-made approach allows for precision in medical treatments and is vital for successful applications in complex biological environments.

Moreover, these polymers' physical and chemical characteristics, including their solubility, charge, and molecular weight, are carefully adjusted to enhance their effectiveness in biological settings. These parameters are essential for ensuring the polymers' proper distribution within the body, facilitating cellular interactions, and maximizing their effectiveness in roles such as drug delivery (17). Polymers engineered for nanobiotechnology are distinctively crafted to meet the demands of living systems, offering safety, efficacy, and responsiveness that are unparalleled by non-medical polymers. This advanced engineering melds materials science with biological science, aiming to overcome the complex challenges faced in contemporary medicine.

To function optimally as nanocarriers in medical treatments, polymers must exhibit critical characteristics that ensure their efficiency in drug delivery at the nanoscale. These properties are pivotal for the encapsulation and protection of therapeutic agents and crucial for successfully navigating the intricate biological environments to target affected sites precisely. A primary requirement for polymers utilized as nanocarriers is biocompatibility. These materials must be inherently non-toxic and should not trigger adverse immune reactions upon administration. Ensuring biocompatibility allows these carriers to operate safely within the body, avoiding detrimental effects on surrounding tissues or inducing unwanted inflammatory responses. This aspect is critical for their acceptance and effectiveness (19).

Biodegradability is another essential feature for effective nanocarriers. They should break down into harmless by-products that the body can readily dispose of. This characteristic is especially important for temporary therapeutic measures, ensuring that the carriers do not persist within the body after accomplishing their drug-delivery tasks, thereby minimizing potential long-term side effects and toxicity (20). Maintaining structural integrity in biological fluids is crucial for nanocarriers until they reach their targets. Polymers must be resilient against the physicochemical challenges encountered in the bloodstream or within cellular environments, including resistance to enzymatic breakdown and stability across diverse pH levels and temperature conditions (21).

The ability to control the release of drugs is fundamental for maintaining therapeutic drug levels at the target site for the intended duration. Polymers used in nanocarriers should facilitate precise control over the release kinetics of the encapsulated drugs, often employing stimuli-responsive mechanisms that activate in response to specific biological signals or environmental changes (21). Advanced targeting is a crucial feature for polymers, enabling them to adhere to and affect diseased cells or tissues specifically. This is accomplished by modifying the polymer surface with ligands or antibodies that recognize and bind to cellular receptors, optimizing therapeutic efficacy and minimizing effects on non-target tissues (19).

Together, these characteristics guarantee that polymers utilized as nanocarriers are efficient in transporting drugs and delivering them in a controlled manner and are safely broken down and eliminated by the body once their therapeutic purpose is fulfilled.

### 4.2.2 Mechanisms of Action

Developing polymer-based nanocarriers that can effectively package and protect biomolecules from premature protease activity during delivery could significantly address therapeutic agents' stability and bioavailability challenges. For this purpose, nanoparticles are engineered for specific properties to allow safe transport and controlled release of the encapsulated bioactive substances, such as proteins, peptides, and nucleic acids. Polymer-based nanocarriers exhibit exceptional efficacy in encapsulating biomolecules. The polymers may then form hydrophobic or hydrophilic interactions with the drug molecules based on their chemical structure, allowing a wide variety of therapeutic agents to be included. Therefore, including various drugs with distinct solubility and stability profiles is crucial.

The process of encapsulation itself ensures protection against premature degradation caused by pH, enzymes, and oxidative stress conditions. This protection safeguards the biological activity within the encapsulated material until it reaches its intended target site (22). Once entrapped, these biomolecules can be protected from the harsh outside environment due to the polymeric matrix. This highlights the significance of safeguarding proteins and nucleic acids, collectively known as biologics, which are often susceptible to degradation by enzymes in the gastrointestinal tract when taken orally or in circulation.

Polymer-based carriers have the ability to enclose these delicate molecules and shield them from enzymatic and hydrolytic breakdown, thereby improving their circulation in the body and increasing their effectiveness in reaching specific tissues (23). Polymeric nanocarriers can be engineered to have controlled release properties by modifying the composition and structure of the polymer. Such systems could react to certain body stimuli, like pH alteration, temperature modification, or many enzymes, for the drug to be secreted only at the site where it will produce its effect. This reduces side effects in the body while increasing therapeutic effectiveness at the target site. Therefore, they are the optimal selection for drug-delivery applications that target specific sites (24).

Nanocarriers must possess biocompatibility and biodegradability to fulfill their intended purpose effectively. The polymers utilized in these systems are predominantly non-toxic and are engineered to gradually break down into harmless substances that can be easily eliminated from the body. This reduces the likelihood of accumulation and long-term toxicity (20). The carriers effectively enclose and safeguard the biomolecules by providing durable protection, shielding encapsulated substances from biological and chemical degradation. Controlled release mechanisms, biocompatibility, and biodegradability features represent key facets in effectively delivering and transporting therapeutic agents, especially when the target is a complex disease.

Polymer-based nanocarriers are engineered to include sophisticated techniques for the optimal transportation and release of therapeutic agents, resulting in a significant enhancement of effectiveness and safety compared to other drug-delivery systems. Both methods of nanocarriers, namely targeted delivery and controlled release, raise drug administration precision.

There are essentially two strategies for a more precise delivery of these therapies: active and passive. The term "active" describes the requirement for conjugation with specific ligands, such as antibodies, peptides, or small molecules, on the surface of the nanocarriers. These ligands, which bind to receptors or antigens on target cells (specifically cancer cells), result in an elevated drug concentration at the disease site while minimizing its effect on healthy tissues. Their work with this strategy highlights improved targeting efficiency and reduced systemic toxicity toward the target. Passive targeting, on the other hand, exploits the enhanced permeability and retention (EPR) effect, a common occurrence in cancer therapy. Polymeric nanocarriers are typically designed with precise dimensions to facilitate their accumulation in tumor tissues, considering tumor vasculature's disordered nature and the lymphatic system's limited drainage capacity. Subhan et al. (2023) discuss how this phenomenon can significantly improve the delivery of anticancer drugs (25).

Controlled release is one of the key aspects of polymer-based nanocarriers that may be under control through pH-responsive, thermo-responsive, or light-responsive systems. The pH-responsive nanocarriers are specifically engineered for release in acidic environments, such as tumor cells or inflamed tissues. Ranjit De et al. (2022) further elaborate that these systems are acid-labile-linked in the polymer structure, and the linkages are destructible in an acidic environment, thus releasing the drug (23). Thermo-responsive systems are based on the use of temperature-change-sensitive polymers. By employing this approach, systems have the capability to focus on elevating tumor temperature levels slightly higher than the average body temperature or react to the influence of an external heat source (26). In their article, Zhang et al. (2021) stated that photoresponsive systems involve photo-sensitive groups integrated into the polymer structure, which specific light wavelengths can trigger to release a drug in a controlled manner (27).

All these systems bring about a fine control of the level of drug release, particularly with the relevance to the localized treatment of cancers—or, literally, with pinpoint precision—in activating photodynamic therapy (PDT). Together, such strategies enable the drug-loaded polymer-based nanocarriers to enhance the drug pharmacokinetics and bio-distribution profoundly. Such a system, in turn, may make it possible for therapeutic agents to be released in a controlled manner right at the site of pathology, ensuring maximum therapeutic efficiency and minimal side effects; potentially, these systems pave the way for more efficient and safe treatment.

## 4.3 DENDRIMERS

### 4.3.1 Structure and Synthesis

Dendrimers, a highly branched and star-shaped macromolecular architecture, are outstanding nanocarriers in drug-delivery applications. The architectural features of dendrimers contribute to their functionality in a big way, and therefore, they have found many therapeutic uses in a very effective manner. Dendrimers have a complex structure with multiple branches and multiple terminal groups. These terminal groups can be modified with various therapeutic agents, enabling the attachment of numerous drug molecules to increase the drug load (28).

Dendrimers are synthesized through a sequential and hierarchical process that involves increasing the size and functionality of the molecule in a tree-like structure. This meticulous procedure enables precise control over the dimensions and shape of the dendrimer, which are critical factors in determining their biological interactions and ability to penetrate biological membranes and enter cells (29). The branches of the dendrimers, which form the internal cavities, can also encapsulate guest molecules. Therefore, the nanoparticles described in the study protect drugs from degradation in sensitive environments until they reach their intended targets. Additionally, these nanoparticles enable the controlled release of the drugs, thus ensuring that therapeutic drug concentration is maintained in the body (30).

The surface of the dendrimers brings even more functional advantages. Chemical modification can be applied to most terminal functional groups to enhance solubility, improve biocompatibility, or attach targeting ligands. When appropriately altered, dendrimers become suitable for therapeutic purposes, such as targeting cancer cells or penetrating the blood–brain barrier (BBB) for drug delivery. Chemists are currently researching dendrimers to enhance the biocompatibility and biodegradability of this type of large molecule. This will result in significantly reduced toxicity in the intended area of application. These properties are demonstrated or improved by incorporating biodegradable linkages or using biodegradable materials within the body (31).

Dendrimers possess a distinctive structure with multiple branches and a specific size. They provide a high level of control due to their chemically adjustable surface. Additionally, they have an internal cavity that can encapsulate a drug in a protected environment. Furthermore, dendrimers have a surface compatible with living organisms and are capable of being broken down naturally. Taken together, all these properties suggest that these agents are more effective as nanocarriers. Dendrimers possess properties that enable efficient drug delivery with minimal side effects and maximum therapeutic impact at specific locations.

Dendrimer synthesis for drug delivery in chemistry has proven to be highly challenging, with numerous problematic factors. The intimate development of innovative solutions is essential for advancing the limits of dendrimer technology in therapeutic applications.

A significant challenge in dendrimer synthesis is the inherent complexity of the process, which involves a stepwise iteration toward the formation of the dendrimer. This requires precise control over the conditions to ensure uniformity and prevent structural defects in each generational layer. This then further adds to the production cost and raises significant issues about scalability. On top of that, it is hard to provide that uniform monodispersity in population. Nevertheless, dendrimers' size and surface functionality vary due to incomplete reactions or side reactions that occur during their synthesis, which ultimately negatively affect their performance in drug-delivery applications (32).

These effective chemical methods will be necessary for the efficient placement of drugs onto the dendrimers and for their modification and attachment of the corresponding functional groups or drugs to their surface. As a result, functionalization is unreliable for therapeutic results; hence, therapeutic outcomes are highly variable (31).

Several sophisticated solutions have been developed for this purpose. Recent technological advancements in automated synthesis techniques have eased the scaling up of dendrimer production, with no need to compromise on the consistency of quality. For these techniques, the conditions and timing of the reaction can be well controlled to realize better yields and uniformity of the final product (33). This is where the most current and one of the best ideas being carried out comes in: the development of orthogonal functionalization strategies to enable the modification of dendrimers at different, non-interfering generational layers. The statement suggests that the functionalization process occurs according to the described control, thereby increasing the capacity of dendrimers to achieve accurate deliveries.

The purification process typically involves dialysis, ultrafiltration, and chromatography to ensure the use of precisely defined dendrimers in subsequent studies and applications. These procedures remove unreacted monomers and their by-products to eliminate batch-to-batch variability (31). Microwave-assisted synthesis has been included in the list of novel reaction conditions used in dendrimer synthesis.

They enhance the synthesis rate and simultaneously decrease unwanted reactions, thereby improving the drug loading efficiency. The conditions are carefully adjusted, primarily to enhance the speed of chemical reactions and achieve precise control over the structure of the dendrimer (32). Therefore, by utilizing these advanced and efficient solutions, significant obstacles in the synthesis process of producing dendrimers for drug delivery can be effectively addressed, thus leading to improved production and facilitating the development of more dependable and potent dendrimer-based therapeutic products.

### 4.3.2 Functionalization and Modification

Hence, modifying dendrimers' surfaces is crucial to augmenting their effectiveness as a drug-delivery system, particularly for precise delivery to specific cells or tissues. Functionalization can occur by attaching various ligands or groups to the surface of the dendrimer. This enables the dendrimer to recognize and bind to a specific receptor or structure on the surface of a target cell.

The surface-modified targeting of the dendrimer works twofold: the targeting of receptors in cells, either cancerous, inflammatory, or infectious type, is only affected by the specific moieties conjugated on the surface of the dendrimer. The dendrimer targets such cells effectively. These may include, but are not limited to, small molecules targeting affinity against a given specific cell surface receptor, such as antibodies, peptides, or even aptamers, depending on the need. Therefore, if used, this approach is likely to have an effective delivery system with less impact only on the cells where the antigen is not targetable, hence fewer side effects and an improved therapeutic index (34).

Surface modification also increases the uptake of cellular dendrimers. The enhanced interaction between modified dendrimers and different components of the cell membrane facilitates more effective endocytosis and uptake, enabling the direct delivery of therapeutic agents to cell targets at elevated levels (35).

Their surface properties can be manipulated in response to specific environmental triggers, such as variations in pH, temperature, or enzymatic activity. Responsive systems are specifically engineered to enable the targeted delivery of drugs at the intended site of action. This design ensures that the drugs are released in a controlled manner, maximizing their therapeutic effectiveness while preventing premature release before the desired time (36). Other designs work on modifying the surface of the dendrimers with hydrophilic polymers, for example, polyethylene glycol (PEG), to improve circulation time and general stability. The modification decreases opsonization and minimizes the likelihood of early clearance by the reticuloendothelial system, thereby ensuring a more extended duration for dendrimers to reach and accumulate at the intended site, as necessary (37).

Lastly, through this surface modification, the inherent toxicity of dendrimers can be reduced by masking or neutralizing charged surface groups that would interact with any biological component adversely. The paramount importance lies in ensuring the safety of dendrimers when utilized for medical purposes (38). Surface functionalization plays a crucial role in optimizing dendrimers as efficient and safe drug carriers that specifically target disease sites, thereby minimizing side effects and maximizing therapeutic efficacy.

Various successful strategies have been applied to conjugate nucleic acids or proteins in the dendrimers, which are aimed at exploiting the specific features that are related to the dendrimers' highly branched structures and functionalizable surfaces. It is important to note that these techniques help to bring about a higher level of stability and delivery efficiency regarding these biomolecules. The nucleic acids or proteins are bound by a covalent bond with dendrimers. Dendrimers can be covalently linked by utilizing the surface reactive functional groups present on the dendrimer, such as amine, carboxyl, or hydroxyl groups. These functional groups allow biomolecules to form stable covalent bonds. Ensuring a robust and enduring bond is crucial for maintaining the bioactivity of the conjugated molecules during delivery (39).

The second approach is bioorthogonal chemistry. Chemical reactions can be utilized without impacting biological systems. Click chemistry, a widely utilized method in bioorthogonal reaction mechanisms, enables the coupling of azide-functionalized proteins or nucleic acids with alkyne-modified dendrimers. It is generally accepted for its specific, efficient, and mild action, thereby ensuring that the biomolecule of interest remains unmodified without any destruction (40). Supramolecular assembly macromolecules, such as proteins and nucleic acids, bond the dendrimers with the aid of hydrogen bonding, electrostatic interactions, and hydrophobic interactions through non-covalent forces.

Another important advantage of this approach is that the conjugates are easily disassembled at the targeted site for the release of therapeutic agents in the reversible modification strategy (41). Additionally, another option is cholesterol conjugation, a method employed to modify dendrimers to enhance the encapsulation of nucleic acids, particularly plasmid DNA, within dendrimer-based complexes. This pathway was further found to highly increase transfection efficiency due to the interaction of the complex with cell membranes, arising from increased cellular uptake (39). In particular, these approaches made dendrimers applicable as carriers for nucleic acids and proteins, ensuring effective delivery to target cells while preserving the integrity and activity of therapeutic biomolecules.

### 4.3.3 Applications of Dendrimers in Nucleic Acid and Protein Delivery

Dendrimers, by their own unique structural and functional characteristics, become the best carriers in applications in gene therapy, often prevailing over all other vectors. Basically, dendrimers refer to a highly branched polymer at the nanometer scale, which possess several advantages and come in handy under the complex requirements of successful genetic material delivery.

They offer a high-density surface in surface functional groups, which are able to link or encapsulate many copies of nucleic acids. This, therefore, significantly improves gene delivery, essentially achieving therapeutic levels of gene expression (42). Apart from that, dendrimers also provide desirable release kinetics because genetic material can be encapsulated either within the interior cavities or covalently attached to the surface. It implies controlled release, essential for maintaining genetic stability, as it provides for sustained release at the target site—hence the ability ever so often tested in other nanoparticle systems (43).

Moreover, dendrimers can easily be functionalized by simply attaching targeting ligands to guide the delivery vehicle to particular cells or tissues. This then increases the specificity of gene therapy and is hence more effective as the off-target effects are reduced, and hence a greater therapeutic index is obtained (44). Another great advantage is the reduction of immunogenicity and toxicity through surface modifications like PEGylation, which would make dendrimers far safer than viral vectors, with strong immune responses and possible safety issues (42).

The said synthetic versatility allows dendrimers to reach high control in terms of precise size, shape, and surface chemistry. Thereby, it can be adapted to several gene therapy applications. In addition, dendrimer synthesis has been well-evidenced to readily allow for scaling up, while some biological vectors may exhibit production difficulties at large scales (43).

Furthermore, dendrimers allow greater control over particle properties and behaviors in biological media compared to other non-viral delivery systems, such as liposomes and polymeric nanoparticles. The material easily condenses DNA, stabilizes it against nucleases' degradation, and hence promotes cellular uptake and endosomal escape, thereby is of particular use in the gene delivery process. Though viral vectors are of high efficacy, dendrimers serve as a secure alternative with less immunogenicity to fill in some of the key limitations associated with the viral systems.

The recent trends toward the use of dendrimers for vaccine delivery increasingly point to very bright fortunes for the application of this new technology in this line. Very importantly, advancement strides have been registered to enhance the effectiveness of dendrimers and broaden the applicability range of vaccine formulations. An incredibly striking development was the design of dendrimers to exhibit multivalency, whereby many copies of either antigens or adjuvants are placed on their surface. In general, with this vaccination approach, the immunogenicity of vaccines will sharply increase due to the enhanced presentation of antigens to immune cells, in such a way strengthening and making the immune response more sustainable (45).

Additionally, dendrimers are being engineered for targeted delivery and controlled release. They can be directed at particular immune cells, such as dendritic cells, which are obligatory in the induction of immune responses. They are capable of being programmed to release their cargo of dendrimers in such a graded way, so there is antigenic exposure for the boosted immune response of longer periods and, possibly, fewer shots (28). The co-delivery of antigens and adjuvants within a single dendrimer system is another innovative approach. In this strategy, the vaccines are more potent as they guarantee that both parts get to the immune cells simultaneously, yielding better activation of the innate and adaptive immune responses (46).

There is increasing interest in the suitability of dendrimers for the delivery of nucleic acid vaccines. Such vaccines, which need protection against degradation until they reach the interior of the target cells, can show their suitability in the flexible and protective environment afforded by dendrimers. This capacity enables the dendrimers to entrap and protect those extremely labile or sensitive molecules when they are delivered (45). With an impact on the formulation, dendrimers help make a more stable formulation of the vaccine with ease in facilitation for storage and transportation. Due to their synthetic creation, they allow scaling up for the production of dendrimers. This is very important in many vaccines and across the globe for economic and sufficient responses.

All these emerging trends underline the sharper role that dendrimers are expected to play in the development of more novel and more effective systems of vaccine delivery, thus pointing toward a revolution in world health.

### 4.3.4 Challenges and Solutions

One of the major challenges of using dendrimers in delivery and therapeutic applications is the need to devise an innovative solution to make the process successful and free of side effects.

The complex synthesis of dendrimers, which requires precise, multistep processes, can be time-consuming and costly. Critical synthesis needs to be uniform in size and structure with consistent therapeutic outcomes (47). Further, the cytotoxicity expressed by some dendrimers is caused by surface charge and the high density of surface groups. These characteristics may act upon cellular membranes, thereby causing disruption or inducing cytotoxic responses (33). Moreover, the surface and core groups of these dendrimers are the determinants of stability and solubility. Biological fluids would act on any alteration made to the dendrimers and would determine their ability to act as a drug-delivery agent. Lastly, the regulatory standard of dendrimer synthesis to the level required can be met for clinical use and commercial production is cumbersome due to its complex synthesis route (31).

To overcome these challenges, several innovative solutions have been developed. The progress in automated synthesis, along with the investigation of novel chemical methodologies, has enabled the cost-effective production of dendrimers on a large scale (47). Dendrimers that are surface-modified, either by PEGylation or by other biocompatible coatings, bestow a remarkable reduction in toxicity. It functions by effectively blocking the groups of reactivity that are ready to be part of unwanted reactions with cellular components (33).

The addition of dendrimers, made from biocompatible materials, can enhance the stability and solubility of composite nanoparticles and hybrid systems in other biomaterials. These help in controlling the release profiles of such encapsulated drugs for their delivery and therapeutic effectiveness (37). Such development is encouraged to be engaged at an earlier stage by regulatory bodies while, in parallel, conducting preclinical studies comprehensively so as not to cause any potential regulatory hurdles.

It is thus very much essential for collaborative efforts—particularly with academic and industrial partners—to facilitate the transition from laboratory to market dendrimer-based therapies. These approaches, placed in a dynamic and ever-evolving landscape of dendrimer use in therapeutic applications, seek to highlight the potential of these nanomaterials in overcoming different life-threatening challenges posed by drug-delivery systems.

## 4.4 POLYMERIC MICELLES

### 4.4.1 Formation and Structural Characteristics

The self-assembly process of the polymeric micelles basically engenders the structural and functional properties that these micelles acquire during the whole process of assembly, something that will have a dramatic effect on the overall drug-delivery efficiency of the system.

Polymeric micelles represent core–shell-type structures formed from the spontaneous interaction of the hydrophobic core with hydrophilic block copolymers in an aqueous environment. The hydrophobic core acts as a storage area for hydrophobic drugs, enhancing the solubility and stability of the pharmaceutical compound. This is particularly beneficial for compounds that would otherwise have low solubility.

This type of encapsulation results in enhanced bioavailability within the body, allowing for more efficient utilization of the administered drug (48). In addition, micelles are stabilized by steric stabilization effects caused by the hydrophilic shell. This is crucial in preventing the micelles from aggregating in the bloodstream. Such stabilization leads to the circulation time of drug-loaded micelles being increased by quite a great percentage to levels where their residency in the bloodstream is guaranteed for longer intervals.

This strategy will prevent the prompt removal of micelles by the mononuclear phagocyte system (MPS) or renal filtration, allowing a greater number of micelles to reach the desired target sites (49). Additionally, hydrophilic outer shells can be surface-modified and functionalized to target specific tissues or cells. Covalent conjugation of targeting ligands includes antibodies or peptides to attain the targeting of cell types or receptors.

The targeted approach suggests a solution not only to losing efficiency but also to the ill effect that originates from the wide distribution of pharmaceuticals within the body (50). In addition, various self-assembled polymeric micelles are supposed to be responsive to environmental triggers such as pH, temperature, and enzymatic conditions. Such stimuli might cause structural changes to the micelles with controlled drug release from the encapsulating medium at the site of action.

For instance, this capacity has been exploited in the therapy of diseases such as cancer, whose tumor microenvironment presents peculiar conditions that can be exploited for therapy (48). Altogether, the self-assembly of polymeric micelles significantly increases their ability in drug delivery, that is, through better solubility of drugs, circulation time, and targeted and controlled drug release.

The design of core–shell polymeric micelles supports drug encapsulation and their release, while assisting further in the development and optimization of the most efficacious and safe drug-delivery system. This hydrophobic core within the polymeric micelles allows it to be a very effective means for both water-insoluble drug encapsulation and drug protection from the physiological aqueous environment and degradation. This characteristic of paramount importance increases the solubility and stability of drugs as it increases transport via the circulation system to the site of required action (51).

The hydrophilic shell in the polymeric micelles stabilizes the structure of the micelle in the bloodstream, thus preventing premature aggregation and uptake by the reticuloendothelial system; hence, it has controlled release. This shell can be further engineered to specifically react to specific environmental stimuli at the target location, such as variations in pH, temperature, or even certain types of enzymes. This enables controlled release of drugs at the right place and time for the encapsulated drug (52).

Polymeric micelles enhance therapeutic effectiveness by significantly improving the solubility and bioavailability of poorly soluble drugs, while also exerting control over their release. This facilitates the provision of an improved pharmacokinetic profile, as well as more precise targeting, enabling more accurate dosage and better management of side effects typically caused by elevated levels of unbound drugs in the body (53).

The core–shell structure also allows for flexibility in functionalization because there are plenty of locations for functionalization in both the core and the shell. This process enables the attachment of particular ligands or signaling molecules to the micelle, guiding the micelles toward specific cells or tissues. As a result, the treatment becomes more targeted and selective (54).

### 4.4.2 Chemical Composition and Functionalization

Block copolymers are among the most important classes of amphiphilic copolymers due to the special structural features they provide, which lead to improvements in the functional diversity of polymeric micelles and flexibility in designing targeted drug-delivery systems.

Block copolymers are structurally versatile, enabling them to easily undergo self-assembly into micelles with distinct core–shell architectures in aqueous solutions by covalently linking two or more polymer blocks of different properties. This versatility allows for designing specific types of micelles—for example, the balance between hydrophobic and hydrophilic constituents can be varied to effectively protect both kinds of drugs (55).

Core–shell-structured polymeric micelles, formed from diblock copolymers, solubilize poorly soluble drugs more effectively and provide protection from premature degradation by sensitive biological environments. The hydrophobic core entraps a hydrophobic drug, while the hydrophilic shell stabilizes the micelle in aqueous environments, thus preventing the aggregation of micelles and promoting extended circulation time in the bloodstream (56).

Block copolymers can be tuned so that they have functional groups with tissue or cell affinity. Meanwhile, the polymeric micelles can actively accumulate at specific sites in the body due to the presence of ligands or antibodies located on the surface of their outer shell, thereby decreasing the drug's side effects and increasing the efficacy of the encapsulated drug (57).

Stimuli-responsive blocks can also be introduced in copolymers so that the prepared micelles can respond to environmental changes, such as pH, temperature, or enzymatic activity. This response triggers the controlled release of encapsulated drugs at the target site, which is crucial for achieving precise therapeutic outcomes and reducing systemic drug exposure (58).

Finally, micelles of different sizes and morphologies can be designed to optimize drug-loading capacities and release kinetics. This flexibility ensures that therapeutic drugs maintain effective levels over time in the target area (59).

The most recent progress in the development of targeting ligands and stimuli-responsive elements for polymeric micelles has greatly enhanced their functionality as a precision drug-delivery system. Approaches toward improved specificity and controlled drug release have the potential effectiveness in increasing therapeutic efficacy with reduced side effects.

Recent advances have armed the polymeric micelles with very sophisticated targeting ligands that can bring about firm binding to disease-related biomarkers. It is also possible to further modify the surface of the micelles with antibodies, peptides, or small molecules, and thus obtain the possibility of their site-selective targeting toward either the cancer cells, inflamed tissues, or any other pathologically altered location. This consequently magnifies the accumulation and retention of therapeutic agents at targeted foci, hence increasing their treatment effectiveness (60).

Moreover, to provide control over stimuli, responsive groups can be incorporated within the core or corona micellar structures, resulting in the controlled release of drugs in response to specific internal or external stimuli, such as changes in pH, temperature, redox conditions, or specific enzymes that are mostly available in inflamed tissue or tumor environments. The guarantee that this mechanism provides is that the drugs are only being released at the right time and the right site of action. This improves their efficiency while reducing systemic toxicity (61).

Further advances have been made in creating polymeric micelles that respond to more than one stimulus at the same time, thereby controlling drug release. Within this modern framework, signals can originate from both internal factors (such as pH and redox) and external factors (such as light and temperature), enabling accurate and focused administration of drugs within the dynamic homeostatic ranges of the human body (62).

This can also allow for the use of cleavable linkages within the micelle structure to increase control over drug release. These links, designed to be responsive to the chemical conditions of the tumor or infection site, ensure precise and localized control over the release of drugs, thereby delivering medication directly to the specific area where it is most necessary (63).

Besides, polymeric micelles have been adapted to the application of PDT. Compared to delivering photosensitizers independently, this system enables more effective delivery and activation of photosensitizers in tumor tissues when exposed to light. As a result, the efficacy of PDT is enhanced without causing damage to the surrounding normal tissues (64).

These developments suggest that the modern platform of polymeric micelles is going to change in a way that could offer even better drug delivery in many therapeutic settings with an enhanced efficacy and safety profile.

### 4.4.3 Applications in Therapeutics

Polymeric micelles have been utilized to enhance the delivery of siRNA and proteins, capitalizing on their unique properties to improve therapeutic efficacy and target accuracy.

One of the leading advances includes conjugation with cell-penetrating peptides (CPPs) and polymeric micelles. These CPPs further potentiate the intracellular uptake of macromolecules, such as siRNA and proteins, allowing for better escape of the macromolecule from endosome compartments and, hence, successful gene silencing or protein function inside the target cell. It has improved the remarkably effective delivery, both for systemic and localized applications, including those to challenging targets such as the brain and inflamed tissues (65).

Another important advancement is the development of cross-linkable polyion complex micelles from polypeptide-based ABC-triblock copolymers. These micelles are better stabilized and can be engineered to give off their payload only in response to certain cellular triggers, thereby enabling greater and more site-specific siRNA delivery with fewer off-targeting effects (66).

The advantage of being protein-based micelles also includes the capacity to carry therapeutic proteins. These micelles have been designed to exploit natural protein properties, which include biocompatibility and penetration through biological barriers. The micelles facilitate the precise release and delivery of these proteins, as they operate specifically at the intended location. These particles have therefore been discovered to be highly beneficial in the treatment of various diseases that require targeted delivery and controlled release (67).

Furthermore, polymeric micelles have been designed to have a multifunctional and stimuli-responsive design. That is, these are designed for the co-packaging of proteins and siRNA at the same time, thus having stimuli-responsive drug release. The dual functionality of this approach enhances therapeutic effectiveness, particularly for diseases with complex origins like cancer, where precise and controlled action is crucial (68).

The interaction between polymeric micelles and siRNA has been extensively studied, leading to the development of optimized designs. Often supported by molecular dynamics simulations, these studies help in fine-tuning the micelle formulations so that the required therapeutic response is obtained with minimal side effects (69).

Polymeric micelles have emerged as revolutionary tools for overcoming biological barriers in drug delivery, because of their unique structural characteristics and the innovative use of block copolymers. The core–shell architecture of polymeric micelles significantly enhances the solubility of hydrophobic drugs, which is a major challenge in pharmaceutical development. By encapsulating these drugs within their hydrophobic core, micelles improve not only their solubility but also their bioavailability, facilitating their passage through biological membranes and enhancing therapeutic efficacy (68).

Polymeric micelles can be engineered to target specific tissues or cells. By modifying their surface with ligands that recognize and bind to specific receptors on the target cells, this targeting capability not only improves delivery efficiency but also minimizes the systemic side effects of drugs, ensuring that they act precisely at a site where needed without affecting other parts of the body (70).

Polymeric micelles possess the remarkable capability to traverse challenging biological barriers, including the BBB. Their small size and potential for surface modification make them ideal for delivering therapeutic agents to the brain, a region traditionally difficult to access due to the protective nature of the BBB (71).

Incorporating stimuli-responsive elements into the micelle structure enables controlled drug release in response to specific biological or environmental triggers, such as changes in pH, temperature, or the presence of specific enzymes. This capability ensures that drugs are released only in the desired location and at the right time, enhancing both treatment effectiveness and safety (71).

Lastly, polymeric micelles can optimize the pharmacokinetics of drugs. By altering the interaction dynamics between the drug and the biological environment, they can prolong the circulation time of drugs in the bloodstream, enhance tissue penetration, and reduce rapid clearance, thus maximizing the therapeutic window of the drugs (67).

### 4.4.4 Challenges and Future Directions

Polymeric micelles have emerged as promising delivery vehicles for drugs, yet their development and application face significant challenges that need to be addressed to fully realize their potential.

One of the primary hurdles is the stability of polymeric micelles in a biological environment. Factors such as temperature, pH, and shear forces in the bloodstream can compromise the structural integrity of the micelles, leading to premature drug release and reduced efficacy. Additionally, the safety profile of polymeric micelles is a concern, particularly regarding their interactions with non-target cells. These toxicological challenges must be rigorously evaluated in both preclinical and clinical settings to prevent adverse effects on the body. Furthermore, the regulatory landscape for nanomedicines is evolving, posing challenges due to the complexity of their design and novel drug-delivery mechanisms. Clear guidelines and standards are necessary to facilitate clinical trials and market approval (72).

Research is progressing toward the development of multifunctional polymeric micelles that can carry multiple types of therapeutic agents, such as drugs, genes, and proteins. This approach aims to improve treatment efficacy by allowing combined therapy modalities (73). Additionally, efforts are being made to create “smart” polymeric micelles that respond to specific biological stimuli for targeted drug release. These include pH-responsive, temperature-sensitive, and redox-responsive micelles that can precisely release their payload at the disease site (74). Moreover, there is a push to expand the applications of polymeric micelles beyond drug delivery to include roles in diagnostics, imaging, and as therapeutic agents themselves. This broadened scope could revolutionize how diseases are diagnosed and treated, making therapies more efficient and less invasive (75).

These insights highlight the ongoing evolution in the field of polymeric micelles, underscoring the complex challenges they face and the innovative directions they are heading toward in medical applications.

## 4.5 HYDROGELS

### 4.5.1 Hydrogel Formation and Properties

Selecting cross-linking techniques for hydrogel formation involves considering several critical factors to ensure the hydrogels meet the desired functional and performance criteria for specific applications. One critical factor is biocompatibility. The cross-linking agents and processes must not compromise the biocompatibility of the hydrogel, especially for medical applications. It is essential to select cross-linking techniques that do not release toxic substances or cause adverse reactions in biological environments (76).

Mechanical properties are also paramount. The strength, elasticity, and responsiveness of the hydrogel to mechanical stress are significantly influenced by the cross-linking density and the nature of the cross-linking bonds. Techniques that allow for precise control over cross-link density are preferred to tailor the mechanical properties according to specific needs (77). The degradation rate is crucial for applications where the hydrogel needs to biodegrade. The selected cross-linking technique should allow for the incorporation of degradable linkages that respond to environmental triggers or biological activities, enabling controlled degradation (78).

The choice between physical and chemical cross-linking techniques depends on the application. Physical cross-linking, being reversible and generally less toxic, is suitable for milder conditions. In contrast, chemical cross-linking provides more stable and durable networks suitable for demanding environments (79). Lastly, responsiveness to stimuli is important in many advanced applications. Hydrogels need to respond to specific stimuli, such as pH, temperature, and enzymes. Selecting a cross-linking technique that can incorporate responsive elements into the hydrogel network is essential for developing smart hydrogel systems (80).

The physical and chemical properties of hydrogels significantly influence their performance as drug-delivery systems, playing a critical role in how they encapsulate, protect, and release therapeutic agents. Hydrogels possess a highly hydrophilic network structure, enabling them to absorb and retain substantial amounts of water. This swelling capacity is crucial for drug release rates; hydrogels that swell rapidly and extensively tend to release drugs quickly, while those with restricted swelling provide a slower, more controlled release. This feature is vital for designing hydrogels that match the specific therapeutic goals of drug-delivery systems (81).

The porosity of hydrogels affects both the loading and subsequent release of therapeutic agents. By tailoring the size and connectivity of the pores within hydrogels, it is possible to control the diffusion rates of drugs, which is essential for achieving sustained or targeted drug release (82). The mechanical strength of hydrogels is particularly important in applications where the hydrogel must withstand physiological stresses without degrading or releasing the drug prematurely. Ongoing research aims to improve the mechanical robustness of hydrogels while maintaining their biocompatibility and effective drug-release properties (83).

For biomedical applications, it is critical that hydrogels are biodegradable, and they decompose into non-toxic by-products easily processed by the body. The rate of degradation influences the drug release rates and is a key property of hydrogels used in temporary therapeutic applications (84). Modern hydrogels are often designed to respond to specific stimuli, such as changes in pH, temperature, or the presence of certain enzymes. These stimuli-responsive hydrogels can change their network structure in response to external conditions, triggering drug release as required by specific physiological triggers (83).

Collectively, these properties play a fundamental role in defining the efficacy of hydrogels as drug-delivery systems. It is crucial to consider each property carefully during the design and development stages to ensure that the hydrogels meet the desired therapeutic outcomes effectively.

### 4.5.2 Modification for Biomedical Applications

The latest innovations in hydrogel technology are propelling their adaptability and efficiency, especially in fields like drug delivery and tissue engineering. These advancements are enhancing the hydrogels' ability to respond to environmental stimuli, making them smarter and more responsive to specific physiological conditions. Advancements have led to the development of hydrogels that can react to various environmental stimuli, such as temperature, pH, and biological signals. These hydrogels adapt their physical properties—like swelling or mechanical strength—in response to external conditions, making them ideal for smart drug-delivery systems that release therapeutic agents precisely when and where needed (85).

Another significant innovation involves integrating hydrogels with other responsive materials, like nanoparticles or conductive polymers. This combination enhances the hydrogels' sensitivity to electrical or magnetic fields, opening new possibilities for controlled drug release and applications in regenerative medicine (86). The introduction of dynamic cross-linking in hydrogels allows for reversible modifications in their network structure in response to environmental stimuli. This adaptability not only provides self-healing properties but also adjustable mechanical features, making these hydrogels more robust and versatile for use in biomedical fields (87).

There is an increasing focus on developing hydrogels that can specifically respond to biomolecules or changes in the biochemical environment. For instance, glucose-responsive hydrogels that can autonomously regulate insulin release in response to blood sugar levels are being developed for diabetes management, demonstrating the potential for personalized medicine applications (88). Advancements in hydrogel technology have also led to the development of hydrogels capable of undergoing controlled shape changes in response to environmental stimuli. These shape-morphing hydrogels are being explored for use in applications such as soft robotics, wearable devices, and minimally invasive medical devices that can adapt their shape to meet specific needs (89).

Biocompatibility and degradation are essential factors meticulously managed in hydrogel design, particularly for the delivery of nucleic acids and proteins. These properties ensure safety and efficacy through innovative material design and formulation approaches.

For hydrogels used in nucleic acid and protein delivery, the choice of material is paramount. Materials like alginate, chitosan, and hyaluronic acid, which are naturally derived, are commonly used because of their inherent biocompatibility. These materials tend to provoke a minimal immune response upon administration. To further enhance biocompatibility, hydrogel surfaces can be chemically modified. Techniques such as PEGylation are applied to decrease protein adsorption and immune recognition, which helps extend the circulating life of the hydrogels and minimizes potential cytotoxic effects.

The degradation rate of hydrogels is crucially engineered to align with the therapeutic release profile of the payload. This is controlled by choosing appropriate cross-linking agents that can respond to specific environmental triggers like pH changes, temperature shifts, or the presence of certain enzymes. Additionally, certain hydrogels are designed to degrade in response to enzymes present in target tissues. For instance, gelatin-based hydrogels can be engineered to degrade through collagenase action, which is particularly beneficial for applications in tissue engineering.

Recent advancements have led to the development of nanohydrogels specifically designed for nucleic acid delivery. These biocompatible nanohydrogels are synthesized with predefined sizes and high positive zeta potential, enhancing their interaction with negatively charged DNA and RNA molecules. They are crafted to ensure minimal cell toxicity, an essential feature for safe *in vivo* applications (90).

High-strength biohydrogels have also been developed for delivering nucleic acids in treatments for conditions such as intervertebral disc degeneration. These hydrogels demonstrate superior biocompatibility and biodegradability, which is crucial for reducing immunogenicity and enabling integration with host tissues (91). Furthermore, injectable DNA hydrogels are being explored for localized drug delivery and immunotherapy, due to their exceptional biocompatibility. These hydrogels can be enzymatically cross-linked and are designed to be responsive to biological stimuli, ensuring precise degradation and payload release at the necessary time (92).

### 4.5.3 Role in Nucleic Acid and Protein Delivery

Hydrogels designed as sustained-release systems for nucleic acid and protein delivery have seen significant advancements, with improvements focused on optimizing their physical and chemical properties to enhance therapeutic outcomes.

Hydrogels are intricately engineered to provide controlled and sustained release of nucleic acids and proteins. This is facilitated by their network structure, which allows the gradual diffusion of therapeutic molecules. By fine-tuning the cross-linking density and the polymer composition, release rates can be precisely managed to meet specific therapeutic needs (93). Incorporating stimuli-responsive materials into hydrogel formulations enables these systems to release their payloads in response to specific biological signals, such as changes in pH, temperature, or enzyme concentrations. This targeted release approach not only helps to reduce side effects but also enhances the efficacy of treatments (94).

For sustained release, hydrogels are typically made from biocompatible and biodegradable materials to ensure safety in medical applications. These materials break down into non-toxic by-products that the body can easily metabolize or excrete, aligning with natural healing processes (95). Hydrogels provide protective environments that safeguard nucleic acids and proteins from enzymatic degradation and denaturation within the biological milieu. This protection is crucial for maintaining the functional integrity of these biomolecules until they reach their target sites (96).

Recent research has focused on developing innovative hydrogel systems that are capable of effectively loading and releasing large biomolecules, such as proteins and nucleic acids. Techniques such as embedding nanoparticles or employing hybrid systems that combine different polymeric materials have significantly enhanced the functional capabilities of these hydrogels (97).

Hydrogels are proving indispensable in the realms of tissue engineering and regenerative medicine, offering a versatile platform that supports tissue repair and regeneration across various tissues, including cardiac, bone, and neural tissues. Their unique properties are particularly suited to supporting cell growth and tissue formation, making them ideal for a wide range of medical applications.

Hydrogels provide a three-dimensional scaffold that closely mimics the natural extracellular matrix, which is essential for supporting cell adhesion, proliferation, and differentiation. This scaffolding is crucial for tissue regeneration as it offers not only structural support but also biochemical cues that facilitate tissue development and integration with surrounding tissues (98). Hydrogels can be engineered to release growth factors, cytokines, or other bioactive molecules in a controlled manner. This capability is vital for directing the tissue regeneration process, such as promoting angiogenesis in cardiac tissue engineering or osteogenesis in bone regeneration (99).

The materials used to construct hydrogels are generally biocompatible and designed to degrade at a rate that coincides with new tissue formation. This gradual degradation allows the hydrogel scaffold to be replaced by natural extracellular matrix materials as new tissue forms, which is critical for maintaining the integrity and functionality of the regenerating area (100). Advanced hydrogels can respond to environmental stimuli such as temperature, pH, or mechanical stress, which is particularly beneficial in dynamic tissue environments like the heart. For instance, hydrogels that react to mechanical stimuli can enhance the integration and functionality of engineered cardiac tissue, adapting to the unique demands of the application (101).

Hydrogels can be tailored in terms of mechanical properties, degradation rates, and biochemical functionalities to meet the specific needs of different tissue types. This level of customizability allows for the development of specialized hydrogels optimized for particular applications, such as nerve regeneration or skin repair (102).

### 4.5.4 Current Limitations and Research Trends

Hydrogels are pivotal in biomedical research, particularly in drug delivery, tissue engineering, and regenerative medicine. Despite their potential, the development of hydrogels encounters various challenges, with ongoing research aimed at addressing these limitations.

A significant challenge with hydrogels is their mechanical strength and stability. Traditional hydrogels often have poor mechanical properties, limiting their use in load-bearing applications where durability is crucial. Enhancing their mechanical strength without losing biocompatibility and bioactivity is a major focus area in current hydrogel research (101). While hydrogels offer potential for controlled drug release, precisely managing release rates and targeting specific tissues or cells remains challenging. Developing sophisticated release mechanisms is critical to ensuring that drugs are delivered efficiently—at the right time, in the right place, and in the correct doses (103).

Controlling the biodegradation rate of hydrogels to match the tissue regeneration pace is essential. Materials that degrade at a predictable and appropriate rate are needed to correspond with new tissue formation, optimizing therapeutic outcomes (104). Research is increasingly focusing on double-network hydrogels, which integrate two polymer networks. This approach enhances mechanical properties while maintaining biocompatibility and hydration, making these materials suitable for high-stress biomedical applications (104).

The development of injectable hydrogels that can be administered in a minimally invasive manner is gaining traction. These hydrogels combine robustness with flexibility, allowing for delivery through a syringe and opening new avenues for localized therapy applications (105). There is a continuous effort to develop hydrogels that respond to specific biological stimuli, such as pH changes, temperature shifts, or enzymatic activities. These responsive systems are particularly promising for creating effective drug-delivery systems that adapt to the body's signals, optimizing therapeutic effects.

## 4.6 COMPARATIVE ANALYSIS OF POLYMER-BASED NANOCARRIERS

### 4.6.1 Efficacy in Delivery of Nucleic Acids and Proteins

Evaluating the efficacy of polymer-based nanocarriers for the delivery of nucleic acids and proteins involves assessing several critical criteria that ensure both the efficiency of delivery and the safety of the carrier system. These criteria include encapsulation efficiency, release profile, targeting efficiency, biocompatibility and toxicity, and physicochemical stability. Nanocarriers must efficiently encapsulate nucleic acids or proteins to protect these therapeutic molecules from degradation until they reach their target. This ensures the agents are functional upon arrival at the site of action, which is crucial for the effectiveness of the treatment. Factors such as the capacity to load a high percentage of the therapeutic agent and the stability of these agents within the carrier are key measures of encapsulation efficiency (16).

The release kinetics of nucleic acids and proteins from nanocarriers must be controlled and predictable. A sustained release that aligns with the therapeutic needs of the treatment is essential for effective therapy.

The release profile should ensure that the therapeutic agents maintain their integrity and are active upon release, which is vital for the success of the treatment (13). The ability of nanocarriers to specifically target certain cells or tissues greatly impacts their efficacy. Enhancements such as targeting ligands or modifications that promote receptor-mediated uptake can significantly improve delivery specificity. This targeted approach reduces off-target effects and increases therapeutic efficacy by ensuring that the therapeutic agents are delivered precisely where needed (106).

Nanocarriers must be biocompatible and exhibit minimal toxicity, especially for clinical applications. It is crucial that the nanocarrier does not elicit significant immune responses or cause adverse side effects. Biocompatibility and minimal toxicity are fundamental for the acceptance and success of nanocarriers in medical treatments (107). The physical and chemical stability of nanocarriers under physiological conditions is essential. They must resist the body's varied environments without premature breakdown or aggregation, which could compromise their effectiveness and safety. Stability under physiological conditions ensures that nanocarriers can deliver their payload effectively and safely (106).

These criteria are essential for evaluating the efficacy of polymer-based nanocarriers in drug delivery, highlighting the importance of balancing effective drug delivery with the safety of the carrier system.

### 4.6.2 Biocompatibility and Safety Profiles

Evaluating the biocompatibility and safety profiles of dendrimers, polymeric micelles, and hydrogels is crucial for determining their suitability for medical applications, especially in drug delivery and tissue engineering. Each of these nanocarriers has unique properties that influence their use in clinical settings.

Dendrimers, characterized by their highly branched, tree-like structures, can be precisely engineered for various medical applications. Their biocompatibility often depends on surface modifications to mitigate the inherent toxicity associated with their cationic surface groups. Techniques like PEGylation or using biodegradable materials help enhance their safety profiles. Despite these modifications, dendrimers' biocompatibility needs careful evaluation due to their potential cytotoxicity, which can vary based on their composition and the specific application (33).

Known for their excellent biocompatibility, polymeric micelles are often made from biodegradable materials such as poly lactic acid-co-glycolic acid or PEG. Their core–shell structure is advantageous for encapsulating drugs, reducing premature interactions with the biological environment, and enhancing delivery efficiency. Although they generally have a favorable safety profile, their stability and potential immunogenic responses must be considered to ensure their safe use in medical scenarios (108).

Hydrogels are highly valued for their biocompatibility due to their hydrophilic nature and ability to mimic the extracellular matrix, which supports cell adhesion and growth. Commonly utilized in tissue engineering, hydrogels can be designed to degrade at rates that coincide with tissue regeneration, enhancing their integration and functionality. Their safety is typically high, provided that the polymers used are non-toxic and the degradation products are harmless (32).

Dendrimers often require significant surface modification to ensure biocompatibility and safety. Polymeric micelles usually offer a more straightforward safety profile but may need specific enhancements for particular applications. Hydrogels provide excellent biocompatibility naturally, but they may need adjustments in their physical properties to meet particular clinical needs.

Ultimately, the choice among dendrimers, polymeric micelles, and hydrogels for a specific medical application depends on balancing these factors against the desired therapeutic outcomes. The decision involves considering each material's unique advantages and limitations in terms of biocompatibility, safety, and the specific needs of the application.

### 4.6.3 Technological and Economic Considerations

When evaluating the technological and economic factors for producing nanocarriers, it is essential to consider several critical aspects to ensure efficient, cost-effective, and compliant manufacturing processes.

Advanced manufacturing techniques such as high-pressure microfluidization are critical for ensuring high-quality, reproducible production of nanocarriers. These methods must be scalable and capable of maintaining consistent particle size and distribution, which is crucial for the efficacy and safety of nanocarriers. Such approaches support scalability and help achieve uniformity in nanocarrier quality (109). The choice of polymers and their chemical properties is pivotal. Materials must be selected for their stability and ability to form nanocarriers with the desired release characteristics, as well as for their compatibility with industrial-scale synthesis processes. This selection ensures that the nanocarriers function effectively in their intended applications (110).

As production scales up, maintaining stringent quality control becomes increasingly challenging but essential. Effective methods are needed to monitor parameters like particle size, stability, and drug loading capacity consistently throughout the manufacturing process. High standards of quality control are vital for the reliability and performance of nanocarriers (111). The cost of raw materials and their availability on a large scale significantly influence the economic viability of nanocarrier production. Economies of scale may reduce costs, but the initial investment required for high-quality, medical-grade materials can be substantial. Strategic sourcing and procurement are crucial for managing material costs effectively.

The efficiency of the production process directly affects the overall cost of nanocarriers. Techniques that minimize waste and energy consumption while maximizing output are preferred. Continuous processing modes are increasingly explored as alternatives to batch processes, offering better efficiency and lower costs due to consistent production flow. Complying with regulatory standards is critical but can entail significant costs. Expenses associated with obtaining approval from regulatory bodies like the FDA are considerable, especially for products intended for clinical use. Ensuring compliance in the early stages of product development is crucial to avoid costly revisions and delays in commercialization.

These technological and economic considerations are fundamental in navigating the complex landscape of nanocarrier production, balancing the need for innovative solutions with cost management and regulatory compliance to successfully bring these products to market.

## 4.7 FUTURE PERSPECTIVES AND EMERGING TRENDS

The field of polymer-based nanocarriers is on the cusp of transformative advancements that promise to revolutionize applications in drug delivery, tissue engineering, and broader medical contexts. These innovations range from smart polymers to high-performance materials, each bringing us closer to more efficient and personalized medical solutions.

Smart polymers are at the forefront of polymer science due to their ability to respond to environmental stimuli like pH, temperature, and light (Figure 4.2).

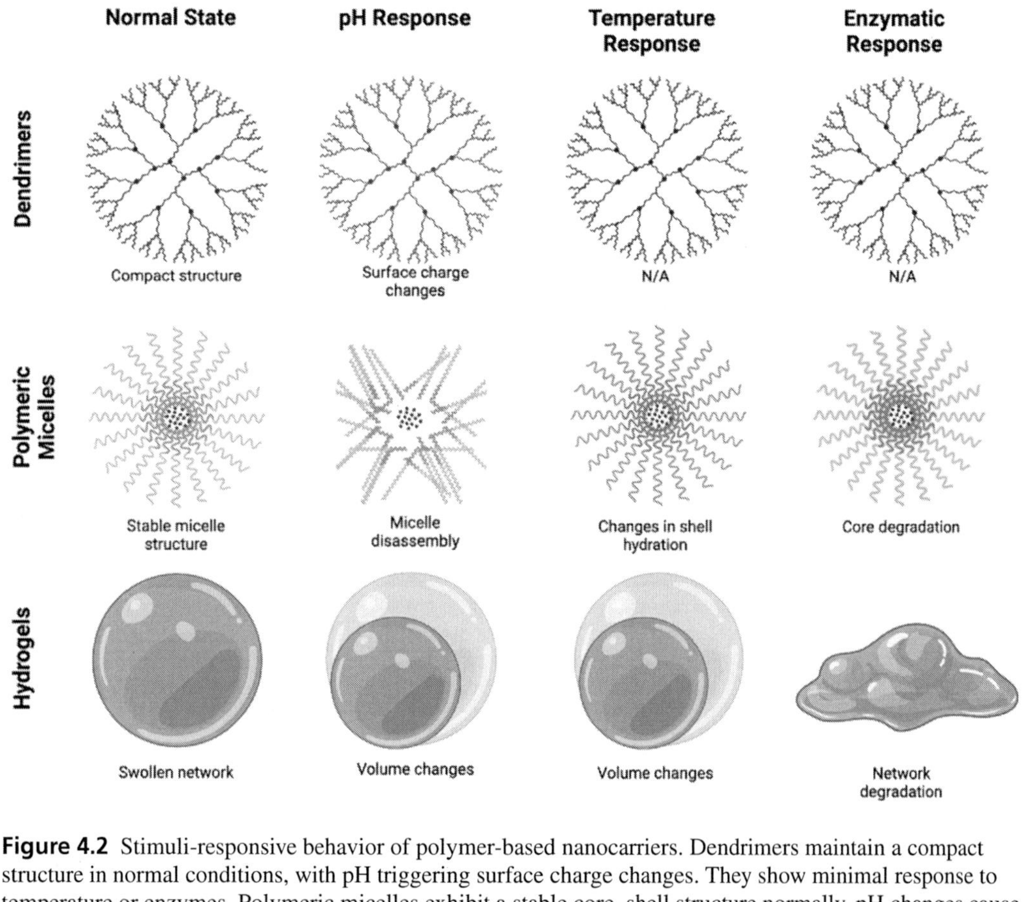


**Figure 4.2** Stimuli-responsive behavior of polymer-based nanocarriers. Dendrimers maintain a compact structure in normal conditions, with pH triggering surface charge changes. They show minimal response to temperature or enzymes. Polymeric micelles exhibit a stable core–shell structure normally. pH changes cause disassembly, temperature affects shell hydration, and enzymes degrade the core, each facilitating controlled drug release. Hydrogels, initially in a swollen state, respond to all stimuli: pH and temperature induce volume changes, while enzymes cause network degradation.

This responsiveness allows for precise control over drug release and targeting, optimizing treatment efficacy and minimizing the side effects typically associated with conventional drug-delivery systems. Such capabilities are crucial in enhancing the precision of medical treatments (112). The development of biodegradable polymers is crucial to addressing the safety profiles of nanocarriers. These polymers decompose into non-toxic by-products, mitigating long-term toxicity concerns and reducing environmental impacts. Research in this area, such as that conducted by Tewari et al. (2022), focuses on ensuring that nanocarriers are not only effective in their therapeutic applications but also safe for long-term use (19).

To enhance the delivery of drugs across complex biological barriers, such as the BBB, researchers are developing high-performance polymers. These materials are designed to overcome physical and biochemical obstacles, ensuring that treatments can reach targeted areas effectively. The work by Ramasundaram et al. (2022) exemplifies efforts to tailor these polymers for specific medical needs, improving access to crucial therapy areas (112). The convergence of polymer-based nanocarriers with other nanotechnologies, such as metal nanoparticles and quantum dots, is fostering the creation of hybrid systems with dual functionalities. These systems, which combine therapeutic and diagnostic capabilities (theranostics), are being explored by researchers like Fernández and Orozco (113). Furthermore, the development of polymer nanocomposites that incorporate graphene or carbon nanotubes is enhancing crucial properties for drug delivery and tissue engineering (114). The use of layer-by-layer assembly techniques is advancing the construction of multifunctional nanocarriers. This method allows for precise control over drug release mechanisms and targeted delivery capabilities, making it a key technique for developing more effective drug-delivery systems (67). These advancements not only promise to enhance targeted and personalized treatment plans, which are tailored to the specific genetic makeup and disease conditions of patients, but also aim to reduce adverse side effects and improve patient compliance. In the field of regenerative medicine, polymer-based nanocarriers are important in providing scaffolds for tissue regeneration and delivering growth factors or genes that promote healing (115).

Collectively, these developments in polymer science and nanotechnology are poised to reshape the future of medicine, ensuring that treatments are not only more effective but also safer, more efficient, and tailored to individual patient needs.

## 4.8 CONCLUSION

In conclusion, polymer-based nanocarriers, including dendrimers, polymeric micelles, and hydrogels, have demonstrated substantial promise across various sectors of healthcare, such as drug delivery, tissue engineering, and regenerative medicine. Advances in polymer science and nanotechnology have not only augmented the functionality of these carriers but have also significantly improved their biocompatibility and safety profiles, making them increasingly viable for a wide range of biomedical applications.

These nanocarriers are now engineered with advanced functionalities that enable them to respond to environmental stimuli and precisely target specific cells or tissues. Such targeting capabilities markedly enhance the precision and efficacy of therapeutic delivery. Alongside these advanced features, significant improvements in polymer chemistry have boosted the biocompatibility and safety of these carriers, which are essential for their use in sensitive medical treatments.

However, scaling up the production of polymer-based nanocarriers presents substantial technological and economic challenges. Addressing these challenges is essential to ensuring the cost-effectiveness and manufacturability of these carriers on an industrial scale. Moreover, the integration of polymer-based nanocarriers with other nanotechnologies has ushered in innovative approaches to drug delivery, offering opportunities for synergistic effects that could lead to enhanced treatment outcomes.

The impact of these advancements on healthcare is profound, paving the way for more effective, targeted, and personalized treatment options. These developments underscore the critical need for ongoing research and innovation in the fields of polymer science and nanotechnology. By continuing to push the boundaries of these disciplines, we can fully unlock the potential of polymer-based nanocarriers to significantly improve patient outcomes and thus shape the future of medicine.

## REFERENCES

1. Napagoda M, Witharana S. Nanotechnology in drug delivery. In: Witharana S, Napagoda MT, editors. *Nanotechnology in Modern Medicine [Internet]*. Springer Nature Singapore; 2023. p. 47–73. Available from: https://doi.org/10.1007/978-981-19-8050-3_3

2. Crossen SL, Goswami T. Nanoparticulate carriers for drug delivery. *Journal of Pharmaceutical and Biopharmaceutical Research [Internet]*. 2022 Apr 20;4(1). Available from: www.syncsci.com/journal/JPBR/article/view/JPBR.2022.01.001

3. Cai X, Jin M, Yao L, He B, Ahmed S, Safdar W, et al. Physicochemical properties, pharmacokinetics, toxicology and application of nanocarriers. *Journal of Materials Chemistry B [Internet]*. 2023;11(4):716–33. Available from: https://doi.org/10.1039/D2TB02001G

4. Kumar V, Nair SC. Nano lipid carriers as a promising drug delivery carrier for neurodegenerative disorders—an overview of recent advances. *Recent Patents on Biotechnology*. 2024;18(1):2–21.

5. Chamundeeswari M, Jeslin J, Verma ML. Nanocarriers for drug delivery applications. *Environmental Chemistry Letters [Internet]*. 2019;17(2):849–65. Available from: https://doi.org/10.1007/s10311-018-00841-1

6. Madgula K, Peddada LM, Pattathil SD. Biologically synthesized nanocarriers for targeted drug delivery applications. In: Pratap Singh R, RB Singh K, Singh J, Adetunji CO, editors. *Nanotechnology for Drug Delivery and Pharmaceuticals [Internet]*. Academic Press; 2023. p. 43–70. Available from: www.sciencedirect.com/science/article/pii/B9780323953252000092

7. Ho LWC, Liu Y, Han R, Bai Q, Choi CHJ. Nano–cell interactions of non-cationic bionanomaterials. *Accounts of Chemical Research [Internet]*. 2019 Jun 18;52(6):1519–30. Available from: https://doi.org/10.1021/acs.accounts.9b00103

8. Shanmugam H, Rengarajan C, Nataraj S, Sharma A. Interactions of plant food bioactives-loaded nano delivery systems at the nano-bio interface and its pharmacokinetics: An overview. *Food Frontiers*. 2022;3(2):256–75.

9. Tiwari S, Gidwani B, Namdeo P, Tripathi A, Pandey RK, Shukla SS, et al. Polymeric nanoparticles as drug delivery system: Basic concepts and applications. *Nanoparticles and Nanocarriers Based Pharmaceutical Formulations*. 2022;26: 26–54.

10. Gouthami K, Lakshminarayana L, Faniband B, Veeraraghavan V, Bilal M, Bhargava RN, et al. Introduction to polymeric nanomaterials. In: Ali N, Bilal M, Khan A, Nguyen TA, Gupta RK, editors. *Smart Polymer Nanocomposites [Internet]*. Elsevier; 2023. p. 3–25. Available from: www.sciencedirect.com/science/article/pii/B9780323916110000086

11. Zhu M, Wang X, Zhou Y, Li M, Wang Y, Wang Y, et al. Breast tumor-targeted drug delivery via polymer nanocarriers: Endogenous and exogenous strategies. *Journal of Applied Polymer Science [Internet]*. 2023;140(31):e54227. Available from: https://onlinelibrary.wiley.com/doi/abs/10.1002/app.54227

12. Mashal M, Attia N, Grijalvo S, Eritja R, Puras G, Pedraz JL. Stability of polymeric cationic niosomes and their plasmid DNA-based complexes as gene delivery carriers. *Drug Delivery [Internet]*. 2023 Dec 31;30(1):2219420. Available from: https://doi.org/10.1080/10717544.2023.2219420

13. Poudel S, Napit PR, Briski KP, Mattheolabakis G. Oral delivery of nucleic acids with passive and active targeting to the intestinal tissue using polymer-based nanocarriers. *Pharmaceutics [Internet]*. 2021;13(7):1075. Available from: www.mdpi.com/1999-4923/13/7/1075

14. Korovkina O, Polyakov D, Korzhikov-Vlakh V, Korzhikova-Vlakh E. Stimuli-responsive polypeptide nanoparticles for enhanced DNA delivery. *Molecules [Internet]*. 2022;27(23):8495. Available from: www.mdpi.com/1420-3049/27/23/8495

15. Liu J, Bao X, Kolesnik I, Jia B, Yu Z, Xing C, et al. Enhancing the in vivo stability of polycation gene carriers by using PEGylated hyaluronic acid as a shielding system. *BIO Integration*. 2022;3(3):103–11.

16. Yuan W, Li H. Polymer-based nanocarriers for therapeutic nucleic acids delivery. In: Andronescu E, Grumezescu AM, editors. *Nanostructures for Drug Delivery [Internet]*. Elsevier; 2017. p. 445–60. Available from: www.sciencedirect.com/science/article/pii/B9780323461436000142

17. Yoon H. Polymer nanomaterials in biomedicine. *International Journal of Molecular Sciences [Internet]*. 2023;24(8):7480. Available from: www.mdpi.com/1422-0067/24/8/7480

18. Sathasivam T, Gugler MC, Janarthanan P. Polymers in nanomedicine. In: Arivarasan VK, Loganathan K, Janarthanan P, editors. *Nanotechnology in Medicine [Internet]*. Springer International Publishing; 2021. p. 175–98. Available from: https://doi.org/10.1007/978-3-030-61021-0_10

19. Tewari AK, Upadhyay SC, Kumar M, Pathak K, Kaushik D, Verma R, et al. Insights on development aspects of polymeric nanocarriers: The translation from bench to clinic. *Polymers (Basel) [Internet]*. 2022;14(17). Available from: www.mdpi.com/2073-4360/14/17/3545

20. Niculescu AG, Grumezescu AM. Polymer-based nanosystems—a versatile delivery approach. *Materials [Internet]*. 2021;14(22):6812. Available from: www.mdpi.com/1996-1944/14/22/6812

21. Muhammad K, Zhao J, Gao B, Feng Y. Polymeric nano-carriers for on-demand delivery of genes via specific responses to stimuli. *Journal of Materials Chemistry B [Internet]*. 2020;8(42):9621–41. Available from: https://doi.org/10.1039/D0TB01675F

22. Markwalter C, Wang L, Sharaf O, Padakanti P, Esposito M, Wilson B, et al. Polymeric nanocarriers co-encapsulating PET probes and protein therapeutics. *Precision Nanomedicine*. 2022;6(1):994–1006.

23. De R, Song YH, Mahata MK, Lee KT. pH-responsive polyelectrolyte complexation on upconversion nanoparticles: A multifunctional nanocarrier for protection, delivery, and 3D-imaging of therapeutic protein. *Journal of Materials Chemistry B [Internet]*. 2022;10(18):3420–33. Available from: https://doi.org/10.1039/D2TB00246A

24. Ramírez-Acosta CM, Cifuentes J, Castellanos MC, Moreno RJ, Muñoz-Camargo C, Cruz JC, et al. pH-responsive, cell-penetrating, core/shell magnetite/silver nanoparticles for the delivery of plasmids: Preparation, characterization, and preliminary in vitro evaluation. *Pharmaceutics*. 2020;12(6):561.

25. Subhan MA, Parveen F, Filipczak N, Yalamarty SSK, Torchilin VP. Approaches to improve EPR-based drug delivery for cancer therapy and diagnosis. *Journal of Personalized Medicine*, 2023, 13(3), 389.

26. de Oliveira Cardoso VM, Ferreira LMB, Comparetti EJ, Sampaio I, Ferreira NN, Miranda RR, et al. Stimuli-responsive polymeric nanoparticles as controlled drug delivery systems. In: Gajbhiye V, Gajbhiye KR, Hong S, editors. *Stimuli-Responsive Nanocarriers*. Elsevier; 2022. p. 87–117.

27. Zhang X, Zhang M, Wu M, Yang L, Liu R, Zhang R, et al. Precise controlled target molecule release through light-triggered charge reversal bridged polysilsesquioxane nanoparticles. *Polymers (Basel)*. 2021;13(15):2392.

28. Mishra S. Hyperbranched nanostructure drug delivery carrier: Dendrimer. *Nanoscience and Nanotechnology-Asia*. 2023;13(1):20–5.

29. Rastogi V, Yadav P, Porwal M, Sur S, Verma A. Dendrimer as nanocarrier for drug delivery and drug targeting therapeutics: A fundamental to advanced systematic review. *International Journal of Polymeric Materials and Polymeric Biomaterials*. 2024;73(4):310–32.

30. Sharma S. Dendrimers in nanomedicine: History, concept and properties of dendrimers. In: *Dendrimers in Nanomedicine*. CRC Press; 2021. p. 41–51.

31. Aravind M, Kumar SP, Begum ASK. An overview of dendrimers as novel carriers in drug delivery. *Research Journal of Pharmacy and Technology*. 2023;16(4):2051–6.

32. Wang J, Li B, Qiu L, Qiao X, Yang H. Dendrimer-based drug delivery systems: History, challenges, and latest developments. *Journal of Biological Engineering*. 2022;16(1):18.

33. Li X, Naeem A, Xiao S, Hu L, Zhang J, Zheng Q. Safety challenges and application strategies for the use of dendrimers in medicine. *Pharmaceutics [Internet].* 2022;14(6):1292. Available from: www.mdpi.com/1999-4923/14/6/1292

34. Singh N, Kumar V. An overview on surface modified dendrimers: Current direction and future perspectives. *Advanced Chemicobiology Research.* 2023;30–41.

35. Jeevanandam J, Tan KX, Rodrigues J, Danquah MK. Target-specific delivery and bioavailability of pharmaceuticals via Janus and dendrimer particles. *Pharmaceutics.* 2023;15(6):1614.

36. Mahmoudi A, Sadi KS, Malaekeh-Nikouei B. Surface engineered dendrimers as novel option for enhanced pharmaceutical and biomedical potential. In: *Dendrimer-Based Nanotherapeutics.* Elsevier; 2021. p. 225–52.

37. Zenze M, Daniels A, Singh M. Dendrimers as modifiers of inorganic nanoparticles for therapeutic delivery in cancer. *Pharmaceutics.* 2023;15(2):398.

38. Prajapati SK, Jain A. Dendrimers for advanced drug delivery. *Advanced Biopolymeric Systems for Drug Delivery.* 2020;339–60.

39. Golkar N, Samani SM, Tamaddon AM. Cholesterol-conjugated supramolecular assemblies of low generations polyamidoamine dendrimers for enhanced EGFP plasmid DNA transfection. *Journal of Nanoparticle Research.* 2016;18:1–20.

40. Gupta K, Singh S, Gupta K, Khan N, Sehgal D, Haridas V, et al. A bioorthogonal chemoenzymatic strategy for defined protein dendrimer assembly. *ChemBioChem.* 2012;13(17):2489–94.

41. Bae J, Kim H, Kim G, Song J, Kim H. Dendrimer-like supramolecular assembly of proteins with a tunable size and valency through stepwise iterative growth. *Advanced Science.* 2021;8(24):2102991.

42. Shetty A, Chikhaliwala P, Suryawanshi J, Chandra S. Understanding physico-chemical interactions of dendrimers with guest molecules for efficient drug and gene delivery. *Current Pathobiology Reports.* 2021;9:57–70.

43. Chis AA, Dobrea CM, Rus LL, Frum A, Morgovan C, Butuca A, et al. Dendrimers as non-viral vectors in gene-directed enzyme prodrug therapy. *Molecules.* 2021;26(19):5976.

44. Baswar D, Devi A, Mishra A. Dendrimers in gene delivery. In: *Dendrimers in Nanomedicine.* CRC Press; 2021. p. 187–99.

45. Kisakova LA, Apartsin EK, Nizolenko LF, Karpenko LI. Dendrimer-mediated delivery of DNA and RNA vaccines. *Pharmaceutics.* 2023;15(4):1106.

46. Bober Z, Bartusik-Aebisher D, Aebisher D. Application of dendrimers in anticancer diagnostics and therapy. *Molecules.* 2022;27(10):3237.

47. Li H, Zha S, Li H, Liu H, Wong K, All AH. Polymeric dendrimers as nanocarrier vectors for neurotheranostics. *Small.* 2022;18(45):2203629.

48. Kuperkar K, Patel D, Atanase LI, Bahadur P. Amphiphilic block copolymers: Their structures, and self-assembly to polymeric micelles and polymersomes as drug delivery vehicles. *Polymers (Basel).* 2022;14(21):4702.

49. Wang H, Ullah A. Synthesis and evaluation of thermoresponsive renewable lipid-based block copolymers for drug delivery. *Polymers (Basel).* 2022;14(17):3436.

50. Yoshizaki Y, Yamasaki M, Nagata T, Suzuki K, Yamada R, Kato T, et al. Drug delivery with hyaluronic acid-coated polymeric micelles in liver fibrosis therapy. *ACS Biomaterials Science & Engineering.* 2023;9(6):3414–24.

51. Cheng X, Li Q, Sun X, Ma Y, Xie H, Kong W, et al. Well-defined shell-sheddable core-crosslinked micelles with pH and oxidation dual-response for on-demand drug delivery. *Polymers (Basel)*. 2023;15(9):1990.

52. Varga N, Bélteki R, Juhász Á, Csapó E. Core-shell structured PLGA particles having highly controllable ketoprofen drug release. *Pharmaceutics*. 2023;15(5):1355.

53. Mehandole A, Walke N, Mahajan S, Aalhate M, Maji I, Gupta U, et al. Core–shell type lipidic and polymeric nanocapsules: The transformative multifaceted delivery systems. *AAPS PharmSciTech*. 2023;24(1):50.

54. Vambhurkar G, Jain N, Srinivasarao DA, Famta P, Singh SB, Srivastava S. Drug solubilization and drug release from polymeric micelles. In: *Polymeric Micelles: Principles, Perspectives and Practices*. Springer; 2023. p. 87–109.

55. Arora V, Bhandari DD, Puri R, Khatri N, Dureja H. Synthesis, self-assembly, and functional chemistry of amphiphilic block copolymers. In: *Polymeric Micelles: Principles, Perspectives and Practices*. Springer; 2023. p. 1–25.

56. Chen Z, Deng S, Tian L. Advances in the preparation and application of polymeric micelles. In: *Eighth International Conference on Energy Materials and Electrical Engineering (ICEMEE 2022)*. SPIE; 2023. p. 45–54.

57. Ali A, Bhadane R, Asl AA, Wilén CE, Salo-Ahen O, Rosenholm JM, et al. Functional block copolymer micelles based on poly (jasmine lactone) for improving the loading efficiency of weakly basic drugs. *RSC Advances*. 2022;12(41):26763–75.

58. Liu T, Lang M. Preparation and characterization of novel functional tri-block copolymer for constructing temperature/redox dual-stimuli responsive micelles. *Journal of Macromolecular Science, Part A*. 2022;59(8):513–25.

59. He Y, Tang Y, Zhang Y, MacFarlane L, Shang J, Shi H, et al. Published under an exclusive license by AIP Publishing. *Applied Physics Reviews*. 2022;9(2).

60. Jin GW, Rejinold NS, Choy JH. Multifunctional polymeric micelles for cancer therapy. *Polymers (Basel)*. 2022;14(22):4839.

61. Jain KK, Kadam AY, Tomar Y, Singhvi G. Stimuli-sensitive polymeric micelles for biomedical applications. *Polymeric Micelles: Principles, Perspectives and Practices*. 2023;125–54.

62. Mahajan D, Suares D, Rachana R, Shetty S. Types of polymeric micelles for controlled drug release. In: Singh SK, Gulati M, Mutalik S, Dhanasekaran M, Dua K, editors. *Polymeric Micelles: Principles, Perspectives and Practices [Internet]*. Springer Nature Singapore; 2023. p. 69–86. Available from: https://doi.org/10.1007/978-981-99-0361-0_4

63. Arafa KK, El-Sherbiny IM. Polymeric micelles with cleavable links for drug delivery. In: Kesharwani P, Greish K, editors. *Polymeric Micelles for Drug Delivery [Internet]*. Woodhead Publishing; 2022. p. 515–30. (Woodhead Publishing Series in Biomaterials). Available from: www.sciencedirect.com/science/article/pii/B9780323898683000203

64. de Santana WMOS, Pochapski DJ, Pulcinelli SH, Fontana CR, Santilli CV. Polymeric micelles–mediated photodynamic therapy. In: Kesharwani P, editor. *Nanomaterials for Photodynamic Therapy [Internet]*. Woodhead Publishing; 2023. p. 105–39. (Woodhead Publishing Series in Biomaterials). Available from: www.sciencedirect.com/science/article/pii/B9780323855952000037

65. Kanazawa T. Cell-penetrating peptide-conjugated polymer micelle for pDNA/siRNA delivery. In: *Cell-Penetrating Peptides [Internet]*. John Wiley & Sons, Ltd; 2023. p. 285–313. Available from: https://onlinelibrary.wiley.com/doi/abs/10.1002/9783527835997.ch16

66. Capelôa L, Yazdi M, Zhang H, Chen X, Nie Y, Wagner E, et al. Cross-linkable polyion complex micelles from polypept(o)ide-based ABC-triblock copolymers for siRNA delivery. *Macromolecular Rapid Communications [Internet]*. 2022;43(12):2100698. Available from: https://onlinelibrary.wiley.com/doi/abs/10.1002/marc.202100698

67. Mustafai A, Zubair M, Hussain A, Ullah A. Recent progress in proteins-based micelles as drug delivery carriers. *Polymers (Basel) [Internet]*. 2023;15(4):836. Available from: www.mdpi.com/2073-4360/15/4/836

68. Ashique S, Garg A, Hussain A, Bhatt S, Agrawal A, Mishra, N. Polymeric micelles in the delivery of proteins. In: Singh SK, Gulati M, Mutalik S, Dhanasekaran M, Dua K, editors. *Polymeric Micelles: Principles, Perspectives and Practices [Internet]*. Springer Nature Singapore; 2023. p. 231–51. Available from: https://doi.org/10.1007/978-981-99-0361-0_12

69. Marquet F, Stojceski F, Grasso G, Patrulea V, Danani A, Borchard G. Characterization of the interaction of polymeric micelles with siRNA: A combined experimental and molecular dynamics study. *Polymers (Basel) [Internet]*. 2022;14(20):4409. Available from: www.mdpi.com/2073-4360/14/20/4409

70. Jain KK, Kadam AY, Tomar Y, Singhvi G. Stimuli-sensitive polymeric micelles for biomedical applications. In: Singh SK, Gulati M, Mutalik S, Dhanasekaran M, Dua K, editors. *Polymeric Micelles: Principles, Perspectives and Practices [Internet]*. Springer Nature Singapore; 2023. p. 125–54. Available from: https://doi.org/10.1007/978-981-99-0361-0_7

71. Kaur J, Gulati M, Kapoor B, Jha NK, Gupta PK, Gupta G, et al. Advances in designing of polymeric micelles for biomedical application in brain related diseases. *Chemico-Biological Interactions [Internet]*. 2022;361:109960. Available from: www.sciencedirect.com/science/article/pii/S000927972200165X

72. Thirugnanasambandham I, Aasaithambi K, Nizam I, Kuppusamy G. Stability of polymeric micelles and their regulatory status. In: Singh SK, Gulati M, Mutalik S, Dhanasekaran M, Dua K, editors. *Polymeric Micelles: Principles, Perspectives and Practices [Internet]*. Springer Nature Singapore; 2023. p. 277–94. Available from: https://doi.org/10.1007/978-981-99-0361-0_15

73. Wang C, Wang Z, Zhao X, Yu F, Quan Y, Cheng Y, et al. DOX loaded aggregation-induced emission active polymeric nanoparticles as a fluorescence resonance energy transfer traceable drug delivery system for self-indicating cancer therapy. *Acta Biomaterialia [Internet]*. 2019;85:218–28. Available from: www.sciencedirect.com/science/article/pii/S1742706118307451

74. Pérez SE, Haidar ZS. Polymeric micelles: Précis for past, present and future. In: Kesharwani P, Greish K, editors. *Polymeric Micelles for Drug Delivery [Internet]*. Woodhead Publishing; 2022. p. 3–14. (Woodhead Publishing Series in Biomaterials). Available from: www.sciencedirect.com/science/article/pii/B9780323898683000197

75. Chen Z, Deng S, Tian L. Advances in the preparation and application of polymeric micelles. In: Palanisamy T, Han LB, editors. *Eighth International Conference on Energy Materials and Electrical Engineering (ICEMEE 2022) [Internet]*. SPIE; 2023. p. 1259809. Available from: https://doi.org/10.1117/12.2672920

76. Bhattacharjee P, Ahearne M. Significance of crosslinking approaches in the development of next generation hydrogels for corneal tissue engineering. *Pharmaceutics [Internet]*. 2021;13(3):319. Available from: www.mdpi.com/1999-4923/13/3/319

77. Gao Y, Peng K, Mitragotri S. Covalently crosslinked hydrogels via step-growth reactions: Crosslinking chemistries, polymers, and clinical impact. *Advanced Materials [Internet]*. 2021;33(25):2006362. Available from: https://onlinelibrary.wiley.com/doi/abs/10.1002/adma.202006362

78. Zhang Z, He C, Rong Y, Ren H, Wang T, Zou Z, et al. A fast and versatile cross-linking strategy via o-phthalaldehyde condensation for mechanically strengthened and functional hydrogels. *National Science Review*. 2021 Apr 1;8(4):nwaa128.

79. Maitra J, Shukla VK, Kumar Shukla V. Cross-linking in hydrogels—a review. *American Journal of Polymer Science [Internet]*. 2014;2014(2):25–31. Available from: https://journal.sapub.org/ajps

80. Mark T, Elia G. *Crosslinkable Hydrogel Compositions [Internet]*. Available from: https://lens.org/074-385-178-151-629

81. Ai Y, Lin Z, Zhao W, Cui M, Qi W, Huang R, et al. Nanocellulose-based hydrogels for drug delivery. *Journal of Materials Chemistry B [Internet]*. 2023;11(30):7004–23. Available from: https://dx.doi.org/10.1039/D3TB00478C

82. Kumar A, Patra P, Nutan B, Bhalani DV, Chandel AKS. Polymeric hydrogels as drug delivery vehicles. *Preprints (Basel) [Internet]*. 2023 Jun. Available from: https://doi.org/10.20944/preprints202306.1118.v1

83. Thang NH, Chien TB, Cuong DX. Polymer-based hydrogels applied in drug delivery: An overview. *Gels [Internet]*. 2023;9(7):523. Available from: www.mdpi.com/2310-2861/9/7/523

84. Akhtar S, Singh G, Singh SP, Kumar S, Malik JK. Current scenario of hydrogel as drug delivery system. *South Asian Research Journal of Pharmaceutical Sciences*. 2023 Mar 15;5(2):33–7.

85. Serpico L, Dello Iacono S, Cammarano A, De Stefano L. Recent advances in stimuli-responsive hydrogel-based wound dressing. *Gels [Internet]*. 2023;9(6):451. Available from: www.mdpi.com/2310-2861/9/6/451

86. Garcia-Torres J. Hybrid hydrogels with stimuli-responsive properties to electric and magnetic fields. In: Popa L, Ghica MV, Dinu-Pîrvu CE, editors. *Hydrogels [Internet]*. IntechOpen; 2022. Available from: https://doi.org/10.5772/intechopen.102436

87. Howard E, Li M, Kozma M, Zhao J, Bae J. Self-strengthening stimuli-responsive nanocomposite hydrogels. *Nanoscale [Internet]*. 2022;14(48):17887–94. Available from: https://doi.org/10.1039/D2NR05408F

88. Lin Z, Ding J, Chen X, He C. pH- and temperature-responsive hydrogels based on tertiary amine-modified polypeptides for stimuli-responsive drug delivery. *Chemistry: An Asian Journal [Internet]*. 2023;18(8):e202300021. Available from: https://aces.onlinelibrary.wiley.com/doi/abs/10.1002/asia.202300021

89. Liu X, Gao M, Chen J, Guo S, Zhu W, Bai L, et al. Recent advances in stimuli-responsive shape-morphing hydrogels. *Advanced Functional Materials [Internet]*. 2022;32(39):2203323. Available from: https://onlinelibrary.wiley.com/doi/abs/10.1002/adfm.202203323

90. Eswaran L, Kazimirsky G, Byk G. New biocompatible nanohydrogels of predefined sizes for complexing nucleic acids. *Pharmaceutics [Internet]*. 2023;15(2):332. Available from: www.mdpi.com/1999-4923/15/2/332

91. Chen T, Qian Q, Makvandi P, Zare EN, Chen Q, Chen L, et al. Engineered high-strength biohydrogel as a multifunctional platform to deliver nucleic acid for ameliorating intervertebral disc degeneration. *Bioactive Materials [Internet]*. 2023;25:107–21. Available from: www.sciencedirect.com/science/article/pii/S2452199X23000105

92. Wang Q, Qu Y, Zhang Z, Huang H, Xu Y, Shen F, et al. Injectable DNA hydrogel-based local drug delivery and immunotherapy. *Gels [Internet]*. 2022;8(7):400. Available from: www.mdpi.com/2310-2861/8/7/400

93. Ke L, Wu YL, Tian H. Preparation and evaluation of supramolecular hydrogels for localized sustained gene delivery. In: Tian H, Chen, X, editors. *Gene Delivery [Internet].* Springer Singapore; 2021. p. 1–16. Available from: https://doi.org/10.1007/978-981-33-6198-0_13-1

94. Ali S, Ranjha NM, Ahmad B, Khan AA, Hassan FU, Aziz T, et al. Sustained release of drug facilitated through chemically crosslinked polyvinyl alcohol-gelatin (PVA-GE) hydrogels. A sustainable biomedical approach. *Polish Journal of Chemical Technology [Internet].* 2023;25(2):56–65. Available from: https://doi.org/10.2478/pjct-2023-0017

95. Nalbadis A, Trutschel ML, Lucas H, Luetzkendorf J, Meister A, Mäder K. Selection and incorporation of siRNA carrying non-viral vector for sustained delivery from gellan gum hydrogels. *Pharmaceutics [Internet].* 2021;13(10):1546. Available from: www.mdpi.com/1999-4923/13/10/1546

96. del Olmo JA, Martínez VS, González RP, Alonso JM. Sustained drug release from biopolymer-based hydrogels and hydrogel coatings. In: Popa L, Ghica MV, Dinu-Pîrvu CE, editors. *Hydrogels [Internet].* IntechOpen; 2022. Available from: https://doi.org/10.5772/intechopen.103946

97. Rafael D, Guerrero M, Marican A, Arango D, Sarmento B, Ferrer R, et al. Delivery systems in ocular retinopathies: The promising future of intravitreal hydrogels as sustained-release dcaffolds. *Pharmaceutics [Internet].* 2023;15(5):1484. Available from: www.mdpi.com/1999-4923/15/5/1484

98. Amirthalingam S, Rajendran AK, Moon YG, Hwang NS. Stimuli-responsive dynamic hydrogels: Design, properties and tissue engineering applications. *Materials Horizons [Internet].* 2023;10(9):3325–50. Available from: https://doi.org/10.1039/D3MH00399J

99. Syed Mohamed SMD, Welsh GI, Roy I. Renal tissue engineering for regenerative medicine using polymers and hydrogels. *Biomaterials Science [Internet].* 2023;11(17):5706–26. Available from: https://doi.org/10.1039/D3BM00255A

100. Jurczak P, Lach S. Hydrogels as scaffolds in bone-related tissue engineering and regeneration. *Macromolecular Bioscience [Internet].* 2023;23(11):2300152. Available from: https://onlinelibrary.wiley.com/doi/abs/10.1002/mabi.202300152

101. Fang W, Yang M, Wang L, Li W, Liu M, Jin Y, et al. Hydrogels for 3D bioprinting in tissue engineering and regenerative medicine: Current progress and challenges. *International Journal of Bioprinting.* 2023;9:207–38. AccScience Publishing.

102. Li X, Xu M, Geng Z, Liu Y. Functional hydrogels for the repair and regeneration of tissue defects. *Frontiers in Bioengineering and Biotechnology [Internet].* 2023;11:1190171. Available from: www.frontiersin.org/journals/bioengineering-and-biotechnology/articles/10.3389/fbioe.2023.1190171

103. Lin KT, Wang A, Nguyen AB, Iyer J, Tran SD. Recent advances in hydrogels: Ophthalmic applications in cell delivery, vitreous substitutes, and ocular adhesives. *Biomedicines [Internet].* 2021;9(9):1203. Available from: www.mdpi.com/2227-9059/9/9/1203

104. Huang M, Huang Y, Liu H, Tang Z, Chen Y, Huang Z, et al. Hydrogels for the treatment of oral and maxillofacial diseases: Current research, challenges, and future directions. *Biomaterials Science [Internet].* 2022;10(22):6413–46. Available from: https://doi.org/10.1039/D2BM01036D

105. Almawash S, Osman SK, Mustafa G, El Hamd MA. Current and future prospective of injectable hydrogels—design challenges and limitations. *Pharmaceuticals [Internet].* 2022;15(3):371. Available from: www.mdpi.com/1424-8247/15/3/371

106. Sadeqi Nezhad M. Poly (beta-amino ester) as an in vivo nanocarrier for therapeutic nucleic acids. *Biotechnology and Bioengineering [Internet].* 2023;120(1):95–113. Available from: https://analyticalsciencejournals.onlinelibrary.wiley.com/doi/abs/10.1002/bit.28269

107. Singh M. Assessing nucleic acid: Cationic nanoparticle interaction for gene delivery. In: Narayanan K, editor. *Bio-Carrier Vectors: Methods and Protocols [Internet]*. Springer US; 2021. p. 43–55. Available from: https://doi.org/10.1007/978-1-0716-0943-9_4

108. Parshad B, Yadav P, Kerkhoff Y, Mittal A, Achazi K, Haag R, et al. Dendrimer-based micelles as cyto-compatible nanocarriers. *New Journal of Chemistry [Internet]*. 2019;43(30):11984–93. Available from: https://doi.org/10.1039/C9NJ02612F

109. Alkanawati MS, Wurm FR, Thérien-Aubin H, Landfester K. Large-scale preparation of polymer nanocarriers by high-pressure microfluidization. *Macromolecular Materials and Engineering [Internet]*. 2018;303(1):1700505. Available from: https://onlinelibrary.wiley.com/doi/abs/10.1002/mame.201700505

110. Förster N, Schmidt S, Vana P. Tailoring confinement: Nano-carrier synthesis via Z-RAFT Star Polymerization. *Polymers (Basel) [Internet]*. 2015;7(4):695–716. Available from: www.mdpi.com/2073-4360/7/4/695

111. Singh R, Rhee HW. Chapter 24 – Industrial scale-up: Lab to commercial-scale manufacturing of POSS composites. In: Thomas S, Somasekharan L, editors. *Polyhedral Oligomeric Silsesquioxane (POSS) Polymer Nanocomposites [Internet]*. Elsevier; 2021. p. 499–516. Available from: www.sciencedirect.com/science/article/pii/B9780128213476000196

112. Ramasundaram S, Sobha S, Saravanakumar G, Oh TH. Recent advances in biomedical applications of polymeric nanoplatform assisted with two-photon absorption process. *Polymers (Basel) [Internet]*. 2022;14(23):5134. Available from: www.mdpi.com/2073-4360/14/23/5134

113. Fernández M, Orozco J. Advances in functionalized photosensitive polymeric nanocarriers. *Polymers (Basel) [Internet]*. 2021;13(15):2464. Available from: www.mdpi.com/2073-4360/13/15/2464

114. Shin M, Lim J, Park Y, Lee JY, Yoon J, Choi JW. Carbon-based nanocomposites for biomedical applications. *RSC Advances [Internet]*. 2024;14(10):7142–56. Available from: https://doi.org/10.1039/D3RA08946K

115. Zhang W, Hou Y, Yin S, Miao Q, Lee K, Zhou X, et al. Advanced gene nanocarriers/scaffolds in nonviral-mediated delivery system for tissue regeneration and repair. *Journal of Nanobiotechnology [Internet]*. 2024;22(1):376. Available from: https://doi.org/10.1186/s12951-024-02580-8

# 5 Inorganic Nanocarriers

## *Gold Nanoparticles, Iron Nanoparticles, Silica Nanoparticles, Quantum Dots*

*Juan C. Cruz and Luis H. Reyes*

### 5.1 INTRODUCTION

Inorganic nanocarriers have gained significant importance in biomedical applications because of their unique, tailorable characteristics, which are impossible in organic forms. These features include better stability, facile functionalization, and distinctive optical, magnetic, and electronic properties. Some of the significant nanocarriers are gold nanoparticles (AuNPs), iron nanoparticles (FeNPs), silica nanoparticles (SiNPs), and quantum dots (QDs).

AuNPs have outstanding optical properties due to surface plasmon resonance (SPR), allowing them to absorb and scatter light. Such a property is essential for imaging and diagnostics. Moreover, attaching different ligands and biomolecules to their surfaces has increased biocompatibility and augmented targeting specificity in their application in drug delivery systems. Initial studies done by El-Sayed et al. in 2005 illustrated that through light, the AuNPs could be converted to heat with their selective propensity to destroy cancer cells, making them multifunctional for simultaneous imaging and therapy (1).

Superparamagnetic FeNPs, especially superparamagnetic iron oxide nanoparticles (SPIONs), have revolutionized magnetic resonance imaging (MRI) due to their better magnetic behavior. SPIONs offer significant enhancement in contrast to MRI, thus improving the diagnosis of diseases. Their magnetism is applied in magnetic hyperthermia to cause localized heating for cancer treatment. This application was introduced by Weissleder et al. in 1990. Studies are underway and are still investigating the potential of the same SPIONs used as vectors for targeted drug delivery and theranostics (2).

The SiNPs are reputed to be biocompatible. They offer an extensive surface area. In addition, they provide a simple approach to their surface functionalization. Such properties make them excellent for use in the delivery of drugs as the pores can be loaded with therapeutic agents and designed for a specific target. Furthermore, high drug payload ability in mesoporous silica nanoparticles (MSNs) was shown to enhance therapeutic efficacy by Slowing et al. in 2008, making it one of the forces compelling a broad interest in research applications in cancer therapy and other medical fields (3).

QDs are very promising for bioimaging and diagnostics, owing to their size-tunable fluorescence and high photostability. These semiconductor nanoparticles yield very bright and long-lasting signals, which are critical for prolonged imaging experiments. The groundbreaking work in 1998 by Chan and Nie opened the way for broad applications of QDs in monitoring cellular processes and identifying disease markers. Studies have been conducted with the hope of reducing QD toxicity and establishing biocompatible coatings in pursuit of safer medical applications (4).

Generally, the distinctive features, from the optical features of AuNPs and the magnetic capabilities of FeNPs to their large surface area and biocompatibility of SiNPs, along with QDs' fluorescence, make them outstanding candidate materials for many biomedical applications. These functions render advanced imaging, precise drug delivery, and a novel approach toward therapeutics, with the possibility of moving the field of nanomedicine further ahead.

In the field of drug delivery, inorganic nanocarriers such as AuNPs, FeNPs, SiNPs, and QDs offer several advantages over their organic counterparts, including polymers and liposomes. These advantages are primarily a result of their inherent physical and chemical properties, which make them efficient in delivery and, most importantly, stable. Furthermore, inorganic nanocarriers have unique surface properties and high surface-to-volume ratios that contribute to their delivery efficiency. For example, the AuNPs can be modified highly precisely at the surface to attach specific targeting ligands and drugs, thus enhancing their delivery precision to particular cells or tissues. Research such as that carried out by Jain et al. in 2012 has already shown that once conjugated with antibodies, AuNPs can target and effectively deliver drugs specifically into cancer cells (5). Among the organic nanocarriers, premature drug release and low targeting selectivity compared to their inorganic counterparts prevent high delivery efficiency.

Another critical advantage is the stability of inorganic nanocarriers under physiological conditions. Materials such as iron oxide and silica maintain the structure of the nanocarriers and are resistant to enzymatic degradation. FeNPs, in particular, can retain their magnetic properties and structural stability, which would be required for their use in MRI and target drug delivery. Biodegradation of organic carriers, such as liposomes, due to enzymes or instability in the blood can reduce the therapeutic effect.

In the case of individual types of inorganic nanocarriers, their functional integrity is increased by enhancing chemical stability and resistance to oxidation by AuNPs during circulation and at the target site.

DOI: 10.1201/9781003473183-5

Core–shell structures and surface coatings are recent innovations that further stabilize AuNPs and add new functionalities (6). The magnetic properties and stability of FeNPs, especially SPIONs, make them a widely popular product. Coatings with dextran or polyethylene glycol (PEG) make FeNPs more biocompatible and help them not to aggregate, a stabilization strategy that is less frequently needed for organic carriers because they are not intrinsically stable against aggregation, unlike inorganic materials (7).

Thus, SiNPs are very stable and possess a high surface area, which makes them very appropriate for making drug delivery systems. This is explained by the porous structure of mesoporous SiNPs, which ensures their effectiveness in high drug loading and the controlled release of substantial drug amounts. In this regard, several studies indicate that MSNs are stable and functional across various biological environments and, in general, outperform many other organic carriers for sustained release and protection against premature degradation (8).

Lastly, QDs are helpful for their unique optical properties and stability, making them applicable to imaging applications. As they are inorganic by constitution, they resist photobleaching and chemical degradation, which otherwise ensures reliable performance in studies where imaging is carried out over a long period of time. Organic fluorescent dyes, on the contrary, are relatively resistant to these situations and render poor performance in terms of prolonged imaging scenarios (9).

Inorganic nanocarriers are much superior to other existing drug delivery systems because of their outstanding physicochemical properties. Problems of stability, loading capacity, and targeted delivery are generally encountered in traditional systems like liposomes and polymer-based carriers. Inorganic nanocarriers, such as AuNPs, IONPs, SiNPs, and QDs, have inherently stable structures, versatile surface chemistry, and multifunctional abilities.

One of the most significant advantages of inorganic nanocarriers is enhanced stability in biological environments. Traditional carriers are easily degradable, making them weak and prone to leakage and degradation, thus causing premature drug release. The inorganic nanoparticles are more stable, do not get damaged easily under physiological conditions, such as enzymatic degradation, and do not disrupt their structure. For instance, SiNPs are well known for their robust inert matrix, which sustains the delivery of drugs without causing premature drug release (10).

In this way, inorganic nanocarriers offer controlled drug release kinetics and targeted delivery. By functionalizing the AuNPs with different types of targeting ligands, they can specifically target specific cells or tissues through either ligand–receptor interactions or physical entrapment. This would minimize off-target impacts and enhance therapeutic effectiveness. By way of an example, Dykman and Khlebtsov in 2012 showed that AuNPs could deliver chemotherapeutics directly into cancer cells so that systemic toxicity is reduced (11).

Moreover, inorganic nanocarriers are inherently multifunctional and allow simultaneous imaging, therapy, and diagnostics. For example, QDs present particular optical properties that have been applied in fluorescence imaging, and they can be easily modified to carry therapeutic agents. These very features make tracking drug delivery in real-time possible, along with monitoring therapeutic outcomes (12)

Another crucial benefit, therefore, is the high drug-loading capacity of inorganic nanoparticles due to their large surface area-to-volume ratio. Large pore volumes of mesoporous SiNPs allow significant amounts of drugs to be encapsulated, providing high therapeutic payloads compared to traditional carriers. This will be essential in effective cancer therapy, for which potent doses of medicines are necessary (3).

Furthermore, inorganic nanocarriers enhance the delivery and stability of nucleic acids and proteins. These are critical but challenging areas in biomedicine due to the very nature of the instability of the biomolecules themselves and their susceptibility to degradation by nucleases and proteases. For example, SiNPs create a protective microenvironment that entraps the nucleic acid inside the porous structure and protects against nucleases, ensuring that the nucleic acid in its original form reaches the target site. Proteins can be similarly immobilized on or within inorganic nanocarriers to maintain their intact structure and functionality, thereby eluding the effects of denaturation and aggregation (13).

Cellular uptake for therapeutic molecules is further enhanced using inorganic nanocarriers. AuNPs can be functionalized by cell-penetrating peptides, which improve cellular uptake through endocytosis and provide therapeutic molecules access to intracellular compartments. Due to their small size and tunable surface properties, such nanoparticles can cross biological barriers and reach the target cells much more effectively than classical carriers (14).

Another important property is controlled release; for instance, MSNs can be designed to unload the drug in response to certain stimuli, like pH changes, redox conditions, or enzyme presence. This allows one to tune the kinetics of the release to the site of action to leave the least possible room for systemic side effects while at the same time achieving the highest possible therapeutic efficacy (3).

Last but not least, targeted delivery—be it gene therapy or protein-based therapy—the relevant delivery of the target to the affected cells will define the fate of failure and success of these interventions. For example, IONPs have been used to deliver siRNA to cancer cells with significant gene-silencing effects (15).

Therefore, such inorganic nanocarriers provide critical advancement over traditional systems of drug delivery that serve to increase therapeutic success and decrease side effects owing to their properties of stability, controlled and targeted delivery, multifunctionality, and high drug-loading capacity. In doing so, they will be essential in devising new and more effective therapeutic strategies for many diseases.

## 5.2 FUNDAMENTALS OF INORGANIC NANOCARRIERS

### 5.2.1 Basic Concepts and Properties

Several fundamental material properties are considered to have a profound influence on the effectiveness of inorganic nanocarriers in their biomedical applications. These directly depend on how nanocarriers act in biological environments, establishing stability, biocompatibility, drug-loading capacity, and the potential to effectively target specific cells or tissues.

The size and shape of inorganic nanoparticles play critical roles in biodistribution, cellular uptake, and clearance from the organism. Generally, nanoparticles range from 1 to 100 nanometers in size. Smaller nanoparticles are better at tissue penetration and less likely to be recognized and cleared by the immune system. Spherical nanoparticles are generally internalized by cells more readily than particles that are rod-shaped or irregularly shaped. Huang et al. in 2010 demonstrated that spherical AuNPs have a higher cellular uptake than their rod-shaped counterparts (16).

At the heart of interaction with biological systems lies surface chemistry. Coating modifications of the surface to be biocompatible, like PEG, helps reduce opsonization and recognition by the mononuclear phagocyte system (MPS) to increase the circulation time of the nanoparticle. Moreover, the functionalization of nanoparticles with targeting ligands (antibodies or peptides) can make drug delivery more selective and enable binding to a specific cell type. For instance, in 2011, Choi and co-authors showed that PEGylated IONPs significantly increased circulation time *in vivo* (17).

The core properties necessary for both imaging and therapy are magnetic and optical. Among the SPIONs, there are strong magnetic properties, and they are applied as a contrast agent in MRI. AuNPs have unique optical properties, such as SPR, that make them suitable for photothermal therapy and imaging. Similarly, QDs are preferable because of their size-tunable fluorescence, thus making them one of the most useful in bioimaging. These characteristics enhance the multifunctionality of the inorganic nanocarriers, therefore making them applicable in theranostics (18).

Notably, drug-loading capacity is an essential characteristic of drug release kinetics. Mesoporous SiNPs offer a large surface area and pore volume that can contain a high drug load. The release kinetics may be tuned with the possibility of changing surface chemistry or using stimuli-responsive materials that may provide a controlled release to minimize systemic side effects and maximize therapeutic effectiveness. Vallet-Regí in 2007 reviewed the potential of MSNs for controlled drug release (19).

Biocompatibility and toxicity are mainstays for *in vivo* applications. While both gold and silica are considered biocompatible, the safety of both components has been evaluated and documented to a high level. However, the long-term implications and toxicity, particularly of heavy metal-containing nanoparticles like cadmium found in QDs, must be studied. Designing safety and surface modification strategies is critical to minimizing potential adverse effects (20).

The common essential characteristics that make an inorganic material suitable for use as a nanocarrier bring about the delivery of therapeutic agents in a biocompatible manner and remain stable in biological environments. Structural stability is one crucial requirement; inorganic materials are, per se, reputed to maintain their structural integrity under physiological conditions, hence preventing premature degradation and fulfilling the objectives of successful delivery to the target site. For example, SiNPs possess excellent chemical and physical stress resistance properties applicable for sustained drug release (10).

The surface functionalization ability allows modification to attach different chemical groups, ligands, or biomolecules for the specificity of nanocarriers, thus further enhancing the efficiency of drug delivery. AuNPs can be functionalized with various thiol-containing molecules to significantly customize their surface properties (5). Controlled release mechanisms are typically accomplished using inorganic materials, such as mesoporous SiNPs. These nanoparticles can be altered to adjust the size and surface chemistry of their pores, enabling precise control over the rate at which drugs are released. Stimuli-responsive release mechanisms—pH- and temperature-responsive—are additional significant enhancements to develop novel targeted drug delivery.

Basic biocompatibility is a requirement. Materials such as gold and silica are well validated for their compatibility with biological systems, which makes it extremely important to establish safety in drug delivery applications. Minimal toxicity and excellent compatibility with biological systems have been demonstrated for AuNPs (16) (Figure 5.1).

Multifunctionality enhances strategies for treatment. Inorganic nanoparticles can be designed to be multifunctional in terms of imaging, therapy, and diagnostics, all within the same platform. This is demonstrated in QDs, which have unique optical properties that offer the possibility of imaging combined with the delivery of a therapeutic agent. This added advantage is not available with organic carriers. All of these allow for more extensive options in diagnostics and therapy, such as MRI, photothermal therapy, and fluorescence imaging, among others.

### 5.2.2 Mechanisms of Action

Various advanced interaction mechanisms between inorganic nanocarriers and cellular environments bring out the effective delivery of therapeutic cargo. This occurs mainly due to the physical and chemical properties of the nanoparticles, such as size, shape, surface charge, and functionalization. In general, this delivery process consists of cellular uptake, intracellular trafficking, and finally, cargo release.

#### *5.2.2.1 Uptake*

Interactions between the inorganic nanocarriers and cells mainly start with their uptake through several endocytic pathways. The size and surface properties of the nanoparticles control the specific pathway employed. Smaller particles, 10–50 nm, are usually internalized more effectively through clathrin-mediated endocytosis, while larger ones >200 nm may be taken up in the cells by phagocytosis. Research by Chithrani et al. in 2006 reflected that AuNPs of different sizes follow varied endocytic pathways for entering the cell (21).

#### *5.2.2.2 Intracellular Trafficking*

Once the inorganic nanocarriers have entered the cell, they pass through the endosomes and lysosomes. If nanoparticles are prepared to show a response to pH changes, then acidic pH in these organelles would lead to the release of the cargo. For instance, pH-responsive MSNs may dissolve under acidic conditions, thus resulting in drug release in cancer cells, as demonstrated by Jiang et al. in 2013 (22).

#### *5.2.2.3 Escape from Endosomes*

To prevent their degradation inside the lysosomes, many inorganic nanocarriers are designed to escape from the endosomes and release their payload into the cytoplasm. This can occur through the proton sponge effect, where nanoparticles induce osmotic swelling and rupture of the endosome. For instance, cationic AuNPs can increase the concentration of protons in endosomes, thereby disrupting these vesicles and releasing their content into the cytoplasm (23).

#### *5.2.2.4 Targeting Specific Organelles*

Some inorganic nanocarriers are specifically designed to deliver their cargo to particular organelles like the nucleus or mitochondria. Functionalizing nanoparticles with nuclear localization signals (NLS) or mitochondrial targeting sequences (MTS) enables their targeted delivery to these critical organelles. For example, AuNPs conjugated with NLS peptides have demonstrated effective delivery of therapeutic agents directly into the nucleus (24).

The inorganic nanocarriers come close to cellular environments during endocytosis, intracellular trafficking, endosomal escape, and targeted delivery to specific organelles. These mechanisms could be assembled to ensure that the therapeutic cargo is effectively taken to suitable intracellular locations for maximum treatment.

Other studies about inorganic nanocarriers include the mechanisms by which targeted and controlled release of drugs is done to increase the precision and efficacy of drug delivery while cutting back on the side effects to yield more therapeutic outcomes.

#### *5.2.2.5 Passive Targeting*

It works through the enhanced permeability and retention (EPR) effect, where nanoparticles accumulate preferentially at the tumor sites due to leaky vasculature for cancerous growth. This increases the drug concentration at the tumor site, thereby increasing therapeutic efficacy, as postulated by Maeda et al. in the year 2000 (25).

#### *5.2.2.6 Active Targeting*

In the functionalization of nanocarrier surfaces, specific ligands such as antibodies, peptides, or small molecules with high binding affinities for receptors typically overexpressed on target cells can be used. This

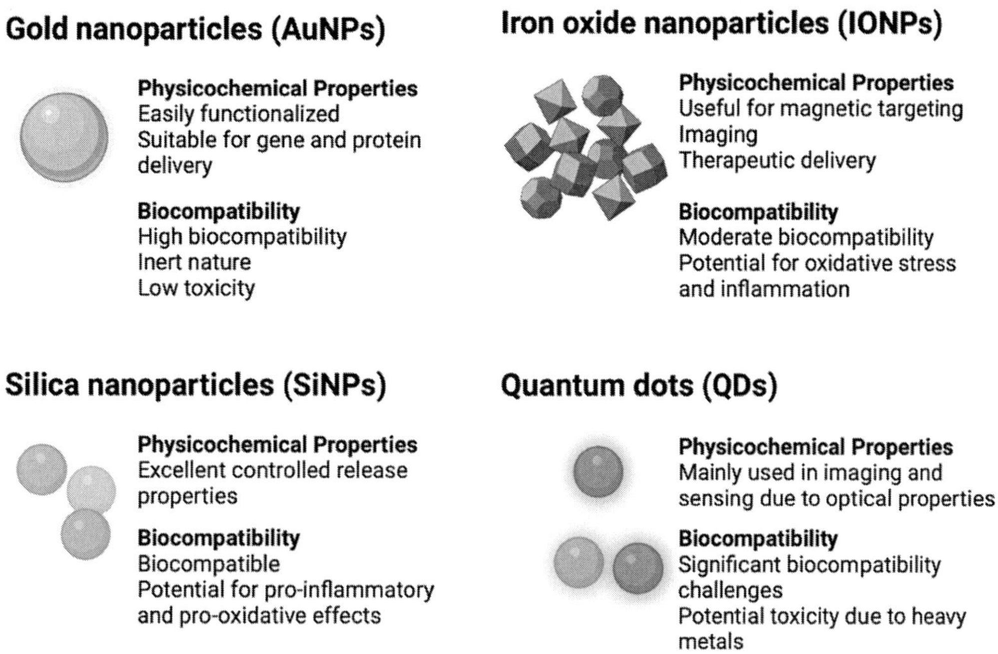


**Figure 5.1** Key properties of inorganic nanocarriers. Gold nanoparticles (AuNPs): Surface plasmon resonance illustrated by oscillating electron cloud and electric field. Iron oxide nanoparticles (IONPs): Magnetic properties demonstrated by alignment under external magnetic field. Silica nanoparticles (SiNPs): Varying pore sizes affect loading capacity, crucial for drug delivery applications. Quantum dots (QDs): Size-dependent optical properties, with smaller particles emitting higher-energy (bluer) light.

allows drug delivery to have greater specificity for diseased cells, for example, in the targeted delivery of AuNPs conjugated with folic acid to cancer cells (26).

#### *5.2.2.7 Magnetic Targeting*

It makes use of magnetic nanoparticles, like iron oxide, which can be directed to a place in the body by an outside magnetic field; this method is most useful in tumor targeting (7).

#### *5.2.2.8 pH-Responsive Release*

Drugs bearing nanocarriers are engineered to release in response to the acidic levels of either tumors or intracellular compartments. MSNs with pH-responsive coatings are the best example. They dissolve in the acid surrounding the environment, hence releasing the drug (27).

#### *5.2.2.9 Temperature-Responsive Release*

This uses temperature-sensitive nanoparticles, which release drugs due to variations in local temperatures and are commonly used along with hyperthermia treatments. For example, AuNPs convert absorbed light into heat, thus releasing the drug in its vicinity (28).

#### *5.2.2.10 Redox-Responsive Release*

Li et al.in their study in 2020 demonstrated that high levels of glutathione cleave the disulfide linkages in the nanoparticles and release the drug; they also demonstrated the exact mechanisms used in enzyme-responsive releases (29). For instance, matrix metalloproteinase-sensitive nanoparticles release drugs in the presence of MMPs, a method that Yao et al. (2018) investigated in targeted cancer therapy (30).

As such, select targeting and well-thought-out overall mechanisms, followed by controlled release mechanisms through inorganic nanocarriers, assure effective and safe delivery of drugs toward overall medical treatment.

## 5.3 GOLD NANOPARTICLES

### 5.3.1 Synthesis and Properties

AuNPs are produced through various methods, which significantly impact their characteristics and render them valuable for a wide range of biomedical uses. Commonly used methods include chemical reduction, seed-mediated growth, green synthesis, and physical processes such as laser ablation.

#### *5.3.1.1 Chemical Reduction*

Out of all the techniques used, the most common one has been reducing gold salts, such as HAuCl4, with agents like sodium citrate, sodium borohydride, or ascorbic acid. By adjusting the reducing agent's concentration, the temperature, and the reaction time during synthesis, AuNPs can develop customized sizes and shapes. In this process, the particles are usually produced spherically, although other shapes, such as rods or cubes, can be realized based on the nature of the reaction parameters. Turkevich et al. first described the essential background for this method of chemical reduction in 1951 with the citrate reduction method (31).

#### *5.3.1.2 Seed-Mediated Growth*

This process involves the formation of small seed particles that are synthesized from gold and then act as nucleation sites for further growth. It is a method in which control over nanoparticles' size and shape is readily possible. Surfactants, like cetyltrimethylammonium bromide (CTAB), enable the production of AuNPs in rod shapes with adjustable aspect ratios by adjusting the concentrations of the seed and growth solutions. Murphy et al. (2008) showed the versatility of this method in producing AuNPs of different shapes and sizes (6).

#### *5.3.1.3 Green Synthesis*

Green synthesis is an environmentally friendly method as it uses biological entities, that is, plant extracts, bacteria, or fungi, as reducing and stabilizing agents, giving biocompatible nanoparticles. The characteristics of the AuNPs synthesized this way can vary considerably depending on the biological source. For example, in 2004, Shankar et al. synthesized stable and biocompatible AuNPs with an aqueous leaf extract of *Azadirachta indica* (neem) (32).

#### *5.3.1.4 Physical Methods*

Techniques such as laser ablation take advantage of high-energy laser pulses to form bulk gold AuNPs without chemical agents, leading to pure nanoparticles. By changing the laser parameters, the size and shape of the AuNPs can be controlled. It was found that the laser ablation for synthesizing AuNPs showed a highly clean synthetic for getting chemically uncontaminated nanoparticles (33).

Each different synthesis route will impact the size, shape, surface chemistry, and stability of the AuNPs and, therefore, their potential for biomedical applications. Chemical reduction typically results in spherical nanoparticles with good dimensional uniformity, while seed-mediated growth allows for the synthesis of various morphologies (such as rods, cubes, or plates). Green synthesis generally yields biocompatible nanoparticles of different shapes, whereas physical methods produce pure nanoparticles suitable for sensitive applications.

The unique optical properties are one of the most outstanding features of AuNPs in their biomedical applications because they derive from SPR. SPR occurs when the conduction electrons at the surface of AuNPs oscillate in response to light, facilitating strong light absorption and scattering (34).

#### *5.3.1.5 Imaging and Diagnostics*

Intense light scattering and absorption phenomena associated with imaging and diagnostics greatly aid AuNPs in imaging applications, for example, dark-field microscopy gives high contrast while imaging biological specimens. Additionally, AuNPs may also be functionalized for targeting specific biomarkers in the early diagnosis of the disease. Jain et al. imaged cancer cells using AuNPs to provide contrast in the images (35).

#### *5.3.1.6 Photothermal Therapy*

Because AuNPs can absorb light and convert it to heat, they can be used in this method, which kills explicitly cancerous cells without affecting the surrounding tissues by just focusing directly on the targeted cancer cells. In a study by Hirsch et al., it was demonstrated that, in principle, tumor photothermal therapy with AuNPs is feasible (36).

#### *5.3.1.7 Biosensing*

Owing to the sensitive response of SPR to environmental conditions, it was readily applied to the sensing of biological entities. The SPR peak shifts in terms of color or optical features upon binding biomolecules to the AuNPs. In 1997, Elghanian et al. generated a colorimetric DNA detection approach that used AuNPs, and in this study, for the first time, this method was shown to be sensitive and specific (37).

#### *5.3.1.8 Drug Delivery*

The optical properties of AuNPs can be used in controlled drug delivery; the drug molecules are released upon photoexcitation at specific wavelengths of light. Such methods offer accurate control of drug release sites and schedules. For example, Fomina et al., in 2012, focused on studying controlled drug release using light-triggered AuNPs (38).

The choice of synthesis method plays a crucial role in determining the properties and functionality of nanoparticles. Additionally, the SPR of AuNPs allows for their use in various biological applications, thus contributing to advancements in nanomedicine. This includes innovations in imaging, therapy, sensing, and drug delivery.

### 5.3.2 Functionalization and Biomedical Applications

Surface modification of AuNPs becomes essential in targeted therapy as it imparts specificity, stability, and biocompatibility. The process entails the direct modification of AuNPs with various ligands, functional groups, or biomolecules to create nanoparticle conjugates that can specifically target cells or tissues. This results in the effective release and delivery of therapeutic agents.

#### *5.3.2.1 Thiol-Based Functionalization*

A common way of functionalizing is through thiol (–SH) groups conjugated on the surface and binding to gold strongly in monolayer form. This allows the molecular attachment of molecules such as drugs, peptides, and antibodies. Thiolated PEG improves biocompatibility and prolongs circulation time by preventing protein adsorption and reducing clearance by the MPS. Gref et al. discovered that PEGylation exponentially improved nanoparticle pharmacokinetics (39).

#### *5.3.2.2 Ligand Conjugation*

A ligand refers to an entity that binds to a specific site of a material. Examples of ligands that can be conjugated on AuNPs to allow them to be delivered directly to the target site include antibodies, peptides, and small molecules. Targeting ligands that bind specifically to overexpressed receptors on target cells can accomplish this by either covalent or non-covalent attachment. One effective example is the folic acid conjugate to target cancer cells overexpressing folate receptors; such constructs will take up the same in higher amounts. For example, Mansoori et al. demonstrated the targeting of cancer cells with folic acid-conjugated AuNPs (40).

#### *5.3.2.3 Aptamer Functionalization*

Aptamers are short, single-stranded DNA or RNA molecules that bind highly to their target molecules through unique three-dimensional structures. The functionalization of AuNPs with aptamers provides a high degree of targeting. Aptamer-functionalized AuNPs can be used for molecular targeting of a range of biomolecules, including proteins and cell surface receptors. Song et al. showed improved targeting and the therapeutic impact of these aptamer-functionalized AuNPs in cancer treatment (41).

#### *5.3.2.4 Peptide Functionalization*

Peptides can be used to modify the surface of AuNPs for targeting and therapeutic applications. Cell-penetrating peptides, such as the trans-activator of transcription (TAT) peptide, have been effectively used to improve the cellular uptake of AuNPs, while targeting peptides, such as RGD (arginine–glycine–aspartic acid), can be used to target integrins overexpressed in cancer cells. Studies report that such peptide-conjugated AuNPs have been applied to imaging and targeted drug delivery. Qin et al. used RGD-modified AuNPs for targeted cancer therapy (42).

To summarize, the surface of AuNPs can be modified for targeted therapies using methods such as thiol-based functionalization, ligand conjugation, aptamer functionalization, and peptide functionalization. All these add to the specificity, stability, and therapeutic effectiveness of the AuNPs in targeted therapy. The high surface area, ease of functionalization, and biocompatibility make functionalized AuNPs potent drug and gene delivery platforms. Their main applications are summarized as follows:

#### *5.3.2.5 Drug Delivery Targeting*

The nanoparticle can be engineered to deliver drugs specifically into target cells, thereby reducing off-target effects while improving therapeutic efficacy. The surface of nanoparticles can be functionalized with a targeting ligand such as antibodies, peptides, or small molecules that specifically bind to specific receptors on the target cell. For example, AuNPs conjugated with antibodies showed efficient delivery of chemotherapeutic drugs specifically to cancer cells while being less toxic to normal tissues. AuNPs have also been found to be effective as carriers of doxorubicin, a DNA-interactive chemotherapy drug, against prostate cancer (43).

#### *5.3.2.6 Gene Transfection*

AuNPs have been extensively applied as carriers in supplying the genetic material for the transfection of DNA, RNA, or siRNA into cellular systems. The positive charge on the surfaces of the AuNPs can facilitate the formation of complexes with negatively charged nucleic acids. This interaction protects nucleic acids from degradation and enhances their uptake by cells. Additionally, AuNPs can be modified with CPPs or other substances to improve the transportation of genes. For example, AuNPs, when functionalized with the TAT peptide, can improve siRNA delivery into cancer cells (44).

#### *5.3.2.7 Theranostics*

Theranostic applications are perfectly applied to AuNPs because of therapy–diagnosis integration. That substantial optical property allows imaging techniques to be carried out; for example, in dark-field microscopy and surface-enhanced Raman spectroscopy (SERS), while their surface is capable of carrying therapeutic agents. This dual functionality can thus enable *in vivo* tracking of drug delivery and therapeutic effects. For instance, Kumar and Lim used AuNPs in 2023 to connect two strategies: photothermal therapy and imaging of cancer cells (45).

#### *5.3.2.8 Controlled Drug Release*

AuNPs can be synthesized to release drugs in the presence of specific stimuli, such as light or pH enzymes. One example is AuNPs that release drugs when exposed to near-infrared light; hence, the user can decide where and when the drugs should be released. This makes drug delivery more efficient and reduces side effects in an intelligent manner. Vines et al. (2019) reviewed the light-triggered anticancer drug delivery from AuNPs (46).

Essentially, the key uses of AuNPs in drug and gene delivery encompass targeted drug delivery, gene transfection, theranostics, and controlled drug release. AuNPs possess remarkable functionalization flexibility and distinctive optical properties, rendering them highly effective instruments in cutting-edge biomedical applications.

### 5.3.3 Challenges in Clinical Translation

The principle underlying the toxicity of AuNPs is based on such imperative criteria as size, surface chemistry, dosage, and accumulation in biological tissues. Despite their high potential and efficiency in various biomedical applications, all of the abovementioned factors may also promote negative biological responses, which in turn must be avoided.

#### *5.3.3.1 Size and Shape*

The size and shape of AuNPs significantly influence their biological interactions and toxicity. Smaller particles of AuNPs (<10 nm) can easily pass through biological membrane barriers and accumulate in tissues. This could result in toxicity in cells. Larger nanoparticles, however, are less capable of penetrating cells but may still produce injury based on bulk and surface interaction. Pan et al. showed that smaller AuNPs (2 nm) had a much higher level of cytotoxicity than the larger ones. In connection with this, size is a governing factor in establishing the toxicity of these metallic nanoparticles (47).

#### *5.3.3.2 Surface Chemistry*

AuNPs are highly engineered toward surface functionalization, and thus, they dramatically influence biocompatibility and interaction with biological systems. Functional groups and coatings, such as PEG, reduce toxicity by preventing protein adsorption and immune recognition. However, certain surface modifications may evoke adverse immune responses or cellular stress. For instance, Alkilany and Murphy (2010) found that uncoated AuNPs had a significantly higher cytotoxic effect than PEGylated ones in one *in vitro* study (48).

#### *5.3.3.3 Dose and Duration of Exposure*

Both the dose and the duration of exposure have a significant impact on the toxicity of AuNPs. They can induce oxidative stress, inflammation, and cell death with high concentrations and prolonged exposure. Some reports have shown that repeated or high-dose administration of AuNPs may lead to the accumulation of AuNPs in organs with potential toxicity. For example, intravenous treatment of AuNPs was found to cause significant accumulation in the liver and spleen, thereby leading to histopathological changes and toxicity in mice (49).

#### *5.3.3.4 Bioaccumulation and Biodistribution*

Generally, the main concerns with using AuNPs are their biodistribution and long-term fate within the body. Potential toxic effects of persistent AuNPs in organs include damage to organs such as the liver, spleen, kidneys, and lungs. In addition to the size effect, the clearance of AuNPs is also related to their surface properties, with smaller AuNPs generally being slowly cleared. Another biodistribution and clearance study on AuNPs by Khlebtsov and Dykman in 2011 pointed out the need for a comprehensive investigation of these aspects within preclinical studies (50).

In essence, major toxicity concerns about AuNPs include size and shape, surface chemistry, dosage, and potential bioaccumulation. Addressing these issues through responsible design and thorough preclinical testing is essential for the clinically safe translation of AuNP-based therapies.

In clinical translation, AuNPs suffer from many key regulatory challenges. The primary one is based on the fact that the area of nanomaterials is inherently complex, so safety and efficiency tests must be rigid. Significant and vital regulatory challenges are standardization, toxicity evaluation, long-term safety, and manufacturing consistency.

#### *5.3.3.5 Standardization*

Lack of standardization in protocols for characterization and evaluation is one of the crucial regulatory challenges for AuNP technologies. Indeed, the wide variation in size, shape, surface modification, and synthesis method detracts from such standardizations in AuNP technologies. Regulatory bodies like the Food and Drug Administration (FDA) and the European Medicines Agency (EMA) stress that there should be a comprehensive characterization of nanomaterials, including their physicochemical properties, stability, and purity. A study by Sharifi et al. (2022) elaborated on the need for standardized methods to provide reproducibility and reliability in the characterization of nanomaterials (51).

#### *5.3.3.6 Toxicity Evaluation*

A full toxicity assessment is an absolute must for regulatory approval of AuNPs. It should include both *in vitro* and *in vivo* studies assessing cytotoxicity, genotoxicity, immunotoxicity, and long-term effects. These biological interactions and the potential for unexpected toxic effects can be pretty complex. Regulatory agencies need a lot of data to prove that AuNPs do not pose undue risks to patients. As previously stated, a literature review by Fadeel et al. (2010) reported the need for the interdisciplinary development of robust toxicity testing frameworks for nanomaterials (52).

#### *5.3.3.7 Long-Term Safety*

Determining the long-term safety of AuNPs becomes somewhat complicated as these nanoparticles have the potential for bioaccumulation and prolonged retention within the body. Long-term studies are required to assess the chronic effects and delayed toxicity potential. Various regulatory agencies demand detailed pharmacokinetics, biodistribution, and excretion studies of AuNPs to understand the long-term behavior of this nanoparticle in biological systems. Indeed, the necessity for long-term safety studies was noted about the risks that can ensue from the persistent nature of the AuNPs by Khlebtsov and Dykman (2011) (50).

#### *5.3.3.8 Manufacturing Consistency*

The manufacturing consistency of AuNPs is critical for their successful clinical translation. Variations in synthesis methods and inconsistent batch-to-batch methods may affect the quality and performance of the synthesized AuNPs. Regulatory agencies call for strict measures about quality control processes that are put into action to ensure that developed nanoparticles meet pre-defined specifications. The formulation of protocols for good manufacturing practices for nanomaterials is essential to satisfy these regulatory standards. Nanoparticle scaling-up is not easy. According to Paliwal et al. (2014), it was not easy to scale up the production of nanomaterials without decreasing product quality and reproducibility (53).

#### *5.3.3.9 Regulatory Pathways*

The regulatory pathways for nanomedicines are complex; specific guidelines must be followed strictly about documentation. Regulatory agencies have developed some guidelines for developing and approving nanomedicines, but the dynamic nature of this field makes it a challenging task. Besides, all these challenges could be tackled when researchers, regulators, and industries work collaboratively to enable the clinical translation of AuNPs. On the other hand, issuing guidance for nanomaterials in drug products is one of the FDA's missions in applying these technologies to new drug product development (54).

In short, there are multiple challenges from a regulatory perspective in the clinical use of AuNPs: standardization, comprehensive toxicology evaluation, long-term safety assessment, manufacturing consistency, and complex regulatory pathways. Addressing such challenges with collaborative efforts and proper testing is key to the successful clinical translation of AuNP-based therapies.

## 5.4 IRON NANOPARTICLES

### 5.4.1 Magnetic Properties and Synthesis

The size, shape, crystallinity, and surface coating of FeNPs in synthesis have a significant impact on their magnetic properties. These characteristics play a crucial role in determining how these nanoparticles interact with magnetic fields and their suitability for various biomedical applications.

#### *5.4.1.1 Size and Shape*

The size and shape of FeNPs are critical to their magnetic properties. Nanoparticles with a size less than 20 nm typically exhibit superparamagnetic characteristics. Consequently, they lack remanence and coercivity, indicating that they do not maintain magnetization once the external magnetic field is removed. In this sense, one of the most critical aspects of developing FeNPs is the control of the synthesis conditions, which include temperature, time of reaction, and precursor concentration. For instance, Peng et al. (2006) demonstrated that, under appropriate control, reaction parameters in the thermal decomposition could lead to monodisperse FeNPs with improved magnetic properties (55).

#### *5.4.1.2 Crystallinity*

Another factor that determines the magnetic behavior of the FeNPs is crystallinity. Better crystallinity often results in better magnetic properties as there are fewer surface defects and the magnetic domains are more aligned. This can be improved during the synthesis process through, for example, the annealing temperature or time. For instance, Park et al. (2005) showed that nanoparticles synthesized at high temperatures have increased crystallinity, increased magnetic saturation, and decreased coercivity (56).

#### *5.4.1.3 Surface Coating*

The surface coating is a crucial step toward enhancing the stability and interaction of FeNPs with biological environments. The surface coatings of nanoparticles serve to prevent aggregation and maintain colloidal stability. These coatings can consist of silica, polymers, or surfactants. Additionally, they may possess functional groups that can be further modified. Foglia et al. (2017) demonstrated that silica-coated FeNPs retained superparamagnetic properties and, when dispersed in aqueous solutions, were biocompatible and stable (57).

#### *5.4.1.4 Synthesis Method*

The magnetic characteristics of the FeNP that is produced are influenced by the specific method used for synthesis, which can be coprecipitation, thermal decomposition, hydrothermal, or solvothermal synthesis. Coprecipitation can result in the formation of nanoparticles with varying sizes and magnetic properties (58). In contrast, thermal decomposition offers a more effective method for controlling the size and shape of particles, resulting in enhanced magnetic properties. Hydrothermal and solvothermal techniques can prepare high-quality, highly crystalline nanoparticles with superior magnetic properties. In a review by Laurent et al. (2009), different synthesis methods and their impact on the magnetic properties of FeNPs were discussed, as were their advantages and drawbacks with the techniques described (59).

The magnetic properties of FeNPs are significantly influenced by manipulating various factors, including size, shape, crystallinity, surface coating, and the choice of synthesis method. Optimizing these conditions is essential for achieving FeNPs with tailored magnetic properties, which are necessary for their application in specific biomedical contexts.

The distinctive magnetic properties of FeNPs play a role in various biomedicine applications of MRI, targeted drug delivery, hyperthermia therapy, biosensing, and MPI. These all depend on the ability of FeNPs to react to externally applied magnetic fields for better diagnostic and curative functions.

MRI is carried out by using SPIONs as contrast agents. Their high magnetic properties allow for an improved contrast of images from MRI by reducing the relaxation times of hydrogen protons located nearby, producing the best possible quality, highly detailed images of tissues and organs. Such an application greatly benefits detecting tumors, inflammation, and vascular diseases. The research by Weissleder et al. (1990) looked at how well SPIONs could improve the contrast of an MRI scan to find metastases in the lymph nodes (60).

#### *5.4.1.5 Targeted Drug Delivery*

The magnetic properties of FeNPs offer the development and exploitation of a targeted drug delivery system by applying an external magnetic field to direct FeNPs to desired body locations and concentrate therapeutic agents at the target site with minimal systemic side effects. In this manner, treatment efficacy is immense, especially in cancer therapy. Sun et al. (2008) reported using magnetically guided FeNPs for targeted delivery of anticancer drugs with increased therapeutic effects in tumor models (18).

#### *5.4.1.6 Hyperthermia Treatment*

Magnetic hyperthermia uses FeNPs for generating localized heating under exposure to an alternating magnetic field. The heat generated is then able to kill cancer cells or improve the effects of chemotherapy and radiotherapy. This treatment will be helpful because it will act on the specific zones without affecting normal tissues around them. Jordan et al. (1999) have presented how this magnetic hyperthermia using FeNPs can have a high potential for effective cancer therapy using a reliable technique (61).

#### *5.4.1.7 Biosensing*

FeNPs are also used for biosensing. To make them work, they are attached to specific ligands or antibodies that allow them to detect and capture biomolecules like proteins, nucleic acids, or pathogens. Additionally, FeNP magnetic responses enable magnetism-based detection systems to quickly and sensitively detect such targets. A research report by Natarajan et al. (2024) focusing on the functionalization of FeNPs for sensitive biomolecule detection has confirmed the high potential of such NPs in diagnostic techniques (62).

#### *5.4.1.8 Magnetic Particle Imaging*

MPI is a new imaging modality based on direct imaging of magnetization, which FeNPs carry. In contrast to MRI, MPI is completely radiation-free and does not depend on the relaxation time of protons, making it much more secure for these reasons. Due to its exceptional sensitivity and specificity, MPI can effectively capture real-time images of the *in vivo* distribution of FeNPs. The technique was initially proposed by Gleich and Weizenecker (2005), and its significant value for medical applications was subsequently demonstrated (63).

The various applications highlight the significance of the magnetic properties of FeNPs in improving diagnostic and therapeutic capabilities, thereby significantly contributing to the advancement of nanomedicine.

### 5.4.2 Functionalization and Applications

Functionalizing FeNPs for biocompatibility is a crucial strategy for their safe and efficient use in various biomedical applications. This involves strategies to enhance their compatibility with biological systems.

Coating the surface is one of the vital approaches to reducing FeNPs' toxicity and enhancing stability. Commonly used materials are PEG, with PEGylation involving the attachment of PEG chains to the surface of the nanoparticle to give it solubility and reduce immunogenicity, thus prohibiting elimination by the reticuloendothelial system (64). Dendrimers, on their part, are branched polymers used for functionalization with numerous ligands that have better biocompatibility and targeting efficiency, providing controlled surface functionality with precision (65). An alternative way is the silica shell, which makes them chemically inert and biocompatible; this can also be further functionalized with biomolecules for targeted delivery (66).

The original surface ligands on FeNPs are exchanged for biocompatible ones to have higher stability in biological environments and lower potential toxicity. Due to their property of forming a stable bilayer similar to the cell membrane, phospholipids decrease immunogenicity (67). The biopolymers chitosan and alginate coat FeNPs to make the interface biocompatible and biodegradable, thereby making the nanoparticles more stable in the surrounding environment (68).

Encapsulation surrounds FeNPs with a biocompatible matrix, enhancing biocompatibility and enabling controlled therapeutic agent release. Liposomes are vesicles capable of entrapping FeNPs to offer a stable and biocompatible environment. They can further be functionalized with targeting ligands to achieve site-specific drug delivery (69). Polymeric micelles, formed from amphiphilic block copolymers, encapsulate the FeNPs and improve solubility and stability in biological fluids (70).

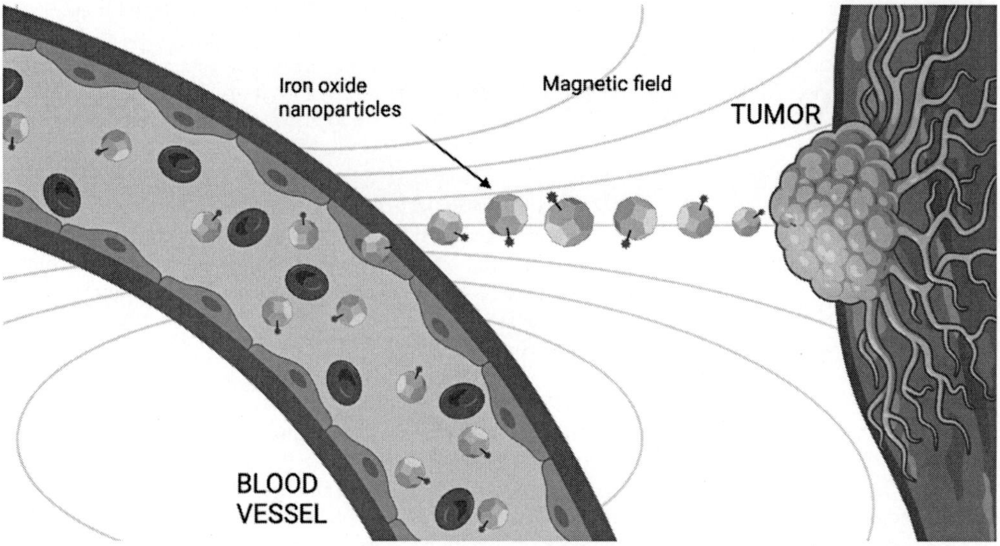


**Figure 5.2** Magnetic targeting of iron oxide nanoparticles for tumor treatment. Innovative use of iron oxide nanoparticles in targeted cancer therapy. An external magnetic field guides these nanoparticles through a blood vessel, directing them towards a tumor site. As the nanoparticles flow alongside red blood cells, the magnetic force concentrates them on the tumor's location. This approach demonstrates the potential of magnetically guided drug delivery to enhance treatment efficacy by increasing local drug concentrations at the tumor while minimizing systemic exposure, thus potentially reducing side effects in cancer therapy.

These functionalization strategies are very important for the successful application of FeNPs in the biomedical arena, ensuring their safe and effective use in the clinical setting.

The characteristic magnetic properties of FeNPs find utility in targeted drug delivery and hyperthermia treatments. In a magnetic targeting system, an external magnetic field is applied to target FeNPs toward sites of interest in the body to increase the precision of drug delivery and lower systemic side effects. This can ensure high local drug concentrations at the target site, leading to huge improvements in treatment outcomes and consequent side effect minimization. For instance, targeted delivery of chemotherapeutics using FeNPs has been shown to yield promising results in reducing tumor sizes with minimal side effects (15). FeNPs can also enhance MRI contrast for better imaging of tissues to monitor targeted delivery processes (71) (Figure 5.2).

In hyperthermia applications, FeNPs generate localized heat under the action of an alternating magnetic field used in the selective ablation of cancer cells. This method can enhance the effectiveness of conventional therapies such as chemotherapy and radiotherapy. The heat generated by FeNPs increases the permeability of tumor cells to chemotherapeutic agents, improving drug uptake and treatment effectiveness (72). Clinical trials have demonstrated the potential of magnetic hyperthermia as an adjunct therapy for treating some types of cancer, highlighting its significant advantages in cancer treatment (73).

Generally, with these advanced functionalization and application techniques, it is evident that FeNPs are potential game changers regarding the treatment of cancer, making huge strides in nanomedicine.

### 5.4.3 Challenges and Opportunities

FeNPs have increasingly been recognized for their potential in various biomedical applications. However, challenges pertaining to their biodegradability and clearance from biological systems act as great obstacles to the safe use and efficiency of these FeNPs.

Most of the time, FeNPs are chemically stable, thus making them persist for a long time inside biological systems. The inert nature of some coatings applied to increase the performance of the coating can slow down the natural degradation processes and cause the possibility of environmental accumulation and ecological risks (7). The nanoparticles can accumulate in the water systems and in the soil, where they may impact different organisms, disrupting ecosystems (74). Moreover, the non-biodegradable FeNPs can accumulate in critical organs like the liver, spleen, and lungs, which may cause long-term toxicity. This accumulation may lead to oxidative stress, inflammation, and cellular damage (75).

The clearance of FeNPs also poses a significant problem. They are quickly taken up by reticuloendothelial system (RES), especially in the liver and spleen, creating a potential for toxicity in these organs (76). Moreover, their size often causes the FeNPs to avoid renal clearance, which has been efficient only for particles that are less than 5–6 nm, therefore increasing the circulation time in the blood and the potential for toxicity (77). Even desirable surface modifications, such as PEGylation for increased biocompatibility and longer circulation times, may impede the clearance processes, thus increasing their bioaccumulation (78).

In view of these limitations, several strategies have been designed, including developing biodegradable coatings, which break down into non-toxic products, toward enhanced clearance from the body. Materials such as dextran, albumin, and other biodegradable polymers are being explored to make this process easier (79). Similar considerations would include better engineering of size and shape for FeNPs to enhance renal clearance; for example, rod-shaped particles often exhibit higher clearance than sphere-shaped particles with the same volume and surface area characteristics (64). Surface functionalization with ligands that enhance clearances through specific cellular pathways would also be an improvement. For example, macrophage-targeted ligands may be more successful in enhancing cellular internalization of FeNPs, followed by lysosomal degradation via phagocytosis (80).

Magnetic targeting technology utilizes the magnetic properties of FeNPs to direct them to specific sites within the body to improve treatment specificity and delivery. Some disadvantages exist, however: The efficiency of magnetic targeting depends significantly on the strength and gradient of the external magnetic field. The use of high-field strengths to guide FeNPs into deep tissue targets can, however, cause technical difficulties and safety issues as these would exert sufficient force on the particles. Furthermore, FeNPs could aggregate in the presence of a magnetic field, leading to less efficient targeting and potentially causing blockages in the vascular system (81). On the other hand, for the same reason, there is an urgent need to ensure that FeNPs are retained specifically at the target site and cause minimal effects on surrounding tissues (82).

Recent technology in magnetic targeting will try to improve such a drawback: gradient magnetic fields have provided the opportunity for enhanced precision of the targeting with FeNPs via improvements in the spatial dispersion of the field (83). New nanoparticle coatings, which would respond predictably to magnetic fields without causing the particles to aggregate, would improve target accuracy. For instance, polymers that become thermosensitive and therefore change their properties with temperature changes caused by a magnetic field have been shown to hold much promise in a number of recent studies (84). In addition, with aggregation inhibition and improvements in targeting accuracy, the method can reduce off-target effects and obtain high specificity during its integration with other methods like ligand-based targeting. Such multimodal approaches combine the advantages of magnetic and biological targeting mechanisms to improve the specificity of targeting (85). The size and shape of the nanoparticles must be well-designed for effective responsiveness to magnetism and subsequent biodistribution. According to these studies, rod-shaped nanoparticles have been found to align much better with the field orientation compared to spherical nanoparticles and hence effectively allow for better targeting (86).

In summary, challenges of biodegradability and clearance are of importance for the safe and effective use of FeNPs in any biomedical application. Continued innovation in magnetic targeting technologies, including magnetic field design, strategies for functionalizing nanoparticles, and multimodal targeting, holds great promise for overcoming present limitations and increasing the clinical applicability of these advanced materials.

## 5.5 SILICA NANOPARTICLES

### 5.5.1 Synthesis and Characterization

SiNPs have evoked great interest in nanotechnology because they are versatile and have a broad application spectrum, mainly in drug delivery systems. The basic issues with using SiNPs are size and porosity; these really vary in terms of their performance in different applications. Therefore, a lot of care is taken in synthesizing these with exact control over structural parameters. Sol–gel is one of the techniques most frequently used in the synthesis of SiNPs. It involves the hydrolysis and condensation of silicon alkoxides, such as tetraethyl orthosilicate (TEOS). By controlling factors such as pH, temperature, and reactant concentration, researchers can finely modulate the size and porosity of the nanoparticles, as given by the seminal work by Stöber, Fink, and Bohn in 1968 (87). The other method is the microemulsion method, where SiNPs are synthesized in micelle confines; surfactant concentration also dictates the size of the micelles and, hence, the size of the nanoparticles produced. This method is known to produce SiNPs that are uniform and monodisperse (88). Organic or inorganic materials are used in template-assisted methods of shaping SiNPs. Most commonly, surfactants like CTAB are used for synthesizing the corresponding MSNs as templates. The template concentration and type may further be adjusted to fine-tune nanoparticle porosity and size (89).

The burning of silicon precursors in a flame during flame synthesis provides a scalable approach to SiNP production, where the flame temperature, precursor concentration, and residence time are just some of the many critical parameters for nanoparticle size control (90).

A few factors affecting the size and porosity of SiNPs are briefly outlined here. Higher concentrations of silicon precursors usually result in the formation of larger nanoparticles because of the increased nucleation and growth rates. The pH and temperature of the reaction mixture can also affect hydrolysis and condensation in the sol-gel process. Besides, the concentration of the surfactant used in the microemulsion and template-assisted methods is very important for the size and porosity of SiNPs. Modification of the surface of SiNPs must provide drug loading and release sites. Organic molecules with amine groups, thiols, or carboxyl groups increase the drug-loading capability of the nanoparticles. Aminated SiNPs create electrostatic interactions with negatively charged drug molecules, increasing loading efficiencies (19). Coating with biocompatible polymers, such as PEG, chitosan, or poly(lactic-co-glycolic acid) (PLGA), not only maximizes the potential for drug loading but can also adjust drug release rates. The hydrophilic environments created by the polymer coatings help stabilize hydrophobic drugs, which would otherwise undergo aggregation. By incorporating a slowly degrading matrix into the coat, drug release is prolonged (91). Inorganic coatings, for instance, calcium phosphate or gold, may also add drug-loading enhancements and serve as controlled release mechanisms through the gradual dissolution or disintegration of the coating in many forms (8). Increasing the number of available binding sites on the SiNP in surface modification causes an increase in the drug-loading capacity. These can also be designed to respond to specific environmental stimuli, such as pH and temperature, for controlled release. For instance, acidic cleavage of polymer-coated SiNPs allows their degradation within tumor tissue, providing targeted drug release (92). In addition, functionalization of SiNPs with biocompatible materials decreases toxicity and increases biocompatibility, hence improving their blood circulation time (93). By coating SiNPs with ligands like antibodies or peptides, it is possible to further improve targeted delivery and reduce unintended side effects (94). This necessitates careful SiNP synthesis and rational surface modification to be carried out to realize the potential of SiNPs in drug delivery. In this way, not only do these approaches refine the structural features of the nanoparticles, but they also improve their drug loading, release properties, and overall biocompatibility. The same confirms that SiNPs are a very versatile platform for advanced therapeutic applications.

### 5.5.2 Biomedical Applications

SiNPs have gained significant interest in nanotechnology for their potential in drug delivery systems due to their unique properties, such as their large surface area, tunable pore size, and ease of surface functionalization. These characteristics render SiNPs ideal for applications in sustained and controlled drug release. SiNPs feature a mesoporous structure that is conducive to the encapsulation of a wide variety of drugs within their pores. By adjusting the pore size, researchers can effectively load different drug molecules and control their release rate. Tailoring the pore diameter enables regulation of the drug's diffusion rate, providing sustained release over prolonged periods. This ability to modulate release kinetics is crucial in applications like chronic disease management, where maintaining consistent drug levels is essential (19).

Additionally, surface functionalization of SiNPs allows for further control over drug release. Coating SiNPs with PEG creates a hydrophilic barrier that moderates the release of hydrophobic drugs, yielding a controlled release profile. Moreover, employing pH-responsive or temperature-sensitive polymers can achieve stimulus-triggered drug release, activated by specific physiological conditions (8). Another approach involves the encapsulation of drugs within SiNPs or the creation of core-shell structures where the core contains the drug and the shell controls its release. This method provides an added layer of control, safeguarding the drug from premature release and ensuring its delivery at the intended site (95).

In cancer therapy, SiNPs are utilized to deliver chemotherapeutic agents in a controlled manner. Doxorubicin-loaded SiNPs, for example, provide sustained release of the drug, maintaining therapeutic levels over an extended time frame, thereby reducing dosing frequency and minimizing side effects. This approach enhances treatment efficacy and patient compliance (3). SiNPs are also employed in the controlled release of antibiotics. Encapsulating antibiotics within SiNPs achieves prolonged release profiles, helping maintain effective drug concentrations at infection sites and reducing the risk of resistance development (96).

SiNPs are highly valued in diagnostics and imaging due to their unique optical properties, high surface area, and capability to be functionalized with various biomolecules, making them efficient contrast agents and probes for different imaging techniques. In fluorescence imaging, SiNPs can be doped with fluorescent dyes or QDs, enhancing fluorescence signals, while the silica matrix protects the dyes from photobleaching. This property is particularly useful for stable and long-lasting imaging applications. For instance, SiNPs doped with rhodamine B have been effectively used for *in vivo* imaging of tumors, offering detailed visualization of cancerous tissues (97).

For MRI, FeNPs coated with silica shells serve as contrast agents, enhancing soft tissue visualization and detecting abnormalities with improved clarity. The silica coating's biocompatibility ensures safe *in vivo* usage (98). In computed tomography (CT) imaging, modifying SiNPs with elements such as gold or bismuth enhances X-ray attenuation, providing superior contrast compared to traditional agents and allowing detailed visualization of vascular structures and tumors (99).

SiNPs also excel in molecular diagnostics. They can be functionalized with antibodies, enzymes, or nucleic acids to create sensitive and specific biosensors for detecting various analytes, including proteins, DNA, and small molecules. These capabilities make SiNPs valuable for disease diagnosis and monitoring. For example, SiNP-based biosensors developed for cancer biomarker detection offer rapid and accurate diagnostics (100). Furthermore, the functionalization flexibility of SiNPs facilitates the development of multiplexed assays, enabling simultaneous detection of multiple biomarkers or pathogens in a single test. This high-throughput capability is invaluable in clinical diagnostics and research, thereby enhancing the efficiency of medical testing (101).

In summary, SiNPs play a pivotal role in controlled drug release, diagnostics, and imaging, offering enhanced drug delivery, contrast, and diagnostic accuracy across a range of medical applications. Their ability to be meticulously tailored and functionalized underscores their versatility and immense potential for advancing healthcare technologies.

### 5.5.3 Challenges and Regulatory Aspects

SiNPs are being increasingly used in biomedical applications due to their unique properties; however, their long-term biocompatibility and toxicity are critical concerns that need to be thoroughly addressed to ensure their safe application.

The potential for chronic toxicity arises when SiNPs accumulate in organs like the liver, spleen, and lungs, leading to adverse effects such as inflammation, fibrosis, and oxidative stress, potentially causing organ damage and dysfunction. This accumulation highlights the need for careful consideration of SiNPs' long-term presence in the body (102). SiNPs can elicit an immune response, with the potential to cause immunotoxicity. The immune system may recognize SiNPs as foreign particles, initiating an inflammatory response that could lead to tissue damage and possibly the onset of autoimmune diseases (103). Studies have raised concerns about SiNPs' ability to cause DNA damage, which could lead to mutations and potentially contribute to carcinogenesis. The potential for SiNPs to induce oxidative DNA damage and interfere with the cell cycle underscores the importance of genotoxicity assessments in their evaluation (104).

To improve SiNPs' safety profile, surface modification techniques are employed. Coating SiNPs with biocompatible polymers such as PEG can minimize protein adsorption and immune recognition, thereby enhancing circulation time and reducing the risk of an adverse immune response (105). Developing biodegradable SiNPs that can be safely metabolized and excreted from the body could mitigate long-term toxicity risks. Researchers are exploring biodegradable materials for SiNPs that degrade into non-toxic byproducts to enhance safety (96). Thorough preclinical testing, including long-term *in vivo* studies, is crucial for assessing the safety and biocompatibility of SiNPs. Such studies should comprehensively evaluate the potential for chronic toxicity, immunotoxicity, and genotoxicity to ensure SiNPs' suitability for human use (106).

Standardizing the manufacturing processes of SiNPs is crucial for their reproducibility, scalability, and regulatory approval. Batch-to-batch variability, scalability challenges, and establishing robust quality control measures are major hurdles in the standardized production of SiNPs. Batch variability often results from differences in synthesis parameters such as reactant concentrations, temperature, and reaction time, which can alter SiNPs' properties like size, shape, and surface characteristics. Implementing Good Manufacturing Practices (GMP) can help minimize such variability by ensuring consistent, controlled production processes (107). Scaling up SiNPs' production while maintaining quality is difficult, as laboratory synthesis methods may not translate directly to industrial scales. Techniques such as flame synthesis and continuous flow reactors are promising for large-scale production but require further optimization to ensure consistent nanoparticle quality across different production scales (108).

Standardized characterization techniques are essential for the consistent quality assessment of SiNPs. Techniques like dynamic light scattering (DLS), transmission electron microscopy (TEM), and Brunauer–Emmett–Teller (BET) analysis should be standardized to provide reliable data on nanoparticle properties (109). Furthermore, establishing clear regulatory frameworks is critical for ensuring the safe production and application of SiNPs. Regulatory guidelines are evolving, with agencies like the FDA and EMA developing robust standards that address the unique challenges posed by nanomaterials, including SiNPs (110).

In summary, addressing the challenges of long-term biocompatibility, toxicity, and standardization in the manufacturing of SiNPs is crucial for leveraging their full potential in biomedical applications. Through strategic modifications, comprehensive testing, and adherence to regulatory standards, SiNPs can be safely integrated into future medical technologies.

## 5.6 QUANTUM DOTS

### 5.6.1 Optical Properties and Synthesis

QDs are nanocrystals made of semiconductors that possess distinctive optical characteristics, specifically photoluminescence. These attributes make QDs extremely beneficial for biological imaging purposes. The photoluminescent characteristics of QDs arise from quantum confinement processes, wherein the size of the material determines its electrical properties. QDs, when stimulated by a light source, produce light at distinct wavelengths depending on their dimensions, composition, and surface properties. The ability of QDs to adjust their emission wavelength allows them to produce a wide range of colors, spanning from ultraviolet to near-infrared. This versatility makes QDs valuable for many imaging applications in biomedicine (111, 112).

The exceptional brightness and photostability of QDs enhance the field of biomedical imaging. Contrary to conventional organic dyes, QDs have a lower tendency to lose their color when exposed to light, which enables longer imaging sessions and repeated exposure without a noticeable decrease in signal strength. This characteristic is especially advantageous in the context of live-cell imaging and long-term tracking research. Moreover, the capacity to combine QDs with biomolecules like antibodies and peptides enables precise visualization of particular cellular structures and processes (113, 114).

The production of QDs entails techniques that meticulously regulate their dimensions, morphology, and surface composition to attain the intended optical characteristics. The most prevalent technique for producing QDs is colloidal synthesis, which involves breaking down precursor materials in the presence of surfactants that restrict particle growth. This process provides monodisperse QDs with high quantum yields. Additional techniques encompass molecular beam epitaxy and chemical vapor deposition, which provide meticulous manipulation of QD composition and structure, albeit at a higher level of intricacy and expense (115).

The size of QDs is inherently connected to their optical properties as a result of the quantum confinement effect (116). As the QD size decreases, the difference in energy between the valence and conduction bands grows, causing the emission wavelength to shift toward the blue end of the spectrum (117). The ability to adjust the properties of QDs enables the creation of QDs with precise emission characteristics that are well-suited for a range of applications, including bioimaging and photovoltaic devices (118).

Size-dependent features play a vital role in enhancing the performance of QDs in various environments. Smaller QDs generally exhibit greater surface-to-volume ratios, which in turn affect their chemical reactivity and stability (119). Surface modifications, such as ligand attachment, are frequently used to enhance solubility, biocompatibility, and functionality (120). These alterations not only provide stability to the QDs but also enable precise adjustment of their optical characteristics to fulfill certain demands.

Several synthesis techniques have been devised to create QDs with accurate control over their size. Hot-injection techniques, such as the fast injection of precursors into a heated surfactant solution, facilitate the controlled formation and development of QDs (121). Solvothermal and hydrothermal procedures provide alternative techniques, frequently yielding high-quality QDs with consistent size distributions (122). In addition, continuous-flow microreactors have become a potential method for producing QDs on a large scale. They offer improved control over reaction conditions and enhance repeatability (123).

### 5.6.2 Biomedical Applications and Functionalization

In the biomedical field, QDs have emerged as a versatile instrument, particularly for the delivery of targeted medications (124). Bioconjugation with biological molecules, including nucleic acids, antibodies, and peptides, frequently leads to the functionalization of QDs, thereby increasing their utility (125). The covalent attachment of biomolecules is facilitated by the modifiable surface chemistry of QDs, which can be amended to include functional groups such as carboxyl, amine, and thiol groups (126).

Peptide-functionalized QDs are an exceptional choice for precision drug delivery due to their exceptional specificity in targeting cell receptors. For instance, RGD peptides can be conjugated with QDs to target integrin receptors that are overexpressed in tumor cells. This conjugation ensures that malignant tissues are precisely targeted, thereby significantly reducing off-target effects. In the same vein, antibodies, which are renowned for their exceptional affinity and specificity for antigens, are an additional popular option for QD functionalization. Drug delivery and imaging can be facilitated by antibody-conjugated QDs, which can precisely target specific proteins or cells (127, 128).

Additionally, QDs can be attached to nucleic acids, including DNA and RNA, for the purpose of biosensing and gene therapy. DNA-functionalized QDs have been implemented to monitor and visualize genetic material within cells, thereby providing valuable insights into molecular cell processes. Additionally, these bioconjugated QDs have the capacity to facilitate the delivery of therapeutic nucleic acids to specific cells, thereby offering potential remedies for genetic disorders (129).

The biomedical imaging, sensing, and therapy fields have been transformed by the superior optical properties of QDs, which include high luminosity, photostability, and size-tunable emission spectra. Both *in vitro* and *in vivo* imaging applications are implemented with QDs. Due to their brilliant and stable fluorescence, they are the most suitable for the labeling and tracing of biomolecules, tissues, and cells over extended periods. QDs have been effectively implemented to visualize and monitor the behavior of cancer cells in real time (130).

In the discipline of biosensing, fluorescent instruments, such as QDs, are capable of detecting a wide range of biological molecules with high sensitivity and specificity. When conjugated with molecules that recognize specific analytes, QDs are capable of detecting biomarkers associated with a variety of diseases. This is crucial for the early diagnosis of the disease (131).

Drug delivery and photodynamic therapy are currently being investigated as therapeutic applications for QDs. Their ability to be functionalized with therapeutic agents and target-specific molecules facilitates the precise delivery of pharmaceuticals to diseased cells, thereby reducing systemic toxicity (132). Upon light activation in photodynamic therapy, QDs have the capacity to directly eradicate cancer cells by generating reactive oxygen species (133). This method offers a treatment option that is less invasive and has fewer adverse effects than conventional therapies.

In addition, the multifunctionality of QDs enables theranostics, a therapeutic and diagnostic approach that integrates their use (134). QDs are capable of administering therapeutic agents and diagnosing and monitoring the progression of diseases, thereby offering a comprehensive solution for disease management (135).

### 5.6.3 Challenges in Clinical Use

There are substantial obstacles to the clinical application of QDs as a result of environmental concerns and uncertainty regarding their toxicity (136). Numerous QDs are contaminated with heavy metals, including cadmium, lead, and selenium, which present toxicity hazards to living organisms. These metals have the potential to leach out of the QDs under physiological conditions, which could result in cytotoxicity and genotoxicity. For example, the release of cadmium ions from CdSe QDs has been demonstrated to result in oxidative stress, inflammation, and cell mortality, which pose significant risks for *in vivo* applications (137). In addition to their direct cytotoxicity, there are significant apprehensions regarding the long-term environmental consequences of QDs. The potential for bioaccumulation and the persistence of these substances in the environment raises concerns regarding their ecological toxicity. QDs have the potential to infiltrate the food chain and affect a variety of organisms by contaminating water sources and soil once they are released (138). Consequently, there is an urgent requirement to create safer QDs that are either more biodegradable or reduced in toxicity.

To reduce the toxicity of QDs, researchers have investigated a variety of strategies, such as the use of core–shell structures. In these structures, a less toxic shell material, such as zinc sulfide (ZnS), encases the toxic core, thereby reducing the discharge of harmful ions. Surface modifications with biocompatible polymers and ligands can also improve the biocompatibility of QDs. For instance, the interaction between biological tissues and QDs is diminished by encapsulating them with PEG, thereby reducing their toxicity.

The clinical efficacy of QDs is significantly influenced by their stability and aggregation in biological environments. QDs are susceptible to aggregation as a result of interactions with proteins and other biomolecules under physiological conditions, which can substantially impact their optical properties and biodistribution (12). Aggregation can result in the rapid clearance of QDs from the body or their unintended accumulation in non-target tissues, thereby compromising their efficacy and safety, as well as a decrease in their fluorescence intensity and stability (4).

There are numerous methods for improving the stability and averting the aggregation of QDs in biological environments. Surface functionalization with hydrophilic and biocompatible molecules, such as PEG, is an effective strategy that creates a steric barrier to prevent aggregation. The colloidal stability of QDs is preserved in physiological conditions through PEGylation, which guarantees consistent performance in therapeutic and imaging applications (139). Another effective approach to safeguarding QDs from the biological milieu is to encapsulate them in biocompatible matrices, or liposomes. These encapsulation techniques establish a physical barrier that protects QDs from aggregation and degradation while at the same time maintaining their optical properties (9). Furthermore, the durability of QDs in biological fluids can be improved by utilizing amphiphilic polymers to establish micelle-like structures around them (140). The integration of QDs with biological systems is enhanced using biomimetic coatings that mimic the cell membrane, which is a promising approach (141). These coatings not only enhance the biocompatibility and targeting efficacy of QDs but also improve their stability, rendering them more suitable for clinical applications.

## 5.7 COMPARATIVE ANALYSIS OF INORGANIC NANOCARRIERS

### 5.7.1 Comparison of Physicochemical Properties

Inorganic nanocarriers' efficacy in the delivery of nucleic acids and proteins: a comparative analysis of their physicochemical properties. Inorganic nanoparticles, such as AuNPs, IONPs, SiNPs, and QDs, possess unique physicochemical properties that significantly influence their effectiveness in the delivery of nucleic acids and proteins. AuNPs are biocompatible, readily functional, and display unique plasmonic features through which effective conjugation with biomolecules like DNA, RNA, and proteins can be done. Their size and shape can be strictly controlled during synthesis to achieve improved cellular internalization and stability (142). IONPs are superparamagnetic and, hence, are used for MRI studies and magnetic targeting in drug delivery. Their surface could be derivatized with polymers or ligands to increase biocompatibility and reduce aggregation. The high potency of IONPs in gene delivery arises from their ability to form complexes with nucleic acids and to assist endosomal escape (143). SiNPs have a high surface area for easy functionalization and a porous structure. They offer a versatile platform for the loading and release of biomolecules for both hydrophilic and hydrophobic substance encapsulation. SiNPs can be engineered to degrade under certain conditions, making it possible to release therapeutic agents in a controlled manner (144). QDs possess unique optical properties like size-tunable emission and high photostability; hence, they have a wide application window in bioimaging and biosensing (145). However, a major drawback that remains with nanomaterials is their potential toxicity due to the release of heavy metals. Surface modifications, such as coating with biocompatible polymers, would be imperative for overcoming these drawbacks and rendering them safe for use in biomedical applications. The differences among these physicochemical properties are one of the key guiding factors toward the suitability of each nanoparticle type for specific biomedical applications. AuNPs are preferable for use because they are biocompatible, easy to functionalize, and have the best suitability for gene and protein delivery. IONPs are useful in applications such as magnetic targeting, imaging, and the delivery of therapeutics. SiNPs have largely passed that stage and are now used with excellent controlled release properties. QDs are mainly used in imaging and sensing, because of their optical properties.

Controlled release and stability are critical parameters for inorganic nanoparticles to be effectively applied in a biomedical setting. In comparison to other types of inorganic nanocarriers, the SiNPs show good stability and controlled release profiles. Because of their porous structure, hydrogels are able to easily encapsulate a variety of therapeutic agents, and surface modifications and environmental triggers like pH and temperature changes can control how quickly these agents release (146). AuNPs are known to be very stable because they are practically inert and do not oxidize easily. Functionalization with thiol-containing ligands can further increase the stability of AuNPs in biological environments. Responsive linkers can be designed to release their therapeutic agents at a site of interest based on specific triggers, such as changes in redox potential or the presence of enzymes (147). IONPs are stable under physiological conditions and can be functionalized for improvement of biocompatibility, such as decreasing aggregation; however, their release characteristics are less controlled compared to SiNPs and AuNPs. On the other hand, together with surface modifications, polymer coatings are also exploited to achieve not only stability but also the controlled release of conjugated biomolecules (148). QDs have problems related to their stability and potential toxicity. QDs can be modified using a wide range of biocompatible materials, such as PEG, to offer improved stability and reduce aggregation. QDs do not serve practical purposes for the controlled release of therapeutic agents; however, their most important application in imaging and sensing benefits from the stability provided by such surface modifications (149). In general, silica NPs embody the most synergistic combination of stability and controlled release features—making them the best candidate for most biomedical applications. Added to this is the huge degree of stability and freedom in their functionalization with AuNPs, whereas IONPs find a preference in applications with special magnetic properties. Following this are QDs, which, despite possibly having some toxicity, are still useful for imaging and sensing due to their distinctive optical properties (Figure 5.3).

### 5.7.2 Biocompatibility and Safety Profiles

The safety and biocompatibility of inorganic nanocarriers, like AuNPs, IONPs, SiNPs, and QDs, depend significantly on the type of composition, size, surface chemistry, and interaction with biological systems. The high biocompatibility of AuNPs derives primarily from their inert nature and low toxicity.

They are immensely stable and can be easily functionalized with biocompatible coatings, like PEG, to further increase their biocompatibility and reduce immune responses (150). IONPs, on the other hand, have moderate biocompatibility, mainly due to the surface modifications. Bare IONPs cause oxidative stress and

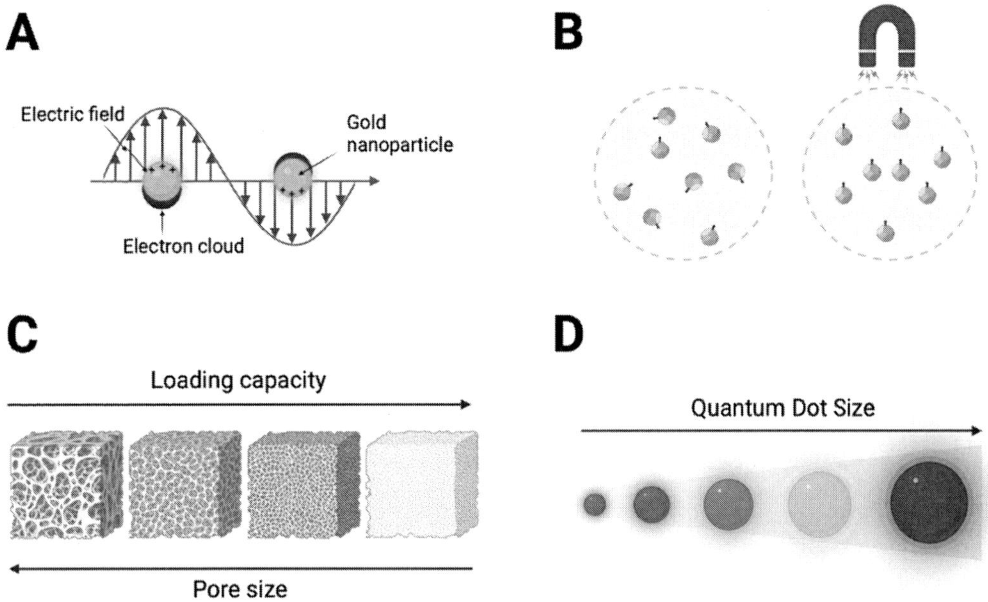


**Figure 5.3** Key properties of inorganic nanocarriers. (A) Gold nanoparticles (AuNPs): They exhibit the surface plasmon resonance effect, showing the oscillation of the electron cloud in response to an electric field, which is key to their optical properties. (B) Iron oxide nanoparticles (IONPs): They respond to an external magnetic field, with nanoparticles aligning along the field lines, which is crucial for magnetic targeting and imaging applications. (C) Silica nanoparticles (SiNPs): They have a relationship between pore size and loading capacity. As pore size decreases from left to right, the available space for cargo loading increases, highlighting their potential for controlled drug delivery. (D) Quantum dots (QDs): They demonstrate size-dependent optical properties, where smaller quantum dots emit higher-energy (bluer) light and larger ones emit lower-energy (redder) light, essential for their use in imaging and sensing applications.

inflammation, but after coating them with some biocompatible polymers—like dextran or PEG—this safety profile improves significantly. Coating prevents aggregation and reduces interactions with proteins and cells, all of which make IONPs suitable for MRI and drug delivery purposes (151). SiNPs are biocompatible, especially if their surfaces are modified with agents that confer improved biocompatibility. Porous structures provide effective drug loading and controlled release, while surface functionalization with amino or carboxyl groups enhances its interaction with biological systems. However, there are doubts about some of the potential adverse effects of SiNPs, particularly in terms of long-term exposure, such as pro-inflammatory and pro-oxidative effects (152). QDs have shown distinct optical properties while, at the same time, causing significant biocompatibility challenges due to potential toxicity. Often containing heavy metals like cadmium and selenium, these metals can leach out of QDs and damage cellular machinery. Surface coatings with biocompatible materials like PEG or silica would reduce these, but long-term safety and the possibility of bioaccumulation remain a concern (153).

Each type of inorganic nanocarrier has specific toxicity concerns that must be addressed to ensure safe biomedical applications. Among the primary concerns with gold nanocarriers are immune response and cytotoxicity at higher concentrations. Mitigation strategies involve biocompatible polymers, such as PEG, which reduce protein adsorption and recognition by the immune system (154). On the other hand, IONPs cause oxidative stress and inflammatory responses because of their iron content, which catalyzes the formation of ROS. Coating IONPs with biocompatible polymers such as dextran, PEG, or silica tends to moderate this effect through a barrier that avoids direct interaction with biological tissues and reduces the release of iron ions (155). More generally, $SiO_2$ nanoparticles are considered to be biocompatible, but under high doses and prolonged exposure, they can trigger inflammation and oxidative stress. Surface functionalization with amino or carboxyl groups, or the use of mesoporous structures, decreases these adverse effects. Finally, the biodegradability of silica materials may be varied to minimize long-term toxicity (156). Highly toxic QDs containing heavy metals are considered serious as the leaching of cadmium or selenium ions from QDs causes cellular damage and severe toxic effects. Surface modifications with thick silica shells or crosslinked polymers produce a robust coating, which can reduce the release of toxic ions and provide better biocompatibility. However, long-term studies are needed to fully understand QDs' environmental and health impacts (157).

In summary, although individual inorganic nanoparticles present unique advantages and challenges related to biocompatibility and safety, adequate surface modification associated with attention to their physicochemical properties can ease toxicity concerns and support their safe and effective use in biomedical applications.

### 5.7.3 Technological and Economic Considerations

One of the many important technological obstacles to increasing the production of inorganic nanocarriers, such as AuNPs, IONPs, SiNPs, and QDs, is ensuring that the size, shape, and surface properties of the particles remain consistent during large-scale syntheses. Nucleation and growth processes to synthesize AuNPs need to be very precisely controlled for uniformity, which dramatically becomes difficult in the industry on a large scale (158). For IONPs, ensuring high magnetic purity and stability while preventing oxidation and aggregation during the scale-up process is complex. Usually, the synthesis process goes through two steps: coprecipitation and thermal decomposition. Good control over these steps is required for the reproducibility of the process. In addition, surface coating with improved biocompatibility and functionalization increases the complexity of applying the coatings uniformly over large batches (159). Given the challenges posed by the pore size and functionalization of porous SiNP, it is crucial to maintain precise control over parameters such as pH, temperature, and reactant concentration during the synthesis process. This is necessary to ensure the reproducibility of both pore size and surface chemistry. These concerns, which are also mentioned in the literature, pertain to maintaining consistent results between batches and preventing clumping during storage and handling (160). Because QDs are composite, containing potentially toxic elements, careful control of the synthesis conditions is necessary to prevent the release of heavy metals. High-temperature processes, like organometallic synthesis, have to be carefully scaled to maintain the optical properties and stability of the QDs. Such surface coatings are also a significant challenge in providing safety and efficacy, free from toxicity and functional interference (161). These general challenges in scaling up nanoparticle production processes were extensively reviewed in 2023 by Fernandes et al. (162).

Economic considerations play a crucial role in the selection of inorganic nanocarriers for commercialization, with a specific focus on drug delivery. The cost of raw materials, synthesis, functionalization, and scalability are key to these nanocarriers' commercial feasibility. This is because AuNPs, although highly biocompatible and easily functionalized, are very expensive due to the expensive nature of the element itself. These result in the material's limited applicability, except when the benefits somehow surpass the cost in most commercial applications. The IONPs are relatively cheaper in terms of raw materials as iron is abundant and way cheaper than gold. On the other hand, high costs accrue from the synthesis process and the high level of surface modification needed for better biocompatibility and functionality. It is cardinal that the commercial values of the nanoparticles be balanced with cost, performance, and safety (163). On the other hand, SiNPs are relatively low-cost materials that offer great flexibility in both functionalization and controlled release properties. SiNPs are economically attractive for drug delivery applications, as compared to other materials, because silica is cheap. SiNPs have the benefit of simplicity and cost-effectiveness in the preparation of materials through the sol–gel synthesis methods. However, the cost of functionalization for ensuring batch consistency may just add to these expenses (164). Although there are unique optical properties of the QDs, they are, in most cases, prohibitively expensive due to their synthesis at high temperatures using complex procedures. They also need extended purification and surface modifications to prevent toxicity. Besides, the high production cost and stringent safety testing of the synthesis make QDs uneconomic in drug delivery (165). Therefore, in making this choice of inorganic nanocarriers for commercial drug delivery applications, the technological challenges as well as economic factors should be critically studied. High-performance AuNPs are expensive; IONPs and SiNPs provide economical solutions with moderate performance. Technologically advanced QDs pose a major economic problem in their commercial use. In the end, it all depends on specific application requirements, such as cost, performance, and the balance between those and safety.

## 5.8 FUTURE PERSPECTIVES AND EMERGING TRENDS

### 5.8.1 Advances in Material Science and Engineering

Recent advances in material science have opened up several new frontiers that have the potential to bring about radical changes in the functionality of inorganic nanocarriers. One of the most significant developments in this regard is the ability to engineer multifunctional nanoparticles capable of delivering multiple functionalities at the same time. For example, AuNPs have been designed for use as diagnostic and therapeutic agents in a concept known as theranostics. The theranostic AuNPs will offer both functional imaging and therapeutic capabilities on the same platform, which is otherwise bulky and less efficient with current technologies (166). Also very promising are hybrid nanoparticles that are based on several inorganic materials. These hybrid structures combine the unique properties of each component, resulting in a synergistic effect that enhances overall performance. As an example, the silica-coated IONPs use the magnetic properties of iron and the biocompatibility and functionalization flexibility of silica to make them highly effective for drug targeting and MRI applications (167).

A further important development is associated with the use of surface modification techniques in an attempt to improve nanoparticle stability, biocompatibility, and target specificity. Adding PEG to the surfaces of nanoparticles has been shown to lower opsonization and increase blood circulation time, which makes drug delivery systems work better (168). New synthesis methods are critical for developing more effective and safer inorganic nanocarriers. These are generally the deficiencies, i.e., the traditional synthesis approaches are not reproducible, burdensome to the environment, and not scalable. However, the recent advances are trying to serve them all at once. Therefore, currently, it is almost on the brink of moving from basic research to practical applications.

One technique is to use green chemistry rules for nanoparticle synthesis. Green synthetic routes are based on biological agents like plant extracts, bacteria, and fungi in the eco-friendly synthesis of nanoparticles. This not only reduces the amount of toxic chemicals but also, in most cases, leads to nanoparticles with enhanced biocompatibility and less toxicity (169). Furthermore, microfluidic synthesis methods have been developed to provide controlled regulation over the size, shape, and composition of the nanoparticle. Microfluidic systems offer the highest level of homogeneity and reproducibility, which are very important in clinical applications. These systems also allow rapid screening of different synthesis conditions, thus accelerating development toward optimized nanocarriers (170).

Incorporation of these novel synthetic approaches can dramatically improve the safety profile of inorganic nanoparticles. For example, control over the size and surface characteristics of QDs makes them less potentially cytotoxic and more amenable to *in vivo* imaging applications. In the same way, magnetic features and coatings of FeNPs can be carefully adapted to mitigate side effects and improve therapeutic efficacy (171). These new synthesis methods in material science bring promising advancements for functionality and safety improvement in inorganic nanocarriers. That is, paramount developments would be realized in the potential of nanotechnology for applications in biomedicine and industry, consequently heralding new opportunities for diagnosis and treatments in medicine and beyond.

### 5.8.2 Integration with Other Nanotechnologies

The synergistic combination of inorganic nanocarriers with other nanotechnologies is likely to become a new paradigm in drug delivery. One such synergy is the production of nanocomposites, which blend nanoparticles with several organic and inorganic materials, providing an exclusive platform for multifunctional systems. As an example, drug loading into AuNPs incorporated with polymeric materials may enhance controlled release and targeting, hence potentiating the overall therapeutic index (172).

In addition, the hybrid systems resulting from combining these inorganic nanoparticles with other nanostructures, such as liposomes or dendrimers, may yield much better results. For example, liposome-coated QDs may combine the benefits of liposomal encapsulation, such as biocompatibility and prolonged circulation time, with the unique optical properties of QDs, thereby enabling simultaneous delivery and imaging. Such systems can be designed to respond to specific triggers, for instance, pH or temperature, in a way that encourages drug release only at sites of interest (173).

Computational modeling and machine learning are part of the transformation to in silico design and optimization of inorganic nanocarriers. Such capability allows the prediction and adjustment of nanoparticle properties toward desired therapeutic effects with a higher degree of precision and efficiency (174). On the other hand, machine learning further personalizes nanomedicine. By incorporating patient-specific information, such as genetic profiles and disease markers, machine learning will predict the behavior of different nanoparticle formulations in every individual patient. This personalized strategy ensures that treatment is in line with the needs of the individual patient, maximizing effectiveness while minimizing side effects (175).

With this, the convergence of inorganic nanocarriers with other forms of nanotechnology and further use of computational modeling and machine learning stand to seriously impact advances in drug delivery. All these novel breakthroughs will likely bring more precision, efficiency, and personalized treatments than ever before, ushering in a new era for nanomedicine.

### 5.8.3 Impact on Healthcare and Regulatory Issues

Advancements in inorganic nanocarriers have the potential to revolutionize several areas, mainly in the diagnosis and targeted therapy of healthcare. AuNPs, for instance, are being explored for the enhancement of imaging techniques and direct drug delivery to cancerous cells. Due to their high surface area, they are able to attach several ligands, enabling multiple druggings and leading to preferred target delivery with minimum side effects. The next is that they possess optical properties, which advance imaging and photothermal treatment, which is a dual application useful both in diagnosis and treatment (176).

Yet another example is FeNPs, specifically when focusing on MRI with respect to cancer hyperthermia treatment, due to their magnetic properties that facilitate better MRI contrast, hence ensuring an early and correct diagnosis. In addition, FeNPs can give out localized heat once subjected to an alternate magnetic field. This selectively kills the cancer cells and spares the surrounding healthy tissue (177).

However, several regulatory challenges have to be overcome before inorganic nanocarriers can be widely implemented in clinical applications. Among these is the comprehensive safety profiling that has to be done. This unpredictability in interactions with biological systems and other potential effects related to the specific properties of nanoparticles requires intensive toxicological testing. Long-term investigations should be carried out to study biodistribution, metabolism, and excretion, as well as any possible chronic effects (178).

Equally important, from a regulatory point of view, is standardization in the manufacturing processes because consistency in both nanoparticle synthesis and functionalization is a prerequisite for the reproducibility and reliability of clinical applications. Regulatory agencies need to establish guidelines for nanoparticle characterization and quality control that encompass the definition of parameters, size, shape, surface charge, and purity of nanoparticles (179).

One of the challenges is the regulatory classification of nanoparticle-based products. Traditional regulatory frameworks are possibly not fully fitted for most nanomedicines that typically reveal hybrid characteristics of drugs, biologics, and medical devices. It could help make clear guidelines and definitions that could enable easier regulatory approval and also assure that such products meet the necessary safety and efficacy standards (180).

On the other hand, developments in inorganic nanocarriers may strongly change healthcare, delivering new opportunities for improved diagnostics, targeted therapy, and precision medicine. Nevertheless, the successful translation of these technologies into clinical practice will depend on addressing the regulatory challenges with regard to their safety, manufacturing, and classification.

## 5.9 CONCLUSION

Inorganic nanocarriers, including AuNPs, FeNPs, SiNPs, and QDs, have shown significant promise in advancing various biomedical applications due to their unique properties. Each type of nanoparticle brings distinct advantages and faces specific challenges in medical treatment delivery. AuNPs are highly valued for their exceptional optical properties, particularly SPR, making them ideal for imaging and diagnostics. Their ability to be functionalized with various ligands and biomolecules enhances biocompatibility and targeting specificity in drug delivery systems. Moreover, AuNPs are effective in photothermal therapy, where light-induced heat is used to destroy cancer cells.

IONPs, known for their superparamagnetic properties, have transformed MRI by improving contrast and diagnostic accuracy. These magnetic properties also enable targeted drug delivery and hyperthermia treatment, where localized heating selectively kills cancer cells. SiNPs offer high surface areas and porous structures that enhance drug delivery mechanisms. Their biocompatibility allows for loading with therapeutic agents and controlled release, increasing therapeutic efficacy in applications like cancer therapy.

QDs are recognized for their size-tunable fluorescence and high photostability, making them valuable in bioimaging and diagnostics. Their bright, long-lasting signals are crucial for extended imaging experiments. However, the toxicity of QDs, primarily due to their heavy metal content, remains a significant concern.

The analysis of AuNPs, FeNPs, SiNPs, and QDs highlights that each nanoparticle type excels in different biomedical applications:

- *AuNPs:* Particularly effective in imaging and therapy, they play a key role in theranostics, which combines diagnosis and treatment on a single platform.
- *FeNPs:* Most suitable for magnetic applications like MRI and magnetically guided targeted drug delivery.
- *SiNPs:* Ideal for drug delivery due to their high surface area, biocompatibility, and controlled release capabilities.
- *QDs:* Excellent for imaging applications owing to their superior brightness and photostability, although their potential toxicity requires careful handling.

These findings are crucial for the broader field of nanomedicine, demonstrating the versatility and specific advantages of different inorganic carriers. The ability to customize nanoparticles facilitates precise site-specific drug delivery, leading to more potent therapeutics with fewer side effects. Additionally, advancements in synthesis and functionalization techniques aim to address existing issues like toxicity and stability, expanding the clinical applicability of these nanomaterials.

Future research should prioritize optimizing the synthesis and functionalization of inorganic nanocarriers to maximize therapeutic efficacy while minimizing toxicity. Developing hybrid systems that combine various nanoparticles could harness their individual strengths to address complex medical conditions more effectively. Furthermore, advancements in computational modeling and machine learning could accelerate the design and optimization of nanoparticles, enhancing their efficacy and safety in clinical applications.

## REFERENCES


1. El-Sayed IH, Huang X, El-Sayed MA. Surface plasmon resonance scattering and absorption of anti-EGFR antibody conjugated gold nanoparticles in cancer diagnostics: Applications in oral cancer. *Nano Letters [Internet]*. 2005 May [cited 2024 Aug 12];5(5):829–34. Available from: https://pubs.acs.org/doi/abs/10.1021/nl050074e

2. Weissleder R, Elizondo G, Wittenberg J, Rabito CA, Bengele HH, Josephson L. Ultrasmall superparamagnetic iron oxide: Characterization of a new class of contrast agents for MR imaging. *Radiology*. 1990;175(2):489–93.

3. Slowing II, Vivero-Escoto JL, Wu CW, Lin VSY. Mesoporous silica nanoparticles as controlled release drug delivery and gene transfection carriers. *Advanced Drug Delivery Reviews*. 2008 Aug 17;60(11):1278–88.

4. Chan WCW, Nie S. Quantum dot bioconjugates for ultrasensitive nonisotopic detection. *Science (1979) [Internet]*. 1998 Sep 25 [cited 2024 Aug 12];281(5385):2016–8. Available from: www.science.org/doi/10.1126/science.281.5385.2016

5. Jain PK, Huang X, El-Sayed IH, El-Sayed MA. Noble metals on the nanoscale: Optical and photothermal properties and some applications in imaging, sensing, biology, and medicine. *Accounts of Chemical Research [Internet]*. 2008 Dec 16 [cited 2024 Aug 12];41(12):1578–86. Available from: https://pubs.acs.org/doi/abs/10.1021/ar7002804

6. Murphy CJ, Gole AM, Stone JW, Sisco PN, Alkilany AM, Goldsmith EC, et al. Gold nanoparticles in biology: Beyond toxicity to cellular imaging. *Accounts of Chemical Research [Internet]*. 2008 Dec 16 [cited 2024 Aug 12];41(12):1721–30. Available from: https://pubs.acs.org/doi/abs/10.1021/ar800035u

7. Gupta AK, Gupta M. Synthesis and surface engineering of iron oxide nanoparticles for biomedical applications. *Biomaterials*. 2005 Jun 1;26(18):3995–4021.

8. Tang F, Li L, Chen D. Mesoporous silica nanoparticles: Synthesis, biocompatibility and drug delivery. *Advanced Materials [Internet]*. 2012 Mar 22 [cited 2024 Aug 12];24(12):1504–34. Available from: https://onlinelibrary.wiley.com/doi/full/10.1002/adma.201104763

9. Gao X, Cui Y, Levenson RM, Chung LWK, Nie S. In vivo cancer targeting and imaging with semiconductor quantum dots. *Nature Biotechnology [Internet]*. 2004 Jul 18 [cited 2024 Aug 12];22(8):969–76. Available from: www.nature.com/articles/nbt994

10. Mai WX, Meng H. Mesoporous silica nanoparticles: A multifunctional nano therapeutic system. *Integrative Biology [Internet]*. 2013 Jan 1 [cited 2024 Aug 12];5(1):19–28. Available from: https://doi.org/10.1039/c2ib20137b

11. Dykman L, Khlebtsov N. ChemInform abstract: Gold nanoparticles in biomedical applications: Recent advances and perspectives. *ChemInform [Internet]*. 2012 Jun 12 [cited 2024 Aug 12];43(24). Available from: https://onlinelibrary.wiley.com/doi/full/10.1002/chin.201224275

12. Smith AM, Duan H, Rhyner MN, Ruan G, Nie S. A systematic examination of surface coatings on the optical and chemical properties of semiconductor quantum dots. *Physical Chemistry Chemical Physics [Internet].* 2006 Aug 16 [cited 2024 Aug 12];8(33):3895–903. Available from: https://pubs.rsc.org/en/content/articlehtml/2006/cp/b606572b

13. Zhang J, Campbell RE, Ting AY, Tsien RY. Creating new fluorescent probes for cell biology. *Nature Reviews Molecular Cell Biology [Internet].* 2002 Dec 1 [cited 2024 Aug 12];3(12):906–18. Available from: www.nature.com/articles/nrm976

14. Mirkin CA, Letsinger RL, Mucic RC, Storhoff JJ. A DNA-based method for rationally assembling nanoparticles into macroscopic materials. *Nature [Internet].* 1996 Aug 15 [cited 2024 Aug 12];382(6592):607–9. Available from: www.nature.com/articles/382607a0

15. Pankhurst QA, Connolly J, Jones SK, Dobson J. Applications of magnetic nanoparticles in biomedicine. *Journal of Physics D: Applied Physics [Internet].* 2003 Jun 18 [cited 2024 Aug 12];36(13):R167. Available from: https://iopscience.iop.org/article/10.1088/0022-3727/36/13/201

16. Huang X, Jain PK, El-Sayed IH, El-Sayed MA. Gold nanoparticles: Interesting optical properties and recent applications in cancer diagnostics and therapy. *Nanomedicine [Internet].* 2007 Oct [cited 2024 Aug 12];2(5):681–93. Available from: www.tandfonline.com/doi/abs/10.2217/17435889.2.5.681

17. Soo Choi H, Liu W, Misra P, Tanaka E, Zimmer JP, Itty Ipe B, et al. Renal clearance of quantum dots. *Nature Biotechnology [Internet].* 2007 Sep 23 [cited 2024 Aug 12];25(10):1165–70. Available from: www.nature.com/articles/nbt1340

18. Sun C, Lee JSH, Zhang M. Magnetic nanoparticles in MR imaging and drug delivery. *Advanced Drug Delivery Reviews.* 2008 Aug 17;60(11):1252–65.

19. Vallet-Regí M, Balas F, Arcos D. Mesoporous materials for drug delivery. *Angewandte Chemie (International Edition) [Internet].* 2007 Oct 8 [cited 2024 Aug 12];46(40):7548–58. Available from: https://onlinelibrary.wiley.com/doi/full/10.1002/anie.200604488

20. Lewinski N, Colvin V, Drezek R. Cytotoxicity of nanoparticles. *Small [Internet].* 2008 Jan 18 [cited 2024 Aug 12];4(1):26–49. Available from: https://onlinelibrary.wiley.com/doi/full/10.1002/smll.200700595

21. Chithrani BD, Ghazani AA, Chan WCW. Determining the size and shape dependence of gold nanoparticle uptake into mammalian cells. *Nano Letters [Internet].* 2006 Apr [cited 2024 Aug 12];6(4):662–8. Available from: https://pubs.acs.org/doi/abs/10.1021/nl052396o

22. Jiang W, Kim BYS, Rutka JT, Chan WCW. Nanoparticle-mediated cellular response is size-dependent. *Nature Nanotechnology [Internet].* 2008 Mar 2 [cited 2024 Aug 12];3(3):145–50. Available from: www.nature.com/articles/nnano.2008.30

23. Li CH, Chan MH, Chang YC, Hsiao M. Gold nanoparticles as a biosensor for cancer biomarker determination. *Molecules [Internet].* 2023 Jan 2 [cited 2024 Aug 12];28(1):364. Available from: www.mdpi.com/1420-3049/28/1/364/htm

24. Ding Y, Jiang Z, Saha K, Kim CS, Kim ST, Landis RF, et al. Gold nanoparticles for nucleic acid delivery. *Molecular Therapy [Internet].* 2014 [cited 2024 Aug 12];22(6):1075. Available from: www.cell.com/molecular-therapy-family/molecular-therapy/fulltext/S1525-0016%2816%2930696-7

25. Maeda H, Wu J, Sawa T, Matsumura Y, Hori K. Tumor vascular permeability and the EPR effect in macromolecular therapeutics: A review. *Journal of Controlled Release.* 2000 Mar 1;65(1–2):271–84.

26. Chauhan VP, Stylianopoulos T, Boucher Y, Jain RK. Delivery of molecular and nanoscale medicine to tumors: Transport barriers and strategies. *Annual Review of Chemical and Biomolecular Engineering [Internet].* 2011 Jul 15 [cited 2024 Aug 12];2:281–98. Available from: www.annualreviews.org/content/journals/10.1146/annurev-chembioeng-061010-114300

27. Zhou L, Liu G, Wang Y, Liu J, Zhang Y, Ma Y. AuNP and ssDNA capped mesoporous silica nanoparticles for laser controlled drug release. *RSC Advances [Internet].* 2019 Oct 28 [cited 2024 Aug 12];9(60):34958–62. Available from: https://pubs.rsc.org/en/content/articlehtml/2019/ra/c9ra07404j

28. Gupta N, Malviya R. Understanding and advancement in gold nanoparticle targeted photothermal therapy of cancer. *Biochimica et Biophysica Acta (BBA)—Reviews on Cancer.* 2021 Apr 1;1875(2):188532.

29. Li R, Peng F, Cai J, Yang D, Zhang P. Redox dual-stimuli responsive drug delivery systems for improving tumor-targeting ability and reducing adverse side effects. *Asian Journal of Pharmaceutical Sciences.* 2020 May 1;15(3):311–25.

30. Yao Q, Kou L, Tu Y, Zhu L. MMP-responsive "smart" drug delivery and tumor targeting. *Trends in Pharmacological Sciences [Internet].* 2018 Aug 1 [cited 2024 Aug 12];39(8):766–81. Available from: www.cell.com/article/S0165614718301007/fulltext

31. Turkevich J, Stevenson PC, Hillier J. A study of the nucleation and growth processes in the synthesis of colloidal gold. *Discussions of the Faraday Society [Internet].* 1951 Jan 1 [cited 2024 Aug 12];11(0):55–75. Available from: https://pubs.rsc.org/en/content/articlehtml/1951/df/df9511100055

32. Shankar SS, Rai A, Ahmad A, Sastry M. Rapid synthesis of Au, Ag, and bimetallic Au core–Ag shell nanoparticles using Neem (Azadirachta indica) leaf broth. *Journal of Colloid and Interface Science.* 2004 Jul 15;275(2):496–502.

33. Vinod M, Jayasree RS, Gopchandran KG. Synthesis of pure and biocompatible gold nanoparticles using laser ablation method for SERS and photothermal applications. Current Applied Physics. 2017 Nov 1;17(11):1430–8.

34. El-Sayed MA. Some interesting properties of metals confined in time and nanometer space of different shapes. *Accounts of Chemical Research [Internet].* 2001 [cited 2024 Aug 12];34(4):257–64. Available from: https://pubs.acs.org/doi/abs/10.1021/ar960016n

35. Jain PK, Lee KS, El-Sayed IH, El-Sayed MA. Calculated absorption and scattering properties of gold nanoparticles of different size, shape, and composition: Applications in biological imaging and biomedicine. *Journal of Physical Chemistry B [Internet].* 2006 Apr 13 [cited 2024 Aug 12];110(14):7238–48. Available from: https://pubs.acs.org/doi/abs/10.1021/jp057170o

36. Hirsch LR, Stafford RJ, Bankson JA, Sershen SR, Rivera B, Price RE, et al. Nanoshell-mediated near-infrared thermal therapy of tumors under magnetic resonance guidance. *Proceedings of the National Academy of Sciences of the United States of America [Internet].* 2003 Nov 11 [cited 2024 Aug 12];100(23):13549–54. Available from: www.pnas.org/doi/abs/10.1073/pnas.2232479100

37. Elghanian R, Storhoff JJ, Mucic RC, Letsinger RL, Mirkin CA. Selective colorimetric detection of polynucleotides based on the distance-dependent optical properties of gold nanoparticles. *Science (1979) [Internet].* 1997 Aug 22 [cited 2024 Aug 12];277(5329):1078–81. Available from: www.science.org/doi/10.1126/science.277.5329.1078

38. Fomina N, Sankaranarayanan J, & Almutairi A. (2012). Photochemical mechanisms of light-triggered release from nanocarriers. *Advanced Drug Delivery Reviews*, 64(11): 1005–1020. https://doi.org/10.1016/j.addr.2012.02.006

39. Gref R, Lück M, Quellec P, Marchand M, Dellacherie E, Harnisch S, et al. "Stealth" corona-core nanoparticles surface modified by polyethylene glycol (PEG): Influences of the corona (PEG chain length and surface density) and of the core composition on phagocytic uptake and plasma protein adsorption. *Colloids and Surfaces. B, Biointerfaces.* 2000 Oct 1;18(3–4):301–13.

40. Ali Mansoori G, Brandenburg KS, Shakeri-Zadeh A. A comparative study of two folate-conjugated gold nanoparticles for cancer nanotechnology applications. *Cancers [Internet].* 2010 Nov 18 [cited 2024 Aug 12];2(4):1911–28. Available from: www.mdpi.com/2072-6694/2/4/1911/htm

41. Song S, Wang L, Li J, Fan C, Zhao J. Aptamer-based biosensors. *TrAC Trends in Analytical Chemistry.* 2008 Feb 1;27(2):108–17.

42. Qin W, Chandra J, Abourehab MAS, Gupta N, Chen ZS, Kesharwani P, et al. New opportunities for RGD-engineered metal nanoparticles in cancer. *Molecular Cancer [Internet].* 2023;22(1):87. Available from: https://doi.org/10.1186/s12943-023-01784-0

43. Giráldez-Pérez RM, Grueso E, Montero-Hidalgo AJ, Luque RM, Carnerero JM, Kuliszewska E, et al. Gold nanosystems covered with doxorubicin/DNA complexes: A therapeutic target for prostate and liver cancer. *International Journal of Molecular Sciences [Internet].* 2022 Dec 1 [cited 2024 Aug 12];23(24):15575. Available from: www.mdpi.com/1422-0067/23/24/15575/htm

44. Elizarova TN, Antopolsky ML, Novichikhin DO, Skirda AM, Orlov AV, Bragina VA, et al. A straightforward method for the development of positively charged gold nanoparticle-based vectors for effective siRNA delivery. *Molecules [Internet].* 2023 Apr 1 [cited 2024 Aug 12];28(8):3318. Available from: www.mdpi.com/1420-3049/28/8/3318/htm

45. Kumar PPP;, Lim DK, Prabhakaran P, Kumar P, Lim DK. Photothermal Effect of Gold Nanoparticles as a Nanomedicine for Diagnosis and Therapeutics. *Pharmaceutics* 2023, Vol 15, Page 2349 [Internet]. 2023 Sep 19 [cited 2024 Aug 12];15(9):2349. Available from: www.mdpi.com/1999-4923/15/9/2349/htm

46. Vines JB, Yoon JH, Ryu NE, Lim DJ, Park H. Gold nanoparticles for photothermal cancer therapy. *Frontiers in Chemistry [Internet].* 2019 Apr 5 [cited 2024 Aug 12];7(APR):432792. Available from: www.frontiersin.org

47. Pan Y, Neuss S, Leifert A, Fischler M, Wen F, Simon U, et al. Size-dependent cytotoxicity of gold nanoparticles. *Small [Internet].* 2007 Nov 5 [cited 2024 Aug 12];3(11):1941–9. Available from: https://onlinelibrary.wiley.com/doi/full/10.1002/smll.200700378

48. Alkilany AM, Murphy CJ. Toxicity and cellular uptake of gold nanoparticles: What we have learned so far? *Journal of Nanoparticle Research [Internet].* 2010 Sep 6 [cited 2024 Aug 12];12(7):2313–33. Available from: https://link.springer.com/article/10.1007/s11051-010-9911-8

49. Chen YS, Hung YC, Liau I, Huang GS. Assessment of the in vivo toxicity of gold nanoparticles. *Nanoscale Research Letters [Internet].* 2009 May 8 [cited 2024 Aug 12];4(8):858–64. Available from: https://link.springer.com/article/10.1007/s11671-009-9334-6

50. Khlebtsov N, Dykmana L. Biodistribution and toxicity of engineered gold nanoparticles: A review of in vitro and in vivo studies. *Chemical Society Reviews [Internet].* 2011 Feb 22 [cited 2024 Aug 12];40(3):1647–71. Available from: https://pubs.rsc.org/en/content/articlehtml/2011/cs/c0cs00018c

51. Sharifi S, Mahmoud NN, Voke E, Landry MP, Mahmoudi M. Importance of standardizing analytical characterization methodology for improved reliability of the nanomedicine literature. *Nano-micro Letters [Internet].* 2022 Dec 1 [cited 2024 Aug 12];14(1):1–15. Available from: https://link.springer.com/article/10.1007/s40820-022-00922-5

52. Fadeel B, Garcia-Bennett AE. Better safe than sorry: Understanding the toxicological properties of inorganic nanoparticles manufactured for biomedical applications. *Advanced Drug Delivery Reviews.* 2010 Mar 8;62(3):362–74.

53. Paliwal R, Babu RJ, Palakurthi S. Nanomedicine scale-up technologies: Feasibilities and challenges. *AAPS PharmSciTech [Internet].* 2014 Nov 27 [cited 2024 Aug 12];15(6):1527–34. Available from: https://link.springer.com/article/10.1208/s12249-014-0177-9

54. Food and Drug Administration. *Guidance for Industry Considering Whether an FDA-Regulated Product Involves the Application of Nanotechnology. Contains Nonbinding Recommendations.* 2014 [cited 2024 Aug 12]; Available from: www.fda.gov/RegulatoryInformation/Guidances/ucm257698.htm

55. Peng S, Wang C, Xie J, Sun S. Synthesis and stabilization of monodisperse Fe nanoparticles. *Journal of the American Chemical Society [Internet].* 2006 Aug 23 [cited 2024 Aug 12];128(33):10676–7. Available from: https://pubs.acs.org/doi/abs/10.1021/ja063969h

56. Park J, An K, Hwang Y, Park JEG, Noh HJ, Kim JY, et al. Ultra-large-scale syntheses of monodisperse nanocrystals. *Nature Materials [Internet].* 2004 Nov 28 [cited 2024 Aug 12];3(12):891–5. Available from: www.nature.com/articles/nmat1251

57. Foglia S, Ledda M, Fioretti D, Iucci G, Papi M, Capellini G, et al. In vitro biocompatibility study of sub-5 nm silica-coated magnetic iron oxide fluorescent nanoparticles for potential biomedical application. *Scientific Reports [Internet].* 2017 Apr 19 [cited 2024 Aug 12];7(1):1–13. Available from: www.nature.com/articles/srep46513

58. Ramírez-Acosta CM, Cifuentes J, Cruz JC, Reyes LH. Patchy core/shell, magnetite/silver nanoparticles via green and facile synthesis: Routes to assure biocompatibility. *Nanomaterials [Internet].* 2020 Sep 17 [cited 2024 Aug 13];10(9):1857. Available from: www.mdpi.com/2079-4991/10/9/1857/htm

59. Laurent S, Forge D, Port M, Roch A, Robic C, Vander Elst L, et al. Magnetic iron oxide nanoparticles: Synthesis, stabilization, vectorization, physicochemical characterizations, and biological applications. *Chemical Reviews [Internet].* 2009 Apr 14 [cited 2024 Aug 12];110(4):2574. Available from: https://pubs.acs.org/doi/full/10.1021/cr900197g

60. Weissleder R, Elizondo G, Wittenberg J, Lee AS, Josephson L, Brady TJ. Ultrasmall superparamagnetic iron oxide: An intravenous contrast agent for assessing lymph nodes with MR imaging. *Radiology [Internet].* 1990 May 1 [cited 2024 Aug 12];175(2):494–8. Available from: https://pubs.rsna.org/doi/10.1148/radiology.175.2.2326475

61. Jordan A, Scholz R, Wust P, Fähling H, Felix R. Magnetic fluid hyperthermia (MFH): Cancer treatment with AC magnetic field induced excitation of biocompatible superparamagnetic nanoparticles. *Journal of Magnetism and Magnetic Materials* 1999 Jul 1;201(1–3):413–9.

62. Natarajan B, Kannan P, Guo L. Metallic nanoparticles-based multicolor visual detection of biomolecules—sensing mechanism-based on the change in various colors. *Chinese Journal of Structural Chemistry.* 2024 May 22;100349.

63. Gleich B, Weizenecker J. Tomographic imaging using the nonlinear response of magnetic particles. *Nature [Internet].* 2005 Jun 30 [cited 2024 Aug 12];435(7046):1214–7. Available from: www.nature.com/articles/nature03808

64. Moghimi SM, Hunter AC, Murray JC. Long-circulating and target-specific nanoparticles: Theory to practice. *Pharmacological Reviews [Internet].* 2001 Jun 1;53(2):283. Available from: http://pharmrev.aspetjournals.org/content/53/2/283.abstract

65. Tomalia DA, Reyna LA, Svenson S. Dendrimers as multi-purpose nanodevices for oncology drug delivery and diagnostic imaging. *Biochemical Society Transactions [Internet]*. 2007 Feb 1 [cited 2024 Aug 12];35(1):61–7. Available from: https://portlandpress.com/biochemsoctrans/article-abstract/35/1/61/64036/Dendrimers-as-multi-purpose-nanodevices-for?redirectedFrom=fulltext

66. Lu AH, Salabas EL, Schüth F. Magnetic nanoparticles: Synthesis, protection, functionalization, and application. *Angewandte Chemie International Edition [Internet]*. 2007 Feb 12 [cited 2024 Aug 12];46(8):1222–44. Available from: https://onlinelibrary.wiley.com/doi/full/10.1002/anie.200602866

67. Torchilin VP. Recent advances with liposomes as pharmaceutical carriers. *Nature Reviews Drug Discovery [Internet]*. 2005 Feb [cited 2024 Aug 12];4(2):145–60. Available from: www.nature.com/articles/nrd1632

68. Donadel K, Felisberto MDV, Fávere VT, Rigoni M, Batistela NJ, Laranjeira MCM. Synthesis and characterization of the iron oxide magnetic particles coated with chitosan biopolymer. *Materials Science and Engineering: C*. 2008 May 1;28(4):509–14.

69. Allen TM, Cullis PR. Liposomal drug delivery systems: From concept to clinical applications. *Advanced Drug Delivery Reviews*. 2013 Jan 1;65(1):36–48.

70. Gaucher G, Dufresne MH, Sant VP, Kang N, Maysinger D, Leroux JC. Block copolymer micelles: Preparation, characterization and application in drug delivery. *Journal of Controlled Release*. 2005 Dec 5;109(1–3):169–88.

71. Lee JH, Huh YM, Jun YW, Seo JW, Jang JT, Song HT, et al. Artificially engineered magnetic nanoparticles for ultra-sensitive molecular imaging. *Nature Medicine [Internet]*. 2006 Dec 24 [cited 2024 Aug 12];13(1):95–9. Available from: www.nature.com/articles/nm1467

72. Maier-Hauff K, Rothe R, Scholz R, Gneveckow U, Wust P, Thiesen B, et al. Intracranial thermotherapy using magnetic nanoparticles combined with external beam radiotherapy: Results of a feasibility study on patients with glioblastoma multiforme. *Journal of Neuro-oncology [Internet]*. 2007 Jan 14 [cited 2024 Aug 12];81(1):53–60. Available from: https://link.springer.com/article/10.1007/s11060-006-9195-0

73. Hergt R, Dutz S. Magnetic particle hyperthermia—biophysical limitations of a visionary tumour therapy. *Journal of Magnetism and Magnetic Materials* 2007 Apr 1;311(1):187–92.

74. Hu X, Li D, Gao Y, Mu L, Zhou Q. Knowledge gaps between nanotoxicological research and nanomaterial safety. *Environment International*. 2016 Sep 1;94:8–23.

75. Arami H, Khandhar A, Liggitt D, Krishnan KM. In vivo delivery, pharmacokinetics, biodistribution and toxicity of iron oxide nanoparticles. *Chemical Society Reviews [Internet]*. 2015 Nov 16 [cited 2024 Aug 12];44(23):8576–607. Available from: https://pubs.rsc.org/en/content/articlehtml/2015/cs/c5cs00541h

76. Zhang Y, Kohler N, Zhang M. Surface modification of superparamagnetic magnetite nanoparticles and their intracellular uptake. *Biomaterials*. 2002 Apr 1;23(7):1553–61.

77. Longmire M, Choyke PL, Kobayashi H. Clearance properties of nano-sized particles and molecules as imaging agents: Considerations and caveats. *Nanomedicine [Internet]*. 2008 Oct [cited 2024 Aug 12];3(5):703–17. Available from: www.tandfonline.com/doi/abs/10.2217/17435889.3.5.703

78. Jinhao GAO, Hongwei GU, Bing XU. Multifunctional magnetic nanoparticles: Design, synthesis, and biomedical applications. *Accounts of Chemical Research [Internet]*. 2009 Aug 18 [cited 2024 Aug 12];42(8):1097–107. Available from: https://pubs.acs.org/doi/abs/10.1021/ar9000026

79. Bobo D, Robinson KJ, Islam J, Thurecht KJ, Corrie SR. Nanoparticle-based medicines: A review of FDA-approved materials and clinical trials to date. *Pharmaceutical Research [Internet].* 2016 Oct 1 [cited 2024 Aug 12];33(10):2373–87. Available from: https://link.springer.com/article/10.1007/s11095-016-1958-5

80. Portilla Y, Mulens-Arias V, Daviu N, Paradela A, Pérez-Yagüe S, Barber DF. Interaction of Iron Oxide Nanoparticles with Macrophages Is Influenced Distinctly by "Self" and "Non-Self" Biological Identities. *ACS Applied Materials & Interfaces [Internet].* 2023 Aug 2 [cited 2024 Aug 12];15(30):35906–26. Available from: https://pubs.acs.org/doi/full/10.1021/acsami.3c05555

81. Dobson J. Magnetic nanoparticles for drug delivery. *Drug Development Research [Internet].* 2006 Jan 1 [cited 2024 Aug 12];67(1):55–60. Available from: https://onlinelibrary.wiley.com/doi/full/10.1002/ddr.20067

82. Alexiou Ch, Schmid R, Jurgons R, Bergemann Ch, Arnold W, Parak FG. Targeted tumor therapy with "magnetic drug targeting": therapeutic efficacy of ferrofluid bound mitoxantrone. *Ferrofluids. Lecture Notes in Physics.* 2002 [cited 2024 Aug 12];233–51. Available from: https://link.springer.com/chapter/10.1007/3-540-45646-5_12

83. Frey NA, Peng S, Cheng K, Sun S. Magnetic nanoparticles: Synthesis, functionalization, and applications in bioimaging and magnetic energy storage. *Chemical Society Reviews [Internet].* 2009 Aug 19 [cited 2024 Aug 12];38(9):2532–42. Available from: https://pubs.rsc.org/en/content/articlehtml/2009/cs/b815548h

84. Reddy LH, Arias JL, Nicolas J, Couvreur P. Magnetic nanoparticles: Design and characterization, toxicity and biocompatibility, pharmaceutical and biomedical applications. *Chemical Reviews [Internet].* 2012 Nov 14 [cited 2024 Aug 12];112(11):5818–78. Available from: https://pubs.acs.org/doi/abs/10.1021/cr300068p

85. Veiseh O, Gunn JW, Zhang M. Design and fabrication of magnetic nanoparticles for targeted drug delivery and imaging. *Advanced Drug Delivery Reviews.* 2010 Mar 8;62(3):284–304.

86. Roca AG, Gutiérrez L, Gavilán H, Fortes Brollo ME, Veintemillas-Verdaguer S, Morales MDP. Design strategies for shape-controlled magnetic iron oxide nanoparticles. *Advanced Drug Delivery Reviews.* 2019 Jan 1;138:68–104.

87. Stöber W, Fink A, Bohn E. Controlled growth of monodisperse silica spheres in the micron size range. *Journal of Colloid and Interface Science.* 1968 Jan 1;26(1):62–9.

88. Arriagada FJ, Osseo-Asare K. Synthesis of nanosize silica in aerosol OT reverse microemulsions. *Journal of Colloid and Interface Science.* 1995 Mar 1;170(1):8–17.

89. Zhao D, Feng J, Huo Q, Melosh N, Fredrickson GH, Chmelka BF, et al. Triblock copolymer syntheses of *Mesoporous silica* with periodic 50 to 300 angstrom pores. *Science (1979) [Internet].* 1998 Jan 23 [cited 2024 Aug 12];279(5350):548–52. Available from: www.science.org/doi/10.1126/science.279.5350.548

90. Yue R, Meng D, Ni Y, Jia Y, Liu G, Yang J, et al. One-step flame synthesis of hydrophobic silica nanoparticles. *Powder Technology.* 2013 Feb 1;235:909–13.

91. Popat A, Hartono SB, Stahr F, Liu J, Qiao SZ, Lu GQ. *Mesoporous silica* nanoparticles for bioadsorption, enzyme immobilisation, and delivery carriers. *Nanoscale [Internet].* 2011 Jul 7 [cited 2024 Aug 12];3(7):2801–18. Available from: https://pubs.rsc.org/en/content/articlehtml/2011/nr/c1nr10224a

92. Rosenholm JM, Meinander A, Peuhu E, Niemi R, Eriksson JE, Sahlgren C, et al. Targeting of porous hybrid silica nanoparticles to cancer cells. *ACS Nano [Internet].* 2009 Jan [cited 2024 Aug 12];3(1):197–206. Available from: https://pubs.acs.org/doi/abs/10.1021/nn800781r

93. Huang X, Teng X, Chen D, Tang F, He J. The effect of the shape of mesoporous silica nanoparticles on cellular uptake and cell function. *Biomaterials.* 2010 Jan 1;31(3):438–48.

94. Lu J, Liong M, Zink JI, Tamanoi F. *Mesoporous silica* nanoparticles as a delivery system for hydrophobic anticancer drugs. *Small [Internet].* 2007 Aug 3 [cited 2024 Aug 12];3(8):1341–6. Available from: https://onlinelibrary.wiley.com/doi/full/10.1002/smll.200700005

95. Zhang L, Gu FX, Chan JM, Wang AZ, Langer RS, Farokhzad OC. Nanoparticles in medicine: Therapeutic applications and developments. *Clinical Pharmacology and Therapeutics [Internet].* 2008 May 1 [cited 2024 Aug 12];83(5):761–9. Available from: https://onlinelibrary.wiley.com/doi/full/10.1038/sj.clpt.6100400

96. He Q, Zhang Z, Gao Y, Shi J, Li Y. Intracellular localization and cytotoxicity of spherical *Mesoporous silica* nano- and microparticles. *Small [Internet].* 2009 Dec 4 [cited 2024 Aug 12];5(23):2722–9. Available from: https://onlinelibrary.wiley.com/doi/full/10.1002/smll.200900923

97. Qhobosheane M, Santra S, Zhang P, Tan W. Biochemically functionalized silica nanoparticles. *Analyst [Internet].* 2001 Jan 1 [cited 2024 Aug 12];126(8):1274–8. Available from: https://pubs.rsc.org/en/content/articlehtml/2001/an/b101489g

98. Iqbal MZ, Ma X, Chen T, Zhang L, Ren W, Xiang L, et al. Silica-coated super-paramagnetic iron oxide nanoparticles (SPIONPs): A new type contrast agent of T1 magnetic resonance imaging (MRI). *Journal of Materials Chemistry. B [Internet].* 2015 Jun 25 [cited 2024 Aug 12];3(26):5172–81. Available from: https://pubs.rsc.org/en/content/articlehtml/2015/tb/c5tb00300h

99. Park JH, Von Maltzahn G, Zhang L, Schwartz MP, Ruoslahti E, Bhatia SN, et al. Magnetic iron oxide nanoworms for tumor targeting and imaging. *Advanced Materials [Internet].* 2008 May 5 [cited 2024 Aug 12];20(9):1630–5. Available from: https://onlinelibrary.wiley.com/doi/full/10.1002/adma.200800004

100. Mladenović M, Jarić S, Mundžić M, Pavlović A, Bobrinetskiy I, Knežević NŽ. Biosensors for cancer biomarkers based on *Mesoporous silica* nanoparticles. *Biosensors.* 2024 *[Internet].* 2024 Jun 30 [cited 2024 Aug 12];14(7):326. Available from: www.mdpi.com/2079-6374/14/7/326/htm

101. Hormsombut T, Mekjinda N, Kalasin S, Surareungchai W, Rijiravanich P. *Mesoporous silica* nanoparticles-enhanced microarray technology for highly sensitive simultaneous detection of multiplex foodborne pathogens. *ACS Applied Bio Materials [Internet].* 2024 Apr 15 [cited 2024 Aug 12];7(4):2367–77. Available from: https://pubs.acs.org/doi/full/10.1021/acsabm.4c00005

102. Murugadoss S, Lison D, Godderis L, Van Den Brule S, Mast J, Brassinne F, et al. Toxicology of silica nanoparticles: An update. *Archives of Toxicology [Internet].* 2017 Jun 1 [cited 2024 Aug 12];91(9):2967–3010. Available from: https://link.springer.com/article/10.1007/s00204-017-1993-y

103. Napierska D, Thomassen LCJ, Rabolli V, Lison D, Gonzalez L, Kirsch-Volders M, et al. Size-dependent cytotoxicity of *Monodisperse silica* nanoparticles in human endothelial cells. *Small [Internet].* 2009 Apr 6 [cited 2024 Aug 12];5(7):846–53. Available from: https://onlinelibrary.wiley.com/doi/full/10.1002/smll.200800461

104. Magdolenova Z, Collins A, Kumar A, Dhawan A, Stone V, Dusinska M. Mechanisms of genotoxicity. A review of in vitro and in vivo studies with engineered nanoparticles. *Nanotoxicology [Internet].* 2014 May [cited 2024 Aug 12];8(3):233–78. Available from: www.tandfonline.com/doi/abs/10.3109/17435390.2013.773464

105. Zhang H, Dunphy DR, Jiang X, Meng H, Sun B, Tarn D, et al. Processing pathway dependence of amorphous silica nanoparticle toxicity: Colloidal vs pyrolytic. *Journal of the American Chemical Society [Internet].* 2012 Sep 26 [cited 2024 Aug 12];134(38):15790–804. Available from: https://pubs.acs.org/doi/abs/10.1021/ja304907c

106. Oberdörster G, Stone V, Donaldson K. Toxicology of nanoparticles: A historical perspective. *Nanotoxicology [Internet].* 2007 Mar [cited 2024 Aug 12];1(1):2–25. Available from: www.tandfonline.com/doi/abs/10.1080/17435390701314761

107. European Medicines Agency. *Multidisciplinary: Nanomedicines.* European Medicines Agency (EMA) [Internet]. 2024 [cited 2024 Aug 12]. Available from: www.ema.europa.eu/en/human-regulatory-overview/research-and-development/scientific-guidelines/multidisciplinary-guidelines/multidisciplinary-nanomedicines

108. Kammler HK, Pratsinis SE. Scaling-up the production of nanosized $SiO_2$-particles in a double diffusion flame aerosol reactor. *Journal of Nanoparticle Research [Internet].* 1999 [cited 2024 Aug 12];1(4):467–77. Available from: https://link.springer.com/article/10.1023/A:1010080004637

109. Yeasmin MN, Sultana M, Siddika A, Tabassum S, Ullah SM, Bashar MS. Structural, optical, and morphological characterization of silica nanoparticles prepared by sol-gel process. *Journal of the Turkish Chemical Society Section A: Chemistry [Internet].* 2022 Nov 30 [cited 2024 Aug 12];9(4):1323–34. Available from: https://dergipark.org.tr/en/pub/jotcsa/issue/72258/1071086

110. Lynch I, Lee RG. In support of the inclusion of data on nanomaterials transformations and environmental interactions into existing regulatory frameworks. *Managing Risk in Nanotechnology. Innovation, Technology and Knowledge Management [Internet].* 2016 [cited 2024 Aug 12];145–69. Available from: https://link.springer.com/chapter/10.1007/978-3-319-32392-3_9

111. Costa MCF, Echeverrigaray SG, Andreeva DV., Novoselov KS, Neto AHC. Two-dimensional quantum dots: From photoluminescence to biomedical applications. *Solids [Internet].* 2022 Oct 19 [cited 2024 Aug 12];3(4):578–602. Available from: www.mdpi.com/2673-6497/3/4/37/htm

112. Singh S, Raina D, Rishipathak D, Babu KR, Khurana R, Gupta Y, et al. Quantum dots in the biomedical world: A smart advanced nanocarrier for multiple venues application. *Archiv der Pharmazie (Weinheim) [Internet].* 2022 Dec 1 [cited 2024 Aug 12];355(12):2200299. Available from: https://onlinelibrary.wiley.com/doi/full/10.1002/ardp.202200299

113. Podder S. Fluorescent quantum dots, a technological marvel for optical bio-imaging: A perspective on associated in vivo toxicity. *Application of Quantum Dots in Biology and Medicine: Recent Advances [Internet].* 2022 Jan 1 [cited 2024 Aug 12];143–63. Available from: https://link.springer.com/chapter/10.1007/978-981-19-3144-4_8

114. Ulag S, Gunduz O. Quantum dot-based nanomaterials for diagnostic and therapeutic applications. *Nanotechnology in the Life Sciences [Internet].* 2022 [cited 2024 Aug 12];429–53. Available from: https://link.springer.com/chapter/10.1007/978-3-031-12658-1_15

115. Gavalajyan PS. Photoluminescence spectra in GaAs biconical quantum dot. *SPIE [Internet].* 2024 Apr 5 [cited 2024 Aug 12];12991:129910X. Available from: https://ui.adsabs.harvard.edu/abs/2024SPIE12991E..0XG/abstract

116. Ali MA, Chun DM, Ibrahim EMM, Abd-Elrahim AG. Optical and temperature-dependent electrical and dielectric properties of ultrasound-synthesized CdS quantum dots. *Physica Scripta [Internet].* 2023 Oct 16 [cited 2024 Aug 12];98(11):115943. Available from: https://iopscience.iop.org/article/10.1088/1402-4896/ad000c

117. Liang KL, Kuo WH, Lin CC, Fang YH. The size-dependent photonic characteristics of colloidal-quantum-dot-enhanced micro-LEDs. *Micromachines (Basel) [Internet].* 2023 Mar 1 [cited 2024 Aug 12];14(3):589. Available from: www.mdpi.com/2072-666X/14/3/589/htm

118. Zhang P, Li S, Li D, Ren L, Qin Z, Jiang L, et al. Quantum size-dependent luminescence and nonlinear optical properties of silicon quantum dots/$SiO_2$ multilayer. *Optics and Laser Technology.* 2023 Jan 1;157:108706.

119. Baruah JM, Narayan J. Semiconductor quantum dots as smart biocompatible imaging agents. *AIP Conference Proceedings [Internet]*. 2019 Oct 29 [cited 2024 Aug 12];2162(1). Available from: https://pubs.aip.org/aip/acp/article-abstract/2162/1/020004/889074/Semiconductor-quantum-dots-as-smart-biocompatible?redirectedFrom=fulltext

120. Li J, Fan J, Wu R, Li N, Lv Y, Shen H, et al. Biomolecular surface functionalization and stabilization method to fabricate quantum dots nanobeads for accurate biosensing detection. *Langmuir [Internet]*. 2022 Apr 26 [cited 2024 Aug 12];38(16):4969–78. Available from: https://pubs.acs.org/doi/abs/10.1021/acs.langmuir.2c00392

121. Hassanabadi E, Latifi M, Gualdrón-Reyes AF, Masi S, Yoon SJ, Poyatos M, et al. Ligand & band gap engineering: Tailoring the protocol synthesis for achieving high-quality CsPbI3 quantum dots. *Nanoscale [Internet]*. 2020 Jul 9 [cited 2024 Aug 12];12(26):14194–203. Available from: https://pubs.rsc.org/en/content/articlehtml/2020/nr/d0nr03180a

122. Kavgacı M, Kalmış HV, Eskalen H. Synthesis of fluorescent carbon quantum dots with hydrothermal and solvothermal method application for anticounterfeiting and encryption. *International Journal of Innovative Engineering Applications [Internet]*. 2023 Jul 6 [cited 2024 Aug 12];7(1):32–8. Available from: https://dergipark.org.tr/en/pub/ijiea/issue/78156/1182009

123. Vikram A, Kumar V, Ramesh U, Balakrishnan K, Oh N, Deshpande K, et al. A Millifluidic reactor system for multistep continuous synthesis of InP/ZnSeS nanoparticles. *ChemNanoMat [Internet]*. 2018 Sep 1 [cited 2024 Aug 12];4(9):943–53. Available from: https://onlinelibrary.wiley.com/doi/full/10.1002/cnma.201800160

124. Singh PK, Singh S, Sachan K, Verma V, Garg S. Recent development and advancement in quantum dots in pharmaceutical and biomedical fields for the delivery of drugs. *Current Nanoscience [Internet]*. 2023 May 18 [cited 2024 Aug 12];20(4):425–35. Available from: www.eurekaselect.com/article/131858

125. Saravanan V, Tippavajhala VK. Quantum dots: Targeted and traceable drug delivery system. *Research Journal of Pharmacy and Technology [Internet]*. 2022 Dec 1 [cited 2024 Aug 12];15(12):5895–902. Available from: www.researchgate.net/publication/368616830_Quantum_Dots_Targeted_and_Traceable_Drug_Delivery_System

126. Sowjanya V, Saravanan G, Venkateswarlu S. Quantum dots conjugation and its advancement in biomedical applications. *Carbon and Graphene Quantum Dots for Biomedical Applications*. 2023 Jan 1;57–74.

127. Janu L, Stanisavljevic M, Krizkova S, Sobrova P, Vaculovicova M, Kizek R, et al. Electrophoretic study of peptide-mediated quantum dot-human immunoglobulin bioconjugation. *Electrophoresis [Internet]*. 2013 Sep 1 [cited 2024 Aug 12];34(18):2725–32. Available from: https://onlinelibrary.wiley.com/doi/full/10.1002/elps.201300088

128. Green CM, Hastman DA, Mathur D, Susumu K, Medintz IL, Diaz SA. Parameters guiding the self-assembly of quantum dots and DNA origami by peptide-PNA. *Proceedings of the IEEE Conference on Nanotechnology*. 2021 Jul 28:448–50.

129. Rahmani P, Goodlad M, Zhang Y, Li Y, Ye T. One-step ligand-exchange method to produce quantum dot-DNA conjugates for DNA-directed self-assembly. *ACS Applied Materials & Interfaces [Internet]*. 2022 Oct 26 [cited 2024 Aug 12];14(42):47359–68. Available from: https://pubs.acs.org/doi/abs/10.1021/acsami.2c10580

130. Mondal S, Pan A. Quantum dots in biosensing, bioimaging, and drug delivery. *Application of Quantum Dots in Biology and Medicine: Recent Advances [Internet]*. 2022 Jan 1 [cited 2024 Aug 12];165–90. Available from: https://link.springer.com/chapter/10.1007/978-981-19-3144-4_9

131. Rajamanickam K, Rajamanickam K. Application of quantum dots in bio-sensing, bio-imaging, drug delivery, anti-bacterial activity, photo-thermal, photo-dynamic therapy, and optoelectronic devices. *Quantum Dots – Recent Advances, New Perspectives and Contemporary Applications [Internet]*. 2022 Sep 16 [cited 2024 Aug 12]; Available from: www.intechopen.com/chapters/83606

132. Wu X, Yang J, Xing J, Lyu Y, Zou R, Wang X, et al. Using host–guest interactions at the interface of quantum dots to load drug molecules for biocompatible, safe, and effective chemo-photodynamic therapy against cancer. *Journal of Materials Chemistry. B [Internet]*. 2023 Jun 7 [cited 2024 Aug 12];11(22):4855–64. Available from: https://pubs.rsc.org/en/content/articlehtml/2023/tb/d3tb00592e

133. Jovanović S, Marković Z, Budimir M, Prekodravac J, Zmejkoski D, Kepić D, et al. Lights and dots toward therapy—carbon-based quantum dots as new agents for photodynamic therapy. *Pharmaceutics [Internet]*. 2023 Apr 6 [cited 2024 Aug 12];15(4):1170. Available from: www.mdpi.com/1999-4923/15/4/1170/htm

134. Phafat B, Bhattacharya S. Quantum dots as theranostic agents: Recent advancements, surface modifications, and future applications. *Mini-Reviews in Medicinal Chemistry [Internet]*. 2022 Apr 6 [cited 2024 Aug 12];23(12):1257–72. Available from: www.eurekaselect.com/article/122304

135. Walton-Raaby M, Woods R, Kalyaanamoorthy S. Investigating the theranostic potential of graphene quantum dots in Alzheimer's disease. *International Journal of Molecular Sciences [Internet]*. 2023 Jun 1 [cited 2024 Aug 12];24(11):9476. Available from: www.mdpi.com/1422-0067/24/11/9476/htm

136. Derfus AM, Chan WCW, Bhatia SN. Probing the cytotoxicity of semiconductor quantum dots. *Nano Letters [Internet]*. 2004 Jan [cited 2024 Aug 12];4(1):11–8. Available from: https://pubs.acs.org/doi/abs/10.1021/nl0347334

137. Hardman R. A toxicologic review of quantum dots: Toxicity depends on physicochemical and environmental factors. *Environmental Health Perspectives [Internet]*. 2006 Feb [cited 2024 Aug 12];114(2):165–72. Available from: https://ehp.niehs.nih.gov/doi/10.1289/ehp.8284

138. Tsoi KM, Dai Q, Alman BA, Chan WCW. Are quantum dots toxic? Exploring the discrepancy between cell culture and animal studies. *Accounts of Chemical Research [Internet]*. 2013 Mar 19 [cited 2024 Aug 12];46(3):662–71. Available from: https://pubs.acs.org/doi/abs/10.1021/ar300040z

139. Pellegrino T, Manna L, Kudera S, Liedl T, Koktysh D, Rogach AL, et al. Hydrophobic nanocrystals coated with an amphiphilic polymer shell: A general route to water soluble nanocrystals. *Nano Letters [Internet]*. 2004 Apr [cited 2024 Aug 12];4(4):703–7. Available from: https://pubs.acs.org/doi/abs/10.1021/nl035172j

140. Chen ZL, Lin Y, Yu XJ, Zhu DL, Guo SW, Zhang JJ, et al. Preparation of monodisperse hydrophilic quantum dots with amphiphilic polymers. *ACS Applied Materials & Interfaces [Internet]*. 2017 Nov 22 [cited 2024 Aug 12];9(46):39901–6. Available from: https://pubs.acs.org/doi/abs/10.1021/acsami.7b09557

141. Singh KR, Nayak V, Sabui P, Mallick S, Singh J, Singh RP. Bioinspired quantum dots: Promising nanosystems for biomedical application. *Nano-Structures & Nano-Objects*. 2022 Oct 1;32:100921.

142. Nash JA, Manning MD, Gulyuk AV., Kuznetsov AE, Yingling YG. Gold nanoparticle design for RNA compaction. *Biointerphases [Internet]*. 2022 Nov 1 [cited 2024 Aug 12];17(6):61001. Available from: https://pubs.aip.org/avs/bip/article-abstract/17/6/061001/2835386/Gold-nanoparticle-design-for-RNA-compaction?redirectedFrom=fulltext

143. Rajan SSR, Thomas J, Francis D, Daniel EC. Effective gene delivery using size dependant nano core-shell in human cervical cancer cell lines by magnetofection. *PLoS One [Internet]*. 2023 Sep 1 [cited 2024 Aug 12];18(9):e0289731. Available from: https://journals.plos.org/plosone/article?id=10.1371/journal.pone.0289731

144. Bobbarjang N, Kasi P, Bandlapalli C, Maddileti MR, Y PR, Maccha KS. Targeted drug delivery to the lungs using mesoporous silica nanoparticles. *Future Journal of Pharmaceuticals and Health Sciences [Internet]*. 2024 Apr 5 [cited 2024 Aug 12];4(2):1–11. Available from: https://pharmasprings.com/fjphs/article/view/589

145. Chen Q, Huang Z, Wang Q, Hu Y, Tang H, Wen R, et al. Novel synthesis of Mn: ZnSe@ZnS core–shell quantum dots based on photoinduced fluorescence enhancement. *Spectrochimica acta. Part A, Molecular and Biomolecular Spectroscopy*. 2021 Mar 5;248:119099.

146. Yu J, Dan N, Eslami SM, Lu X. State of the art of silica nanoparticles: An overview on biodistribution and preclinical toxicity studies. *The AAPS Journal [Internet]*. 2024 Mar 21 [cited 2024 Aug 12];26(3):1–13. Available from: https://link.springer.com/article/10.1208/s12248-024-00906-w

147. Truong LB, Medina-Cruz D, Mostafavi E. Gold nanoparticles for delivery of nucleic acid constructs for cancer treatment. *Gold Nanoparticles for Drug Delivery*. 2024 Jan 1;141–65.

148. Arango D, Cifuentes J, Puentes PR, Beltran T, Bittar A, Ocasión C, et al. Tailoring magnetite-nanoparticle-based nanocarriers for gene delivery: Exploiting CRISPRa potential in reducing conditions. *Nanomaterials [Internet]*. 2023 May 31 [cited 2024 Aug 12];13(11):1782. Available from: www.mdpi.com/2079-4991/13/11/1782/htm

149. Gil HM, Booth Z, Price TW, Lee J, Naylor-Adamson L, Avery M, et al. Impact of surface ligand on the biocompatibility of InP/ZnS quantum dots with platelets. *Small [Internet]*. 2024 Mar 1 [cited 2024 Aug 12];20(12):2304881. Available from: https://onlinelibrary.wiley.com/doi/full/10.1002/smll.202304881

150. Bhatt CS, Rajavel A, Parimi DS, Sella RN, Murugaiyan J, Suresh AK. Sustainable exscalar monodispersed gold nanoparticles with enhanced dispersion stability and biocompatibility for theragnostics. *ACS Applied Nano Materials [Internet]*. 2023 Jul 14 [cited 2024 Aug 12];6(13):12548–59. Available from: https://pubs.acs.org/doi/abs/10.1021/acsanm.3c02407

151. Schemberg J, El Abbassi A, Lindenbauer A, Chen LY, Grodrian A, Nakos X, et al. Synthesis of biocompatible superparamagnetic iron oxide nanoparticles (SPION) under different microfluidic regimes. *ACS Applied Materials & Interfaces [Internet]*. 2022 Oct 26 [cited 2024 Aug 12];14(42):48011–28. Available from: https://pubs.acs.org/doi/full/10.1021/acsami.2c13156

152. Bragato C, Mazzotta R, Persico A, Bengalli R, Ornelas M, Gomes F, et al. Biocompatibility analysis of bio-based and synthetic silica nanoparticles during early Zebrafish development. *International Journal of Molecular Sciences [Internet]*. 2024 May 1 [cited 2024 Aug 12];25(10):5530. Available from: www.mdpi.com/1422-0067/25/10/5530/htm

153. Naylor-Adamson L, Price TW, Booth Z, Stasiuk GJ, Calaminus SDJ. Quantum dot imaging agents: Haematopoietic cell interactions and biocompatibility. *Cells [Internet]*. 2024 Feb 18 [cited 2024 Aug 12];13(4):354. Available from: www.mdpi.com/2073-4409/13/4/354/htm

154. de Oliveira JM, Silva DPD, Floresta LRS, Rocha GG, Almeida LIM, Dias EHV., et al. Tuning biocompatibility and bactericidal efficacy as a function of doping of gold in ZnO nanocrystals. *ACS Omega [Internet]*. 2024 May 21 [cited 2024 Aug 12];9(20):21904–16. Available from: https://pubs.acs.org/doi/full/10.1021/acsomega.3c09680

155. Paramo L, Jiménez-Chávez A, Medina-Ramirez IE, Böhnel HN, Escobar-Alarcón L, Esquivel K. Biocompatibility evaluation of $TiO_2$, $Fe_3O_4$, and $TiO_2/Fe_3O_4$ nanomaterials: Insights into potential toxic effects in erythrocytes and HepG2 cells. *Nanomaterials [Internet]*. 2023 Nov 1 [cited 2024 Aug 12];13(21):2824. Available from: www.mdpi.com/2079-4991/13/21/2824/htm

156. Romano M, González Gómez MA, Santonicola P, Aloi N, Offer S, Pantzke J, et al. Synthesis and characterization of a biocompatible nanoplatform based on silica-embedded SPIONs functionalized with polydopamine. *ACS Biomaterials Science & Engineering [Internet]*. 2023 Jan 9 [cited 2024 Aug 12];9(1):303–17. Available from: https://pubs.acs.org/doi/abs/10.1021/acsbiomaterials.2c00946

157. Paydary P, Larese-Casanova P. Water chemistry influences on long-term dissolution kinetics of CdSe/ZnS quantum dots. *Journal of Environmental Sciences*. 2020 Apr 1;90:216–33.

158. Khanal BP, Zubarev ER. Gram-scale synthesis of isolated monodisperse gold nanorods. *Chemistry—A European Journal [Internet]*. 2019 Jan 28 [cited 2024 Aug 12];25(6):1595–600. Available from: https://onlinelibrary.wiley.com/doi/full/10.1002/chem.201805571

159. Ovejero JG, Gallo-Cordova A, Roca AG, Morales MP, Veintemillas-Verdaguer S. Reproducibility and scalability of magnetic nanoheater synthesis. *Nanomaterials [Internet]*. 2021 Aug 1 [cited 2024 Aug 12];11(8):2059. Available from: www.mdpi.com/2079-4991/11/8/2059/htm

160. Baba YD, Chiacchia M, Patwardhan SV. A novel method for understanding the mixing mechanisms to enable sustainable manufacturing of bioinspired silica. *ACS Engineering Au [Internet]*. 2022 Feb 15 [cited 2024 Aug 12];3(1):17–27. Available from: https://pubs.acs.org/doi/full/10.1021/acsengineeringau.2c00028

161. Tao CL, Ma J, Wei C, Xu D, Xie Z, Jiang Z, et al. Scalable synthesis of high-quality core/shell quantum dots with suppressed blinking. *Advanced Optical Materials* [Internet]. 2023 Sep 1 [cited 2024 Aug 12];11(17):2300533. Available from: https://onlinelibrary.wiley.com/doi/full/10.1002/adom.202300533

162. Fernandes C, Jathar M, Sawant BKS, Warde T. Scale-up of nanoparticle manufacturing process. In: *Pharmaceutical Process Engineering and Scale-up Principles*. Springer; 2023. p. 173–203.

163. Hammed A, Voronov A, Pryor S. Scalable and sustainable synthesis of superparamagnetic iron oxide nanoparticles using new tubular electrochemical system. *Journal of Applied Electrochemistry [Internet]*. 2021 Jul 1 [cited 2024 Aug 12];51(7):1093–100. Available from: https://link.springer.com/article/10.1007/s10800-021-01560-2

164. Gulin-Sarfraz T, Jonasson S, Wigenstam E, Von Haartman E, Bucht A, Rosenholm JM. Feasibility study of *Mesoporous silica* particles for pulmonary drug delivery: Therapeutic treatment with dexamethasone in a mouse model of airway inflammation. *Pharmaceutics [Internet]*. 2019 Apr 1 [cited 2024 Aug 12];11(4):149. Available from: www.mdpi.com/1999-4923/11/4/149/htm

165. Ahmad A, Ganguly P. Quantum dots in drug delivery: The artificial atom. *International Journal of Chemical and Environmental Sciences*. 2023 Aug 28;4(4):22–30.

166. Kawasaki R, Kondo K, Miura R, Yamana K, Isozaki H, Shimada R, et al. Theranostic agent combining fullerene nanocrystals and gold nanoparticles for photoacoustic imaging and photothermal therapy. *International Journal of Molecular Sciences [Internet]*. 2022 May 1 [cited 2024 Aug 12];23(9):4686. Available from: www.mdpi.com/1422-0067/23/9/4686/htm

167. Pourmadadi M, Rahmani E, Shamsabadipour A, Mahtabian S, Ahmadi M, Rahdar A, et al. Role of iron oxide ($Fe_2O_3$) nanocomposites in advanced biomedical applications: A state-of-the-art review. *Nanomaterials [Internet]*. 2022 Nov 2 [cited 2024 Aug 12];12(21):3873. Available from: www.mdpi.com/2079-4991/12/21/3873/htm

168. Xie X, Yue T, Gu W, Cheng WY, He L, Ren WY, et al. Recent advances in *Mesoporous silica* nanoparticles delivering siRNA for cancer treatment. *Pharmaceutics [Internet]*. 2023 Oct 17 [cited 2024 Aug 12];15(10):2483. Available from: www.mdpi.com/1999-4923/15/10/2483/htm

169. Niveditha M, Shelar VR, Bhalerao BM. Green synthesis: An alternative sustainable route for nanotechnology. *Journal of Advances in Biology & Biotechnology [Internet]*. 2024 Jun 26 [cited 2024 Aug 12];27(7):1054–69. Available from: https://journaljabb.com/index.php/JABB/article/view/1065

170. Khizar S, Zine N, Errachid A, Jaffrezic-Renault N, Elaissari A. Microfluidic-based nanoparticle synthesis and their potential applications. *Electrophoresis [Internet]*. 2022 Apr 1 [cited 2024 Aug 12];43(7–8):819–38. Available from: https://onlinelibrary.wiley.com/doi/full/10.1002/elps.202100242

171. Narayanaswamy V, Jagal J, Al-Omari IA, Khurshid H, Haider M, Obaidat IM, et al. Investigating antitumor therapeutic efficacy using magnetic hyperthermia of $Fe_3O_4$ nanoparticles. *Journal of Physics. Conference Series [Internet]*. 2024 Apr 1 [cited 2024 Aug 12];2751(1):012002. Available from: https://iopscience.iop.org/article/10.1088/1742-6596/2751/1/012002

172. Shahid N, Erum A, Hanif S, Malik NS, Tulain UR, Syed MA. Nanocomposite hydrogels—a promising approach towards enhanced bioavailability and controlled drug delivery. *Current Pharmaceutical Design [Internet]*. 2023 Dec 29 [cited 2024 Aug 12];30(1):48–62. Available from: www.eurekaselect.com/article/136835

173. Ateeq H, Zia A, Husain Q, Khan MS. Role of inorganic nanocomposite materials in drug delivery systems. *Synthesis and Applications of Nanomaterials and Nanocomposites. Composites Science and Technology*. 2023 [cited 2024 Aug 12];171–95. Available from: https://link.springer.com/chapter/10.1007/978-981-99-1350-3_7

174. Dawoud MHS, Fayez AM, Mohamed RA, Sweed NM. Optimization of nanovesicular carriers of a poorly soluble drug using factorial design methodology and artificial neural network by applying quality by design approach. *Pharmaceutical Development and Technology [Internet]*. 2021 Nov 26 [cited 2024 Aug 12];26(10):1035–50. Available from: www.tandfonline.com/doi/abs/10.1080/10837450.2021.1980009

175. Barbosa RM, Lima CC, Oliveira FF, Câmara GBM, Viseras C, Moura TFAL, et al. New machine learning approach for the optimization of nano-hybrid formulations. *Nanomanufacturing [Internet]*. 2022 Jul 18 [cited 2024 Aug 12];2(3):82–97. Available from: www.mdpi.com/2673-687X/2/3/7/htm

176. Simelane NWN, Abrahamse H. Actively targeted photodynamic therapy in multicellular colorectal cancer spheroids via functionalised gold nanoparticles. *Artificial Cells, Nanomedicine, and Biotechnology [Internet]*. 2024 Dec 31 [cited 2024 Aug 12];52(1):309–20. Available from: www.tandfonline.com/doi/abs/10.1080/21691401.2024.2357693

177. Goel D, Santhiya D, Kumar S, Kumar Mahapatro A. Synthesis of mesoporous core shell magnetite bioactive glass nanoparticles for magnetic hyperthermia treatment of cancer. *ChemistrySelect [Internet]*. 2024 Jan 26 [cited 2024 Aug 12];9(4):e202302119. Available from: https://onlinelibrary.wiley.com/doi/full/10.1002/slct.202302119

178. Zielińska A, Costa B, Ferreira MV., Miguéis D, Louros JMS, Durazzo A, et al. Nanotoxicology and nanosafety: Safety-by-design and testing at a glance. *International Journal of Environmental Research and Public Health [Internet]*. 2020 Jun 28 [cited 2024 Aug 12];17(13):4657. Available from: www.mdpi.com/1660-4601/17/13/4657/htm

179. Soares S, Sousa J, Pais A, Vitorino C. Nanomedicine: Principles, properties, and regulatory issues. *Frontiers in Chemistry [Internet]*. 2018 Aug 1 [cited 2024 Aug 12];6(AUG):356901. Available from: www.frontiersin.org

180. Đorđević S, Gonzalez MM, Conejos-Sánchez I, Carreira B, Pozzi S, Acúrcio RC, et al. Current hurdles to the translation of nanomedicines from bench to the clinic. *Drug Delivery and Translational Research [Internet]*. 2021 Jul 23 [cited 2024 Aug 12];12(3):500–25. Available from: https://link.springer.com/article/10.1007/s13346-021-01024-2

# 6 Carbon-Based Nanocarriers

## *Carbon Nanotubes, Graphene Oxide, Fullerenes, and Carbon Dots*

*Cristian F. Rodríguez, Paula Guzmán-Sastoque, Coryna Rodriguez-Bazurto, Juan C. Rojas-Hernández, Luis H. Reyes, and Juan C. Cruz*

## ABSTRACT

Carbon-based nanocarriers—graphene, carbon nanotubes (CNTs), carbon dots (CDs), and fullerenes—represent transformative platforms at the nexus of nanotechnology and biomedicine. This chapter extensively reviews their historical evolution, groundbreaking discoveries, and sophisticated synthesis methodologies, detailing top-down approaches like mechanical exfoliation and bottom-up processes such as chemical vapor deposition. Emphasizing their unique physicochemical properties, including graphene's unparalleled mechanical strength and conductivity, CNTs' remarkable flexibility, CDs' sustainable production and tunable fluorescence, and fullerenes' versatile structural adaptability, the chapter underscores their revolutionary implications for medicine. Biomedical applications spotlight the robust potential of these nanomaterials in targeted drug and gene delivery, precise bioimaging, therapeutic peptide immobilization, and innovative therapies such as photothermal and photodynamic treatments. While their biocompatibility, biodegradability, and potential toxicity remain critical challenges, ongoing advancements in surface functionalization and nanocarrier engineering offer promising strategies to mitigate risks. Addressing these challenges is vital for translating laboratory breakthroughs into clinical realities. This comprehensive overview underscores how carbon-based nanocarriers are poised to redefine therapeutic paradigms, driving forward the future of personalized medicine.

## 6.1 INTRODUCTION

In 1959, Professor Richard P. Feynman delivered a seminal lecture titled "There's Plenty of Room at the Bottom," marking the inception of nanomaterial design and analysis (see Figure 6.1) [1]. Since then, the field of nanomaterials has experienced remarkable growth, with millions of articles investigating the design, synthesis, and characterization of these materials [2]. Among the various nanomaterials being developed, carbon-based nanomaterials have garnered significant interest due to their exceptional properties [3–5]. These properties, which include high electrical conductivity, remarkable mechanical strength, excellent thermal stability, and unique optical characteristics, have revolutionized multiple fields [4, 6, 7]. Notable examples of carbon nanomaterials include graphene, carbon dots (CDs), carbon nanotubes (CNTs), and fullerenes [8–13].

The journey of carbon-based nanomaterials began long before their current prominence. Graphene, for instance, can trace its origins to the mid-19th century with early studies on graphite oxide [14]. However, the true breakthrough came in 2004 when Sir Andre Geim and Sir Kostya Novoselov successfully isolated monolayer graphene, earning them the Nobel Prize in Physics in 2010 (see Figure 6.1) [15–17]. This discovery sparked an explosion of research into graphene's extraordinary properties. Similarly, CNTs were discovered in 1991 by Sumio Iijima, although their existence was noted much earlier by Soviet scientists in the 1950s [18, 19]. The cylindrical structure of CNTs, formed by rolling graphene sheets, endowed them with exceptional strength and electrical properties, making them a focal point in nanotechnology research [20]. CDs were discovered more recently, in 2004, as a by-product during the purification of single-walled CNTs [21]. Their small size, excellent biocompatibility, and tunable fluorescence have made them attractive alternatives to traditional quantum dots (QDs), particularly in biomedical imaging and sensing [22]. Fullerenes were first discovered in 1985 by Harold Kroto, Robert Curl, and Richard Smalley while studying carbon clusters formed by vaporizing graphite [23]. This discovery, which also won a Nobel Prize in Chemistry in 1996, revealed the unique spherical structure of $C_{60}$, a molecule that has since become a symbol of nanotechnology's potential [24] (see Figure 6.1).

The exceptional properties of carbon-based nanomaterials are central to their applications in various fields, particularly in biomedicine [6]. Graphene and CNTs exhibit high electrical conductivity, far surpassing traditional materials like copper [25]. This makes them ideal for use in electronics, biosensors, and energy storage devices where efficient charge transfer is crucial [25]. The mechanical strength of graphene and CNTs is unparalleled. Graphene is known for its tensile strength, which is 200 times greater than that of steel, despite being just one atom thick [26, 27]. CNTs also exhibit extraordinary mechanical strength and flexibility, making them suitable for reinforcement materials and as nanocarriers capable of withstanding physiological conditions [28]. Both graphene and CNTs possess excellent thermal conductivity, with graphene's thermal conductivity being approximately ten times higher than that of copper [29]. This property is invaluable in applications requiring efficient heat dissipation, such as in high-performance electronics. CDs

DOI: 10.1201/9781003473183-6

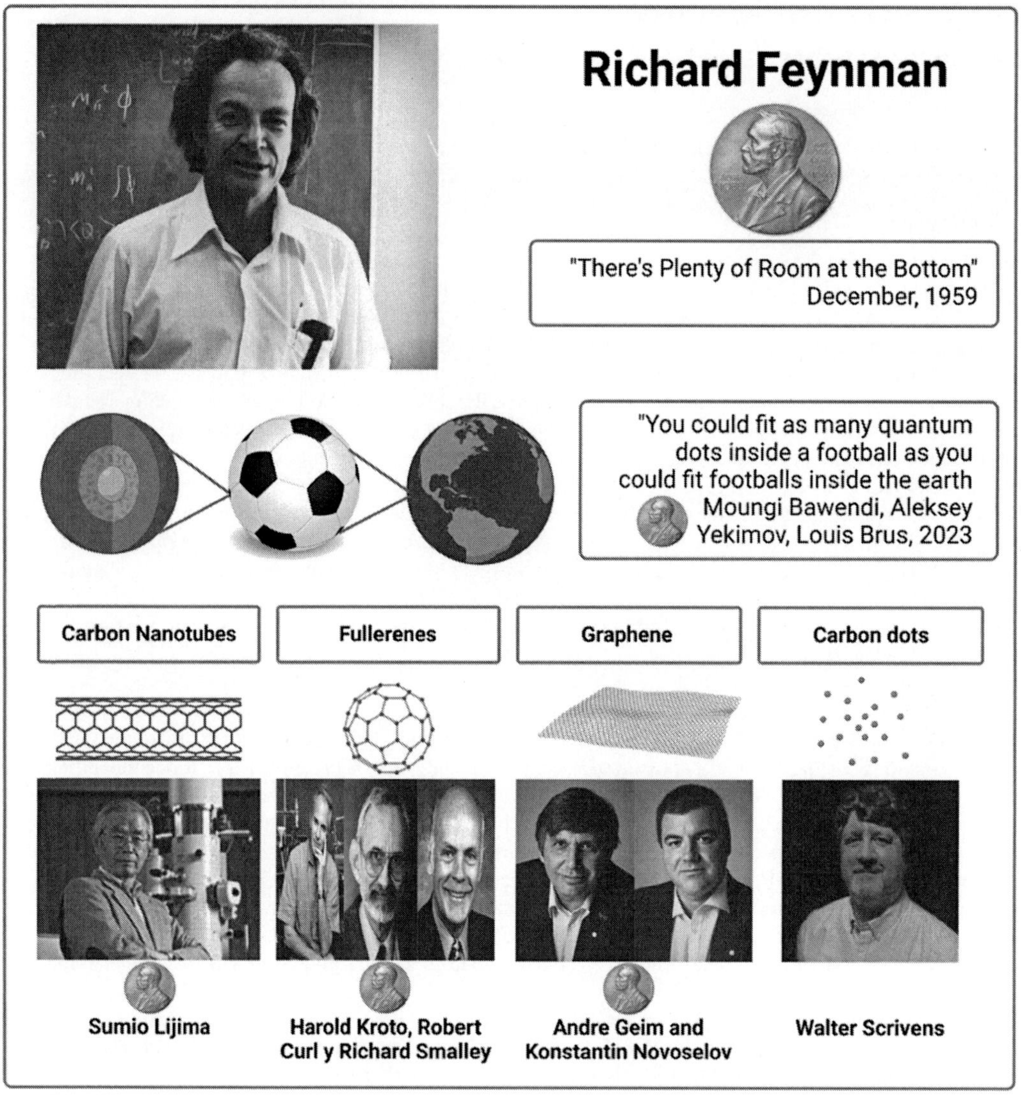


**Figure 6.1** Key milestones in the development and recognition of carbon-based nanomaterials. The top section honors Professor Richard P. Feynman, whose 1959 lecture "There's Plenty of Room at the Bottom" laid the groundwork for the field of nanomaterial science. The figure also showcases significant contributions by Professor Sumio Iijima, who discovered CNTs in 1991; Professors Harold Kroto, Robert Curl, and Richard Smalley, who discovered fullerenes in 1985; Professors Sir Andre Geim and Sir Kostya Novoselov, who isolated monolayer graphene in 2004; and Professor Walter Scrivens, who discovered CDs in 2004. These materials are characterized by their small size and high surface area. As Professor Moungi Bawendi mentioned in his Nobel Prize lecture on CDs, "You could fit as many quantum dots inside a football as you could fit footballs inside the Earth."

are particularly noted for their unique optical properties. CDs exhibit tunable fluorescence, which is highly desirable for bioimaging and diagnostic applications [30].

Carbon-based nanomaterials have shown great promise in the biomedical field due to their ability to interact with biological systems at the molecular level [6]. The ultra-thin structure, high surface area, and amenability for functionalization of graphene make it an excellent candidate for drug delivery systems, particularly for delivering nucleic acids (NAs) and peptides [31]. Its biocompatibility and ability to be chemically modified allow for targeted delivery and controlled release of therapeutic agents [32]. CNTs have been extensively explored as carriers for DNA, RNA, and proteins due to their needle-like structure, which facilitates cellular uptake [33]. Their ability to protect genetic material from degradation and their high surface area for drug loading make them promising tools for gene therapy and cancer treatment [33]. CDs, with their biocompatibility and fluorescence properties, are ideal for bioimaging and biosensing [34]. Their small size allows for easy penetration into cells, and their surface can be functionalized to enhance specificity for targeted delivery of therapeutic molecules [35]. Fullerenes' ability to undergo various chemical modifications allows them to serve as versatile nanocarriers for drug and gene delivery [36]. Their unique structure enables them to protect NAs from degradation, facilitate targeted delivery, and act as photodynamic agents in cancer therapy [37].

The advantages of carbon-based nanomaterials over traditional materials in nanomedicine are evident in their superior performance in drug delivery, gene therapy, and diagnostics [38]. Their scalability, biocompatibility, and ability to be functionalized for specific applications make them indispensable tools in the development of next-generation medical treatments [39, 40]. However, the translation of these materials from the laboratory to clinical use requires overcoming challenges related to scalability, biocompatibility, and environmental impact [41]. This chapter will delve into the history, synthesis methods, properties, and biomedical applications of graphene, CNTs, CDs, and fullerenes. It will also address the challenges and future directions for research in this field, providing a comprehensive understanding of how these nanomaterials are poised to revolutionize medicine.

## 6.2 GRAPHENE AS A NANOCARRIER

### 6.2.1 History and Development

The isolation of graphene is the result of multiple scientific contributions over an extended period. This process can be traced back to the mid-19th century, starting with the work of Sir Benjamin Collins Brodie, who served as President of the Royal Society from 1858 to 1861 [42, 43] (see Figure 6.2). Brodie studied the synthesis of crystalline graphitic acid and patented a method for the purification of graphite [44, 45]. He was the first to describe the formation of graphite oxide in 1855 [46–48]. In 1859, he published his subsequent work, where he presented the Brodie method for the reaction of graphite with potassium chlorate in fuming nitric acid [45, 49, 50].

Brodie's results were supported and confirmed in 1865 by Gottschalk in the publication "Beiträge zur Kenntnis der Graphitsäure." With the development of X-ray diffraction (XRD) techniques, Hofmann and collaborators were able to analyze graphite oxide, producing the first widely accepted structural findings [51].

In 1947, P. R. Wallace predicted the electronic properties of graphene, including its unique electronic structure and the linear dispersion relation [52–54] (see Figure 6.2). Wallace's predictions sparked interest in the potential production of nanomaterials with these remarkable properties.

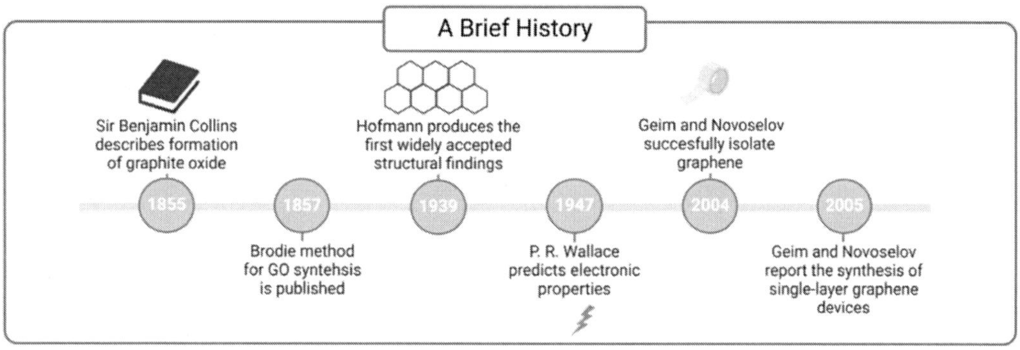


**Figure 6.2** Timeline that outlines the key historical milestones leading to the isolation of graphene. The journey began in 1855 with Sir Benjamin Collins Brodie, who first described the formation of graphite oxide and later developed the Brodie method for its synthesis. In 1939, Hofmann produced the first widely accepted structural findings of graphite oxide. P. R. Wallace's 1947 predictions about graphene's electronic properties further fueled interest in this material. The timeline culminates with Sir Andre Geim and Sir Kostya Novoselov's groundbreaking work in 2004, where they successfully isolated graphene using the mechanical exfoliation method, followed by their report on the synthesis of single-layer graphene devices in 2005. These contributions collectively paved the way for the development of graphene as a revolutionary material in nanotechnology.

This scientific race reached its pinnacle in 2004 when Sir Andre Geim and Sir Kostya Novoselov successfully isolated graphene [15] (see Figure 6.2). They employed a technique known as mechanical exfoliation, commonly referred to as the "Scotch tape method," to peel off flakes of graphite from a block of highly oriented pyrolytic graphite [55]. Among these flakes, they discovered several that were only a single atom thick, effectively isolating monolayer graphene. This groundbreaking discovery challenged the prevailing assumption that a single layer of graphene would be too fragile to exist independently.

In 2005, Geim and Novoselov reported the creation of single-layer graphene devices, marking a significant milestone in the field of nanotechnology [16]. Their work not only demonstrated the practical potential of graphene but also laid the foundation for future research and applications in various technological fields.

### 6.2.2 Structural and Physical Properties

Graphene, a single layer of carbon atoms arranged in a two-dimensional honeycomb or hexagonal lattice, is an allotrope of carbon just one atom thick [56–58] (see Figure 6.3). Each carbon atom in this lattice forms three σ bonds with its neighboring atoms through $sp^2$ hybridization, which involves the mixing of one *s* orbital and two *p* orbitals (px and py) to generate in-plane bonds [59]. These σ bonds create a robust and stable planar network. Meanwhile, the unhybridized *pz* orbital extends perpendicular to the plane, forming π bonds. It is these π bonds that are chiefly responsible for graphene's exceptional electrical conductivity, as they allow electrons to move freely across the lattice [60].

The interatomic distance between neighboring carbon atoms in graphene is approximately 0.142 nm, significantly smaller than in many other carbon-based materials, contributing to its remarkable mechanical strength and elasticity [59, 61]. Despite being ultra-thin, at approximately 0.35 nm, graphene remains exceptionally stable due to the strength of its carbon–carbon covalent bonds [59, 62, 63]. This structural integrity enables it to withstand external forces without breaking or altering its atomic arrangement. Additionally, the slight twisting of the lattice allows for force redistribution across the plane, enhancing its flexibility without compromising stability.

A fascinating feature of graphene is its edge configuration, which can exist in two distinct forms: zigzag or armchair, depending on the arrangement of the carbon atoms [62, 64] (see Figure 6.3). Zigzag-edged graphene exhibits metallic characteristics and facilitates electron conduction [65, 68]. In contrast, armchair-edged graphene can exhibit either metallic or semiconducting properties, depending on its structural arrangement. This tunable electronic behavior opens up exciting possibilities for graphene's application in advanced electronics [69–72].

Graphene is also extraordinarily lightweight, with a density of approximately 0.77 mg/m², which makes it highly attractive for applications where weight reduction is crucial, such as in aerospace engineering and

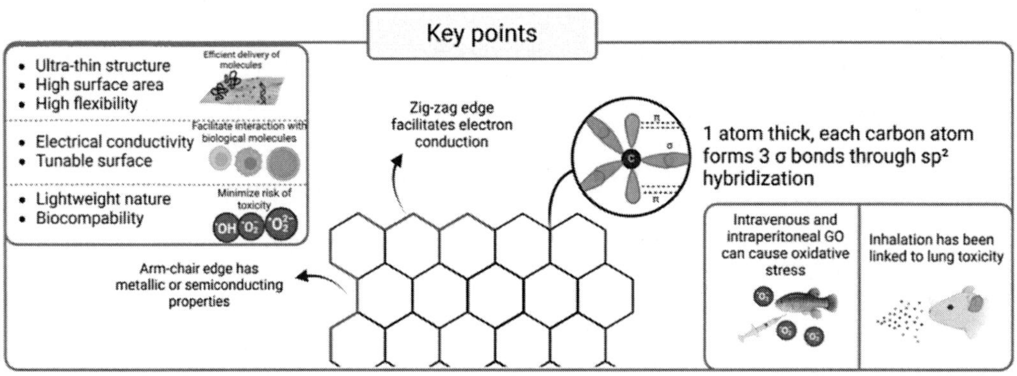


**Figure 6.3** Key points about graphene's structure and properties. Graphene, a single layer of carbon atoms arranged in a hexagonal lattice, is characterized by its ultra-thin structure, high surface area, and flexibility. Each carbon atom forms three σ bonds through $sp^2$ hybridization, creating a robust planar network, while the π bonds contribute to its exceptional electrical conductivity. The zigzag edge facilitates electron conduction, while the armchair edge can exhibit metallic or semiconducting properties. Graphene's lightweight nature, high tensile strength, and biocompatibility make it an ideal material for drug delivery systems, allowing for efficient molecule delivery and minimizing toxicity risks. Additionally, graphene's electron mobility is remarkably high, supporting applications in bio-sensing and precision drug delivery. However, potential toxicity concerns, particularly with GO, such as oxidative stress and lung toxicity upon inhalation, are noted.

portable electronics. Furthermore, graphene has an exceptional tensile strength of 125 GPa, making it one of the strongest known materials [56, 73]. This combination of low density and high tensile strength has generated significant interest in graphene as a potential material for diverse applications.

Equally remarkable are graphene's thermal properties. It boasts a thermal conductivity of around 5000 W/mK, which is approximately ten times higher than that of copper [74, 75]. This exceptional thermal conductivity makes graphene an ideal candidate for thermal management in high-performance electronics and cooling systems, where efficient heat dissipation is crucial [76].

Perhaps one of the most groundbreaking characteristics of graphene is its ultra-high electron mobility, which has been measured at approximately $2 \times 10^5$ $cm^2/V*s$ over 140 times greater than that of silicon, the convectional semiconductor material used in electronics [56]. This extraordinary electron mobility arises from graphene's $sp^2$ hybridization and the resulting delocalized $\pi$-bond network, which allows electrons to traverse the material with minimal resistance [46, 56] (see Figure 6.3).

Graphene's unique combination of properties makes it highly promising for drug delivery systems. Its ultra-thin structure, high surface area, and flexibility allow for efficient loading and delivery of therapeutic molecules. The exceptional tensile strength and stability of graphene ensure that it can withstand various physiological conditions without degradation during transport. Moreover, graphene's remarkable electrical conductivity and tunable surface chemistry facilitate interaction with biological molecules, allowing for controlled and target release. Additionally, its lightweight nature and biocompatibility minimize the risk of toxicity, while its high electron mobility supports potential applications in biosensing and precision drug delivery systems. These properties collectively make graphene an ideal candidate for advanced drug delivery applications, enabling the efficient and precise administration of therapeutic peptides and NAs.

### 6.2.3 Synthesis Methods

Various methods and approaches have been developed to synthesize graphene nanosheets, which can be categorized into two main strategies: top-down and bottom-up methods [55, 77–79]. The top-down methods involve breaking down bulk graphite into individual graphene layers through techniques such as mechanical exfoliation, electrochemical exfoliation, chemical oxidation-reduction processes, and liquid-phase exfoliation [55, 80]. In contrast, bottom-up methods synthesize graphene from carbon-containing precursors, often through chemical vapor deposition (CVD), thermal pyrolysis, direct organic synthesis, and epitaxial growth [55, 79, 81] (see Figure 6.4).

One of the most historically significant top-down methods is mechanical exfoliation, pioneered by Andre Geim and Konstantin Novoselov, who were awarded the Nobel Prize for "groundbreaking experiments regarding the two-dimensional material graphene" [82]. This method, often referred to as the Scotch tape method, involves using adhesive tape to peel off single-layer graphene from highly oriented pyrolytic graphite [55, 80]. While this technique produces high-quality graphene with excellent structural integrity, its low yield and limited scalability restrict its use primarily in laboratory research [83]. The method relies on weak van der Waals forces between graphite layers, which enable the peeling and sliding of graphene sheets, although these forces also limit the technique's utility for mass production [83, 84] (see Figure 6.4).

Ball milling is another widely used top-down approach for graphene production. This method applies shear forces through grinding media, typically metallic balls, to break down graphite into graphene flakes [77, 81]. Recent innovations have sought to make ball milling more environmentally friendly, such as by using ethanol to remove surfactants [81]. However, a major drawback of this method is the potential contamination of the graphene with metallic impurities from the grinding media [85]. The size and quality of the resulting graphene depend on specific milling parameters and media, whether the milling is conducted under wet or dry conditions [86] (see Figure 6.4).

Ultrasonication is frequently employed in liquid-phase exfoliation of van der Waals materials like graphene [87]. This method leverages cavitation to create high-quality graphene dispersions. However, ultrasonication is often used in conjunction with other methods due to its low exfoliation rates and variable flake size [88]. Recent advances have improved graphene yields by optimizing parameters such as cavitation dose, sonication time, and temperature control, although the technique remains limited by its inability to scale efficiently [89–92] (see Figure 6.4).

One of the most effective techniques for large-scale graphene production is the chemical oxidation-reduction method [83]. This approach involves the oxidative exfoliation of graphite to produce graphene oxide (GO), which is then reduced to form reduced graphene oxide (rGO) [84]. GO serves as an important precursor due to its straightforward synthesis and scalability. The oxidative process introduces oxygen-containing groups into the graphite structure, which disrupts the van der Waals forces and facilitates exfoliation into single-, double-, or few-layer GO sheets [85, 86].

Several well-established oxidation methods are employed to produce GO, including the Brodie, Staudenmaier, Hofmann, Hummers, and Tour methods [93]. These processes typically utilize strong

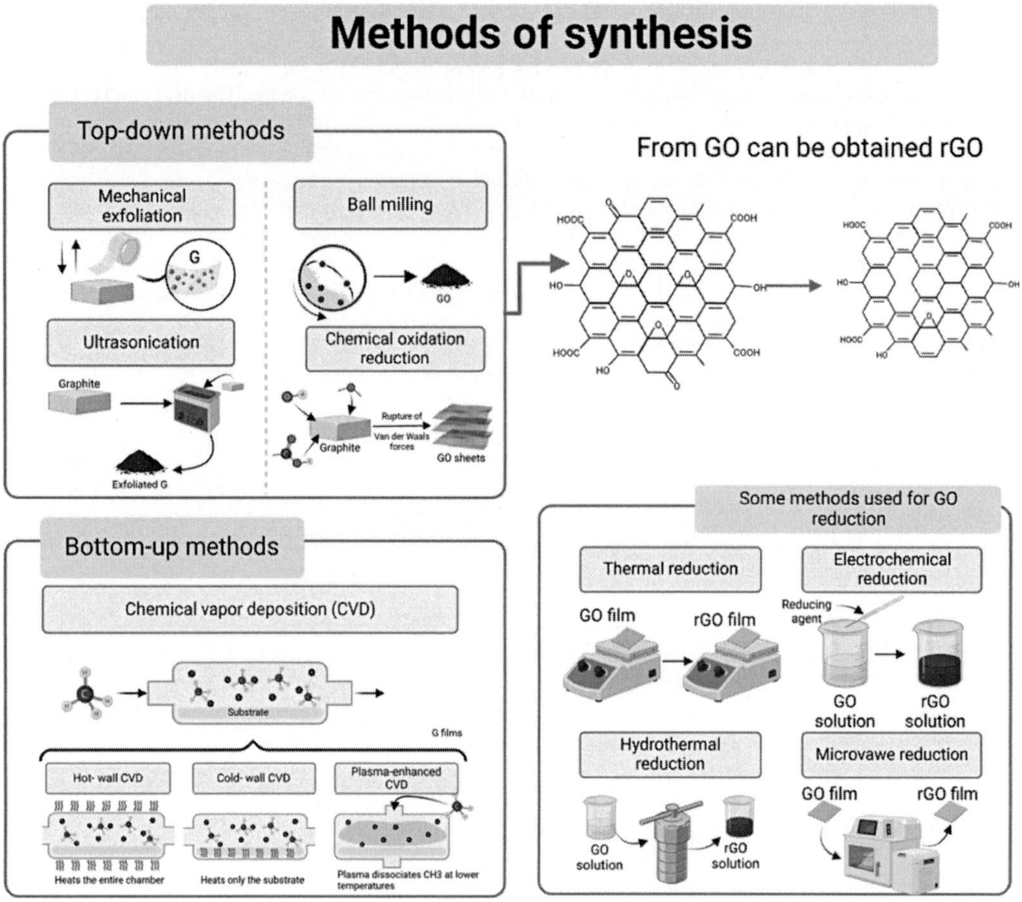


**Figure 6.4** Overview of graphene synthesis methods, categorized into top-down and bottom-up approaches. Top-down methods, such as mechanical exfoliation, ball milling, and chemical oxidation-reduction, break down bulk graphite into graphene layers. Bottom-up methods, like CVD, synthesize graphene from carbon precursors. The figure also illustrates various reduction methods for converting GO into rGO, highlighting the diverse techniques used to optimize graphene for different applications.

acids like sulfuric, phosphoric, or nitric acid, in conjunction with oxidizing agents such as potassium permanganate or sodium nitrate [94]. Over time, these methods have been refined to enhance safety, minimize toxic by-products, and improve efficiency. The Hummers method, developed in 1958, is one of the most widely used because of its relatively quick reaction time and improved safety over older methods, although it still produces nitrogen-based toxic gases [95]. In 2010, the Tour method further improved upon Hummers' by replacing sodium nitrate with phosphoric acid, resulting in higher oxidation levels, fewer defects, and improved safety, making it more suitable for industrial-scale GO production [96].

Once GO is synthesized, it can be reduced to graphene using a variety of techniques, including thermal, chemical, electrochemical, hydrothermal, microwave, and photoreduction methods [55, 78]. The primary goal of these reduction processes is to remove the oxygen groups introduced during oxidation and repair the structural defects, thereby restoring graphene's intrinsic properties [55, 97, 98]. Each reduction method offers different advantages depending on the intended application. For instance, thermal reduction, typically performed at temperatures above 1000°C, effectively removes oxygen groups and restores the graphitic structure [99]. However, this process can cause significant mass loss, resulting in wrinkled and defect-rich graphene sheets. In contrast, chemical reduction is more cost-effective and scalable, utilizing reducing agents like hydrazine or environmentally friendly alternatives such as ascorbic acid to produce rGO with fewer [100].

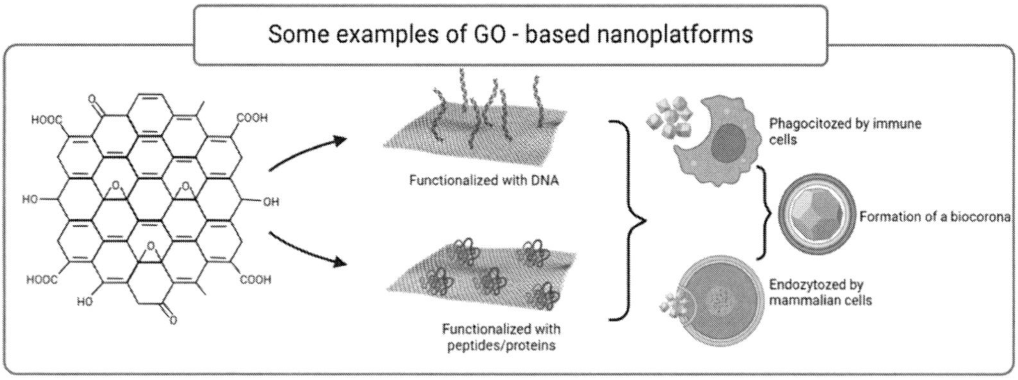


**Figure 6.5** Examples of GO-based nanoplatforms, illustrating the functionalization of GO with DNA and peptides/proteins. The figure shows how these functionalized nanoplatforms can interact with biological systems: they can be phagocytosed by immune cells, leading to the formation of a biocorona, or be endocytosed by mammalian cells. These interactions highlight the versatility of GO in gene and peptide delivery systems, demonstrating its potential in advanced biomedical applications due to its strong binding capabilities and protective properties.

While the chemical oxidation-reduction method shows great potential for large-scale graphene production, ongoing research is focused on optimizing these processes to reduce costs, improve safety, and enhance the quality of the final graphene product. Particular attention is being given to developing greener reduction methods that minimize environmental impact and reduce reliance on hazardous chemicals.

In terms of bottom-up methods, CVD stands out as one of the most commonly used approaches for graphene synthesis [101]. In CVD, carbon atoms are deposited onto a substrate through the decomposition of gaseous hydrocarbons, such as methane ($CH_4$), at high temperatures [102]. These carbon atoms adsorb onto the substrate surface and self-assemble into a graphene layer through chemical reactions. CVD is favored for its ability to produce large-area, single-layer, and few-layer graphene films at relatively low cost with high scalability, making it a crucial technology for both research and industrial applications [101]. Multiple factors, such as temperature, pressure, and gas flow rates, influence the final properties of the graphene, including thickness, grain size, and defect density [103] (see Figure 6.4).

CVD methods can be classified based on their heating configurations, pressure conditions, wall/substrate arrangements, precursor nature, gas flow state, and activation mechanisms [104]. Hot-wall CVD, one of the most common approaches, heats the entire reaction chamber, ensuring uniform temperature distribution, which facilitates the growth of high-quality graphene [101]. This method is particularly effective for synthesizing large-area, monolayer graphene, although careful control of gas ratios and pressures is required to minimize the formation of multilayer graphene [105]. In contrast, cold-wall CVD heats only the substrate, while the chamber remains at room temperature. This technique reduces energy consumption and allows for faster heating and cooling cycles, making it more cost-efficient while also minimizing side reactions that could introduce impurities into the graphene film [114, 106].

Plasma-enhanced CVD (PECVD) is another variation of the process, in which plasma is used to dissociate the hydrocarbon gases at lower temperatures compared to traditional thermal CVD [114, 107]. The plasma enhances the reactivity of the gases, allowing for more precise control over the graphene nanostructure and faster growth rates [108]. PECVD is especially useful for growing graphene on nonmetal substrates, eliminating the need for the post-synthesis transfer that is often required when using metal substrates.

### 6.2.4 Graphene-Based Nanomaterials in Biomedical Research

Graphene-based nanomaterials (GBNs), encompassing GO, rGO, and graphene quantum dots (GQDs), have garnered significant attention in biomedical research, particularly within the domain of gene therapy [109–111]. These materials, derived from graphene or graphite, exhibit unique physicochemical properties—such as tunable layer thickness, surface chemistry, and nanoscale dimensions—that render them suitable for various medical applications [112]. GO, for instance, is synthesized through the oxidation of graphite, introducing oxygen-containing functional groups (e.g., hydroxyl, epoxy, and carboxyl groups) that enhance its dispersibility in polar solvents. This increased dispersibility facilitates chemical modifications, positioning GO as a promising platform for drug delivery and gene therapy [113].

However, the oxidative process disrupts the $sp^2$ hybridized carbon network of GO, diminishing its mechanical, electrical, and thermal properties [114]. To mitigate these issues, GO is often reduced to rGO through thermal, chemical, or UV treatments [115]. This reduction partially restores its conjugated structure, improving electrical conductivity while preserving its biocompatibility [116, 117]. Due to its capacity for adsorbing polymers and small molecules through non-covalent interactions (e.g., π–π stacking and van der Waals forces), rGO has shown promise in applications such as cell culture, biosensing, and tissue engineering.

The internalization of GO into mammalian cells typically occurs via endocytosis, with macropinocytosis, clathrin-mediated endocytosis, and caveolin-mediated endocytosis representing the most common pathways [118]. The specific route of cellular entry is contingent upon various factors, including particle size, surface functionalization, and the type of cell involved [119]. Larger GO particles are predominantly internalized by immune cells through phagocytosis, whereas smaller particles are internalized by non-phagocytic cells via clathrin- or caveolin-mediated mechanisms [120]. Surface modifications such as polyethylene glycol (PEG) coating or conjugation with targeting molecules (e.g., hyaluronic acid (HA)) can significantly enhance uptake efficiency and cell specificity, thereby improving GO's utility as a vehicle for gene and drug delivery [121]. Notably, GO's interactions with proteins in biological environments often result in the formation of a "biocorona." This biocorona can either facilitate or impede cellular uptake, adding an additional layer of complexity to the optimization of GO-based delivery systems [122–124].

#### *6.2.4.1 Graphene-Based Nanomaterials as Gene Delivery Platforms*

GBNs have emerged as compelling alternatives to traditional viral gene delivery systems, due to their distinctive physicochemical and optical properties [125]. Their two-dimensional structure, biocompatibility, and ease of chemical functionalization make them highly effective platforms for developing gene delivery vectors [126] (see Figure 6.5). One of the primary advantages of GBNs is their ability to engage in strong π–π stacking interactions and van der Waals forces, which enable the efficient binding of NAs while offering protection against enzymatic degradation [125, 126]. For instance, Alibolandi et al. [127] demonstrated that GO could effectively shield single-stranded DNA (ssDNA) from enzymatic degradation by developing a dextran-coated GO platform (GO-DEX). Moreover, when functionalized with the AS1411 aptamer—targeting nucleolin to enhance cellular uptake—the platform achieved a high curcumin loading capacity (~29%) via π–π stacking [128].

A notable breakthrough in gene delivery using GBNs is the development of nitrogen-doped graphene quantum dots (NGQDs) by Ahn et al. [129]. This approach gained particular relevance during the COVID-19 pandemic, where NGQDs were utilized as transfection agents for messenger RNA (mRNA) and plasmid DNA (pDNA). By using polyethyleneimine (PEI) and citric acid (CA) as precursors, the researchers synthesized positively charged NGQDs that formed stable electrostatic complexes with negatively charged NAs [129–131]. These NGQDs demonstrated transfection efficiencies comparable to, and in some cases exceeding, those of commercially available lipid nanoparticles (LNPs)—the latter being widely used in mRNA vaccines for COVID-19. Moreover, NGQDs exhibited superior stability at room temperature and lower cytotoxicity compared to LNPs, suggesting their potential as versatile and safer alternatives in gene therapy applications [132, 133].

#### *6.2.4.2 Peptide Immobilization on Graphene-Based Nanomaterials for Biomedical Applications*

Peptide immobilization on GBNs has also emerged as a promising strategy for biomedical applications, leveraging both covalent and non-covalent interactions [134, 135]. Non-covalent interactions such as π–π stacking, electrostatic forces, van der Waals interactions, and hydrophobic effects enable peptides with aromatic amino acids to bind to graphene's basal planes [136]. This non-covalent immobilization preserves the intrinsic properties of graphene, rendering it particularly suitable for sensitive applications such as biosensing [137]. For GO, the presence of oxygen-rich functional groups enhances its biocompatibility, facilitating peptide interaction through electrostatic forces and hydrogen bonding [138, 139]. Environmental factors like pH and ionic strength are crucial in determining the extent and stability of these interactions, highlighting the importance of careful experimental optimization.

In contrast, covalent immobilization offers a more robust method for attaching peptides to GBNs, particularly in applications requiring long-term stability [140, 141]. Cross-linkers such as 1-ethyl-3-(3-dimethylaminopropyl) carbodiimide (EDC) and N-hydroxysuccinimide (NHS) enable the formation of stable amide bonds between GO's carboxyl groups and peptides' amino groups [142, 143]. This covalent attachment not only ensures strong binding but also maintains stability under harsh environmental conditions, making it advantageous for diagnostic devices and enzyme immobilization [144]. For example, functionalization with PEG can enhance the hydrophilicity of GO, improving the solubility of peptide–graphene conjugates in aqueous environments while minimizing the nonspecific adsorption of other biomolecules [145].

The antimicrobial properties of peptide–GO conjugates have been particularly promising. For instance, Kumar et al. [142] developed an antimicrobial surface by covalently attaching the peptide nisin onto a sulfonated polyetheretherketone (SPEEK) substrate coated with GO. This SPEEK–GO–nisin composite exhibited potent antibacterial activity against *Staphylococcus aureus*, forming a 27 mm inhibition zone. The covalent attachment of nisin, facilitated by EDC cross-linking, not only enhanced antibacterial properties but also improved resistance to bacterial adhesion and biofilm formation. Compared to SPEEK–nisin and SPEEK–GO, the composite exhibited superior performance, making it a viable option for implantable devices in dental and orthopedic applications where both antimicrobial activity and biocompatibility are critical.

#### *6.2.4.3 Biocompatibility of GBNs in Gene Therapy and Peptide Delivery*

GBNs are garnering significant interest in biomedical applications, particularly for gene therapy and peptide delivery, due to their unique physicochemical properties [146]. Their large surface area, mechanical strength, and ease of functionalization make them highly attractive for delivering therapeutic molecules [146]. However, the biocompatibility of GBNs remains a critical concern, especially regarding their interactions with biological systems and potential toxicity. Thus, a thorough understanding of GBN biocompatibility is essential to optimize their use in gene therapy and peptide delivery while minimizing adverse effects.

GBNs can enter the body via multiple routes, including inhalation, intravenous injection, oral administration, and dermal exposure [142]. Once inside, their biodistribution and accumulation are influenced by several factors such as particle size, surface functionalization, and the biological environment they encounter [147]. For instance, protein corona formation on GBNs can modulate their interaction with cells and tissues, which, in turn, affects how they distribute across different organs and their eventual clearance from the body [142, 148]. This biodistribution is crucial for gene therapy and peptide delivery, where precise targeting and minimal off-target effects are key to therapeutic success.

Toxicity is another significant concern when considering the biomedical applications of GBNs. The size, oxidation state, and surface chemistry of graphene-based materials significantly influence their toxicity profiles [142, 149–151]. For example, graphene and GO have been shown to induce varying degrees of toxicity, particularly at higher concentrations [152–157]. Inhalation of graphene has been linked to lung toxicity in animal models, manifesting as local inflammation, mitochondrial damage, and compromised lung function [142, 158, 159]. Similarly, intravenous and intraperitoneal administration of GO can result in systemic toxicities such as oxidative stress and inflammation, primarily in organs like the liver and spleen [142, 160]. These observations underscore the need to carefully consider both dosage and exposure duration when utilizing GBNs in therapeutic contexts [161, 162].

In gene therapy and peptide delivery, the ability of GBNs to cross biological barriers is particularly important [142]. For instance, PEGylated GO has demonstrated the capacity to traverse the blood–brain barrier (BBB), which is a key requirement for central nervous system (CNS) gene therapy. Nevertheless, concerns about neurotoxicity persist [163, 164]. Accumulation of high levels of graphene in the brain can lead to oxidative stress and neuronal damage [165]. Similarly, the blood–testis barrier and the blood–epididymis barrier present challenges for reproductive health. Studies have shown that larger GO particles can reduce sperm motility and induce DNA damage, although these effects may be reversible over time [18, 166].

The surface functionalization of GBNs is crucial for enhancing biocompatibility in gene therapy and peptide delivery [18, 167]. Strategies such as PEGylation have been effective in reducing toxicity and improving the biocompatibility of GBNs [13]. PEGylated GBNs are more efficiently excreted through renal and fecal pathways, reducing long-term retention in critical organs. This property is particularly beneficial in gene therapy, where precise and safe delivery of therapeutic molecules is paramount. Furthermore, surface modifications can improve the targeting of specific tissues while reducing off-target effects and systemic toxicity, thereby enhancing the therapeutic potential of GBNs [18, 168, 169].

## 6.3 CNT AS NANOCARRIERS

### 6.3.1 History and Development

The discovery of CNTs is often credited to Iijima and Ichihashi in 1991 [170] when they published transmission electron microscopy (TEM) images showcasing CNTs with three, five, and seven concentric graphene shells [164]. However, the existence of CNTs was first reported in 1952 by Radushkevich and Lukyanovich [171], though their work was largely ignored due to their affiliation with the Soviet Union. The publication by Radushkevich and Lukyanovich in 1952 showcased multi-walled carbon nanotubes (MWCNTs) with a diameter of 50 nm [18, 165, 172]. Moreover, as early as 1889–1890, some research groups documented the growth of carbon filaments, which is now understood to be related to CNTs [165, 173, 174].

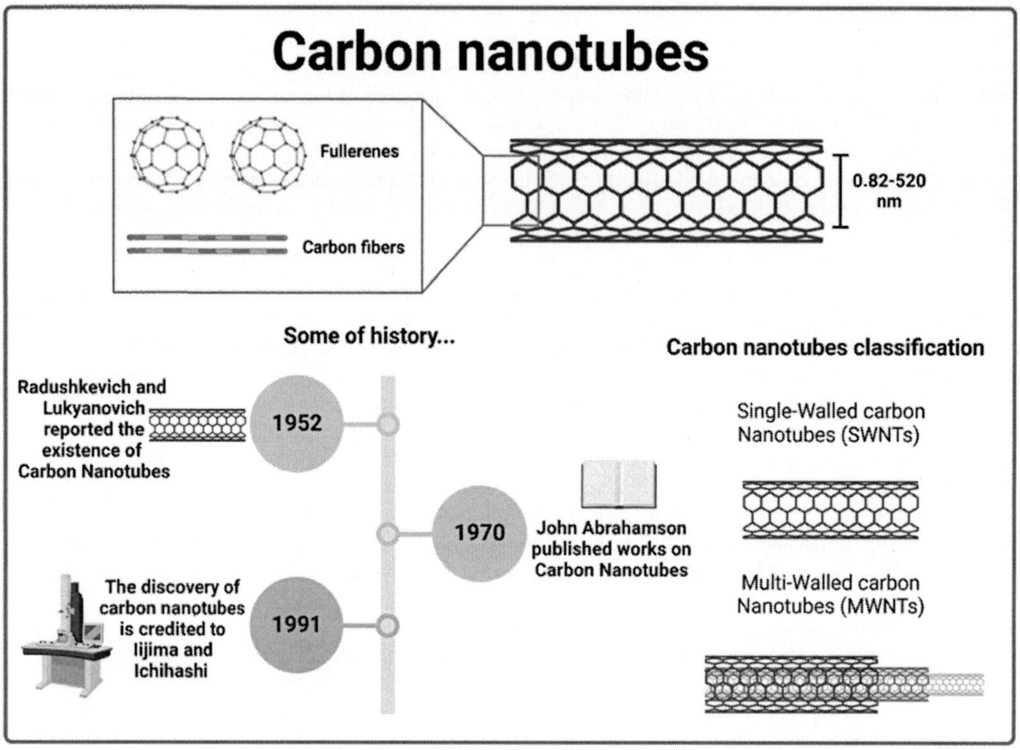


**Figure 6.6** Historical timeline and structural overview of CNTs. The timeline traces the discovery and documentation of CNTs, beginning with Radushkevich and Lukyanovich's 1952 report on MWCNTs, followed by John Abrahamson's 1970 work, and culminating in Iijima and Ichihashi's 1991 transmission electron microscopy (TEM) images that are widely credited with bringing CNTs to global attention. The figure also illustrates the cylindrical structure of CNTs, formed by rolling graphene sheets, and highlights the range in diameters of SWNTs and MWNTs, which exhibit remarkable thermal conductivity, flexibility, and strength.

In 1970, John Abrahamson also published works on CNTs, describing them as carbon fibers capable of growing under a low-pressure nitrogen atmosphere [18, 175; see Figure 6.6].

### 6.3.2 Structural and Physical Properties

CNTs are derived from both carbon fibers and fullerenes [18]. They are characterized by their cylindrical structure, formed by rolling graphene sheets. The diameter of CNTs can range from 0.82 to 520 nm, depending on whether they are SWNTs or MWNTs [13, 20]. One of the most outstanding properties of CNTs is their thermal conductivity [176]. Other notable properties include flexibility, strength, and high rigidity [176]. Generally, Young's modulus increases as the number of walls in the CNTs increases [176].

### 6.3.3 Synthesis Methods

#### *6.3.3.1 Laser Ablation*

Laser ablation is a technique that involves vaporizing graphite with a laser inside a quartz reactor [177]. A high-energy laser vaporizes graphite and a metal catalyst in an inert gas atmosphere at high temperatures. As the carbon plasma cools, it forms CNTs [178] (see Figure 6.7). First introduced in 1996 by Thess et al. [179] for the synthesis of SWNTs, this method involved vaporizing a carbon–nickel–cobalt mixture at 1200°C. It achieved a high yield of over 70%, producing nearly uniform SWNTs that self-organized into "ropes" containing 100 to 500 nanotubes. Key factors influencing this process include temperature, laser wavelength, power, pressure, and fluid dynamics near the carbon target [20]. While this method yields nanotubes of superior purity and quality, it also alters their structure and chemical composition, making it essential to carefully control the aforementioned parameters.

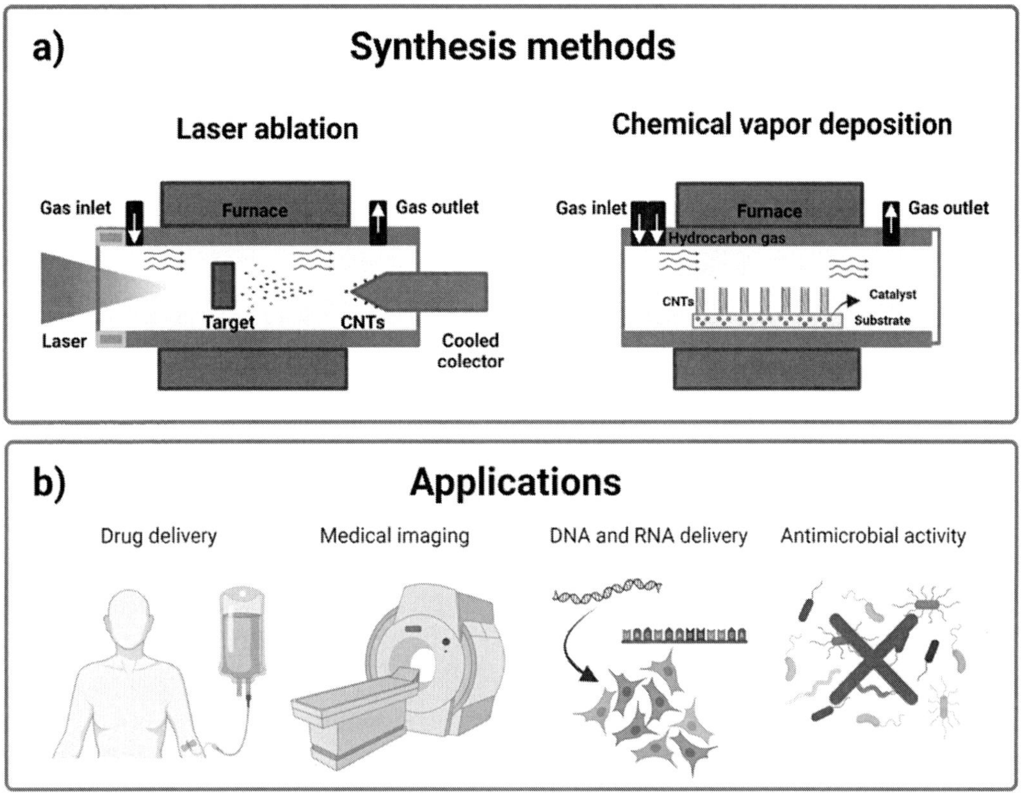


**Figure 6.7** Synthesis methods and applications of CNTs. The top panels illustrate two key methods for CNT synthesis: Laser ablation, where a high-energy laser vaporizes a graphite target in an inert gas atmosphere, leading to the formation of CNTs as the carbon plasma cools; and CVD, in which hydrocarbon gas decomposes over a metal catalyst at high temperatures, resulting in the growth of CNTs on a substrate. The bottom panels highlight the diverse applications of CNTs in nanomedicine, including drug delivery, medical imaging (MRI), gene therapy, and antimicrobial activity.

#### *6.3.3.2 Chemical Vapor Deposition*

In 1996, CVD was proposed as a method for synthesizing CNTs. CVD involves the decomposition of a hydrocarbon gas at high temperatures in the presence of a metal catalyst (see Figure 6.7). This process causes carbon atoms to deposit onto a substrate, resulting in the formation of CNTs [178]. This technique can achieve synthesis product yields of up to 90% with high purity [180]. One of the major challenges is large-scale synthesis while maintaining high-quality CNTs. Researchers have investigated the impact of substrates on the morphology and structure of CNTs using this method. It was found that CVD with bulk silica as the substrate can produce aligned CNTs with high purity and density [181], with CNT lengths reaching approximately 2 mm. However, a drawback of this technique is the high consumption of precursors. To address this, the use of inductively coupled plasma CVD (ICP-CVD) has been proposed. This method, which involves synthesis in the presence of plasma, results in crystalline CNTs. Overall, one of the most important factors to consider when using this technique is the size of the catalyst [182].

#### *6.3.3.3 Catalytic Decomposition of Methane*

Catalytic decomposition of methane (CDM) is a widely used technique for synthesizing CNTs [183]. This process is characterized by being an endothermic reaction that allows the production of carbon nanostructures and hydrogen in a single step [184]. The most commonly used catalysts are cobalt (Co) and nickel (Ni), although they are limited to operating at temperatures below 650°C [184]. For applications requiring higher temperatures, iron-based (Fe) catalysts can be employed [184]. Overall, CDM is a more cost-effective technique that does not produce $CO_2$ emissions [185]. As with other synthesis methods, the choice of catalyst is crucial in determining the quality and growth of CNTs [178].

#### *6.3.3.4 Arc Discharge*

This technique involves using two high-purity graphite electrodes placed less than 1 mm apart to generate an electric arc [178]. The reaction takes place at low pressure in a controlled environment with an inert or reactive gas [178]. The arc is manually produced between the electrodes for less than 1 minute, causing the graphite or graphite-metal electrode (anode) to sublimate and deposit carbon onto the cathode electrode [186]. In this method, voltage and current values are crucial for controlling the quantity of CNTs produced. Although it is one of the simpler methods, it typically yields short nanotubes of varying sizes [178, 187].

### 6.3.4 Applications in NA and Protein Delivery

CNTs have multiple applications in nanomedicine, including [188] (see Figure 6.7):

- *Drug Delivery:* CNTs can be used for the delivery of anticancer drugs such as doxorubicin and paclitaxel. Their surfaces can be modified to enhance therapeutic efficacy.
- *Medical Imaging:* They can serve as contrast agents in magnetic resonance imaging (MRI).
- *Therapeutic Uses:* CNTs are used in photothermal therapy and for the delivery of biomolecules.

Besides, CNTs have other applications related to NA and protein delivery:

#### *6.3.4.1 DNA and RNA Delivery*

In 2004, Pantarotto et al. [189] first reported the use of CNTs for delivering biologically active molecules, such as NAs. In their study, CNTs were functionalized with ammonium groups (f-CNT) and associated with pDNA through electrostatic interactions. These f-CNTs possess notable characteristics, including the ability to penetrate cell membranes, low cytotoxicity, and efficient cellular uptake [189, 190].

It has been demonstrated that SWNTs can protect DNA from enzymatic cleavage and protein interference [191]. This results in enhanced intracellular biostability, even for mRNA, if it is the material being delivered, and a higher capacity for self-administration [191]. An example of this is the use of SWNTs for delivering an mRNA vaccine for HIV-1, offering a valuable alternative by avoiding the use of viral vectors [192].

Overall, the most attractive features of CNTs for NA delivery include their stability, biocompatibility, high surface area, and substantial loading capacity [193]. Additionally, the primary use of CNTs in genetic material delivery has involved f-CNTs for pDNA delivery, utilizing modifications such as PEIs, polyamidoamine hybrids, cationic glycopolymers, and ethylenediamine [194].

#### *6.3.4.2 Therapeutic Protein Transport*

CNTs have a needle-like morphology, which facilitates their internalization into cells, along with a very high surface area, making them highly effective nanocarriers for protein administration and delivery [194]. For example, they were used to deliver streptavidin, a protein with a very high molecular weight (approximately 60 kDa) that has potent anticancer activity but cannot easily penetrate cell membranes. By conjugating it with an SWNT-biotin, improved delivery was achieved [194].

Another application related to therapeutic proteins and effects derivative from CNTs use is tissue regeneration. CNTs also hold the potential for bone tissue regeneration. Specifically, MWCNTs can be oxidized and functionalized with enzymes such as rhBMP-2 (recombinant bone morphogenetic protein 2). This is significant in relation to their interaction with proteins, as it stimulates alkaline phosphatase, Cbfa1 (alpha-1 subunit of the core-binding factor), and COLIA1 (type I collagen) [195].

However, there are still some limitations concerning how the size, chemical rigidity of the surface, and charge density of CNTs affect their cellular uptake and internalization mechanisms [196]. Therefore, it is crucial to consider the size of CNTs, the choice of catalysts, and the surface functionalizations, especially when they are intended for transporting proteins that do not easily enter cells on their own but have significant therapeutic potential.

#### *6.3.4.3 Environmental Applications of CNTs*

CNTs can positively impact the environment by adsorbing heavy metals from water. For example, carboxylated MWCNTs have been used to adsorb Cr(VI) [197]. CNTs have also been employed to adsorb various metals, including Cu(II), Cr(III), Cr(VI), Cd(II), Pb(II), Ni(II), Hg(II), Mn(VII), and Au(III). The adsorption capacity depends on the specific metal and can range from 18 mg/g to 449 mg/g.

#### 6.3.4.4 Other Applications

##### 6.3.4.4.1 Antimicrobial Activity of SWCNTs

SWCNTs exhibit significant antimicrobial activity, with their size playing a crucial role in deactivating microorganisms. Smaller CNTs have a higher surface-to-volume ratio, enhancing their binding to microbial cell walls or membranes and improving their effectiveness. The antimicrobial mechanism involves disrupting the cellular membrane, metabolic processes, and morphology of microorganisms. Exposure to smaller CNTs has been shown to enhance pDNA and RNA, as well as induce the efflux of cytoplasmic materials, indicating their potential in targeted antibiotic delivery.

##### 6.3.4.4.2 Molecular Dynamics of CNTs and Lipid Bilayers

Recent studies using coarse-grained molecular dynamics (CG-MD) and density functional theory (DFT) simulations have explored the interaction of SWCNTs with lipid bilayers. These simulations reveal that SWCNTs, including amino-functionalized and pristine variants, diffuse into lipid bilayers, with free energy differences indicating a favorable insertion mechanism. Accurate simulations require consideration of CNTs' physical properties, such as unsaturated bonds and dipole moments, which are influenced by nanotube chirality and diameter [198].

#### 6.3.4.5 Challenges and Risks Associated with CNTs

Despite their benefits, CNTs also have certain disadvantages and pose potential risks:

- *Oxidative Stress and Toxicity:* The nanotoxicity of CNTs generally depends on their length, electronic structure, and surface functionalization [199]. The toxicity of CNTs is influenced by factors such as their type, length, metal impurities, solubilizing agents, and surface functionalization [200]. CNTs can induce oxidative stress, inflammation, and fibrosis, promote angiogenesis, and generate free radicals [201, 202]. The induction of oxidative stress can lead to adverse effects such as necrosis, inhibition of cell growth, and DNA damage [203].
- *Genotoxicity:* Genotoxicity of CNTs has been evaluated using various assays, including the comet assay, γH2AX assay, micronucleus assay, and HPRT mutation assay. These tests have been conducted on different cell lines, such as A549, BEAS-2B, SAEC, THP-1, HK-2, V79, 3-KT, and MDA-435 [204]. However, the results have been inconclusive and highly variable.
- *Occupational Exposure:* Studies on the effects of occupational exposure to CNTs have shown that such exposure can lead to increased levels of biomarkers like malondialdehyde (MDA) and 4-hydroxy-2-hexenal (4-HHE) [164]. Additionally, exposure to aerosols containing MWCNTs can alter gene expression profiles related to cell cycle regulation, apoptosis, and cell proliferation [164].

## 6.4 CARBON DOTS

### 6.4.1 History and Discovery

The exploration of QDs has been a central focus of nanotechnology since its discovery in the early 1980s [205]. These semiconductor nanoparticles are notable for their size-dependent properties, a result of quantum confinement, where the movement of electrons is restricted within a nanoscale volume [206–208]. This confinement leads to the formation of discrete energy levels, giving rise to unique optical and electronic behaviors [209–211]. Unlike traditional materials, where properties are dictated primarily by chemical composition, the size of QDs determines their characteristics, such as the ability to emit light at different wavelengths [212–214]. This tunable emission makes them highly valuable across various fields, from advanced display technologies to medical imaging [215–217].

The journey toward understanding QDs began with Aleksey Ekimov's pioneering work in the early 1980s [218, 219]. While experimenting with copper chloride nanoparticles embedded in glass, Yekimov observed size-dependent quantum effects for the first time [219]. He demonstrated that changing the size of these nanoparticles affected the color of the glass, marking a key observation of quantum phenomena in nanostructures [220, 221]. Louis Brus built upon this discovery by showing that similar effects could be observed in colloidal particles suspended in liquid, establishing QDs as a foundational material in nanotechnology [222–224].

A significant breakthrough came in 1993 when Moungi Bawendi developed a synthesis method for producing highly uniform QDs [225–227]. This innovation marked a turning point in the field, as it enabled the creation of nearly monodisperse particles, facilitating more precise and reliable applications. Bawendi's method played a crucial role in the commercialization of QDs, particularly in display technologies such as QD light-emitting diode (QLED) screens, which offer superior brightness and color accuracy compared to traditional displays [228–230]. QDs have also found significant applications in medical imaging, particularly in oncology, where their distinct optical properties enhance the accuracy of diagnostic and surgical procedures. In recognition of their groundbreaking contributions, the 2023 Nobel Prize in Chemistry was awarded to Yekimov, Brus, and Bawendi, underscoring the transformative impact of QDs on both scientific research and technology [231].

In 2004, a new class of nanomaterials emerged with the accidental discovery of CDs during the electrophoretic purification of SWNTs [21]. Researchers noticed unexpected fluorescence from carbon-based nanoparticles, sparking widespread interest in these environmentally friendly alternatives to traditional QDs [232]. Unlike QDs, which are often composed of heavy metals like cadmium or lead, CDs consist primarily of carbon, making them more biocompatible and less toxic [233].

CDs exhibit quantum confinement effects similar to traditional QDs, giving them size-dependent optical properties [234–236]. However, they offer additional advantages such as high conductivity, excellent water solubility, and the ability to be synthesized from renewable sources like food waste, aligning with global efforts toward sustainability [237, 238]. In 2006, researchers improved the photoluminescent (PL) properties of CDs through surface passivation techniques, such as coating the nanoparticles with PEG [239]. This advancement expanded their use in bioimaging and optoelectronics, where their biocompatibility and stable luminescence offer promising alternatives to conventional imaging agents [34, 240, 241].

Today, CDs are being explored across a range of applications. In technology, they are utilized in light-emitting devices and energy storage systems, valued for their high conductivity and customizable optical properties [242–244]. In medicine, CDs have become essential tools as fluorescent probes for imaging, providing a nontoxic alternative to traditional imaging agents. Their ability to be derived from sustainable sources further enhances their appeal in environmental sensing and pollution control [233, 245, 246].

The discovery and development of both QDs and CDs represent pivotal milestones in nanotechnology. The recognition of QDs with the 2023 Nobel Prize highlights their revolutionary role in electronics and medicine, while the rise of CDs points to a future where nanomaterials are not only innovative but also sustainable. As research continues to refine the synthesis and applications of these materials, they are expected to play an increasingly vital role in addressing global challenges, from improving energy efficiency to advancing medical diagnostics.

### 6.4.2 Synthesis Methods

#### *6.4.2.1 Top-Down Approach*

The synthesis of CDs can be achieved through a variety of techniques, broadly categorized into top-down and bottom-up approaches [247]. Top-down methods involve the fragmentation of larger carbon-based nanomaterials—such as graphene, GO, or CNTs—into smaller CDs [248]. When graphene or GO serves as the precursor material, the resultant nanomaterials are often referred to as GQDs, highlighting their unique structural properties [249].

In the top-down synthesis paradigm, the primary goal is to reduce the size of preexisting nanomaterials through energy-intensive or chemical processes, leading to the formation of CDs with tailored properties [248]. Laser ablation exemplifies a top-down technique that is particularly effective when working with graphene or GO [250, 251]. This method involves directing high-energy laser pulses at the carbon material, which induces rapid, localized heating, causing the material to fragment into nanoscale particles that form CDs [251]. The type of laser used significantly influences the synthesis outcomes; for instance, femtosecond lasers, with their ultrafast pulses, minimize thermal diffusion, thereby producing smaller, more uniform CDs with fewer defects [252, 253]. Conversely, nanosecond lasers, which generate greater thermal effects, often result in a broader size distribution of the synthesized dots. The crystallinity of the precursor material, such as highly crystalline graphene, also plays a crucial role in producing uniform nanoparticles with desirable properties [254].

Oxidative cleavage, or oxidative cutting, is another widely adopted top-down method, particularly effective for synthesizing GQDs from nanomaterials like graphene, GO, or CNTs [255]. This technique involves breaking the carbon–carbon bonds within these materials using strong oxidizing agents such as sulfuric acid ($H_2SO_4$) or nitric acid ($HNO_3$). Oxidative cleavage is favored for its ability to produce GQDs with specific desirable properties, including fluorescence and high solubility, which are essential for applications in bioimaging and sensing [248, 255]. While commonly applied to graphene and GO, this method's versatility extends to other carbon-based nanomaterials, such as fullerenes and CNTs, enabling their breakdown into smaller QDs.

Both solvothermal and hydrothermal methods exhibit versatility, being applicable in both top-down and bottom-up approaches. In a top-down context, these methods involve subjecting nanomaterials like GO to high temperatures and pressures, facilitating their breakdown into nanoscale GQDs [256]. The key distinction between these techniques lies in the reaction medium: the hydrothermal method utilizes an aqueous environment, while the solvothermal method employs an organic solvent [257]. By meticulously adjusting parameters such as temperature, pressure, and reaction duration, researchers can finely control the size and photoluminescence properties of the resulting GQDs [258]. These methods are particularly valued for their capacity to produce GQDs with controlled sizes and optical properties, which are critical for applications in fields like optoelectronics and bioimaging. Additionally, the ability to modify the surface chemistry of GQDs by introducing functional groups during synthesis further enhances the adaptability of these methods, allowing for the creation of GQDs with tailored properties [259].

The microwave-assisted method, which is also applicable in both top-down and bottom-up approaches, represents an innovative technique that addresses some limitations of traditional methods, particularly in terms of reaction time [260]. In a top-down context, microwave irradiation offers rapid and uniform heating, which significantly accelerates the breakdown of carbon nanomaterials like GO into GQDs [261–263]. This method is advantageous for its efficiency in producing GQDs with consistent size distribution and enhanced PL properties. The rapid heating provided by microwave-assisted synthesis not only reduces the overall synthesis time but also improves energy efficiency, making it a more sustainable option for large-scale production [260, 262], this technique has gained popularity for its ability to quickly produce high-quality GQDs, suitable for a wide range of technological applications [259].

#### *6.4.2.2 Bottom-Up Approach*

The synthesis of CDs through bottom-up approaches has garnered significant attention due to the ability of these methods to manipulate the fundamental properties of the resulting nanomaterials. These approaches typically involve the use of small organic molecules, chosen for their carbon-rich content and functional groups that facilitate carbonization [264, 265]. The choice of precursor, solvent, and synthesis technique plays a pivotal role in dictating the optical, structural, and electronic properties of the CDs [266]. Recent studies, such as those conducted by Ding et al. [267] and Jiang et al. [268], have illustrated how variations in these parameters can lead to substantial differences in the size, emission properties, and elemental composition of CDs, thereby influencing their suitability for specific applications, including PL devices and bioimaging.

CA is among the most frequently utilized organic molecules in CD synthesis due to its propensity for thermal decomposition, which leads to the formation of graphitic carbon structures with high quantum yields [269, 270]. Similarly, tartaric acid is another widely used precursor known for producing CDs with favorable PL properties [271, 272]. Other notable examples include triammonium citrate and naphthalenediamine, both of which provide a rich carbon source and contribute to CDs with specific optical and structural characteristics [273–275].

Recent advancements have highlighted the use of aromatic molecules in CD synthesis, particularly for generating multicolor emissions [266]. Aromatic compounds such as ortho-phenylenediamine (oPDA), meta-phenylenediamine (mPDA), and para-phenylenediamine (pPDA) have been extensively investigated for their ability to produce CDs that emit light across various wavelengths. The conjugated structures of these aromatic molecules facilitate the formation of CDs with tunable emission properties [266].

Jiang et al. [268] conducted a comprehensive study on the synthesis of CDs using different phenylenediamine isomers (oPDA, mPDA, and pPDA) under identical solvothermal conditions (12 hours at 180°C). Their findings demonstrated that the emission colors of the CDs could be precisely controlled based on the precursor used: mPDA-derived CDs exhibited blue emission, oPDA-derived CDs showed green emission, and pPDA-derived CDs produced red emission when excited at 365 nm. This variation in emission color was attributed to differences in particle size, with average diameters of 6.0 nm for mPDA-CDs, 8.2 nm for oPDA-CDs, and 10.0 nm for pPDA-CDs [268]. The differences in particle size influenced quantum confinement effects and electronic properties, which in turn affected the emission wavelengths.

Furthermore, the nitrogen content in these CDs varied significantly, with 3.69% nitrogen in mPDA-CDs, 7.32% in oPDA-CDs, and 15.57% in pPDA-CDs [268]. This variation in nitrogen content influenced the surface chemistry and electronic states of the CDs, further contributing to differences in their PL characteristics. The ability to produce CDs with tailored emission spectra is particularly valuable in applications such as light-emitting devices and bioimaging, where distinct colors are required for different purposes.

Incorporating nitrogen into the CD structure is a common strategy to enhance the photoluminescence and overall performance of CDs [276–278]. Small molecules like urea and ethylenediamine (EDA) are frequently employed as nitrogen-rich precursors [279–281]. These molecules not only provide carbon but also introduce nitrogen atoms into the carbon lattice during synthesis, which can significantly improve the quantum yield and modulate the electronic properties of the CDs [276, 279]. Such enhancements make nitrogen-doped CDs particularly effective in applications like sensing and imaging.

Additionally, other nitrogen-containing molecules, such as thiourea and ammonium citrate, have been used to achieve similar effects [279, 282]. The incorporation of nitrogen atoms within the carbon structure often results in CDs with brighter fluorescence, improved stability under varying environmental conditions, and enhanced optical properties, further expanding their potential applications.

The choice of solvent in the synthesis of CDs plays a pivotal role in dictating the optical and structural properties of the resulting nanomaterials. During solvothermal synthesis, the solvent's polarity significantly influences the size and distribution of $sp^2$-conjugated domains within the CDs [266]. Highly polar solvents, such as water, typically yield CDs with blue PL [266]. This is because the high polarity of water hinders the aggregation of conjugated structures, thereby limiting the formation of extensive $sp^2$-conjugated networks that are necessary for red-shifted emissions [266]. Conversely, solvents with lower polarity, such as dimethylformamide (DMF), promote the development of larger conjugated domains, resulting in CDs that exhibit red-shifted emissions [266]. This solvent-dependent tuning of the CDs' emission properties underscores the importance of solvent selection in the controlled synthesis of CDs with desired optical characteristics.

In a study conducted by Ding et al. [267], CDs were synthesized using L-glutamic acid and oPDA as precursors in the presence of various solvents, including formamide, DMF, ethanol, mixtures of DMF and formamide, and aqueous solutions of sulfuric acid ($H_2SO_4$) at different concentrations. The emission colors of the resulting CDs, when excited by UV light at 365 nm, varied significantly depending on the solvent used.

The study reveals that the variation in the emission colors of synthesized CDs is primarily influenced by the solvents' impact on carbonization and dehydration processes during the solvothermal reaction. High-polarity solvents, such as formamide, restrict the growth of $sp^2$-conjugated domains, resulting in smaller CDs with blue emissions due to the larger energy gaps associated with their reduced size. In contrast, low-polarity solvents like DMF and ethanol support the formation of larger $sp^2$-conjugated structures, leading to red-shifted emissions as the energy gap decreases [267].

Additionally, the study emphasizes the crucial role of solvent boiling points and catalytic properties, particularly in sulfuric acid solutions, in accelerating the carbonization process [267]. This acceleration promotes the development of larger CDs, further shifting the emission spectrum towards the red or near-infrared regions [267]. The systematic tuning of CDs' optical properties through solvent selection underscores the importance of the reaction environment in controlling the size, graphitization degree, and surface states of the resulting particles [267]. Thus, the study demonstrates that by adjusting solvent conditions, the emission color of CDs can be precisely tuned across the visible spectrum to near-infrared, providing a versatile approach for customizing the PL properties of CDs for specific applications [267].

The method chosen for the synthesis of CDs plays a crucial role in determining the properties of the final product. For instance, the study by Crista et al. [283] investigated the impact of different synthesis techniques—namely hydrothermal, microwave, and calcination—on CDs synthesized from CA and urea. In their experiments, the same quantities of CA (0.75 g) and urea (0.25 g) were employed for all synthesis methods, allowing for a direct comparison of the outcomes.

## Table 6.1 Emission colors and wavelengths of carbon dots synthesized using L-glutamic acid and o-phenylenediamine as precursors in various solvents [266]

| Solvent | Emission Color | Emission Wavelength (nm) |
|---|---|---|
| Formamide | Blue | 443 |
| DMF/formamide (1:6 vol/vol) | Green | 515 |
| DMF/formamide (3:2 vol/vol) | Yellow | 544 |
| DMF | Yellow-orange | 571 |
| Ethanol | Orange | 594 |
| $H_2SO_4/H_2O$ (2.8 mol/L) | Red-orange | 640 |
| $H_2SO_4/H_2O$ (6.1 mol/L) | Deep red | 715 |
| $H_2SO_4/H_2O$ (7.3 mol/L) | Dark red | 745 |

The emission colors, observed under UV light excitation at 365 nm, show significant variation depending on the solvent used, highlighting the influence of the solvent on the optical properties of the CDs.

During the hydrothermal synthesis, the CA and urea mixture was dissolved in water and heated at 200°C for 2 hours in a Teflon-lined reactor [283]. This method resulted in the lowest yield of 1.5%. In contrast, the microwave-assisted synthesis involved mixing CA and urea in water, followed by exposure to microwave irradiation at 700 W for 5 minutes. This approach produced a significantly higher yield of 31.6%. Lastly, in the calcination-assisted synthesis, the mixture of CA and urea was heated in an oven at 200°C for 2 hours, resulting in a moderate yield of 26.1% [283].

The fluorescence quantum yield (QYFL) of the CDs varied notably between the different synthesis methods. Calcination-assisted synthesis produced CDs with the highest QYFL at 29.3%, closely followed by the microwave-assisted method, which achieved a QYFL of 25.1% [283]. In contrast, the CDs synthesized via the hydrothermal method exhibited a significantly lower QYFL of 3.7%, indicating that this method is less effective in terms of fluorescence efficiency [283].

Elemental analysis further revealed differences in the composition of the CDs. Hydrothermal synthesis resulted in the highest oxygen incorporation at 28.8%, but the lowest nitrogen content at 9.1%, which likely contributed to its reduced fluorescence properties [283]. On the other hand, CDs produced by the microwave-assisted method displayed a more balanced elemental composition, with higher nitrogen content at 13.1% and oxygen content at 26.9%. Calcination-assisted CDs had the highest nitrogen content at 13.5%, along with the lowest oxygen content at 24.7% [283].

A life cycle assessment (LCA) was conducted to evaluate the sustainability of the three synthesis methods. The microwave-assisted synthesis was identified as the most sustainable approach, considering both its high yield and relatively high QYFL [283]. In contrast, hydrothermal synthesis was deemed the least sustainable due to its low yield and QYFL, which resulted in higher environmental impacts. The calcination-assisted method, while showing promise in terms of achieving a high QYFL, was associated with slightly higher environmental impacts compared to microwave-assisted synthesis [283].

The findings discussed underscore the importance of careful selection of synthesis parameters in the bottom-up production of CDs. By varying the organic precursors, solvents, and synthesis techniques, researchers can fine-tune the optical and structural properties of CDs, enabling their application in a wide range of fields [267, 268, 283]. The work of Ding et al. [267] and Jiang et al. [268] demonstrates that through controlled synthesis, it is possible to achieve tailored emission spectra, optimized quantum yields, and enhanced environmental sustainability. This highlights the potential of bottom-up methods not only to produce high-quality CDs but also to push the boundaries of their functionality and application (see Figure 6.8).

### 6.4.3 Applications in NA and Peptide Delivery

Carbon dots (CDs) possess excellent biocompatibility and versatile surface properties that allow for functionalization, particularly through surface passivation [8, 9, 34]. These characteristics make CDs highly attractive for applications in NA and peptide delivery [285, 286]. One approach to enhance the functionalization potential of CDs involves using a nitrogen source during synthesis to passivate the surface and introduce amine groups, which facilitate subsequent functionalization [287, 288].

An illustrative example is the study by Wang et al. [289], where the authors synthesized CDs using a microwave-assisted one-step process, with CA as the carbon source and tryptophan (Trp) as both the nitrogen source and passivation agent. This method produced CDs with uniform size, excellent water solubility, biocompatibility, and high quantum yield. The synthesized CDs were then functionalized with PEI to form Cdots@PEI complexes, which were utilized to deliver Survivin siRNA into human gastric cancer cells (MGC-803). The Cdots@PEI complex effectively delivered the siRNA, resulting in a 6.1% downregulation of Survivin protein expression. This siRNA-mediated silencing of Survivin induced cell cycle arrest in the G1 phase and increased apoptosis in the cancer cells.

The efficacy and biocompatibility of carbon dots (Cdots) have been extensively studied and validated through various *in vivo* models, with promising implications for cancer therapy. A significant study by Wu et al. [290] demonstrated the synthesis and application of polyethyleneimine-passivated carbon dots (PEI-Cdots) for lung cancer treatment.

In this study, CDs were synthesized by combining 10 ml of glycerol with 0.5 g of PEI in 6 ml of 10 mM phosphate buffer. The mixture was then subjected to microwave pyrolysis at 550 W for 5 minutes. During this process, glycerol acted as the carbon source, undergoing carbonization to form the carbon dot core, while PEI served as a passivating agent. This passivation not only stabilized the CDs but also introduced functional groups essential for further modifications, such as conjugation with folic acid and complexation with siRNA [290].

The resulting PEI-Cdots exhibited strong photoluminescence, emitting visible blue light at 460 nm when excited at 360 nm. This property renders them highly suitable for bioimaging applications, particularly in real-time monitoring of therapeutic effects and tumor progression. The nanoparticles were effectively

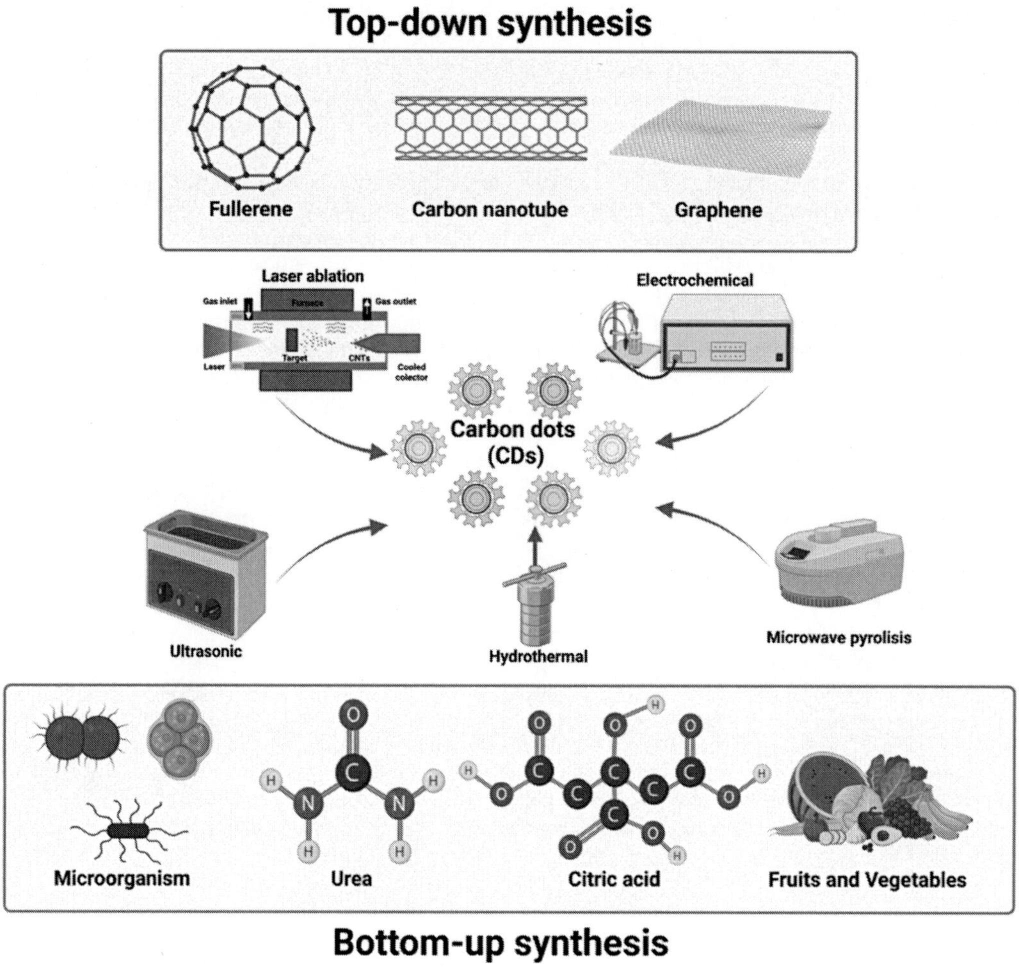


**Figure 6.8** Schematic representation of top-down and bottom-up synthesis methods for CDs. Top-down methods involve the fragmentation of larger carbon-based nanomaterials such as fullerenes, CNTs, and graphene into smaller CDs through processes like laser ablation and oxidative cleavage. Bottom-up methods utilize small organic molecules (e.g., urea, CA) and natural sources (e.g., fruits and vegetables) to form CDs via hydrothermal, solvothermal, and microwave-assisted techniques. These approaches allow for the control of size, photoluminescence, and other properties of the synthesized CDs, enabling their application in various fields.

**Source: Adapted from ref. [283].**

visualized in lung cancer cells (H460) using fluorescence microscopy, highlighting their potential as bioimaging agents [290].

Beyond imaging, the PEI-Cdots demonstrated significant therapeutic potential. The CDs efficiently complexed with siRNA, releasing it within the reductive intracellular environment characteristic of cells with high glutathione levels. *In vitro* studies revealed a notable reduction in lung cancer cell viability, with a 70% decrease observed 72 hours post-treatment with PEI-Cdots/siRNA complexes. Additionally, gene expression analysis confirmed the effective silencing of cyclin B1 and EGFR genes, with sustained effects lasting up to 48 hours [290].

The biocompatibility of the synthesized PEI-Cdots was confirmed through cytotoxicity assays, where cell viability remained above 95% in control cells, indicating minimal cytotoxicity [290]. Both normal fibroblast and lung cancer cells exhibited high viability post-treatment, reinforcing the nontoxic nature of these nanoparticles.

*In vivo*, studies further supported the therapeutic efficacy of PEI-Cdots. When delivered via aerosol inhalation, the PEI-Cdots/siRNA complexes effectively targeted lung tumors, resulting in a significant reduction in tumor size over a 10-day period. Bioluminescence imaging indicated a 50% decrease in tumor intensity compared to pre-treatment levels, evidencing successful gene silencing and tumor inhibition [290]. The selective accumulation of PEI-Cdots in lung cancer tissues, corroborated by both bioimaging and tumor size reduction, underscores their potential as a multifunctional theragnostic agent for lung cancer therapy.

In addition, to the biocompatibility the use of CDs in biomedical research has proven advantageous due to their unique properties that facilitate both gene delivery and cellular imaging [8, 9]. Among these, the autofluorescence of CDs stands out as a particularly significant benefit, especially in the context of *in vitro* studies [212]. This property not only enhances the efficiency of gene delivery but also simplifies the experimental processes involved in evaluating the delivery and colocalization of nanoparticles within cells. As a result, CDs have become invaluable tools in nanobiotechnology.

Wang et al. [291] provide a compelling example of the efficacy of CDs in gene delivery and cell imaging. They developed a simple and environmentally friendly method to synthesize hyaluronic acid-based carbon dots (HA-Cds) using HA as a carbon source and PEI as passing through microwave-assisted pyrolysis. This method is notable for its lack of required additives, which contributes to its green chemistry appeal. Moreover, the resulting HA-Cds exhibit significantly lower cytotoxicity compared to PEI alone, making them safer for biological applications, particularly in gene therapy and *in vivo* experiments [291].

One of the most remarkable outcomes of this study is the HA-Cds' ability to achieve up to 50 times higher gene transfection efficiency compared to traditional PEI, even in the presence of serum. This enhanced efficiency is crucial because serum typically poses a challenge for nonviral gene delivery systems. Additionally, the HA-Cds demonstrate strong and tunable fluorescence under UV light, allowing them to function as both gene delivery vehicles and bioimaging agents [291]. The HA component of the CDs provides further advantages by targeting CD44 receptors, which are overexpressed on various tumor cells, thereby increasing the specificity and potential efficacy of gene delivery treatments.

Furthermore, the negative charge on the HA-Cds imparts a "stealth effect," minimizing nonspecific interactions with serum proteins and thereby enhancing their biocompatibility and serum tolerance. The CDs predominantly enter cells through clathrin-mediated endocytosis, an efficient pathway that facilitates the precise delivery of genetic material to the cell nucleus.

The autofluorescence of the CDs played a crucial role in the authors' experiments, particularly in the fluorescence microscopy images of HeLa and HepG2 cells incubated with HA-Cds [291]. This autofluorescence enabled real-time monitoring of the gene delivery process and facilitated the tracking of nanoparticle colocalization within cells. Consequently, the multifunctionality of HA-Cds, combining gene delivery and imaging capabilities, represents a significant advancement in nanobiotechnology, making *in vitro* tests more straightforward and effective.

#### *6.4.3.1 Applications in Peptide and Proteins Delivery*

The application of CDs in peptide and protein delivery has gained significant attention due to their unique combination of fluorescence and biocompatibility [9, 30, 34, 212]. Among the various types of CDs, protein-derived CDs have shown particular promise in targeted delivery applications, especially within the context of cancer research [292]. These protein-derived CDs not only retain the inherent fluorescence of CDs but also inherit the biological functions of the parent proteins, thereby facilitating both targeted delivery and imaging [292].

A prominent example of this is the use of transferrin-derived carbon dots (Tf-CDs) for selective targeting of cancer cells. Strickland et al. [292] demonstrated that Tf-CDs effectively target cancer cells by combining the fluorescent properties of CDs with the targeting capabilities of transferrin. In their study, Tf-CDs were synthesized using transferrin as the carbon source, resulting in CDs that maintained the fluorescence of traditional CDs while incorporating the targeting functionality of transferrin [292]. This is especially significant given transferrin's natural affinity for the CD71 receptor, a well-established cancer biomarker.

The dual functionality of Tf-CDs enables not only efficient gene delivery but also the simultaneous imaging of cells, thereby allowing for the *in vitro* evaluation of nanoparticle delivery and their localization within cells. The study also introduced a novel synthesis method for Tf-CDs that preserves the functional groups of the protein, ensuring that the CDs maintain their affinity for CD71 receptors [292]. This selective binding triggers transferrin-mediated endocytosis, a mechanism through which nanoparticles are internalized into the cell via clathrin-coated pits.

Furthermore, the fluorescent properties of Tf-CDs, with an emission peak at 413 nm under UV excitation, facilitate the visualization of these nanoparticles within cells [292]. This fluorescence simplifies the experimental procedures required to assess the effectiveness of nanoparticle delivery. The specificity of

Tf-CDs in targeting cancer cells is particularly advantageous in cancer research, where precise delivery of therapeutic agents is crucial. The targeting ability of Tf-CDs was confirmed through flow cytometry and fluorescence microscopy, both of which demonstrated effective binding and internalization by CD71-expressing cells. In contrast, cells with suppressed CD71 expression exhibited significantly lower fluorescence, underscoring the specificity of Tf-CDs in targeting cancer cells [292].

While using peptides as precursors to synthesize CDs has shown promising results in targeted delivery, an alternative approach involves synthesizing CDs first and then functionalizing them with peptides to enhance their targeting capabilities [287]. This method allows for precise control over the surface properties of the CDs, which can be tailored for specific applications, such as nuclear targeting.

A compelling example of this approach is demonstrated in the study by Song et al. [293] The researchers developed an innovative method for nuclear-targeted imaging by functionalizing CDs with the trans-activator of transcription (TAT) peptide. This functionalization enables the CDs to selectively target the cell nucleus, enhancing their utility in both one-photon and two-photon fluorescence imaging.

The CDs were synthesized using Trp and formic acid, resulting in CDs with a high QYFL. To functionalize these CDs with the TAT peptide, a coupling reaction was employed [293]. The CDs were first activated by mixing them with benzotriazol-1-yl-oxytripyrrolidinophosphonium hexafluorophosphate (PyBOP) and 1-hydroxybenzotriazole hydrate (HOBt) in DMF, which activates the carboxyl groups on the CD surface [293]. The TAT peptide, which contains free amino groups on the N-terminal, was then conjugated to the CDs using DIPEA (N,N-diisopropylethylamine) to facilitate the formation of stable amide bonds [293].

After the reaction, the TAT-CD conjugates were purified through a series of washes and were finally isolated and stored for future use. The successful functionalization of CDs with TAT peptides is crucial, as it allows the CDs to translocate across the nuclear membrane, enabling high-contrast imaging of the cell nucleus [293]. The resultant TAT-CDs were highly effective in one-photon and two-photon fluorescence imaging, providing clear and detailed images of the cell nucleus [293]. This dual functionality of TAT-CDs makes them particularly valuable for applications in cellular imaging, with potential extensions to disease diagnosis and therapeutic monitoring.

#### *6.4.3.2 Biocompatibility*

CDs have emerged as a promising tool in nanomedicine, particularly for peptide and gene delivery applications [290, 291, 293, 294]. Their unique physicochemical properties—including high biocompatibility, versatile surface functionalization capabilities, and straightforward synthesis methods—position them as strong candidates for therapeutic use [9, 34]. The structural core of CDs, composed of graphitic, amorphous, or polymeric materials, inherently possesses biocompatible characteristics, thereby minimizing the risk of generating harmful by-products under physiological conditions [295]. Additionally, CDs exhibit low chemical reactivity, capacity for biodegradation—especially in the case of polymeric CDs—and minimal bioaccumulation, particularly for particles smaller than 6 nm [295]. These attributes contribute to their generally low propensity for inducing significant or prolonged inflammatory responses in biological systems.

The biocompatibility of CDs is a critical consideration in their application for gene and peptide delivery [290, 294]. Surface functionalization can enhance the targeting specificity of CDs; however, it is crucial to avoid the introduction of toxic chemical groups during this process [287]. Interestingly, in some cases, the attachment of ligands with known toxic side effects has been observed to paradoxically increase bioavailability while simultaneously reducing toxicity [295]. However, this phenomenon requires further investigation to fully understand the underlying mechanisms. Despite the widespread recognition of the high biocompatibility of CDs, certain conditions—such as repeated dosing, prolonged light exposure, the use of particles larger than 6 nm, high positive surface charges, and densely localized charge areas—can lead to adverse biological effects [295, 296].

A comprehensive evaluation of the toxicological profile of CDs is essential to ensure their safe application in biomedical contexts. Toxicity is often dose-dependent, with most studies reporting low toxicity at concentrations below 200 μg/mL, maintaining 80% cell viability over 24 hours [295]. However, it is important to note that the majority of toxicity studies have focused on single-dose exposures, whereas clinical applications may necessitate multiple dosing regimens. Research has shown that repeated dosing can trigger immune system activation, compromise cell membrane integrity, and alter hepatic clearance, as evidenced by animal studies [295]. These findings highlight the importance of conducting thorough toxicological assessments that consider the long-term effects of repeated exposure to CDs.

In the context of gene and peptide delivery, the size of CDs is a key determinant of their biocompatibility and toxicity [295]. CDs smaller than 6 nm are typically excreted via the renal system, resulting in minimal toxicity [295, 297, 298]. In contrast, larger particles may persist in the body for extended periods, potentially eliciting immune responses or undergoing enzymatic degradation [295]. Furthermore, the surface charge of CDs significantly influences their interaction with biological systems. While positively charged CDs are particularly effective in gene delivery due to their strong affinity for negatively charged NAs, they also present a higher risk of toxicity, as evidenced by increased cellular stress and morphological alterations.

## 6.5 FULLERENES

### 6.5.1 History and Discovery

#### *6.5.1.1 Background and Interest in Non-Benzene Aromatics*

Interest in non-benzenoid aromatics and the search for molecules with three-dimensional electron delocalization over highly symmetric surfaces led to the synthesis of molecules such as corannulene. These molecules serve as examples of this class of compounds.

#### *6.5.1.2 Discovery of Fullerenes*

The history of fullerenes began in 1970 with Harold Kroto and David R. M. Walton, who were working with $H\text{-}(CC)_nCN$ molecules [23]. These molecules showed microwave spectra consistent with their linear dipolar structure. Kroto hypothesized that it was possible to simulate the explosion of a red giant star to obtain linear clusters similar to cyanopolyynes. In 1984, Kroto visited Robert Curl, who was experimenting with clusters by using a laser to vaporize solids. This collaboration led Kroto to suggest vaporizing graphite [299].

In 1985, a stable species of $C_{60}$ was identified in vaporized graphite by Harold Kroto, Robert Curl, and Richard E. Smalley. This molecule was named "buckminsterfullerene" by Kroto, in honor of Buckminster Fuller. $C_{60}$ is notable for having 60 carbon atoms positioned at the vertices of truncated regular icosahedrons and possesses 12,498 resonance forms, more than benzene. The discovery of fullerenes stemmed from studies on carbon nucleation, which simulated processes occurring in red giant stars [23]. In 1995, $C_{60}$ was named "Molecule of the Year" by Science magazine.

#### *6.5.1.3 Nobel Prize and Subsequent Discoveries*

In 1996, Kroto, Curl, and Smalley were awarded the Nobel Prize in Chemistry for their discovery of fullerenes. The identification of fullerenes also led to the discovery of other carbon forms, such as CNTs and graphene [300].

#### *6.5.1.4 Macroscopic Production of Fullerenes*

The macroscopic production of $C_{60}$ and $C_{70}$ was developed by Krätschmer and colleagues, who used an electric arc between two graphite electrodes to produce carbon soot in a helium atmosphere. This method yielded approximately 3–5% by weight of $C_{60}$ and $C_{70}$ [301, 302].

In more detail, Krätschmer's experiment involved evaporating graphite electrodes at 100 torr in a helium atmosphere. The resulting soot was collected and scraped from the surfaces inside the apparatus. The $C_{60}$ present in the soot was then dissolved in benzene and dried using gentle heating. An alternative separation method involved the sublimation of $C_{60}$ by heating the soot to 400°C in an inert environment [302].

Today, there is extensive interest in fullerenes, with thousands of research articles published on the advancements in fullerene research [303; see Figure 6.9].

### 6.5.2 Structural and Optical Properties

Natural allotropes of carbon include diamond, graphite, and amorphous carbon. In general, the bonding structure and atomic orientation that carbon adopts within the structure of an allotrope determine its properties. As mentioned earlier, the discovery of $C_{60}$ in 1985 marked a significant turning point in the history of carbon allotropes. $C_{60}$ is a spherical cage composed of 60 carbon atoms forming a truncated icosahedron with an average diameter of 0.72 nm, containing 12 pentagons and 20 hexagons of bonded carbon [304].

Currently, there are families of carbon cage structures such as $C_{30}$, $C_{50}$, $C_{70}$, $C_{72}$, $C_{76}$, $C_{84}$, and $C_{540}$. However, $C_{60}$ remains notable for its unique properties, especially its ability to destroy viruses when irradiated, as it can cleave biological molecules like DNA and RNA [304].

Some properties of fullerenes include derivatives of $C_{60}$ exhibiting a certain level of solubility, have a low density (1.65 g/cm$^3$) compared to diamond (3.51 g/cm$^3$), are stable at high temperatures, and can react with nucleophiles [305]. Additionally, fullerenes are borderless, uncharged, and have no unpaired electrons. They are excellent conductors of electricity and heat. Notably, $C_{60}$ readily accepts and donates electrons [306].

Fullerenes also exhibit solubility in a wide range of solvents, making them suitable for use in solutions for coatings and electrodes. Furthermore, fullerenes can be functionalized to extend their applications [306].

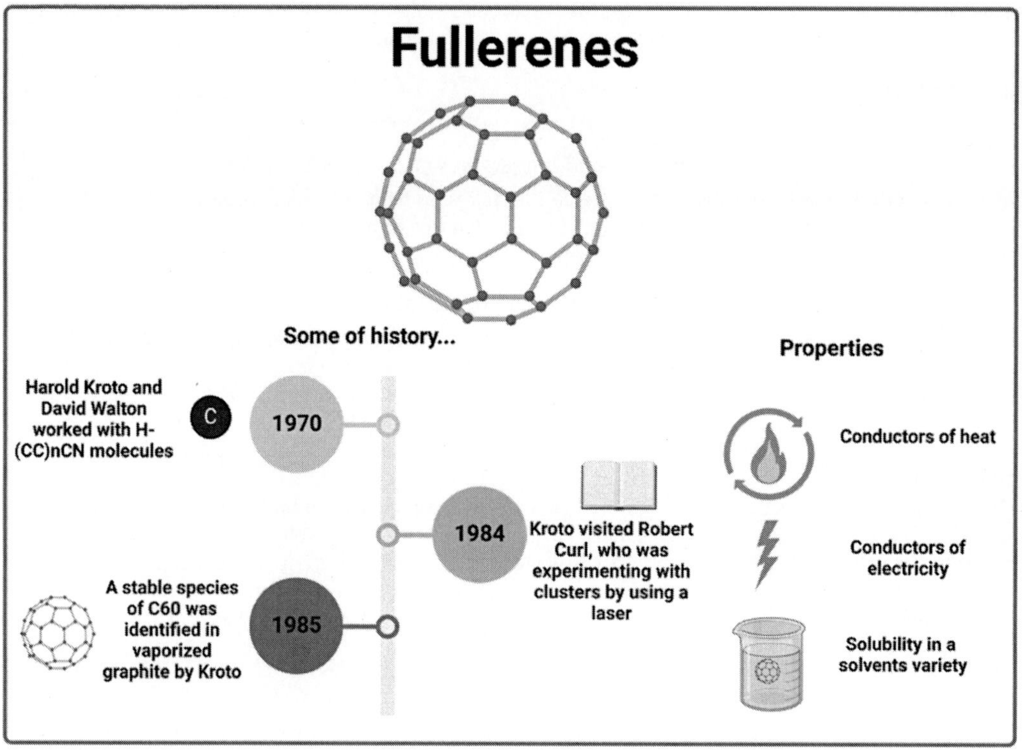


**Figure 6.9** Key milestones in the discovery and development of fullerenes. The timeline highlights significant events, beginning in 1970 with Harold Kroto and David R. M. Walton's work on H-$(CC)_n$CN molecules, leading to the hypothesis of simulating red giant star explosions to obtain linear clusters similar to cyanopolyynes. In 1984, Kroto collaborated with Robert Curl, who was experimenting with laser vaporization of solids, which led to the suggestion of vaporizing graphite. In 1985, this collaboration resulted in the identification of a stable species of $C_{60}$, known as buckminsterfullerene, in vaporized graphite by Kroto, Curl, and Richard E. Smalley. The properties of $C_{60}$, including its structure with 60 carbon atoms and its unique resonance forms, are depicted. The figure also notes the ability of fullerenes to conduct heat and electricity and their solubility in various solvents.

### 6.5.3 Methods of Synthesis

The synthesis of fullerenes, particularly $C_{60}$, has been a central focus of research since their discovery in 1985 by Kroto et al. [23]. Over the years, various methods have been developed and refined to produce these unique carbon structures, each presenting distinct advantages and challenges [307, 308]. One of the earliest and most widely used techniques involves the vaporization of a carbon source, typically graphite [309]. In this process, high temperatures are used to vaporize graphite in the presence of an inert gas, such as helium, leading to the condensation of carbon atoms into fullerene molecules like $C_{60}$ and $C_{70}$. While this method was instrumental in the initial discovery and mass production of fullerenes, it is not without limitations [309]. The efficiency of this process is generally low, with yields often below 1%, and it requires extreme conditions, including temperatures around 1300°C and pressures of approximately 1 kbar [309]. Additionally, the process typically results in a mixture of different fullerenes, necessitating extensive purification efforts, which can be both time-consuming and costly.

In contrast, chemical synthesis offers a more controlled approach to fullerene production [309–311]. This method involves assembling smaller carbon-based precursors through a series of reactions that ultimately form the fullerene structure [311]. A significant advantage of chemical synthesis is the ability to produce specific fullerene isomers or derivatives that are challenging to obtain through vaporization techniques [309, 311]. However, this approach is often hampered by low yields and the complexity of the reactions involved, making it less practical for large-scale production [309]. The precise conditions required for these reactions further complicate the process, rendering it time-consuming and expensive.

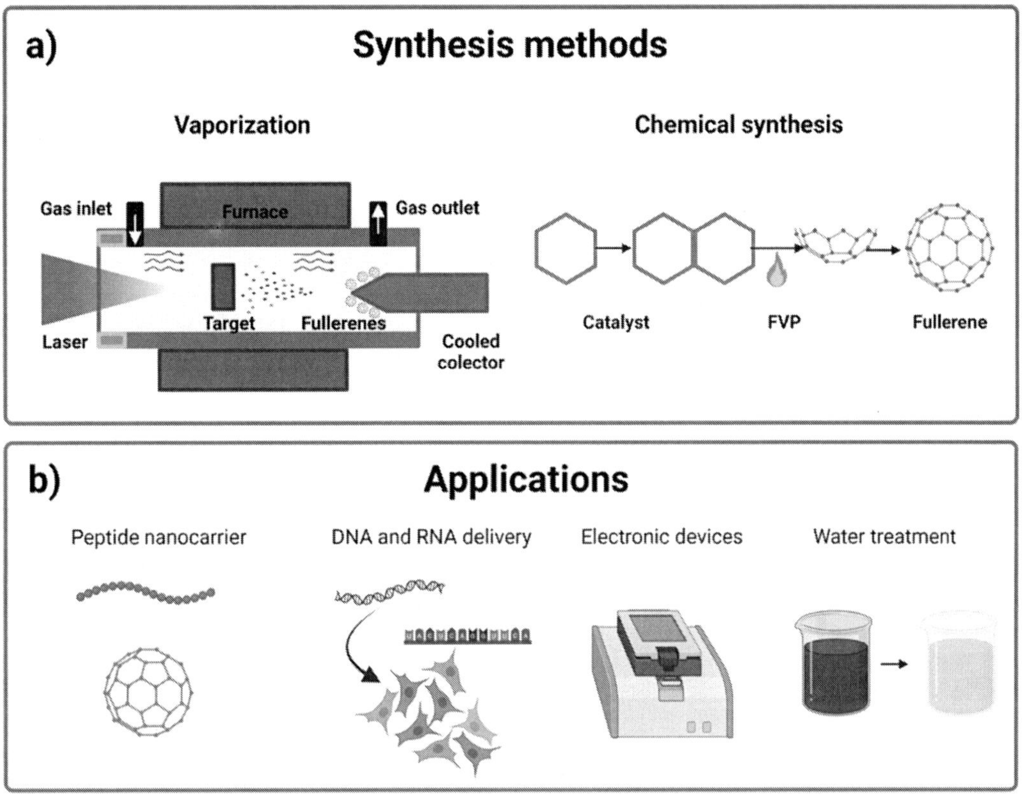


**Figure 6.10** Methods of fullerene synthesis and their applications in drug and gene delivery. (A) Synthesis methods: The vaporization method (left) involves the use of a high-temperature furnace to vaporize a carbon target, typically graphite, in the presence of an inert gas, leading to the formation of fullerene molecules. The chemical synthesis method (right) depicts the catalytic formation of fullerenes from smaller carbon-based precursors via flash vacuum pyrolysis (FVP), facilitating the controlled production of specific fullerene isomers and derivatives. (B) Applications: Fullerenes are utilized in various fields, including as peptide nanocarriers, in DNA and RNA delivery, in electronic devices, and in water treatment processes, demonstrating their versatility in both biological and industrial applications.

Beyond these traditional methods, researchers have explored several innovative approaches to fullerene synthesis [309]. For instance, Viñes and Görling have proposed a theoretical method that utilizes supported metallic nanoparticles as templates to catalyze the formation of fullerenes from simple precursors such as methane and ethylene [309, 312]. This approach could potentially reduce the high temperatures typically required for fullerene synthesis, as the nanoparticles facilitate the dehydrogenation of carbon precursors. However, while promising, this method remains largely theoretical and requires further development before it can be considered viable for large-scale production. Another novel approach, developed by Chen and Lou, involves the reduction of $CO_2$ to produce fullerenes [309, 313]. This method offers an environmentally friendly alternative by using a greenhouse gas as a carbon source. Nonetheless, the current yields are low, and the process demands extremely high pressures (around 100 MPa) and temperatures (approximately 700°C), limiting its practicality (see Figure 6.10).

### 6.5.4 Fullerenes as Nanocarriers

Fullerenes, particularly $C_{60}$, have garnered significant attention in the field of drug and gene delivery due to their unique structural and chemical properties [314, 315]. The spherical, hollow structure of $C_{60}$, combined with its ability to undergo various chemical modifications, makes it an attractive candidate for developing multifunctional nanocarriers [316]. This versatility is especially advantageous in the delivery of peptides and NAs, where targeted delivery, protection from degradation, and controlled release are critical factors [317].

One of the primary reasons fullerenes are interesting for peptide and NA delivery is their ability to be easily functionalized. The surface of the $C_{60}$ molecule can be modified to attach various functional groups, including peptides, NAs, and other biologically active molecules [316–318]. This multifunctionalization enables fullerenes to serve as platforms that can carry multiple therapeutic agents simultaneously, enhancing the efficacy of treatment. For example, peptides can be conjugated to fullerenes to facilitate targeted delivery to specific cells or tissues, while NAs can be attached to deliver genetic material directly into cells, where it can influence gene expression.

The conjugation of peptides to fullerenes is particularly important in the design of nanoplatforms for targeted delivery [319, 320]. Peptides can be selected for their ability to recognize and bind to specific cell surface receptors, thereby directing the fullerene-NA complex to specific cell types [319, 320]. This targeting capability reduces off-target effects and enhances the therapeutic efficacy of the delivered NAs. Additionally, the unique structure of fullerenes allows for the formation of stable complexes with NAs, protecting them from enzymatic degradation during delivery [315–317]. This protection is crucial for ensuring that the NAs remain intact and functional once they reach their target cells [321].

Furthermore, the use of fullerenes in gene delivery is supported by their ability to cross biological membranes, including the cell membrane and the nuclear envelope [322]. This property is essential for delivering NAs into the nucleus, where they can effectively influence gene expression.

#### *6.5.4.1 Fullerenes as Peptides Nanocarriers*

Fullerene $C_{60}$ nanocrystals have emerged as a promising nanocarrier for the delivery of therapeutic peptides, particularly in the field of antiviral treatments. The study by Dostalova et al. [323] demonstrates the potential of fullerenes in enhancing the antiviral activity of peptides derived from maximin H5, an antimicrobial peptide originally isolated from the Asian toad *Bombina maxima*. The unique structure of fullerene $C_{60}$ allows for extensive functionalization, making it an ideal platform for conjugating biologically active molecules such as peptides and NAs. In this study, the authors employed two methods to activate the surface of fullerene $C_{60}$ nanocrystals: oxidation with nitric acid (F-$HNO_3$) and modification with trimesic acid (F-TMA). These methods introduced reactive groups on the fullerene surface, facilitating the covalent or non-covalent attachment of peptides [323].

The peptides utilized in this study were derivatives of maximin H5, where the aspartic acid at position 11 was substituted with various amino acids, including alanine, glycine, histidine, asparagine, valine, or tyrosine. These mutations were designed to investigate the impact of amino acid substitution on the antiviral efficacy of the peptides when conjugated with fullerene $C_{60}$. The conjugation process involved mixing the synthesized peptides with the activated fullerene nanocrystals in DMF, resulting in the formation of peptide-fullerene conjugates. The conjugates were then purified to remove any unbound peptides and residual reagents [323].

The results of the study indicated that the antiviral activity of the maximin H5 derivatives was significantly enhanced when they were bound to fullerene $C_{60}$ nanocrystals. Notably, the peptides conjugated to fullerene nanocrystals modified with trimesic acid (F-TMA) exhibited higher binding efficiency and greater antiviral effects compared to those bound to nitric acid-activated fullerenes (F-$HNO_3$). Among the peptide variants, those with asparagine (H5N), valine (H5V), or tyrosine (H5Y) substitutions demonstrated the most potent antiviral activity against bacteriophage k, a model virus [323]. The study highlights the importance of selecting appropriate surface modifications for fullerenes to optimize their performance as nanocarriers, as the enhanced surface chemistry provided by trimesic acid modification significantly improved the interaction between the peptides and the fullerene $C_{60}$ nanocrystals [323].

#### *6.5.4.2 Fullerenes as Acid Nucleic Nanocarriers*

Fullerene $C_{60}$ has emerged as a highly versatile nanocarrier in the realm of gene delivery, primarily due to its unique structural properties and capacity for functionalization [321]. The study by Uritu et al. [324] explores the potential of $C_{60}$ fullerenes conjugated with PEI and PEG for the efficient delivery of DNA into cells. The $C_{60}$ fullerene serves as a stable and hydrophobic core, which is functionalized through the nucleophilic addition of PEI and PEG-PEI, resulting in dendrimer-like structures. These functionalizations were carried out under ambient conditions in nonpolar solvents, followed by purification steps including dialysis and lyophilization, yielding stable $C_{60}$-PEI and $C_{60}$-PEG-PEI conjugates [324].

The $C_{60}$-PEI and $C_{60}$-PEG-PEI conjugates are particularly noteworthy for their ability to form stable polyplexes with both linear and pDNA. The PEI component provides multiple positive charges, facilitating strong interactions with the negatively charged phosphate groups of DNA, which is essential for the formation of stable polyplexes. The addition of PEG enhances the solubility and biocompatibility of these conjugates, further improving their potential as gene-delivery vehicles. The study demonstrated that the resulting polyplexes exhibit colloidal stability, with particle sizes ranging from 50 to 200 nm, and maintain a positive charge at higher nitrogen-to-phosphorus (N/P) ratios, a characteristic that is critical for efficient cellular uptake [324].

In terms of biocompatibility, the $C_{60}$-PEG-PEI conjugates showed significant advantages, promoting cell proliferation by up to 200% compared to control samples [324]. This high level of cytocompatibility is attributed to the presence of PEG chains, which mitigate the cytotoxic effects commonly associated with PEI. Consequently, the $C_{60}$-PEG-PEI conjugates offer a safer alternative for gene delivery applications, particularly in scenarios where maintaining cell viability is paramount.

The transfection efficiency of these conjugates was also evaluated using HEK 293T cells transfected with a plasmid encoding an enhanced yellow fluorescent protein (EYFP). The $C_{60}$-PEI polyplexes achieved transfection efficiencies exceeding 25% at optimal N/P ratios, outperforming commercial transfection reagents like SuperFect®. Although the $C_{60}$-PEG-PEI conjugates exhibited lower transfection efficiency, around 6% at higher N/P ratios, their superior biocompatibility makes them an attractive option for applications where reduced cytotoxicity is a priority [324]. This study highlights the potential of $C_{60}$ fullerenes as effective and adaptable nanocarriers for gene delivery, demonstrating the benefits of integrating fullerene's structural capabilities with functional polymers like PEI and PEG to address complex challenges in gene therapy

#### *6.5.4.3 Enhancing Nanocarrier Functionality through Fullerene Conjugation*

The fullerenes have the possibility of being conjugated to other nanomaterials and in this way potentially the properties of hybrid nanocarriers. The study by Nag et al. [325] presents a novel approach to developing nanocarrier systems by combining fullerenes ($C_{60}$), QDs, and peptides to create a highly efficient platform for visualizing membrane potentials in living cells. This research emphasizes the potential of integrating different nanomaterials to enhance the functionality and efficiency of nanocarriers, particularly for applications in bioimaging and biosensing.

Fullerene $C_{60}$, a carbon-based nanomaterial, was employed in this study due to its exceptional electron-accepting properties, which make it ideal for facilitating electron transfer (ET) processes when conjugated with QDs. In the designed bioconjugate system, $C_{60}$ acts as an electron acceptor, playing a crucial role in quenching the PL of the QDs upon the depolarization of the cellular membrane. This quenching is integral to the system's ability to function as a membrane potential sensor [325].

The synthesis of the QD-peptide-$C_{60}$ bioconjugates involved a meticulous multi-step process. The first step was the covalent attachment of $C_{60}$ to a peptide via EDC chemistry. The peptides used in the study, designated as JBD1, JBD2, and JBD3, were designed with specific sequences rich in alanine and leucine to promote helix formation, which is essential for their insertion into the plasma membrane. Each peptide contained a lysine residue that served as the attachment point for the $C_{60}$ moiety. The QDs, serving as the central scaffold, were coated with either dihydrolipoic acid (DHLA) or DHLA-PEG750-OMe ligands, which provided colloidal stability and facilitated the assembly of the peptide-$C_{60}$ conjugates on their surface.

The functionalization process was carefully controlled to achieve precise assembly of the peptide-$C_{60}$ conjugates onto the QD surface. The number of peptide-$C_{60}$ units per QD, referred to as valence, was a critical parameter that directly influenced the photophysical properties of the bioconjugate. The study demonstrated that the efficiency of quenching of the QD photoluminescence by $C_{60}$ was dependent on this valence, with maximal quenching observed when approximately 60 peptide-$C_{60}$ units were assembled around each QD. Additionally, the choice of ligand coating on the QD significantly impacted the quenching efficiency, with DHLA-PEG750-OMe-coated QDs exhibiting greater efficiency compared to those coated with DHLA alone. This difference is likely due to reduced electrostatic shielding and better exposure of the $C_{60}$ to the QD surface in the PEGylated system [325].

The QD-peptide-$C_{60}$ bioconjugates were successfully utilized as sensors for visualizing changes in membrane potential in living cells. Upon membrane depolarization, these bioconjugates exhibited significant quenching of QD photoluminescence, with the extent of quenching correlating with the proximity of the $C_{60}$ to the QD surface. This system was also effective in dual-channel imaging experiments, where it was used in combination with conventional voltage-sensitive dyes (VSDs) to provide complementary readouts of membrane potential changes, thereby enhancing the versatility and applicability of the nanocarrier.

Moreover, the study extended the application of the QD-peptide-$C_{60}$ system to *in vivo* imaging, particularly for monitoring cortical brain function in live mice. The bioconjugates demonstrated superior photostability and larger fluorescence changes compared to traditional VSDs, making them highly suitable for real-time imaging of brain activity with improved signal strength and temporal resolution. These findings underscore the potential of combining fullerenes with other nanomaterials, such as QDs and peptides, to create multifunctional nanocarriers with enhanced capabilities for studying complex biological processes [325].

This research exemplifies the significant advantages of integrating fullerenes into nanocarrier systems, particularly when combined with other nanomaterials to address specific challenges in bioimaging and biosensing. The QD-peptide-$C_{60}$ bioconjugate system not only showcases the potential for more powerful and versatile nanocarriers but also paves the way for future innovations in the development of next-generation nanomaterials for biomedical applications.

#### *6.5.4.4 Biocompatibility of Fullerenes*

The biocompatibility of fullerenes, particularly $C_{60}$, has been extensively studied, revealing their low acute toxicity alongside potential long-term risks. While pristine $C_{60}$ fullerenes generally exhibit minimal cytotoxic effects on human and animal cells *in vitro* and demonstrate low acute toxicity *in vivo*, the long-term implications of their persistence in biological systems remain inadequately understood [326, 327].

In their comprehensive investigation, Halenova et al. [328] examined the biocompatibility of water-soluble pristine $C_{60}$ fullerenes using both *in vitro* and *in vivo* models. Their research focused on the effects of these fullerenes on various blood components, including erythrocytes, platelets, and coagulation factors, to evaluate their suitability as safe nanomaterials for biomedical applications. The study revealed that $C_{60}$ fullerenes did not induce hemolysis in red blood cells, indicating a high degree of compatibility with erythrocyte membranes. However, prolonged exposure to higher concentrations of $C_{60}$ fullerenes was found to enhance platelet aggregation, suggesting a potential prothrombotic effect. Furthermore, while standard coagulation times remained largely unaffected, $C_{60}$ fullerenes were observed to inhibit thrombin activity, indicating potential anticoagulant properties.

The findings of Halenova et al. [328] provide critical insights into the intricate interactions between $C_{60}$ fullerenes and the hemostatic system. Although the study underscores the generally favorable biocompatibility of these nanomaterials, it also highlights specific risks related to platelet function and coagulation, which necessitate further investigation.

The safety profile of fullerenes is significantly influenced by surface modifications and environmental conditions. For instance, Rancan et al. [329] reported that exposure to UVA or UVB light renders fullerenes cytotoxic, leading to UV-dose-dependent cell membrane damage. This observation underscores the importance of considering environmental factors in safety assessments [327, 329].

Surface functionalization plays a pivotal role in determining the toxicity of fullerenes. Chemically modified fullerenes, such as polyhydroxylated fullerenes (fullerenols), tend to produce fewer reactive oxygen species (ROS) and, consequently, exhibit lower cytotoxicity compared to their unmodified counterparts [326, 327]. Nevertheless, even these modified variants can generate singlet oxygen under UV or visible light, potentially leading to phototoxicity and oxidative stress in cells. Interestingly, at subcytotoxic levels, fullerenols may function as scavengers of hydroxyl radicals, highlighting their dual role as both protective and potentially harmful agents [326, 327].

The ability of fullerenes to penetrate cell membranes is a critical determinant of their biocompatibility [330, 331]. Cationic fullerene derivatives, which exhibit greater ease of cellular entry, tend to be more toxic than their neutral or anionic counterparts. Research indicates that fullerene nanoparticles, once in systemic circulation, preferentially localize in lipid-rich regions such as cell membranes, where they may induce oxidative damage [330]. Although transient lung inflammation has been observed in rodent studies following high-dose exposure, these effects typically do not culminate in long-term pathological changes in tissues

## 6.6 CONCLUSIONS

The exploration of carbon-based nanomaterials as nanocarriers for peptides and NAs has unveiled a range of promising applications that leverage the unique properties of these materials [5, 6, 31]. The distinct structural, electrical, mechanical, and optical characteristics of graphene, CNTs, CDs, and fullerenes have opened new avenues in biomedical research, particularly in the fields of gene therapy, drug delivery, and biosensing [22, 58, 180].

Graphene and its derivatives have demonstrated exceptional potential due to their high surface area, mechanical strength, and tunable surface chemistry [53, 58]. These properties enable efficient loading and targeted delivery of therapeutic molecules, making graphene an ideal candidate for advanced drug delivery systems. The various synthesis methods of a top-down approach, such as the chemical oxidation-reduction, have enabled the scalable production of high-quality graphene materials, further enhancing their applicability in biomedical research [55].

CNTs stand out for their remarkable electrical conductivity, thermal stability, and mechanical strength [180]. These properties make CNTs highly effective for delivering NAs and proteins, protecting them from enzymatic degradation and ensuring their efficient cellular uptake [332]. Advances in synthesis methods, such as CVD and Laser Ablation, have improved the quality and yield of CNTs, while functionalization strategies have enhanced their biocompatibility and targeting efficiency.

Carbon dots have emerged as versatile nanomaterials with excellent biocompatibility, PL properties, and ease of functionalization [8, 22, 212]. Their quantum confinement effects and tunable optical properties make them highly suitable for both gene and peptide delivery, as well as real-time imaging applications [30]. The flexibility in their synthesis, achieved through both top-down and bottom-up approaches, allows for precise control over their size and emission characteristics, enabling their use in a wide range of biomedical applications [22, 212].

Fullerenes, particularly $C_{60}$, offer a unique combination of structural stability and the ability to undergo various chemical modifications [37, 315]. These properties make fullerenes effective nanocarriers for peptides and NAs, providing protection from degradation and enabling targeted delivery to specific cells or tissues [37, 316]. The ability to conjugate fullerenes with other nanomaterials, such as QDs, has further enhanced their functionality, making them valuable tools in bioimaging and biosensing [37, 316].

Despite the significant progress made in utilizing carbon-based nanomaterials for biomedical applications, several challenges remain. The biocompatibility of these materials, although generally favorable, requires further investigation, particularly concerning long-term effects and potential toxicity. Surface functionalization plays a crucial role in mitigating toxicity, but the impact of repeated dosing and prolonged exposure in clinical settings is not yet fully understood.

Looking forward, the development of hybrid nanomaterials that combine the strengths of different carbon-based nanomaterials presents an exciting opportunity for enhancing the efficacy and safety of therapeutic applications. Furthermore, advancements in synthesis techniques that allow for the large-scale production of high-quality nanomaterials will be critical in translating laboratory findings into clinical practices.

In conclusion, carbon-based nanomaterials hold immense potential for revolutionizing the fields of drug delivery, gene therapy, and biomedical imaging. However, realizing their full potential will require continued interdisciplinary research focused on overcoming the current challenges and optimizing these materials for safe and effective use in human health. The ongoing innovation in this field promises to lead to new therapeutic strategies that could significantly improve the precision and efficacy of medical treatments, ultimately contributing to better patient outcomes.

## REFERENCES

1. O. Eshel, "Reading Winnicott into nano-psychoanalysis: 'there's plenty of room at the bottom,'" *Psychoanal Inq*, vol. 33, no. 1, pp. 36–49, Jan. 2013, doi: 10.1080/07351690.2013.743803.

2. M. Jafari and H. R. Zarghami, "Measuring nanotechnology development through the study of the dividing pattern between developed and developing countries during 2000–2014," *J Nanopart Res*, vol. 18, no. 7, p. 180, Jul. 2016, doi: 10.1007/s11051-016-3431-0.

3. S. J. Malode, S. Pandiaraj, A. Alodhayb, and N. P. Shetti, "Carbon nanomaterials for biomedical applications: progress and outlook," *ACS Appl Bio Mater*, vol. 7, no. 2, pp. 752–777, Feb. 2024, doi: 10.1021/acsabm.3c00983.

4. A. T. Krasley, E. Li, J. M. Galeana, C. Bulumulla, A. G. Beyene, and G. S. Demirer, "Carbon nanomaterial fluorescent probes and their biological applications," *Chem Rev*, vol. 124, no. 6, pp. 3085–3185, Mar. 2024, doi: 10.1021/acs.chemrev.3c00581.

5. N. Parvin, V. Kumar, S. W. Joo, and T. K. Mandal, "Emerging trends in nanomedicine: carbon-based nanomaterials for healthcare," *Nanomaterials*, vol. 14, no. 13, p. 1085, Jun. 2024, doi: 10.3390/nano14131085.

6. M. Shin, J. Lim, Y. Park, J.-Y. Lee, J. Yoon, and J.-W. Choi, "Carbon-based nanocomposites for biomedical applications," *RSC Adv*, vol. 14, no. 10, pp. 7142–7156, 2024, doi: 10.1039/D3RA08946K.

7. C. L. Brito, J. V. Silva, R. V. Gonzaga, M. A. La-Scalea, J. Giarolla, and E. I. Ferreira, "A review on carbon nanotubes family of nanomaterials and their health field," *ACS Omega*, vol. 9, no. 8, pp. 8687–8708, Feb. 2024, doi: 10.1021/acsomega.3c08824.

8. J. Wang et al., "Multifunctional carbon dots for biomedical applications: diagnosis, therapy, and theranostic," *Small*, vol. 20, no. 3, Jan. 2024, doi: 10.1002/smll.202303773.

9. H. Chen, K. Luo, C. Xie, and L. Zhou, "Nanotechnology of carbon dots with their hybrids for biomedical applications: a review," *Chem Eng J*, vol. 496, p. 153915, Sep. 2024, doi: 10.1016/j.cej.2024.153915.

10. C. Pesado-Gómez et al., "Fullerenes: historical background, novel biological activities versus possible health risks," *Coord Chem Rev*, vol. 501, p. 215550, Feb. 2024, doi: 10.1016/j.ccr.2023.215550.

11. M. A. Islam et al., "Recent achievement of graphene in biomedicine: advancements by integrated microfluidics system and conventional techniques," *Sensors Int*, vol. 5, p. 100293, 2024, doi: 10.1016/j.sintl.2024.100293.

12. G. Bae, H. Cho, and B. H. Hong, "A review on synthesis, properties, and biomedical applications of graphene quantum dots (GQDs)," *Nanotechnology*, vol. 35, no. 37, p. 372001, Sep. 2024, doi: 10.1088/1361-6528/ad55d0.

13. N. Anzar, R. Hasan, M. Tyagi, N. Yadav, and J. Narang, "Carbon nanotube—a review on synthesis, properties and plethora of applications in the field of biomedical science," *Sensors Int*, vol. 1, p. 100003, 2020, doi: 10.1016/j.sintl.2020.100003.

14. C. Galande et al., "Science and engineering of graphene oxide," *Particle & Particle Systems Characterization*, vol. 31, no. 6, pp. 619–638, Jun. 2014, doi: 10.1002/ppsc.201300232.

15. K. S. Novoselov et al., "Electric field effect in atomically thin carbon films," *Science (1979)*, vol. 306, no. 5696, pp. 666–669, Oct. 2004, doi: 10.1126/science.1102896.

16. A. K. Geim and K. S. Novoselov, "The rise of graphene," *Nat Mater*, vol. 6, no. 3, pp. 183–191, Mar. 2007, doi: 10.1038/nmat1849.

17. Y. Hancock, "The 2010 Nobel Prize in physics—ground-breaking experiments on graphene," *J Phys D Appl Phys*, vol. 44, no. 47, p. 473001, Nov. 2011, doi: 10.1088/0022-3727/44/47/473001.

18. J. Chen, S. Wei, and H. Xie, "A brief introduction of carbon nanotubes: history, synthesis, and properties," *J Phys Conf Ser*, vol. 1948, no. 1, p. 012184, Jun. 2021, doi: 10.1088/1742-6596/1948/1/012184.

19. S. Iijima and T. Ichihashi, "Single-shell carbon nanotubes of 1-nm diameter," *Nature*, vol. 363, no. 6430, pp. 603–605, Jun. 1993, doi: 10.1038/363603a0.

20. S. Rathinavel, K. Priyadharshini, and D. Panda, "A review on carbon nanotube: an overview of synthesis, properties, functionalization, characterization, and the application," *Mater Sci Eng B*, vol. 268, p. 115095, Jun. 2021, doi: 10.1016/j.mseb.2021.115095.

21. X. Xu et al., "Electrophoretic analysis and purification of fluorescent single-walled carbon nanotube fragments," *J Am Chem Soc*, vol. 126, no. 40, pp. 12736–12737, Oct. 2004, doi: 10.1021/ja040082h.

22. B. Wang, G. I. N. Waterhouse, and S. Lu, "Carbon dots: mysterious past, vibrant present, and expansive future," *Trends Chem*, vol. 5, no. 1, pp. 76–87, Jan. 2023, doi: 10.1016/j.trechm.2022.10.005.

23. H. W. Kroto, J. R. Heath, S. C. O'Brien, R. F. Curl, and R. E. Smalley, "C60: buckminsterfullerene," *Nature*, vol. 318, no. 6042, pp. 162–163, Nov. 1985, doi: 10.1038/318162a0.

24. R. E. Smalley, "Discovering the fullerenes (Nobel lecture)," *Angew Chem Int Ed Eng*, vol. 36, no. 15, pp. 1594–1601, Aug. 1997, doi: 10.1002/anie.199715941.

25. A. R. Urade, I. Lahiri, and K. S. Suresh, "Graphene properties, synthesis and applications: a review," *JOM*, vol. 75, no. 3, pp. 614–630, Mar. 2023, doi: 10.1007/s11837-022-05505-8.

26. V. Khanna, V. Kumar, and S. A. Bansal, "Mechanical properties of aluminium–graphene/carbon nanotubes (CNTs) metal matrix composites: advancement, opportunities and perspective," *Mater Res Bull*, vol. 138, p. 111224, Jun. 2021, doi: 10.1016/j.materresbull.2021.111224.

27. Y. Li, S. Wang, Q. Wang, and M. Xing, "A comparison study on mechanical properties of polymer composites reinforced by carbon nanotubes and graphene sheet," *Compos B Eng*, vol. 133, pp. 35–41, Jan. 2018, doi: 10.1016/j.compositesb.2017.09.024.

28. J.-P. Salvetat et al., "Mechanical properties of carbon nanotubes," *Appl Phys A Mater Sci Process*, vol. 69, no. 3, pp. 255–260, Sep. 1999, doi: 10.1007/s003390050999.

29. J. Che, T. Çagin, and W. A. Goddard, "Thermal conductivity of carbon nanotubes," *Nanotechnology*, vol. 11, no. 2, pp. 65–69, Jun. 2000, doi: 10.1088/0957-4484/11/2/305.

30. F. Yan, Z. Sun, H. Zhang, X. Sun, Y. Jiang, and Z. Bai, "The fluorescence mechanism of carbon dots, and methods for tuning their emission color: a review," *Microchim Acta*, vol. 186, no. 8, p. 583, Aug. 2019, doi: 10.1007/s00604-019-3688-y.

31. S. K. Debnath and R. Srivastava, "Drug delivery with carbon-based nanomaterials as versatile nanocarriers: progress and prospects," *Front Nanotechnol*, vol. 3, Apr. 2021, doi: 10.3389/fnano.2021.644564.

32. C. Liao, Y. Li, and S. C. Tjong, "Graphene nanomaterials: synthesis, biocompatibility, and cytotoxicity," *Int J Mol Sci*, vol. 19, no. 11, p. 3564, Nov. 2018, doi: 10.3390/ijms19113564.

33. K. Umemura, "Hybrids of nucleic acids and carbon nanotubes for nanobiotechnology," *Nanomaterials*, vol. 5, no. 1, pp. 321–350, Mar. 2015, doi: 10.3390/nano5010321.

34. H. Li et al., "Recent advances in carbon dots for bioimaging applications," *Nanoscale Horiz*, vol. 5, no. 2, pp. 218–234, 2020, doi: 10.1039/C9NH00476A.

35. W. Zhang et al., "Carbon dots: a future blood–brain barrier penetrating nanomedicine and drug nanocarrier," *Int J Nanomed*, vol. 16, pp. 5003–5016, Jul. 2021, doi: 10.2147/IJN.S318732.

36. M. Mohajeri, B. Behnam, and A. Sahebkar, "Biomedical applications of carbon nanomaterials: drug and gene delivery potentials," *J Cell Physiol*, vol. 234, no. 1, pp. 298–319, Jan. 2019, doi: 10.1002/jcp.26899.

37. H. Kazemzadeh and M. Mozafari, "Fullerene-based delivery systems," *Drug Discov Today*, vol. 24, no. 3, pp. 898–905, Mar. 2019, doi: 10.1016/j.drudis.2019.01.013.

38. K. Brindhadevi, H. A. L. Garalleh, A. Alalawi, E. Al-Sarayreh, and A. Pugazhendhi, "Carbon nanomaterials: types, synthesis strategies and their application as drug delivery system for cancer therapy," *Biochem Eng J*, vol. 192, p. 108828, Mar. 2023, doi: 10.1016/j.bej.2023.108828.

39. P. R. Riley and R. J. Narayan, "Recent advances in carbon nanomaterials for biomedical applications: a review," *Curr Opin Biomed Eng*, vol. 17, p. 100262, Mar. 2021, doi: 10.1016/j.cobme.2021.100262.

40. S. Zheng, Y. Tian, J. Ouyang, Y. Shen, X. Wang, and J. Luan, "Carbon nanomaterials for drug delivery and tissue engineering," *Front Chem*, vol. 10, Sep. 2022, doi: 10.3389/fchem.2022.990362.

41. K. P. Loh, D. Ho, G. N. C. Chiu, D. T. Leong, G. Pastorin, and E. K. Chow, "Clinical applications of carbon nanomaterials in diagnostics and therapy," *Adv Mater*, vol. 30, no. 47, Nov. 2018, doi: 10.1002/adma.201802368.

42. R. Travers and V. Sayag-Boukris, "Sir Benjamin Collins Brodie (1783–1862)—a pioneer rheumatologist," *Int J Rheum Dis*, vol. 18, no. 5, pp. 587–590, Jun. 2015, doi: 10.1111/1756-185X.12646.

43. "Sir Benjamin Brodie, F. R. S. (1783–1862)," *Notes Rec R Soc Lond*, vol. 19, no. 1, pp. 42–52, Jun. 1964, doi: 10.1098/rsnr.1964.0004.

44. "XIII. On the atomic weight of graphite," *Philos Trans R Soc Lond*, vol. 149, pp. 249–259, Dec. 1859, doi: 10.1098/rstl.1859.0013.

45. V. Dhinakaran, M. Lavanya, K. Vigneswari, M. Ravichandran, and M. D. Vijayakumar, "Review on exploration of graphene in diverse applications and its future horizon," *Mater Today Proc*, vol. 27, pp. 824–828, 2020, doi: 10.1016/j.matpr.2019.12.369.

46. M. Salvatore, G. Carotenuto, S. De Nicola, C. Camerlingo, V. Ambrogi, and C. Carfagna, "Synthesis and characterization of highly intercalated graphite bisulfate," *Nanoscale Res Lett*, vol. 12, no. 1, p. 167, Dec. 2017, doi: 10.1186/s11671-017-1930-2.

47. L. Chen et al., "Modifying graphite oxide nanostructures in various media by high-energy irradiation," *RSC Adv.*, vol. 4, no. 2, pp. 1025–1031, 2014, doi: 10.1039/C3RA46203J.

48. A. Lerf, "Graphite oxide story—from the beginning till the graphene hype," in *Graphene Oxide*, Wiley, 2016, pp. 1–35, doi: 10.1002/9781119069447.ch1.

49. S. Gambhir, R. Jalili, D. L. Officer, and G. G. Wallace, "Chemically converted graphene: scalable chemistries to enable processing and fabrication," *NPG Asia Mater*, vol. 7, no. 6, pp. e186–e186, Jun. 2015, doi: 10.1038/am.2015.47.

50. A. Jiříčková, O. Jankovský, Z. Sofer, and D. Sedmidubský, "Synthesis and applications of graphene oxide," *Materials*, vol. 15, no. 3, p. 920, Jan. 2022, doi: 10.3390/ma15030920.

51. U. Hofmann and R. Holst, "Über die Säurenatur und die Methylierung von Graphitoxyd," *Ber dtsch Chem GesA and B Ser*, vol. 72, no. 4, pp. 754–771, Apr. 1939, doi: 10.1002/cber.19390720417.

52. P. R. Wallace, "The band theory of graphite," *Phys Rev*, vol. 71, no. 9, pp. 622–634, May 1947, doi: 10.1103/PhysRev.71.622.

53. T. Barkan, "Graphene: the hype versus commercial reality," *Nat Nanotechnol*, vol. 14, no. 10, pp. 904–906, Oct. 2019, doi: 10.1038/s41565-019-0556-1.

54. O. V. Prezhdo, "Graphene—the ultimate surface material," *Surf Sci*, vol. 605, no. 17–18, pp. 1607–1610, Sep. 2011, doi: 10.1016/j.susc.2011.05.009.

55. N. Kumar et al., "Top-down synthesis of graphene: a comprehensive review," *FlatChem*, vol. 27, p. 100224, May 2021, doi: 10.1016/j.flatc.2021.100224.

56. Z. Zhen and H. Zhu, "Structure and properties of graphene," in *Graphene*, Elsevier, 2018, pp. 1–12, doi: 10.1016/B978-0-12-812651-6.00001-X.

57. V. B. Mbayachi, E. Ndayiragije, T. Sammani, S. Taj, E. R. Mbuta, and A. Ullah Khan, "Graphene synthesis, characterization and its applications: a review," *Results Chem*, vol. 3, p. 100163, Jan. 2021, doi: 10.1016/j.rechem.2021.100163.

58. F. Zhang et al., "Recent advances on graphene: synthesis, properties and applications," *Compos Part A Appl Sci Manuf*, vol. 160, p. 107051, Sep. 2022, doi: 10.1016/j.compositesa.2022.107051.

59. S. Trivedi, K. Lobo, and H. S. S. Ramakrishna Matte, "Synthesis, properties, and applications of graphene," in *Fundamentals and Sensing Applications of 2D Materials*, Elsevier, 2019, pp. 25–90, doi: 10.1016/B978-0-08-102577-2.00003-8.

60. Y.-W. Son, M. L. Cohen, and S. G. Louie, "Energy gaps in graphene nanoribbons," *Phys Rev Lett*, vol. 97, no. 21, p. 216803, Nov. 2006, doi: 10.1103/PhysRevLett.97.216803.

61. P. Delhaes, Ed., *Graphite and Precursors*, CRC Press, 2000, doi: 10.1201/9781482296921.

62. S. K. Tiwari, S. Sahoo, N. Wang, and A. Huczko, "Graphene research and their outputs: status and prospect," *J Sci Adv Mater Dev*, vol. 5, no. 1, pp. 10–29, Mar. 2020, doi: 10.1016/j.jsamd.2020.01.006.

63. A. Santhiran, P. Iyngaran, P. Abiman, and N. Kuganathan, "Graphene synthesis and its recent advances in applications—a review," *C J Carbon Res (Basel)*, vol. 7, no. 4, p. 76, Nov. 2021, doi: 10.3390/c7040076.

64. G. Yang, L. Li, W. B. Lee, and M. C. Ng, "Structure of graphene and its disorders: a review," *Sci Technol Adv Mater*, vol. 19, no. 1, pp. 613–648, Dec. 2018, doi: 10.1080/14686996.2018.1494493.

65. P. Pachfule, D. Shinde, M. Majumder, and Q. Xu, "Fabrication of carbon nanorods and graphene nanoribbons from a metal–organic framework," *Nat Chem*, vol. 8, no. 7, pp. 718–724, Jul. 2016, doi: 10.1038/nchem.2515.

66. J. Liu et al., "Toward cove-edged low band gap graphene nanoribbons," *J Am Chem Soc*, vol. 137, no. 18, pp. 6097–6103, May 2015, doi: 10.1021/jacs.5b03017.

67. "Graphene: zig-zag filters," *NPG Asia Mater*, May 2009, doi: 10.1038/asiamat.2009.189.

68. J.-J. He, X.-H. Yan, Y.-D. Guo, C.-S. Liu, Y. Xiao, and L. Meng, "The electron transport properties of zigzag graphene nanoribbons with upright standing linear carbon chains," *Solid State Commun*, vol. 227, pp. 28–32, Feb. 2016, doi: 10.1016/j.ssc.2015.11.013.

69. T. Enoki, S. Fujii, and K. Takai, "Zigzag and armchair edges in graphene," *Carbon N Y*, vol. 50, no. 9, pp. 3141–3145, Aug. 2012, doi: 10.1016/j.carbon.2011.10.004.

70. L. Yang, C.-H. Park, Y.-W. Son, M. L. Cohen, and S. G. Louie, "Quasiparticle energies and band gaps in graphene nanoribbons," *Phys Rev Lett*, vol. 99, no. 18, p. 186801, Nov. 2007, doi: 10.1103/PhysRevLett.99.186801.

71. R. M. Jacobberger et al., "Direct oriented growth of armchair graphene nanoribbons on germanium," *Nat Commun*, vol. 6, no. 1, p. 8006, Aug. 2015, doi: 10.1038/ncomms9006.

72. J. Kim et al., "Distinguishing zigzag and armchair edges on graphene nanoribbons by X-ray photoelectron and Raman spectroscopies," *ACS Omega*, vol. 3, no. 12, pp. 17789–17796, Dec. 2018, doi: 10.1021/acsomega.8b02744.

73. M. A. Torkaman-Asadi and M. A. Kouchakzadeh, "Atomistic simulations of mechanical properties and fracture of graphene: a review," *Comput Mater Sci*, vol. 210, p. 111457, Jul. 2022, doi: 10.1016/j.commatsci.2022.111457.

74. P. V. Polyakova and J. A. Baimova, "Mechanical properties of graphene networks under compression: a molecular dynamics simulation," *Int J Mol Sci*, vol. 24, no. 7, p. 6691, Apr. 2023, doi: 10.3390/ijms24076691.

75. H. Kashani, Y. Ito, J. Han, P. Liu, and M. Chen, "Extraordinary tensile strength and ductility of scalable nanoporous graphene," *Sci Adv*, vol. 5, no. 2, Feb. 2019, doi: 10.1126/sciadv.aat6951.

76. H. Li, S. Xiao, H. Yu, Y. Xue, and J. Yang, "A review of graphene-based films for heat dissipation," *New Carbon Mater*, vol. 36, no. 5, pp. 897–908, Oct. 2021, doi: 10.1016/S1872-5805(21)60092-6.

77. C. V. Kumar and A. Pattammattel, "Synthetic routes to graphene preparation from the perspectives of possible biological applications," in *Introduction to Graphene*, Elsevier, 2017, pp. 17–44, doi: 10.1016/B978-0-12-813182-4.00002-7.

78. N. Liu, Q. Tang, B. Huang, and Y. Wang, "Graphene synthesis: method, exfoliation mechanism and large-scale production," *Crystals (Basel)*, vol. 12, no. 1, p. 25, Dec. 2021, doi: 10.3390/cryst12010025.

79. F. Liu et al., "Synthesis of graphene materials by electrochemical exfoliation: recent progress and future potential," *Carbon Energy*, vol. 1, no. 2, pp. 173–199, Dec. 2019, doi: 10.1002/cey2.14.

80. Q. Abbas et al., "Graphene synthesis techniques and environmental applications," *Materials*, vol. 15, no. 21, p. 7804, Nov. 2022, doi: 10.3390/ma15217804.

81. A. L. Olatomiwa, T. Adam, S. C. B. Gopinath, S. Y. Kolawole, O. H. Olayinka, and U. Hashim, "Graphene synthesis, fabrication, characterization based on bottom-up and top-down approaches: an overview," *J Semiconduct*, vol. 43, no. 6, p. 061101, Jun. 2022, doi: 10.1088/1674-4926/43/6/061101.

82. E. Gerstner, "Nobel Prize 2010: Andre Geim & Konstantin Novoselov," *Nat Phys*, vol. 6, no. 11, pp. 836–836, Nov. 2010, doi: 10.1038/nphys1836.

83. A. A. MOOSA and M. S. ABED, "Graphene preparation and graphite exfoliation," *Turk J Chem*, vol. 45, no. 3, pp. 493–519, Jun. 2021, doi: 10.3906/kim-2101-19.

84. V. Singh, D. Joung, L. Zhai, S. Das, S. I. Khondaker, and S. Seal, "Graphene based materials: past, present and future," *Prog Mater Sci*, vol. 56, no. 8, pp. 1178–1271, Oct. 2011, doi: 10.1016/j.pmatsci.2011.03.003.

85. C. K. Chua, Z. Sofer, B. Khezri, R. D. Webster, and M. Pumera, "Ball-milled sulfur-doped graphene materials contain metallic impurities originating from ball-milling apparatus: their influence on the catalytic properties," *Phys Chem Chem Phys*, vol. 18, no. 27, pp. 17875–17880, 2016, doi: 10.1039/C6CP03004A.

86. N. Zhang, Y. Mao, S. Wu, and W. Xu, "Effects of the ball milling process on the particle size of graphene oxide and its application in enhancing the thermal conductivity of wood," *Forests*, vol. 13, no. 8, p. 1325, Aug. 2022, doi: 10.3390/f13081325.

87. A. V. Tyurnina et al., "Ultrasonic exfoliation of graphene in water: a key parameter study," *Carbon N Y*, vol. 168, pp. 737–747, Oct. 2020, doi: 10.1016/j.carbon.2020.06.029.

88. P. Turner, M. Hodnett, R. Dorey, and J. D. Carey, Controlled inertial cavitation as a route to high yield liquid phase exfoliation of graphene, Sep. 2018. arXiv.org. https://arxiv.org/abs/1809.07630

89. M. Telkhozhayeva et al., "Higher ultrasonic frequency liquid phase exfoliation leads to larger and monolayer to few-layer flakes of 2D layered materials," *Langmuir*, vol. 37, no. 15, pp. 4504–4514, Apr. 2021, doi: 10.1021/acs.langmuir.0c03668.

90. P. Turner, M. Hodnett, R. Dorey, and J. D. Carey, "Controlled sonication as a route to in-situ graphene flake size control," *Sci Rep*, vol. 9, no. 1, p. 8710, Jun. 2019, doi: 10.1038/s41598-019-45059-5.

91. A. Ručigaj, J. G. Connell, M. Dular, and B. Genorio, "Influence of the ultrasound cavitation intensity on reduced graphene oxide functionalization," *Ultrason Sonochem*, vol. 90, p. 106212, Nov. 2022, doi: 10.1016/j.ultsonch.2022.106212.

92. A. Kaur et al., "Temperature as a key parameter for graphene sono-exfoliation in water," *Ultrason Sonochem*, vol. 90, p. 106187, Nov. 2022, doi: 10.1016/j.ultsonch.2022.106187.

93. J. G. S. Moo, B. Khezri, R. D. Webster, and M. Pumera, "Graphene oxides prepared by Hummers', Hofmann's, and Staudenmaier's methods: dramatic influences on heavy-metal-ion adsorption," *ChemPhysChem*, vol. 15, no. 14, pp. 2922–2929, Oct. 2014, doi: 10.1002/cphc.201402279.

94. O. Jankovský et al., "Synthesis procedure and type of graphite oxide strongly influence resulting graphene properties," *Appl Mater Today*, vol. 4, pp. 45–53, Sep. 2016, doi: 10.1016/j.apmt.2016.06.001.

95. H. Yu, B. Zhang, C. Bulin, R. Li, and R. Xing, "High-efficient synthesis of graphene oxide based on improved Hummers method," *Sci Rep*, vol. 6, no. 1, p. 36143, Nov. 2016, doi: 10.1038/srep36143.

96. D. C. Marcano et al., "Improved synthesis of graphene oxide," *ACS Nano*, vol. 4, no. 8, pp. 4806–4814, Aug. 2010, doi: 10.1021/nn1006368.

97. M. U. Khan and M. A. Shaida, "Reduction mechanism of graphene oxide including various parameters affecting the C/O ratio," *Mater Today Commun*, vol. 36, p. 106577, Aug. 2023, doi: 10.1016/j.mtcomm.2023.106577.

98. A. Zhou, J. Bai, W. Hong, and H. Bai, "Electrochemically reduced graphene oxide: preparation, composites, and applications," *Carbon N Y*, vol. 191, pp. 301–332, May 2022, doi: 10.1016/j.carbon.2022.01.056.

99. C.-Y. Su et al., "Highly efficient restoration of graphitic structure in graphene oxide using alcohol vapors," *ACS Nano*, vol. 4, no. 9, pp. 5285–5292, Sep. 2010, doi: 10.1021/nn101691m.

100. Y. Harima et al., "Electrochemical reduction of graphene oxide in organic solvents," *Electrochim Acta*, vol. 56, no. 15, pp. 5363–5368, Jun. 2011, doi: 10.1016/j.electacta.2011.03.117.

101. M. Saeed, Y. Alshammari, S. A. Majeed, and E. Al-Nasrallah, "Chemical vapour deposition of graphene—synthesis, characterisation, and applications: a review," *Molecules*, vol. 25, no. 17, p. 3856, Aug. 2020, doi: 10.3390/molecules25173856.

102. S. Kataria et al., "Chemical vapor deposited graphene: from synthesis to applications," *Phys Status Sol (A)*, vol. 211, no. 11, pp. 2439–2449, Nov. 2014, doi: 10.1002/pssa.201400049.

103. X. Yang, G. Zhang, J. Prakash, Z. Chen, M. Gauthier, and S. Sun, "Chemical vapour deposition of graphene: layer control, the transfer process, characterisation, and related applications," *Int Rev Phys Chem*, vol. 38, no. 2, pp. 149–199, Apr. 2019, doi: 10.1080/0144235X.2019.1634319.

104. J. Wang et al., "A review of graphene synthesis at low temperatures by CVD methods," *New Carbon Mater*, vol. 35, no. 3, pp. 193–208, Jun. 2020, doi: 10.1016/S1872-5805(20)60484-X.

105. Y. H. Miao, C., Zheng, C., Liang, O., & Xie, *Physics and Applications of Graphene: Experiments*, IntechOpen, 2011.

106. Q. H. Chang et al., "Few-layer graphene direct deposition on Ni and Cu foil by cold-wall chemical vapor deposition," *J Nanosci Nanotechnol*, vol. 12, no. 8, pp. 6516–6520, Aug. 2012, doi: 10.1166/jnn.2012.5432.

107. Z. Bo, Y. Yang, J. Chen, K. Yu, J. Yan, and K. Cen, "Plasma-enhanced chemical vapor deposition synthesis of vertically oriented graphene nanosheets," *Nanoscale*, vol. 5, no. 12, p. 5180, 2013, doi: 10.1039/c3nr33449j.

108. K. Ostrikov, "Colloquium: reactive plasmas as a versatile nanofabrication tool," *Rev Mod Phys*, vol. 77, no. 2, pp. 489–511, Jun. 2005, doi: 10.1103/RevModPhys.77.489.

109. H. Bao et al., "Chitosan-functionalized graphene oxide as a nanocarrier for drug and gene delivery," *Small*, vol. 7, no. 11, pp. 1569–1578, Jun. 2011, doi: 10.1002/smll.201100191.

110. H. Hu, C. Tang, and C. Yin, "Folate conjugated trimethyl chitosan/graphene oxide nanocomplexes as potential carriers for drug and gene delivery," *Mater Lett*, vol. 125, pp. 82–85, Jun. 2014, doi: 10.1016/j.matlet.2014.03.133.

111. Y. He et al., "Enhanced chemotherapy efficacy by co-delivery of shABCG2 and doxorubicin with a pH-responsive charge-reversible layered graphene oxide nanocomplex," *J Mater Chem B*, vol. 3, no. 31, pp. 6462–6472, 2015, doi: 10.1039/C5TB00923E.

112. R. Imani, F. Mohabatpour, and F. Mostafavi, "Graphene-based nano-carrier modifications for gene delivery applications," *Carbon N Y*, vol. 140, pp. 569–591, Dec. 2018, doi: 10.1016/j.carbon.2018.09.019.

113. M. Hoseini-Ghahfarokhi et al., "Applications of graphene and graphene oxide in smart drug/gene delivery: is the world still flat?," *Int J Nanomed*, vol. 15, pp. 9469–9496, Nov. 2020, doi: 10.2147/IJN.S265876.

114. D. N. Trivedi and N. V. Rachchh, "Graphene and its application in thermoplastic polymers as nano-filler—a review," *Polymer (Guildf)*, vol. 240, p. 124486, Feb. 2022, doi: 10.1016/j.polymer.2021.124486.

115. A. Pandey and P. Chauhan, "Functionalized graphene nanomaterials: next-generation nanomedicine," in *Functionalized Carbon Nanomaterials for Theranostic Applications*, Elsevier, 2023, pp. 3–18, doi: 10.1016/B978-0-12-824366-4.00020-0.

116. J.-H. Jiang, J. Pi, H. Jin, and J.-Y. Cai, "Functional graphene oxide as cancer-targeted drug delivery system to selectively induce oesophageal cancer cell apoptosis," *Artif Cells Nanomed Biotechnol*, vol. 46, no. supl. 3, pp. 297–307, Nov. 2018, doi: 10.1080/21691401.2018.1492418.

117. R. Muñoz, D. P. Singh, R. Kumar, and A. Matsuda, "Graphene oxide for drug delivery and cancer therapy," in *Nanostructured Polymer Composites for Biomedical Applications*, Elsevier, 2019, pp. 447–488, doi: 10.1016/B978-0-12-816771-7.00023-5.

118. B. Dąbrowski, A. Żuchowska, and Z. Brzózka, "Graphene oxide internalization into mammalian cells—a review," *Colloids Surf B Biointerfaces*, vol. 221, p. 112998, Jan. 2023, doi: 10.1016/j.colsurfb.2022.112998.

119. M. Azizi-Lalabadi, H. Hashemi, J. Feng, and S. M. Jafari, "Carbon nanomaterials against pathogens; the antimicrobial activity of carbon nanotubes, graphene/graphene oxide, fullerenes, and their nanocomposites," *Adv Colloid Interface Sci*, vol. 284, p. 102250, Oct. 2020, doi: 10.1016/j.cis.2020.102250.

120. M. V. Baranov, M. Kumar, S. Sacanna, S. Thutupalli, and G. van den Bogaart, "Modulation of immune responses by particle size and shape," *Front Immunol*, vol. 11, Feb. 2021, doi: 10.3389/fimmu.2020.607945.

121. H. Zhao et al., "Graphene-based nanomaterials for drug and/or gene delivery, bioimaging, and tissue engineering," *Drug Discov Today*, vol. 22, no. 9, pp. 1302–1317, Sep. 2017, doi: 10.1016/j.drudis.2017.04.002.

122. J. Shannahan, "The biocorona: a challenge for the biomedical application of nanoparticles," *Nanotechnol Rev*, vol. 6, no. 4, pp. 345–353, Aug. 2017, doi: 10.1515/ntrev-2016-0098.

123. L. S. Franqui et al., "Interaction of graphene oxide with cell culture medium: evaluating the fetal bovine serum protein corona formation towards in vitro nanotoxicity assessment and nanobiointeractions," *Mater Sci Eng C*, vol. 100, pp. 363–377, Jul. 2019, doi: 10.1016/j.msec.2019.02.066.

124. E. Quagliarini, D. Pozzi, F. Cardarelli, and G. Caracciolo, "The influence of protein corona on graphene oxide: implications for biomedical theranostics," *J Nanobiotechnol*, vol. 21, no. 1, p. 267, Aug. 2023, doi: 10.1186/s12951-023-02030-x.

125. B. Liu, S. Salgado, V. Maheshwari, and J. Liu, "DNA adsorbed on graphene and graphene oxide: fundamental interactions, desorption and applications," *Curr Opin Colloid Interface Sci*, vol. 26, pp. 41–49, Dec. 2016, doi: 10.1016/j.cocis.2016.09.001.

126. K. Chauhan, E. Cho, and D.-E. Kim, "Graphene oxide and nucleic acids," in *Handbook of Chemical Biology of Nucleic Acids*, Springer Nature Singapore, Singapore, 2022, pp. 1–31, doi: 10.1007/978-981-16-1313-5_62-1.

127. M. Alibolandi, M. Mohammadi, S. M. Taghdisi, M. Ramezani, and K. Abnous, "Fabrication of aptamer decorated dextran coated nano-graphene oxide for targeted drug delivery," *Carbohydr Polym*, vol. 155, pp. 218–229, Jan. 2017, doi: 10.1016/j.carbpol.2016.08.046.

128. B. Mao, L. Cheng, S. Wang, J. Zhou, and L. Deng, "Combat biofilm by bacteriostatic aptamer-functionalized graphene oxide," *Biotechnol Appl Biochem*, vol. 65, no. 3, pp. 355–361, May 2018, doi: 10.1002/bab.1631.

129. M. Ahn, J. Song, and B. H. Hong, "Facile synthesis of N-doped graphene quantum dots as novel transfection agents for mRNA and pDNA," *Nanomaterials*, vol. 11, no. 11, p. 2816, Oct. 2021, doi: 10.3390/nano11112816.

130. L. Feng et al., "Polyethylene glycol and polyethylenimine dual-functionalized nano-graphene oxide for photothermally enhanced gene delivery," *Small*, vol. 9, no. 11, pp. 1989–1997, Jun. 2013, doi: 10.1002/smll.201202538.

131. H. Kim, R. Namgung, K. Singha, I.-K. Oh, and W. J. Kim, "Graphene oxide–polyethylenimine nanoconstruct as a gene delivery vector and bioimaging tool," *Bioconjug Chem*, vol. 22, no. 12, pp. 2558–2567, Dec. 2011, doi: 10.1021/bc200397j.

132. J. H. Kim, F. Marks, and J. D. Clemens, "Looking beyond COVID-19 vaccine phase 3 trials," *Nat Med*, vol. 27, no. 2, pp. 205–211, Feb. 2021, doi: 10.1038/s41591-021-01230-y.

133. J. Gustafsson, G. Arvidson, G. Karlsson, and M. Almgren, "Complexes between cationic liposomes and DNA visualized by cryo-TEM," *Biochim Biophys Acta (BBA)—Biomembr*, vol. 1235, no. 2, pp. 305–312, May 1995, doi: 10.1016/0005-2736(95)80018-B.

134. J. Zhang et al., "Graphene oxide as a matrix for enzyme immobilization," *Langmuir*, vol. 26, no. 9, pp. 6083–6085, May 2010, doi: 10.1021/la904014z.

135. Y. Wang, Z. Zhu, H. Zhang, J. Chen, B. Tang, and J. Cao, "Investigation on the conformational structure of hemoglobin on graphene oxide," *Mater Chem Phys*, vol. 182, pp. 272–279, Oct. 2016, doi: 10.1016/j.matchemphys.2016.07.032.

136. K. Chaudhary, K. Kumar, P. Venkatesu, and D. T. Masram, "Protein immobilization on graphene oxide or reduced graphene oxide surface and their applications: influence over activity, structural and thermal stability of protein," *Adv Colloid Interface Sci*, vol. 289, p. 102367, Mar. 2021, doi: 10.1016/j.cis.2021.102367.

137. S. Palanisamy, B. Unnikrishnan, and S.-M. Chen, "An amperometric biosensor based on direct immobilization of horseradish peroxidase on electrochemically reduced graphene oxide modified screen printed carbon electrode," *Int J Electrochem Sci*, vol. 7, no. 9, pp. 7935–7947, Sep. 2012, doi: 10.1016/S1452-3981(23)17966-1.

138. E. I. Biru, M. I. Necolau, A. Zainea, and H. Iovu, "Graphene oxide–protein-based scaffolds for tissue engineering: recent advances and applications," *Polymers (Basel)*, vol. 14, no. 5, p. 1032, Mar. 2022, doi: 10.3390/polym14051032.

139. S. Goenka, V. Sant, and S. Sant, "Graphene-based nanomaterials for drug delivery and tissue engineering," *J Control Release*, vol. 173, pp. 75–88, Jan. 2014, doi: 10.1016/j.jconrel.2013.10.017.

140. J. Shen et al., "Covalent attaching protein to graphene oxide via diimide-activated amidation," *Colloids Surf B Biointerfaces*, vol. 81, no. 2, pp. 434–438, Dec. 2010, doi: 10.1016/j.colsurfb.2010.07.035.

141. G. Hernández-Cancel, D. Suazo-Dávila, A. J. Ojeda-Cruzado, D. García-Torres, C. R. Cabrera, and K. Griebenow, "Graphene oxide as a protein matrix: influence on protein biophysical properties," *J Nanobiotechnol*, vol. 13, no. 1, p. 70, Dec. 2015, doi: 10.1186/s12951-015-0134-0.

142. S. R. Kumar, C.-C. Hu, T. T. T. Vi, D. W. Chen, and S. J. Lue, "Antimicrobial peptide conjugated on graphene oxide-containing sulfonated polyetheretherketone substrate for effective antibacterial activities against *Staphylococcus aureus*," *Antibiotics*, vol. 12, no. 9, p. 1407, Sep. 2023, doi: 10.3390/antibiotics12091407.

143. E. Ravelo-Nieto et al., "Unlocking cellular barriers: silica nanoparticles and fullerenol conjugated cell-penetrating agents for enhanced intracellular drug delivery," *Front Bioeng Biotechnol*, vol. 11, May 2023, doi: 10.3389/fbioe.2023.1184973.

144. S. Joshi et al., "A review on peptide functionalized graphene derivatives as nanotools for biosensing," *Microchim Acta*, vol. 187, no. 1, p. 27, Jan. 2020, doi: 10.1007/s00604-019-3989-1.

145. S. S. Banerjee, N. Aher, R. Patil, and J. Khandare, "Poly(ethylene glycol)-prodrug conjugates: concept, design, and applications," *J Drug Deliv*, vol. 2012, pp. 1–17, May 2012, doi: 10.1155/2012/103973.

146. G. Hernández-Cancel, D. Suazo-Dávila, A. J. Ojeda-Cruzado, D. García-Torres, C. R. Cabrera, and K. Griebenow, "Graphene oxide as a protein matrix: influence on protein biophysical properties," *J Nanobiotechnol*, vol. 13, no. 1, p. 70, Dec. 2015, doi: 10.1186/s12951-015-0134-0.

147. R. Patil, P. Bahadur, and S. Tiwari, "Dispersed graphene materials of biomedical interest and their toxicological consequences," *Adv Colloid Interface Sci*, vol. 275, p. 102051, Jan. 2020, doi: 10.1016/j.cis.2019.102051.

148. K. Bhattacharya et al., "Biological interactions of carbon-based nanomaterials: from coronation to degradation," *Nanomedicine*, vol. 12, no. 2, pp. 333–351, Feb. 2016, doi: 10.1016/j.nano.2015.11.011.

149. I. M. Adjei, B. Sharma, and V. Labhasetwar, "Nanoparticles: cellular uptake and cytotoxicity," *Adv Exp Med Biol*, 2014, 811, pp. 73–91, doi: 10.1007/978-94-017-8739-0_5.

150. J. Ma et al., "Crucial role of lateral size for graphene oxide in activating macrophages and stimulating pro-inflammatory responses in cells and animals," *ACS Nano*, vol. 9, no. 10, pp. 10498–10515, Oct. 2015, doi: 10.1021/acsnano.5b04751.

151. F. F. Contreras-Torres et al., "Differential cytotoxicity and internalization of graphene family nanomaterials in myocardial cells," *Mater Sci Eng C*, vol. 73, pp. 633–642, Apr. 2017, doi: 10.1016/j.msec.2016.12.080.

152. A. N. Ghulam, O. A. L. dos Santos, L. Hazeem, B. Pizzorno Backx, M. Bououdina, and S. Bellucci, "Graphene oxide (GO) materials—applications and toxicity on living organisms and environment," *J Funct Biomater*, vol. 13, no. 2, p. 77, Jun. 2022, doi: 10.3390/jfb13020077.

153. N. Krasteva et al., "Aminated graphene oxide as a potential new therapy for colorectal cancer," *Oxid Med Cell Longev*, vol. 2019, pp. 1–15, Mar. 2019, doi: 10.1155/2019/3738980.

154. M. Bangeppagari, S. H. Park, R. R. Kundapur, and S. J. Lee, "Graphene oxide induces cardiovascular defects in developing zebrafish (*Danio rerio*) embryo model: in-vivo toxicity assessment," *Sci Total Environ*, vol. 673, pp. 810–820, Jul. 2019, doi: 10.1016/j.scitotenv.2019.04.082.

155. T. T. Liao et al., "Dose-dependent cytotoxicity evaluation of graphite nanoparticles for diamond-like carbon film application on artificial joints," *Biomed Mater*, vol. 12, no. 1, p. 015018, Jan. 2017, doi: 10.1088/1748-605X/aa52ca.

156. K. Srikanth, L. S. Sundar, E. Pereira, and A. C. Duarte, "Graphene oxide induces cytotoxicity and oxidative stress in bluegill sunfish cells," *J Appl Toxicol*, vol. 38, no. 4, pp. 504–513, Apr. 2018, doi: 10.1002/jat.3557.

157. L. Xu et al., "Graphene quantum dots in alveolar macrophage: uptake-exocytosis, accumulation in nuclei, nuclear responses and DNA cleavage," *Part Fibre Toxicol*, vol. 15, no. 1, p. 45, Dec. 2018, doi: 10.1186/s12989-018-0279-8.

158. M. Ema, M. Gamo, and K. Honda, "A review of toxicity studies of single-walled carbon nanotubes in laboratory animals," *Regulat Toxicol Pharmacol*, vol. 74, pp. 42–63, Feb. 2016, doi: 10.1016/j.yrtph.2015.11.015.

159. K. Krajnak et al., "Exposure to graphene nanoparticles induces changes in measures of vascular/renal function in a load and form-dependent manner in mice," *J Toxicol Environ Health A*, vol. 82, no. 12, pp. 711–726, Jun. 2019, doi: 10.1080/15287394.2019.1645772.

160. S. Syama, W. Paul, A. Sabareeswaran, and P. V. Mohanan, "Raman spectroscopy for the detection of organ distribution and clearance of PEGylated reduced graphene oxide and biological consequences," *Biomaterials*, vol. 131, pp. 121–130, Jul. 2017, doi: 10.1016/j.biomaterials.2017.03.043.

161. Z. Tang et al., "Mechanisms of oxidative stress, apoptosis, and autophagy involved in graphene oxide nanomaterial anti-osteosarcoma effect," *Int J Nanomed*, vol. 13, pp. 2907–2919, May 2018, doi: 10.2147/IJN.S159388.

162. S. Gurunathan, M.-H. Kang, M. Jeyaraj, and J.-H. Kim, "Differential cytotoxicity of different sizes of graphene oxide nanoparticles in Leydig (TM3) and Sertoli (TM4) cells," *Nanomaterials*, vol. 9, no. 2, p. 139, Jan. 2019, doi: 10.3390/nano9020139.

163. M. C. P. Mendonça et al., "PEGylation of reduced graphene oxide induces toxicity in cells of the blood–brain barrier: an in vitro and in vivo study," *Mol Pharm*, vol. 13, no. 11, pp. 3913–3924, Nov. 2016, doi: 10.1021/acs.molpharmaceut.6b00696.

164. N. Grobert, "Carbon nanotubes—becoming clean," *Mater Today*, vol. 10, no. 1–2, pp. 28–35, Jan. 2007, doi: 10.1016/S1369-7021(06)71789-8.

165. M. Baldrighi, M. Trusel, R. Tonini, and S. Giordani, "Carbon nanomaterials interfacing with neurons: an in vivo perspective," *Front Neurosci*, vol. 10, Jun. 2016, doi: 10.3389/fnins.2016.00250.

166. O. Akhavan, E. Ghaderi, E. Hashemi, and E. Akbari, "Dose-dependent effects of nanoscale graphene oxide on reproduction capability of mammals," *Carbon N Y*, vol. 95, pp. 309–317, Dec. 2015, doi: 10.1016/j.carbon.2015.08.017.

167. T. A. Tabish et al., "Biocompatibility and toxicity of graphene quantum dots for potential application in photodynamic therapy," *Nanomedicine*, vol. 13, no. 15, pp. 1923–1937, Aug. 2018, doi: 10.2217/nnm-2018-0018.

168. H. Shen et al., "PEGylated graphene oxide-mediated protein delivery for cell function regulation," *ACS Appl Mater Interfaces*, vol. 4, no. 11, pp. 6317–6323, Nov. 2012, doi: 10.1021/am3019367.

169. K. Yang, J. Wan, S. Zhang, Y. Zhang, S.-T. Lee, and Z. Liu, "In vivo pharmacokinetics, long-term biodistribution, and toxicology of PEGylated graphene in mice," *ACS Nano*, vol. 5, no. 1, pp. 516–522, Jan. 2011, doi: 10.1021/nn1024303.

170. S. Iijima, "Helical microtubules of graphitic carbon," *Nature*, vol. 354, no. 6348, pp. 56–58, Nov. 1991, doi: 10.1038/354056a0.

171. L.V. Radushkevich, V.M. Lukyanovich, "O strukture ugleroda obrazujucegosja pri termiceskom razlozenii okisi ugleroda na zeleznom kontakte," *Z Fiz Khimii*, vol. 26, pp. 88–95, 1952. (Translated into L.V. Radushkevich, V.M. Lukyanovich, "About the structure of carbon formed by thermal decomposition of carbon monoxide on iron substrate," *Russ J Phys Chem A*, vol. 26, pp. 88–95, 1952.)

172. M. Paramsothy, "70th year anniversary of carbon nanotube discovery—focus on real-world solutions," *Nanomaterials*, vol. 13, no. 24, p. 3162, Dec. 2023, doi: 10.3390/nano13243162.

173. J. Xu, J. P. Donohoe, and C. U. Pittman, "Preparation, electrical and mechanical properties of vapor grown carbon fiber (VGCF)/vinyl ester composites," *Compos Part A Appl Sci Manuf*, vol. 35, no. 6, pp. 693–701, Jun. 2004, doi: 10.1016/j.compositesa.2004.02.016.

174. H. Terrones and M. Terrones, "Curved nanostructured materials," *New J Phys*, vol. 5, pp. 126–126, Oct. 2003, doi: 10.1088/1367-2630/5/1/126.

175. J. Abrahamson, P. G. Wiles, and B. L. Rhoades, "Structure of carbon fibres found on carbon arc anodes," *Carbon N Y*, vol. 37, no. 11, pp. 1873–1874, Jan. 1999, doi: 10.1016/S0008-6223(99)00199-2.

176. J. Abraham, S. Thomas, and N. Kalarikkal, Eds., *Handbook of Carbon Nanotubes*, Springer International Publishing, Cham, 2022, doi: 10.1007/978-3-030-91346-5.

177. P. Darvishi, S. A. Mousavi, A. Mahmoudi, and D. Nayeri, "A comprehensive review on the removal of antibiotics from water and wastewater using carbon nanotubes: synthesis, performance, and future challenges," *Environ Sci (Camb)*, vol. 9, no. 1, pp. 11–37, 2023, doi: 10.1039/D1EW00912E.

178. A. Szabó, C. Perri, A. Csató, G. Giordano, D. Vuono, and J. B. Nagy, "Synthesis methods of carbon nanotubes and related materials," *Materials*, vol. 3, no. 5, pp. 3092–3140, May 2010, doi: 10.3390/ma3053092.

179. A. Thess et al., "Crystalline ropes of metallic carbon nanotubes," *Science (1979)*, vol. 273, no. 5274, pp. 483–487, Jul. 1996, doi: 10.1126/science.273.5274.483.

180. S. Rathinavel, K. Priyadharshini, and D. Panda, "A review on carbon nanotube: an overview of synthesis, properties, functionalization, characterization, and the application," *Mater Sci Eng B*, vol. 268, p. 115095, Jun. 2021, doi: 10.1016/j.mseb.2021.115095.

181. S. Xie, W. Li, Z. Pan, B. Chang, and L. Sun, "Carbon nanotube arrays," *Mater Sci Eng A*, vol. 286, no. 1, pp. 11–15, Jun. 2000, doi: 10.1016/S0921-5093(00)00657-2.

182. S. V. Sawant, A. W. Patwardhan, J. B. Joshi, and K. Dasgupta, "Boron doped carbon nanotubes: synthesis, characterization and emerging applications—a review," *Chem Eng J*, vol. 427, p. 131616, Jan. 2022, doi: 10.1016/j.cej.2021.131616.

183. G. Song et al., "Catalytic decomposition of methane for controllable production of carbon nanotubes and high purity H2 over LTA zeolite-derived Ni-based yolk-shell catalysts," *Chem Eng J*, vol. 474, p. 145643, Oct. 2023, doi: 10.1016/j.cej.2023.145643.

184. D. Torres, J. L. Pinilla, M. J. Lázaro, R. Moliner, and I. Suelves, "Hydrogen and multiwall carbon nanotubes production by catalytic decomposition of methane: thermogravimetric analysis and scaling-up of Fe–Mo catalysts," *Int J Hydrogen Energy*, vol. 39, no. 8, pp. 3698–3709, Mar. 2014, doi: 10.1016/j.ijhydene.2013.12.127.

185. S. K. Saraswat and K. K. Pant, "Ni–Cu–Zn/MCM-22 catalysts for simultaneous production of hydrogen and multiwall carbon nanotubes via thermo-catalytic decomposition of methane," *Int J Hydrogen Energy*, vol. 36, no. 21, pp. 13352–13360, Oct. 2011, doi: 10.1016/j.ijhydene.2011.07.102.

186. R. Sharma, A. K. Sharma, and V. Sharma, "Synthesis of carbon nanotubes by arc-discharge and chemical vapor deposition method with analysis of its morphology, dispersion and functionalization characteristics," *Cogent Eng*, vol. 2, no. 1, p. 1094017, Dec. 2015, doi: 10.1080/23311916.2015.1094017.

187. J. Yan, Z. Fan, and L. Zhi, "Functionalized carbon nanotubes and their enhanced polymers," in *Polymer Science: A Comprehensive Reference*, Elsevier, 2012, pp. 439–478, doi: 10.1016/B978-0-444-53349-4.00217-X.

188. C. Zhang, L. Wu, M. de Perrot, and X. Zhao, "Carbon nanotubes: a summary of beneficial and dangerous aspects of an increasingly popular group of nanomaterials," *Front Oncol*, vol. 11, Jul. 2021, doi: 10.3389/fonc.2021.693814.

189. D. Pantarotto et al., "Functionalized carbon nanotubes for plasmid DNA gene delivery," *Angew Chem Int Ed*, vol. 43, no. 39, pp. 5242–5246, Oct. 2004, doi: 10.1002/anie.200460437.

190. W. Cheung, F. Pontoriero, O. Taratula, A. M. Chen, and H. He, "DNA and carbon nanotubes as medicine," *Adv Drug Deliv Rev*, vol. 62, no. 6, pp. 633–649, Apr. 2010, doi: 10.1016/j.addr.2010.03.007.

191. Y. Wu, J. A. Phillips, H. Liu, R. Yang, and W. Tan, "Carbon nanotubes protect DNA strands during cellular delivery," *ACS Nano*, vol. 2, no. 10, pp. 2023–2028, Oct. 2008, doi: 10.1021/nn800325a.

192. Y. Xu et al., "Short carbon nanotube-based delivery of mRNA for HIV-1 vaccines," *Biomolecules*, vol. 13, no. 7, p. 1088, Jul. 2023, doi: 10.3390/biom13071088.

193. H. Zare et al., "Carbon nanotubes: smart drug/gene delivery carriers," *Int J Nanomed*, vol. 16, pp. 1681–1706, Mar. 2021, doi: 10.2147/IJN.S299448.

194. K. H. Son, J. H. Hong, and J. W. Lee, "Carbon nanotubes as cancer therapeutic carriers and mediators," *Int J Nanomed*, vol. 11, pp. 5163–5185, Oct. 2016, doi: 10.2147/IJN.S112660.

195. N. W. S. Kam, Z. Liu, and H. Dai, "Carbon nanotubes as intracellular transporters for proteins and DNA: an investigation of the uptake mechanism and pathway," *Angew Chem Int Ed*, vol. 45, no. 4, pp. 577–581, Jan. 2006, doi: 10.1002/anie.200503389.

196. K. Castillo, A. Tasset, M. Marinkovic, and A. Foote, "Discrete multiwalled carbon nanotubes for versatile intracellular transport of functional biomolecular complexes," *C J Carbon Res (Basel)*, vol. 10, no. 2, p. 37, Apr. 2024, doi: 10.3390/c10020037.

197. Z. Peng et al., "Advances in the application, toxicity and degradation of carbon nanomaterials in environment: a review," *Environ Int*, vol. 134, p. 105298, Jan. 2020, doi: 10.1016/j.envint.2019.105298.

198. S. H. Tabari, Y. Jamali, and R. Poursalehi, "Multi-scale simulation of carbon nanotubes interactions with cell membrane: DFT calculations and molecular dynamic simulation," *Proc Mater Sci*, vol. 11, pp. 423–427, 2015, doi: 10.1016/j.mspro.2015.11.025.

199. T. Jiang et al., "Toxicity of single-walled carbon nanotubes (SWCNTs): effect of lengths, functional groups and electronic structures revealed by a quantitative toxicogenomics assay," *Environ Sci Nano*, vol. 7, no. 5, pp. 1348–1364, 2020, doi: 10.1039/D0EN00230E.

200. B. O. Murjani, P. S. Kadu, M. Bansod, S. S. Vaidya, and M. D. Yadav, "Carbon nanotubes in biomedical applications: current status, promises, and challenges," *Carbon Letters*, vol. 32, no. 5, pp. 1207–1226, Aug. 2022, doi: 10.1007/s42823-022-00364-4.

201. C. Zhang, L. Wu, M. de Perrot, and X. Zhao, "Carbon nanotubes: a summary of beneficial and dangerous aspects of an increasingly popular group of nanomaterials," *Front Oncol*, vol. 11, Jul. 2021, doi: 10.3389/fonc.2021.693814.

202. M. Barbarino and A. Giordano, "Assessment of the carcinogenicity of carbon nanotubes in the respiratory system," *Cancers (Basel)*, vol. 13, no. 6, p. 1318, Mar. 2021, doi: 10.3390/cancers13061318.

203. M. R. Chetyrkina, F. S. Fedorov, and A. G. Nasibulin, "In vitro toxicity of carbon nanotubes: a systematic review," *RSC Adv*, vol. 12, no. 25, pp. 16235–16256, 2022, doi: 10.1039/D2RA02519A.

204. P. Møller, R. S. Wils, E. Di Ianni, C. A. T. Gutierrez, M. Roursgaard, and N. R. Jacobsen, "Genotoxicity of multi-walled carbon nanotube reference materials in mammalian cells and animals," *Mutat Res/Rev Mutat Res*, vol. 788, p. 108393, Jul. 2021, doi: 10.1016/j.mrrev.2021.108393.

205. A. M. Wagner, J. M. Knipe, G. Orive, and N. A. Peppas, "Quantum dots in biomedical applications," *Acta Biomater*, vol. 94, pp. 44–63, Aug. 2019, doi: 10.1016/j.actbio.2019.05.022.

206. S. Dua, P. Kumar, B. Pani, A. Kaur, M. Khanna, and G. Bhatt, "Stability of carbon quantum dots: a critical review," *RSC Adv*, vol. 13, no. 20, pp. 13845–13861, 2023, doi: 10.1039/D2RA07180K.

207. S. Kargozar, S. J. Hoseini, P. B. Milan, S. Hooshmand, H. Kim, and M. Mozafari, "Quantum dots: a review from concept to clinic," *Biotechnol J*, vol. 15, no. 12, Dec. 2020, doi: 10.1002/biot.202000117.

208. A. Deshmukh, J. Bhaiswar, V. Kapse, and D. Meghe, "Quantum dots: a brief review," in *Quantum Dots*, Elsevier, 2023, pp. 41–66, doi: 10.1016/B978-0-323-85278-4.00001-5.

209. N. Le, M. Zhang, and K. Kim, "Quantum dots and their interaction with biological systems," *Int J Mol Sci*, vol. 23, no. 18, p. 10763, Sep. 2022, doi: 10.3390/ijms231810763.

210. F. P. García de Arquer, D. V. Talapin, V. I. Klimov, Y. Arakawa, M. Bayer, and E. H. Sargent, "Semiconductor quantum dots: technological progress and future challenges," *Science (1979)*, vol. 373, no. 6555, Aug. 2021, doi: 10.1126/science.aaz8541.

211. F. Antolini and L. Orazi, "Quantum dots synthesis through direct laser patterning: a review," *Front Chem*, vol. 7, Apr. 2019, doi: 10.3389/fchem.2019.00252.

212. M. Jorns and D. Pappas, "A review of fluorescent carbon dots, their synthesis, physical and chemical characteristics, and applications," *Nanomaterials*, vol. 11, no. 6, p. 1448, May 2021, doi: 10.3390/nano11061448.

213. C. Giansante, "Surface chemistry impact on the light absorption by colloidal quantum dots," *Chem Euro J*, vol. 27, no. 58, pp. 14358–14368, Oct. 2021, doi: 10.1002/chem.202102168.

214. J. Sobhanan, J. V. Rival, A. Anas, E. Sidharth Shibu, Y. Takano, and V. Biju, "Luminescent quantum dots: synthesis, optical properties, bioimaging and toxicity," *Adv Drug Deliv Rev*, vol. 197, p. 114830, Jun. 2023, doi: 10.1016/j.addr.2023.114830.

215. C. Matea et al., "Quantum dots in imaging, drug delivery and sensor applications," *Int J Nanomed*, vol. 12, pp. 5421–5431, Jul. 2017, doi: 10.2147/IJN.S138624.

216. Z. Zahed et al., "Recent advances in fluorescence nanoparticles 'quantum dots' as gene delivery system: a review," *Int J Biol Macromol*, vol. 254, p. 127802, Jan. 2024, doi: 10.1016/j.ijbiomac.2023.127802.

217. M. Wojnicki and V. Hessel, "Quantum materials made in microfluidics—critical review and perspective," *Chem Eng J*, vol. 438, p. 135616, Jun. 2022, doi: 10.1016/j.cej.2022.135616.

218. A. I. Ekimov, A. A. Onushchenko, "Quantum size effect in three-dimensional microscopic semiconductor crystals," *Z Eksperimental I Teoretich Fiziki* (in Russian), vol. 34, pp. 363–366, 30 September 1981.

219. "Nanocrystals in their prime," *Nat Nanotechnol*, vol. 9, no. 5, pp. 325–325, May 2014, doi: 10.1038/nnano.2014.101.

220. A. I. Ekimov, Al. L. Efros, and A. A. Onushchenko, "Quantum size effect in semiconductor microcrystals," *Solid State Commun*, vol. 56, no. 11, pp. 921–924, Dec. 1985, doi: 10.1016/S0038-1098(85)80025-9.

221. A. I. Ekimov, A. Onushchehko, A. G. Plyukhin, and A. Efros, "Size quantization of exciton and determination of their energy spectrum parameters in CuCl," *J Exp Theor Phys*, vol. 88, pp. 1491–1501, Sep. 1985.

222. R. Rossetti, S. Nakahara, and L. E. Brus, "Quantum size effects in the redox potentials, resonance Raman spectra, and electronic spectra of CdS crystallites in aqueous solution," *J Chem Phys*, vol. 79, no. 2, pp. 1086–1088, Jul. 1983, doi: 10.1063/1.445834.

223. L. E. Brus, "A simple model for the ionization potential, electron affinity, and aqueous redox potentials of small semiconductor crystallites," *J Chem Phys*, vol. 79, no. 11, pp. 5566–5571, Dec. 1983, doi: 10.1063/1.445676.

224. R. Rossetti, J. L. Ellison, J. M. Gibson, and L. E. Brus, "Size effects in the excited electronic states of small colloidal CdS crystallites," *J Chem Phys*, vol. 80, no. 9, pp. 4464–4469, May 1984, doi: 10.1063/1.447228.

225. C. B. Murray, D. J. Norris, and M. G. Bawendi, "Synthesis and characterization of nearly monodisperse CdE (E = sulfur, selenium, tellurium) semiconductor nanocrystallites," *J Am Chem Soc*, vol. 115, no. 19, pp. 8706–8715, Sep. 1993, doi: 10.1021/ja00072a025.

226. C. B. Murray, C. R. Kagan, and M. G. Bawendi, "Synthesis and characterization of monodisperse nanocrystals and close-packed nanocrystal assemblies," *Ann Rev Mater Sci*, vol. 30, no. 1, pp. 545–610, Aug. 2000, doi: 10.1146/annurev.matsci.30.1.545.

227. B. O. Dabbousi et al., "(CdSe)ZnS core–shell quantum dots: synthesis and characterization of a size series of highly luminescent nanocrystallites," *J Phys Chem B*, vol. 101, no. 46, pp. 9463–9475, Nov. 1997, doi: 10.1021/jp971091y.

228. G. J. Supran et al., "QLEDs for displays and solid-state lighting," *MRS Bull*, vol. 38, no. 9, pp. 703–711, Sep. 2013, doi: 10.1557/mrs.2013.181.

229. W. Mei et al., "High-resolution, full-color quantum dot light-emitting diode display fabricated via photolithography approach," *Nano Res*, vol. 13, no. 9, pp. 2485–2491, Sep. 2020, doi: 10.1007/s12274-020-2883-9.

230. D. Qi et al., "Finely controlled synthesis of $Zn_{1-x}Mg_xO$ nanoparticles with uniform size distribution used as electron transport materials for red QLEDs," *ACS Appl Electron Mater*, vol. 4, no. 4, pp. 1875–1881, Apr. 2022, doi: 10.1021/acsaelm.2c00088.

231. L. Manna and T. W. Odom, "Profile of Alexei I. Ekimov, Louis E. Brus, and Moungi G. Bawendi: 2023 Nobel laureates in chemistry," *Proc Nat Acad Sci*, vol. 121, no. 29, Jul. 2024, doi: 10.1073/pnas.2410357121.

232. K. Kasirajan, M. Karunakaran, and H. K. Choi, "Synthesis of environmentally-friendly carbon quantum dots from orange juice for selective detection of $Fe^{3+}$ ions, antibacterial activity, and bioimaging applications," *J Environ Chem Eng*, vol. 12, no. 5, p. 113535, Oct. 2024, doi: 10.1016/j.jece.2024.113535.

233. N. Khansili, "Review on synthesis and properties of carbon dots with emphasis to combat environmental toxins," *Environ Adv*, vol. 16, p. 100542, Jul. 2024, doi: 10.1016/j.envadv.2024.100542.

234. S. Sarkar, D. Banerjee, U. K. Ghorai, N. S. Das, and K. K. Chattopadhyay, "Size dependent photoluminescence property of hydrothermally synthesized crystalline carbon quantum dots," *J Lumin*, vol. 178, pp. 314–323, Oct. 2016, doi: 10.1016/j.jlumin.2016.05.033.

235. Y. Liu, H. Huang, W. Cao, B. Mao, Y. Liu, and Z. Kang, "Advances in carbon dots: from the perspective of traditional quantum dots," *Mater Chem Front*, vol. 4, no. 6, pp. 1586–1613, 2020, doi: 10.1039/D0QM00090F.

236. M. K. Barman and A. Patra, "Current status and prospects on chemical structure driven photoluminescence behaviour of carbon dots," *J Photochem Photobiol C Photochem Rev*, vol. 37, pp. 1–22, Dec. 2018, doi: 10.1016/j.jphotochemrev.2018.08.001.

237. M. J. Molaei, "Principles, mechanisms, and application of carbon quantum dots in sensors: a review," *Anal Methods*, vol. 12, no. 10, pp. 1266–1287, 2020, doi: 10.1039/C9AY02696G.

238. N. Oladzadabbasabadi, M. A. Dheyab, A. M. Nafchi, M. Ghasemlou, E. P. Ivanova, and B. Adhikari, "Turning food waste into value-added carbon dots for sustainable food packaging application: a review," *Adv Colloid Interface Sci*, vol. 321, p. 103020, Nov. 2023, doi: 10.1016/j.cis.2023.103020.

239. Y.-P. Sun et al., "Quantum-sized carbon dots for bright and colorful photoluminescence," *J Am Chem Soc*, vol. 128, no. 24, pp. 7756–7757, Jun. 2006, doi: 10.1021/ja062677d.

240. C. Dias et al., "Biocompatibility and bioimaging potential of fruit-based carbon dots," *Nanomaterials*, vol. 9, no. 2, p. 199, Feb. 2019, doi: 10.3390/nano9020199.

241. A. N. Emam, S. A. Loutfy, A. A. Mostafa, H. Awad, and M. B. Mohamed, "Cyto-toxicity, biocompatibility and cellular response of carbon dots–plasmonic based nano-hybrids for bioimaging," *RSC Adv*, vol. 7, no. 38, pp. 23502–23514, 2017, doi: 10.1039/C7RA01423F.

242. Z. Jiang, L. Guan, X. Xu, E. Wang, and C. Wang, "Applications of carbon dots in electrochemical energy storage," *ACS Appl Electron Mater*, vol. 4, no. 11, pp. 5144–5164, Nov. 2022, doi: 10.1021/acsaelm.2c01152.

243. Y. Zhai et al., "Carbon dots as new building blocks for electrochemical energy storage and electrocatalysis," *Adv Energy Mater*, vol. 12, no. 6, Feb. 2022, doi: 10.1002/aenm.202103426.

244. D. Huang et al., "Carbon dots-decorated carbon-based metal-free catalysts for electrochemical energy storage," *Small*, vol. 17, no. 4, Jan. 2021, doi: 10.1002/smll.202002998.

245. T. Rasheed, "Carbon dots as potential greener and sustainable fluorescent nanomaterials in service of pollutants sensing," *TrAC Trends Anal Chem*, vol. 158, p. 116841, Jan. 2023, doi: 10.1016/j.trac.2022.116841.

246. C. Long, Z. Jiang, J. Shangguan, T. Qing, P. Zhang, and B. Feng, "Applications of carbon dots in environmental pollution control: a review," *Chem Eng J*, vol. 406, p. 126848, Feb. 2021, doi: 10.1016/j.cej.2020.126848.

247. D. Ozyurt, M. Al Kobaisi, R. K. Hocking, and B. Fox, "Properties, synthesis, and applications of carbon dots: a review," *Carbon Trends*, vol. 12, p. 100276, Sep. 2023, doi: 10.1016/j.cartre.2023.100276.

248. D. Nagarajan, D. Gangadharan, and S. Venkatanarasimhan, "Synthetic strategies toward developing carbon dots via top-down approach," in *Carbon Dots in Analytical Chemistry*, Elsevier, 2023, pp. 1–13, doi: 10.1016/B978-0-323-98350-1.00016-5.

249. A. Kaur, K. Pandey, R. Kaur, N. Vashishat, and M. Kaur, "Nanocomposites of carbon quantum dots and graphene quantum dots: environmental applications as sensors," *Chemosensors*, vol. 10, no. 9, p. 367, Sep. 2022, doi: 10.3390/chemosensors10090367.

250. R. M. Altuwirqi, "Graphene nanostructures by pulsed laser ablation in liquids: a review," *Materials*, vol. 15, no. 17, p. 5925, Aug. 2022, doi: 10.3390/ma15175925.

251. A. Khayal et al., "Advances in the methods for the synthesis of carbon dots and their emerging applications," *Polymers (Basel)*, vol. 13, no. 18, p. 3190, Sep. 2021, doi: 10.3390/polym13183190.

252. L. Yang et al., "The fabrication of micro/nano structures by laser machining," *Nanomaterials*, vol. 9, no. 12, p. 1789, Dec. 2019, doi: 10.3390/nano9121789.

253. A. Kaczmarek et al., "Luminescent carbon dots synthesized by the laser ablation of graphite in polyethylenimine and ethylenediamine," *Materials*, vol. 14, no. 4, p. 729, Feb. 2021, doi: 10.3390/ma14040729.

254. L. Wang et al., "Gram-scale synthesis of single-crystalline graphene quantum dots with superior optical properties," *Nat Commun*, vol. 5, no. 1, p. 5357, Oct. 2014, doi: 10.1038/ncomms6357.

255. A. Vibhute, T. Patil, R. Gambhir, and A. P. Tiwari, "Fluorescent carbon quantum dots: synthesis methods, functionalization and biomedical applications," *Appl Surf Sci Adv*, vol. 11, p. 100311, Oct. 2022, doi: 10.1016/j.apsadv.2022.100311.

256. M. Xu et al., "Hydrothermal/solvothermal synthesis of graphene quantum dots and their biological applications," *Nano Biomed Eng*, vol. 5, no. 2, Apr. 2013, doi: 10.5101/nbe.v4i3.p65-71.

257. M. Li, T. Chen, J. J. Gooding, and J. Liu, "Review of carbon and graphene quantum dots for sensing," *ACS Sens*, vol. 4, no. 7, pp. 1732–1748, Jul. 2019, doi: 10.1021/acssensors.9b00514.

258. D. Ghosh, K. Sarkar, P. Devi, K.-H. Kim, and P. Kumar, "Current and future perspectives of carbon and graphene quantum dots: from synthesis to strategy for building optoelectronic and energy devices," *Renew Sustain Energy Rev*, vol. 135, p. 110391, Jan. 2021, doi: 10.1016/j.rser.2020.110391.

259. S. Tajik et al., "Carbon and graphene quantum dots: a review on syntheses, characterization, biological and sensing applications for neurotransmitter determination," *RSC Adv*, vol. 10, no. 26, pp. 15406–15429, 2020, doi: 10.1039/D0RA00799D.

260. M. Nazar, M. Hasan, B. Wirjosentono, B. A. Gani, and C. E. Nada, "Microwave synthesis of carbon quantum dots from Arabica coffee ground for fluorescence detection of $Fe^{3+}$, $Pb^{2+}$, and $Cr^{3+}$," *ACS Omega*, vol. 9, no. 18, pp. 20571–20581, May 2024, doi: 10.1021/acsomega.4c02254.

261. G. Jeong et al., "Microwave-assisted synthesis of multifunctional fluorescent carbon quantum dots from A4/B2 polyamidation monomer sets," *Appl Surf Sci*, vol. 542, p. 148471, Mar. 2021, doi: 10.1016/j.apsusc.2020.148471.

262. A. O. Adeola, M. P. Duarte, and R. Naccache, "Microwave-assisted synthesis of carbon-based nanomaterials from biobased resources for water treatment applications: emerging trends and prospects," *Front Carbon*, vol. 2, Jul. 2023, doi: 10.3389/frcrb.2023.1220021.

263. S. Yalshetti et al., "Microwave-assisted synthesis, characterization and in vitro biomedical applications of *Hibiscus rosa-sinensis* Linn.-mediated carbon quantum dots," *Sci Rep*, vol. 14, no. 1, p. 9915, Apr. 2024, doi: 10.1038/s41598-024-60726-y.

264. Y. Choi, Y. Choi, O. Kwon, and B. Kim, "Carbon dots: bottom-up syntheses, properties, and light-harvesting applications," *Chem Asian J*, vol. 13, no. 6, pp. 586–598, Mar. 2018, doi: 10.1002/asia.201701736.

265. D. Qu and Z. Sun, "The formation mechanism and fluorophores of carbon dots synthesized via a bottom-up route," *Mater Chem Front*, vol. 4, no. 2, pp. 400–420, 2020, doi: 10.1039/C9QM00552H.

266. Z. Zhu et al., "Red carbon dots: optical property regulations and applications," *Mater Today*, vol. 30, pp. 52–79, Nov. 2019, doi: 10.1016/j.mattod.2019.05.003.

267. H. Ding, J. Wei, P. Zhang, Z. Zhou, Q. Gao, and H. Xiong, "Solvent-controlled synthesis of highly luminescent carbon dots with a wide color gamut and narrowed emission peak widths," *Small*, vol. 14, no. 22, May 2018, doi: 10.1002/smll.201800612.

268. K. Jiang et al., "Red, green, and blue luminescence by carbon dots: full-color emission tuning and multicolor cellular imaging," *Angew Chem*, vol. 127, no. 18, pp. 5450–5453, Apr. 2015, doi: 10.1002/ange.201501193.

269. C. F. Rodríguez, P. Guzmán-Sastoque, C. Muñoz-Camargo, L. H. Reyes, J. F. Osma, and J. C. Cruz, "Enhancing magnetic micro- and nanoparticle separation with a cost-effective microfluidic device fabricated by laser ablation of PMMA," *Micromachines (Basel)*, vol. 15, no. 8, p. 1057, Aug. 2024, doi: 10.3390/mi15081057.

270. J. Guo et al., "Solvent-engineered synthesis of multicolor emissive carbon dots from citric acid and cyanamide," *Ind Eng Chem Res*, vol. 63, no. 20, pp. 9050–9057, May 2024, doi: 10.1021/acs.iecr.4c00136.

271. M. Xu et al., "Hydrophilic and blue fluorescent N-doped carbon dots from tartaric acid and various alkylol amines under microwave irradiation," *Nanoscale*, vol. 7, no. 38, pp. 15915–15923, 2015, doi: 10.1039/C5NR04209G.

272. H. Wang et al., "Carbon dots with positive surface charge from tartaric acid and *m* -aminophenol for selective killing of Gram-positive bacteria," *J Mater Chem B*, vol. 9, no. 1, pp. 125–130, 2021, doi: 10.1039/D0TB02332A.

273. Q. Chang et al., "Full color fluorescent carbon quantum dots synthesized from triammonium citrate for cell imaging and white LEDs," *Dyes Pigm*, vol. 193, p. 109478, Sep. 2021, doi: 10.1016/j.dyepig.2021.109478.

274. F. Yan et al., "Carbon dots serve as an effective probe for the quantitative determination and for intracellular imaging of mercury(II)," *Microchim Acta*, vol. 183, no. 5, pp. 1611–1618, May 2016, doi: 10.1007/s00604-016-1788-5.

275. Y. An et al., "Red emission carbon dots prepared by 1,4-diaminonaphthalene for light-emitting diode application and metal ion detection," *Materials*, vol. 14, no. 16, p. 4716, Aug. 2021, doi: 10.3390/ma14164716.

276. Y. Xu et al., "Nitrogen-doped carbon dots: a facile and general preparation method, photoluminescence investigation, and imaging applications," *Chem Euro J*, vol. 19, no. 7, pp. 2276–2283, Feb. 2013, doi: 10.1002/chem.201203641.

277. Z. Yang et al., "Nitrogen-doped, carbon-rich, highly photoluminescent carbon dots from ammonium citrate," *Nanoscale*, vol. 6, no. 3, pp. 1890–1895, 2014, doi: 10.1039/C3NR05380F.

278. Y. Park, Y. Kim, H. Chang, S. Won, H. Kim, and W. Kwon, "Biocompatible nitrogen-doped carbon dots: synthesis, characterization, and application," *J Mater Chem B*, vol. 8, no. 39, pp. 8935–8951, 2020, doi: 10.1039/D0TB01334J.

279. Y. Zhu, L. Yan, M. Xu, Y. Li, X. Song, and L. Yin, "Difference between ammonia and urea on nitrogen doping of graphene quantum dots," *Colloids Surf A Physicochem Eng Asp*, vol. 610, p. 125703, Feb. 2021, doi: 10.1016/j.colsurfa.2020.125703.

280. N. Jing, M. Tian, Y. Wang, and Y. Zhang, "Nitrogen-doped carbon dots synthesized from acrylic acid and ethylenediamine for simple and selective determination of cobalt ions in aqueous media," *J Lumin*, vol. 206, pp. 169–175, Feb. 2019, doi: 10.1016/j.jlumin.2018.10.059.

281. K. Wang, C. Geng, F. Wang, Y. Zhao, and Z. Ru, "Urea-doped carbon dots as fluorescent switches for the selective detection of iodide ions and their mechanistic study," *RSC Adv*, vol. 11, no. 44, pp. 27645–27652, 2021, doi: 10.1039/D1RA04558J.

282. H. Zhou, Y. Ren, Z. Li, W. He, and Z. Li, "Selective detection of $Fe^{3+}$ by nitrogen–sulfur-doped carbon dots using thiourea and citric acid," *Coatings*, vol. 12, no. 8, p. 1042, Jul. 2022, doi: 10.3390/coatings12081042.

283. D. M. A. Crista, J. C. G. Esteves da Silva, and L. Pinto da Silva, "Evaluation of different bottom-up routes for the fabrication of carbon dots," *Nanomaterials*, vol. 10, no. 7, p. 1316, Jul. 2020, doi: 10.3390/nano10071316.

284. B. De and N. Karak, "Recent progress in carbon dot–metal based nanohybrids for photochemical and electrochemical applications," *J Mater Chem A Mater*, vol. 5, no. 5, pp. 1826–1859, 2017, doi: 10.1039/C6TA10220D.

285. Z. Wang, Z. Wang, and F. Wu, "Carbon dots as drug delivery vehicles for antimicrobial applications: a minireview," *ChemMedChem*, vol. 17, no. 13, Jul. 2022, doi: 10.1002/cmdc.202200003.

286. G. Calabrese et al., "Carbon dots: an innovative tool for drug delivery in brain tumors," *Int J Mol Sci*, vol. 22, no. 21, p. 11783, Oct. 2021, doi: 10.3390/ijms222111783.

287. F. Yan, Y. Jiang, X. Sun, Z. Bai, Y. Zhang, and X. Zhou, "Surface modification and chemical functionalization of carbon dots: a review," *Microchim Acta*, vol. 185, no. 9, p. 424, Sep. 2018, doi: 10.1007/s00604-018-2953-9.

288. L. Li and T. Dong, "Photoluminescence tuning in carbon dots: surface passivation or/and functionalization, heteroatom doping," *J Mater Chem C Mater*, vol. 6, no. 30, pp. 7944–7970, 2018, doi: 10.1039/C7TC05878K.

289. Q. Wang, C. Zhang, G. Shen, H. Liu, H. Fu, and D. Cui, "Fluorescent carbon dots as an efficient siRNA nanocarrier for its interference therapy in gastric cancer cells," *J Nanobiotechnol*, vol. 12, no. 1, p. 58, Dec. 2014, doi: 10.1186/s12951-014-0058-0.

290. Y.-F. Wu et al., "Multi-functionalized carbon dots as theranostic nanoagent for gene delivery in lung cancer therapy," *Sci Rep*, vol. 6, no. 1, p. 21170, Feb. 2016, doi: 10.1038/srep21170.

291. H.-J. Wang, J. Zhang, Y.-H. Liu, T.-Y. Luo, X. He, and X.-Q. Yu, "Hyaluronic acid-based carbon dots for efficient gene delivery and cell imaging," *RSC Adv*, vol. 7, no. 25, pp. 15613–15624, 2017, doi: 10.1039/C7RA01417A.

292. S. Strickland, M. Jorns, L. Fourroux, L. Heyd, and D. Pappas, "Cancer cell targeting via selective transferrin receptor labeling using protein-derived carbon dots," *ACS Omega*, vol. 9, no. 2, pp. 2707–2718, Jan. 2024, doi: 10.1021/acsomega.3c07744.

293. Y. Song, X. Li, S. Cong, H. Zhao, and M. Tan, "Nuclear-targeted of TAT peptide-conjugated carbon dots for both one-and two-photon fluorescence imaging," *Colloids Surf B Biointerfaces*, vol. 180, pp. 449–456, Aug. 2019, doi: 10.1016/j.colsurfb.2019.05.015.

294. A. Selva Sharma and N. Y. Lee, "Comprehensive review on fluorescent carbon dots and their applications in nucleic acid detection, nucleolus targeted imaging and gene delivery," *Analyst*, vol. 149, no. 16, pp. 4095–4115, 2024, doi: 10.1039/D4AN00630E.

295. A. Truskewycz et al., "Carbon dot therapeutic platforms: administration, distribution, metabolism, excretion, toxicity, and therapeutic potential," *Small*, vol. 18, no. 16, Apr. 2022, doi: 10.1002/smll.202106342.

296. P. Gao et al., "Renal clearing carbon dots-based near-infrared fluorescent super-small nanoprobe for renal imaging," *Carbon N Y*, vol. 201, pp. 805–814, Jan. 2023, doi: 10.1016/j.carbon.2022.09.052.

297. B. Tian et al., "Renal-clearable nickel-doped carbon dots with boosted photothermal conversion efficiency for multimodal imaging-guided cancer therapy in the second near-infrared biowindow," *Adv Funct Mater*, vol. 31, no. 26, Jun. 2021, doi: 10.1002/adfm.202100549.

298. Y. Su, S. Liu, Y. Guan, Z. Xie, M. Zheng, and X. Jing, "Renal clearable hafnium-doped carbon dots for CT/fluorescence imaging of orthotopic liver cancer," *Biomaterials*, vol. 255, p. 120110, Oct. 2020, doi: 10.1016/j.biomaterials.2020.120110.

299. M. Sharon, *History of Nanotechnology: From Prehistoric to Modern Times*, Wley-Scrivener, Beverly, MA, 2019.

300. N. Martín, "The legacy of Sir Harold W. Kroto: fullerenes and beyond," *Chem*, vol. 5, no. 4, pp. 733–738, Apr. 2019, doi: 10.1016/j.chempr.2019.03.015.

301. M. V. Putz, Ed., *New Frontiers in Nanochemistry: Concepts, Theories, and Trends*, Apple Academic Press, Palm Bay, FL, 2020.

302. W. Krätschmer, L. D. Lamb, K. Fostiropoulos, and D. R. Huffman, "Solid C60: a new form of carbon," *Nature*, vol. 347, no. 6291, pp. 354–358, Sep. 1990, doi: 10.1038/347354a0.

303. R. Bawa, G. Audette, and I. Rubinstein, *Handbook of Clinical Nanomedicine: Nanoparticles, Imaging, Therapy and Clinical Applications*, Pan Stanford Publishing, Singapore, 2020.

304. G. T. Hermanson, *Bioconjugate Techniques*, Elsevier, 2008, doi: 10.1016/B978-0-12-370501-3.X0001-X.

305. P. Bhakta and B. Barthunia, "Fullerene and its applications: a review," *J Indian Acad Oral Med Radiol*, vol. 32, no. 2, p. 159, 2020, doi: 10.4103/jiaomr.jiaomr_191_19.

306. M. Paukov, C. Kramberger, I. Begichev, M. Kharlamova, and M. Burdanova, "Functionalized fullerenes and their applications in electrochemistry, solar cells, and nanoelectronics," *Materials*, vol. 16, no. 3, p. 1276, Feb. 2023, doi: 10.3390/ma16031276.

307. A. Ramazani, M. A. Moghaddasi, A. Mashhadi Malekzadeh, S. Rezayati, Y. Hanifehpour, and S. W. Joo, "Industrial oriented approach on fullerene preparation methods," *Inorg Chem Commun*, vol. 125, p. 108442, Mar. 2021, doi: 10.1016/j.inoche.2021.108442.

308. G. Peters and M. Jansen, "A new fullerene synthesis," *Angew Chem Int Ed Eng*, vol. 31, no. 2, pp. 223–224, Feb. 1992, doi: 10.1002/anie.199202231.

309. M. Mojica, J. A. Alonso, and F. Méndez, "Synthesis of fullerenes," *J Phys Org Chem*, vol. 26, no. 7, pp. 526–539, Jul. 2013, doi: 10.1002/poc.3121.

310. H. Singh and M. Srivastava, "Fullerenes: synthesis, separation, characterization, reaction chemistry, and applications—a review," *Energy Sources*, vol. 17, no. 6, pp. 615–640, Nov. 1995, doi: 10.1080/00908319508946108.

311. L. T. Scott, "Methods for the chemical synthesis of fullerenes," *Angew Chem Int Ed*, vol. 43, no. 38, pp. 4994–5007, Sep. 2004, doi: 10.1002/anie.200400661.

312. F. Viñes and A. Görling, "Template-assisted formation of fullerenes from short-chain hydrocarbons by supported platinum nanoparticles," *Angew Chem Int Ed*, vol. 50, no. 20, pp. 4611–4614, May 2011, doi: 10.1002/anie.201006588.

313. C. Chen and Z. Lou, "Formation of C60 by reduction of $CO_2$," *J Supercrit Fluids*, vol. 50, no. 1, pp. 42–45, Aug. 2009, doi: 10.1016/j.supflu.2009.04.008.

314. E. Kılınç, "Fullerene $C_{60}$ functionalized $\gamma$-$Fe_2O_3$ magnetic nanoparticle: synthesis, characterization, and biomedical applications," *Artif Cells Nanomed Biotechnol*, vol. 44, no. 1, pp. 298–304, Jan. 2016, doi: 10.3109/21691401.2014.948182.

315. F. Moussa, "[60]Fullerene and derivatives for biomedical applications," in *Nanobiomaterials*, Elsevier, 2018, pp. 113–136, doi: 10.1016/B978-0-08-100716-7.00005-2.

316. A. Bratovcic, "Biomedical application of nanocomposites based on fullerenes-C60," 2023, pp. 107–117, doi: 10.1007/978-3-031-34721-4_12.

317. J. Conyers, "Biomedical applications of functionalized fullerene-based nanomaterials," *Int J Nanomed*, p. 261, Nov. 2009, doi: 10.2147/IJN.S5964.

318. M. Kumar and K. Raza, "C60-fullerenes as drug delivery carriers for anticancer agents: promises and hurdles," *Pharm Nanotechnol*, vol. 5, no. 3, Jan. 2018, doi: 10s.2174/2211738505666170301142232.

319. E. I. Pochkaeva et al., "Fullerene derivatives with amino acids, peptides and proteins: from synthesis to biomedical application," *Prog Solid State Chem*, vol. 57, p. 100255, Mar. 2020, doi: 10.1016/j.pro gsolidstchem.2019.100255.

320. D. Pantarotto, "Synthesis and biological properties of fullerene-containing amino acids and peptides," *Mini Rev Med Chem*, vol. 4, no. 7, 2012, doi: 10.2174/1389557043403602.

321. E. Nakamura and H. Isobe, "In vitro and in vivo gene delivery with tailor-designed aminofullerenes," *Chem Record*, vol. 10, no. 5, pp. 260–270, Oct. 2010, doi: 10.1002/tcr.201000021.

322. I. Andreev et al., "Penetration of fullerene $C_{60}$ derivatives through biological membranes," *Fuller Nanotub Carbon Nanostruct*, vol. 16, no. 2, pp. 89–102, Mar. 2008, doi: 10.1080/15363830701885831.

323. S. Dostalova et al., "Antiviral activity of fullerene C60 nanocrystals modified with derivatives of anionic antimicrobial peptide maximin H5," *Monatsh Chem Chem Monthly*, vol. 147, no. 5, pp. 905–918, May 2016, doi: 10.1007/s00706-016-1675-0.

324. C. M. Uritu et al., "Hybrid fullerene conjugates as vectors for DNA cell-delivery," *J Mater Chem B*, vol. 3, no. 12, pp. 2433–2446, 2015, doi: 10.1039/C4TB02040E.

325. O. K. Nag et al., "Quantum dot–peptide–fullerene bioconjugates for visualization of in vitro and in vivo cellular membrane potential," *ACS Nano*, vol. 11, no. 6, pp. 5598–5613, Jun. 2017, doi: 10.1021/acsnano.7b00954.

326. S. T. Zuckerman and W. J. Kao, "Nanomaterials and biocompatibility: carbon nanotubes and fullerenes," in *Nanotechnology in Drug Delivery*, Springer New York, New York, NY, 2009, pp. 229–266, doi: 10.1007/978-0-387-77668-2_8.

327. S. Fiorito, A. Serafino, F. Andreola, A. Togna, and G. Togna, "Toxicity and biocompatibility of carbon nanoparticles," *J Nanosci Nanotechnol*, vol. 6, no. 3, pp. 591–599, Mar. 2006, doi: 10.1166/jnn.2006.125.

328. T. Halenova et al., "Evaluation of the biocompatibility of water-soluble pristine C60 fullerenes in rabbit," *Bionanoscience*, vol. 10, no. 3, pp. 721–730, Sep. 2020, doi: 10.1007/s12668-020-00762-w.

329. F. Rancan et al., "Cytotoxicity and photocytotoxicity of a dendritic C60 mono-adduct and a malonic acid C60 tris-adduct on Jurkat cells," *J Photochem Photobiol B*, vol. 67, no. 3, pp. 157–162, Jul. 2002, doi: 10.1016/S1011-1344(02)00320-2.

330. S. Goodarzi, T. Da Ros, J. Conde, F. Sefat, and M. Mozafari, "Fullerene: biomedical engineers get to revisit an old friend," *Mater Today*, vol. 20, no. 8, pp. 460–480, Oct. 2017, doi: 10.1016/j.mattod.2017.03.017.

331. S. Bosi et al., "Hemolytic effects of water-soluble fullerene derivatives," *J Med Chem*, vol. 47, no. 27, pp. 6711–6715, Dec. 2004, doi: 10.1021/jm0497489.

332. D. F. Chowdhury, "Carbon nanotube for drug delivery and controlled release," in *Comprehensive Biotechnology*, Elsevier, 2019, pp. 615–627, doi: 10.1016/B978-0-444-64046-8.00312-8.

# 7 Virus-Inspired Nanocarriers

*Ankita Alice Singh and Seema Kohli*

## ABBREVIATIONS

| | |
|---|---|
| APC | antigen-presenting cell |
| BHK-21 | Baby Hamster Kidney fibroblast cells |
| CCMV | cowpea chlorotic mottle virus |
| DC | dendritic cell |
| EYFP | enhanced yellow fluorescent protein |
| NLS | nuclear localization signals |
| QS | quorum sensing |
| VINP | virus-inspired nanoparticle |
| VLP | Virus-like particle |

## 7.1 INTRODUCTION

Viruses are known to be non-reactive when they have not been attached to a host. They are unicellular organisms that tend to replicate by attaching to the host's cell and interfering with their DNA. With their evolution, viruses have been able to bypass the host's immune system effectively. To get enough amount of time for cell infection, they take advantage of amphoteric charges which ultimately increase their systemic circulation time [1, 2].

The very first step for the viral infection is the process of attachment. The transformation of viral-envelope proteins occurs because of polyvalent ligand–receptor interactions which further allow the viruses to enter and penetrate the cell [3]. The host's cell nucleus plays an important role for some viruses in the replication step. Viruses are capable of specifically targeting the nucleus due to the presence of nuclear localization signals (NLS), which enables the virus to enter and release their genomes for replication [4].

Additionally, communication between the viral surface cellular components and their microenvironment leads to contributory viral infection steps. It is interesting to know that the coat protein and structural morphology of the viruses play a unique role in crossing various physiological barriers and attaching to host cells during the whole infection process [5]. Different types of bioengineered vectors have been developed for nano-cargo delivery in biomedical applications which was inspired by the efficient viral infection process [6]. For example, high transfection efficiency of adenoviral and lentiviral vectors has been employed to treat multiple diseases [7], such as hemophilia [8], cystic fibrosis [9], and tumors [10].

Bacteriophages seem to have a greater ability to target bacterial biofilms. Biofilm formation is a survival mechanism for bacteria and fungi to adapt to their living environment (Table 7.1), especially in a hostile environment. The high numbers of bacteria present inside the biofilms actually facilitate the actions of bacteriophages by allowing rapid and efficient infection of the host and consequent amplification of the bacteriophagic action. Bacteriophages are known to produce (or to be able to induce) enzymes that degrade the extracellular matrix. They are also able to infect persister cells, remaining dormant within them, but re-activating when they become metabolically active. It is perhaps unsurprising that bacteriophages as the natural predators of bacteria have the ability to target this common form of bacterial life.

## 7.2 WHAT ARE VIRUS-INSPIRED NANOPARTICLES

It is a technology where naturally occurring nanomaterials derived from plant viruses, bacteriophages, mammalian viruses, etc. are used in drug delivery systems. The development of virus-inspired nanoparticles (VINPs) or virus-like particles (VLPs) and their genome-free versions are rapidly growing for their applications in nanomedicines.

Naturally, viruses act as nucleic acid carriers, as they carry and protect their biomolecules and release them according to their suitable environment (host's cellular environment); this particular property is widely utilized for drug delivery [7].

Active agents can be loaded by different processes such as encapsulation, infusion, absorption, or conjugation to the interior or exterior surfaces of the coat protein interfaces. This can be done using different amalgamations of chemistries and with attachment to numerous functional groups provided by the cellular protein structure [7]. Some important parameters such as reaching the target site of action, protection of the cargo (drug) in the proteinaceous matrix, and reversible binding of the functional group as well as active molecules are made possible due to the abovementioned features.

DOI: 10.1201/9781003473183-7

## Table 7.1 Biofilm: Formation, dispersal and the risk of dissemination

| | |
|---|---|
| Mechanism of biofilm formation | The formation of a biofilm consists of various different phases such as reversible attachment, irreversible attachment, colonization, maturation and dispersion. Microorganisms living inside a biofilm have distinct mechanisms that enable initial surface attachment, the development of a colony formation and ecosystem, and successive separation from the biofilm. The attachment of microorganisms to a surface can be accelerated by factors such as increased shear forces, bacterial mobility, and electrostatic interactions between the microorganism and surface. In a state of "reversible attachment," there is thought to be equilibrium between attached and free-floating microorganisms. However, there are traits of the microbial cell surface that facilitate the attachment procedure to the surface, inclusive of flagella, pili, fimbriae, and glycocalyx [11]. |
| Molecular signaling of biofilm | As the cell populace density of the maturing biofilm varies, the gene expression of cells within the biofilm is managed by a process known as quorum sensing (QS). Through this system, bacteria release chemical indicators called auto-inducers, which are constitutively produced and increase in concentration as the density of the biofilm increases. Some integral factors that are required for the survival of microorganisms within biofilm are lined up by physiological processes which include motility, sporulation, and release of virulence factors, all these changes take place when the concentration of the auto-inducers reaches a critical threshold which further leads to modifications in gene expression [12, 13]. Molecules called acyl-homoserine lactones and oligopeptide molecules are released by gram-negative bacteria and gram-positive bacteria, respectively [12]. |
| Dispersal or detachment of bacteria from a biofilm and the risk of dissemination | A severe ultimatum gives rise to the host, when parts of the biofilm sheds and are capable of forming colonies on new sites; this approach is known as biofilm dispersal which further stimulates distribution within the host [14]. Biofilm dispersal is a process that occurs towards the end of the biofilm life cycle; during this phase, the cells that were once part of a complex, relatively static, slow-growing micro-populace within the biofilm become distinguished, frequently highly motile microorganisms [15]. It is important to note that these diffused cells are different from bacteria that slough off from the biofilm and are truly specialized cells; furthermore, these dispersed cells are capable of initiating biofilm growth by attaching to new surfaces. |

### 7.3 RISKS AND CHALLENGES

Some important challenges are given below:

1. Clinical false-positive rate of some detection methods is still high;
2. In the virus tracking process, there is less adaptability of quantum dots (QDs) owing to larger size, flicker, and possible interference with virus function; and
3. Real-time *in vivo* tracking of a single virus needs further refinement. In the future, smaller, non-toxic, and chemically stable nanomaterials are needed to improve the efficiency and accuracy of detection, and monitoring of virus infections to curb mortalities.

Although there are still many safety risks involved in using the viruses directly as a carriage, some of which include high immunogenicity, wide viral tropism, and pathogenicity, remarkably restricting their further implementations [16]. Currently, the booming evolution of nanotechnology has resulted in unparalleled strategies for drug delivery [17]. Solid lipid nanoparticles [18], inorganic nanoparticles [19], and polymeric nanoparticles (NPs) [20] are some of the traditional nanocarriers that have been developed for drug delivery.

Despite countless attempts, limitations are still multifarious. Initially, the carrier must be able to escape the immune system's clearance efficaciously, then traverse the biological hurdles or (barricades) and cytomembrane while avoiding endosomal entanglement.

Furthermore, the nanocarrier requires outstanding stability on the outer side of the cell, as well as it must target and set free the payload inside the nucleus successfully.

Potentially, virus-inspired nanocarriers or nanosystems (VINs) have been established which have shown a certain level of distinctive supremacy to overcome these challenges in biomedicine, especially for drug delivery [21, 22].

Numerous nanocarriers have been developed to overcome challenges that occur due to a number of barriers during drug delivery. With an increased understanding of viral infection, researchers have been inspired to mimic the viral structures and functions while designing a drug carrier. To improve the bioavailability of drugs, the VINPs have been highly optimized based on the specific viral features.

## 7.4 ADVANTAGES OF VIRUS-INSPIRED NANOPARTICLES AND BACTERIOPHAGE-BASED SYSTEMS

VINPs offer the following distinctive advantages as nanocarriers:

- First, due to their morphological consistency, increased biocompatibility, hydrophilic property, easy functionalization, and high absorption efficiency, VINPs have been effortlessly utilized. These nanocarriers can easily pass through the host's immune barriers because their physiochemical properties can be edited and they can deal with the physiological surroundings during drug delivery [23, 24].
- Other than this, they have been used as a delivering agent for genes; with this delivery property, VINs can be employed to deliver a wide range of pharmaceutical or therapeutic drugs [25].
- There is enough space in the hollow structure, in which a therapeutic product can be encapsulated, which further may result in improved water solubility, loading efficiency, and biocompatibility of the drug.
- Better bioavailability and higher cellular uptake efficiency can be improved by implementing a virus-like design of VINs [26].
- Bacteriophages, which were discovered by Frederick Twort and Felix d'Herelle, respectively in 1915 and 1917, have been widely employed over antibiotic therapy to treat human bacterial infections (due to their ability to attack and kill bacteria).
- Some of the important advantages of phage therapy over antibiotic therapy are that bacteriophages are effective against a wide range of pathogenic bacteria (multi-drug-resistant bacteria), microbial flora does not get affected due to bacteriophage specificity (only the targeted pathogenic bacteria gets killed by phage).
- Rate of mutation in bacteriophages is significantly much higher than that of bacteria, because of which phages can respond quickly to the appearance of phage-resistant bacterial mutants. Phage therapy-related side effects are very rare when compared with antibiotic therapy. Developing the cost of new antibiotics is costlier than that of phage therapy [27, 28].

Bacteriophages have some of the most desirable properties and important advantages that make them enthralling applicants for tackling antibiotic resistance in bacteria. Some of the most important advantages are given as follows:

- Bacteriophages are environmentally friendly.
- Isolation and identification of suitable bacteriophage for phage therapy are rapid, relatively easy, and cost-effective.
- Bacterial resistance against phage develops about ten times slower than the resistance to that of antibiotics [29].
- Under very harsh environmental conditions, phages tend to replicate continuously and they remain infective until the host bacterial population density has been significantly reduced [30].

- Once the bacteria get lysed by the lytic phage they will not be able to regain their viability; by contrast, antibiotic therapy, in which the targeted bacteria might not get killed, facilitates the development of antibiotic resistance [31].

Apart from these generalized advantages, there are some specific advantages that phage therapy has. These are outlined below:

- Phages are considered to be self-dosing or auto-dosing because they are capable of increasing in number in the presence of host bacteria (depending on the host population density) and are not able to multiply in its absence and thus ultimately get eliminated from the human/animal.
- Bacteriophages are highly specific for their host cell; it is another advantage of phage therapy over antibiotics. They target specific bacterial strains without affecting the normal beneficial bacterial microflora of the body. Therefore, the risk of secondary infections is reduced.
- Phages are inherently non-toxic as phages consist mostly of nucleic acids and proteins. Most of the phages exhibit a relatively narrow host range which limits the number of bacterial types with which selection for specific phage-resistance mechanisms can occur. Thus, the chances of resistance associated with phage therapy are low. It is because when phages infect and kill bacteria, the mechanism they use is different from those of antibiotics; specific antibiotic resistance mechanisms do not translate into mechanisms of phage resistance [32].
- Bacteriophages are non-toxic to humans; thus, there are no severe side effects, and are suitable for use in humans as phages do not infect eukaryotic cells [29].

## 7.5 APPLICATIONS

VINPs have a wide range of applications in the field of biomedicine. They have been continuously employed as delivery agents successfully. Some of the major applications have been discussed as follows.

### 7.5.1 As a Carrier of Genes and Nucleic Acids

Viruses are natural carriers of genetic material (genes, nucleic acid). Because VINs do not have their own genome, with the help of the encapsulation method nucleic acid can be encapsulated in it and can be widely employed as a gene-delivering agent. According to a study, viruses are used as vectors to transport genetic material in about 67% of gene therapy clinical trials [33].

### 7.5.2 Plant-Derived VLPs

They have been used for the delivery of mRNA. In one study, cowpea chlorotic mottle virus (CCMV) was assessed for the delivery of heterologous mRNA encoding enhanced yellow fluorescent protein (EYFP) to mammalian (Baby Hamster Kidney fibroblast (BHK-21) cells). It appears that through the encapsulation method, RNA cargo was stabilized into the plant viral capsid. Transfection was done on cells using lipofectamine; it was observed that VLPs successfully delivered and released nucleic acids into the cytoplasm of the BHK-21 cells which further facilitates EYFP expression [34].

Immune cells like dendritic cells (DCs) and macrophages take up VLPs effectively due to VLPs' (most VLPs') immunogenic nature. This particular property can be employed for the delivery of therapeutic nucleic acids to innate immune cells in cancer treatment.

### 7.5.3 Vaccines and Immunotherapy

As VLPs cannot replicate nor infect in the absence of the viral genome, they do not compromise on safety and efficiency. On the other hand, VLPs can mimic the conformation of viruses (native) utilizing their intrinsic immunogenicity. VLPs are powerful and successful platforms for vaccines [35].

Bacteriophage and plant viruses are unable to replicate within mammalian cells (with or without their viral genome), which provides a secondary level of safety; these have been developed as heterologous nanoparticle display platforms, whereas VLPs have been developed as subunit vaccines.

When exposed to the host's immune system (mammalian origin or not), VLPs are taken up by antigen-presenting cells (APCs), stimulating a strong and effective immune response. Thus, VLPs can be strongly exploited as a carrier to deliver and present epitopes for vaccine designs.

Through various studies, it has been observed that bacteriophages have a wide range of therapeutic properties. Some important applications of bacteriophages have been discussed in Table 7.2.

## Table 7.2 Applications of bacteriophage

| | |
|---|---|
| Phage therapy | In 1919, phages were first used as a therapeutic agent in human, and that was the time when they were just discovered [36]. Phage therapy started back in 1896, when the existence of antibacterial activity against *Vibrio cholera,* the causative agent of cholera, was first reported by Ernest Hankin. Cholera was considered one of the deadliest perils humans had faced. Frederick Twort, in 1915, hypothesized that antibacterial activity could be due to the phage, but he did not follow his discovery; therefore, in 1917, bacteriophages were discovered by Fe'lix d'He'relle. In 1925, d'He'relle drew attention toward phage therapy by reporting the treatment of plague (four types) by anti-plague phages [37]. The Eliava institute in Tbilisi Georgia is considered the pioneer in this regard where phage therapy is extensively studied and applied [38]. |
| Phages as antibacterial agent | Phages are prominent candidates for exploitation as antibacterial agents as they are the natural killers of bacteria. Phages have multiple intrinsic characteristics which make them suitable candidates for such applications. Phages neither can incorporate their DNA into the genome of eukaryotic cells nor can replicate in such cells. They are highly specific in their bactericidal potential. Some phages are strain specific and they normally target a single bacterial species, which means that there is little or no effect on the natural microbiota of the patient/animal. This is a powerful advantage over the many broad range antimicrobials (including antibiotics) which are commonly used nowadays. Besides, there have been no reports of harmful effects on eukaryotic cells by purified phages. Phages are non-toxic in nature because humans are exposed to phages in the natural environment every day without any adverse side effects. Phages multiply in the presence of the target bacteria and are not able to multiply in its absence and thus ultimately get eliminated from the human/animal, which indicates that the phages are considered to be self-dosing. Phages also have the aptness to lyse bacteria present in a biofilm, mucous membrane, and suppurative wounds. These are bacterial niches, which are unsuitable for treatment with antibiotics [39–41]. |
| Phage display | In 1985, the concept of phage display was first introduced [42]. Phage display was possibly the first phage applied as a tool of modern biotechnology [43]. Phage display is a unique molecular technique used for synthesizing polypeptides with novel characteristics. In this technique, the DNA that encodes the protein is fused with phage coat protein gene and the desired protein is expressed on the surface of the phage particle or bacteriophage. The most extensively used phage in phage display technique is filamentous phage M13 of *Escherichia coli*; other than this, phages like lambda and T7 phage have also been widely used in phage display system [44].<br>For the purpose of screening and isolation of peptides, phage libraries can be used; these peptide molecules are highly specific and have affinity for target proteins. These highly specific peptides can be applied in drug designing as reagents for interpretation of molecular recognition and they also reduce the limitations for receptors. These peptides then further can be used as therapeutic agents by inhibiting the interaction between receptor and ligand or acting as agonist [45].<br>In addition, phage display can be used for various applications which include mapping of epitopes, in vaccine design, in study of protein–protein interactions, in determination of specificity of enzymes and inhibitors, screening for anticancer peptide and protein, and in the screening for receptors [43]. |

*(continued)*

## Table 7.2 (Continued)

| | |
|---|---|
| Phage typing | Bacteriophages infect and demolish only one or a few types of bacteria, which makes them highly specific to their host. This specificity of phage is very helpful and widely used for bacterial cells which allows them to be used for the typing of bacterial strain, a technique known as phage typing, and for the detection of pathogenic bacteria [46]. In phage typing, specific phages use sensitivity patterns for precisely identifying the microbial strains. If the phages that are bound to bacteria are detected by specific bacterial antibodies, then the sensitivity of the detection could be increased [47]. For the purpose of detection of unknown bacterial strain, the lawn of unknown bacterial strains is provided with various different phages. Once the phages start infecting and lysing the specific bacterial strains, formation of a plaque (clear zone) appears on the lawn, which makes it easy to identify the specific bacterial strain. Additionally, there are certain other techniques that can be employed, which are also widely used for the detection of pathogenic bacteria, for example, use of phages that deliver reporter genes (e.g., Lux) specifically or use of green fluorescent proteins that would express after infecting the bacteria [44]. |
| Phages used as vehicle for the delivery of vaccines | Bacteriophages have been widely used as vehicles for the delivery of vaccines. Bacteriophages can be employed in two different ways. Firstly, phages can be used directly, in which they transport the vaccine antigen that is displayed on their surface. Secondly, in DNA vaccine, in this the gene that is responsible for the synthesis of vaccine antigen, it is first incorporated into the phage genome, then this phage would act as a vehicle for the delivery of DNA vaccines [48]. Phages can be constructed using phage display techniques, that further would display the antigenic peptide on their surface [44]. In some studies, whole phage particles that displayed antigenic peptides have been used as vaccines in animal models [49]. |
| Phages used in targeted gene delivery | Bacteriophages have been potentially and widely used as specific gene delivery vectors [50]. The phages that are used for the purpose of targeted gene delivery is similar to that for using phages for DNA vaccine delivery in which the DNA inside is protected by phage coat. This phage coat protects the DNA from degradation after it has been injected. The ability of phages to display foreign proteins on their surfaces allows them to target specific cell types which is essential for successful gene therapy [44]. For the purpose of display targeting and processing molecules on the surfaces of phages, phage display and artificial covalent conjugation are widely used [51, 52]. |
| Phages therapy in plants | For the elimination of plant pathogens, bacteriophages have been widely used as a bio-control. Phage-mediated bio-controls of plant pathogens such as *Xanthomonas pruni* associated with the infection of cabbage, peppers, and peaches. *Ralstonia solanacearum* causes infection in tobacco. They have been successfully used against *Xanthomonas campestris*, which causes spots on tomatoes and against *Pseudomonas talaasi,* which causes blotch of mushrooms [53]. |
| | The application of phage therapy in plants is still in the early stages. For the use of phage therapy, new phages are being discovered and constantly been utilized in many different ways. Treatment of diseased crop with the help of phage therapy has provided promising results, including the treatment of some antibiotic-resistant infections involved in bacterial blight on soyabeans. A number of phages have been approved by the United States Food and Drug Administration (FDA) for use on crops meant for human consumption. To control bacterial pathogens in environmental, medical settings, and food processing and also in the expectation of reducing product loss and production cost, many companies and organizations are concentrated on the discovery, isolation, and marketing of bacteriophage products [54]. |

## Table 7.2 (Continued)

| | |
|---|---|
| Phage therapy in animals | Bacteriophages are widely used for the treatment and elimination of infection in chickens, calves, pigs, and lambs, caused by *E. coli* and *Salmonella*. In 1983, experiment performed by Smith and Huggins proves that a single dose of phage R is very effective in preventing death in mice due to septicemia caused by *E. coli*; this infection cannot be treated with multiple doses of antibiotics. They eventually used bacteriophages for the treatment of infections caused by *E. coli* (entero toxigenic) in neonatal lambs, calves, and pigs [55]. Animal studies have shown that phage therapy works in animal system. However, many experiments with animals recommend that the therapeutic application of phages in animals have some advantages over antibiotics. |
| Phages in food industry | Food-borne diseases are the contamination of food that are caused by many pathogenic bacteria [56]. Currently, food-borne pathogens like *Listeria* and *Salmonella*, followed by *E. coli* and *Campylobacter jejuni*, are the major cause of death. Bacteriophages can be used effectively and safely to eradicate the food-borne pathogenic bacteria that contaminate the food [57]. For example, *Listeria* phage P100 (under the commercial name Listex P100) was developed to eradicate biofilms present in processed meat products and on factory working surfaces, which has already been authorized in the United States by the Department of Agriculture [58]. Other commercial bacteriophage products have been targeted for other pathogenic species, for example, *S. enterica* or *E. coli* (Salmofresh™ and ScoShield™, respectively) [59]. |
| Eradication of biofilm by using phages | Formation of biofilm is an important strategy that is adopted by the bacteria for the survival purpose. During this strategy, microorganisms secrete a huge amount of extracellular polymer. Microorganisms aggregate to form biofilm either on living or non-living surfaces. Normally, environmental conditions stimulate most of the bacteria to form biofilm. In humans, formation of biofilms causes severe diseases, and the biofilm is resistant to antibiotics and host immune system [29]. An alternative therapy is needed to control biofilm-associated diseases, as biofilms are resistant to antimicrobial drugs [60].<br>Nevertheless, phages are furnished with enzymes (e.g., exopolysaccharide (EPS) depolymerase) outside of the capsid that degrade the extracellular polymeric substance and disseminate bacterial biofilms, allowing the phage to enter the bacteria embedded within the EPS matrix. Once the lytic cycle is completed, the phage progeny is released, propagate the dispersal of the biofilm through the removal of biofilm embedded within bacteria in following layers [61]. For the penetration of dense biofilms, high doses of antibiotics are typically required to observe any inhibition of bacterial growth, yet complete eradication of biofilm is rare and at the end of antibiotic treatments re-growth of colonies begin [62]. |

### 7.6 CONCLUSION

Nanosized viruses when encapsulated with proteins or nucleic acids act as one of the best carriers to transport these cargos across various barriers of the host's immune system. VNPs have exceptional applications in the field of nanomedicine, as they are safe, effective, biodegradable, and biocompatible. On the other hand, phage therapy is an exciting, rediscovered field that will surely provide countless benefits to science, agriculture, veterinary science, and medicine, including offering a possible solution to counteract the increased prevalence of antibiotic-resistant pathogens. The potential of combining antibiotic therapy and phage therapy, the use of phage cocktails, or the use of phage protein products may be the best areas for successful phage treatment of infections. Considering the utility and diverse applications of phage therapy, this area of research is extremely necessary.

Bacteriophages possess unique properties and show considerable promise in the control of biofilms. Bacteriophages do not appear to naturally produce substantial quantities of antibiotic-like antibacterial agents, that is, relatively small extracellular factors that can bind to and then inhibit the metabolism of target bacteria.

Plant VNPs offer biocompatibility, efficacy, low cost, easy manufacturing, and optimal cargo loading capacity. Plant VNPs offer biodegradability, genetic engineering customization, and superior modification efficacy over synthetic nanomaterials, which rely on chemical synthesis and are persistent in the body.

## REFERENCES


1. Figueroa SM, Fleischmann D and Goepferich A, Biomedical nanoparticle design: What we can learn from viruses. *J. Controlled Release*, 2021, **329**, 552–569.
2. Blanco E, Shen H and Ferrari M, Principles of nanoparticle design for overcoming biological barriers to drug delivery. *Nat. Biotechnol.*, 2015, **33**, 941–951.
3. Fleischmann D, Figueroa SM, Beck S, et al., Adenovirus-mimetic nanoparticles: Sequential ligand–receptor interplay as a universal tool for enhanced in vitro/in vivo cell identification. *ACS Appl. Mater. Interfaces*, 2020, **12**, 34689–34702.
4. Mudhakir D and Harashima H, Learning from the viral journey: How to enter cells and how to overcome intracellular barriers to reach the nucleus. *AAPS J.*, 2009, 11, 65–77.
5. Steinmetz NF, Biological and evolutionary concepts for nanoscale engineering: Viruses as natural nanoparticles have great potential for a wide range of nanoscale products. *EMBO Rep.*, 2019, **20**, e48806.
6. Ng J, Barral S, Barrigon CD, et al., Gene therapy restores dopamine transporter expression and ameliorates pathology in iPSC and mouse models of infantile parkinsonism. *Sci. Transl. Med.*, 2021, **13**, eaaw1564.
7. Bulcha JT, Wang Y, Ma H, et al. Viral vector platforms within the gene therapy landscape. *Sig. Transduct. Target Ther.*, 2021, **6**, 53.
8. Stephens CJ, Lauron EJ, Kashentseva E, et al., Long-term correction of hemophilia B using adenoviral delivery of CRISPR/Cas9. *J. Control Release.*, 2019, Mar 28, **298**, 128–141. doi:10.1016/j.jconrel.2019.02.009. Epub 2019 Feb 13.
9. Zabner J, Couture LA, Gregory RJ, et al., Adenovirus-mediated gene transfer transiently corrects the chloride transport defect in nasal epithelia of patients with cystic fibrosis. *Cell*, 1993, 75, 207–216.
10. Schmid M, Ernst P, Honegger A, et al., Adenoviral vector with shield and adapter increases tumor specificity and escapes liver and immune control. *Nat. Commun.*, 2018, 9, 1–16.
11. Donlan RM, Biofilms and device-associated infections. *Emerg. Infect. Dis.*, 2001, **7**, 277–281.
12. Lindsay D and von Holy A, Bacterial biofilms within the clinical setting: what healthcare professionals should know. *J. Hosp. Infect.*, 2006, **64**, 313–325.
13. Mangwani N, Dash HR, Chauhan A, et al., Bacterial quorum sensing: functional features and potential applications in biotechnology. *J. Mol. Microbiol. Biotechnol.*, 2012, **22**, 215–227.
14. Donelli G, Vascular catheter-related infection and sepsis. *Surg. Infect. (Larchmt)*, 2006, **7** (Suppl. 2), s-25–s-27.
15. McDougald D, Rice SA, Barraud N, et al., Should we stay or should we go: mechanisms and ecological consequences for biofilm dispersal. *Nat. Rev. Microbiol.*, 2012, **10**, 39–50.
16. Zuber G, Dauty E, Nothisen M, et al., Towards synthetic viruses. *Adv. Drug Delivery Rev.*, 2001, **52**, 245–253.
17. Mitchell MJ, Billingsley MM, Haley RM, et al., Engineering precision nanoparticles for drug delivery. *Nat. Rev. Drug Discovery*, 2021, **20**, 101–124.

18. Hou X, Zaks T, Langer R, et al., Lipid nanoparticles for mRNA delivery. *Nat. Rev. Mater.*, 2021, **6**. doi:10.1038/s 41578-021-00358-0.

19. Huo S, Jiang Y, Jiang Z, et al., Stable and oxidant responsive zwitterionic nanoclusters. *Nanoscale*, 2018, **10**, 7382–7386.

20. Manzari MT, Shamay Y, Kiguchi H, et al., Targeted drug delivery strategies for precision medicines. *Nat. Rev. Mater.*, 2021, **6**, 351–370.

21. Shan WJ, Chen RH, Zhang Q, et al., Improved stable indocyanine green (ICG)-mediated cancer optotheranostics with naturalized hepatitis B core particles. *Adv. Mater.*, 2018, **30**, 1707567.

22. Perrault SD and Shih WM, Virus-inspired membrane encapsulation of DNA nanostructures to achieve in vivos. *ACS Nano.*, 2014, **8**, 5132–5140.

23. Elzoghby AO, Samy WM and Elgindy NA, Protein-based nanocarriers as promising drug and gene delivery systems. *J. Control Release*, 2012, 161, 38–49. doi:10.1016/j.jconrel.2012.04.036.

24. Samandoulgou I, Hammami R, Rayas RM, et al, Stability of secondary and tertiary structures of virus-like particles representing noroviruses: Effects of pH, ionic strength, and temperature and implications for adhesion to surfaces. *Appl. Environ. Microbiol.*, 2015, **81**, 7680–7686.

25. Lostale-Seijo I and Montenegro J, Synthetic materials at the forefront of gene delivery. *Nat. Rev. Chem.*, 2018, **2**, 258–277.

26. Ni R, Zhou JL, Hossain N, et al., Virus-inspired nucleic acid delivery system: Linking virus and viral mimicry. *Adv. Drug Delivery Rev.*, 2016, **106**, 3–26.

27. Lu Y, Chan W, Ko BY, et al., Assessing sequence plasticity of a virus-like nanoparticle by evolution toward a versatile scaffold for vaccines and drug delivery. *Proc. Natl. Acad. Sci. U. S. A.*, 2015, **112**, 12360–12365.

28. *Gene Therapy Clinical Trials Worldwide*. 2019. www.abedia.com/wiley/, accessed April 30, 2020.

29. Parasion S, Kwiatek M, Gryko R, et al., Bacteriophages as an alternative strategy for fighting biofilm development. *Pol. J. Microbiol.*, 2014, **63**, 137–145.

30. Schmelcher M and Loessner MJ, Application of bacteriophages for detection of foodborne pathogens. *Bacteriophage*, 2014, 4, e28137.

31. Stratton CW, Dead bugs don't mutate: susceptibility issues in the emergence of bacterial resistance. *Emerg. Infect. Dis.* 2003, **9**, 10–16. doi:10.3201/eid0901.020172.

32. Loc-Carrillo C and Abedon ST, Pros and cons of phage therapy. *Bacteriophage*, 2011, **1**, 111–114.

33. Matsuzaki S, Rashel M, Uchiyama J, et al., Bacteriophage therapy: a revitalized therapy against bacterial infectious diseases. *J. Infect. Chemother.*, 2005, **11**(5), 211–219, 2-s2.0-27644460427. https://doi.org/10.1007/s10156-005-0408-9.

34. Matsuzaki S, Yasuda M, Nishikawa H, et al., Experimental protection of mice against lethal *Staphylococcus aureus* infection by novel bacteriophage $\phi$MR11, *J. Infect. Dis.*, 2003, 187(4), 613–624, 2-s2.0-0037442798. https://doi.org/10.1086/374001.

35. Chroboczek J, Szurgot I and Szolajska E, Virus-like particles as vaccine. *Acta Biochim. Pol.*, 2014, **61**, 531–539.

36. Summers WC, Bacteriophage discovered. In: *Felix d'Herelle and the Origins of Molecular Biology*. Yale University Press, 1999, pp. 47–59.

37. Hermos JA, Garcia JL and Garcia P, Taking aim on bacterial pathogens: from phage therapy to enzybiotics. *Curr. Opin. Microbiol.*, 2007, **10**(5), 461–472.

38. Sulakvelidze A and Kutter E, Bacteriophage therapy in humans. In: Kutter E, Sulakvelidze A, editors. *Bacteriophages: Biology and Application*. Boca Raton, FL: CRC Press; 2005, pp. 381–436.

39. Slopek S, Weber-Dabrowska B, Dabrowski M, et al., Results of bacteriophage treatment of suppurative bacterial infections in the years 1981–1986. *Arch. Immunol. Ther. Exp. (Warsz)*, 1987, 35(5), 569–583.

40. Fischetti VA, Novel method to control pathogenic bacteria on human mucous membranes. *Ann. N. Y. Acad. Sci.*, 2003, **987**, 207–214.

41. Olson ME, Ceri H, Morck DW, Read RR, et al., Biofilm bacteria: formation and comparative susceptibility to antibiotics. *Can. J. Vet. Res.*, 2002, 66(2), 86–92.

42. Smith GP, Filamentous fusion phage: novel expression vectors that display cloned antigens on the virion surface. *Science*, 1985, 228(4705), 1315–1317.

43. Krylov VN, Tolmachova TO and Akhverdyan, VZ, DNA homology in species of bacteriophages active on *Pseudomonas aeruginosa*. *Arch. Virol.*, 1993, **131**, 141–151.

44. Clark JR and March JB, Bacteriophages and biotechnology: vaccines, gene therapy and antibacterials. *Trends Biotechnol.,* 2006, 24, 212–218.

45. Sidhu SS, Phage display in pharmaceutical biotechnology. *Curr. Opin. Biotechnol.*, 2000, 11(6), 610–616.

46. Gill J and Abedon ST, Bacteriophage ecology and plants. *Virology J.*, 2012, 9, 9.

47. Watson BB and Eveland WC, The application of the phage fluorescent antiphage staining system in the specific identification of Listeria monocytogenes. I. Species specificity and immunofluorescent sensitivity of *Listeria monocytogenes* phage observed in smear preparations. *J. Infect. Dis.*, 1965, 115(4), 363–369.

48. Clark JR and March JB, Bacterial viruses as human vaccines? *Expert Rev. Vaccines*, 2004, 3(4), 463–476.

49. Irving MB, Pan O and Scott JK, Random-peptide libraries and antigen fragment libraries for epitope mapping and the development of vaccines and diagnostics. *Curr. Opin. Chem. Biol.*, 2001, 5(3), 314–324.

50. Barry MA, Dower WJ and Johnston SA, Toward cell-targeting gene therapy vectors: selection of cell-binding peptides from random peptide presenting phage libraries. *Nat. Med.*, 1996, 2(3), 299–305.

51. Larocca D, Witte A, Johnson W, et al., Targeting bacteriophage to mammalian cell surface receptors for gene delivery. *Hum. Gene Ther.*, 1998, **9**(16), 2393–2399.

52. Larocca D, Kassner PD, Witte A, et al., Gene transfer to mammalian cells using genetically targeted filamentous bacteriophage. *FASEB J.*, 1999, **13**(6), 727–734.

53. Hertwig S, Hammerl JA, Appel B, et al,, Post-harvest application of lytic bacteriophages for biocontrol of foodborne pathogens and spoilage bacteria. *Berl. Munch Tierarztl. Wochenschr.*, 2013, **126**(9–10), 357–369.

54. Meaden S and Koskella B, Exploring the risks of phage application in the environment. *Front. Microbiol.*, 2013, **4**, 358.

55. Smith HW and Huggins MB, Effectiveness of phages in treating experimental *Escherichia coli* diarrhea in calves, piglets and lambs. *J. Gen. Microbiol.*, 1983, 129, 2659–2675.

56. Endersen L, O'Mahony J, Hill C, et al., Phage therapy in the food industry. *Ann. Rev. Food Sci. Technol.*, 2014, 5, 327–349.

57. Mahony J, McAuliffe O, Ross RP, et al., Bacteriophages as biocontrol agents of food pathogens. *Curr. Opin. Biotechnol.*, 2011, 22, 157–163.

58. Fister S, Robben C, Witte AK, et al., Influence of environmental factors on phage–bacteria interaction and on the efficacy and infectivity of phage P100. *Front. Microbiol.*, 2016, **7**, 1152. doi:10.3389/fmicb.2016.01152.

59. Gutiérrez D, Rodríguez-Rubio L, Martínez B, et al., Bacteriophages as weapons against bacterial biofilms in the food industry. *Front. Microbiol.*, 2016, 7, 825. doi:10.3389/fmicb.2016.00825.

60. Azeredo J and Sutherland I, The use of phages for the removal of infectious biofilms. *Curr. Pharmaceut. Biotechnol.*, 2008, 9(4), 261.

61. Abedon ST, Ecology of anti-biofilm agents I: antibiotics versus bacteriophages. *Pharmaceuticals (Basel)*, 2015, **8**, 525–558.

62. Anwar H, Strap JL, Chen K, et al., Dynamic interactions of biofilms of mucoid *Pseudomonas aeruginosa* with tobramycin and piperacillin. *Antimicrob. Agents Chemother.*, 1992, 36, 1208–1214.

# 8 Nucleic Acid Nanocarriers, DNA Nanostructures, RNA Nanostructures, Hybrid Nucleic Acid Systems

*Komal Parmar and Jayvadan Patel*

## 8.1 INTRODUCTION

Recent years have seen the development of potent and innovative nucleic acid-based nanocarriers that are tailored to specific functions such as drug and gene delivery, vaccine delivery, photosensitization, vaccine transport, biosensing, preventing drug resistance, and acting as carriers and logic gates (Burgahn et al. 2019, Kedmi et al. 2018, Aji Alex et al. 2017, Zhou et al. 2017, Kim et al. 2017a, 2017b, Chen et al. 2013). Specifically, the application of nucleic acid-based nanocarriers as nano-pharmaceutical carriers has garnered significant interest due to their capacity to surmount numerous current constraints. They adhere closely to the complementary base-pairing concept, which makes it simple to design a targeted nanocarrier by varying the base's kind or order (Wang et al. 2020, Sarett et al. 2017). Commercial DNA and RNA synthesizers make it simple and quick to generate nucleic acids. As a result, nucleic acid-based nanocarriers have emerged as very promising nanomaterials for application in the area of nanomedicine. Different types of nucleic acids—mRNA, siRNA, miRNA, siRNA, gRNA, aptamer, and plasmid—can be used. Moreover, nucleic acid nanocarriers can be combined with other nano-systems to produce multiple-level combined nano-pharmaceutical systems (Picchetti et al. 2023, El Moukhtari et al. 2023). Figure 8.1 gives a glimpse of the types and properties of nucleic acid-based nanocarriers.

Following the Watson–Crick pairing principle, the ATCGU nucleotides are used to self-assemble the nucleic acids facilitated by extremely strict base pairing, thereby resulting in a biomacromolecule. Nucleic acids have atomic-level accuracy and the capacity to properly self-assemble as each atom and link are perfectly specified. They give nearly infinite sequences that can be used to fabricate specific nanostructures and allow programmable interactions between molecules (Seeman and Sleiman 2017). The spatial addressability of nucleic acid nanocarriers can locate the position of molecules by adjusting the sequence. Thus, nucleic acid nanostructures and aptamers enable the exact integration of functional molecules like folic acid, bovine serum albumin, and aptamers into specific nucleic acid nanocarrier sites (Li et al. 2023, Wang et al. 2021). Chemical synthesis is a viable method for producing them with a high yield and minimal quality fluctuation between batches. Additionally, phosphate backbones of nucleic acid which have negative charge can chelate with metal ions and form $\pi$-stacking. This can result in desired pharmacokinetics and function of the nanocarriers.

## 8.2 DNA NANOSTRUCTURES

DNA molecules, the building blocks of life, exhibit a variety of biological properties, including specificity, low toxicity, customization, and biocompatibility. In strict adherence to the base-pairing principle, double-stranded DNA hybridizes to form an antiparallel structure resembling a double helix. As a result, by altering the kind or order of bases, desired nucleic acid structures can be created (Zhao et al. 2019). When it comes to its applications, DNA nanostructure design is crucial. The father of DNA nanotechnology, Ned Seeman, hypothesized in the 1980s that guest molecules may be crystallized using DNA as a framework (Seeman 1982). DNA nanocarriers are essential for a variety of chemical processes, bioimaging, and targeted delivery (Kang et al. 2024, Duangrat et al. 2020). They fall into two categories: self-assembled DNA and nanoparticles coated with DNA. Self-assembled DNA mainly includes a tile-based approach, DNA origami, and wireframe method. Seeman originally proposed the concept of creating periodic two- (2D) or three-dimensional (3D) lattices and geometric shapes by combining branched DNA junctions with sticky-end cohesion. Later, Sa-Ardyen et al. constructed a class of branched complexes known as double-crossover molecules by joining two double helices through strand exchange. In addition to simplifying the sequence design, perfect sequence symmetry in the tile guarantees perfect geometric symmetry, which eliminates any unforeseen distortions in the DNA nanostructure and permits the growth of 2D DNA arrays to bigger sizes. These include triple-crossover molecules, three-helix and six-helix bundles, four-helix, eight-helix, and twelve-helix planar tiles, parallelogram DNA junctions, cross-shaped DNA tiles, and triangular and three-point star motifs (Figure 8.2). Each of these examples demonstrates how DNA-based self-assembly is adaptable and programmable. DNA origami facilitates the bottom-up self-assembly of predetermined DNA nanomaterial. The fundamental idea behind DNA origami design is that several short synthetic DNA strands attach to various DNA scaffold locations and fold the material into a predetermined shape. A hexagonal, square, or honeycomb lattice can be formed from the cross-section of DNA helices. The collection of vertices and faces that make up a wireframe mesh can be used to represent both 2D and 3D geometries in simple text format. Various DNA-based nanostructures are reported for their application in theranostics.



DOI: 10.1201/9781003473183-8

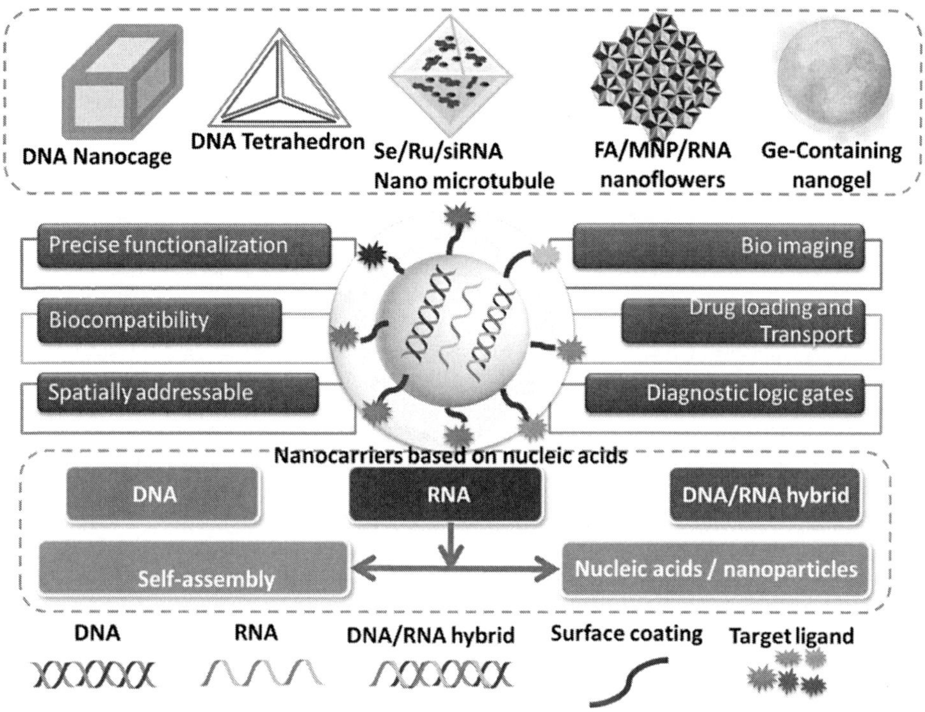


**Figure 8.1** Nucleic acid nanocarriers.

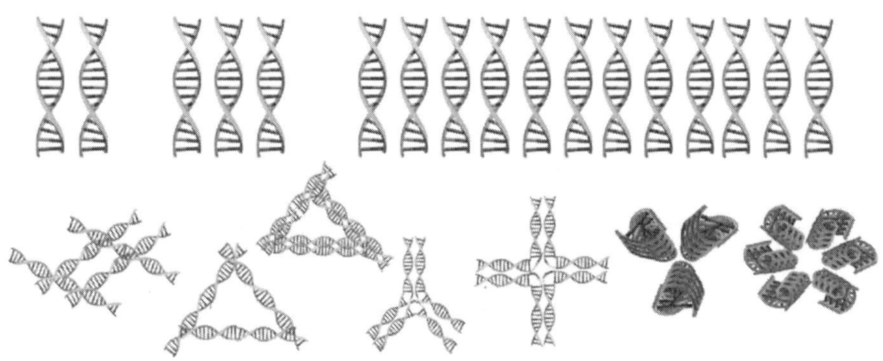

**Figure 8.2** Various shapes of DNA assemblies.

DNA nanoparticles were exploited for the co-delivery of paclitaxel and doxorubicin for cancer management. Targeting agent mucin 1 (*MUC-1*) aptamer was conjugated on the DNA nanocarrier (Wang et al. 2023). In another work, tetrahedral DNA nanostructures were loaded with paclitaxel. The cytotoxicity of formulation and pure drug alone on non-small cell lung cancer (NSCLC) cells (A549) and the PTX-resistant cell line (A549/T) was determined using a cell count kit-8 (CCK-8) assay. Drug-loaded DNA nanostructures exerted strong lethality on both cell lines with negligible drug resistance (Xie et al. 2018). Targeting ligand (2-[3-(1,3-carboxy propyl)-ureido] pentanedioic acid)-modified DNA origami nanostructure loaded with doxorubicin for the selective delivery of anticancer drug to prostate-specific membrane antigen+ cancer cell line Lymph Node Carcinoma of the Prostate (LNCaP) was reported by Ge and co-researchers. The cytotoxicity of the formulation was found to be higher in the cancer cell line (Ge et al. 2020). DNA nanostructures named nano cockleburs inspired by cockleburs with aptamer covering were fabricated and loaded with doxorubicin. DNA sequences employed were S1, S2, S3, and S4 DNA strands. The aptamer-nano cockleburs showed improved cellular internalization and selective binding when compared to DNA nano cockleburs without aptamer covering. Flow cytometry demonstrated that the aptamer-nano cockleburs might have significant binding affinity to MCF-7 cells whether in binding buffer or cell culture medium. In addition, the aptamer-nano cocklebur-doxorubicin showed increased cytotoxicity when compared with non-aptamer-nano cocklebur-doxorubicin in MCF7 cells target cells (Sun et al. 2020). Self-assembled DNA aptamer-grafted photoresponsive hyperbranched polymer was used for the targeted and photoresponsive drug delivery for doxorubicin. The doxorubicin-loaded nano assembly exhibited selective photo triggered cytotoxicity toward cancer cells, indicating its promising therapeutic effect as a smart drug delivery system (Yang et al. 2018). Aptamer-functionalized albumin-based nanoparticles for effective drug delivery and targeted cancer therapy of doxorubicin were developed. Aptamer-modified nanoparticles show higher cellular uptake and stronger cell inhibitory efficacy against cancerous MCF-7 cells as compared with the nanoparticles without aptamer modification. In addition, drug loaded aptamer-functionalized nanoparticles can induce more significant down-regulation of Bcl-2 and PCNA as well as up-regulation of pRB, PARP, and Bax in MCF-7 cells compared with unmodified nanoparticles, indicating the aptamer modification can induce cell apoptosis more effectively. Besides, aptamer-modified nanoparticles display a significantly improved capability in up-regulating p16, p21, and E-cadherin, and down-regulating EpCAM, vimentin, Snail, MMP-9, CD44, and CD133, thereby implying the favorable effects of drug delivery on the prevention of tumor progression and metastasis (Xu et al. 2018). DNA origami nanostructure loaded with doxorubicin was developed for customizable chemotherapeutic drug release. 2D and 3D scaffolded DNA origami nanostructures were prepared and characterized. Drug aggregation mechanisms and spectral changes were observed to be linked to pH, magnesium, and drug concentration which further was found to be influencing drug release profiles from the nanostructures (Ijäs et al. 2021). Apurinic/apyrimidinic endonuclease/redox effector factor 1 (*APE1*) responsive DNA prism was developed as a nano vehicle for doxorubicin for targeted therapy. Cellular uptake and drug release were investigated in a human cervical cancer cell HeLa and human embryonic kidney cell HEK-293 T. Due to overexpression of *APE1* on these cancer cells, it was observed that DNA prism could selectively release the trapped doxorubicin in response to APE1 activity in cancer cells, thus, providing a new targeted drug therapy for precision medicine (Liang et al. 2022). Doxorubicin and DNA nanocomplexes were prepared and investigated against lymphoma. Extended circulation of drug–DNA nano complexes, while avoiding off-target accumulation in the lung and being cleared from the liver, resulted in rapid accumulation in tumors and lowered cardiotoxicity. Tumor growth of EL4-challenged C57BL/6 mice (syngeneic model) and OPM2-challenged NSG mice (human xenograft model) was efficiently inhibited by the nano complexes with enhanced overall survival, in comparison with free drug (Mohammad et al. 2023). ATP-responsive Y-shaped DNA structure-capped mesoporous silica nanoparticles loaded with doxorubicin were fabricated for the targeted and controlled release in cancer therapy. An amine-modified MUC1 aptamer was covalently anchored on the surface of carboxyl-functionalized mesoporous nanoparticles. ATP aptamers (ATP1 and ATP2 aptamers) were immobilized on the surface of mesoporous nanoparticles through partial hybridization with the MUC1 aptamer. The *in vitro* cytotoxicity and the *in vivo* antitumor efficacy of the nanoformulation showed greater cytotoxicity than the nanoparticles decorated with scrambled ATP aptamers in C26 and MCF-7 cell lines (MUC1+). The biodistribution and in vivo anticancer efficacy on the C26 tumor-bearing mice indicated that the nanoformulation had a higher tumor accumulation and superior tumor growth inhibitory effect compared to a free drug (Bagheri et al. 2021).

## 8.3 RNA NANOSTRUCTURES

RNA is essential to many important biological processes in nature. These activities include the storage of information in messenger RNA (mRNA) and the control of gene expression by RNA interference (RNAi) elements such as small interfering RNA (siRNA) and microRNA (miRNA).

This yields a wider range of potential capabilities and provides fascinating functional advantages over DNA. The fact that RNA is biocompatible is one of its best qualities; unlike DNA, which typically elicits a significant immune response when found in the cytoplasm, most RNA sequences are generally well tolerated. This benefit gives RNA-based nanostructures the potential to develop into biocompatible therapeutics that interact with biological mechanisms. Protective strategies such as chemical modifications that guarantee RNA stability and enable nuclease escape are necessary as RNA is fragile and prone to chemical and enzymatic destruction (Roberts et al. 2020). Chemical changes can be done via chemical synthesis or enzymatically during *in vitro* transcription process (Flamme et al. 2019). Further, functionalization of RNA can help to target the RNA nanostructures. RNA can be functionalized in two ways: covalently, by conjugating functional groups either internally or at the 5′ and/or 3′ ends of the RNA sequence, or non-covalently, by charged contacts or annealing with other oligonucleotides (Grijalvo et al. 2018, Alberti et al. 2016).

The development of optimized RNA nanostructures in combination with pharmaceuticals, targeting ligands, and imaging tracers for use in nanomedicine as treatments, imaging tracers, or theranostic agents has advanced significantly during the past 20 years. Recently, nanoengineered polymeric RNA nanoparticles were reported for targeted cancer therapy. Polymeric siRNA nanoparticles were fabricated employing a rolling circle transcription process and terminated using biopolymers poly-L-lysine, poly-L-glutamate, and hyaluronic acid. Hyaluronic acid-based nanoparticles showed promising results in the orthotopic breast cancer model due to the CD44 receptor targeting and optimized nanostructure. *In vivo*, siRNA delivery with protein expression inhibition differed by up to approximately 20-fold (Kim et al. 2024). In another similar study, polymeric nanoparticles of siRNA were prepared and targeted for various non-Hodgkin's lymphoma (NHL) cell lines (Daudi, Raji, Ramos, and Toledo cells) via receptor-mediated endocytosis. The engineered nanoparticles showed almost 10-fold enhanced NHL-specific intracellular delivery and induced significantly *in vitro* anticancer effects (Kwak et al. 2022). Hyaluronan decorated lipid nanoparticles encasing siRNA were developed using a layer-by-layer technique with an aim of targeting the luciferase gene to the reporter luciferase-expressing glioblastoma cell line, GL261. Lipids employed were 1,2-dioleyloxy-3-dimethylaminopropane (DODMA), (6Z,9Z,28Z,31Z)-heptatriaconta-6,9,28,31-tetraen-19-yl 4-(dimethylamino)butanoate (MC3), and 1,2-dioleoyl-3-dimethylammonium-propane (DODAP). The results obtained demonstrated gene transfection *in vitro* successfully (Passos et al. 2024). Ligand-tethered nanoparticles for the targeted delivery of RNA for the treatment of liver fibrosis were developed using a one-pot, two-step modular synthetic method. The lipid employed for the fabrication of nanoparticles was AA-lipidoid AA-T3A-C12. In a preclinical model of liver fibrosis, AA-T3A-C12 enables ~65% silencing of heat shock protein 47, a therapeutic target primarily expressed by activated fibroblasts, which is twofold more potent than MC3 (Food and Drug Administration (FDA)-approved ionizable lipid), leading to significantly reduced collagen deposition and liver fibrosis (Han et al. 2023).

Doxorubicin-conjugated siRNA lipid nanoparticles were fabricated for combination cancer therapy. Co-delivery strategy via conjugation of doxorubicin to the surface of siRNA-loaded lipid nanoparticles. Optimized lipid nanoparticles enabled potent knockdown of anti-apoptotic protein B-cell lymphoma 2 (Bcl-2) and efficient delivery of a drug into the nucleus of Burkitts' lymphoma (Raji) cells, leading to effective inhibition of tumor growth in a mouse model of lymphoma (Butowska et al. 2023). Ultra-thermostable RNA nanoparticles of paclitaxel were developed with the aim of improving solubilizing and high-yield loading efficiency of the drug. The aqueous solubility of paclitaxel is increased by a factor of 32,000 when RNA nanoparticles are employed as carriers. Intravenous infusions of RNA-paclitaxel nanoparticles with a particular cancer-targeting ligand significantly suppress the growth of breast cancer in mice while causing almost little toxicity or immunological reactions. Paclitaxel doses equivalent to the reported $LD_{50}$ do not result in any fatalities (Guo et al. 2020). Multivalent rubber-like RNA nanoparticles were developed for targeted co-delivery of paclitaxel and miRNA with an aim to silence the drug efflux transporter and liver cancer drug resistance. The rubber-like RNA nanoparticles allow for enhanced targeting ability to the hepatocellular carcinoma tumors. This was demonstrated to be associated with alleviated enhanced permeability and retention effect and transcytosis effect as well as the receptor-mediated endocytosis by hepatocyte targeting ligands. $IC_{50}$ was determined to be 460 nM (Wang et al. 2020). RNA pyramids were developed for the light-triggered controlled release of paclitaxel enclosed in the RNA nanocages. The light-triggered release of paclitaxel or fluorophore Cy5 was achieved by incorporation of photocleavable spacers into the RNA nanoparticles. Upon irradiation with ultraviolet light, cargos were rapidly released (within 5 min). *In vitro*, treatment of breast cancer cells with the RNA nanoparticles enclosing photocleavable paclitaxel showed higher cytotoxicity as compared to RNA nanoparticles without the photocleavable spacer (Xu et al. 2019).

## 8.4 NUCLEIC ACID HYBRIDS

Nucleic acid hybrid nano-systems can be produced by conjugation of DNA nanostructures with organic or inorganic materials. This has been suggested as a way to increase specificity and stability, reduce immunological response, or facilitate distribution by alternative pathways. DNA nanostructures' versatility allows for the easy incorporation of many functional elements into a wide diversity of hybrid entities. The formation of nucleic acid amphiphiles makes DNA more hydrophobic and is a popular method for improving cellular uptake of therapeutic nucleic acids. It also provides a means of delivering medications that are both hydrophilic and hydrophobic. An aqueous solution of DNA amphiphiles can form a variety of nanostructures, such as micelles, fibers, vesicles, and nanosheets (Ye et al. 2022). These conjugates are frequently used in drug delivery, biosensing, diagnostics, and nanoreactors; further functionalization has been demonstrated to improve targeting and biocompatibility (Gubu et al. 2023, Obuobi and Škalko-Basnet 2020). Cholesterol functionalization is observed to be the most versatile conjugation with DNA nanostructures. Functionalization can help the DNA nanoparticles to self-assemble into a variety of structures. In one such study, the aggregation of cholesterol-modified DNA nanoparticles was studied using a combined approach of non-denaturing polyacrylamide gel electrophoresis, dynamic light scattering, confocal microscopy, and atomistic molecular dynamics simulations. Results demonstrated that aggregation of cholesterol-tagged single strand DNA (ssDNA) is sequence-dependent, while for assembled DNA constructs, the number and position of the cholesterol tags are the dominating factors. Molecular dynamics simulations of cholesterol-modified ssDNA exhibited that the nucleotides wrap around the hydrophobic moiety, protecting it from the environment (Ohmann et al. 2019). In another study, cellular interactions of DNA nano bundles that were altered with one and three cholesterol anchors were evaluated using flow cytometry and confocal microscopy. Surface-bound DNA, which is nuclease accessible, was separated from internalized DNA by nuclease digestion. When compared to unaltered DNA, three cholesterol anchors were found to boost cellular interaction by as much as ten times. Within 24 hours, the bundles were effectively endocytosed (Whitehouse et al. 2019).

DNA nanoparticles are also explored to form a variety of conjugates with inorganic materials such as gold and silver (Liu et al. 2024). In one study, gold–DNA nanosunflowers were fabricated for efficient gene silencing with controllable transformation. The prepared sunflower-like nanostructures exhibited strong near-infrared absorption and photothermal conversion ability. When exposed to near-infrared absorption, the large-sized nanostructure could break down and produce ultrasmall nanoparticles altered with c-myc oncogene silencing sequence, which could directly target the cell nucleus. Furthermore, by concurrently modulating the preincubation period with the self-assembled nanostructure (*in vitro* and *in vivo*) and near-infrared irradiation time point, the controlled gene silencing effect could be achieved (Huo et al. 2019). Recently, double-stranded DNA-functionalized silver nanoparticles containing poly(acrylic acid) as the passivating ligand which can store elastic energy releasable through a stimulus of heat, light, or pH was reported. Heating or light excitation to blunt end-stacking-induced assemblies of double-stranded DNA-functionalized silver nanoparticles causes terminal opening of the double-stranded DNA and entropic repulsion to overcome the attraction force, restoring individual double-stranded DNA-functionalized silver nanoparticles. Furthermore, a pH change results in charge association/dissociation of the carboxyl groups in poly(acrylic acid), setting up an electrostatic force switch for the controlled actuation of double-stranded DNA-functionalized silver nanoparticles (Kong et al. 2022). Table 8.1 summarizes recent hybrid nucleic acid-based nanostructures employed for theranostic applications.

## Table 8.1 Hybrid nucleic acid nano-systems for theranostic applications

| DNA Nanostructure | Application | Reference |
|---|---|---|
| Platinum-based DNA intercalator | Targets and blocks epidermal growth factor receptor in cancer therapy | Wu et al. 2019 |
| Magnesium-stabilized multifunctional DNA nanoparticles | Tumor-targeted chemotherapy | Zhao et al. 2018 |
| Aptamer | Doxorubicin targeted drug delivery to prostate/breast cancer cells | Yaghoobi et al. 2022, Taghdisi et al. 2018, Abnous et al. 2018 |
| Oligonucleotide-conjugated multi-color carbon dots | Intracellular recognition of biological events | Srivastava et al. 2020 |
| Insulin-loaded nucleic acid–metal organic framework nanoparticle | Intracellular delivery | Wang et al. 2019 |
| Peptide nucleic acid-based zirconium coordination nanoparticles | Drug delivery | Öztürk et al. 2023 |
| Core–shell DNA-cholesterol nanoparticles | Lysosomolytic activity in African trypanosomes | Knieß et al. 2022 |
| Carbon nanoparticles with DNA aptamers | Colorimetric, fluorescent, and aggregation-based assays | Khramtsov et al. 2018 |
| Nano-biosensor amplified with DNA-templated silver nanoclusters | Detection of carcinoembryonic antigen | Wang et al. 2020, Xu et al. 2020 |
| Aptamer–gold nanoparticle | Detection of prostate-specific antigen | Li et al. 2021 |
| Dual-aptamers functionalized gold nanoprobe | Classification of different subtypes of breast cancer | Wu et al. 2022 |
| Cellular prion protein $PrP^C$ aptamer conjugated-gold nanoparticles | Targeted delivery of doxorubicin to colorectal cancer cells | Go et al. 2021 |
| Chitosan–gold nanoparticles coated with nucleolin aptamer | Delivery of doxorubicin | Khademi et al. 2020 |

## 8.5 FUTURE CONSIDERATIONS

The chapter highlights various nucleic acid-based nanocarriers in different areas of theranostics. The nanocarriers discussed include RNA-based nanocarriers, DNA-based nanocarriers, and hybrid nucleic acid nanocarriers. Due to the great programmability of nucleic acids, their sequence design can produce a wide variety of complex nanostructures. This characteristic has broad applications in material science, biomedicine, drug delivery, and diagnostics. Despite the substantial contributions of various investigators, clinical practice is still in its infancy, and much work remains.

## REFERENCES

Abnous K, Danesh NM, Ramezani M, Charbgoo F, Bahreyni A, Taghdisi SM. Targeted delivery of doxorubicin to cancer cells by a cruciform DNA nanostructure composed of AS1411 and FOXM1 aptamers. *Expert Opin Drug Deliv.* 2018;15(11):1045–1052. doi:10.1080/17425247.2018.1530656.

Aji Alex MR, Nehate C, Veeranarayanan S, Kumar DS, Kulshreshtha R, Koul V. Self assembled dual responsive micelles stabilized with protein for co-delivery of drug and siRNA in cancer therapy. *Biomaterials.* 2017;133:94–106. doi:10.1016/j.biomaterials.2017.04.022.

Alberti E, Zampakou M, Donghi D. Covalent and non-covalent binding of metal complexes to RNA. *J Inorg Biochem.* 2016 Oct;163:278–291. doi:10.1016/j.jinorgbio.2016.04.021.

Bagheri E, Alibolandi M, Abnous K, Taghdisi SM, Ramezani M. Targeted delivery and controlled release of doxorubicin to cancer cells by smart ATP-responsive Y-shaped DNA structure-capped mesoporous silica nanoparticles. *J Mater Chem B.* 2021;9(5):1351–1363. doi:10.1039/d0tb01960g.

Burgahn T, Garrecht R, Rabe KS, Niemeyer CM. Solid-phase synthesis and purification of protein-DNA origami nanostructures. *Chemistry.* 2019;25(14):3483–3488. doi:10.1002/chem.201805506.

Butowska K, Han X, Gong N, El-Mayta R, Haley RM, Xue L, Zhong W, Guo W, Wang K, Mitchell MJ. Doxorubicin-conjugated siRNA lipid nanoparticles for combination cancer therapy. *Acta Pharm Sin B.* 2023;13(4):1429–1437. doi:10.1016/j.apsb.2022.07.011.

Chen Q, Liu H, Lee W, Sun Y, Zhu D, Pei H, Fan C, Fan X. Self-assembled DNA tetrahedral optofluidic lasers with precise and tunable gain control. *Lab Chip.* 2013;13(17):3351–3354. doi:10.1039/c3lc50629k.

Duangrat R, Udomprasert A, Kangsamaksin T. Tetrahedral DNA nanostructures as drug delivery and bioimaging platforms in cancer therapy. *Cancer Sci.* 2020;111(9):3164–3173. doi:10.1111/cas.14548.

El Moukhtari SH, Garbayo E, Amundarain A, Pascual-Gil S, Carrasco-León A, Prosper F, Agirre X, Blanco-Prieto MJ. Lipid nanoparticles for siRNA delivery in cancer treatment. *J Control Release.* 2023;361:130–146. doi:10.1016/j.jconrel.2023.07.054.

Flamme M, McKenzie LK, Sarac I, Hollenstein M. Chemical methods for the modification of RNA. *Methods.* 2019;161:64–82. doi:10.1016/j.ymeth.2019.03.018.

Ge Z, Guo L, Wu G, Li J, Sun Y, Hou Y, Shi J, Song S, Wang L, Fan C, Lu H, Li Q. DNA Origami-enabled engineering of ligand-drug conjugates for targeted drug delivery. *Small.* 2020;16(16):e1904857. doi:10.1002/smll.201904857.

Go G, Lee CS, Yoon YM, Lim JH, Kim TH, Lee SH. $PrP^C$ aptamer conjugated-gold nanoparticles for targeted delivery of doxorubicin to colorectal cancer cells. *Int J Mol Sci.* 2021;22(4):1976. doi:10.3390/ijms22041976.

Grijalvo S, Alagia A, Jorge AF, Eritja R. Covalent strategies for targeting messenger and non-coding RNAs: an updated review on siRNA, miRNA and antimiR conjugates. *Genes (Basel).* 2018;9(2):74. doi:10.3390/genes9020074.

Gubu A, Zhang X, Lu A, Zhang B, Ma Y, Zhang G. Nucleic acid amphiphiles: synthesis, properties, and applications. *Mol Ther Nucleic Acids.* 2023;33:144–163. doi:10.1016/j.omtn.2023.05.022.

Guo S, Vieweger M, Zhang K, Yin H, Wang H, Li X, Li S, Hu S, Sparreboom A, Evers BM, Dong Y, Chiu W, Guo P. Ultra-thermostable RNA nanoparticles for solubilizing and high-yield loading of paclitaxel for breast cancer therapy. *Nat Commun*. 2020;11(1):972. doi:10.1038/s41467-020-14780-5.

Han X, Gong N, Xue L, Billingsley MM, El-Mayta R, Shepherd SJ, Alameh MG, Weissman D, Mitchell MJ. Ligand-tethered lipid nanoparticles for targeted RNA delivery to treat liver fibrosis. *Nat Commun*. 2023;14(1):75. doi:10.1038/s41467-022-35637-z.

Huo S, Gong N, Jiang Y, Chen F, Guo H, Gan Y, Wang Z, Herrmann A, Liang XJ. Gold-DNA nanosunflowers for efficient gene silencing with controllable transformation. *Sci Adv*. 2019;5(10):eaaw6264. doi:10.1126/sciadv.aaw6264.

Ijäs H, Shen B, Heuer-Jungemann A, Keller A, Kostiainen MA, Liedl T, Ihalainen JA, Linko V. Unraveling the interaction between doxorubicin and DNA origami nanostructures for customizable chemotherapeutic drug release. *Nucleic Acids Res*. 2021;49(6):3048–3062. doi:10.1093/nar/gkab097.

Kang H, Yang Y, Wei B. Synthetic molecular switches driven by DNA-modifying enzymes. *Nat Commun*. 2024;15(1):3781. doi:10.1038/s41467-024-47742-2.

Kedmi R, Veiga N, Ramishetti S, Goldsmith M, Rosenblum D, Dammes N, Hazan-Halevy I, Nahary L, Leviatan-Ben-Arye S, Harlev M, Behlke M, Benhar I, Lieberman J, Peer D. A modular platform for targeted RNAi therapeutics. *Nat Nanotechnol*. 2018;13(3):214–219. doi:10.1038/s41565-017-0043-5.

Khademi Z, Lavaee P, Ramezani M, Alibolandi M, Abnous K, Taghdisi SM. Co-delivery of doxorubicin and aptamer against Forkhead box M1 using chitosan-gold nanoparticles coated with nucleolin aptamer for synergistic treatment of cancer cells. *Carbohydr Polym*. 2020;248:116735. doi:10.1016/j.carbpol.2020.116735.

Khramtsov P, Kropaneva M, Kalashnikova T, Bochkova M, Timganova V, Zamorina S, Rayev M. Highly stable conjugates of carbon nanoparticles with DNA aptamers. *Langmuir*. 2018;34(35):10321–10332. doi:10.1021/acs.langmuir.8b01255.

Kim E, Zwi-Dantsis L, Reznikov N, Hansel CS, Agarwal S, Stevens MM. One-pot synthesis of multiple protein-encapsulated DNA flowers and their application in intracellular protein delivery. *Adv Mater*. 2017a;29(26):10.1002/adma.201701086. doi:10.1002/adma.201701086.

Kim H, Jeong J, Kim D, Kwak G, Kim SH, Lee JB. Bubbled RNA-based cargo for boosting RNA interference. *Adv Sci (Weinh)*. 2017b;4(8):1600523. doi:10.1002/advs.201600523.

Kim M, Lee JS, Kim W, Lee JH, Jun BH, Kim KS, Kim DE. Aptamer-conjugated nano-liposome for immunogenic chemotherapy with reversal of immunosuppression. *J Control Release*. 2022;348:893–910. doi:10.1016/j.jconrel.2022.06.039.

Kim T, Han HS, Yang K, Kim YM, Nam K, Park KH, Choi SY, Park HW, Choi KY, Roh YH. Nanoengineered polymeric RNA nanoparticles for controlled biodistribution and efficient targeted cancer therapy. *ACS Nano*. 2024;18(11):7972–7988. doi:10.1021/acsnano.3c10732.

Knieß R, Leeder WM, Reißig P, Geyer FK, Göringer HU. Core-shell DNA-cholesterol nanoparticles exert lysosomolytic activity in African trypanosomes. *Chembiochem*. 2022;23(20):e202200410. doi:10.1002/cbic.202200410.

Kong L, Wang L, Shi Y, Peng L, Liang X, Wang G, Mitomo H, Takarada T, Ijiro K, Maeda M. DNA-functionalized silver nanoparticles in an alcoholic solvent for environment-dictated multimodal actuation. *ACS Appl Nano Mater*. 2022;5(8): 10321–10330.

Kwak E, Kim T, Yang K, Kim YM, Han HS, Park KH, Choi KY, Roh YH. Surface-functionalized polymeric siRNA nanoparticles for tunable targeting and intracellular delivery to hematologic cancer cells. *Biomacromolecules*. 2022;23(6):2255–2263. doi:10.1021/acs.biomac.1c01497.

Li M, Zhang W, Zhang Y. Aptamer–gold nanoparticle-signal probe bioconjugates amplify electrochemical signal for the detection of prostate specific antigen. *Anal Methods*. 2021;13:4150–4156. doi:10.1039/D1AY01175H.

Li X, Wang S, Gao Q, Li N, Dong S, Gao Y, Wang Z, Zhang B, He X. MiRNA-520a-3p combined with folic acid conjugated Fe2O3@PDA multifunctional nanoagents for MR imagine and antitumor gene-photothermal therapy. *Nanotechnology*. 2023;34(37). doi:10.1088/1361-6528/acd5d9.

Liang M, Li N, Liu F, Zeng N, Yu C, Li S. Apurinic/apyrimidinic endonuclease triggered doxorubicin-releasing DNA nanoprism for target therapy. *Cell Cycle*. 2022;21(24):2627–2634. doi:10.1080/15384101.2022.2108567.

Liu B, Duan H, Liu Z, Liu Y, Chu H. DNA-functionalized metal or metal-containing nanoparticles for biological applications. *Dalton Trans*. 2024;53(3):839–850. doi:10.1039/d3dt03614f.

Mohammad SN, Choi YS, Chung JY, Cedrone E, Neun BW, Dobrovolskaia MA, Yang X, Guo W, Chew YC, Kim J, Baek S, Kim IS, Fruman DA, Kwon YJ. Nanocomplexes of doxorubicin and DNA fragments for efficient and safe cancer chemotherapy. *J Control Release*. 2023;354:91–108. doi:10.1016/j.jconrel.2022.12.048.

Obuobi S, Škalko-Basnet N. Nucleic acid hybrids as advanced antibacterial nanocarriers. *Pharmaceutics*. 2020;12(7):643. doi:10.3390/pharmaceutics12070643.

Ohmann A, Göpfrich K, Joshi H, Thompson RF, Sobota D, Ranson NA, Aksimentiev A, Keyser UF. Controlling aggregation of cholesterol-modified DNA nanostructures. *Nucleic Acids Res*. 2019;47(21):11441–11451. doi:10.1093/nar/gkz914.

Öztürk Ö, Lessl AL, Höhn M, Wuttke S, Nielsen PE, Wagner E, Lächelt U. Peptide nucleic acid-zirconium coordination nanoparticles. *Sci Rep*. 2023;13(1):14222. doi:10.1038/s41598-023-40916-w.

Passos Gibson V, Tahiri H, Gilbert C, Yang C, Phan QT, Banquy X, Hardy P. Modulating the nature of ionizable lipids and number of layers in hyaluronan-decorated lipid nanoparticles for in vitro delivery of RNAi. *Pharmaceutics*. 2024;16(4):563. doi:10.3390/pharmaceutics16040563.

Picchetti P, Volpi S, Rossetti M, Dore MD, Trinh T, Biedermann F, Neri M, Bertucci A, Porchetta A, Corradini R, Sleiman H, De Cola L. Responsive nucleic acid-based organosilica nanoparticles. *J Am Chem Soc*. 2023;145(42):22896–22902. doi:10.1021/jacs.3c00393.

Roberts TC, Langer R, Wood MJA. Advances in oligonucleotide drug delivery. *Nat Rev Drug Discov*. 2020;19(10):673–694. doi:10.1038/s41573-020-0075-7.

Sa-Ardyen P, Vologodskii AV, Seeman NC. The flexibility of DNA double crossover molecules. *Biophys J*. 2003 Jun;84(6):3829–3837. doi: 10.1016/S0006-3495(03)75110-8. PMID: 12770888; PMCID: PMC1302964.

Sarett SM, Werfel TA, Lee L, Jackson MA, Kilchrist KV, Brantley-Sieders D, Duvall CL. Lipophilic siRNA targets albumin in situ and promotes bioavailability, tumor penetration, and carrier-free gene silencing. *Proc Natl Acad Sci U S A*. 2017;114(32):E6490–E6497. doi:10.1073/pnas.1621240114.

Seeman NC. Nucleic acid junctions and lattices. *J Theor Biol*. 1982;99(2):237–47. doi:10.1016/0022-5193(82)90002-9.

Seeman NC, Sleiman HF. DNA nanotechnology. *Nat Rev Mater*. 2017;3:17068.

Srivastava I, Misra SK, Bangru S, Boateng KA, Soares JANT, Schwartz-Duval AS, Kalsotra A, Pan D. Complementary oligonucleotide conjugated multicolor carbon dots for intracellular recognition of biological events. *ACS Appl Mater Interfaces*. 2020;12(14):16137–16149. doi:10.1021/acsami.0c02463.

Sun S, Cheraga N, Jiang HN, Xiao QR, Gao PC, Wang Y, Wei YY, Wang XW, Jiang Y. Bioinspired DNA nanococklebers for targeted delivery of doxorubicin. *Colloids Surf B Biointerfaces*. 2020;186:110733. doi:10.1016/j.colsurfb.2019.110733.

Taghdisi SM, Danesh NM, Ramezani M, Yazdian-Robati R, Abnous K. A novel AS1411 aptamer-based three-way junction pocket DNA nanostructure loaded with doxorubicin for targeting cancer cells in vitro and in vivo. *Mol Pharm*. 2018;15(5):1972–1978. doi:10.1021/acs.molpharmaceut.8b00124.

Wang D, Liu Q, Wu D, He B, Li J, Mao C, Wang G, Qian H. Isothermal self-assembly of spermidine-DNA nanostructure complex as a functional platform for cancer therapy. *ACS Appl Mater Interfaces*. 2018;10(18):15504–15516. doi:10.1021/acsami.8b03464. Erratum in: *ACS Appl Mater Interfaces*. 2020;12(26):30019. doi:10.1021/acsami.0c10222.

Wang H, Ellipilli S, Lee WJ, Li X, Vieweger M, Ho YS, Guo P. Multivalent rubber-like RNA nanoparticles for targeted co-delivery of paclitaxel and MiRNA to silence the drug efflux transporter and liver cancer drug resistance. *J Control Release*. 2021;330:173–184. doi:10.1016/j.jconrel.2020.12.007.

Wang J, Guo X, Liu R, Guo J, Zhang Y, Zhang W, Sang S. Detection of carcinoembryonic antigen using a magnetoelastic nano-biosensor amplified with DNA-templated silver nanoclusters. *Nanotechnology*. 2020;31(1):015501. doi:10.1088/1361-6528/ab4506.

Wang J, Li Y, Nie G. Multifunctional biomolecule nanostructures for cancer therapy. *Nat Rev Mater*. 2021;6(9):766–783. doi:10.1038/s41578-021-00315-x.

Wang J, Zhang T, Li X, Wu W, Xu H, Xu XM, Zhang T. DNA nanobarrel-based drug delivery for paclitaxel and doxorubicin. *Chembiochem*. 2023;24(19):e202300424. doi:10.1002/cbic.202300424.

Wang S, Chen Y, Wang S, Li P, Mirkin CA, Farha OK. DNA-functionalized metal-organic framework nanoparticles for intracellular delivery of proteins. *J Am Chem Soc*. 2019;141(6):2215–2219. doi:10.1021/jacs.8b12705.

Whitehouse WL, Noble JE, Ryadnov MG, Howorka S. Cholesterol anchors enable efficient binding and intracellular uptake of DNA nanostructures. *Bioconjug Chem*. 2019;30(7):1836–1844. doi:10.1021/acs.bioconjchem.9b00036.

Wu T, Liu J, Liu M, Liu S, Zhao S, Tian R, Wei D, Liu Y, Zhao Y, Xiao H, Ding B. A nanobody-conjugated DNA nanoplatform for targeted platinum-drug delivery. *Angew Chem Int Ed Engl*. 2019;58(40):14224–14228. doi:10.1002/anie.201909345.

Wu Y, Feng Y, Li X. Classification of breast cancer by a gold nanoparticle based multicolor fluorescent aptasensor. *J Colloid Interface Sci*. 2022;611:287–293. doi:10.1016/j.jcis.2021.12.039.

Xie X, Shao X, Ma W, Zhao D, Shi S, Li Q, Lin Y. Overcoming drug-resistant lung cancer by paclitaxel loaded tetrahedral DNA nanostructures. *Nanoscale*. 2018;10(12):5457–5465. doi:10.1039/c7nr09692e.

Xu C, Li H, Zhang K, Binzel DW, Yin H, Chiu W, Guo P. Photo-controlled release of paclitaxel and model drugs from RNA pyramids. *Nano Res*. 2019;12(1):41–48. doi:10.1007/s12274-018-2174-x.

Xu X, Ji J, Chen P, Wu J, Jin Y, Zhang L, Du S. Salt-induced gold nanoparticles aggregation lights up fluorescence of DNA-silver nanoclusters to monitor dual cancer markers carcinoembryonic antigen and carbohydrate antigen 125. *Anal Chim Acta*. 2020;1125:41–49. doi:10.1016/j.aca.2020.05.027.

Yaghoobi E, Zavvar T, Ramezani M, Alibolandi M, Rahimzadeh Oskuei S, Zahiri M, Alinezhad Nameghi M, Abnous K, Taghdisi SM. A multi-storey DNA nanostructure containing doxorubicin and AS1411 aptamer for targeting breast cancer cells. *J Drug Target*. 2022;30(10):1106–1112. doi:10.1080/1061186X.2022.2094387.

Yang L, Sun H, Liu Y, Hou W, Yang Y, Cai R, Cui C, Zhang P, Pan X, Li X, Li L, Sumerlin BS, Tan W. Self-assembled aptamer-grafted hyperbranched polymer nanocarrier for targeted and photoresponsive drug delivery. *Angew Chem Int Ed Engl.* 2018;57(52):17048–17052. doi:10.1002/anie.201809753.

Ye J, Cao Y, Lu X, Wu F, Liu N, Dong Y, Shi Q. Self-assembly of DNA-organic hybrid amphiphiles by frame-guided assembly strategies. *Giant.* 2022; 11: 100113. doi:10.1016/j.giant.2022.100113.

Zhao H, Yuan X, Yu J, Huang Y, Shao C, Xiao F, Lin L, Li Y, Tian L. Magnesium-stabilized multifunctional DNA nanoparticles for tumor-targeted and pH-responsive drug delivery. *ACS Appl Mater Interfaces.* 2018;10(18):15418–15427. doi:10.1021/acsami.8b01932.

Zhao Y, Dai X, Wang F, Zhang X, Fan C, Liu X. Nanofabrication based on DNA nanotechnology. *Nanotoday.* 2019;26:123–148.

Zhou X, Li H, He M, Yin X, Yao D, Xiao S, Liang H. Photoresponsive spherical nucleic acid: spatiotemporal control of the assembly circuit and intracellular microRNA release. *Chem Commun (Camb).* 2017;54(1):106–109. doi:10.1039/c7cc07932j.

# 9 Protein-Based Nanocarriers

## *Protein Cages, Self-Assembling Peptides, Fusion Proteins*

*Nishant B. Chopade, Prashant K. Deshmukh, and Mahesh P. More*

## 9.1 INTRODUCTION

Delivery of therapeutic agents is a challenging field and course of interest from a material background. Targeted delivery to different organs, systemic delivery, oral, topical, nasal, ocular, and so many interests of anatomical sites were investigated. The oral route of administration was the preferred route over intravenous, considering side effects and other limitations like instability of therapeutic agents *in vivo*. Transportation of any pharmaceutical compound is governed by the carrier system, which helps to deliver it at the appropriate site and maintains biological characteristics. The carrier designed for the delivery of a therapeutic agent may articulate the physical or chemical characteristics to enhance its efficacy and delivery. Many delivery systems were commercialized starting from lipid-based nanoparticles for the delivery of RNA and liposomes for the delivery of doxorubicin. Various nanomaterials like chitosan, sodium alginate, and gelatin from natural origin while Eudragit, polylactic acid, polylactide-co-glycolide (PLGA), polycaprolactone (PCL), etc. from synthetic sources were demonstrated for the delivery of therapeutic agents [1–3]. From inorganic sources, inorganic materials like mesoporous alumina, silica, and metal-based nanomaterials like silver and gold were demonstrated for the delivery. The carbon-based nanomaterial acts as an excellent material for the delivery of a variety of therapeutic agents. The unique structural characteristics of carbon-based nanomaterials like graphene, carbon nanotubes, diamonds, and fullerenes can deliver the therapeutic agent at the targeted site. Toxicity may be a limitation to use as a carrier for the delivery of therapeutic agents. These nanomaterials can cross biological barriers and protect the therapeutic agents from exogenous and endogenous degradation pathways [4–6].

Naturally derived materials are preferred as carriers for the delivery of therapeutic agents. Mostly natural proteins are focused carriers for the delivery or modified version of synthetics peptides derived from proteins that have similar characteristics. The major precursors used for the designing proteins as carriers are amino acids, which are formed by self-assembly or aggregation through polymerization [7]. The building blocks of proteins are formed by the condensation polymerization process in the presence of monomers (amino acids). The process of protein synthesis is naturally accelerated in living cells which are mediated in the presence of RNA and ribosomes known as translation. The variation in the sequence of amino acids and changes in the length and dissimilarities make diverse groups of proteins. The proteins perform multiple functions like enzymes, bioregulators, biotransporters, carriers for natural components, hormones, etc., based on dissimilarities. Proteins also have therapeutic potential or show synergistic activity with other therapeutic agents [8].

The proteins can be obtained from animals, plants, microorganisms, fungi, etc. The protein nanomaterial is biodegradable, non-toxic, biocompatible, and metabolized easily inside the body with easy elimination [9]. From animal sources, protein-based nanomaterials like gelatin, albumin, casein, elastin, whey, and silk are extracted, whereas zein, gliadin, and soybean are derived from plant sources. The properties of protein nanocarriers and their application for the delivery of chemotherapeutic agents are highlighted in Table 9.1. The protein cages are new, diverse types of proteins considered to be attractive candidates for delivery of drugs.

### 9.1.1 Computational Studies for the Development of Protein Self-Assembly

Proteins have multiple functions in a cellular environment. Computational design helps to understand the symmetric protein complexes and provides intriguing knowledge on the shapes and solvent interaction besides the cellular environment. The subunits obtained from the larger proteins can be studied using a computational environment to create an individual set of functionalized protein complexes. Similarly, a short sequence of peptides has also been defined from the self-assembled large peptides. The design of nanocages has emerged recently due to its unique characteristics for the delivery of enzymes, drugs, and catalysts for biochemical reactions. The solubility, pore dimensions, solute diffusion, and solvent interaction are important characteristics considered for computation design space. The symmetric structures were designed in the computational environment by considering the natural building blocks of proteins. Tunable sizes of cages were limited by subunit arrangement with respect to natural protein blocks.



DOI: 10.1201/9781003473183-9

## Table 9.1 Summary of protein nanocarrier and its properties for the delivery of therapeutic agents

| Nanoparticle System | Physicochemical Properties | Therapeutic Agent |
|---|---|---|
| HSA | Abundant and negative plasma protein; high affinity to particular receptors on cells; low opsonization | Paclitaxel, aspirin, doxorubicin, ganciclovir, azidothymidine, pilocarpine, obidoxime, cvclodextrins. insulin |
| Gelatin | The main constructive component of connective tissues, skin and bones; production by acid and alkaline hydrolysis process of collagen | Hydrocortisone, 5-fluorouracil, paclitaxel, tizanidine, gatifloxacin, fluconazole, doxorubicin, amphotericin B, didanosine, isoniazide |
| Fibrin | Generated during blood coagulation cascade; as a scaffold to cell growth; tissue regeneration and immobilization of growth factors and drugs | Tacrolimus, ciprofloxacin, fluconazole |
| Elastin | Undergo an inverse phase transition at a certain temperature | Rapamycin, BMP-2, BMP-14, dexamethasone phosphate |
| Collagen | The major component of the extracellular matrix; hydrogel formation without the addition of a cross-linker | Adriamycin, dactinomycin, estradiol |
| Silk | Produced by certain insects and spiders; as a support to enzyme immobilization; easy processing to different morphologies | Insulin, paclitaxel, methotrexate, doxorubicin |
| Casein | The main protein of milk; gel swelling; considerable stability to temperature; easy degradation by stomach protease and suitable to treat stomach diseases | Celecoxib, cisplatin, vitamin D, mitoxantrone, flutamide, vinblastine. Alfuzosin hydrochloride |
| Whey | Gel-forming and emulsifying properties | Caffeine, doxorubicin, chlorothiazide |
| Giladin | Extracted from wheat gluten; low aqueous solubility; a mucoadhesive protein | Cyclophosphamide, vitamin E, alltrans-retinoic acid, clarithromycin, omeprazole, carbazole |
| Zein | Extracted from maize seeds; **G**IBBERELLIC ACID INSENSITIVE (GAI), **R**EPRESSOR OF GAI (RGA), and **S**CARECROW (SCR) (GRAS) protein; hydrophobic and insoluble in aqueous solution; slow enzymatic digestion | Metformin, daidzin, vitamin D3 |
| Soy protein | The most dominant storage of plant proteins isolated from soybean | Vitamin B12 |

*Source:* Reprint with permission from ref. [9].
BMP, bone morphogenetic protein.

The computational design can help to identify the modifiable domain within the existing protein structure and help to re-engineer them to produce novel nanostructures. The single protein fold can be responsible for the representation of the conformation space of peptides, whereas the protein fold can augment new functionality from the design space [10]. Nature produces highly symmetric larger proteins and defines the assembly structure similar to viral proteins (VPs), actin filaments, clathrin scaffolds, and S-layer proteins. Viral capsids are one of the best examples to understand and design new protein structures, which are stable, highly organized nanostructures. The protein capsids consist of genomes and building blocks which are very limited in number but show stable nanostructure. The building blocks of Epstein-Barr virus were encapsulated with ferritin or encapsulation-like self-assembled peptides, and surfaces were anchored with gp350 ectodomain [11]. Naturally, oligomerization of domain via fusion creates new sets of nanostructures, for example, polyhedra. The tetrahedral cages of proteins are formed by fusion between di- and trimerization domains. The major requirement in designing the tetrahedral structure was the domains must consist of geometric restrictions. The PDB structures initiated and ended with the α-helix of di- and trimerizing proteins acting as helical linkers during nanostructure formation. The fusing takes place between dimer to trimer and dimer to dimer to get the cages and filaments. Initially, the process forms a heterogeneous structure based on its shape and size. Based on the position and nature of the linker, it provides helicity and limits flexibility by locking the subunits at specific positions. The resulting formation of tetrahedral cages was monodisperse in nature [12].

Protein interface design (PID) is an alternative approach based on protein–protein interaction (PPI) to generate a self-assembled nanostructure. The advantage over the natural oligomerization process is that this design interface can use alternative synthetic domain subunits instead of natural ones. The PID produces polydisperse nanostructures, which may be due to oligomerizing domains reducing conformational flexibility [13]. The PID utilizes a principle similar to the de novo synthesis protocol, which has greater selectivity between PPI and produces novel nanostructures using computational design. The computational design generates as much as possible conformations and is useful to interpret based on the hydrogen bonding and solvation energy. The scoring based on hydrogen bonding and solvation energy is considered a better quantification unit in a computational environment. The scoring function works on empirical and physical attributes of the protein molecules simulated in a computational environment. The conformational space utilizes many algorithms such as Monte Carlo simulation, linear programming, dead-end elimination, etc. Other software packages such as Rosetta, orbit are used for self-assembled protein nanostructures [10].

### 9.1.2 Protein and Peptide Self-Assembly Mechanism

The creation of an organized structure from the disordered system to present a unique organizational behavior and function is referred to as a self-assembled structure [14]. The complex nanostructures formed by self-assembly structures are considered an essential component of the life cycle. The hierarchical structure formation is a group of events formed in the living cells. The hydrogen-bonded interaction creates complex nanostructures that are secondary in nature [15, 16]. The self-assembled structure of proteins can be modified using both chemical and genetic methods to create a hollow interior cavity. The hollow interior cavity can be used to load therapeutic cargo. The external surface can be engineered to provide binding affinity for ligands for target-based therapy. The internal spaces available in the assembled structure are useful for creating channels for solvent interaction and the release of payload molecules [17].

The self-assembly is a bottom-up approach to create nanostructured carriers for the delivery of therapeutic agents. Spontaneously organizing molecules with the assistance of free energy is a practical approach to molecular self-assembly formation. The various energy levels were accounted for, such as hydrogen bonding, van der Waals forces for attraction, aromatic interaction, etc., to measure the self-assembly [18]. The intra- and intermolecular forces between the precursor molecules affect the arrangement of nanostructures. For the fabrication of carriers for pharmaceuticals, synthetic polypeptides derived from natural amino acids act as monomers for the self-assembly structure. The synthetic polypeptides are biocompatible, biodegradable, and have high physicochemical stability in biological environment, making them suitable precursors for self-assembly formation. The choice of starting material in self-assembly fabrication changes the characteristics of the property of nanostructures. Either self-assembled peptide acts as a carrier or anchor for active targeting to cell-specific receptors. Simultaneously, the peptide shows both characteristic features as it acts as a therapeutic agent or can be used as a delivery carrier [19].

The precursor amino acids are a primary structure or monomer in the design of self-assembly. It is also a site for modification to make specific structures for end applications. Based on the surface charge, hydrophilicity/hydrophobicity, size, shape, polarity, etc., the characteristics of side chains differed. The change in the sequence of the amino acids, the number of amino acids, and the type of amino acids may change the self-assembled nanostructure. These changes were reflected in secondary structure and physicochemical properties. Figure 9.1 explores the different hydrophobic amino acids, their architecture in peptide self-assembly, and possible applications [20]. The exploration of self-assembly was classified based on the physicochemical properties of the precursor used. Figure 9.1 highlights the categorization of hydrophilic and hydrophobic amino acids with their surface charge behavior encased in the structure formation. The hydrophobic amino acids are further categorized into aliphatic and aromatic subgroups for understanding interaction. The general hydrophobic environment is represented by aliphatic amino acids, while aromatic represents $\pi$–$\pi$ interaction during the conjugation. These interactions are revamped during folding while uncharged species participate in hydrogen bonding. The charged residues shift their surface charges based on the pKa and pH of the medium to either positive or negative charge and interact to form a charged species. The interaction forms a beta-sheet structure that is specifically useful for biological applications. The alpha-helices are used in sensing and drug delivery and act as support for cells during growth/scaffolding. The peptide amphiphiles and aromatic-directed assembly structure are used specifically in non-biologicals like nanoreactors or templating [20].

The mechanism of peptide and protein self-assembly involves several key processes. Firstly, the primary structure of peptides, which consists of amino acids linked by peptide bonds, plays a crucial role in self-assembly properties. Specific amino acid sequences can either promote or inhibit this process. Similarly,

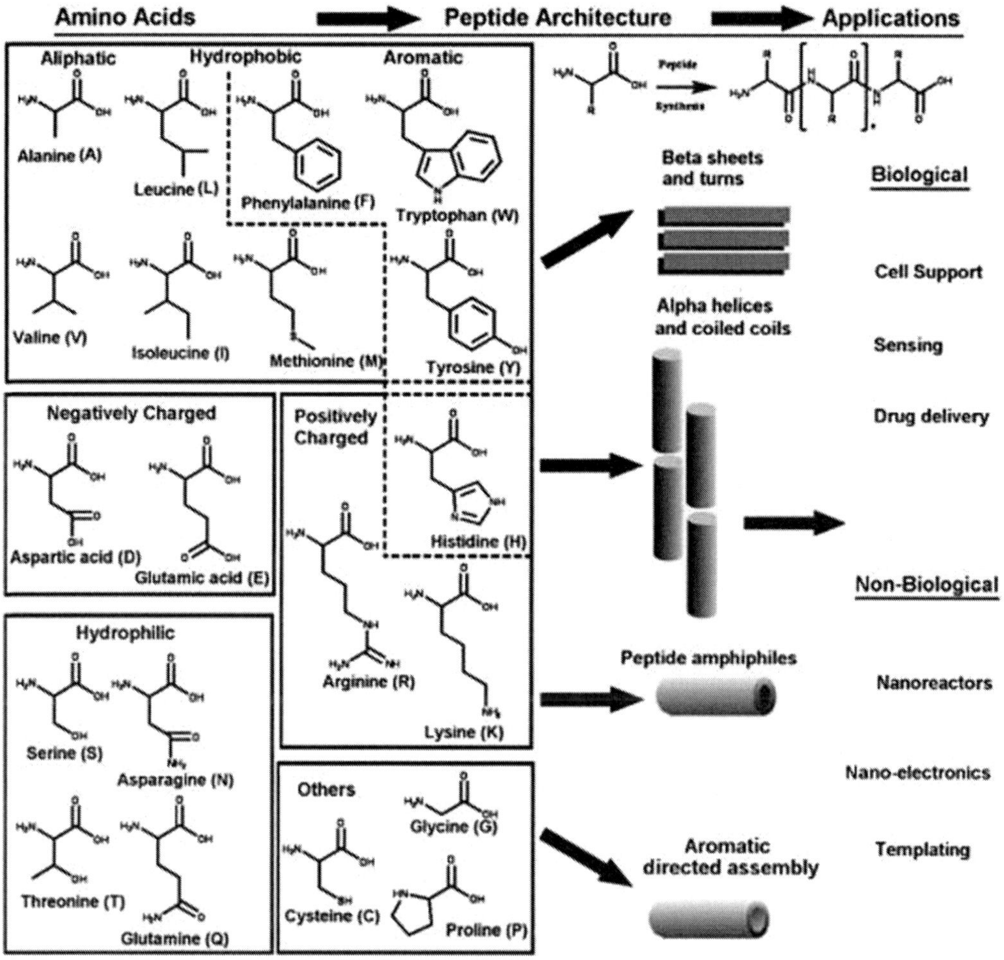


**Figure 9.1** Different hydrophobic amino acids used for peptide designing and four main classes of possible architecture with applications.

*Source:* **Adapted with permission from ref. [20].**

proteins, also composed of amino acids linked through peptide bonds, rely on interactions between specific residues to drive their self-assembly.

Secondary structures, such as alpha-helices or beta-sheets, often arise in peptides and proteins from hydrogen bonding interactions among backbone amide groups. These structures enhance stability and promote self-assembly by providing a framework that influences the overall physical attributes of the proteins. Hydrophobic interactions, involving hydrophobic amino acid residues, drive peptides to shield their hydrophobic regions from water, leading to the formation of stable aggregates. In proteins, nonpolar residues cluster together to minimize water contact, promoting stable structure formation. Additionally, charge-shielding effects can increase hydrophobic contacts during salt-triggered self-assembly, with aromatic residues contributing through $\pi$–$\pi$ or hydrophobic interactions, each exhibiting distinct patterns.

Hydrogen bonds, with bond strengths of 10–40 kJ/mol, are crucial in peptide self-assembly. These interactions stabilize secondary structures and contribute to forming specific nanostructures. In peptides, hydrogen bonds are selective and strong, enabling the formation of stable self-assembled structures. van der Waals forces, though weak, also contribute to the stability of self-assembled peptide and protein structures by facilitating close packing through slight changes in electron distribution.

Electrostatic interactions between charged amino acid residues can attract or repel, influencing self-assembly. These interactions, although weak and non-directional, are significantly affected by pH and ionic strength. Adjusting these conditions can regulate peptide self-assembly, as demonstrated by the formation of nanostructures through electrostatic interactions for drug delivery systems. Aromatic amino acids in peptides can undergo π–π stacking, where aromatic side chains align and stack, contributing to the stability and organization of self-assembled structures. This interaction is resilient in water and prominent in organic solvents, driving directional growth.

Environmental conditions, such as pH and ionic strength, significantly affect peptide self-assembly by altering electrostatic interactions and hydrogen bonding. Additionally, self-assembling peptides can respond to external stimuli like temperature, light, or specific ions, which can trigger conformational changes, promoting or disrupting self-assembly. Proteins containing cysteine residues can form disulfide bonds that stabilize their structures by linking different parts of the protein, maintaining structural integrity. Electrostatic interactions between charged amino acid residues also influence protein self-assembly by attracting or repelling, affecting the overall structure. Many proteins consist of multiple subunits that form a functional quaternary structure. The arrangement and interactions of these subunits contribute to the stability and function of the protein assembly. In living organisms, chaperone proteins assist in the correct folding and assembly of other proteins, preventing misfolding and aggregation, and ensuring proteins adopt their native, functional structures. Changes in environmental conditions, such as pH and temperature, can influence protein self-assembly, leading to denaturation or aggregation of proteins [21–24].

### 9.1.3 Structure, Properties, and Types of Self-Assembled Proteins and Peptides

The self-assembled nature of peptides was formed by amino acid residues and converted to large or short peptides based on the number of amino acids. The short peptides consist of less than six amino acid residues and can be synthesized easily, economically, and highly bioactively [25]. The self-assembled peptides can be converted to any shape based on the arrangement of amino acid residues, like twisted nanofibers, nanotubes, nanowires, nanovesicles, nanotapes, nanocrystals, necklaces, nanofibers, etc. The self-assembled structure forms secondary, tertiary, as well as quaternary structures based on the requirement. The self-assembled structures respond to pH, temperature, ionic strength, etc., converted to desirable shapes.

Peptides can load a wide range of hydrophilic and hydrophobic chemical entities and promote smart as well as targeted drug delivery. The self-assembly reaction does not require any toxic cross-linker and is reversible in nature once it reaches the target location and releases the payload. The release mechanism is stimuli sensitive and can be triggered using pH, enzymes, ions, temperature, or external triggers like magnetic effect, heat, laser, etc.

## 9.2 PROTEIN NANOCAGES

Protein nanocages (PNCs) have attracted significant interest because of their diverse biological roles. These include viral capsids that store and transport nucleic acids, ferritin that helps in iron mineralization, DNA-binding proteins from starved cells (Dps) that protect DNA, heat shock proteins (HSPs) that prevent protein aggregation, bacterial microcompartments that encapsulate enzymes, and clathrin that is involved in endocytosis. PNCs are often constructed using monomeric subunits to create strong nanostructures that can be likened to molecular Lego sets. Cage-like proteins possess three distinct structural characteristics: Firstly, all of them possess remarkably symmetrical structures and interfaces, resulting in beneficial interaction with biomolecules. Viral capsids have the role of encapsulating and safeguarding the viral genome. The assembly of viral capsids mostly relies on a network of weak, non-covalent contacts, metal coordination facilitated by metal ions or metallocofactors, and occasionally covalent connections that occur either between or within subunits. Furthermore, PNC architectures possess distinct exterior and interior surfaces, as well as interfaces between subunits and within subunits. Additionally, the PNCs exhibit exceptional durability and can endure considerable alterations at the genetic or chemical level [26, 27].

The development of PNC with various functions motivates us to modify and manipulate them for varied objectives. One prominent aspect of catalysis is a key focus on the natural usage of the PNCs. The nanostructure is used to create artificial nanoreactors that can carry out certain catalytic activities. Efficient catalytic biomaterials have been constructed using these nanoreactors at various size scales. PNCs commonly contain guest molecules within their interiors for purposes such as storage, sequestration, protection, and repurposing. Utilizing these nanocages as nanocarriers has proven to be an effective method for incorporating a diverse range of non-native cargos. The limited volume within the cages serves as a crucial limitation in determining the capacity for packing and the size of cargo that can be packaged, such as the "head-full" genome packaging mechanism observed in certain bacteriophages like P22. A similar characteristic was also demonstrated by polymers to restrict the growth of synthetic nanomaterials. Moreover, protein cages undergo self-assembly to create a repeated arrangement of a restricted set of constituent units, which plays a crucial role in their biological recognition. By employing this framework in conjunction with the modification of protein subunits, it has become possible to utilize these structures for surface presentation, such as for the purpose of directing peptides and immunogenic epitopes. The utilization of PNC architecture has emerged as a robust framework for biomimetic design and engineering. The modifications were tested with physical, chemical, or genetic modification, which are beneficial in a variety of applications [28].

Viruses are resilient entities capable of enduring environmental stresses, yet they possess the sensitivity to perceive and respond to alterations in cellular signals. Viruses have served as a source of inspiration for creating a variety of self-assembling PNCs derived from natural sources. Viral capsids can be modified using a bioengineering approach to modulate their functional characteristics. For instance, bacteriophages work as peptide displays to help serve as a template for nanofabrication. The filamentous phage formed after the nanofabrication process anchored on the surfaces and acted as immunogens [29, 30].

### 9.2.1 Types of PNCs

The PNCs can be categorized into two groups: virus-like particles (VLPs) and non-viral protein nanocages (NVPNs). VLPs, which make up the primary category of PNCs, exhibit a wide range of variations in terms of their structures and dimensions. Although VLPs share a similar structure to viruses, they are non-infectious due to the absence of viral genes. NVPNs, however, have no connection to viral particles. NVPNs are created through the spontaneous arrangement of protein monomers. The outcome of the process is heavily influenced by the surface properties and interaction with adjacent subunits [31].

#### *9.2.1.1 Non-Viral PNCs*

Proteins must be able to self-assemble into shapes resembling similar capsules useful in non-viral protein nanostructures. NVPNs are formed by the assembly of a number of protein monomers, resulting in a complex structure that is precisely defined, symmetric, and homogeneous in nature. The non-viral particles have a size of 10–100 nm and are found in various prokaryotes and eukaryotes. The structural features are crucial for vital cellular functions, such as mineral storage, iron homeostasis regulation, chaperone activity to safeguard other proteins during high temperatures, DNA protection against oxidative harm, nucleic acid transportation, and catalytic assistance in enzymatic reactions [31, 32]. The four major types of NVPNs are described in the following section.

##### *9.2.1.1.1 Ferritin*

Ferritin is a ubiquitous protein found in living organisms, serving a fundamental role in iron metabolism. Composed of 24 subunits that self-assemble into a hollow spherical nanocage, ferritin acts as an intracellular iron storehouse [33]. Ferritins are widely distributed proteins present in all living species, including yeast. The physiological function of ferritin is associated with the metabolism of nonheme iron and has the ability to store up to 4500 iron ions per molecule. The term "ferritin" is derived from the Latin word "ferratus," which translates to "covered with iron." From a physiological perspective, it is crucial to encapsulate Fe(III) within a protein shell to prevent excessive iron accumulation in the cytoplasm. High levels of iron can stimulate the formation of reactive oxygen species (ROS), which in turn can activate growth factor signaling pathways and induce DNA alterations that contribute to the development of tumors [34]. The structure of ferritin, internal dimensions, interface, and helix conformations are represented in Figure 9.2.

Scientists are exploring applications ranging from magnetic resonance imaging (MRI) contrast agents to drug delivery vehicles, leveraging ferritin's natural ability to encapsulate therapeutic agents. Proteins belonging to the ferritin family exhibit remarkable thermodynamic stability, enabling them to withstand elevated temperatures (up to 85°C) and extreme pH conditions (8.5–9). This robustness allows these proteins to remain structurally unchanged for extended periods, even in challenging ionic environments. Moreover, their biodegradability and non-immunogenic nature are noteworthy characteristics, rendering them suitable for a variety of applications in medical diagnostics. Moreover, ferritin nanocages have been investigated as nanoreactors for catalytic reactions, showcasing the protein's versatility in diverse fields, from medicine to nanotechnology [35].

Ferritin's capacity to bind to various tumor markers establishes it as a crucial factor in numerous medical applications, particularly in tumor diagnosis. The H chain of ferritin (Hfn) features an arginylglycylaspartic acid (RGD) domain at its N-terminus, characterized by the sequence -Cys, Phe, Cys, Asp, Gly, Arg, Cys, Asp, Cys-. This RGD domain, denoted by the "C4" symbol indicating the presence of four cysteine residues in the domain, forms the central core of the ferritin protein and performs important functions for interaction with other components. The RGD domain exhibits genetic variations across different species, making it adaptable for diverse applications. Research indicates that Hfn-c4-RGD, where Hfn is integrated with the C4 domain, demonstrates an affinity for targeting THP-1 (acute monocyte leukemia). Recent research demonstrated that the mutant form of Hfn may exhibit superior performance compared to its wild type. These findings underscore the potential of engineered ferritin constructs for targeted cellular interactions in biomedical contexts [36].

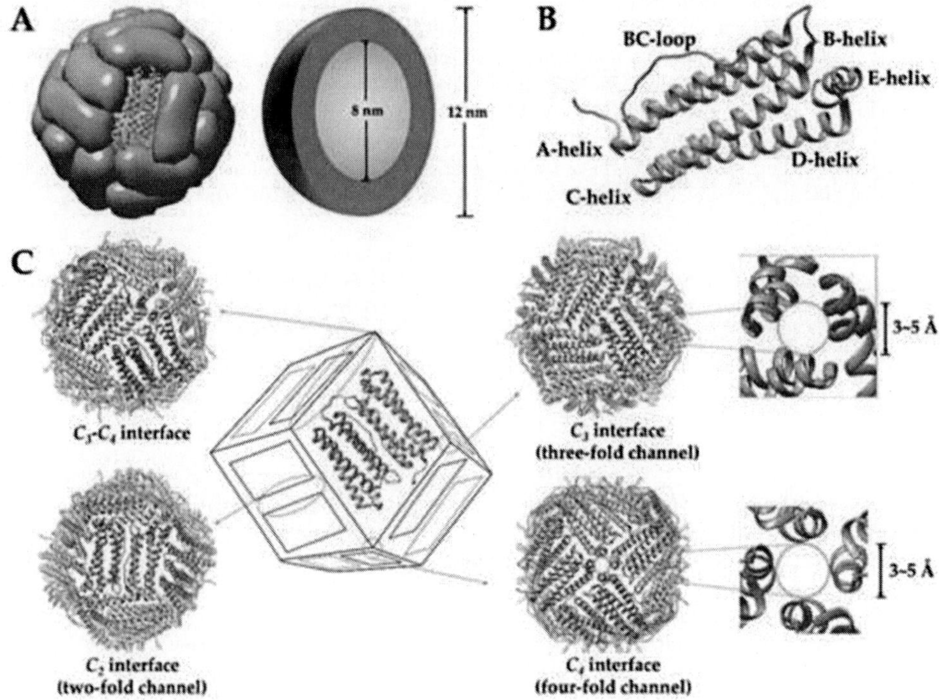


**Figure 9.2** Structure of ferritin nanocages (A), subunit structure (B), inter-subunit interface and pore size (C).

*Source:* **Published under open access CCBY 4.0 [33].**

Immunizations based on ferritin have garnered a lot of attention because of their safety and effectiveness. Because ferritin nanocages are small (10–12 nm), dendritic cells (DCs) can easily absorb them on the surface and migrate directly towards the lymphatic system. The interaction and delivery enhance the humoral and cellular immune response. Ferritin-based vaccines have been shown to be very effective as a result of non-immunogenic effects and useful to demonstrate therapeutics for cancer and autoimmune illnesses in addition to infectious diseases [37].

##### *9.2.1.1.2 Heat Shock Proteins*

HSPs form a widespread class of protein family expressed in various living cells and play a crucial role in responding to stress. HSPs are self-assembling proteins that are particularly intriguing because of their strong ability to bind to certain peptides or proteins. The HSP family preserves the three-dimensional structures of proteins facing unfavorable denaturing conditions. Within the spectrum of HSPs, classified into subclasses based on structures like small HSPs (sHSPs), medium HSPs, and large HSPs. One notable subtype of sHSPs is derived from *Methanococcus jannaschii* (MjHSP). MjHSP exhibits the ability to self-assemble into PNCs, forming an octahedral structure with 24 identical subunits. This type of nanocage was characterized by a weight of 16.5 kDa and outer (12 nm) and inner (6.5 nm) dimensions, respectively. The sHSPs share similarities with a typical ferritin nanocage. However, it distinguishes itself by featuring a multiple number of 3 nm pores on its surface, facilitating the transport of small molecules across the cage. Moreover, this structure boasts significant thermodynamic stability, aligning with the stability observed in ferritin nanostructures. These distinctive properties make MjHSP and its nanocage structure promising candidates for various applications, including potential roles in drug delivery and nanotechnology [38].

The HSP can effectively boost immunological responses against tumors, either by acting as carriers for antigens or by activating natural immune stimulants. HSPs are molecules available inside the cell and increase their expression when the cells are exposed to stressful conditions. During complex formation with other protein units, it forms stable spherical structures. The subsequent molecules attach to the interior center cavity of the chaperone structure. The HSPs are upregulated in various forms of cancer, such as breast, lung, prostate, and gastrointestinal malignancies. The HSP70 and HSP90 are the most efficacious and extensively employed types of HSPs in immunotherapy. The tumor antigens retrieved from tumors are combined with HSPs and utilized for immunization, yielding encouraging outcomes [39, 40].

#### *9.2.1.1.3 Chaperonins*

Chaperonins, essential molecular chaperones, play a fundamental role in maintaining cellular proteostasis. The bacterial chaperonin (GroEL) is involved in protein folding within the cellular environment while its cofactor GroES accelerates the process. The structure of GroEL is characterized by two stacked rings, each composed of seven identical subunits, forming a cylindrical complex. This architecture creates a central cavity capable of encapsulating unfolding or partially folded proteins. GroES (attached to one end of GroEL) forms a lid that encapsulates the folding protein within the central cavity. Unlike the majority of other protein cages, chaperonins possess sizable openings that permit the entry and exit of proteins, synthetic macromolecules, and inorganic nanoparticles from their cavities [41–43]. The primary function of chaperonins is to facilitate correct protein folding by providing a secluded environment, shielding the folding protein from potential misfolding or aggregation. The challenging task for chaperonins is to verify the folding process of newly synthesized folded proteins and assist in the refolding of denatured or misfolded proteins, particularly during cellular stress conditions like heat shock. By preventing misfolding and aggregation, chaperonins contribute to the maintenance of cellular homeostasis [38, 44].

Chaperonins are not only crucial for cellular functions but also hold significance in biotechnological and biomedical applications. The unique structure of chaperonins also acts as carriers for cargo, such as therapeutic proteins or enzymes, offering stability and protection during transport. The potential for targeted delivery is enhanced by modifying the exterior of chaperonins and introducing ligands or antibodies that enable precise delivery to the targeted site. Additionally, their applications extend to protein production and purification processes in biotechnology [42, 44].

#### *9.2.1.1.4 Encapsulins*

Encapsulins are a type of PNCs with an icosahedral structure, exhibiting sizes between 24 and 42 nm and diverse triangulation numbers (T1, T3, or T4). The encapsulins are formed from the shell proteins by self-assembly and are composed of 60–240 amino acids. The capsids are similar to the Hong Kong 97 (HK97) phage-like fold [45]. The encapsulins are classified into four distinct groups, drawing distinctions based on variations in their sequence, operon layout, overall structure, and the mechanism through which encapsulation occurs [46, 47]. Encapsulins name is derived from the distinctive properties associated with high payload capacity and self-assembled formation with particular cargo proteins. The self-assembly forms via selective cargo-loading mechanisms that rely on targeting domains (TDs) or targeting peptides (TPs) located at the N- or C-terminus of each cargo protein [45, 48].

For the encapsulation process, the TP is located at the C-terminus of encapsulins. The flexible liner was used to attach TPs and encapsulate. The linker is characterized by a high content of glycine and proline and spans approximately 10–50 residues in length. The presence of glycine and proline helps reduce the stearic interaction between cargo protein and shell. The optimum interaction increases cargo-loading capacity. Due to the higher mobility associated with the cargo proteins, they are less resolved in encapsulins, which results in the creation of a flexible network structure. It is noteworthy that shell assembly does not depend on cargo protein loading; encapsulin shells accelerate self-assembly without cargo proteins. The shell self-assembly process helps to understand the mechanism of interaction and expression at the targeted site. The self-assembly of shell formation helps to understand the nature of cargo being loaded and its interaction with TPs [48]. Khaleeq et al. [49] studied that modifying the encapsuling scaffold for the delivery of SAR-CoV-2 immunogen presents a significant advancement in the field of vaccine development. The study explores the use of encapsulin, a self-assembling protein extracted from *Myxococcus xanthus*, as a scaffold for the multivalent delivery of a monomeric receptor-binding domain derivative (mRBD). The approach involves designing 180 copies of mRBD on the encapsulin scaffold, which enhances the binding affinity for conformation-specific antibodies. The stability of the scaffold demonstrates long-term stability in the presence of encapsuling. The long-term stability is a crucial attribute for vaccine distribution and storage, particularly in varied global conditions. *In vivo* studies conducted on mice showed that the immunogen induced high titers of neutralizing antibodies after a single immunization. Notably, these titers were significantly boosted, by 42-fold, following a subsequent immunization. The increase in neutralizing antibody titers displays the high efficiency of vaccine and long-term protection from the viral infection. The developed immunogen was also effective against several variants of concern (VOC), including the α, β, and δ variants, with IC50s comparable to the wild-type virus. However, like other SARS-CoV-2 vaccines, the immunogen showed a notable reduction in infection in the challenge study in the Omicron variant. The study underscores the potential of encapsulin-based vaccine platforms in enhancing immune responses through multivalent display mechanisms [49].

The inherent efficiency and modularity of this native cargo-loading process position encapsulins as exceptional protein carriers with versatile applications, including targeted drug delivery vehicles, vaccine platforms, bionanoreactors, and many more.

##### *9.2.1.1.5 β-Lactoglobulin*

The β-lactoglobulin (BLG) is found in many animal species, and this is one of the types of whey protein extracted from cow or sheep milk. The structurally globular in shape, small in size and consists of a β-sheet having eight strands long and antiparallel to each other, a 3-turn structure, with α-helix on the outside. The disadvantage of using BLG as a carrier for drug delivery is the solvent accessibility ligand associated with it. Alternatively, the problem has been partially addressed through several methods, such as introducing an additional layer, such as pectin, to safeguard the ligand during the delivery process. The assembly of BLG into nanostructures has also been enhanced to augment the carrier potential [50, 51].

##### *9.2.1.1.6 Albumin*

Albumin is the predominant protein found in plasma, and other forms of albumins have been identified, such as ovalbumin (OVA), human serum albumin (HSA), bovine serum albumin (BSA), and rat serum albumin (RSA). Albumin possesses the potential to function as a carrier for nanodrugs due to its biodegradable nature, lack of toxicity, ready availability, low cost, practical purification process, and ability to stimulate an immune response. The albumin exhibits stability over a wide pH range (4–9) and possesses unique ligand-binding properties attributed to its highly charged nature. Notably, albumins are preferentially taken up by inflamed tissues and, surprisingly, tumors. The unique features of albumin make it acceptable as a universal carrier for site-specific delivery applications. Assembling albumins into precise nanostructures provides additional therapeutic benefits, such as increased tolerance, absence of unwanted interactions with serum, and a high payload [36].

Additionally, the amino acid residues can be utilized for physical interaction and adsorption of drug molecules on the surface. Recently, the US Food and Drug Administration (FDA) has granted approval for various therapy regimens that utilize albumin for clinical treatment. Furthermore, the levels of albumin receptors, including Gp18, Gp30, Gp60, and Secreted Protein Acidic and Rich in Cysteine (SPARC), are excessively increased in cancer cells. Albumin-based drug carriers can partially target tumors by interacting with albumin receptors and albumin. Another approach for drug delivery involves the synthesis of hydrogels using albumin. Recently, researchers have demonstrated cross-linkable albumin-based hydrogels (CABH) by covalent attachment in the presence of N-hydroxysuccinimide (NHS). The hydrogels release the payload only in an acidic environment, which is considered to be advantageous for acidic tumor targeting [52].

#### *9.2.1.2 Viral PNC*

Viral PNCs, also called VLPs, consist of precise structural protein assemblies that exhibit excellent specificity and efficiency. The exterior viral structure is composed of repetitive arrangements of viral coat proteins (VCPs), while the inner core consists of genetic material. The viral protein is structurally either icosahedral or helical in shape. The VLPs resemble similar capsid or protein coats of viruses but lack genetic materials. The process of viral transduction is facilitated by self-assembling protein structures known as capsids. The interaction between capsids and host cells facilitates the transport of genetic material inside the cell. The viral capsids can be explored for numerous applications in industry for chemical functionalization. Viruses possess a natural mechanism to safeguard their genetic material, release it into the extracellular environment, evade the immune system upon encountering target cells through interactions with receptors, and transport the encapsulated nucleic acids into the intended cell compartment.

The protein subunits from the capsids, self-assembled to form particles that have a narrow range of sizes (20–200 nm). The surface is anchored with antigens that facilitate cellular internalization [53]. Researchers harness the capability of viral capsid to carry genetic material by substituting the viral genetic material like RNA that replaces non-viral nucleic acids that encode to the desired protein. The non-viral delivery of capsid not only results in the synthesis of the protein but also prevents viral multiplication. The VLPs act like adjuvant used in vaccine therapy, stimulating immune response but do not cause viral infection. The VLPs can be useful for the delivery of nucleic acid and act as a carrier for vaccines. The AstraZeneca COVID-19 vaccine utilizes a genetically altered chimpanzee adenovirus vector to carry the genetic code for the spike protein (from SARS-CoV-2). The delivery leads to the activation of B- and T-cell responses via the expression of non-viral nucleic acid to protein [54].

The VLPs can be prepared using multiple expression systems followed by purification and extraction of proteins from the virulent part. The production of VLPs can be easily set up in a laboratory with the complete removal of viral genetic material. The process can be easier to attach and deliver multiple therapeutic agents using VLPs. An alternative modification strategy can be the amalgamation of VLPs with Clustered Regularly Interspaced Short Palindromic Repeats (CRISPR)-Cas technology to act as an efficient tool in gene editing techniques. The approach further can be used to co-encapsulate guiding RNA and mRNA encoding for Cas9 within VLPs as shown in Figure 9.3. The surface modification using the Cas9 strategy can be used for transgene introduction or cell surface-specific targeting like CD4+ [55].

Likewise, genetically engineered VLPs with lipophilic amino acids can carry tiny hydrophobic medicines such as the anticancer medication (doxorubicin). By including TPs and PEG chains to enhance immune evasion, additional surface modifications can enhance VLPs and create customized smart delivery systems [56]. Concerning the stability of VLPs and interaction with chemical agents or genes, it has suitable characteristics to deliver chemotherapeutic agents. VLPs consist of self-assembled homologous or heterogeneous VPs, which self-assemble and organize into highly symmetrical, monodisperse, and stable nano- or micro-scale structures. The VLPs can be converted into virus-like nanoparticles (VLNPs) and developed as carrier systems [36]. The surface functional groups can be modified using genetic engineering approaches to enhance their potential in nanomedicine. The alteration of amino and sulfur from lysine and cysteine can be used to modulate the surface characteristics [44, 57].

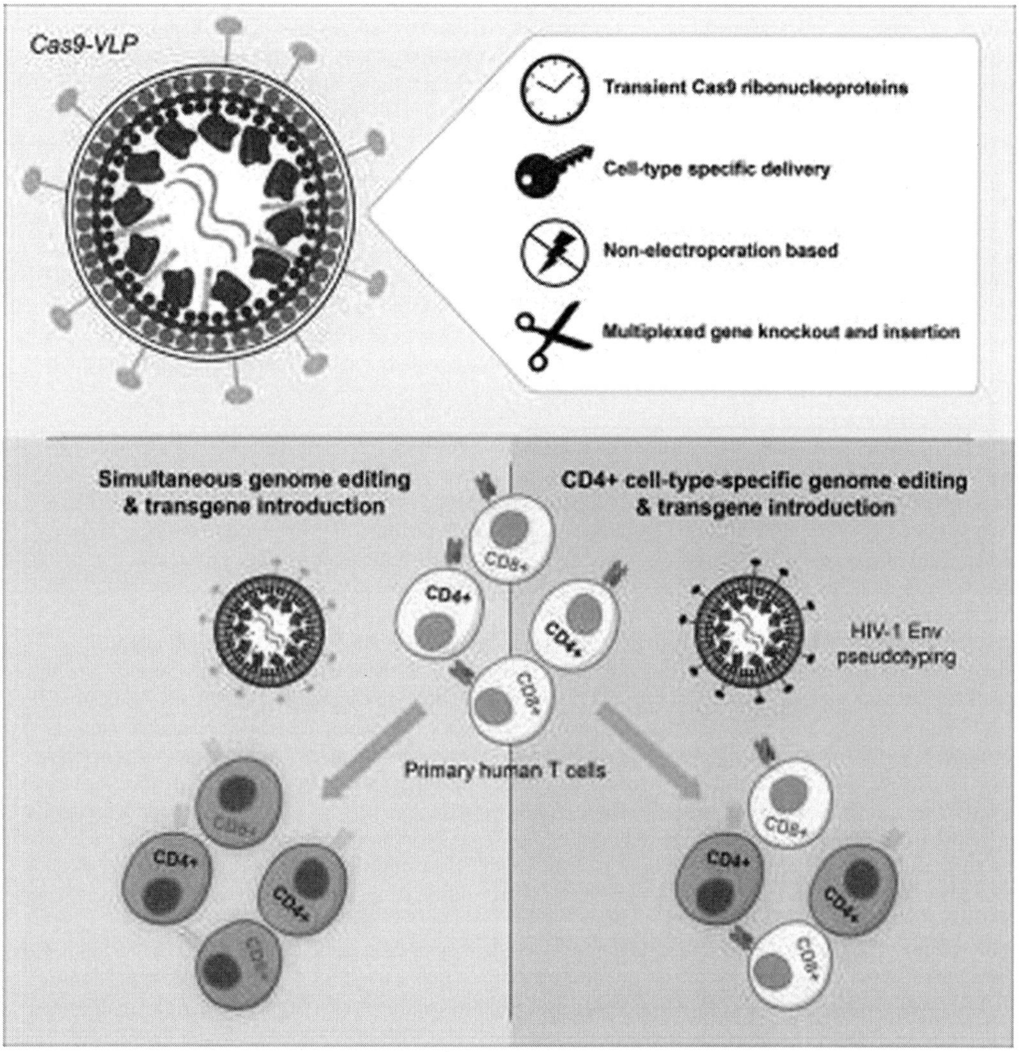


**Figure 9.3** VLPs and Cas9 for transgene introduction and cancer targeting.

*Source:* **Reproduced with permission from ref. [55].**

## 9.3 SELF-ASSEMBLING PEPTIDES

Self-assembly is used to synthesize molecular templates and supramolecular structures through the utilization of nanoengineering methods. Macromolecules spontaneously arrange themselves into nanostructures, including nanofibers, nanotubes, vesicles, helical ribbons, and fibrous scaffolds [58, 59]. Amphiphilic molecules were formed by interaction between polar and nonpolar precursor-containing monomers. The polar hydrophilic portions of amphiphilic molecules self-assemble in aqueous solutions to form special structures, including tubules, vesicles, and micelles. The amphiphiles created via self-assembly prevent direct interaction with the aqueous core; this may be due to nonpolar hydrophobic portions being extended in an outward direction during assembly formation [60].

Self-assembled peptides typically comprise monomer chains or repeating sequences of 8–16 amino acids, which spontaneously reorganize into nanostructures. Due to their tiny size, typical peptides have little ability to remain in the patient's body for extended durations. The modifying sequences within self-assembled peptides can generate structures that exhibit greater stability compared to conventional non-biological materials. The stimuli-responsive self-assembled peptides can be synthesized using the modification approach.

Drug-loaded hydrogels that respond to changes in pH are stimuli-specific responses used to develop using self-assembled peptides. The swellable potential of self-assembled peptide is designed to efficiently transport drugs specifically in acidic tumor with pH below 4.2 [61]. The researchers fabricated biocomposites by utilizing self-assembled peptides and gold nanoparticles. Self-assembled peptides can be designed to reach the target site and release actives based on light as stimuli, which generates a heat-specific response in the presence of gold nanoparticles, which can achieve multimodal therapeutics for cancer management [62]. The peptides and proteins were modified using bioengineering approaches to modulate the target specificity and delivery payload. The bioengineering provides an additional improvement in the precision of nanocomplexes, granting them a competitive edge over other cancer treatments [52].

The functions of self-assembled peptides/proteins are often manifested in three aspects: The objective is to achieve functional improvement by boosting synergistic activity. (1) Polyoxometalates can facilitate the self-assembly of short peptides by hydrophobic and/or π–π interactions, resulting in enhanced stability in their natural environment [63]. (2) Acquisition of novel functionalities that are not present in the individual building blocks. For example, *MUC1* glycopeptide that modifies with Tn glycosylation in the PDTRP domain is responsible for generating a notable immunological response [64]. (3) Enhance functionality by integrating functional molecules, such as incorporating phototherapeutic agents for photodynamic therapy (PDT) and photothermal therapy (PTT) [65].

### 9.3.1 β-Sheet Peptide

The β-sheet, a prevalent naturally occurring motif, stands out as a highly valuable element for the self-assembled peptides. The folding of β-sheets relies on the hydrogen bonds formed between amides in the structure of two neighboring β-strands. Typically, the side chain residues of polypeptides are used in the synthesis and maintenance of β-sheets by engaging in extra electrostatic interactions or π–π stacking. In addition, the aggregation of β-sheets leads to increasingly intricate and hierarchical formations [66]. β-sheet forming peptides constitute a class of peptides inherently inclined to adopt β-sheet secondary structures, characterized by hydrogen-bonded alignments of adjacent peptide strands. Examples of such peptides include beta-amyloid peptides, notorious for their involvement in Alzheimer's disease, and those designed to form β-hairpin structures or hydrophobic zipper patterns, promoting the stacking of β-strands. Additionally, certain antimicrobial peptides, collagen triple-helix peptides, silk fibroin peptides, and self-assembling peptide amphiphiles, with hydrophobic tails and beta-sheet-forming sequences, exhibit a natural propensity to organize into beta-sheet-rich structures. Short dipeptide sequences like diphenylalanine (FF) are known to self-assemble into beta-sheet-rich nanofibers, forming hydrogels. The versatility of β-sheet peptides extends to give rise to diverse nanostructures, including nanotubes, nanoscale-ordered monolayers, and nanoribbons. This unique self-assembly property positions β-sheet peptides as promising candidates for applications in the development and engineering of functional materials with tailored nanostructures for various biomedical and nanotechnological purposes [22, 67].

For example, the peptide QQRFEWEFEQQ, featuring ionizable side chains from Glu and Arg amino acids, exhibits the ability to undergo self-assembly, forming a pH-responsive hydrogel. This peptide transitions into a hydrogel structure under low pH conditions and remains soluble at neutral pH levels. The mechanism behind this behavior lies in the creation of antiparallel β-sheet tapes at lower pH, which can then stack to generate hydrogel nanofibrils. Additionally, at interfaces between air and water, the β-hairpin peptides derived from these sequences demonstrate self-assembly into diverse nanostructures. The fundamental process for the self-assembly of β-hairpins in proteins consists of the arrangement of two β-sheets in antiparallel configurations. A responsive hydrogel was developed using a β-hairpin peptide with the sequence VKVKVKVKVDPPTKVKVKV, with the substance forming in response to increasing pH levels. The mechanism involves the formation of hydrogels derived from the hairpin structure via self-assembly to create β-sheets. The self-assembly accelerated once after the change in the pH value at constant reaction conditions [68–70].

### 9.3.2 α-Helix

The α-helix is a secondary structure that is important in protein- and peptide-like structures. It is defined as the peptide backbone wrapping into a right-handed helix with about 3–6 amino acids per turn. It is among the main structural components present in a section of peptide hydrogels. In contrast to the β-sheet structure, an α-helix is created by a single peptide chain in which the main-chain (amide) components are involved in intramolecular hydrogen bonding. The intrinsic thermodynamic instability of α-helices is responsible for eliminating helical confirmation of linear peptides in solution. Similar simple methodologies were used for the synthesis of α-helix and modification of structural conformation [71, 72]. The instability of α-helical structure at variable temperature conditions, it is necessary to stabilize it before self-assembly. The earlier reports suggest that hydrogen bonding, cross-coupling between the side chain, or salt bridging can help to stabilize the α-helical structures.

### 9.3.3 β-Hairpin

The β-hairpin self-assembling peptides are biomolecules characterized by their capability to spontaneously organize into a specific structural motif known as a β-hairpin. In this arrangement, the peptide chain folds back on itself, forming a hairpin-like structure with two antiparallel beta strands connected by a reverse turn. These peptides often consist of a simultaneous sequence of lipophilic and lipophobic amino acids, contributing to their amphiphilic nature. The self-assembly of β-hairpin peptides was driven by interactions between lipophilic and lipophobic regions, leading to the formation of stable nanostructures. These nanostructures hold promise in various applications, including the design of functional materials, carrier system, and nanoscale materials. As described by Pochan et al., β-hairpin self-assembling peptides initially exhibit a random coil shape at ambient temperature and high pH. However, upon heating, these peptides undergo a structural transformation, adopting a β-hairpin conformation. The modified β-hairpin structure facilitates conversion into a swellable mass-like hydrogel formed by the self-assembly. Notably, the formed hydrogel dissolves entirely upon cooling. The temperature range for hydrogel formation, ranging between 25°C and 60°C, can be modulated by adjusting the lipophilicity of the peptides [73, 74].

### 9.3.4 Surfactant-Like Peptides

In broad terms, surfactant-like peptides consist of hydrophilic heads linked to hydrophobic tails. The hydrophilic head is usually composed of more than one amino acid residue, such as His, Asp, or Glu, and has a surface charge for interaction. The hydrophobic tail typically consists of nonpolar amino acids, which are 3–9 in number, specifically a combination of Ala, Phe, Ile, or Val. These peptides exhibit a propensity to self-assemble, concealing the lipophilic tails from polar head groups once they come in contact with water molecules. The tendency to self-assemble and convert into aggregated conformation tends to form nanostructures like nanotubes and nanocapsules [22]. Similar to lipids and fatty acids, the formation of a polar-to-nonpolar bridge in the supramolecular structure is responsible for isolating the lipophilic tail from water.

Glutamic acid (E) and aspartic acid (D) have a negative charge and hydrophilic properties. While other amino acids based on pKa are positively charged, arginine (R), histidine (H), and lysine (K) also have hydrophilic properties. Conversely, the lipophilic amino acids are isoleucine (I), valine (V), glycine (G), alanine (A), and leucine (L). Surfactant-like peptide design can be achieved by directional organization using Ac-AAAAAAD (A6D), Ac-VVVVVVD (V6D), and positively charged Ac-AAAAAK (A6K) or any other sequence design [75]. The primary structures like nanotubes or nanocapsules formed in surfactant-like peptides were due to self-assembly [76]. The nanoassembly acts on the extracellular membrane in a manner similar to lipid-scavenging micelles. Additionally, a similar peptide known as bola-amphiphilic peptide displays an alternative connection between charged amino acids and a hydrophobic sequence at the center. A difference of hydrophilic head groups in the building block is a major distinction with bola-amphiphilic peptides from surfactant-like peptides [77]. Bola-amphiphilic peptides are characterized by the presence of lipophobic heads with lipophobic groups separated by lipophilic sites at both ends. In contrast, surfactant-like peptides possess only one lipophobic head [78].

### 9.3.5 Amphiphilic Peptides

Amphiphilic peptides, sharing structural similarities with lipid molecules, commonly exhibit the binding between lipophilic long-chain alkyl groups and lipophobic peptides. The interaction between the lipophobic head and lipophilic tail of amphiphilic peptides serves as both intra- and intermolecular driving forces, facilitating the stabilization of various supramolecular structures. Additionally, these peptides can generate nanostructures with diverse morphologies [79, 80]. Amphiphilic peptides are unique self-assembled peptide nanostructures that are attractive due to their ease of usage, adaptability, and biocompatibility. Amphiphilic peptides are simple to design and synthesize for use in biomedical applications. Additionally, amphiphilic peptide-derived nanostructures are biologically active and are crucial for biomedical applications [81].

### 9.3.6 Ionic-Complementary Peptides

The ionic-complementary peptides (ICPs) exhibit a molecular-level self-assembly into nanofibrous hydrogels, comprising alternating arrangements of lipophobic and lipophilic charged amino acids. These polypeptides present a distinctive charge distribution pattern, categorizing them into three types: Type I (–+–+–+–+), Type II (––++––++), and Type IV (––––++++). The self-assembly mechanisms vary depending on the sequence of their charge distribution, which allows the rational design of ICP. The manipulating and combining of different charged species contribute to the overall distribution. Apart from the self-assembled peptide EAK16, RADA16-I, KFE8, etc. were identified and extensively investigated for various applications. The ability of RADA16-I, a classical ionic complementarity peptide, to spontaneously form fibrous hydrogels in aqueous solutions has made it useful in the biological and clinical domains [67]. Nevertheless, the challenge commonly shared among peptides, including RADA16-I, is their instability under low pH conditions. To address this issue, researchers are actively exploring modifications to enhance the stability of ICPs, aiming to broaden their applicability in medical contexts. Moreover, recent findings indicate that hydrogels formed by two ICPs featuring opposite charges demonstrate enhanced biocompatibility with fibroblasts under physiological pH conditions [82].

### 9.3.7 Triple Helical Peptides

A triple helical peptide is a molecular structure characterized by the intertwining of three peptide chains in a helical conformation. This unique arrangement forms a stable and rigid structure, resembling a triple-stranded helix. Typically, triple helical peptides are rich in proline and glycine residues helping to improve the stability of the helical structure. The collagen's triple helical domains have specific structural characteristics resulting from its particular amino acid sequence. The triple helical sequence is made up of $(GXY)_n$ repeating patterns forming a polyproline Type II left-side helix. The location of proline is at the amino acid X position, which is fixed and 4R-hydroxyproline, a post-translational modification of proline that is denoted O in the one-letter amino acid code and Hyp in the three-letter code, is frequently found at the Y position in the exo conformation. Through a network of hydrogen bonds between the amide group of the glycine and the carbonyl group of the residue at the X position, as well as van der Waals contacts, each $(GXY)_n$ chain associates with two other chains [83, 84]. The three chains form a rope-like triple-helix structure with one amino acid staggered between each chain as they interlace in a right-handed helix. Because of its distinct structure, collagen triple helices can be stacked unevenly both laterally and axially to generate fibrils that exhibit an electron microscope-visible periodic banding pattern of 67 nm in length [84].

### 9.3.8 Diphenylalanine

FF peptide refers to a peptide composed of two phenylalanine amino acid residues. Phenylalanine is an aromatic amino acid, and when two such residues are joined together, it forms a dipeptide known as FF. The interaction between the aromatic side chains leads to the formation of self-assembly containing FF peptides into various nanostructures. The FF peptide, sourced from the Alzheimer's beta-amyloid polypeptide, acts as the central recognition [85]. FF is a peptide constituent associated with modification in the other peptides related to the progress of Alzheimer's disease. The identification of the core recognition in Alzheimer's β-amyloid polypeptide studies suggests its role in guiding self-assembly [86]. Following the discovery by Reches and Gazit [87] that FF may generate nanotubes, other studies suggest the FF-based building blocks can possibly generate various functional nanostructures. The functional nanostructures include ordered molecular chains or nanocontainers.

Warren et al. [88] demonstrated the synthesis of polymers capable of mimicking the complex functions of natural biopolymers. Despite advancements in polymer synthesis methods, synthetic polymers often lack the intricate network of non-covalent interactions characteristic of natural proteins. To address this challenge, the researchers utilized a biomimetic di(phenylalanine) acrylamide (FF) monomer to study the role of non-covalent interactions in creating ordered assembled structures. They synthesized amphiphilic copolymers resembling β-sheet-like local structures, which converted to single-chain assemblies upon contact with water. Systematic analysis of these amphiphilic copolymers revealed that hydrophobic forces primarily accelerated the collapse of linear chains. Additionally, hydrogen bonding and aromatic interactions were found to stabilize the local structure, as evidenced by circular dichroism and thioflavin T fluorescence indicating β-sheet-like interactions. Further analysis of phenylalanine (F) and alanine-phenylalanine acrylamide (AF) copolymers showed that positioning the aromatic residue away from the polymer backbone could induce β-sheet-like local structures similar to those observed in FF copolymers. However, the interactions between AF subunits were less stable than those formed by FF. Moreover, incorporating hydrogen-bond-donating hydrophilic monomers disrupted the internal structure formed by FF within the collapsed assemblies.

Warren et al.'s [88] findings provide critical insights into the design principles necessary for incorporating multiple facets of protein-mimetic, higher-order structures within synthetic polymers. This research advances the understanding of how to create synthetic polymers that more closely replicate the complex functionality of natural proteins through strategic non-covalent interactions.

## 9.4 FUSION PROTEIN

A fusion protein is a hybrid molecular entity created by combining genetic sequences of two or more distinct proteins, resulting in a single gene that directs the synthesis of a chimeric protein. This merging of genetic information often serves specific purposes across molecular biology, biotechnology, and medicine. In biotechnology, fusion proteins simplify the purification of target proteins by incorporating known affinity tags or marker proteins, facilitating their isolation from complex mixtures. Additionally, fusion proteins are integral in molecular biology as reporter systems, aiding the study of gene expression, protein localization, and interactions. Furthermore, fusion proteins find applications in medicine, where they are designed for therapeutic use, enhancing stability, half-life, or tissue-specific targeting. Vaccine development benefits from fusion proteins, as they augment immune responses by combining antigens with carrier proteins. Lastly, fusion proteins contribute to functional studies, allowing researchers to gain insights into the roles and regulation of specific proteins. Common fusion partners, such as green fluorescent protein (GFP), glutathione S-transferase (GST), maltose-binding protein (MBP), and histidine tags, are selected based on experimental requirements, collectively providing valuable tools for protein manipulation and analysis in diverse scientific disciplines.

### 9.4.1 GFP Fusion

GFP is a naturally occurring protein derived from the jellyfish *Aequorea victoria*, known for its unique ability to fluoresce and emit green light when exposed to specific wavelengths. This intrinsic fluorescence has made GFP a pivotal tool in molecular and cell biology. As a reporter gene, GFP allows for the real-time monitoring of gene expression and protein localization within living cells. Researchers frequently use GFP as a fusion tag by genetically fusing the GFP gene to target genes, enabling easy visualization and tracking of proteins of interest. GFP's applications in cellular imaging have revolutionized the field, facilitating the observation of dynamic processes such as cell division and organelle movement under fluorescence microscopy. Introducing the GFP gene into the genomes of transgenic organisms, including plants and animals, allows for the study of *in vivo* developmental processes and gene expression patterns. Additionally, GFP is instrumental in studying protein–protein interactions through techniques like fluorescence resonance energy transfer (FRET). Ongoing modification and engineering of GFP have produced a spectrum of variants with diverse colors and enhanced characteristics. These advancements have expanded GFP's utility in various experimental contexts, significantly contributing to our understanding of cellular and molecular biology [89, 90].

The GFP fusion involves the genetic fusion of the GFP gene with the gene encoding a protein of interest, creating a chimeric or hybrid protein. The advanced technique leverages GFP's natural fluorescence properties to visualize and study the expression, localization, and movement of the fused protein within living cells [91]. The GFP fusion process typically begins with the insertion of the GFP gene into the target gene's coding sequence or its attachment at the protein's termini. The resulting fusion protein retains the fluorescent characteristics of GFP, allowing researchers to track the tagged protein in real time. The choice of fusion site can influence the functionality and behavior of the hybrid protein. GFP fusion proteins find broad applications in molecular and cell biology and serve as invaluable tools for studying cellular processes, protein trafficking, and interactions. By observing the fluorescent signals emitted by GFP, researchers can gain insights into the spatial and temporal dynamics of the gene of interest, which expresses specific protein without disrupting cellular processes. This non-invasive approach is particularly useful for live-cell imaging and has revolutionized the field by providing a direct visualization method for studying various biological phenomena [92].

The GFP produces a green color fluorescence excited in the presence of UV light. This distinctive characteristic allows for the intracellular monitoring of the protein using a fluorescent microscope, offering vital insights into the absorption and dispersion of nanoparticles that contain GFP. In 2013, Xia et al. demonstrated a method for creating GFP nanoparticles by attaching different linear and Y-shaped polyacrylamides and polyacrylates to GFP. When the temperature was raised to 50°C, which is higher than the lower critical solution temperature (LCST) of the polymer, the material underwent a transformation and formed micellar nanoparticles. These nanoparticles had a hydrodynamic radius ranging from 12 to 15 nm. Upon reaching room temperature, this process underwent a reversal, causing the particles to dismantle again. UV/Vis and circular dichroism (CD) spectroscopy support the protein structure that remained unchanged following conjugation with the polymer. However, no IVIVC correlation was conducted to assess the particles' suitability for imaging applications [93]. Kim et al. [94] employed a coacervation technique to create nanoparticles by combining surface-modified GFP variants with positively charged polymers. To examine the impact of charge distribution on the external surface of GFP, various mutants were created. These mutants had modified distributions of arginine, lysine, glutamate, or aspartate on the protein's surface. All mutants displayed a comparable secondary structure and demonstrated activity as fluorophores. The process of charge-driven coacervation resulted in the formation of nanoparticles with a size of around 120 nm. These nanoparticles were generated by the interaction of a highly positively charged vinyl pyridine polymer [94].

### 9.4.2 Polyhistidine

Polyhistidine, often referred to as a histidine tag or His-tag is a short peptide sequence comprised of consecutive histidine residues, commonly used as an affinity tag in recombinant protein expression and purification. Typically introduced by fusing the tag to the N- or C-terminus of a target protein during the cloning process, polyhistidine tags serve a crucial role in facilitating the purification of the expressed protein. The affinity purification is achieved through immobilized metal affinity chromatography (IMAC), where the His-tagged protein selectively binds to nickel ions on a column and can be subsequently eluted under mild conditions. This method allows for efficient and specific isolation of the protein of interest from complex mixtures. Beyond purification, polyhistidine tags find utility in protein detection, with antibodies specific to the His-tag employed in techniques such as Western blotting and ELISA. Additionally, polyhistidine tags can be valuable in protein interaction studies, enabling researchers to immobilize His-tagged proteins for the investigation of their interactions with other molecules. While polyhistidine tags offer advantages such as minimal impact on protein structure and cost-effective purification, careful consideration of potential influences on protein function and the design of appropriate control experiments are essential aspects of their use in research and biotechnology [95, 96].

### 9.4.3 GST Fusion Proteins

GST fusion proteins represent a widely utilized strategy in recombinant protein expression and purification. In this approach, the gene encoding the target protein is fused with the GST gene, generating a hybrid construct capable of producing a chimeric protein. This fusion protein is then expressed in a host system, commonly *Escherichia coli*, and subsequently purified using glutathione affinity chromatography. The GST portion selectively binds to glutathione resin, facilitating efficient purification. The method is particularly valuable for isolating soluble and intact proteins. Purified GST fusion proteins find diverse applications in research, serving as essential tools in biochemical studies, structural biology, and functional assays. Moreover, the protein tag is exploited in protein–protein interaction studies, where GST-tagged proteins can be immobilized to capture interacting partners from cellular lysates. The GST fusion system is also employed in the production of antigens for antibody generation, benefiting from the enhanced immunogenicity conferred by the GST tag. In structural biology, GST fusion proteins contribute to crystallization studies, providing a stable and soluble form of the protein for X-ray crystallography. Overall, the GST fusion protein approach is a robust and versatile technique that significantly aids researchers in various aspects of molecular and cell biology [95, 97].

### 9.4.4 MBP Fusion Proteins

MBP fusion proteins involve the fusion of interest or specific protein with MBP, offering a valuable approach in recombinant protein expression and purification. The process begins with the genetic fusion of the MBP gene to the gene encoding the target protein, resulting in a hybrid gene responsible for the production of chimeric protein. The recombinant construct is then introduced into a host expression system, commonly *E. coli*, where it drives the production of the MBP fusion protein. MBP serves as a versatile fusion partner due to its solubility-enhancing properties, aiding in the expression of challenging or insoluble proteins. The purification of MBP fusion proteins typically involves affinity chromatography using amylose resin, due to higher separation for maltose. The interested sequence protein can be subsequently cleaved from MBP using self-cleavable linkage, resulting in the isolation of the untagged target protein. MBP fusion proteins

## Table 9.2 Protein nanocarriers for the delivery of drug and gene

| Sl. No. | Protein Nanocarrier | Nanoparticles from Synthetic Origin/Surface Modification | Therapeutics | Application | References |
|---|---|---|---|---|---|
| 1 | HSA | – | Cu. Zn, gene superoxide dismutase | Sustained release | [101] |
| 2 | HSA | Polyethylenimine (PEI) | pGL3 | BBB cross and targeting cerebellar granule cells | [102] |
| 3 | Thiolated gelatin | Wt-p53 plasmid | Gemcitabine | Tumor growth suppression in pancreatic adenocarcinoma cells | [103] |
| 4 | Peptide amphiphile KRRASVAGK [C12] NH2 | Protein kinase A | Doxorubicin | Extended release up to 7 hours. Breast cancer cells show high efficiency | [104] |
| 5 | F-ΔF dipeptide nanotubes | | Pazopanib | Choroidal neovascularization—sustained release for 35 days | [105] |
| 6 | RADA16 amphiphiles | Heparin-binding sequence domain | Growth factor | Sustained release and provide therapeutic effect post 1 month results to improve cardiac function | [106] |
| 7 | Phosphorylated serine or Arginine-Glycine-Aspartic Acid (RGD) | Hydroxyapatite nucleated | Bone tissue delivery | Improve mineralization, cell adhesion | [107] |

are extensively used in biochemical and structural studies, where the enhanced solubility provided by the MBP tag improves the likelihood of obtaining properly folded and functional proteins [98]. Similar to other fusion proteins, largely concentrated MBP fusion proteins produce inclusion bodies in *E. coli* resulting in the formation of insoluble protein aggregates. The MBP tag's 45 kDa size may have an impact on proteins function as it starts its cleavage potential. Proteolytic cleavage of the tag, when bound to amylose resin, is ineffective, necessitating elution of the fusion protein with free maltose prior to cleavage, unlike in glutathione affinity chromatography. Often, a mild elution with 10 mM maltose is sufficient to preserve protein function and antigenicity [95].

### 9.5 PROTEIN-BASED NANOCARRIER IN DRUG DELIVERY AND GENE DELIVERY

The delivery of chemotherapeutic agents and genes at the targeted site was accomplished with the help of nanocarriers. The protein self-assembled architecture forms a complex structure that has high payload capacity and delivers at the targeted site. The gene delivery uses genetic materials such as RNA, DNA, and nucleic acid to the infected cells and treats them by modifications. The viral vector and non-viral vectors are the two major parts of gene delivery. The viral vectors were immunogenic, non-specific, and costly, considered to be limited for use in therapeutics [99]. Different protein-based nanocarriers such as zein, HSA, gelatin, casein, etc. were effectively used as carrier for the delivery of gene; a few examples are highlighted in Table 9.2. Zein is a type of plant protein extracted from maize seeds. Due to its highly hydrophobic nature, it has slower enzymatic digestion. The positively charged zein at its isoelectric pH interacts with negatively charged plasmid DNA and forms a self-assembly structure. The self-assembly cleaved at reduced pH and released at active inside intracellular pH [100]. A few self-assembled peptides as carriers for therapeutic delivery are described in Table 9.2.

### 9.6 CONCLUSION

The study of protein and peptide self-assembly mechanisms, as outlined in the preceding sections, highlights the intricate interplay of various structural and interactional factors. Understanding the primary and secondary structures of proteins and peptides is crucial, as these foundational elements dictate the subsequent self-assembly processes. Key interactions such as hydrophobic interactions, hydrogen bonding, van der Waals forces, electrostatic interactions, and $\pi$–$\pi$ stacking, along with environmental factors like pH and ionic strength, collectively influence the assembly and stability of these biomolecules. Advancements in computational studies have significantly contributed to our understanding of protein self-assembly, offering insights that guide experimental approaches. The structural diversity and functional properties of self-assembled proteins and peptides have paved the way for innovative applications, particularly in the development of PNCs. Both viral and NVPNs exhibit unique characteristics that make them suitable for various biotechnological applications. The exploration of self-assembling peptides has unveiled numerous structural motifs, including β-sheet peptides, α-helices, β-hairpins, surfactant-like peptides, amphiphilic peptides, ICPs, triple helical peptides, and FF. Each of these motifs offers distinct advantages and functionalities, further expanding the toolkit for designing biomimetic materials and nanostructures. Fusion proteins, especially those involving GFP, have revolutionized molecular and cellular biology by enabling real-time visualization of protein localization and interactions. Other fusion tags like polyhistidine, GST, and MBP fusion proteins have facilitated protein purification and functional studies, enhancing our ability to manipulate and study proteins in various contexts. The development of protein-based nanocarriers for drug and gene delivery underscores the practical implications of understanding protein and peptide self-assembly. These nanocarriers offer targeted delivery, controlled release, and improved therapeutic efficacy, making them invaluable in medical and pharmaceutical applications. In conclusion, the comprehensive study of protein and peptide self-assembly, along with the development of advanced biotechnological tools and nanocarriers, holds immense potential for future research and applications. Continued exploration in this field will undoubtedly lead to further innovations, enhancing our ability to harness the unique properties of proteins and peptides for diverse scientific and medical purposes.

## REFERENCES


1. Bonlawar J, Setia A, Challa RR, Vallamkonda B, Mehata AK, Vaishali, et al. Targeted nanotheranostics: integration of preclinical MRI and CT in the molecular imaging and therapy of advanced diseases. *Nanotheranostics*. 2024;8(3):401–26.

2. Setia A, Vallamkonda B, Challa RR, Mehata AK, Badgujar P, Muthu MS. Herbal theranostics: controlled, targeted delivery and imaging of herbal molecules. *Nanotheranostics*. 2024;8(3):344–79.

3. More MP, Deshmukh PK. Computational studies and biosensory applications of graphene-based nanomaterials: a state-of-the-art review. *Nanotechnology*. 2020;31(43):432001.

4. Suseela MNL, Mehata AK, Vallamkonda B, Gokul P, Pradhan A, Pandey J, et al. Comparative evaluation of liquid-liquid extraction and nanosorbent extraction for HPLC-PDA analysis of cabazitaxel from rat plasma. *Journal of Pharmaceutical and Biomedical Analysis*. 2024;245:116149.

5. Pasala PK, Prasanth DSNBK, Rudrapal M, Challa RR, Ahmad SF, Vallamkonda B, et al. Anti-Parkinson potential of hesperetin nanoparticles: in vivo and in silico investigations. *Natural Product Research*. 2024:1–10.

6. Chakravarthy PSA, Popli P, Challa RR, Vallamkonda B, Singh I, Swami R. Bile salts: unlocking the potential as bio-surfactant for enhanced drug absorption. *Journal of Nanoparticle Research*. 2024;26(4):76.

7. More MP, Patil S, Ghodke S, Patil PO, Jain R, Dandekar P, et al. Development of cross-linked collagen/pullulan ocular film for sustained delivery of Besifloxacin using novel spin-coating technique. *Journal of Materials Research*. 2021;36(16):3278–92.

8. Riba A, Di Nanni N, Mittal N, Arhné E, Schmidt A, Zavolan M. Protein synthesis rates and ribosome occupancies reveal determinants of translation elongation rates. *Proceedings of the National Academy of Sciences*. 2019;116(30):15023–32.

9. Salatin S, Jelvehgari M, Maleki-Dizaj S, Adibkia K. A sight on protein-based nanoparticles as drug/gene delivery systems. *Therapeutic Delivery*. 2015;6(8):1017–29.

10. Aupič J, Lapenta F, Strmšek Ž, Jerala R. Towards designing new nano-scale protein architectures. *Essays in Biochemistry*. 2016;60(4):315–24.

11. Kanekiyo M, Bu W, Joyce MG, Meng G, Whittle JRR, Baxa U, et al. Rational design of an Epstein-Barr virus vaccine targeting the receptor-binding site. *Cell*. 2015;162(5):1090–100.

12. Lai YT, Cascio D, Yeates TO. Structure of a 16-nm cage designed by using protein oligomers. *Science*. 2012;336(6085):1129.

13. Zhang J, Zheng F, Grigoryan G. Design and designability of protein-based assemblies. *Current Opinion in Structural Biology*. 2014;27:79–86.

14. Mendes AC, Baran ET, Reis RL, Azevedo HS. Self-assembly in nature: using the principles of nature to create complex nanobiomaterials. *WIREs Nanomedicine and Nanobiotechnology*. 2013;5(6):582–612.

15. Leite DM, Barbu E, Pilkington GJ, Lalatsa A. Peptide self-assemblies for drug delivery. *Current Topics in Medicinal Chemistry*. 2015;15(22):2277–89.

16. Panda JJ, Chauhan VS. Short peptide based self-assembled nanostructures: implications in drug delivery and tissue engineering. *Polymer Chemistry*. 2014;5(15):4418–36.

17. Sandra F, Khaliq NU, Sunna A, Care A. Developing protein-based nanoparticles as versatile delivery systems for cancer therapy and imaging. *Nanomaterials*. 2019;9(9):1329.

18. Mandal D, Shirazi AN, Parang K. Self-assembly of peptides to nanostructures. *Organic and Biomolecular Chemistry*. 2014;12(22):3544–61.

19. Eskandari S, Guerin T, Toth I, Stephenson RJ. Recent advances in self-assembled peptides: implications for targeted drug delivery and vaccine engineering. *Advanced Drug Delivery Reviews*. 2017;110–1:169–87.

20. Ulijn RV, Smith AM. Designing peptide based nanomaterials. *Chemical Society Reviews*. 2008;37(4):664–75.

21. Li T, Lu XM, Zhang MR, Hu K, Li Z. Peptide-based nanomaterials: self-assembly, properties and applications. *Bioactive Materials*. 2022;11:268–82.

22. Gao Y, Wang L, Zhang X, Zhou Z, Shen X, Hu H, et al. Advances in self-assembled peptides as drug carriers. *Pharmaceutics*. 2023;15(2):482.

23. Wang J, Liu K, Xing R, Yan X. Peptide self-assembly: thermodynamics and kinetics. *Chemical Society Reviews*. 2016;45(20):5589–604.

24. Zhang L, Yao L, Zhao F, Yu A, Zhou Y, Wen Q, et al. Protein and peptide-based nanotechnology for enhancing stability, bioactivity, and delivery of anthocyanins. *Advanced Healthcare Materials*. 2023;12(25):2300473.

25. Marchesan S, Vargiu AV, Styan KE. The Phe-Phe motif for peptide self-assembly in nanomedicine. *Molecules*. 2015;20(11):19775–88.

26. Lv C, Zhang X, Liu Y, Zhang T, Chen H, Zang J, et al. Redesign of protein nanocages: the way from 0D, 1D, 2D to 3D assembly. *Chemical Society Reviews*. 2021;50(6):3957–89.

27. Sutter M, Greber B, Aussignargues C, Kerfeld CA. Assembly principles and structure of a 6.5 MDa bacterial microcompartment shell. *Science*. 2017;356(6344):1293–7.

28. Wang Y, Douglas T. Protein nanocage architectures for the delivery of therapeutic proteins. *Current Opinion in Colloid & Interface Science*. 2021;51:101395.

29. Bhaskar S, Lim S. Engineering protein nanocages as carriers for biomedical applications. *NPG Asia Materials*. 2017;9(4):e371–e371.

30. Uchida M, Klem MT, Allen M, Suci P, Flenniken M, Gillitzer E, et al. Biological containers: protein cages as multifunctional nanoplatforms. *Advanced Materials*. 2007;19(8):1025–42.

31. João J, Prazeres DMF. Manufacturing of non-viral protein nanocages for biotechnological and biomedical applications. *Frontiers in Bioengineering Biotechnology*. 2023;11:1200729.

32. Diaz D, Care A, Sunna A. Bioengineering strategies for protein-based nanoparticles. *Genes*. 2018;9(7):370.

33. Zhang C, Zhang X, Zhao G. Ferritin nanocage: a versatile nanocarrier utilized in the field of food, nutrition, and medicine. *Nanomaterials*. 2020;10(9):1894.

34. Ray PD, Huang BW, Tsuji Y. Reactive oxygen species (ROS) homeostasis and redox regulation in cellular signaling. *Cell Signal*. 2012;24(5):981–90.

35. Mohanty A, Parida A, Raut RK, Behera RK. Ferritin: a promising nanoreactor and nanocarrier for bionanotechnology. *ACS Bio Med Chem Au*. 2022;2(3):258–81.

36. Roudi NE, Saraygord-Afshari N, Hemmaty M. Protein nano-cages: novel carriers for optimized targeted remedy. *F1000Res*. 2017;6:1541.

37. Lee NK, Cho S, Kim IS. Ferritin—a multifaceted protein scaffold for biotherapeutics. *Experimental and Molecular Medicine*. 2022;54(10):1652–7.

38. Rother M, Nussbaumer MG, Renggli K, Bruns N. Protein cages and synthetic polymers: a fruitful symbiosis for drug delivery applications, bionanotechnology and materials science. *Chemical Society Reviews*. 2016;45(22):6213–49.

39. Karimi M, Zangabad PS, Mehdizadeh F, Malekzad H, Ghasemi A, Bahrami S, et al. Nanocaged platforms: modification, drug delivery and nanotoxicity opening synthetic cages to release the tiger. *Nanoscale*. 2017;9(4):1356–92.

40. Zong J, Wang C, Liu B, Liu M, Cao Y, Sun X, et al. Human hsp70 and HPV16 oE7 fusion protein vaccine induces an effective antitumor efficacy. *Oncology Reports*. 2013;30(1):407–12.

41. Bigotti MG, Clarke AR. Chaperonins: the hunt for the group II mechanism. *Archives of Biochemistry and Biophysics*. 2008;474(2):331–9.

42. Gunkel-Grabole G, Sigg S, Lomora M, Lörcher S, G. Palivan C, P. Meier W. Polymeric 3D nano-architectures for transport and delivery of therapeutically relevant biomacromolecules. *Biomaterials Science*. 2015;3(1):25–40.

43. Ishii D, Kinbara K, Ishida Y, Ishii N, Okochi M, Yohda M, et al. Chaperonin-mediated stabilization and ATP-triggered release of semiconductor nanoparticles. *Nature*. 2003;423(6940):628–32.

44. Aljabali AA, Rezigue M, Alsharedeh RH, Obeid MA, Mishra V, Serrano-Aroca Á, et al. Protein-based nanomaterials: a new tool for targeted drug delivery. *Therapeutic Delivery*. 2022;13(6):321–38.

45. Giessen TW, Orlando BJ, Verdegaal AA, Chambers MG, Gardener J, Bell DC, et al. Large protein organelles form a new iron sequestration system with high storage capacity. *eLife*. 2019;8:e46070.

46. Gabashvili AN, Chmelyuk NS, Sarkisova VA, Melnikov PA, Semkina AS, Nikitin AA, et al. *Myxococcus xanthus* encapsulin as a promising platform for intracellular protein delivery. *International Journal of Molecular Sciences*. 2022;23(24):15591.

47. Giessen TW, Silver PA. Widespread distribution of encapsulin nanocompartments reveals functional diversity. *Nature Microbiology*. 2017;2(6):1–11.

48. Jones JA, Benisch R, Giessen TW. Encapsulin cargo loading: progress and potential. *Journal of Materials Chemistry B*. 2023;11(20):4377–88.

49. Khaleeq S, Sengupta N, Kumar S, Patel UR, Rajmani RS, Reddy P, et al. Neutralizing efficacy of encapsulin nanoparticles against SARS-CoV2 variants of concern. *Viruses*. 2023;15(2):346.

50. Ghalandari B, Divsalar A, Saboury AA, Parivar K. β-Lactoglobulin nanoparticle as a chemotherapy agent carrier for oral drug delivery system. *Journal of Iranian Chemical Society*. 2015;12(4):613–9.

51. Torchilin VP. Micellar nanocarriers: pharmaceutical perspectives. *Pharmaceutical Research*. 2007;24(1):1–16.

52. Zhang S, Chen M, Geng Z, Liu T, Li S, Yu Q, et al. Potential application of self-assembled peptides and proteins in breast cancer and cervical cancer. *International Journal of Molecular Sciences*. 2023;24(23):17056.

53. de Oliveira EG, Carvalho JVRP, Botelho BB, da Costa Filho CA, Henriques LR, de Azevedo BL, et al. Giant viruses as a source of novel enzymes for biotechnological application. *Pathogens*. 2022;11(12):1453.

54. Watanabe Y, Mendonça L, Allen ER, Howe A, Lee M, Allen JD, et al. Native-like SARS-CoV-2 spike glycoprotein expressed by ChAdOx1 nCoV-19/AZD1222 vaccine. *ACS Central Science*. 2021;7(4):594–602.

55. Hamilton JR, Tsuchida CA, Nguyen DN, Shy BR, McGarrigle ER, Sandoval Espinoza CR, et al. Targeted delivery of CRISPR-Cas9 and transgenes enables complex immune cell engineering. *Cell Reports*. 2021;35(9):109207.

56. Kaltbeitzel J, Wich PR. Protein-based nanoparticles: from drug delivery to imaging, nanocatalysis and protein therapy. *Angewandte Chemie International Edition*. 2023;62(44):e202216097.

57. Ferrer-Miralles N, Rodríguez-Carmona E, Corchero JL, García-Fruitós E, Vázquez E, Villaverde A. Engineering protein self-assembling in protein-based nanomedicines for drug delivery and gene therapy. *Critical Reviews in Biotechnology*. 2015;35(2):209–21.

58. Dong Y, Chen Y, Zhu D, Shi K, Ma C, Zhang W, et al. Self-assembly of amphiphilic phospholipid peptide dendrimer-based nanovectors for effective delivery of siRNA therapeutics in prostate cancer therapy. *Journal of Controlled Release*. 2020;322:416–25.

59. Wong CK, Qiang X, Müller AHE, Gröschel AH. Self-assembly of block copolymers into internally ordered microparticles. *Progress in Polymer Science*. 2020;102:101211.

60. Lombardo D, Kiselev MA, Magazù S, Calandra P. Amphiphiles self-assembly: basic concepts and future perspectives of supramolecular approaches. *Advances in Condensed Matter Physics*. 2015;2015(1):151683.

61. Zhang J, Lin W, Yang L, Zhang A, Zhang Y, Liu J, et al. Injectable and pH-responsive self-assembled peptide hydrogel for promoted tumor cell uptake and enhanced cancer chemotherapy. *Biomaterials Science*. 2022;10(3):854–62.

62. Daso RE, Banerjee IA. Self-assembled peptide-based biocomposites for near-infrared light triggered drug release to tumor cells. *Biotechnology Journal*. 2020;15(12):2000128.

63. Li J, Chen Z, Zhou M, Jing J, Li W, Wang Y, et al. Polyoxometalate-driven self-assembly of short peptides into multivalent nanofibers with enhanced antibacterial activity. *Angewandte Chemie International Edition English*. 2016;55(7):2592–5.

64. Huang ZH, Shi L, Ma JW, Sun ZY, Cai H, Chen YX, et al. A totally synthetic, self-assembling, adjuvant-free MUC1 glycopeptide vaccine for cancer therapy. *Journal of the American Chemical Society*. 2012;134(21):8730–3.

65. Rong P, Huang P, Liu Z, Lin J, Jin A, Ma Y, et al. Protein-based photothermal theranostics for imaging-guided cancer therapy. *Nanoscale*. 2015;7(39):16330–6.

66. Zhou Y, Li Q, Wu Y, Li X, Zhou Y, Wang Z, et al. Molecularly stimuli-responsive self-assembled peptide nanoparticles for targeted imaging and therapy. *ACS Nano*. 2023;17(9):8004–25.

67. La Manna S, Di Natale C, Onesto V, Marasco D. Self-assembling peptides: from design to biomedical applications. *International Journal of Molecular Sciences*. 2021;22(23):12662.

68. Aggeli A, Bell M, Carrick LM, Fishwick CWG, Harding R, Mawer PJ, et al. pH as a trigger of peptide β-sheet self-assembly and reversible switching between nematic and isotropic phases. *Journal of the American Chemical Society*. 2003;125(32):9619–28.

69. Fan T, Yu X, Shen B, Sun L. Peptide self-assembled nanostructures for drug delivery applications. *Journal of Nanomaterials*. 2017;2017:e4562474.

70. Powers ET, Yang SI, Lieber CM, Kelly JW. Ordered Langmuir–Blodgett films of amphiphilic β-hairpin peptides imaged by atomic force microscopy. *Angewandte Chemie—International Edition*. 2002;41(1):127–30.

71. Hu K, Xiong W, Sun C, Wang C, Li J, Yin F, et al. Self-assembly of constrained cyclic peptides controlled by ring size. *CCS Chemistry*. 2020;2(1):42–51.

72. Liu J, Wang D, Zheng Q, Lu M, Arora PS. Atomic structure of a short α-helix stabilized by a main chain hydrogen-bond surrogate. *Journal of the American Chemical Society*. 2008;130(13):4334–7.

73. Pochan DJ, Schneider JP, Kretsinger J, Ozbas B, Rajagopal K, Haines L. Thermally reversible hydrogels via intramolecular folding and consequent self-assembly of a de novo designed peptide. *Journal of the American Chemical Society*. 2003;125(39):11802–3.

74. Rad-Malekshahi M, Lempsink L, Amidi M, Hennink WE, Mastrobattista E. Biomedical applications of self-assembling peptides. *Bioconjugate Chemistry*. 2016;27(1):3–18.

75. Lee S, Trinh THT, Yoo M, Shin J, Lee H, Kim J, et al. Self-assembling peptides and their application in the treatment of diseases. *International Journal of Molecular Sciences*. 2019;20(23):5850.

76. Castelletto V, Seitsonen J, Ruokolainen J, W. Hamley I. Alpha helical surfactant-like peptides self-assemble into pH-dependent nanostructures. *Soft Matter*. 2021;17(11):3096–104.

77. Mello LR, Aguiar RB, Yamada RY, Moraes JZ, Hamley IW, Alves WA, et al. Amphipathic design dictates self-assembly, cytotoxicity and cell uptake of arginine-rich surfactant-like peptides. *Journal of Materials Chemistry B*. 2020;8(12):2495–507.

78. Li J, Wang J, Zhao Y, Zhou P, Carter J, Li Z, et al. Surfactant-like peptides: from molecular design to controllable self-assembly with applications. *Coordination Chemistry Reviews*. 2020;421:213418.

79. Gupta S, Singh I, Sharma AK, Kumar P. Ultrashort peptide self-assembly: front-runners to transport drug and gene cargos. Frontiers in bioengineering and biotechnology [Internet]. 2020 [cited 2024 Jan 16];8. Available from: www.frontiersin.org/articles/10.3389/fbioe.2020.00504

80. Qiu F, Chen Y, Tang C, Zhao X. Amphiphilic peptides as novel nanomaterials: design, self-assembly and application. *International Journal of Nanomedicine*. 2018;13:5003–22.

81. Zhao C, Chen H, Wang F, Zhang X. Amphiphilic self-assembly peptides: rational strategies to design and delivery for drugs in biomedical applications. *Colloids and Surfaces B: Biointerfaces*. 2021;208:112040.

82. Pal VK, Jain R, Roy S. Tuning the supramolecular structure and function of collagen mimetic ionic complementary peptides via electrostatic interactions. *Langmuir*. 2020;36(4):1003–13.

83. Brodsky B, Ramshaw JAM. The collagen triple-helix structure. *Matrix Biology*. 1997;15(8):545–54.

84. Malcor JD, Mallein-Gerin F. Biomaterial functionalization with triple-helical peptides for tissue engineering. *Acta Biomaterialia*. 2022;148:1–21.

85. Yan X, Zhu P, Li J. Self-assembly and application of diphenylalanine-based nanostructures. *Chemical Society Reviews*. 2010;39(6):1877–90.

86. Görbitz CH. The structure of nanotubes formed by diphenylalanine, the core recognition motif of Alzheimer's β-amyloid polypeptide. *Chemical Communications*. 2006;(22):2332–4.

87. Reches M, Gazit E. Designed aromatic homo-dipeptides: formation of ordered nanostructures and potential nanotechnological applications. *Physical Biology*. 2006;3(1):S10.

88. Warren JL, Dykeman-Bermingham PA, Knight AS. Controlling amphiphilic polymer folding beyond the primary structure with protein-mimetic di(phenylalanine). *Journal of the American Chemical Society*. 2021;143(33):13228–34.

89. Chalfie M. Green fluorescent protein. *Photochemistry and Photobiology*. 1995;62(4):651–6.

90. Yu K, Liu C, Kim BG, Lee DY. Synthetic fusion protein design and applications. *Biotechnology Advances*. 2015;33(1):155–64.

91. Van der Weken H, Cox E, Devriendt B. Rapid production of a chimeric antibody-antigen fusion protein based on 2A-peptide cleavage and green fluorescent protein expression in CHO cells. *mAbs*. 2019;11(3):559–68.

92. Chan KY, Kwong YHH, Szeto, DM. Green fluorescent protein: its development protein engineering, and applications in protein research. *Journal of Young Investigators*. 2022;25(3).

93. Xia Y, Tang S, Olsen BD. Site-specific conjugation of RAFT polymers to proteins via expressed protein ligation. *Chemical Communications*. 2013;49(25):2566–8.

94. Kim SK, Foote MB, Huang L. The targeted intracellular delivery of cytochrome C protein to tumors using lipid-apolipoprotein nanoparticles. *Biomaterials*. 2012;33(15):3959–66.

95. Kimple ME, Brill AL, Pasker RL. Overview of affinity tags for protein purification. *Current Protocols in Protein Science*. 2013;73:Unit-9.9.

96. Sun H, Wang H, Chen Q, Dong W, Gao C, Song H, et al. Cation affinity purification of histidine-tagged proteins. *Applied Microbiology and Biotechnology*. 2023;107(7):2639–51.

97. Gnanasekaran P, Pappu HR. Detection of protein–protein interactions using glutathione-S-transferase (GST) pull-down assay technique. In: Mukhtar S, editor. *Protein–Protein Interactions: Methods and Protocols* [Internet]. New York, NY: Springer US; 2023 [cited 2024 Feb 2]. pp. 111–5. Methods in Molecular Biology. Available from: https://doi.org/10.1007/978-1-0716-3327-4_10

98. Soltau AD, Malisetti N, Winslow PJ, Seelig B. TEV protease cleavage of MBP and his affinity tags in de novo fusion proteins. 2023 Jul [cited 2024 Feb 2]. Available from: http://conservancy.umn.edu/handle/11299/256042

99. Boulaiz H, Marchal JA, Prados J, Melguizo C, Aránega A. Non-viral and viral vectors for gene therapy. *Cellular and Molecular Biology (Noisy-le-Grand, France)*. 2005;51(1):3–22.

100. Regier MC, Taylor JD, Borcyk T, Yang Y, Pannier AK. Fabrication and characterization of DNA-loaded zein nanospheres. *Journal of Nanobiotechnology*. 2012;10(1):44.

101. Mo Y, Barnett ME, Takemoto D, Davidson H, Kompella UB. Human serum albumin nanoparticles for efficient delivery of Cu, Zn superoxide dismutase gene. *Molecular Vision*. 2007;13:746–57.

102. Langiu M, Dadparvar M, Kreuter J, Ruonala MO. Human serum albumin-based nanoparticle-mediated in vitro gene delivery. *PLOS One*. 2014;9(9):e107603.

103. Xu J, Singh A, Amiji MM. Redox-responsive targeted gelatin nanoparticles for delivery of combination wt-p53 expressing plasmid DNA and gemcitabine in the treatment of pancreatic cancer. *BMC Cancer*. 2014;14(1):75.

104. Webber MJ, Newcomb CJ, Bitton R, Stupp SI. Switching of self-assembly in a peptide nanostructure with a specific enzyme. *Soft Matter*. 2011;7(20):9665–72.

105. Veerman C, Rajagopal K, Palla CS, Pochan DJ, Schneider JP, Furst EM. Gelation kinetics of β-hairpin peptide hydrogel networks. *Macromolecules*. 2006;39(19):6608–14.

106. Guo H Dong, Cui G Hong, Yang J Jun, Wang C, Zhu J, Zhang L Sheng, et al. Sustained delivery of VEGF from designer self-assembling peptides improves cardiac function after myocardial infarction. *Biochemical and Biophysical Research Communications*. 2012;424(1):105–11.

107. Rapaport H. Amphiphilic peptides and hydrogel matrices thereof for bone repair [Internet]. US8658763B2, 2014 [cited 2024 Jul 20]. Available from: https://patents.google.com/patent/US8658763B2/en

# 10 Nanocarrier Functionalization Strategies

## *Targeting Ligands, Imaging Agents, Stimuli-Responsive Elements*

*Sampada Satish Shinde, Bhavna Sonule, and Lalit Kumar*

### ABBREVIATIONS

| | |
|---|---|
| API | active pharmaceutical ingredient |
| DDS | drug delivery systems |
| EGFR | epidermal growth factor receptor |
| EPR | enhanced permeability and retention |
| HA | hyaluronic acid |
| NP | nanoparticle |
| PEG | polyethylene glycol |
| PET | positron emission tomography |
| SDDS | Smart Drug Delivery Systems |
| SPECT | single photon emission computed tomography |
| Tf | transferrin |

### 10.1 INTRODUCTION

Nanomedicine is a rapidly emerging field of study that utilizes nanotechnology to provide highly specialized medical interventions for illness prevention, diagnosis, and therapy. Diverse fields, from materials science to nanomedicine, establish nanotechnology as attractive due to its unique physical, optical, and electronic characteristics [1]. Nanotechnology has emerged as a potent tool in the finding of novel strategies to fight against unmet clinical challenges. Nowadays, nanotechnology is being used for a variety of biological purposes and is growing more advanced and developed scientifically. It can be used to treat complex illnesses like cancer or neurological disorders or to diagnose these pathologies early enough to potentially eradicate them before any symptoms manifest [2]. Many types of nanostructures (Table 10.1), including nanoparticles (NPs), nanocomposites, nanotubes, and nanofibers, are effective in the identification and management of a wide range of diseases. Furthermore, these nanostructures are used as transporters or carrier molecules for drugs, vaccines, proteins, genes, and enzymes [3]. Nanocarriers are colloidal NPs that are commonly utilized to deliver chemical compounds, such as medicinal agents, to a particular site. Drug delivery by different nanocarriers, including liposomes, dendritic macromolecules, exosomes, inorganic NPs, viral-like particles, and protein and polypeptide NPs, has been somewhat successful in recent years. Due to the inactive nature of nanocarriers, they are usually considered to be safe mediums, and they provide good biocompatibility [4].

Systemic adverse effects can occasionally arise from the non-specific biological distribution and uncontrolled drug release properties of conventional drug delivery system (DDS). Investigating novel drug delivery technologies is necessary for substantial growth in the pharmaceutical industry as it can have both therapeutic and financial benefits. Furthermore, several drugs are difficult to administer using traditional drug delivery strategies due to a lack of therapeutic activity and many problems such as limited bioavailability, sensitive toxicity, insufficient specificity, and many more. In addition, there are other obstacles to take into account and get beyond, such as enzyme attacks, poor tissue permeability, and access concerns to the target once it gets to the destination cells. Investigations into novel, innovative approaches to drug distribution are necessary [6]. The drawbacks of conventional drug delivery overcome with the development of smart DDSs (SDDSs), which enable payloads to be delivered to specific locations in a spatially controlled manner. In addition to their many other uses, SDDS has the potential to be developed into a smart systems approach that contains imaging and therapeutic chemicals inside of them and has stealth characteristics. Additionally, SDDSs can be utilized to create diagnostic instruments like positron emission tomography (PET) scanning and magnetic resonance imaging-contrast agents for accurate and timely cancer diagnosis [7].

Most of the time, there is a significant distance between the site of drug administration and the expected therapeutic effect. Targeted drug delivery has thus been the focus of extensive research. Combining research from several different fields—such as chemistry, biology, material science, pharmaceutics, and clinical medicine—NP DDS is the outcome of these efforts [8]. The intrinsic properties of NPs, such as size, shape, and morphology, as well as their extrinsic properties from surface changes are crucial for their cellular transport and uptake. The selection of a targeted ligand is influenced by several factors, such as cost, immunogenicity, flexibility, affinity, ease of manufacture, and availability. When preparing NPs with a maximal targeting capacity and a low cost, these factors should be carefully taken into consideration. NPs can be made functional by attaching themselves to ligands of different kinds, such as aptamers, small molecules,

DOI: 10.1201/9781003473183-10

## Table 10.1 Commonly used nanocarriers in SDDS [1–9]

| Nanocarriers | Challenges | Functionalized Strategies | Uses | Method of Preparation |
|---|---|---|---|---|
| Liposomes | Slower blood circulation times, increased drug release, inadequate drug loading, and instability | ***Targeting ligand:*** monoclonal antibodies, antibody fragments, proteins, peptides, vitamins, carbohydrates, and glycoproteins are typically grafted on the liposome | Actively target the tumor site | Solvent injection technique, detergent dialysis, Bangham method, and reverse-phase evaporation are the most commonly used methods |
| | | PEGylated liposomes | PEGylation is required for liposomes to exit the RES. PEGylated liposomes circulate blood for an extended duration of time | |
| | | Radiolabeled liposome | Used to identify the biodistribution of liposomes within the body, as well as to identify the tumor and deliver treatment | |
| Micelles | Crossing the critical micelle concentration (CMC) results in the release of immature drugs. Furthermore, the balance between the micelle and blood may be changed by interactions with blood and the absorption of unimers into plasma protein | • Cross-linked, i.e., a disulfide bond forms between two polymer chains<br>• Ligands are used to decorate the micelle surface<br>• Stimuli response<br>• Multifunctional with co-delivery strategies | Together, doxorubicin administration with ultrasound-assisted tumor imaging | Direct dissolution and the film casting |
| Dendrimers | Lower absorption by cancer cells and quick removal by the immune system | • Chemical alteration, copolymerization with a linear polymer<br>• Hybridization with different nanocarriers | Nucleic acid delivery | Divergent and convergent method |

## Table 10.1 (Continued)

| Nanocarriers | Challenges | Functionalized Strategies | Uses | Method of Preparation |
|---|---|---|---|---|
| Mesoporous silica nanoparticles | Human serum protein (HSA), hemolysis of red blood cells, and phagocytosis by THP-1 macrophages—a cell line generated from human leukemia—all contribute to the half-life of blood circulation | • PEGylation<br>• Polymer-grafted MSNs<br>• Imaging agent<br>• Stimuli-responsive drug delivery | MSN-based theranostic drug delivery in cancer | Soft and hard template method |
| Gold nanoparticle | Unstable in blood and face higher uptake by the RES | • PEGylated gold NPs show enhanced solubility and stability<br>• FA-targeting ligand | To identify the cancer site, gold NPs modified with fluorescently labeled heparin may be utilized | Physical, chemical, and biological method |
| Superparamagnetic iron oxide nanoparticles | Aggregation of SPIONs | • Use of targeting ligand<br>• Stimuli-responsive polymer-coated SPIONs | Theranostic properties | Synthesis techniques that use co-precipitation, thermal breakdown, hydrothermal processes, microemulsion, sonochemical reactions, and microwave assistance synthesis |
| Carbon nanotubes | RES and toxicity | PEGylation is used to reduce toxicity, avoid RES, and increase solubility | Early cancer detection | Chemical vapor deposition and the laser ablation method |
| QDs | RES's non-specific uptake | Different ligands, including peptides, folate, and major proteins (monoclonal antibodies), can be grafted on the surface of QDs to actively target a tumor site | Natural fluorescence for cancer imaging | Top-down processing techniques include molecular beam epitaxy, ion implantation, ebeam lithography, and X-ray lithography; bottom-up techniques produce colloidal QDs through self-assembly in solution after chemical reduction |

peptides, antibodies, glycoproteins, polysaccharides, DNA or RNA, conjugation methods, or characteristics of the target cell [9].

## 10.2 STRATEGIES FOR TARGETING

By using targeted delivery systems, most drugs can be precisely and successfully delivered to tumor cells or tissues rather than normal cells or tissues. Nanotechnology can be utilized to develop such delivery systems. Because of the EPR effect, which allows NPs to assemble at tumor locations, NP-based delivery of drugs has generally acquired more attention. The "targeting fraction" is a key element of targeted DDS; it can bind selectively to particular moieties or receptors at the target region. Furthermore, because of its low drug dosage, great efficacy, and few adverse effects, targeted drug delivery can achieve the goal of personalized therapy. Targeted nanocarriers, through a variety of targeting mechanisms, are widely known to successfully increase the bioavailability and efficacy of drugs. There are two ways by which targeting of NPs may be achieved, namely passive and active targeting [10].

1. Direct targeting
2. Stimuli-based drug targeting
3. Active targeting
4. Passive targeting

### 10.2.1 Direct Targeting

It can appear challenging to provide a medicinal drug directly to the target site using nanocarriers. This is because diseases spread throughout the body's cells and tissues on a systemic level. Rarely does the localized disease site appear [3].

### 10.2.2 Stimuli-Based Drug Targeting

Stimuli-responsive elements in the DDS refer to the material or system that responds to the internal or external factors to release the drug at targeted sites and times. The stimuli-responsive elements can be differentiated into two types, i.e., endogenous (internal) responsive elements and exogenous (external) responsive elements. The internal stimuli are pH gradient, enzymes, and oxidative stress, whereas external stimuli include magnetic fields, temperature, electrical, and photo/light-responsive DDS.

#### *10.2.2.1 Internal Stimuli-Responsive DDS*

##### *10.2.2.1.1 pH Gradient Responsive DDS*

One of the simplest and most used DDS approaches is elements that respond to pH. The pH-responsive DDS must be stable and responsive to well below physiological pH ranges (5.0–6.5, 7.4) for a wide range of biological applications. Because tumor tissues have a little different pH than normal tissue, using pH-responsiveness to target tumors with chemotherapy is a great option [11]. Variations in pH can be utilized for dispersing drugs as it is possible to change a polymer's properties in response to exposure to a certain pH. The most widely utilized method for creating pH-responsive polymeric systems is to include functional groups that ionize at various pH levels in the polymer structure. For example, at neutral pH carboxylic acid carries a negative charge which becomes protonated with the loss of charge at lower pH. Similar to carboxylic acid, tertiary amines can be used; however, they have the exact opposite function of carboxylic acid. The amines are uncharged and hydrophobic in neutral pH and become positively charged when exposed to the acidic pH [12]. Indeed, utilizing changes in the polymer protonation states based on variations in pH is a common strategy in the design of responsive polymers. This pH-responsive behavior can be utilized for controlled drug delivery. For example, if the polymer is used as a drug carrier, it can be designed to remain intact and insoluble in the acidic environment of the stomach. Once it reaches the more basic or neutral environment of the intestines or the targeted tissues, where the pH is high, the polymer undergoes a transition to a water-soluble form. This transformation facilitates the release of the encapsulated drug. The advantage of such systems lies in their ability to respond to specific environmental signals, providing a means for targeted and controlled drug release. This approach has various applications in drug delivery where precise drug delivery is crucial [13]. Sahu et.al developed the pH-responsive chitosan bleomycin biodegradable nanogels for the treatment of breast cancer. Chitosan is pH-responsive biodegradable polymer that can release the bleomycin in a controllable mechanism due to the swelling action as chitosan is hydrophilic and accumulates more quantity of water and swells. Nanogel matrix degrades due to the acidic pH of the skin which releases the bleomycin–chitosan NPs at pH 5.5–6.5. Then due to the characteristics of melanoma and leaky vasculature accumulation of NPs occurs through the enhanced permeability and retention (EPR) effect. In the skin, normal melanin cells have a pH of 7.4, whereas skin cancer melanoma tumor cells have a pH of 5.5–6.5, i.e., slightly more acidic than normal cells. The pH-responsive nanogels observed no release of bleomycin from chitosan NPs at normal melanin cell pH 7.4, whereas it showed the rapid release of bleomycin at tumor cells with pH 5.5–6.5 [13].

#### *10.2.2.1.2 Enzyme-Responsive DDS*

Enzymes are important components of cell control and systemic metabolism, and they are useful markers for a wide range of illness diagnoses [14]. Various disease sites and enzymatic expression dysregulation areas have higher enzyme concentration which can be linked with certain diseases [15]. In cancer, targeted therapies can specifically kill the tumor cells based on enzymatic activation. Also, the specific enzyme of the tumor may control the release of drugs from the NPs [16]. These enzyme-responsive stimuli have various applications in the drug delivery. The phospholipase A2 (PLA2) is one of the enzymes which cleaves the fatty acids of the mammalian cell. This PLA2 is found in some advanced cancer stages including cancers of digestive organs [17–19]. To develop gemcitabine nanovectors for pancreatic cancer therapy, Han et al. [19] developed a dual enzymatic reaction-assisted gemcitabine delivery system. Matrix metalloproteinase-9 (MMP-9) conjugation, detachable polyethylene glycol (PEG), and cathepsin B-cleavable gemcitabine were used in this procedure. Because the pancreatic tumor overexpresses cathepsin B and MMP-9, the overexpression of MMP-9 in the tumor environment broke down the PEG barrier layer, which made it easier for the cancer cells to internalize. The increased cathepsin B further encourages the drug's release from the nanovectors during cellular absorption [34]. Also, Vaghasiya et al. [20] enveloped enzyme-responsive NPs coated with concanavalin-A for lung cancer therapy. Along with MMPs, the mannose receptors are also overexpressed on the tumor cell surface. In their study, they formulated the cisplatin-loaded gelatin NPs which were surface-coated with concanavalin A. The concanavalin A binds with the mannose receptors which gives site-specific cellular internalization of the NPs, after that the subsequently high level of MMP tumor tissue helps to release the cisplatin in a controlled release manner [20].

#### *10.2.2.1.3 Oxidative Stress-Responsive DDS*

Reactive oxygen species (ROS) serve as essential for metabolism and signal transduction. However, excessive ROS production in tissues or cells frequently results in oxidative stress, which has been linked to several illnesses, including inflammation, atherosclerosis, cancer, and aging. Reactive DDSs are gaining more attention from researchers due to the growing requirement for on-demand drug delivery and the advancement of ROS-responsive materials and nanomedicine. In normal pathological and physiological processes, the optimum level of ROS has an important role but the production of ROS at the normal level activates the antioxidant defense system which leads to oxidative stress [21]. All of the molecules that are produced when the body produces oxygen are known as ROS. These compounds include hydroxyl radicals ($HO^-$), superoxide ($O2^-$), and hydrogen peroxide ($H_2O_2$). These molecules can change into one another through a sequence of reactions [22]. These ROS can be generated both endogenously and exogenously by NADPH enzyme-catalyzed reactions or mitochondrial metabolism and exposure to UV light or xenobiotic compounds respectively [21, 23]. The overproduction of these ROS is associated with various diseases including diabetes [24], cancer [25], neurodegenerative diseases [26], and cardiovascular diseases [27]. This method of drug delivery has several applications. For instance, Yin et al. [28] created integrated block copolymer prodrug NPs for a combination of ROS-responsive drug release and tumor oxidative amplification. This is because tumor tissues have higher levels of ROS than normal tissues, which can act as a stimulus to initiate drug release at the targeted site. In this study, to encapsulate β-apache, the researchers synthesized amphiphilic block copolymer prodrugs using PEG and polymerized methacrylate monomers of thioketal-linked camptothecin (CPT). These prodrugs were subsequently self-assembled to form micelles. Once this nanoformulation is administered systemically, β-lapachone can specifically cause an increase in ROS levels by catalyzing the overexpressed quinone oxidoreductase-1 (NQO1) enzyme seen in NADPH cancer cells. Thioketal linkers cleave as a result of the elevated ROS level, releasing the medication that inhibits tumor growth [28]. In another example, Behroozi et al. [29] developed liposomal drug delivery for the treatment of age-related macular degeneration (AMD), as oxidative stress is responsible for the progression of AMD. In this study, they have developed an antioxidant delivery system that is based on the redox-sensitive liposome. As previously discussed, the oxidative stress is high in AMD so the liposome containing di-selenide can disrupt at redox condition and release the drug [29].

#### *10.2.2.2 External Stimuli-Responsive DDS*

##### *10.2.2.2.1 Temperature-Responsive DDS*

Temperature is unique among all the stimuli as it can act as both internal as some tumors are naturally having high temperatures and external as temperature can be applied externally to increase the temperature in the disease site or body [30]. Many polymeric materials can be used as temperature-responsive systems and many studies also confirmed that these thermo-responsive polymers can trigger the release of the drug by slight variation in the temperature [31–34]. By using this temperature as a stimulus, Ruan et al. [36] have developed a mitochondrial temperature-responsive DDS for lung cancer. In the literature, it is mentioned that the mitochondrial temperature difference has been observed between normal cells and cancer cells [35]. The temperature-responsive nanocarriers can prevent doxorubicin efflux and also improve the accumulation in cancer cells and thus mitochondrial targeting [36]. Another application of this stimulus is in the field of pesticides, Yao et al. [37] have developed the nano-pesticide of imidacloprid based on mesoporous silica NPs. Imidacloprid nanocarriers were prepared by using the mesoporous silica NPs as core and coated with paraffin wax. The imidacloprid will only release at high temperature due to the temperature-responsive paraffin wax.

##### *10.2.2.2.2 Magnetic Field-Responsive DDS*

Applying a magnetic field to the target place allows magnetic nanocarriers to be directed noninvasively. Different formulations of magnetic NPs have been developed to diagnose and treat disorders for which traditional therapy has demonstrated poor effectiveness. Although this method has several benefits, such as enhanced magnetic nanocarrier cellular uptake at the disease's site, it also has certain drawbacks. Several factors, including vascular supply, depth of the target tissue, blood flow velocity, field strength, and the physicochemical characteristics of the particles, affect drug delivery efficiency. The blood flow rate may be altered because of the magnetic nanocarrier in the blood vessel [38]. Therapeutic medications and genes are put into the coating shell of magnetic nanocarriers, which are made up of a magnetic core. Targeting molecules can alter the surface of the coating shell [39]. The strength of the magnetic field can determine how well magnetic nanocarriers perform mechanical or thermal effects that are used in hyperthermia theranostics [40]. When magnetization reversal occurs, magnetic particles produce heat. The therapeutic effects of an alternating magnetic field can be enhanced by increasing its frequency and amplitude [35, 60]. This magnetic field-responsive DDS has various applications in the diagnostic and therapeutic area. Kong et al. [41] explained the magnetic field-sensitive lipid–polymer hybrid NPs for breast cancer. Here, the 10 nm magnetic beads and active pharmaceutical ingredient (API) CPT are encapsulated in NPs made via the nanoprecipitation process of lipid–polymer hybrid. For systemic distribution, the hydrophobic drug is contained in a hydrophobic polymeric core, and the lipid coating provides the barriers to prevent drug release in the absence of external stimuli. As soon as the remote radio frequency magnetic field is actuated, it results in the localized heating of $Fe_3O_4$ inside the polymeric core which loses the polymeric matrix and releases the drug for cytotoxicity to cancer cells. Fang et al. [42] prepared the magnetic poly(lactic-co-glycolic) acid (PLGA) microspheres of doxorubicin and coated them with the $\gamma$-$Fe_2O_3$. The $\gamma$-$Fe_2O_3$ was applied electrostatically to the microsphere's surface to guarantee a high level of sensitivity to an external magnetic field that changed with frequency. The drug release from the microspheres was induced by the external alternating current magnetic field, which activated the $\gamma$-$Fe_2O_3$.

##### *10.2.2.2.3 Photo/Light-Responsive DDS*

A photoresponsive drug delivery systems (DDS) refers to a specialized drug delivery platform designed to release therapeutic agents in response to light stimuli. This type of system typically involves incorporating photosensitive materials or molecules into the drug carrier. When exposed to a specific wavelength of light, these materials change their structure or properties, triggering the release of the encapsulated drug at the target site. Based on the light-induced reaction pathways, the photoresponsive DDSs can be differentiated into two types, i.e., direct photoresponsive and indirect photoresponsive. Direct photoresponsive means when the light of a specific wavelength is absorbed by the material, it changes its chemical structure where whereas the indirect photoresponsive stimulus means the light is absorbed by the intermediate signal or the molecule to trigger the transformation [43]. The direct photoresponsive stimuli include UV light [44], visible light [45], and near-infrared (NIR) [46] responsive drug delivery. Indirect light response can also be used as an alternative the direct light which includes light-induced heat reactions, ROS production, gas release, and hypoxia occurrence [43]. Photoresponsive DDSs hold significant promise for enhancing the precision, safety, and efficacy of therapeutic interventions, offering a versatile platform for targeted drug delivery in various biomedical applications. Using a surface coating of Ce6 on the cell membrane and loading doxorubicin into the cell, Gao et al. [47] developed a red blood cell (RBC)-based DDS of doxorubicin. The RBC membrane is disrupted and the medication is released when exposed to 660 nm with low power density singlet oxygen produced by Ce6.

##### *10.2.2.2.4 Electro-Responsive DDS*

An electrically responsive DDS utilizes electrical stimuli to trigger the release of the therapeutic agent from a drug carrier or reservoir. This system includes electro-responsive materials like polymers which change conductivity or swelling in response to the applied electric field, which releases the encapsulated drug. Qu et al. [48] formulated the biocompatible conductive hydrogel of dextran and aniline trimer for localized drug release. Hexamethylene di-isocyanate was used as a crosslinker when dextran and electroreactive aniline trimer were combined to formulate this hydrogel. An on-off, precisely regulated drug release mechanism was developed by controlling the hydrogel with external electrical inputs. Hydrogels exhibited greater drug release when an increased voltage was applied, and decreased drug release when the voltage was removed. Spizzirri et al. [49] developed spherical gelatin and carbon nanotube spherical hybrid hydrogel with sodium methacrylate and N, N-ethylene bisacrylamide for the delivery of diclofenac sodium salt.

### 10.2.3 Passive Targeting

Passive targeting means targeting drugs through the leaky vasculature of a diseased area to cause passive accumulation of drugs. It was discovered that the permeability of vascular endothelial cells increases, and they become leaky under specific pathological conditions such as tumors, infarcts, and inflammation. NPs like liposomes and micelles can accumulate and have therapeutic effects in these vascularly permeable regions. Also known as an EPR effect, this spontaneous or "passive" drug delivery is mostly employed to target cancer. The drug doxil, which is a readily available formulation of doxorubicin embedded in PEG-coated liposomes that circulate throughout time, is an ideal example of EPR-based targeting [50].

### 10.2.4 Active Targeting

Active targeting usually involves the conjugation of one or more targeting moieties to the NP's surface, which interact selectively with antigens or receptors that are either overexpressed or uniquely expressed on tumor cells in comparison to normal tissues. Another benefit of this approach, if internalizing receptors are targeted, is that once the particles enter the tumor's extracellular space, they can be transported into the cells by a particular pathway [51]. Active targeting primarily depends on a biological interaction between the cell target and the ligands on the surface of NPs, which is in contrast to the passive targeting approach. Many biological ligands that support the active targeting of NPs have been found and investigated. The cellular absorption and therapeutic efficacy of drug-containing NPs are enhanced when these biological ligands attach to specific receptors on the surface of the target cells. A higher ligand density is better for improving binding and cellular absorption than a single ligand due to the multivalent effect. To achieve this, ligands belonging to several groups have been employed, such as proteins, polysaccharides, peptides, nucleic acids, and small compounds. These ligands functionalize NPs in two ways generally. Both chemical conjugation and physical adsorption on the NPs following NP formation are options for coupling them with NP components, such as polymers, before NP preparation [2, 4].

#### *10.2.4.1 Ligand-Based Targeting*

TDDS is an effective and easy method. Several targeting ligands have been used in TDDS recently, including folic acid (FA), polysaccharides, peptides, aptamers, and antibodies (as shown in Figure 10.1). They have different structural properties as well as distinct advantages and disadvantages. Their function in differentiating between healthy and unhealthy tissues is important. The markers expressed by targeted tumor cells are precisely recognized by these targeting ligands, which bind to them selectively. Due to this identification, only cells expressing the correct receptor can receive active drugs supplied via targeting ligand-conjugated nanocarriers or small molecules that directly bind to the targeting ligand. Thus, the targeting carriers do not affect normal cells. The effective application of these TDDS has reduced the deleterious side effects and effectively improved the therapeutic efficacy of cytotoxic drugs. This invention is presently being developed on a global scale [52].

##### *10.2.4.1.1 Protein*

For active nanocarrier targeting, proteins like transferrin and human serum albumin can be utilized as targeting ligands.

###### *10.2.4.1.1.1 TRANSFERRIN*

Tf is an 80 kDa glycoprotein that is essential to iron metabolism and is in the role of delivering ferric ions ($Fe^{3+}$). The iron that binds to Tf is called $Fe^{3+}$, and it is produced when $Fe^{2+}$ is oxidized. Tf then interacts with

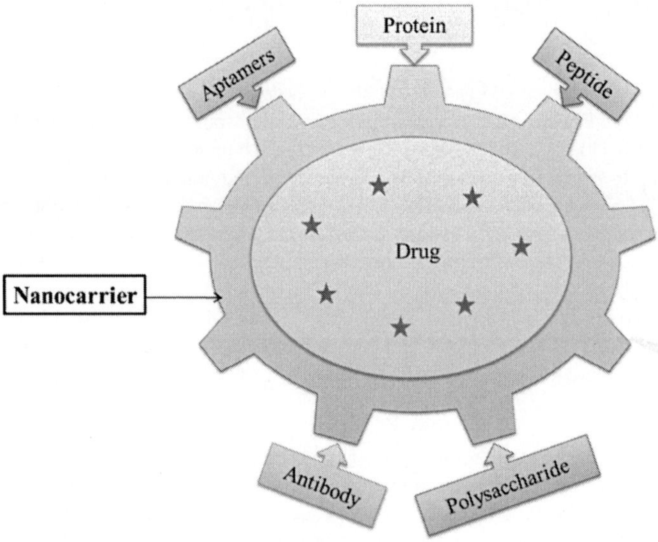


**Figure 10.1** Surface modification of nanocarriers: targeting ligands.

the Tf receptor. The body expresses transferrin receptors widely on a variety of tissues, including endothelial cells in the brain, red blood cells, and several carcinoma cell types. Application of transferring in glioma, brain tumors that overexpress Tf receptors allow more Tf to enter the bloodstream through endocytosis because the tumor requires more iron to keep up with its rapid growth. Iron homeostasis in the brain is further influenced by endocytosis, transcytosis, and paracellular transport. Targeting the Tf receptor on brain endothelial cells using Tf-conjugated NPs (of any substance) may be a viable therapeutic option for treating gliomas due to its function and the overexpression of Tf receptors on human glioma cell lines. Because the NPs can pass through the blood–brain barrier, they can help reduce both the size of the glioma and the rate of cell proliferation [53].

###### *10.2.4.1.1.2 HUMAN SERUM ALBUMIN*

The most abundant protein in plasma is albumin, which makes up over half of all the proteins in the fluid (50%). It serves the essential physiological needs of the body. In particular, the human body uses albumin as an endogenous supply of amino acids and uses it to help keep the plasma osmotic pressure steady. A free sulfhydryl group and eight disulfide linkages provide albumin excellent water solubility (up to 40% w/v at pH 7.4). Consequently, albumin functions as a general transport protein that may produce soluble complexes *in vivo* by binding forces with a variety of insoluble organic compounds and inorganic ions. As an organic biomaterial, albumin is non-immunogenic, biodegradable, and biocompatible. In solid tumors, albumin-coated NPs are taken up by specific cells through a process called transcytosis. It starts when native albumin attaches itself to the 60-kDa glycoprotein receptor on the cell surface (gp60). Furthermore, albumin binding to SPARC (secreted protein acidic and rich in cysteine), an extracellular matrix glycoprotein that is overexpressed in several cancers, constitutes a second pathway [54]. The presence of functional groups on HSA makes it possible to modify them with a variety of targeted ligands, improving DDS targeting and lowering the accumulation of therapeutic API in non-targeted sites [55].

##### *10.2.4.1.2 Polysaccharide*

Hyaluronic acid (HA) is a polymer that can interact with cell surface receptors to facilitate active targeting. Along with collagen, HA is a polysaccharide that is a major part of the extracellular matrix. CD44 is frequently overexpressed on the surface of cancer cells, leading to the hypothesis that it is a distinctive marker of cancer stem cells. HA binds to CD44. At physiological pH, HA is negatively charged. As a result, surface modification of nanocarriers using HA as ligands can reverse the charges on their surfaces [8].

##### *10.2.4.1.3 Peptide*

The peptide, consisting of fewer than 50 amino acids, is called a low-molecular-weight ligand. Peptides play an important role as an effective targeting ligand because they can bind with great affinity and selectivity to receptors inside cells, on the surface of cells, or in the extracellular matrix [56]. Numerous peptide receptors, including the somatostatin analog, vasoactive intestinal peptide, gastrin-related peptide, cholecystokinin, and luteinizing hormone-releasing hormone, are significantly expressed in abundance in specific tumor cells [57]. Peptide 1 (also known as GE11) is a dodecapeptide that is utilized as a cancer cell-targeting peptide. It specifically binds to the EGFR, also known as ErbB1, which is overexpressed in several tumors originating from the epithelium, including breast cancer. Peptide 1 was discovered by comparing a pure human EGFR protein with a phage display peptide library. Hossein-Nejad-Ariani et al. specifically designed a synthetic peptide library of 29 GE11 analogs to target EGFR in triple-negative breast cancer cells. Utilizing flow cytometry and a whole cell-peptide binding experiment, they discovered multiple novel peptide 1 analogs that exhibited increased binding and uptake to triple-negative breast cancer cells. These novel peptide 1 analogs that target EGFR are useful [58].

##### *10.2.4.1.4 Antibody*

The beneficial qualities of antibody-coupled nanocarriers, such as their selectivity and stability toward the biological system, make them a suitable drug-targeting approach. Antibody specificity is responsible for the active targeting of various tumor types. They have the capacity to recognize and distinguish between malignant and non-cancerous cells across several cancer types, with a primary focus on colorectal cancer [59]. Recombinant monoclonal antibodies (mAbs) are most commonly used because they exhibit significant binding coefficient KD in the sub-nanomolar range and superior specificity when interacting naturally with their complementary antigens. By inhibiting ligand–receptor interactions or reducing tumor development, mAbs specifically target overexpressed antigens or receptors in tumors, thereby inducing anticancer effects [4].

##### *10.2.4.1.5 Aptamers*

Aptamers represent a class of nucleic acids that consist of several nucleotides. Two types of aptamers can be distinguished: (1) peptide aptamers, which consist of one or more short variable peptide domains; and (2) DNA, RNA, or XNA (xeno nucleic acid) aptamers, which usually consist of short oligonucleotide strands. While some aptamers are produced specifically for a target, most aptamers are found naturally as the ligand-binding components of riboswitches [60]. The majority of aptamers that are chosen using systematic evolution of ligands by exponential enrichment (SELEX) demonstrate a strong affinity for targets that the range between the micromolar and nanomolar range in terms of their dissociation constant (Kd). Aptamers are better than antibodies as they are not immunogenic, making them useful tools for a range of applications that were initially intended for antibodies [61]. Among the various aptamers most commonly utilized in cancer treatment is AS1411, a single-strand aptamer. Prostate, breast, and lung cancer cell lines are only a few of the human tumor cell lines have been shown to effectively suppress. To enhance the bioactivity of AS1411 and facilitate its efficient cellular transport, Apt-AuNS (aptamer-conjugated gold NPs) nanocarriers were employed [62].

##### *10.2.4.1.6 Small-Molecule Weight Ligand*

Targeting small molecules aims to influence the activity of protein targets, which is the molecular biology basis of cancer. The roles that small molecules targeted chemicals play vary depending on the type of target. Currently, enzymes and receptors are the primary protein targets of small molecule-targeted medicines. While functioning on the receptor, small molecules that target receptors can be classified as either receptor agonists or antagonists. The small-molecule ligands include carbohydrates (fructose, glucose, mannose), vitamin (vitamin B12, FA, riboflavin, biotin, thiamine), and amino acid (glycine, arginine, alanine, tryptophan, lysine). Small molecules with an average molecular weight of less than 1 kDa, or 1 nm, are known as small molecule-targeting ligands and are utilized for targeted administration [63]. Several NPs target FA receptors by receptor-mediated endocytosis because it is well known that these receptors are overexpressed in macrophages and solid tumor cells. In a recent study, Lv et al. [64] used the large gas-filled microbubble (MB) approach to decorate mesoporous silica NPs (MSNs) modified with FA for active targeting. By using local ultrasonic irradiation to break down the gas-filled MB, this approach releases the FA-modified MSNs into the target tissue through the endothelium layer. This method was used in a study to load the hydrophobic medication tanshinone IIA (TAN) into MSNs, which showed a strong power to trigger tumor cell death as well as a high loading capacity. The HeLa and A549 cells exhibited minimal cytotoxicity toward the FA-modified MSNs and MB [8]. The main obstacles faced by small molecule-targeting ligands are their small size, which necessitates a molecular interaction to demonstrate tissue selectivity, and their high mobility, which can cause off-target effects. Nevertheless, small molecules are a great target due to their small size, simplicity in synthesis, and absence of immunogenicity [63].

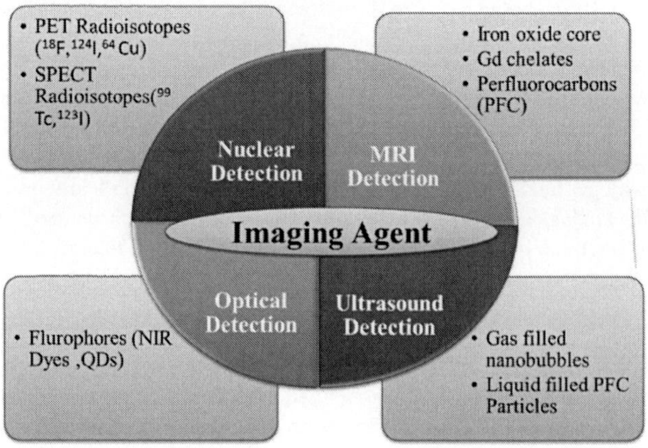


**Figure 10.2** Imaging agents for theranostic approach.

## 10.3 IMAGING AGENT

In theory, specific NPs offer numerous advantages for cellular target in vivo imaging. Most particles have a large loading capacity that allows for the site-specific delivery of large amounts of imaging agents, improving the signal-to-noise ratio. Additionally, the surface of the particle can be specifically designed to provide intrinsic imaging functionality, as is the case with superparamagnetic iron oxide nanoparticles (SPIONs), gold NPs, or quantum dots (QDs) (as shown in Figure 10.2). Early research using antibody-decorated NPs for imaging purposes ran into issues with selective accumulation; extravasation of the NPs from the vasculature, as opposed to cell attachment, was discovered to be the limiting step [65]. While the drug may be chemically modified, the labeling of the drug itself may contain information about the drug's localization. Co-loading drugs with NPs and imaging agents is an additional approach; nevertheless, the drug's and the imaging agent's release profiles can still be different. The use of API and NPs with inherent imaging detectability, such as drugs, iron oxide nanoparticles (ION) particles, gold nanoparticles (AuNP), and chemical exchange saturation transfer (CEST) magnetic resonance imaging (MRI)-detectable NPs, may have the advantage of eliminating the requirement for additional imaging labels [21]. The primary emphasis of Chacko et al.'s research was the targeted nanocarrier for vascular inflammation imaging and treatment [66]. An appropriately chosen label or probe (such as a radionuclide, iron oxide, gadolinium-chelate, gas- or liquid-filled bubble, fluorophore, etc.) is stably bonded to a carrier in a targeted molecular imaging NP probe. The inclusion of masking (such as PEG) and membrane-permeating moieties in a nanocarrier is based upon the needs of both intracellular delivery of the probe and longevity in circulation. Maximum imaging of vascular inflammation can be accomplished with NP probes whose pharmacokinetics support delivery, precise binding to the target, and rapid blood clearance [66]. Different imaging probe detection modalities have different spatial and temporal resolutions, signal penetration depths, imaging probe detection thresholds, and imaging probe availability. The aspects of the modalities that can be used to trace NP imaging probes in vivo and have the most promise for clinical translation when probes become available are briefly discussed in this section [67].

### 10.3.1 Nuclear Detection

Imaging technologies employed in medical diagnosis have always concentrated on either the functional alteration (i.e., areas of malignant growth) or the anatomical structure (i.e., locations of suspicious masses and internal organs) (i.e., single photon emission computed tomography (SPECT), positron emission tomography (PET)). Dual modality imaging is made more potent by combining these two distinct imaging systems into a single unit with more recent imaging technologies. SPECT and/or PET pictures combined with CT and/or MR data provide doctors with a better understanding of the location and changes in tissue. Noninvasive 3D tomographic imaging methods such as SPECT and PET may identify gamma ($\gamma$)-rays from radionuclide-labeled probes placed deep inside living tissue. These methods can identify radiotracers at concentration levels ranging from nanomolar to picomolar ($10^{-9}$–$10^{-12}$ M) in real time with a rapid temporal resolution [66].

### 10.3.2 MRI Detection

Utilizing non-ionizing radiation produced by an electromagnetic field, MRI produces images with superior soft tissue contrast, high temporal resolution, and submillimeter spatial resolution. Usually, it depends on detecting the relaxation characteristics (T1 and T2) of excited hydrogen (1H) nuclei in soft tissue lipids and water [66]. The MR contrast agent promotion has been dominated by gadolinium and iron oxide-based agents, which are used to improve the contrast of T1- and T2-weighted images. Bright signals appear on T1-weighted images when T1 agents primarily shorten T1 relaxation, which is dependent on local water flow. T1 agents impact protons nearby. However, iron oxide-based compounds, also referred to as T2 agents, have a high magnetic sensitivity and cause localized field disruptions. These disruptions in magnetic susceptibility heterogeneity cause signal dephasing in tissues, which ultimately results in a loss of signal via T2 decay, particularly on gradient echo pictures. Generally, the reticuloendothelial system (RES) absorbs non-targeted iron oxide-based agents non-specifically, and the size of the NP as a whole controls the biodistribution of the agent [68].

### 10.3.3 Ultrasound Detection

Ultrasound imaging, also known as sonography or ultrasound scanning, is a diagnostic imaging technique based on ultrasound that is used to see internal organs and muscles, as well as their sizes, structures, and potential diseases or lesions. The internal organs of the body, their movement, and the flow of blood through vessels can all be seen in real-time ultrasound imaging [68]. Among the inorganic NPs used as contrast agents in ultrasonic theranostics are perfluorocarbon NPs, carbon-based nanomaterials (like graphene oxides, carbon nanotubes, carbon dots, and graphene), noble metal materials (like Au, Ag, Pd, and Pt), and silicon-based nanomaterials [69].

### 10.3.4 Optical Detection

Fluorescence detection technologies, which are sensitive and have quick acquisition and processing times, make optical imaging a potent technique for imaging small animals. This allows for real-time, high-throughput imaging. As background tissue autofluorescence, it decreases the ratio of target to nontarget and causes significant photon absorption and scattering. Therefore, *in vivo*, fluorescence imaging with fluorophores that produce visible light (450–650 nm) is still limited. The sensitivity of fluorescent light emitters in the NIR band (650–900 nm) is higher because of improved light transmission and detection at concentrations ranging from picomolar to femtomolar ($10^{12}$–$10^{15}$ M) [66]. Optical imaging is mainly used for the pre-clinical research. Sara F.F. Almeida reported a study of exosomes loaded with D-luciferin and labeled with indocyanine green are used in optical imaging to determine the targetability for lung metastases [70].

## 10.4 CONCLUSION

Nanocarriers, a wonder of modern technology, are crucial for biological applications, especially when it comes to the delivery of chemotherapy drugs. To overcome the side effects of conventional chemotherapy, smart nanocarrier-based DDSs have been developed. The science underlying SDDSs is still being researched, though, as there are still a lot of obstacles to be overcome before they can be used as a viable substitute for chemotherapy in the treatment of cancer. One of the biggest obstacles to an effective SDDS is the toxicity of the nanocarriers. Research has been done to generate new nanocarriers with lower toxicity or to optimize the toxicity of currently available nanocarriers. Targeted drug delivery mainly focused on the increased therapeutic effect and reduced unwanted side effects and drug toxicity. The targeting is mainly achieved by surface modification by using targeting ligands, imaging agents, and stimuli response elements. According to the biological environment and disease state parameters, the nanocarrier strategies are developed to increase the effectiveness of the API. Compared to nontargeted delivery, the ligand-targeted approach of drug delivery provides more flexibility for drug optimization. Imaging technology is used for the diagnostic and therapeutic response of the drug in different disease states.

## REFERENCES


1. Farah FH. Nanocarriers as delivery systems for therapeutics agents Farah Hamad Farah Department of Pharmaceutical Sciences, College of Pharmacy and Health Sciences, Ajman University, Ajman, PO Box 346, United Arab Emirates. *Int J Pharm Sci Res*. 2019;10(8):3487–507.

2. Villaverde G, Baeza A. Targeting strategies for improving the efficacy of nanomedicine in oncology. *Beilstein J Nanotechnol*. 2019;10(1):168–81.

3. Chamundeeswari M, Jeslin J, Verma ML. Nanocarriers for drug delivery applications. *Environ Chem Lett* [Internet]. 2019;17(2):849–65. Available from: https://doi.org/10.1007/s10311-018-00841-1

4. Yang T, Zhai J, Hu D, Yang R, Wang G, Li Y, et al. "Targeting design" of nanoparticles in tumor therapy. *Pharmaceutics*. 2022;14(9):1–25.

5. Hossen S, Hossain MK, Basher MK, Mia MNH, Rahman MT, Uddin MJ. Smart nanocarrier-based drug delivery systems for cancer therapy and toxicity studies: a review. *J Adv Res* [Internet]. 2019;15:1–18. Available from: https://doi.org/10.1016/j.jare.2018.06.005

6. Rana A, Adhikary M, Singh PK, Das BC, Bhatnagar S. "Smart" drug delivery: a window to future of translational medicine. *Front Chem.* 2023;10(January).

7. Wang H, Huang Q, Chang H, Xiao J, Cheng Y. Stimuli-responsive dendrimers in drug delivery. *Biomater Sci.* 2016;4(3):375–90.

8. Yoo J, Park C, Yi G, Lee D, Koo H. Active targeting strategies using biological ligands for nanoparticle drug delivery systems. *Cancers (Basel).* 2019;11(5).

9. Fernández M, Orozco J. Advances in functionalized photosensitive polymeric nanocarriers. *Polymers (Basel).* 2021;13(15).

10. Li J, Wang Q, Xia G, Adilijiang N, Li Y, Hou Z, et al. Recent advances in targeted drug delivery strategy for enhancing oncotherapy. *Pharmaceutics.* 2023;15(9).

11. Raza A, Rasheed T, Nabeel F, Hayat U, Bilal M, Iqbal HMN. Endogenous and exogenous stimuli-responsive drug delivery systems for programmed site-specific release. *Molecules.* 2019;24(6):1–21.

12. Colson YL, Grinstaff MW. Biologically responsive polymeric nanoparticles for drug delivery. *Adv Mater.* 2012 Jul;24(28):3878–86.

13. Sahu P, Kashaw SK, Kushwah V, Sau S, Jain S, Iyer AK. pH-responsive biodegradable nanogels for sustained release of bleomycin. *Bioorg Med Chem.* 2017;25(17):4595–613.

14. Andresen TL, Thompson DH, Kaasgaard T. Enzyme-triggered nanomedicine: drug release strategies in cancer therapy (Invited Review). *Mol Membr Biol.* 2010;27(7):353–63.

15. Fang Z, Shen Y, Gao D. Stimulus-responsive nanocarriers for targeted drug delivery. *New J Chem.* 2021 Mar;45(10):4534–44.

16. Danhier F, Feron O, Préat V. To exploit the tumor microenvironment: passive and active tumor targeting of nanocarriers for anti-cancer drug delivery. *J Control Release.* 2010 Dec;148(2):135–46.

17. Abe T, Sakamoto K, Mita S, Kamohara H, Hirano YI, Kuwahara N, et al. Surgical trauma induces group II phospholipase A2 production by neutrophils at a local site after surgery. *Clin Biochem.* 1998 Mar;31(2):101–6.

18. Shahriari M, Zahiri M, Abnous K, Taghdisi SM, Ramezani M, Alibolandi M. Enzyme responsive drug delivery systems in cancer treatment. *J Control Release.* 2019 Aug;308:172–89.

19. Han H, Valdepérez D, Jin Q, Yang B, Li Z, Wu Y, et al. Dual enzymatic reaction-assisted gemcitabine delivery systems for programmed pancreatic cancer therapy. *ACS Nano.* 2017 Feb;11(2):1281–91.

20. Vaghasiya K, Ray E, Singh R, Jadhav K, Sharma A, Khan R, et al. Efficient, enzyme-responsive, and tumor receptor-targeting gelatin nanoparticles decorated with concanavalin-A for site-specific and controlled drug delivery for cancer therapy. *Mater Sci Eng C Mater Biol Appl.* 2021 Apr;123.

21. Jing L, and Liu B. ROS-responsive drug delivery systems.pdf. *Bioeng Transl Med.* 2016;3:239–51.

22. D'Autréaux B, Toledano MB. ROS as signaling molecules: mechanisms that generate specificity in ROS homeostasis. *Nat Rev Mol Cell Biol.* 2007;8(10):813–24.

23. Nathan C. Specificity of a third kind: reactive oxygen and nitrogen intermediates in cell signaling. *J Clin Invest.* 2003 Mar;111(6):769.

24. Haigis MC, Yankner BA. The aging stress response. *Mol Cell.* 2010 Oct;40(2):333–44.

25. Pelicano H, Carney D, Huang P. ROS stress in cancer cells and therapeutic implications. *Drug Resist Updat.* 2004;7(2):97–110.

26. Andersen JK. Oxidative stress in neurodegeneration: cause or consequence? *Nat Med.* 2004;10 Suppl(7):S18.

27. Dhalla NS, Temsah RM, Netticadan T. Role of oxidative stress in cardiovascular diseases. *J Hypertens.* 2000;18(6):655–73.

28. Yin W, Ke W, Chen W, Xi L, Zhou Q, Mukerabigwi JF, et al. Integrated block copolymer prodrug nanoparticles for a combination of tumor oxidative stress amplification and ROS-responsive drug release. *Biomaterials.* 2019 Mar;195:63–74.

29. Behroozi F, Abdkhodaie MJ, Abandansari HS, Satarian L, Ashtiani MK, Jaafari MR, et al. Smart liposomal drug delivery for treatment of oxidative stress model in human embryonic stem cell-derived retinal pigment epithelial cells. *Int J Pharm.* 2018 Sep;548(1):62–72.

30. Karimi M, Sahandi Zangabad P, Ghasemi A, Amiri M, Bahrami M, Malekzad H, et al. Temperature-responsive smart nanocarriers for delivery of therapeutic agents: applications and recent advances. *ACS Appl Mater Interfaces.* 2016 Aug;8(33):21107–33.

31. Zhou J, Pishko MV, Lutkenhaus JL. Thermoresponsive layer-by-layer assemblies for nanoparticle-based drug delivery. *Langmuir.* 2014 May;30(20):5903–10.

32. Sun F, Wang Y, Wei Y, Cheng G, Ma G. Thermo-triggered drug delivery from polymeric micelles of poly(N-isopropylacrylamide-co-acrylamide)-b-poly(n-butyl methacrylate) for tumor targeting. http://dx.doi.org/101177/0883911514535288. 2014 May;29(4):301–17.

33. Chen G, Chen R, Zou C, Yang D, Chen ZS. Fragmented polymer nanotubes from sonication-induced scission with a thermo-responsive gating system for anti-cancer drug delivery. *J Mater Chem B.* 2014 Feb;2(10):1327–34.

34. Kashyap S, Jayakannan M. Thermo-responsive and shape transformable amphiphilic scaffolds for loading and delivering anticancer drugs. *J Mater Chem B.* 2014 Jun;2(26):4142–52.

35. Wang D, Zhou M, Huang H, Ruan L, Lu H, Zhang J, et al. Gold nanoparticle-based probe for analyzing mitochondrial temperature in living cells. *ACS Appl Bio Mater.* 2019 Jul;2(8):3178–82.

36. Ruan L, Chen J, Du C, Lu H, Zhang J, Cai X, et al. Mitochondrial temperature-responsive drug delivery reverses drug resistance in lung cancer. *Bioact Mater.* 2021 Jul;13:191–9.

37. Yao P, Zou A, Tian Z, Meng W, Fang X, Wu T, et al. Construction and characterization of a temperature-responsive nanocarrier for imidacloprid based on mesoporous silica nanoparticles. *Colloids Surf B Biointerfaces.* 2021 Feb;198.

38. Barakat NS. Magnetically modulated nanosystems: a unique drug-delivery platform. *Nanomedicine.* 2009;4(7):799–812.

39. Lee N, Yoo D, Ling D, Cho MH, Hyeon T, Cheon J. Iron oxide based nanoparticles for multimodal imaging and magnetoresponsive therapy. *Chem Rev.* 2015 Oct;115(19):10637–89.

40. Mosayebi J, Kiyasatfar M, Laurent S. Synthesis, functionalization, and design of magnetic nanoparticles for theranostic applications. *Adv Healthc Mater.* 2017 Dec;6(23).

41. Deok Kong S, Sartor M, Jack Hu CM, Zhang W, Zhang L, Jin S. Magnetic field activated lipid–polymer hybrid nanoparticles for stimuli-responsive drug release. *Acta Biomater.* 2013;9(3):5447.

42. Fang K, Song L, Gu Z, Yang F, Zhang Y, Gu N. Magnetic field activated drug release system based on magnetic PLGA microspheres for chemo-thermal therapy. *Colloids Surf B Biointerfaces*. 2015 Dec;136:712–20.

43. Zhou Y, Ye H, Chen Y, Zhu R, Yin L. Photoresponsive drug/gene delivery systems. *Biomacromolecules*. 2018 Jun;19(6):1840–57.

44. De Gracia Lux C, Lux J, Collet G, He S, Chan M, Olejniczak J, et al. Short soluble coumarin crosslinkers for light-controlled release of cells and proteins from hydrogels. *Biomacromolecules*. 2015 Oct;16(10):3286–96.

45. Lee S, Saito K, Lee HR, Lee MJ, Shibasaki Y, Oishi Y, et al. Hyperbranched double hydrophilic block copolymer micelles of poly(ethylene oxide) and polyglycerol for pH-responsive drug delivery. *Biomacromolecules*. 2012 Apr;13(4):1190–6.

46. Zhu K, Liu G, Hu J, Liu S. Near-infrared light-activated photochemical internalization of reduction-responsive polyprodrug vesicles for synergistic photodynamic therapy and chemotherapy. *Biomacromolecules*. 2017 Aug;18(8):2571–82.

47. Gao M, Hu A, Sun X, Wang C, Dong Z, Feng L, et al. Photosensitizer decorated red blood cells as an ultrasensitive light-responsive drug delivery system. *ACS Appl Mater Interfaces*. 2017 Feb;9(7):5855–63.

48. Qu J, Liang Y, Shi M, Guo B, Gao Y, Yin Z. Biocompatible conductive hydrogels based on dextran and aniline trimer as electro-responsive drug delivery system for localized drug release. *Int J Biol Macromol*. 2019 Nov;140:255–64.

49. Spizzirri UG, Hampel S, Cirillo G, Nicoletta FP, Hassan A, Vittorio O, et al. Spherical gelatin/CNTs hybrid microgels as electro-responsive drug delivery systems. *Int J Pharm*. 2013 May;448(1):115–22.

50. Prasad D, Chauhan H. Key targeting approaches for pharmaceutical drug delivery. *Am Pharm Rev*. 2013;16(6):1–8.

51. Pearce AK, O'Reilly RK. Insights into active targeting of nanoparticles in drug delivery: advances in clinical studies and design considerations for cancer nanomedicine. *Bioconjug Chem*. 2019;30(9):2300–11.

52. Yan S, Na J, Liu X, Wu P. Different targeting ligands-mediated drug delivery systems for tumor therapy. *Pharmaceutics*;2024;248.

53. Koneru T, McCord E, Pawar S, Tatiparti K, Sau S, Iyer AK. Transferrin: biology and use in receptor-targeted nanotherapy of gliomas. *ACS Omega*. 2021;6(13):8727–33.

54. Hashada RA, Ishaka RAH, Geneidia AS, Mansoura S. Methotrexate loading in chitosan nanoparticles at a novel pH: Response surface modeling, optimization and characterization. *Int J Biol Macromol*. 2016;91:630–39, https://doi.org/10.1016/j.ijbiomac.2016.06.014.

55. Li C, Zhang D, Pan Y, Chen B. Human serum albumin based nanodrug delivery systems: recent advances and future perspective. *Polymers (Basel)*. 2023;15(16):1–18.

56. Li ZJ, Cho CH. Peptides as targeting probes against tumor vasculature for diagnosis and drug delivery. *J Transl Med*. 2012;10(Suppl 1):1–9.

57. Das M, Mohanty C, Sahoo SK. Ligand-based targeted therapy for cancer tissue. *Expert Opin Drug Deliv*. 2009;6(3):285–304.

58. Bajracharya R, Song JG, Patil BR, Lee SH, Noh HM, Kim DH, et al. Functional ligands for improving anticancer drug therapy: current status and applications to drug delivery systems. *Drug Deliv* [Internet]. 2022;29(1):1959–70. Available from: https://doi.org/10.1080/10717544.2022.2089296

59. Chaubey P, Momin M, Sawarkar S. Significance of ligand-anchored polymers for drug targeting in the treatment of colonic disorders. *Front Pharmacol.* 2020;10(February):1–14.

60. Kapoor D, Bhatt S, Kumar M, Maheshwari R, Tekade RK. Ligands for targeted drug delivery and applications [Internet]. In: *Basic Fundamentals of Drug Delivery.* Elsevier Inc.; 2018. 307–342. Available from: http://dx.doi.org/10.1016/B978-0-12-817909-3.00008-X

61. Darmostuk M, Rimpelova S, Gbelcova H, Ruml T. Current approaches in SELEX: an update to aptamer selection technology. *Biotechnol Adv* [Internet]. 2014;33(6):1141–61. Available from: http://dx.doi.org/10.1016/j.biotechadv.2015.02.008

62. Kenchegowda M, Rahamathulla M, Hani U, Begum MY, Guruswamyz S, Osmani RAM, et al. Smart nanocarriers as an emerging platform for cancer therapy: a review. *Molecules.* 2022;27(1):1–40.

63. Zhao Z, Ukidve A, Kim J, Mitragotri S. Targeting strategies for tissue-specific drug delivery. *Cell* [Internet]. 2020;181(1):151–67. Available from: https://doi.org/10.1016/j.cell.2020.02.001

64. Lv Y, Cao Y, Li P, Liu J, Chen H, Hu W, Zhang L. Ultrasound-triggered destruction of folate-functionalized mesoporous silica nanoparticle-loaded microbubble for targeted tumor therapy. *Adv Healthcare Mater.* 2017, 6, 1700354. https://doi.org/10.1002/adhm.201700354

65. Richards DA, Maruani A, Chudasama V. Antibody fragments as nanoparticle targeting ligands: a step in the right direction. *Chem Sci* [Internet]. 2016;8(1):63–77. Available from: http://dx.doi.org/10.1039/C6SC02403C

66. Chacko AM, Hood ED, Zern BJ, Muzykantov VR. Targeted nanocarriers for imaging and therapy of vascular inflammation. *Curr Opin Colloid Interface Sci* [Internet]. 2011;16(3):215–27. Available from: http://dx.doi.org/10.1016/j.cocis.2011.01.008

67. Haque ST, Banaszak Holl MM, Chowdhury EH. Strategies to assemble therapeutic and imaging molecules into inorganic nanocarriers. *Front Mater Sci.* 2022;16(3):1–44.

68. Pan D, Lanza GM, Wickline SA, Caruthers SD. Nanomedicine: perspective and promises with ligand-directed molecular imaging. *Eur J Radiol.* 2009;70(2):274–85.

69. Siafaka PI, Okur NÜ, Karantas ID, Okur ME, Gündoğdu EA. Current update on nanoplatforms as therapeutic and diagnostic tools: a review for the materials used as nano theranostics and imaging modalities. *Asian J Pharm Sci.* 2021;16(1):24–46.

70. Almeida SFF, Fonseca A, Sereno J, Ferreira HRS, Lapo-Pais M, Martins-Marques T, et al. Osteosarcoma-derived exosomes as potential PET imaging nanocarriers for lung metastasis. *Small.* 2022;18(49).

# 11 Delivery of Nucleic Acids Using Nanocarriers

## *siRNA and miRNA Delivery, mRNA and DNA Delivery, CRISPR-Cas Systems*

*Paula Guzmán–Sastoque, María Camila Monsalve, Cristian F. Rodríguez, Stiven Castellanos, Juan C. Cruz, and Luis H. Reyes*

## 11.1 INTRODUCTION

The prefix "nano" is derived from the Greek word *nannos*, meaning "very short man." According to ISO standards, nanomaterials are defined as having dimensions between 1 and 100 nm. The history of nanotechnology dates back to the ninth century in Mesopotamia, where artisans used copper and silver (AgNPs) nanoparticles to enhance the luster of ceramics. The Romans also utilized nanocomposites unknowingly around AD 400, as exemplified by the Lycurgus Cup, made from glass containing nanoparticles of a gold–silver alloy. Similarly, the Damascenes and the people of Chichen Itza incorporated nanoparticles into their defensive tools and pigments [1].

In 1857, Michael Faraday reported the synthesis of a colloidal solution of gold nanoparticles (AuNPs). Although he did not use the term "nanoparticles," it is now recognized that he was working with nanomaterials. Additionally, Faraday reduced AuNPs using phosphorus. The manufacturing of silica nanoparticles began in 1940 [1].

The concept of nanotechnology was introduced to the scientific community in 1959 by Nobel laureate physicist Richard Feynman [2], who is now recognized as the father of nanotechnology [3]. The application of nanotechnology in the biomedical field began in the late 1970s, using nanoscale materials ($10^{-9}$ m) to design more effective methods for drug delivery and therapies targeting cardiovascular diseases, skin disorders, cancer, neurodegenerative diseases, and more [4]. In 1980, Eric Drexler, who published the book "Engines of Creation", proposed the potential of a field of molecular nanotechnology.

Nanomaterials have become particularly appealing as carriers for nucleic acids in nanomedicine due to their chemical versatility. They offer the ability to protect genetic material, deliver nucleic acids in a controlled manner, provide high loading capacity, and ensure efficient cellular internalization. Various types of nanocarriers are used for nucleic acid delivery, including liposomes, dendrimers, organic and inorganic nanoparticles, hybrid nanoparticles, quantum dots, and carbon nanotubes (CNTs) [3].

Nucleic acids are promising because they offer greater specificity and often produce more sustained or even curative effects compared to conventional drugs. Their applications extend beyond gene therapy to include diseases without a genetic cause, as well as cancer and viral infections, as demonstrated in various scientific and clinical studies. However, a major limitation of nucleic acids is their low cellular internalization and poor stability *in vivo* [5].

To address these challenges, non-viral strategies, such as those involving nanomaterials, and viral strategies—including retroviruses, adenoviruses, lentiviruses, and parvoviruses—have been developed for nucleic acid delivery. While viral strategies have been widely used, they present limitations, particularly concerning cargo capacity and immunogenicity [5].

This chapter aims to provide a comprehensive overview of the delivery of nucleic acids using nanocarriers, with a focus on different types of genetic materials, including siRNA, miRNA, mRNA, DNA, and CRISPR-Cas systems.

### 11.1.1 Nanocarriers for Nucleic Acid Delivery

#### *11.1.1.1 Types of Nanocarriers*

Nanocarriers play a fundamental role in the delivery of nucleic acids by protecting genetic material from degradation and enhancing cellular uptake. Various types of nanocarriers have been developed, each with distinct physicochemical properties that influence their delivery efficiency, stability, and biocompatibility. Figure 11.1 provides an overview of the major types of nanocarriers used for nucleic acid delivery.

##### *11.1.1.1.1 Lipid-Based Nanocarriers*

The use of lipid-based nanoparticles for nucleic acid delivery is widespread due to the structural similarities between lipids and cell membranes [6, 7]. Cationic lipids, which carry a permanent positive charge, are commonly employed in this context. For instance, 1,2-di-O-octadecenyl-3-trimethylammonium-propane



DOI: 10.1201/9781003473183-11

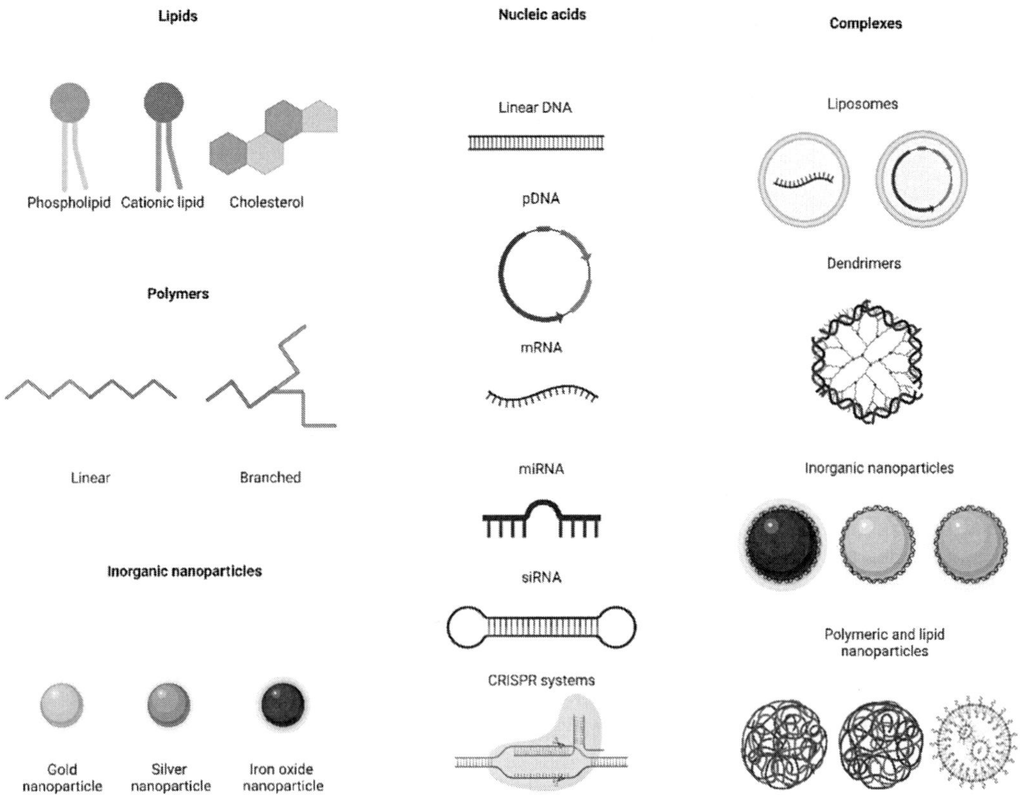


**Figure 11.1** Types of nanocarriers for nucleic acid delivery.

(DOTMA) has been used for mRNA delivery. DOTMA has also been used for DNA delivery in conjunction with unilamellar liposomes, as it facilitates the binding of the complex to the cell membrane [8]. Other reagents include Lipofectin, which is often combined with 1,2-dioleoyl-sn-glycero-3-phosphoethanolamine (DOPE) and 1,2-dioleoyl-3-trimethylammonium-propane (DOTAP) [9, 10].

Another example is BHEM-Cholesterol, which has been utilized for the delivery of CRISPR-Cas9, and ethylphosphatidylcholine (ePC), which incorporates a third alkyloxy group into phosphatidylcholines to neutralize their negative charge [9].

Ionizable lipids are particularly versatile due to their pH sensitivity. This property enhances their interactions with cell membranes and improves the biocompatibility of lipid nanoparticles (LNPs). Notable examples of ionizable lipids include (2S)-2,5-bis(3-aminopropylamino)-N-[2-(dioctadecylamino) acetyl]pentanamide (DOGS; Transfectam), N1-[2-((1S)-1-[(3-aminopropyl)amino]-4-[di(3-aminopropyl) amino]butylcarboxamido)ethyl]-3,4-di[oleyloxy]-benzamide (MVL5), DC-Cholesterol, N4-cholesteryl-spermine (GL67), 1,2-dilinoleyloxy-N,N-dimethyl-3-aminopropane (DLin-DMA), and 2,2-dilinoleyl-4-dimethylaminoethyl-[1,3]-dioxolane (DLin-KC2-DMA) [9].

Another type of lipid that can be incorporated into nanoparticles is biodegradable lipids. These lipids have been shown to be more effective for delivering nucleic acids, particularly mRNA, as they make nanoparticles more tolerable and enable rapid clearance from the liver and plasma. For example, hexa(octan-3-yl) 9,9′,9″,9‴,9″″,9‴″-((((benzene-1,3,5-tricarbonyl)tris(azanediyl))tris(propane-3,1-diyl))tris(azanetriyl)) hexanoate (FTT5) has been effective for *in vivo* mRNA delivery [9].

There are also other LNP formulations that include phospholipids, cholesterol, or PEGylated lipids (PEG-lipids). For instance, cholesterol can be used to increase stability, but it may reduce delivery efficacy. To address this, cholesterol analogs have been proposed to maintain high delivery efficacy and stability. Finally, PEGylated lipids prevent particle aggregation, which increases stability and facilitates conjugation, enhancing the specificity of particle release to the target [7, 9].

##### *11.1.1.1.2 Polymer-Based Nanoparticles*

Polymeric nanoparticles can come in various forms, including solid nanoparticles, polymeric micelles, or polymersomes [11]. The most common form of polymeric nanoparticles are nanocapsules or nanospheres. Polymeric nanoparticles offer structural versatility, biocompatibility, and high transfection efficiency. Polyplexes typically contain cationic polymers to condense nucleic acids. In contrast, polymersomes, or polymeric vesicles, are synthesized using amphiphilic block copolymers [12].

Polymers can be linear, branched, or cross-linked, and they can be either natural or synthetic. Cationic polymers are commonly used for nucleic acid delivery because they can facilitate endosomal escape. Some of the most commonly used cationic polymers include polyethyleneimine (PEI), poly(amidoamine) (PAMAM), poly (β-amino ester) (PBAE), poly-L-lysine, poly-L-arginine, and chitosan [11]. They readily form complexes through electrostatic interactions, allowing them to condense nucleic acids. However, these polymers can be cytotoxic, necessitating modifications to improve their safety. In contrast, non-cationic polymers tend to have higher cytocompatibility but require covalent bonding with nucleic acids for effective delivery. Additionally, temperature- or pH-sensitive polymers can be used to enhance nucleic acid delivery in response to specific stimuli.

Non-cationic polymers, such as poly(ethylene glycol) (PEG), are widely used due to their ability to create a steric barrier that prevents unwanted molecules from binding to what has been conjugated to PEG. Polyesters are another type of polymer frequently used in the synthesis of nanomaterials. Some examples include polylactic acid (PLA) and poly(lactic-co-glycolic acid) (PLGA). These polymers have been employed for the delivery of nucleic acids, such as mRNA and RNAi, to restore tumor suppressors [13] and for gene silencing in cancer cells [14], respectively.

##### *11.1.1.1.3 Inorganic Nanoparticles*

Inorganic nanoparticles are typically made of gold, silica, or iron. Their size and structure can be chemically modified to increase specificity, stability, and delivery efficacy. Notably, iron and AuNPs have been shown to be cytocompatible. However, the immune response they trigger can vary depending on their surface chemistry [12].

AuNPs were proposed in 1990 as a nanomaterial with high potential for molecule delivery because they are chemically inert and non-cytotoxic. Additionally, their surfaces can be modified by functionalizing specific ligands to enhance delivery efficacy [15]. AuNPs typically range in size from 13 to 15 nm, and nucleic acids are covalently conjugated to them via thiol groups [12].

Silica nanoparticles, with diameters ranging from 100 to 250 nm, are utilized for nucleic acid delivery due to their biocompatibility. The conjugation to nucleic acids is generally through non-covalent interactions. If the surface of silica nanoparticles has a negative charge, it is often modified with cationic molecules to facilitate binding [16, 17].

Iron oxide nanoparticles ($Fe_3O_4$) (IONPs) have a highly versatile surface chemistry that facilitates the conjugation of nucleic acids and even CRISPR systems [18, 19]. Besides, other properties of IONPs are biocompatibility, immunogenicity, and highly endosomal escape. Conjugation can be covalent, via thiol exchange, or through electrostatic.interactions

#### *11.1.1.2 Design and Functionalization*

##### *11.1.1.2.1 Surface Modification*

When designing nanomaterials, it is essential to consider that surface modifications are generally needed to ensure efficient nucleic acid delivery, prevent immunogenic responses, achieve high endosomal escape (see Figures 11.2 and 11.3), and maintain chemical inertness and cytocompatibility. Strategies to achieve these goals include PEGylation, the use of calcium ions, and ionizable materials.

PEGylation enhances drug efficacy, reduces undesirable effects, and prevents nanoparticle aggregation. The use of this strategy dates back to the 1970s, when PEG was first conjugated to proteins, leading to improved pharmacokinetics [20]. As a result, PEGylation emerged as a potential alternative to prevent rapid clearance and instability in drug and nucleic acid carriers. It has been found that PEG conjugation to nanoparticles and liposomes increases circulation time and stability. One of the most well-known studies is by Alberto Gabizon, who demonstrated that delivering doxorubicin (DOX) with PEGylated liposomes was more efficient than using non-PEGylated liposomes [20].

Another pioneering study using this strategy was conducted in the 1990's by Yokoyama [21]. In this study, PEG was conjugated to micelles containing adriamycin, demonstrating a significant reduction in tumor growth.

The use of calcium ions has emerged as a strategy to enhance transfection efficiency. Calcium is abundant in the human body and is recognized as a highly biocompatible metal [22]. It can bind to phosphate ions to

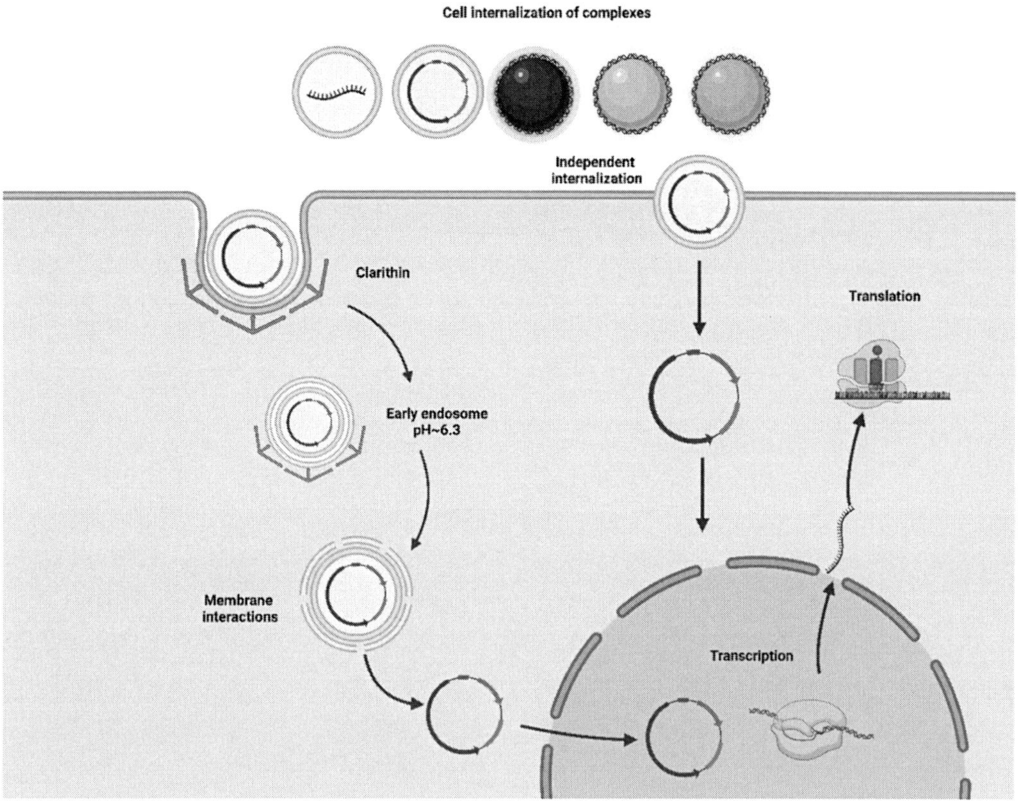


**Figure 11.2** Internalization of a liposome loaded with pDNA via the clathrin-mediated pathway. Initially, the nanocarrier can be internalized either independently or through clathrin-mediated endocytosis, resulting in its encapsulation within an early endosome. Subsequently, due to changes in charge, protonation, or interactions between membranes, the liposome's cargo is released and can reach the nucleus. In this case, the pDNA may encode a protein, which is then transcribed and subsequently translated in the cell's cytoplasm.

form complexes that facilitate the release of nucleic acids. Calcium phosphate-based complexes have been shown to have low cytotoxicity *in vivo*, improve cellular uptake, and effectively reach desired targets [22].

The surface can also be modified by adding a secondary polymer that binds to a polymeric core through copolymerization, adsorptive coating, or covalent bonding with another polymer. Hydrophilic coatings are the most commonly used because they increase biocompatibility and reduce the binding of proteins or other biomolecules. The polymers most frequently used for these types of modifications include PEG, poly(vinyl alcohol) (PVA), poly(acrylic acid), poly(methacrylic acid), poly(acrylamide), and poly(vinylpyrrolidone).

##### *11.1.1.2.2 Functionalization*

The functional groups typically exposed on the surface of the nanocarriers include carboxylate, aliphatic amine, hydroxyl, hydrazide, aldehyde, amide, thiol, aromatic amine, epoxide, chloromethyl, and tosyl-activated bead groups [23].

The coupling of ligands with an amino end to nanocarriers containing a carboxylate group can be performed using N-hydroxysuccinimide (NHS) and EDC (1-ethyl-3-(3-dimethylaminopropyl)carbodiimide) in a two-step reaction, or using only EDC (1-ethyl-3-(3-dimethylaminopropyl)carbodiimide) in a one-step coupling reaction [23].

If the functional group of the nanocarriers is a primary amine, the coupling can be carried out as previously mentioned, with the difference that activation will occur on the ligand if it has a carboxylate group. Additionally, crosslinking agents can be used with either a heterobifunctional compound, which has different reactive groups at each end, or a homobifunctional compound, which has the same reactive groups at each end. Some common homobifunctional crosslinking agents include those with NHS esters, imidoesters, or aldehyde groups at each end [23]. On the other hand, if the nucleic acids have a thiol group and the nanomaterial has

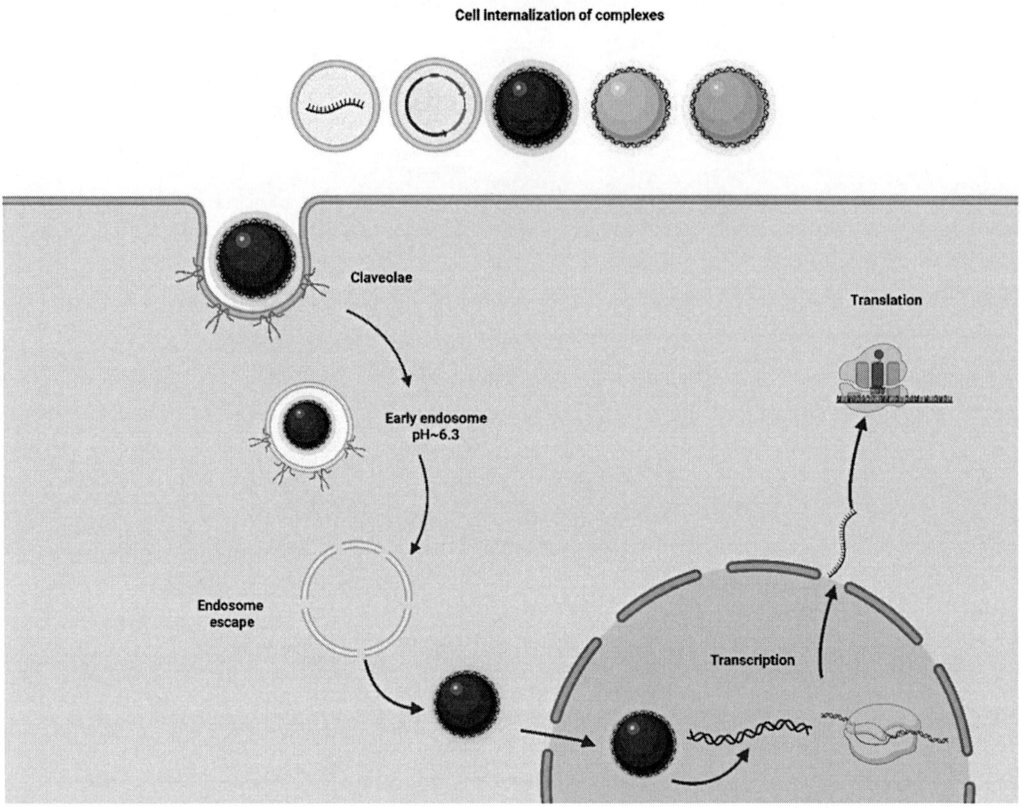


**Figure 11.3** Internalization of an IONP via the caveolae-mediated pathway. Initially, the nanocarrier can be internalized through caveolae-mediated endocytosis, resulting in its encapsulation within an early endosome. Subsequently, the nanoparticle can escape from the endosomes. In this case, the DNA may encode a protein, which is then transcribed and subsequently translated in the cell's cytoplasm.

a primary amine group, a heterobifunctional crosslinker such as sulfo-succinimidyl-4-(N-maleimidomethyl) cyclohexane-1-carboxylate (SMCC) can be used [23]. This crosslinker has an NHS ester on one end, which is reactive to the amine, and a maleimide group on the other end, which is reactive to the thiol [23].

If nanocarriers, particularly nanoparticles, contain an epoxide group, nucleic acids with a thiol can also be attached through a ring-opening reaction under basic conditions [23]. This type of nanoparticle can also be used to create surface aldehydes by opening the epoxide ring through acid hydrolysis and then oxidizing the resulting diols with sodium periodate [23].

##### *11.1.1.3 Clinical Trials*

According to a search on ClinicalTrials.gov, to date, over 563 clinical trials have been proposed involving the use of nanoparticles for drug delivery. Some of these trials include the use of nanocarriers for the delivery of nucleic acids, aiming to treat various pathological conditions (see Table 11.1).

## 11.2 SIRNA AND MIRNA DELIVERY

### 11.2.1 Mechanisms of RNA Interference

siRNA and microRNA (miRNA) are molecules involved in gene inhibition, collectively known as RNA interference (RNAi) (see Figure 11.4). These molecules inhibit gene expression through two mechanisms: miRNA represses the translation of mRNA, while both siRNA and miRNA can induce targeted degradation of mRNA [24]. RNAi molecules function as guides for the silencing of specific sequences, and their discovery was first made in *Caenorhabditis elegans*.

The RNAi canonic pathway begins with the cleavage of double-stranded RNA (dsRNA) into small interfering RNAs (siRNAs) of 21–23 nucleotides by the enzyme Dicer. The dsRNA can originate from

## Table 11.1 Clinical trials using nanocarriers to treat or investigate diseases

| NCT Number | Conditions | Phases |
|---|---|---|
| NCT03020017 | Gliosarcoma I glioblastoma | EARLY PHASE1 |
| NCT04352465 | COVID-19 | PHASE1 I PHASE2 |
| NCT04616872 | Atherosclerosi | PHASE2 I PHASE3 |
| NCT04610567 | COVID-19 | PHASE1 I PHASE2 |
| NCT00470548 | Breast cancer | PHASE1 I PHASE2 |
| NCT01847326 | Recurrent salivary gland cancer I recurrent squamous cell carcinoma of the hypopharynx I recurrent squamous cell carcinoma of the larynx I recurrent squamous cell carcinoma of the lip and oral cavity I recurrent squamous cell carcinoma of the nasopharynx I recurrent squamous cell carcinoma of the oropharynx I recurrent squamous cell carcinoma of the paranasal sinus and nasal cavity I recurrent verrucous carcinoma of the larynx I recurrent verrucous carcinoma of the oral cavity I salivary gland squamous cell carcinoma I tongue cancer | PHASE1 |
| NCT02258659 | Human papilloma virus infection I stage III squamous cell carcinoma of the oropharynx I stage IVA squamous cell carcinoma of the oropharynx I stage IVB squamous cell carcinoma of the oropharynx | PHASE2 |
| NCT02369198 | Malignant pleural mesothelioma I non-small cell lung cancer | PHASE1 |
| NCT04892953 | Lung cancer | PHASE2 |
| NCT04956692 | Non-small-cell lung cancer | PHASE3 |
| NCT04964960 | Lung cancer I lung cancer metastatic I brain cancer I cancer | PHASE2 |
| NCT05501665 | Lung non-small-cell carcinoma I stage IV lung cancer AJCC v8 I stage III lung cancer | PHASE1 I PHASE2 |
| NCT06096844 | Advanced lung non-small-cell carcinoma I stage IIIB lung cancer AJCC v8 I stage IIIC lung cancer AJCC v8 I stage IV lung cancer AJCC v8 | PHASE3 |
| NCT03107182 | HPV-related squamous cell carcinoma I HNSCC | PHASE2 |
| NCT05451043 | Pancreatic cancer I hepatocellular cancer I biliary tract cancer I cholangiocarcinoma | PHASE2 |

viruses or exogenous sources. The siRNAs are then unwound into single strands, and the guide strand is incorporated into the RNA-induced silencing complex (RISC). RISC binds to the target mRNA, where an Argonaute protein (AGO) facilitates the degradation of the mRNA [25].

Regarding the miRNA pathway, it begins with a pre-miRNA that, after being transported to the cytoplasm, is cleaved by Dicer. Similar to the RNAi pathway, a complex is formed with RISC and an Argonaute protein, which is then directed to a P-body. This ultimately results in the repression of mRNA translation by the miRNA [26].

miRNAs are of great interest for clinical applications because they impact metabolism, differentiation, proliferation, and programmed cell death. In fact, miRNAs have been associated with cancer progression, neurodegenerative disorders, and cardiovascular diseases. Additionally, miRNAs can regulate more than one mRNA, allowing their use to target multiple therapeutic pathways [27].

### 11.2.2 Nanocarriers for siRNA Delivery

The most commonly used nanocarriers for siRNA delivery are lipid-based nanoparticles, polymeric nanoparticles, and inorganic material nanoparticles.

#### *11.2.2.1 Lipid-Based Nanoparticles*

Lipid-based nanoparticles are frequently employed for RNAi delivery and are often stabilized with cholesterol or modified lipids, such as PEGylated lipids. These types of nanocarriers are used because the electrostatic interactions and functional groups present favor endosomal escape, which is crucial for effective siRNA delivery [28–30]. Lipid-based nanocarriers offer several advantages, such as biocompatibility, low immunogenicity, and low toxicity, although they may have limitations like low cargo capacity and stability [31]. Typically, cationic lipids are used to enhance electrostatic interactions. The most common

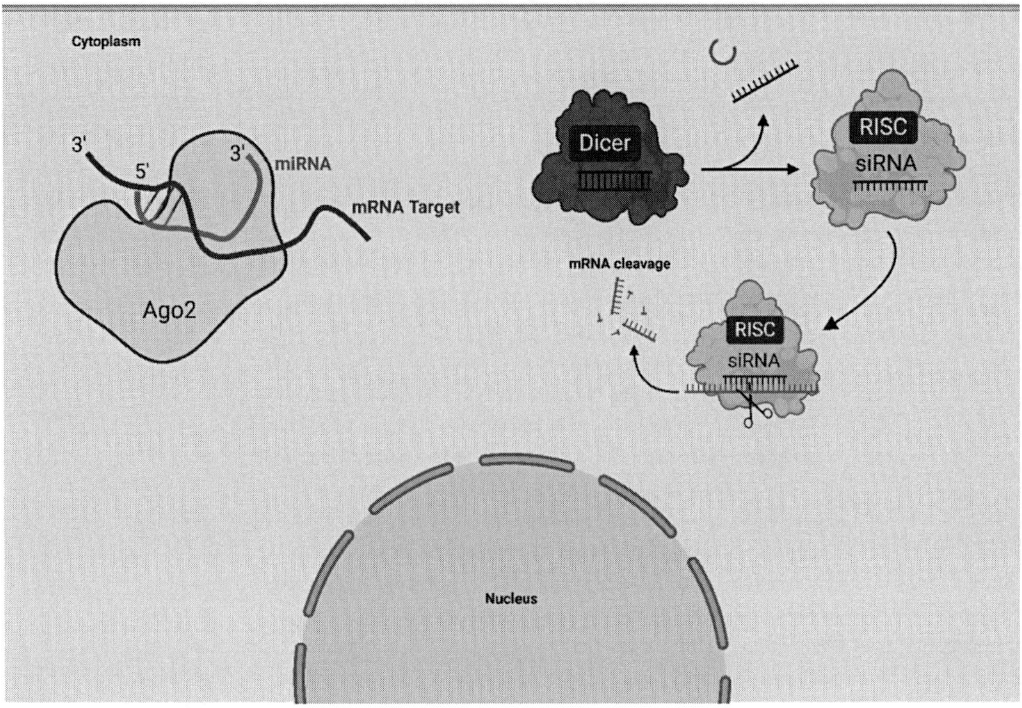


**Figure 11.4** Mechanisms of RNA interference (RNAi) pathways. The figure illustrates the distinct processes of siRNA and miRNA pathways. In the siRNA pathway, double-stranded RNA (dsRNA) is cleaved by the enzyme Dicer into siRNAs, which are then incorporated into the RNA-induced silencing complex (RISC). The RISC–siRNA complex guides the Argonaute protein (Ago2) to cleave the target mRNA. In the miRNA pathway, pre-miRNA is processed by Dicer and incorporated into RISC, where it associates with Ago2. The RISC–miRNA complex binds to target mRNA, leading to translational repression.

administration routes for these nanocarriers are intravenous or intratumoral injections. They are primarily used in cancer therapy due to their efficiency in gene silencing, and they have been predominantly employed in treating myeloid leukemia, colon cancer, lung cancer, cervical cancer, and breast cancer [32]. Commonly used lipid-based nanocarriers include liposomes such as DOPE and DOTAP [31, 32].

#### *11.2.2.2 Polymeric Nanoparticles*

Polymeric nanoparticles are another effective type of nanocarrier for siRNA delivery. These nanoparticles protect siRNA by incorporating ionizable groups and are characterized by their ability to condense siRNA, enhancing delivery efficiency. Polymersomes are the most commonly used polymeric systems, particularly for the treatment of lung, stomach, liver, and breast cancers [33]. These treatments are administered via various routes, including intravenous, parenteral, intratumoral, and subcutaneous injections [33]. Several genes, such as *Orai3*, *PLK1*, and *Bcl-xL*, have been successfully silenced using these systems [31, 32].

#### *11.2.2.3 Inorganic Material Nanoparticles*

Inorganic nanoparticles, including CNTs, liposomes, dendrimers, polymeric micelles, and nanocrystals, are also used for siRNA delivery [33].

##### *11.2.2.3.1 Metal Oxide Nanoparticles*

Metal oxide nanoparticles, such as magnetite nanoparticles, are widely used for the effective delivery of siRNA and to enhance the efficacy of cationic polymers. Manganese oxide ($MnO_2$) nanoparticles, for example, have been proposed for use with PEI in the treatment of *HER2*-positive breast cancer. Silica nanoparticles are also commonly used for siRNA delivery and cancer treatment due to their low immunogenicity and ability to be modified for targeted tumor recognition [33].

##### *11.2.2.3.2 pH-Sensitive Nanoparticles*

Another type of nanoparticle that can be used is pH-sensitive nanoparticles, such as carbonated apatite nanoparticles, which release their payload under acidic endosomal conditions. These nanoparticles are also biodegradable, which is an additional beneficial feature [33].

###### *11.2.2.3.2.1 Factors Influencing siRNA Delivery*

Several factors influence the efficiency of siRNA delivery, including the size of the nanocarrier and the ability to cross the cell membrane. siRNA can be easily recognized by the reticuloendothelial system, so the size of the nanocarrier is crucial for avoiding premature clearance. Effective crossing of the cell membrane is vital for delivery, for which cell-penetrating peptides and cationic polymers like PEI are commonly used [33]. The phase structure of the nanocarrier also plays a significant role in determining the absorption pathway, biodistribution, and delivery mechanisms.

In conclusion, the choice of nanocarrier and its modifications is crucial for optimizing siRNA delivery and enhancing therapeutic outcomes.

### 11.2.3 Nanocarriers for miRNA Delivery

There are two types of miRNA: viral and non-viral. Non-viral nanocarriers are characterized by their lower immunogenicity, high loading capacity, and scalability for industrial production. Examples of non-viral nanocarriers include liposomes, polymeric nanoparticles, iron nanoparticles, AuNPs, and quantum dots [34, 35]. For instance, aluminum nanoparticles with *miR-205* have been used for prostate cancer treatment, and mesoporous silica nanoparticles (MSNs) have been employed to deliver anti-miR-221-octaarginine in combination with temozolomide for glioma treatment [36].

#### *11.2.3.1 Polymeric Nanoparticles*

Polymeric systems with cationic polymers are among the most studied for miRNA delivery. This is because nucleic acids interact with amine groups to form polyplexes. Polymers like PEI, PLGA, and chitosan have also been utilized for this purpose [36].

#### *11.2.3.2 Lipid-Based Nanoparticles*

Lipid-based nanocarriers generally require stabilization with polymers such as PEG to reduce immune recognition and extend circulation time. PEGylation can also impart a slight negative charge to the nanoparticles, preventing their uptake by the reticuloendothelial system (RES) [37, 38]. pH-sensitive PEGylated lipids are often preferred to enhance internalization.

The most commonly used cationic lipoplexes for miRNA and pre-miRNA delivery include those based on DOTMA, DDAB, DOTAP, and DOPE, combined with PEGylated lipids [37]. These formulations are used to treat various conditions such as lung cancer, head and neck squamous cell carcinoma, atherosclerosis, glioblastoma, leukemia, myeloma, and liver cancer [37, 38].

### 11.2.4 Delivery of miRNA Using Nanocarriers: Therapeutic Applications

miRNA-based gene therapies have garnered significant attention due to their capacity to silence multiple genes through mRNA cleavage, translational repression, or DNA methylation [39]. Despite representing only about 3% of human genes, miRNAs regulate approximately 30% of protein-coding genes. They typically bind to the 3′ untranslated regions (3′ UTRs) of mRNAs, leading to translation inhibition or mRNA degradation [39]. miRNA genes can be monocistronic, polycistronic, or intronic [40].

Challenges in miRNA-based therapies include short half-life, off-target effects, vulnerability to blood-borne enzymes, safety concerns, and high production costs. Effective delivery of miRNAs to target cells while minimizing off-target effects remains a critical issue [39]. Non-viral delivery systems have emerged as promising solutions, utilizing bionanotechnology to improve miRNA delivery through inorganic nanoparticles, polymer-derived nanovehicles, and lipid-based vehicles [41].

#### *11.2.4.1 Cancer Therapy*

In cancer treatment, the variability of genes among individuals poses a challenge for developing universal genetic therapies. However, miRNAs can regulate multiple genes post-transcriptionally, a mechanism that can be conserved across patients [39]. Some miRNAs can influence several genes involved in cancer and enhance the chemosensitivity of cancer cells to conventional therapies [41].

OncomiRs and anti-oncomiRs are types of dysregulated miRNAs associated with oncogenic events and cancer development. Key tumor-suppressive miRNA families include *miR-34*, *let-7*, and *miR-200* [34]. Nanocarriers used in cancer therapy can be classified into inorganic, polymer-based, and lipid-based types.

##### *11.2.4.1.1 Inorganic Nanocarriers*

Metal-based nanocarriers, such as AuNPs, are notable for their tunable size, biostability, and surface modulation capabilities, which enhance targeted delivery. AuNPs can combine genetic therapy with photothermal therapy due to their strong optical absorption in the near-infrared region [41].

- *Gold Nanocages:* Functionalized with PEI and PEGylated, these nanocages are used for delivering *miR-181b* in hepatocellular carcinoma. The positive charge from PEI enhances nucleic acid loading, while PEGylation improves stability and biocompatibility [42].
- *Thiol-modified miR-375:* PEGylated AuNPs were used to deliver thiol-modified *miR-375*, successfully targeting hepatocellular carcinoma in mice [43].
- *Gold–Silver–Gold Core–Shell–Shell (CSS) Nanoparticles:* These nanoparticles, activated by near-infrared light, release miRNA mimics and have shown efficacy in delivering *miR-34a* for esophageal cancer treatment [44].
- *IONPs:* Used for magnetofection to deliver pDNA-encoded miRNA-34a, these nanoparticles showed effectiveness in triple-negative breast cancer and non-small-cell lung cancer [45].
- *Graphene Oxide (GO):* Modified with PEI, GO nanocarriers effectively delivered miRNAs like *miR-194-5p* and *miR-125b-5p*, inhibiting tumor growth in intrahepatic cholangiocarcinoma [46].

##### *11.2.4.1.2 Polymer-Based Nanocarriers*

Polymeric nanocarriers, such as those made from PLGA and PEI, offer high biocompatibility and effective delivery.

- *PLGA Nanoparticles:* PLGA, when modified with cationic surfaces, can deliver *miR-34a* effectively, reducing cell proliferation and inducing cell cycle arrest [47].
- *PEI-Based Nanocarriers:* Modified PEI, including hyperbranched and disulfide-linked versions, has been used to deliver *miRNA-135a* and *miR-148b* in glioma and other cancers, demonstrating their improved transfection efficiency and reduced cytotoxicity [48].
- *Elastin-Like Peptides (ELP):* ELP nanocarriers, functionalized with targeting ligands and TAT peptides, showed significant success in delivering *miRNA-34a* and inhibiting cancer cell proliferation [34].
- *Chitosan:* Lactosylated-chitosan-5-fluorouracil nanocarriers demonstrated efficient *miRNA-122* delivery in hepatocellular carcinoma, utilizing receptor-mediated endocytosis [49].

##### *11.2.4.1.3 Lipid-Based Nanocarriers*

Lipid-based nanocarriers are effective due to their biocompatibility and high transfection capacity.

- *Nanostructured Lipid Carriers (NLCs):* Combining solid and liquid lipids, NLCs improve drug-loading capacity and have been used to deliver *miR-375* in hepatocellular carcinoma, showing significant anti-proliferative effects [50].
- *Niosomes:* PEGylated niosomes, formed from synthetic non-ionic surfactants, efficiently deliver *miRNA-15a* and *miRNA-16-1* to prostate cancer cells, inducing apoptosis [51].
- *Quatsomes:* Unilamellar carriers with surfactants and pH-sensitive sterols, quatsomes offer an alternative to liposomes for miRNA delivery.

#### *11.2.4.2 Infectious Diseases*

Infectious diseases caused by bacteria, fungi, viruses, and parasites are responsible for nearly 15 million deaths globally. Major contributors include human immunodeficiency virus (HIV), malaria, tuberculosis, and acute respiratory infections such as COVID-19 [52]. Several of these infections can be managed using miRNAs that either silence genes essential for the proliferation of the infectious agents or reduce localized inflammation.

*Application in Kaposi's Sarcoma-Associated Herpesvirus (KSHV) Treatment:* A study focused on treating KSHV used ZIF-8, which was synthesized and loaded with GCV (9-(1,3-dihydroxy-2-propoxymethyl) guanine), an antiviral that inhibits KSHV DNA polymerase [53]. Additionally, ZIF-8 was loaded with *miR-34a-5p* due to its role in KSHV treatment. The external surface of ZIF-8 was functionalized with PEI, which binds to ZIF-8 on one end and folic acid (FA) on the other, facilitating miRNA loading and targeting of infected cells. *In vitro* experiments demonstrated a significant reduction in the proliferation and migration of KSHV-infected cells, attributed to both the antiviral effect and the downregulation of mRNA from latent and lytic genes due to *miR-34a-5p* inhibition [53].

*Lipid-Based Nanocarriers for Cerebral Malaria:* Lipid-based vehicles have also been investigated for treating cerebral malaria caused by *Plasmodium falciparum* [54]. Researchers developed lipid nanocarriers using a hot-melt homogenization process, combining Precirol (a solid lipid) and Labrasol (a liquid lipid) with Artemether (ARM), an antimalarial drug. Artemether generates free radicals that damage proteins and essential structures of the malaria parasite and exerts anti-inflammatory effects through various mechanisms [54]. The lipid nanocarriers, loaded with *miR-223*, inhibit ICAM-1 (a membrane receptor involved in leukocyte adhesion) by phosphorylating several proteins, thereby reducing inflammation caused by *P. falciparum*. Administered intranasally to bypass the blood–brain barrier, these nanocarriers showed significant effectiveness in a mouse model, improving drug bioavailability, targeting the brain effectively, and demonstrating promise for treating cerebral malaria [54].

### 11.2.5 Delivery of siRNA Using Nanocarriers: Therapeutic Application

siRNA is increasingly recognized as a promising platform for cancer therapy. Despite this potential, the clinical application of these RNA-based therapies faces significant challenges, primarily stemming from their susceptibility to degradation by ubiquitous RNases and their inherent instability in biological environments. These challenges necessitate the development of protective and efficient delivery systems to ensure that the RNA-based therapies can reach and function effectively within target cells [55].

To address the instability and degradation issues associated with siRNA, various nanomaterials—including AuNPs, dendrimers, CNTs, and MSNs—have been engineered as delivery vehicles [56, 57]. These nanocarriers are designed to protect RNA molecules from enzymatic degradation, enhance cellular uptake, and promote endosomal escape, thereby ensuring the effective delivery of therapeutic siRNA to the appropriate intracellular locations [58].

Yang et al. have tackled the specific challenge of delivering siRNA for cancer therapy [59]. Their study introduced a targeted nanomaterial-based drug delivery system, DOX/MSN/Chimeras, designed to overcome chemotherapy resistance in bladder cancer. This system co-delivers DOX and siRNA directly to bladder cancer cells, addressing drug resistance while minimizing the adverse effects typically associated with conventional chemotherapy.

The DOX/MSN/Chimera system integrates several key components critical to the delivery process. Carboxylated mesoporous silica nanoparticles (COOH-MSNs) form the core, providing enhanced biocompatibility and stability. DOX, a widely used chemotherapeutic agent, is loaded into these nanoparticles, which are then coated with bovine serum albumin (BSA) and PEI. BSA stabilizes the nanoparticles and prevents non-specific interactions with non-target cells, while PEI facilitates the attachment of the nucleic acid chimera and enhances cellular uptake by promoting endosomal escape.

Central to this system is the nucleic acid chimera, comprising an aptamer that specifically targets bladder cancer cells and siRNA that silences the PI3K gene, a crucial factor in drug resistance. This dual-function chimera ensures that therapeutic agents are delivered precisely to the cancer cells, thereby reducing off-target effects and improving overall treatment efficacy. The aptamer-siRNA chimera is engineered to bind with high specificity to bladder cancer cells, enhancing the precision of drug delivery and minimizing side effects on healthy tissues [59].

In both *in vitro* and *in vivo* evaluations, the DOX/MSN/Chimera system demonstrated significant improvements in drug delivery specificity, tumor size reduction, and apoptosis promotion in drug-resistant bladder cancer cells. In animal models, this system led to a marked reduction in tumor volume and a significant increase in apoptosis, as indicated by specific biomarkers. Moreover, the system effectively mitigated chemotherapy-induced damage to healthy tissues, addressing a common drawback of conventional chemotherapy [59].

This innovative strategy underscores the potential for personalized and targeted treatments in combating drug-resistant cancers, offering a promising alternative to traditional chemotherapy approaches. The implications of this research extend beyond bladder cancer, as the platform could be adapted for other types of drug-resistant cancers. Future research could explore the clinical translation of this nanoplatform, including its application in other malignancies and its integration into broader cancer treatment regimens.

## 11.3 MRNA AND DNA DELIVERY

### 11.3.1 DNA and mRNA

The process of converting DNA to mRNA is defined by the central dogma of molecular biology. This dogma outlines the flow of genetic information through replication, transcription, and translation. DNA contains genetic information encoded in four nucleotides: adenine (A), guanine (G), cytosine (C), and thymine (T). While DNA can be replicated, it can also be transcribed into mRNA, which serves as a template for protein synthesis in ribosomes. Generally, genetic information is decoded in groups of three nucleotides, each of which forms a codon that corresponds to one of the 20 canonical amino acids [60].

From a research perspective, DNA is preferred for delivery because it is very stable and less susceptible to degradation compared to RNA, which is easily degraded. Clinically, mRNA is advantageous because it performs vital functions without needing to enter the cell nucleus and has a low likelihood of integrating into the genome, making its use safer by reducing the risk of mutations and carcinogenesis [61].

Delivering both DNA and mRNA poses challenges due to their insufficient cellular entry and stability for achieving a therapeutic effect. To overcome these limitations, viral vectors have been employed, but they come with disadvantages such as immunological response, toxicity, and low payload capacity. Therefore, non-viral nanocarriers have been developed as alternatives. These nanocarriers offer several advantages, including the ability to condense DNA and mRNA, which protects them from enzymatic degradation [61]. Additionally, they can be engineered for targeted delivery and designed to collapse upon endosomal escape due to pH sensitivity or the presence of bioreducible bonds, thereby increasing transfection efficiency and payload release [61].

### 11.3.2 Nanocarrier Systems for DNA Delivery

Nanocarrier-based delivery systems have evolved significantly, becoming increasingly sophisticated to ensure efficient delivery of DNA to its target. Currently, there is a variety of such systems, including dendrimers, liposomes, metallic nanoparticles, polymeric nanoparticles, lipid-based nanoparticles, MSNs, and CNTs [62]. Below are some types of nanocarriers used for DNA delivery:

#### *11.3.2.1 Liposomes*

Liposomes are characterized by their amphiphilic properties, high biocompatibility, and a size range of 50–150 nm [12]. One of the advantages of liposomes is their ability to encapsulate DNA within their structure and deliver it into cells via endocytosis.

#### *11.3.2.2 Lipid-Based Nanoparticles*

For DNA delivery, cationic lipids such as DOTMA are commonly used. These lipids facilitate electrostatic interactions with negatively charged DNA. Additionally, ionizable lipids can aid in endosomal escape due to their positive charge at acidic pH [12].

#### *11.3.2.3 Polymeric Nanoparticles*

Polymeric nanocarriers often consist of polycations with ionizable amine groups, which interact with the negatively charged phosphate groups of DNA, promoting condensation. This interaction aids in escaping from endosomes and releasing DNA into the cell [12].

#### *11.3.2.4 Inorganic Nanoparticles*

Inorganic nanoparticles, such as gold, mesoporous silica, and IONPs, are frequently used for DNA delivery. AuNPs and IONPs are considered non-toxic and are often modified to reduce adverse immune responses. DNA can be conjugated to these nanoparticles via thiol groups [12].

Delivery platforms for DNA can also be designed with membrane-penetrating and translocating peptides to facilitate internalization. One example is Buforin II (BUF-II) [18], a 21-amino acid peptide derived from the histone 2A of the Asian toad.

### 11.3.3 Nanocarrier Systems for mRNA Delivery

mRNA technology has rapidly advanced as a transformative approach for treating a wide range of diseases, with mRNA-based COVID-19 vaccines standing out as a recent and impactful example. Despite its promise, mRNA is inherently unstable, being highly susceptible to degradation by nucleases and other enzymes. This vulnerability necessitates the use of protective delivery systems to ensure the integrity and efficacy of mRNA-based therapies. Nanocarriers, particularly those utilizing LNPs as seen in COVID-19 vaccines, are essential in shielding mRNA from enzymatic degradation and facilitating its targeted delivery to cells.

Nanoparticles employed in mRNA delivery are typically categorized into three major types: inorganic, polymeric, and lipid-based nanoparticles. Each type offers unique advantages and presents specific challenges in the context of mRNA delivery.

#### *11.3.3.1 Inorganic Nanoparticles*

Inorganic nanoparticles, such as IONPs, silica nanoparticles, and AuNPs, are valued for their high stability and robustness, which make them effective at protecting mRNA from enzymatic degradation [63]. Furthermore, the surfaces of these nanoparticles can be readily modified with targeting ligands, enabling precise delivery of mRNA to specific tissues or cells [64]. However, a significant limitation of inorganic nanoparticles lies in their biocompatibility. Surface passivation techniques are often employed to mitigate this issue, enhancing their compatibility with biological systems [65]. Additionally, many inorganic nanoparticles are non-biodegradable, raising concerns about their potential accumulation in the body and associated long-term toxicity [66].

#### *11.3.3.2 Polymeric Nanoparticles*

Polymeric nanoparticles, often formed using polymers such as chitosan and PLGA, are widely used due to their inherent biocompatibility and biodegradability [67–70]. These polymers can be chemically modified to introduce reactive functional groups, such as amine or carboxyl groups, which facilitate targeting and the incorporation of therapeutic agents [71]. However, the synthesis of polymeric nanoparticles is highly sensitive to various conditions, posing challenges in achieving consistent reproducibility. Moreover, some synthetic polymers may degrade into toxic byproducts, potentially compromising their biocompatibility [71].

#### *11.3.3.3 Lipid Nanoparticles*

LNPs, including liposomes, have garnered significant attention, especially following their successful application in COVID-19 vaccines. The development of LNPs traces back to the discovery of liposomes by Bangham et al. in 1965 [72]. LNPs are generally well-tolerated by the body, exhibiting lower toxicity compared to other nanoparticle types. Recent advances in microfluidic technology have significantly improved the scalability of LNP production, thereby facilitating their clinical application [73, 74]. Additionally, LNPs can be easily modified with functional groups to enhance their targeting ability, stability, and mRNA encapsulation efficiency. Despite these advantages, challenges remain, such as the potential for immune responses upon repeated administration, including the accelerated blood clearance (ABC) phenomenon [75].

### 11.3.4 Applications and Therapeutic Potential

#### *11.3.4.1 Gene Therapy*

Gene therapy has gained significant relevance in the scientific community, as it can be used to treat diseases such as cancer, HIV/AIDS, diabetes, hemophilia, neurodegenerative disorders, cardiovascular conditions, muscular diseases, and more. The reason gene therapy is utilized is because it allows for the alteration of the expression of a target gene and/or metabolic pathways for therapeutic purposes. In general, gene therapy can be used to edit genome mutations, positively regulate gene expression with plasmid DNA (pDNA), minicircle DNA (mcDNA), and synthetic mRNA [12], and negatively regulate expression using other nucleic acid molecules such as siRNA and miRNA, which have been previously discussed in this chapter.

Dendrimers have been employed with pDNA encoding IL-12, pDNA-p53, and siRNA for various cancer treatments, including ovarian cancer and triple-negative breast cancer, where they are combined with DOX and lycopene. Liposomes have been used to deliver therapeutic genes such as acetylcholinesterase (AChE), the p53 gene, and plasmids. Additionally, lipid-based, polymeric, and inorganic nanoparticles have been utilized for the delivery of plasmids encoding advanced systems like CRISPR-Cas9 [76].

Some of the clinical trials proposed for gene therapy using nanoparticles are shown in Table 11.2.

**Table 11.2 Clinical trials using nanocarriers to do gene therapy**

| NCT Number | Conditions | Phases |
|---|---|---|
| NCT05062980 | Non-small-cell lung cancer | PHASE1 I PHASE2 |
| NCT05703971 | Small-cell lung cancer extensive stage | PHASE1 I PHASE2 |
| NCT00505271 | Breast cancer | PHASE1 I PHASE2 |
| NCT04486833 | Carcinoma, non-small-cell lung | PHASE1 I PHASE2 |
| NCT01437007 | Colorectal cancer with hepatic metastases | PHASE1 |

#### *11.3.4.2 Vaccines*

##### *11.3.4.2.1 DNA Vaccines*

DNA vaccines were proposed in 1990 and are characterized by their industrial scalability, stability, and the ability to generate immune responses while being safe for immunocompromised individuals [77]. The initial DNA vaccines used pDNA to induce protein expression, with ZyCovD against SARS-CoV-2 being the first DNA vaccine approved for human use [77].

These vaccines activate the adaptive immune system by delivering pDNA, which is transcribed and translated to produce antigens that are presented primarily via the major histocompatibility complex class I (MHC-I) pathway [77]. Although antigens can also be presented through the major histocompatibility complex class II (MHC-II) pathway, DNA vaccines typically utilize MHC-I presentation, which activates CD8+ T cells. Additionally, pDNA may contain CpG motifs that activate the innate immune system [78].

Recently, this type of vaccine has been proposed for cancer treatment, as they can encode genetic information against tumor antigens, thereby enhancing effectiveness [79]. For DNA delivery, nanocarriers such as liposomes, polymeric nanoparticles, and AuNPs have been suggested.

##### *11.3.4.2.2 mRNA Vaccines*

The first mRNA vaccines were proposed in 1990, but it was only recently that significant advancements were made in improving their stability, delivery efficacy, and immunogenicity. The main advantages of mRNA vaccines are that they are safe because they do not integrate with the host's DNA, structural modifications to the mRNA can enhance their effectiveness, and their production can be scaled up to an industrial level [80]. Additionally, mRNA vaccines can encode multiple antigens.

Some of the most widely recognized mRNA vaccines globally due to the SARS-CoV-2 pandemic are the Pfizer-BioNTech (BNT162b2) and Moderna (mRNA-1273) vaccines, which feature pseudouridine modifications to increase mRNA stability [81]. The delivery systems used for these vaccines involved LNPs.

**Table 11.3 Clinical trials using nanocarriers and mRNA vaccines**

| NCT Number | Conditions | Phases |
|---|---|---|
| NCT05264974 | Melanoma | PHASE1 |
| NCT05231369 | COVID-19 I SARS-CoV-2 | PHASE1 |
| NCT04847102 | COVID-19 I SARS-CoV-2 | PHASE3 |
| NCT05057182 | COVID 19 vaccine | PHASE4 |
| NCT05168813 | SARS-CoV-2 infection I HIV infections I COVID-19 | PHASE2 I PHASE3 |
| NCT05945485 | Influenza | PHASE1 |
| NCT05525208 | COVID-19 | PHASE2 |
| NCT05755620 | Influenza | PHASE1 |
| NCT04566276 | COVID-19 vaccine | PHASE1 I PHASE2 |
| NCT05876364 | COVID-19 I SARS-CoV-2 | PHASE1 |
| NCT04776317 | COVID-19 | PHASE1 |
| NCT05658523 | COVID-19 | PHASE3 |
| NCT04480957 | SARS-CoV-2 | PHASE1 I PHASE2 |
| NCT05534048 | SARS-CoV-2 infection | PHASE3 |
| NCT05543356 | COVID-19 | PHASE3 |
| NCT05534035 | SARS-CoV-2 infection | PHASE3 |
| NCT05387317 | COVID-19 | PHASE3 |
| NCT04811664 | SARS-CoV-2 infection | PHASE3 |
| NCT05175950 | COVID-19 (healthy volunteers) | PHASE2 |
| NCT04785144 | COVID-19 I COVID-19 immunisation | PHASE1 |
| NCT05516459 | Coronavirus infections |  |
| NCT05057169 | COVID-19 vaccination | PHASE4 |
| NCT05289037 | COVID-19 | PHASE2 |
| NCT04283461 | COVID-19 I COVID-19 immunisation | PHASE1 |
| NCT05515042 | COVID-19 | PHASE2 |
| NCT04889209 | COVID-19 | PHASE1 I PHASE2 |

LNPs proposed for mRNA vaccines often include cationic lipids, which facilitate endosomal escape, and may also contain PEGgylated lipids to enhance stability, prevent aggregation, and extend circulation time [81]. Recently, LNPs have also been proposed for use in mRNA vaccines as cancer immunotherapies to prevent premature degradation of the mRNA [82].

mRNA vaccines that use nanoparticles for delivery have gained prominence in recent years due to the SARS-CoV-2 pandemic, which is why several clinical trials have been reported to date.

## 11.4 CRISPR-CAS SYSTEMS DELIVERY

### 11.4.1 Overview of CRISPR-Cas Technology

Clustered Regularly Interspaced Short Palindromic Repeats (CRISPR) associated with the Cas protein constitute an adaptive antiviral immunity system that was discovered in bacteria and archaea. To trace the origins of CRISPR-Cas systems, we can look back to the discovery of unusual sequences in *Escherichia coli*. In 1987, Ishino et al. [83] at Osaka University published a paper in the *Journal of Bacteriology* reporting the nucleotide sequence of the iap gene in *E. coli*. This gene sequence contained a series of short, repetitive DNA sequences, which we now recognize as the first identified CRISPR sequence. This discovery laid the groundwork for understanding how these sequences function in a prokaryotic immune system.

Fast forward to the twenty-first century, Spanish microbiologist Francisco Martínez Mojica further elucidated the role of these CRISPR sequences in bacterial immunity. He demonstrated that bacteria incorporate fragments of viral DNA, which guide the bacteria in recognizing and cutting foreign genetic elements [84]. These DNA pieces, known as spacers, are stored between CRISPR cassettes, and the entire section of the host genome is known as the CRISPR locus. The host transcribes and processes the CRISPR locus to form CRISPR RNAs (crRNAs). Each crRNA encodes a unique spacer sequence. Concurrently, the organism produces Cas proteins, which then combine with a specific crRNA. This crRNA serves as a guide to silence foreign genetic elements that match the crRNA sequence (see Figure 11.5) [85].

This marked the beginning of our understanding of CRISPR as an adaptive immune system in bacteria and archaea. Building on these foundational discoveries, Jennifer Doudna and Emmanuelle Charpentier made a groundbreaking contribution in 2012 by developing a method to harness CRISPR-Cas9 as a precise genome-editing tool [86]. Their work revolutionized biotechnology, enabling scientists to edit genes with unprecedented accuracy, and earned them the Nobel Prize in Chemistry in 2020.

#### *11.4.1.1 Classification of CRISPR-Cas Systems*

The action of the CRISPR-Cas system can be divided into three main stages: (1) spacer integration, (2) processing of the CRISPR locus transcript and maturation of the crRNA, and (3) DNA or RNAi [87]. It is important to note that this system varies widely, leading to extensive classification (see Table 11.4). Initially, two classes are defined based on evolutionary taxonomy: Class 1 and Class 2. The Cas1–Cas2 complex, which is present in almost all CRISPR-Cas systems, plays a crucial role in the insertion of spacers into the CRISPR cassettes [87].

##### *11.4.1.1.1 Class 1 CRISPR-Cas Systems*

Class 1 CRISPR-Cas systems are characterized by the use of multiple different Cas proteins, while Class 2 systems use a single Cas protein. Each class is further subdivided into types: Class 1 into Types I, III, and IV, and Class 2 into Types II, V, and VI. Among Class 1 systems, Type I is the most common. It employs a multiprotein complex known as the CRISPR-associated complex for antiviral defense (Cascade), which is bound to the crRNA guide. Additionally, a Cas3 endonuclease is involved in cleaving the foreign DNA [88]. In contrast, Type III CRISPR systems use Csm or Cmr complexes to cleave both RNA and DNA and are characterized by the presence of the multidomain Cas10 protein [89].

##### *11.4.1.1.2 Class 2 CRISPR-Cas Systems*

Class 2 CRISPR systems have become widely used for genomic engineering in eukaryotic cells due to their simplicity, associated with a single Cas protein, and their small size, which facilitates delivery into target cells. Among Class 2 systems, Type II is the most utilized in research. The defining feature of Type II is the Cas9 protein, a multidomain protein that integrates recognition, binding, cleavage functions, and crRNA maturation [87]. The Cas9 protein contains two endonuclease domains: the HNH-nuclease domain and the RuvC-like nuclease domain, which together cut both strands of the target DNA, resulting in blunt ends [90]. Additionally, Type II systems include a trans-activating CRISPR RNA (tracrRNA), which pairs with the crRNA to form a guide RNA (gRNA). The Cas9 complex uses the tracrRNA portion as a handle, while the crRNA directs the complex to a matching viral sequence (see Figure 11.6A). Among Type II subtypes, II-A includes an additional gene known as *csn2*, which is associated with spacer integration [87].

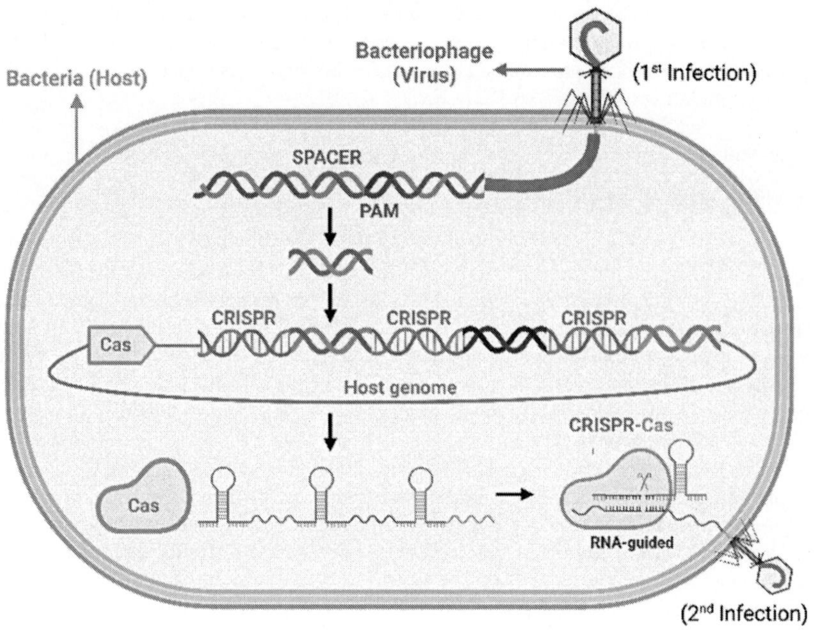


**Figure 11.5** Overview of the CRISPR-Cas system in bacterial immunity. The CRISPR-Cas9 system in bacteria serves as an adaptive immune mechanism. When a bacteriophage infects a bacterium, a fragment of the viral DNA (adjacent to a PAM) is integrated into the host genome at the CRISPR locus as a new spacer. The CRISPR locus is transcribed and processed to generate CRISPR RNAs (crRNA). Each crRNA associates with Cas effector proteins that use crRNAs as guides to silence foreign genetic elements that match the crRNA sequence.

## Table 11.4 CRISPR-Cas systems classification

| Class | Common Characteristics | Type | Common Characteristics | Subtypes |
|---|---|---|---|---|
| 1 | Multiple Cas proteins | I | Cascade complex + Cas3 endonuclease DNA targeting | A–E |
| | | III | Csm/Cmr complex<br>Multidomain Cas10 protein<br>DNA+RNA targeting | A–F |
| | | IV | Lack *cas1* and *cas2* genes<br>No CRISPR cassette detectable in genome<br>*csf1* signature gene | - |
| 2 | Single Cas protein | II | Cas9<br>DNA targeting<br>tracr-RNA | A–C |
| | Used for genomic engineering in eukaryotic cells | V | Cas12<br>DNA targeting | A–I, K, U, CRISPR–CasΦ |
| | | VI | Cas13<br>RNA targeting | |

Type V CRISPR systems are characterized by the presence of an RuvC-like endonuclease domain at the C-terminus of the effector protein (Cas12) and a variable N-terminus. Some effector proteins in Type V systems are capable of acting on single-stranded DNA (ssDNA) and even single-stranded RNA (ssRNA), in addition to their basic double-stranded DNA (dsDNA) functions. The first known Type V system, V-A, features the effector protein Cas12a. This endonuclease cleaves both the non-complementary and complementary strands

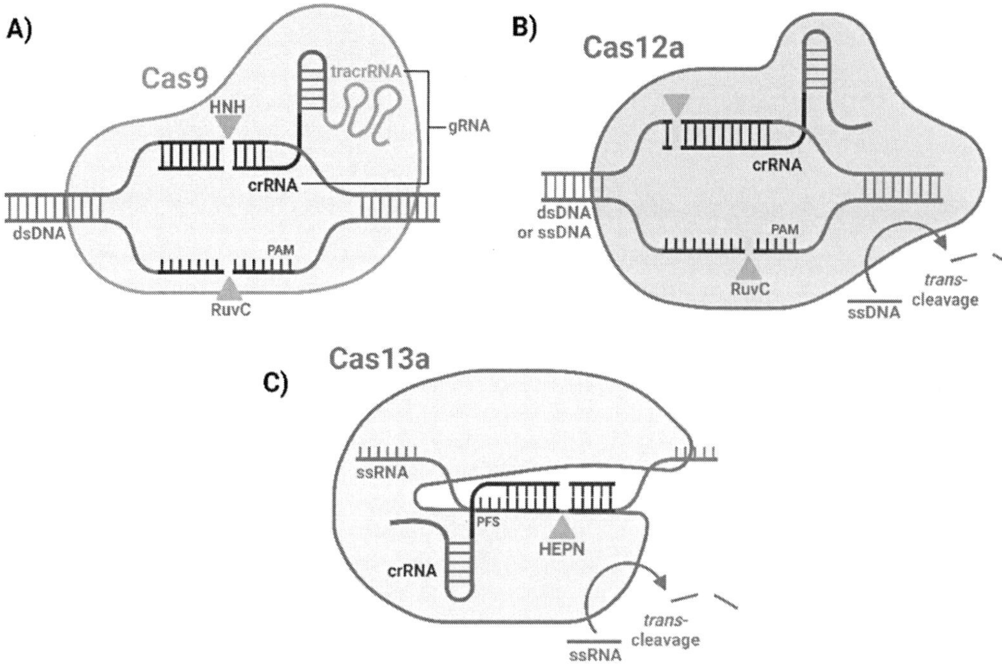


**Figure 11.6** CRISPR-Cas9, -Cas12a and -Cas13a systems. (A) CRISPR-Cas9 systems cleave dsDNA using a tracrRNA:crRNA hybrid (gRNA). The Cas9 uses the tracrRNA portion to hold the gRNA and the crRNA portion to recognize and bind to the target DNA. Cas9 has two nuclease domains: HNH cleaves the strand complementary to the gRNA, and RuvC cleaves the non-target strand. (B) CRISPR-Cas12a systems cleave dsDNA or ssDNA using a crRNA to recognize and bind to the target DNA. Cas12a has a single nuclease domain: RuvC. The single domain can cut both strands in dsDNA leaving a 5-7-nt overhang (sticky ends). Cas12 not only cleaves the target DNA strand (cis-cleavage) but also trans-cleavage of ssDNA indiscriminately. (C) CRISPR-Cas13a systems cleave ssRNA using a crRNA to recognize and bind to the target RNA. Cas13a has two HEPN domains that come together to initiate cis and trans-cleavage of ssRNA.

of the target DNA using a single RuvC domain, a process referred to as *cis*-cleavage. Cas12 systems are also known for their collateral or trans-cleavage activity (see Figure 11.6B) [91].

Type VI CRISPR systems are distinguished by the Cas13 effector protein. Unlike Cas9 and Cas12, Cas13 is an RNA-guided RNA endonuclease, making it particularly suitable for RNA detection, targeting, and editing technologies. The Cas13 protein contains two HEPN (higher eukaryotes and prokaryotes nucleotide-binding) nuclease domains. Guided by crRNA, Cas13 recognizes ssRNA targets that are complementary to the spacer portion of the crRNA. Like Type V, Type VI systems also exhibit both *cis*- and *trans*-cleavage activities (see Figure 11.6C). Additionally, Cas13 is involved in the maturation of crRNA [92].

#### *11.4.1.2 CRISPR-Cas9 and CRISPR-dCas9 Systems*

Type II CRISPR systems, renowned for their simplicity and compact size, have emerged as powerful tools for genome editing, revolutionizing biomedical research. Notably, the CRISPR-Cas9 system has become widely utilized due to its versatility in genetic manipulation. It is employed for correcting specific mutations within the genome, regulating the transcription of target genes, and even switching genes on or off entirely [93].

The CRISPR-Cas9 system consists of two essential components: a gRNA and the CRISPR-associated protein 9 (Cas9). The gRNA binds to the complementary sequence in the target DNA, while Cas9, an endonuclease, creates a double-stranded break (DSB) (see Figure 11.7) [94]. To achieve precise DNA cutting, a protospacer adjacent motif (PAM) must be present adjacent to the target sequence recognized by the gRNA. This 2–5-nucleotide sequence is specific to the bacterial species from which the Cas9 protein is derived.

The DSB induced by the CRISPR-Cas9 system triggers the cell's repair mechanisms to address the cleavage. There are two major pathways for repairing DSBs: non-homologous end joining (NHEJ) and homologous recombination (HDR). NHEJ ligates the DNA ends through random insertion or deletion of DNA, often resulting in gene knockout (KO). In contrast, HDR uses a DNA repair template, allowing

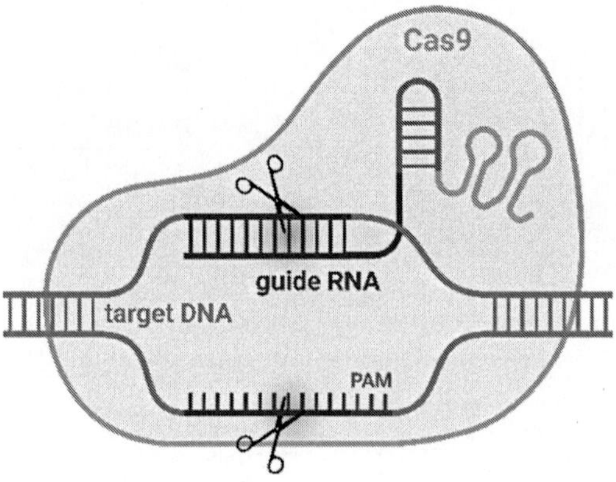


**Figure 11.7** CRISPR-Cas9 system. The Cas9 protein is an endonuclease that creates a DSB at a specific location in the DNA. Cas9 binds to the target sequence specified by the gRNA.

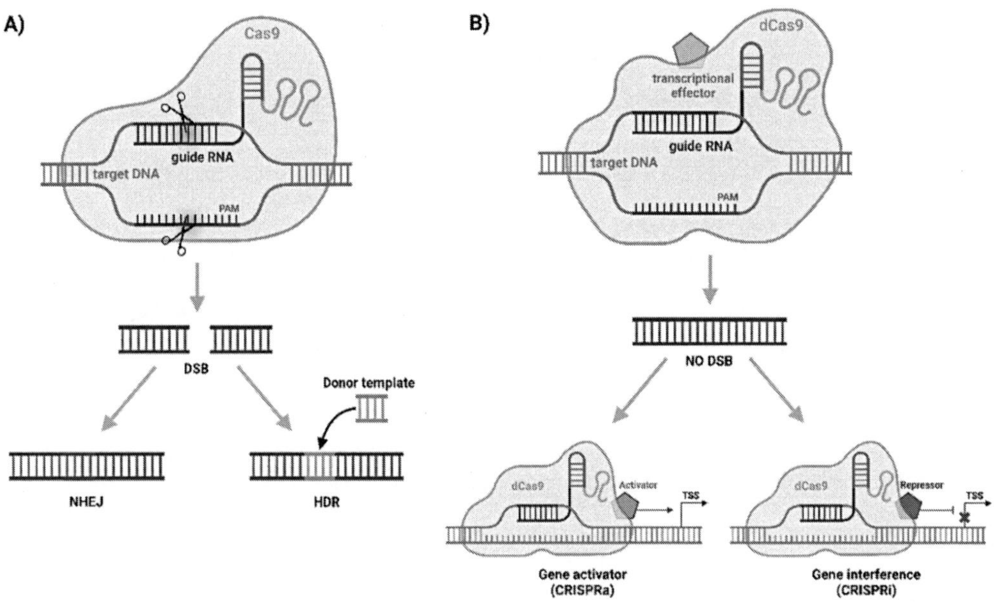


**Figure 11.8** (A) CRISPR-Cas9 system: the Cas9 protein, guided by a gRNA, creates a DSB in the target DNA. The DSB can be repaired by either NHEJ or HDR using a donor template for precise editing. (B) CRISPR-dCas9 system: The dCas9 protein binds to the target DNA without causing a DSB. When combined with a transcriptional effector, it can lead to gene activation (CRISPRa) or gene interference (CRISPRi).

for precise genome editing and facilitating gene knock-in (KI) with specific modifications [95] (see Figure 11.8A). Both types of gene editing are permanent and absolute, functioning as an "on-off switch."

An alternative approach developed for transcriptional regulation allows for temporal and gradual modifications, akin to a "light dimmer." This system, known as CRISPR-dCas9, consists of three main components: a gRNA, a catalytically inactive Cas9 (dCas9) protein, and a transcriptional effector. As previously mentioned, the Cas9 protein has two nuclease domains (HNH and RuvC). To create the deactivated Cas9 protein, a single amino acid in each of the HNH and RuvC domains is mutated, rendering the dCas9 protein unable to perform endonuclease activity. The transcriptional effector can be either a transcriptional activator (CRISPR activation; CRISPRa) or a repressor (CRISPR interference; CRISPRi) (see Figure 11.8B) [96].

In the CRISPR interference (CRISPRi) system, gene silencing is achieved by targeting the catalytically inactive Cas9 (dCas9) protein to the promoter region of a gene. This approach inhibits downstream gene expression by sterically hindering the transcriptional machinery, achieving 60–80% repression on its own. To enhance gene silencing, a transcriptional repressor domain, such as the Kruppel-associated box (KRAB), is incorporated, enabling inducible and reversible gene silencing [97]. Conversely, gene activation or overexpression is accomplished by fusing dCas9 to transcriptional activators such as VP64 and p65, targeting these activators to the promoter and enhancer regions of the target DNA [97].

When a single amino acid is mutated in either the HNH or RuvC domain of the Cas9 effector protein, it behaves as a nickase (nCas9) [90]. This modification results in a single-strand cleavage of the target DNA, depending on which domain is deactivated. This feature is particularly useful in applications such as base editing, where nCas9 introduces precise nucleotide changes without generating double-strand breaks. For instance, nCas9 paired with a cytidine deaminase enzyme has been used to efficiently convert C•G base pairs to T•A in specific genomic regions, facilitating precise gene correction or the creation of specific mutations [98]. Additionally, nCas9-induced single-strand breaks can be used in homology-directed repair (HDR) to guide the insertion of desired DNA sequences with higher fidelity, reducing the risk of unwanted mutations [99].

### 11.4.2 Nanocarrier Strategies for CRISPR Delivery

In Section 11.4.1, we extensively discussed the CRISPR-Cas system and its components. Among the various CRISPR systems, Class 2 systems are of particular interest in biotechnology research due to their simplicity and size advantages. Class 2 CRISPR systems generally consist of a single Cas protein effector and an RNA guide, which can be either a hybrid gRNA (tracrRNA) or just a crRNA.

The CRISPR-Cas system can be delivered into a cell in three main forms: (1) as a ribonucleoprotein (RNP) complex, which consists of the Cas endonuclease protein and the gRNA molecule; (2) as pDNA containing the genetic information for both the Cas protein and the gRNA, or two separate plasmids, each encoding one of these components; and (3) as Cas9 mRNA together with the gRNA molecule (see Figure 11.9) [100]. Each delivery method offers unique advantages and faces specific challenges.

pDNA is more stable than linear DNA constructs, as it resists degradation by exonucleases. It is efficiently transfected into cells and can be easily produced through bacterial amplification. However, pDNA can provoke immune responses due to bacterial backbone elements, and it requires nuclear entry for gene expression, which can reduce efficiency. In contrast, mRNA allows for faster protein expression by bypassing nuclear translocation and transcription, making it more effective. Nevertheless, mRNA has high innate immunogenicity and low chemical stability, resulting in a very short half-life in biological fluids. RNPs, composed of the Cas protein and gRNA, offer immediate gene-editing functionality upon delivery. They provide a rapid onset of activity without the need for transcription or translation, and they reduce off-target effects due to their rapid degradation post-function. However, RNPs are prone to degradation and require protective delivery mechanisms to ensure stability and efficient cellular uptake, which makes their delivery formulation more complex compared to pDNA or mRNA [101].

Due to the immense potential of CRISPR-Cas systems for disease therapy, researchers are highly invested in developing nanocarriers to deliver editing machinery effectively. Specialized delivery vehicles are crucial for several reasons. They protect CRISPR components from degradation by nucleases and clearance by the immune system once they enter the body. They also enhance cellular uptake and facilitate endosomal escape,

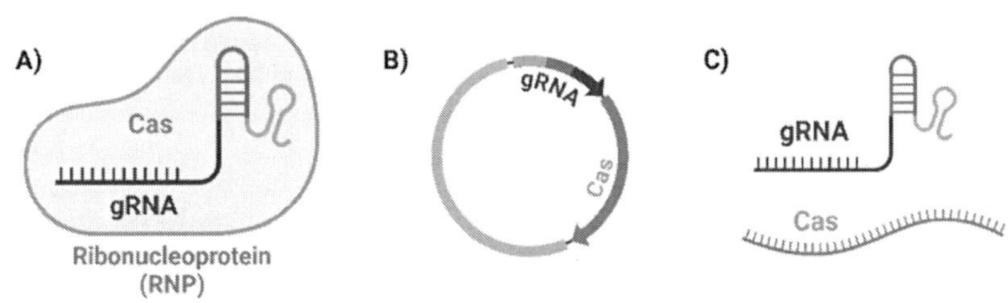


**Figure 11.9** CRISPR-Cas forms for delivery. (A) RNP complex, composed of the effector Cas protein and the gRNA molecule, (B) a plasmid that contains the gRNA sequence and the *cas* gene, (C) the Cas mRNA and the gRNA molecule.

thus improving the likelihood of reaching the target DNA. Additionally, surface modifications can enable targeted delivery to specific cells or tissues. Furthermore, they help reduce the risk of off-target effects, as systemic administration of CRISPR components can increase the chances of unintended modifications in non-target DNA [102].

Various delivery strategies for CRISPR-Cas systems have been developed, including physical methods, viral vectors, and non-viral vectors [103]. While viral vectors are efficient for gene transfection, they present significant challenges, including high risk of immunogenicity, mutagenicity, and limited loading capacity [104]. Consequently, non-viral nanocarriers have gained considerable attention in recent years due to their ease of synthesis, cost-effectiveness, tunability, and non-immunogenicity [104]. These non-viral vectors are further categorized into lipid, polymeric, inorganic, and hybrid types.

#### *11.4.2.1 Lipid Nanoparticles*

LNPs have emerged as one of the most utilized vehicles for nucleic acid delivery due to their ability to traverse physiological barriers, ease of formulation, spontaneous self-assembly, high potency, excellent biocompatibility, large payload capacity, and versatile design options for specific applications [105]. LNPs can be categorized into various subtypes, including solid lipid nanoparticles (SLN), NLC, and liposomes [102]. These nanoparticles generally function as nanospheres that encapsulate therapeutic compounds such as nucleic acids.

They typically comprise four main surface components: ionizable cationic lipids, PEG, zwitterionic phospholipids, and cholesterol [106]. PEG lipids enhance LNP stability, reduce immunogenicity, and prolong circulation time [63]. They also prevent particle aggregation during formulation, storage, and circulation, which helps control particle size [107]. Zwitterionic phospholipids provide structural stability and delivery efficiency. Various phospholipids (also known as helper lipids) can be added to the LNP surface to improve the efficiency and safety of nucleic acid delivery. For instance, the helper lipid dioleoylphosphatidylethanol amine (DOPE) facilitates the fusion of LNPs with cellular membranes, enhancing cellular uptake and fusion with the endosomal membrane, thereby promoting the release of cargo into the cytoplasm [108]. Another helper lipid, distearoylphosphatidylcholine (DSPC), has been used to improve stability [109]. Cholesterol increases particle stability, promotes cellular uptake, and assists in destabilizing the cellular bilayer during fusion [110].

Ionizable cationic lipids with a pKa of around 6.4 are crucial for delivery mechanisms [111]. In acidic environments (pH ~ 3–5), these lipids acquire a positive charge, facilitating the binding of negatively charged RNA molecules during LNP formation [112]. Upon entering the bloodstream, where the pH becomes neutral, these lipids lose their charge, stabilizing the particles and preventing premature release. Once internalized via endocytosis, LNPs are encapsulated in endosomes where the pH decreases, restoring the positive charge on the lipids. This change promotes interaction with the negatively charged endosomal membrane, leading to fusion and subsequent release of the cargo into the cytoplasm [111]. This encapsulation and release strategy is applicable to various CRISPR-Cas systems (see Figure 11.10), although modifications, especially to the helper lipids, are often proposed to enhance gene delivery efficiency or address specific limitations of different CRISPR-Cas forms.

A notable example of LNP-based delivery for CRISPR/Cas9 gene editing involves the proprietary ionizable lipid-based LNP, iLP181. In a proof-of-concept study, Li et al. [113] developed this non-viral delivery system for an anti-cancer CRISPR/Cas9 application targeting the *PLK1* gene. The selected plasmid was encapsulated within the iLP181 LNP formulation. Remarkably, this formulation achieved effective gene editing without requiring plasmid condensation reagents, using an additional incubation step to reassemble and stabilize the LNPs. The iLP181/psgPLK1 complex was efficiently internalized by hepatoma carcinoma cells (HepG2-Luc) through interaction with the ApoE protein, resulting in a 32% knockdown of *PLK1* mRNA *in vitro*. In HepG2-Luc-xenograft tumor-bearing mice, the LNPs demonstrated robust endosomal escape, leading to significant tumor growth inhibition, reduced *PLK1* mRNA levels, and increased tumor cell apoptosis. Furthermore, the iLP181 formulation showed an excellent safety profile *in vivo*, highlighting its potential as a powerful non-viral delivery platform for CRISPR/Cas9 systems and a promising approach for gene editing-based cancer therapies.

In another study, Sun et al. [114] developed several new amino lipids by modifying amino head groups, resulting in stable nanoparticles with CRISPR/Cas9 plasmids, ranging in size from 100 to 200 nm. These nanoparticles exhibited pH-sensitive hemolysis, with minimal activity at pH 7.4 and increased activity at acidic pH, which is crucial for endosomal escape. The ECO and iECO lipids demonstrated the highest gene editing efficiency, with ECO achieving up to 50% GFP suppression in NIH3T3-GFP cells. The study suggests that these amino lipids, particularly ECO and iECO, are promising carriers for CRISPR/Cas9 delivery due to their efficient formulation, low cytotoxicity, and high gene editing efficacy. This makes them potential candidates for gene therapy applications, especially for treating autosomal dominant diseases.

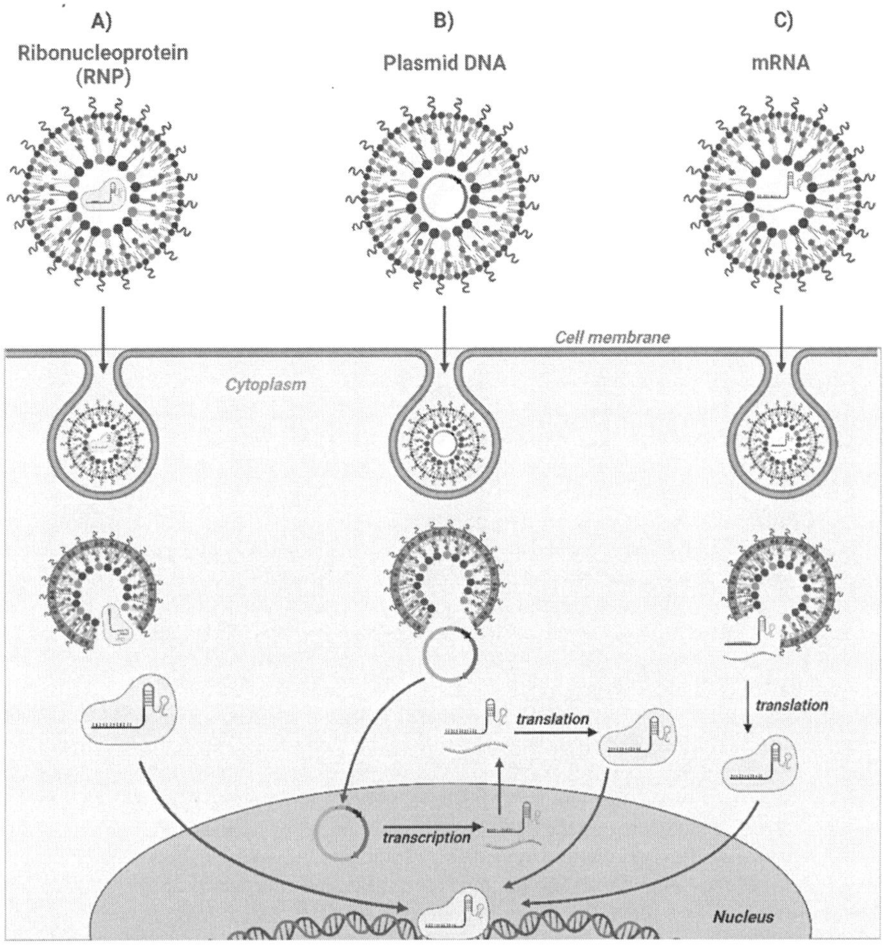


**Figure 11.10** Intracellular CRISPR-Cas delivery using LNPs. (A) RNP complex can enter the nucleus directly and perform the gene editing. (B) Plasmid DNA enters the nucleus where it is transcribed into mRNA. The mRNA exits the nucleus and the Cas mRNA in translated to form the Cas effector protein, together with the gRNA they re-enter the nucleus to perform gene editing. (C) The mRNA form is released into the cytosol and the Cas mRNA is directly translated into the Cas protein. It joins the gRNA to enter the nucleus and perform gene editing.

#### *11.4.2.2 Polymeric Nanoparticles*

Polymeric-based nanoparticles have garnered significant interest in delivering CRISPR-Cas9 systems due to their low immunogenicity and high carrying capacity [115]. Their primary advantage over other non-viral vectors is their potential for chemical diversity and functionalization, which allows for adaptation to various cargo forms and specific targeted editing, along with establishing release triggers associated with known microenvironments. Additionally, they exhibit strong biocompatibility and biodegradability. Common polymeric nanoparticles include PEI, poly-L-lysine, chitosan, and PAMAM-based nanovehicles. Like LNPs, these polymers can pass through cellular membranes and protect the cargo from degradation or clearance. The main challenge with polymeric nanoparticles has been the toxicity associated with cationic vectors, which has been addressed through modifications to the cationic polymeric vector or co-coating with other polymers [115]. Hydrophilic PEG is widely used to coat cationic polymers, enhancing stability, biocompatibility, and biodegradability while prolonging blood circulation [115].

PEI is often regarded as the gold standard for *in vitro* transfection due to its ability to effectively condense pCas9. However, its high density of positive charges can induce cytotoxicity, necessitating modifications for safer *in vivo* use. To address this, researchers have incorporated PEG to enhance biocompatibility and reduce immunogenicity, along with cholesterol to improve cellular uptake [116, 117]. Li et al. [104] developed a

PEG-PC/PEI-CHO/pCas9 system for genomic editing of HeLa tumor cells in vitro and *in vivo*. Polycarbonate (PC) was chosen for its degradable properties, as its breakdown products are inert, reducing potential toxicity from repeated use. The PEG component enhances biocompatibility, while PEI facilitates endosomal escape through the proton sponge effect. This effect occurs when PEI captures incoming protons pumped into the endosome, leading to an influx of counter-ions to maintain electrical neutrality. This osmotic imbalance causes water to flow into the endosome, leading to its rupture and the release of the pCas9 complex into the cytoplasm. The system resulted in significant suppression of tumor growth by inducing apoptosis.

To address the cytotoxicity challenges associated with transfection reagents, researchers have explored biodegradable polycations, such as PBAEs. Gao et al. [118] reported that PBAEs exhibit high biodegradability, biocompatibility, and ease of synthesis. When positively charged, PBAEs effectively condense CRISPR-Cas9 pDNA, protect the cargo from immune responses, facilitate cellular uptake, and ensure cargo release within the cytosol. Hyperbranched polymers, which show higher transfection efficiency compared to linear counterparts, led researchers to develop a copolymer combining poly(amide-amine) (PAMAM) with PBAEs, resulting in the PAMAM-PBAE hyperbranched copolymer (hPPC). PAMAM, a dendrimer vector, efficiently condenses pDNA and protects the cargo. The hPPC vector was tested for delivering a CRISPR/Cas9 recombinant plasmid targeting the *HPV16 E7* oncogene.

Another advancement with polymer-based nanoparticles was achieved by Rui et al. [119], who synthesized a novel class of bioreducible branched delivery vectors, termed reducible branched ester-amine quadpolymers (rBEAQs). These vectors were designed for the co-delivery of Cas9 pDNA and single guide RNA (sgRNA) oligonucleotides. The branched architecture of rBEAQs significantly increased charge density, leading to stronger nucleic acid binding and enhanced compaction. The vector's multiple disulfide bonds enable environmentally triggered release upon entering the cytosol, a reducing environment. The high hydrophobicity of rBEAQs further improves nanoparticle stability and cellular uptake. This optimized delivery method achieved a 40% gene (GFP) KO efficiency, marking the first successful co-delivery of plasmid Cas9 and sgRNA for CRISPR gene editing.

In another advancement with polymer-based nanoparticles, Abbasi et al. [120] developed Cas9 mRNA/sgRNA polyplex micelles (PMs). These PMs are nanoscale structures formed from a combination of PEG and a polycation block copolymer, featuring a core-shell architecture. RNA, such as mRNA, is condensed and protected within the core, while PEG forms a protective outer shell. The polycation segment, poly(N′-(N-(2-aminoethyl)-2-aminoethyl)aspartamide (PAsp(DET)), facilitates interactions and condenses the charged mRNA into the micelle core, while also being biodegradable and aiding in endosomal escape. Simultaneously, PEG reduces inflammatory and immune responses and enhances tissue penetration. The treatment was applied directly to the mouse brain, targeting the stop cassettes of tdTomato expression. Notably, sgRNA stability increased when co-loaded within a single micelle alongside Cas9 mRNA, resulting in significantly higher genome editing efficiency with the PMCas9/sgRNA treatment compared to a mixture of $PM_{Cas9}$ and $PM_{sgRNA}$.

#### *11.4.2.3 Inorganic Nanoparticles*

Inorganic nanoparticles are nanoscale structures composed of inorganic substances, including metals, metal oxides, or semiconductors. These nanovehicles are characterized by their non-toxic, hydrophilic, biocompatible, and highly stable features, making them suitable for various biomedical applications [121]. Their unique physicochemical properties, such as electrochemical and magnetic characteristics, render them highly effective for nucleic acid delivery [122].

These nanoparticles encompass a range of types based on their composition and structure. Carbon-based nanoparticles, such as CNTs, and metal-based nanoparticles, including AuNPs, AgNPs, IONPs/$Fe_3O_4$, zinc oxide (ZnO NPs), and copper oxide (CuO NPs), are prominent examples. Additionally, metal-organic frameworks (MOFs) combine metal ions with organic ligands, while silica-based nanoparticles, particularly MSNs, are notable for their porous structure that enhances functionality [123, 124].

Despite their advantages, inorganic nanoparticles can provoke immune responses and are susceptible to clearance by macrophages. To address these issues, PEG coatings are often applied to reduce immune system recognition and extend circulation time [123]. Given their diverse potential, these inorganic nanoparticles have been explored for their application in delivering CRISPR-Cas9 systems. The following sections will provide key examples of how each type of nanoparticle has been utilized to enhance the precision and efficiency of CRISPR-Cas9 delivery.

##### *11.4.2.3.1 Metal-Based Nanoparticles*

Among the various metals used for nanoparticle synthesis, AuNPs have garnered significant interest due to their chemically inert properties, which help avoid immune system activation [124]. Their ease of synthesis and versatile surface functionalizations further enhance their appeal. The gold surface's strong affinity for thiol groups enables the formation of stable gold–thiol bonds when CRISPR-Cas9 components are modified with thiol groups [125]. This bond is advantageous for CRISPR-Cas9 delivery as it can be cleaved under reductive conditions found in the cytosol, allowing for controlled release of the cargo. However, CRISPR-Cas9 components are still vulnerable to degradation by nucleases and immune responses. To improve stability, co-conjugation with polymers such as PEG and PAMAM has been utilized [125].

Building on these advantages, Lee et al. [126] developed the CRISPR-Gold technology, which enables the direct delivery of Cas9 RNP and donor DNA *in vivo* through local administration, effectively inducing HDR. HDR requires a donor DNA template, and in this system, the AuNP surface is functionalized with thiol-modified oligonucleotides (DNA-thiol), which hybridize with the donor DNA (ssDNA). The Cas9 RNP is then loaded onto this complex via the affinity between the ssDNA and the Cas9 nuclease (see Figure 11.11). The encapsulation of this assembly with the cationic polymer PAsp(DET), which acts as an endosomal disruptor, forms the CRISPR–Gold complex. This polymer enhances cellular uptake and endosomal escape, and once in the cytoplasm, the reductive environment cleaves the gold–thiol bond, releasing the CRISPR–Cas9 complex along with the donor DNA [126].

This system has demonstrated the capability for *in vivo* gene editing, as shown in the Ai9 mouse model with the tdTomato gene, where the stop cassettes were effectively removed, enabling downstream expression of the fluorescent protein [126]. The development of CRISPR-Gold has significantly accelerated the application of AuNPs in the field of gene editing.

Following the success of CRISPR–Gold, IONPs have emerged as another powerful tool for delivering CRISPR-Cas9 components. IONPs, particularly those composed of magnetite ($Fe_3O_4$), are recognized for their unique magnetic properties, biocompatibility, and multifunctionality. At the nanoscale, magnetite exhibits superparamagnetic properties, meaning it shows magnetic properties only in the presence of an external magnetic field and does not retain magnetization when the field is removed [127]. These properties make IONPs highly suitable for targeted gene delivery applications. Unlike AuNPs, IONPs can be guided to specific tissues or cells using an external magnetic field, enhancing the precision of gene editing. This magnetic targeting helps reduce off-target effects and increases the efficiency of CRISPR-Cas9 delivery to the desired location [128]. Recent studies have highlighted the potential of IONPs in facilitating efficient gene editing with minimal cytotoxicity, establishing them as a versatile and effective alternative to gold-based nanoparticles in gene therapy.

In a study by Mohammadi Ghanbarlou et al. [128], superparamagnetic iron oxide nanoparticles (SPIONs) were used to enhance the delivery of a CRISPR/dCas9 activator system into human foreskin fibroblast cells. The SPIONs were coated with PEI, a cationic polymer that facilitates the binding of the CRISPR-Cas9 plasmid through electrostatic interactions and aids in cellular uptake, thus improving the delivery efficiency. The CRISPR/dCas9 system was designed to activate the expression of stem cell-related genes, specifically OCT4 and SOX2, crucial for cell reprogramming and stem cell maintenance. Upon transfection, a 2.5-fold increase in the expression of these genes was observed in fibroblast cells, as confirmed by qRT-PCR analysis. This method demonstrated significantly higher transfection efficiency compared to traditional lipofection, with a transfection rate of 28.2% versus 7.89% in control cells using lipofection. Importantly, magnetofection with SPIONs did not result in significant cytotoxicity, making this approach a promising alternative for CRISPR delivery.

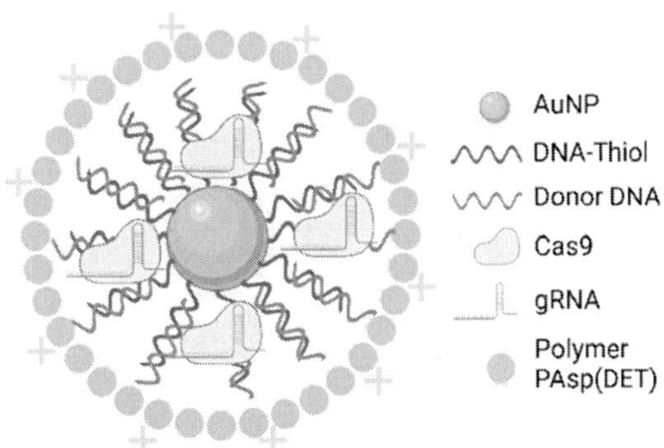


**Figure 11.11** CRISPR–Gold mechanism. The DNA–thiol binds to the gold surface via a gold–thiol bond. The donor DNA (ssDNA) hybridizes with the DNA–thiol strand, and the Cas9 RNP is loaded via its affinity with the ssDNA. The cationic polymer PAsp(DET) encapsulates all components.

#### *11.4.2.3.2 Silica-Based Nanoparticles*

Silica-based nanoparticles, especially MSNs, have become prominent as nanocarriers for CRISPR-Cas9 delivery due to their unique properties. MSNs are known for their highly ordered pore structures, large surface area, and excellent biocompatibility, making them ideal for encapsulating and protecting CRISPR-Cas9 components such as Cas9 protein, gRNA, and donor DNA. These features help shield the cargo from enzymatic degradation and premature immune clearance. Compared to other inorganic nanoparticles, MSNs offer advantages including controllable size, easily modifiable surfaces, and suitability for large-scale production, all while maintaining excellent biocompatibility. Their stability and non-toxic nature have led to FDA approval for a Phase-1 clinical trial, underscoring their potential for clinical applications. Given their negative charge in aqueous conditions, MSNs are often coated with a cationic layer to facilitate the loading of negatively charged nucleic acids, enhancing cellular uptake and promoting endosomal escape [129].

Building on these advantages, Wang et al. developed a novel glutathione (GSH)-responsive silica nanoparticle (SNP) system designed to enhance CRISPR-Cas9 delivery efficiency. These SNPs were engineered with disulfide bond-containing crosslinkers, which enable cargo release upon exposure to the high levels of glutathione in the cytosol, where the disulfide bonds are cleaved. Additionally, imidazole groups were incorporated to facilitate endosomal escape via the proton sponge effect, further improving delivery efficiency. The surface of these SNPs was PEGylated to enhance biocompatibility and circulation time, and they were functionalized with targeting ligands specific to retinal pigment epithelium (RPE) and liver cells. *In vivo* studies demonstrated that these SNPs could efficiently deliver CRISPR-Cas9 RNP, resulting in a significant increase in tdTomato expression in Ai14 mice [130].

### 11.4.3 Targeted Gene Editing

Targeted gene editing has transformed molecular biology by allowing for precise modifications at specific genomic loci. Four principal families of engineered nucleases have been developed for this purpose: zinc-finger nucleases (ZFNs), transcription activator-like effector nucleases (TALENs), CRISPR-Cas, and engineered meganucleases. Despite their structural differences, these platforms share the common ability to introduce site-specific DSBs, which are then repaired by the cell's endogenous DNA repair mechanisms.

Meganucleases are noted for their high specificity but are challenging to re-engineer for new targets [131]. ZFNs and TALENs, while instrumental in the early development of gene editing, are often considered less efficient, labor-intensive, and costly [132]. In contrast, CRISPR/Cas9 has gained prominence for its simplicity, efficiency, and accessibility. Designing sgRNA for CRISPR/Cas9 is more user-friendly, adaptable, and cost-effective compared to the creation of DNA-binding domains required for ZFNs or TALENs [132].

Despite the potential of CRISPR-Cas9 for treating both genetic and non-genetic diseases, a significant challenge remains in minimizing off-target effects [133]. Off-target effects refer to unintended edits at sites other than the intended target, which can have particularly detrimental consequences. In gene therapies aimed at correcting genetic disorders, unintended edits could disrupt other critical genes, leading to new mutations or adverse outcomes. Therefore, developing strategies to enhance specificity and reduce off-target effects is crucial for the safe application of these technologies in medicine.

#### ***11.4.3.1 Off-Target Effects***

Off-target effects represent unintended genetic modifications that occur when engineered nucleases, like CRISPR-Cas9, cleave DNA sequences other than the intended target. The CRISPR-Cas9 system relies on a sgRNA to direct the Cas9 protein to the specific DNA site. The sgRNA is approximately 80 base pairs (bp) long, with only 20 bp dedicated to binding the target DNA. Given the vast size of the human genome, which encompasses around 3 billion bp, a 20 bp guide is not sufficiently complex to ensure uniqueness across the entire genome. Additionally, Cas9 can tolerate up to three mismatches within this 20 bp binding region, further amplifying the risk of off-target sites [134, 135].

The impact of mismatches on off-target effects is influenced by their location relative to the PAM region. Mismatches within the PAM-proximal region (the first 7–12 bp) significantly affect Cas9 binding and R-loop formation, often preventing the induction of a DSB. In contrast, mismatches in the PAM-distal region are more tolerable, allowing for stable Cas9 binding and increasing the risk of DSBs at off-target sites [133, 134]. These off-target effects are sgRNA-dependent, meaning they can be predicted using in silico tools that scan the genome for sequences with partial complementarity to the sgRNA [132]. Such tools assess the likelihood of off-target editing and help minimize unintended modifications.

Furthermore, off-target effects can also arise independently of sgRNA-DNA mismatches. This indicates that factors such as chromatin structure, DNA accessibility, and the physical properties of the Cas9 protein itself might also contribute to off-target activity [131]. Understanding and mitigating these variables is crucial for enhancing the precision and safety of CRISPR-based gene editing applications.

#### *11.4.3.2 Precision in Gene Editing*

Precision in CRISPR-Cas9 gene editing is essential for achieving accurate and predictable genome modifications while minimizing unintended consequences. Several factors influence the effectiveness and accuracy of CRISPR-Cas9 editing, including the sequence specificity of the sgRNA, local chromatin structure, cellular environment, and the duration of Cas9 expression.

*Sequence Specificity of sgRNA:* The sgRNA directs the Cas9 nuclease to the target DNA sequence by binding near the PAM. For precise binding and cleavage, the sgRNA must have a perfect match with the target DNA sequence. Mismatches between the sgRNA and the target DNA, particularly within the seed region (the first 7–12 bp proximal to the PAM), can reduce binding affinity and editing precision. Mismatches in the PAM-distal region are more tolerated but can lead to off-target effects where Cas9 cleaves unintended genomic sites [134, 136].

*Local Chromatin Structure:* The accessibility of the target DNA is influenced by the local chromatin structure, which includes DNA methylation, histone modifications, and nucleosome positioning. Open chromatin regions are more accessible to the CRISPR–Cas9 complex, leading to higher editing efficiency. Conversely, tightly packed chromatin can hinder access and reduce precision [137].

*Cellular Environment and Repair Pathways:* After Cas9 induces a DSB, the cell uses either the error-prone NHEJ or the more precise HDR to repair the break. NHEJ often results in indels that may be unpredictable, whereas HDR can enable precise insertion or correction if a donor template is available. However, HDR is less efficient in mammalian cells [138].

*Duration of Cas9 Expression:* The length of time that Cas9 is expressed in the cell can influence off-target effects. Prolonged Cas9 expression increases the window for potential off-target cleavage. Cas9 binds strongly to target DNA and remains associated for approximately 15 hours after cleavage, delaying repair mechanisms [137]. This highlights the need for transient expression systems or inducible Cas9 variants to minimize off-target effects.

In summary, achieving precision in CRISPR-Cas9 gene editing involves careful consideration of the sgRNA sequence, chromatin accessibility, repair pathways, and expression duration to minimize off-target effects and ensure accurate genetic modifications.

#### *11.4.3.3 Evaluation of Off-Target Effects*

Evaluating off-target effects is a critical aspect of ensuring the safety and efficacy of CRISPR-Cas9 genome editing. Despite the system's precision, CRISPR-Cas9 can induce unintended edits at genomic sites other than the intended target. Detecting and characterizing these off-target effects is essential for optimizing gene editing strategies, particularly in clinical applications where unintended mutations could have harmful consequences [131]. Several methodologies, including both *in vitro* and *in vivo* approaches, have been developed to assess off-target activity. These methods can be broadly categorized into cell-free, cell-based, and *in vivo* techniques.

##### *11.4.3.3.1 Cell-Free Methods*

Cell-free methods reconstitute the nuclease activity *in vitro*, allowing for the identification of off-target sites without the complexities of the cellular environment. One prominent approach is Digenome-seq (digested genome sequencing), which involves incubating purified genomic DNA with the Cas9/sgRNA RNP complex, followed by whole-genome sequencing (WGS) to identify DSBs. Digenome-seq can detect off-targets with a sensitivity as low as 0.1%, making it a powerful tool for off-target analysis [134]. However, it is limited by the high cost associated with the extensive sequencing coverage required and the omission of chromatin state information, which is crucial for accurate detection. Additionally, Digenome-seq cannot distinguish between RNP-independent genomic breaks and those induced by the off-target activity of the nuclease [139].

To address the limitation of chromatin state omission, DIG-seq was developed. This method incorporates chromatin into the analysis by using cell-free chromatin instead of purified DNA, thereby improving the accuracy of off-target detection by accounting for the influence of chromatin structure on Cas9 activity [132]. Further advancement in this area led to the development of Extru-seq, a method that more closely mimics the intracellular environment by mechanically lysing cells to release genomic DNA for Cas9/sgRNA interaction. Extru-seq reduces false-negative rates by integrating the advantages of both cell-free and cell-based methods [140].

To reduce the costs associated with WGS, alternative strategies such as SITE-seq and CIRCLE-seq have been developed. SITE-seq involves tagging, enriching, and sequencing cleaved genomic fragments, thereby reducing sequencing coverage requirements, though its accuracy remains relatively low [141]. CIRCLE-seq, on the other hand, uses circularized DNA to minimize background noise and requires fewer reads, but it still suffers from a high false-positive rate [142, 143].

##### *11.4.3.3.2 Cell-Based Methods*

Given that the cellular context can significantly influence the behavior of CRISPR-Cas9, cell-based methods provide a more accurate representation of off-target effects. WGS in edited cells is the most direct approach, comparing genomic sequences before and after CRISPR-Cas9 editing. However, its application is limited by its low sensitivity for detecting off-target events that occur at low frequencies [144].

To increase sensitivity and reduce costs, techniques such as Cas9 ChIP-seq (chromatin immunoprecipitation followed by sequencing) have been employed. ChIP-seq captures sequences surrounding DSBs and characterizes the genomic landscape of these breaks. Traditional ChIP-seq, using catalytically inactive Cas9 (dCas9) to pull down Cas9 binding sites, has been shown to yield a high rate of false positives, as dCas9 may bind to many genomic locations without inducing cleavage, leading to the identification of numerous sites that are not genuine off-target cleavage sites [132].

GUIDE-seq is a robust cell-based method that integrates double-stranded oligodeoxynucleotides (dsODNs) into DSBs during NHEJ. GUIDE-seq can detect off-target sites with indel frequencies as low as 0.03% [145].

Comparative studies of these methods have shown that while they identify a substantial number of overlapping off-target sites, each method also detects unique sites, suggesting that a combination of approaches may be necessary to fully capture the off-target landscape for certain gRNAs [146].

BLESS (breaks labeling, enrichment on streptavidin, and next-generation sequencing) and BLISS (breaks labeling in situ and sequencing) are techniques that label DSBs in situ. BLESS captures DSBs at the time of cell fixation, which can limit its ability to detect all DSBs, leading to a potential underestimation of off-target events. Despite this limitation, BLESS has a low false-positive rate, making it a reliable method for off-target detection [147]. BLISS, an advancement of BLESS, enhances sensitivity and reduces the number of cells required by using T7-mediated transcription to amplify tagged DNA fragments before sequencing [148].

##### *11.4.3.3.3 In Vivo Detection*

Evaluating off-target effects *in vivo* is crucial for understanding the safety and efficacy of CRISPR-Cas9 in therapeutic applications. DISCOVER-seq is a pioneering method that utilizes the DNA repair protein MRE11 to detect DSBs *in vivo*, offering an advantage over methods like GUIDE-seq, which can exhibit cytotoxicity [146]. DISCOVER-seq relies on ChIP-seq to detect MRE11 binding to DSBs using specific antibodies, making it applicable across various tissue types and cell samples [149]. However, DISCOVER-seq has limitations due to the transient nature of MRE11 binding to DSBs, capturing only those off-targets present during sample preparation, which can result in a higher false-negative rate [146].

To address these limitations, a newer technique called GUIDE-tag was developed. Derived from GUIDE-seq, GUIDE-tag enhances the capture of double-stranded oligodeoxynucleotides (dsODNs) at DSB sites. This improvement is achieved by attaching a monomeric streptavidin (mSA) to the Cas9 protein, which recruits biotinylated dsODNs to the DSB sites during the NHEJ repair process, thereby facilitating their integration and increasing the discovery rate of off-targets [150]. Comparative studies between GUIDE-tag and DISCOVER-seq have demonstrated that GUIDE-tag offers superior sensitivity, providing broader coverage of off-target sites [150].

### 11.4.4 Therapeutic Applications

CRISPR-Cas systems have emerged as powerful tools in biomedical science, enabling precise and efficient genome editing. This technology has rapidly transitioned from a research tool to a potential therapeutic approach for treating a wide range of diseases. The ability to modify specific genes with high accuracy has opened new possibilities in the treatment of genetic disorders, cancer, infectious diseases, and other conditions [151].

CRISPR-Cas9, in particular, has been at the forefront of these developments, offering a versatile platform for gene editing. Its applications extend from correcting genetic mutations that cause hereditary diseases to modifying immune cells for improved cancer therapies. Additionally, the recent exploration of other CRISPR systems, such as Cas12 and Cas13, has expanded the scope of potential therapeutic uses, including targeted DNA and RNA editing [152].

#### *11.4.4.1 Genetic Disorders*

CRISPR-Cas systems have revolutionized the treatment of genetic disorders by enabling precise genome editing, offering hope for conditions previously deemed incurable. The CRISPR-Cas9 system, in particular, has demonstrated remarkable success in treating several genetic diseases, including sickle cell disease (SCD), Duchenne muscular dystrophy (*DMD*), and hemophilia.

##### *11.4.4.1.1 Sickle Cell Disease*

SCD is caused by a single nucleotide mutation in the *HBB* gene, leading to the production of abnormal (sickle) hemoglobin (HbS). Due to the nature of red blood cells (RBCs), which are enucleated and therefore lack DNA, direct gene correction in mature RBCs is not feasible. Consequently, therapeutic interventions must focus on hematopoietic stem cells (HSCs), the progenitors of RBCs, to ensure the correction of the mutation is passed on to the resulting blood cells [153].

CRISPR-Cas9-based approaches for treating SCD can be categorized into two primary strategies: (I) gene correction and (II) fetal hemoglobin (HbF) reactivation. Gene correction aims to directly modify the disease-causing mutation in the *HBB* gene responsible for HbS. This approach relies on HDR, a precise but challenging process, often hindered by competition with the NHEJ pathway, which can introduce mutations. HDR is also restricted to specific cell cycle phases and requires a donor template, adding complexity and potential immune responses [153]. Newer gene editing platforms, such as base editors and prime editors, offer an alternative to HDR. These tools allow for precise single nucleotide changes without inducing DSBs [154, 155]. While base editors can correct some mutations by directly converting specific bases, they are limited and cannot address all types of mutations, such as the T>A transversion in SCD. Prime editing, offers broader editing capabilities, and was tested by Anzalone et al., in the SCD context. They obtained very promising correction results (58%); however, implementation is very complex and requires further optimization [155].

The second strategy, HbF reactivation, is based on the protective effect HbF has due to the reduction in HbS polymerization [156]. This strategy is inspired by naturally occurring mutations of fetal hemoglobin gene (HPFH) which allow continued HbF production [157]. CRISPR-Cas9 can be used to mimic HPFH mutations either by recreating large deletions or introducing single nucleotide changes that disrupt repressor binding sites, such as *BCL11A* [158, 159].

##### *11.4.4.1.2 Duchenne Muscular Dystrophy*

*DMD* is a severe genetic disorder caused by mutations in the *DMD* gene, which encodes the dystrophin protein [160]. Researchers developed a method using CRISPR-Cas9 to delete exons 45–55 of the *DMD* gene in human-induced pluripotent stem cells (hiPSCs), now that approximately 60% of *DMD* patients have mutations in this region. By removing these exons, the study aimed to restore the reading frame of the *DMD* gene, leading to the production of a truncated but functional dystrophin protein, similar to what is observed in individuals with Becker muscular dystrophy (a milder form of muscular dystrophy). They obtained improved membrane integrity and the reassembly of the dystrophin–glycoprotein complex (DGC), both *in vitro* and *in vivo*. Additionally, the study found no significant off-target effects at the top 10 candidate sites, underscoring the precision of the CRISPR-Cas9 approach used in this context [161].

##### *11.4.4.1.3 Hemophilia*

Hemophilia is an X-linked recessive disorder characterized by mutations in coagulation genes, leading to impaired blood clotting. Hemophilia A (HA) results from mutations in the factor VIII (*F8*) gene, while Hemophilia B (HB) is due to mutations in the factor IX (*F9*) gene. These mutations result in a loss of function of the respective coagulation proteins, critically affecting the coagulation cascade and leading to severe bleeding disorders [162].

Hemophilia A is the most common type of hemophilia. CRISPR/Cas9 technology offers a promising alternative to traditional gene replacement strategies by enabling the targeted repair of the endogenous F8 gene. In a recent study, a Hemophilia A mouse model was established with a targeted deletion of the promoter and exon 1 of the F8 gene, mimicking the effects of common human mutations. Researchers used an adeno-associated virus (AAV) vector to deliver the CRISPR/Cas9 components along with a donor DNA template to hepatocytes. This approach facilitated the integration of the missing coding sequence of exon 1 and a promoter into the F8 gene via HDR [163].

The study demonstrated successful restoration of F8 expression in the liver, with significant improvements in coagulation as measured by activated partial thromboplastin time (aPTT). Importantly, the CRISPR/Cas9-mediated repair was shown to improve survival in a tail-clip challenge, a standard test for bleeding disorders. The study highlighted that even with low levels of corrected hepatocytes, there was a marked improvement in clinical outcomes, suggesting the feasibility of this approach for treating severe Hemophilia A [163].

Hemophilia B, though less common than Hemophilia A, presents similar therapeutic challenges. The F9 gene is smaller than F8, making it more amenable to packaging in AAV vectors, a significant advantage for gene therapy. In preclinical studies, CRISPR/Cas9 has been used to correct mutations in the F9 gene by inserting a normal copy in hepatocytes [164]. The study achieved a significant increase in F9 levels, reaching about 10% of normal activity, which is sufficient to alleviate symptoms in mild Hemophilia B. However, the approach faced challenges, including off-target integrations and immune responses against the viral vector and Cas9 protein [164].

*11.4.4.1.4 Clinical Trials*

In general, several well-studied genetic diseases can be targeted for treatment using CRISPR-Cas systems. It is crucial to emphasize that these diseases are primarily monogenic, meaning they are caused by mutations in a single gene. As the CRISPR-Cas system relies on a gRNA to precisely target and modify specific genes, optimizing this technology is essential for addressing single-gene disorders. This focused approach provides a foundational step before considering more complex strategies for targeting multiple genes in clinical applications.

To date, there are 102 clinical trials reported on ClinicalTrials.gov that utilize CRISPR systems. Of these, 30 are specifically targeting genetic diseases (see Table 11.5). The most advanced trials are currently in Phase 3 and focus on treating SCD.

***11.4.4.2 Cancer Therapy***

CRISPR-Cas systems have dramatically impacted cancer research and treatment, particularly in the development of cancer models, identification of drug targets, and direct therapeutic interventions [151].

**Table 11.5 Clinical trials using CRISPR systems to treat or investigate genetic diseases**

| Id | Conditions | Phase |
|---|---|---|
| NCT06025032 | Congenital hearing loss | EARLY PHASE1 |
| NCT05565248 | Diabetes mellitus, Type 1 I glucose metabolism disorders | PHASE1 I PHASE2 |
| NCT05210530 | Diabetes mellitus, Type 1 I glucose metabolism disorders | PHASE1 |
| NCT05514249 | DMD | PHASE1 |
| NCT06379789 | Hemophilia B | PHASE1 I PHASE2 |
| NCT05120830 | Hereditary angioedema | PHASE1 I PHASE2 |
| NCT03855631 | Kabuki syndrome 1 | |
| NCT03872479 | Leber congenital amaurosis 10 I inherited retinal dystrophies | PHASE1 I PHASE2 |
| NCT05805007 | Retinitis pigmentosa | EARLY_PHASE1 |
| NCT05740761 | Rett syndrome | |
| NCT06506461 | Sickle cell disease | PHASE1 |
| NCT06300723 | Sickle cell disease | NA |
| NCT06287086 | Sickle cell disease | NA |
| NCT06287099 | Sickle cell disease | NA |
| NCT05951205 | Sickle cell disease | PHASE3 |
| NCT04819841 | Sickle cell disease | PHASE1 I PHASE2 |
| NCT04774536 | Sickle cell disease | PHASE1 I PHASE2 |
| NCT04443907 | Sickle cell disease | PHASE1 |
| NCT03167450 | Sickle cell disease | |
| NCT03745287 | Sickle cell disease I hemoglobinopathies | PHASE2 I PHASE3 |
| NCT05329649 | Sickle cell disease I hematological diseases | PHASE3 |
| NCT06363760 | Sickle cell disease I transfusion-dependent beta thalassemia | |
| NCT03728322 | Thalassemia | EARLY_PHASE1 |
| NCT06041620 | Thalassemia, beta I thalassemia major | NA |
| NCT05577312 | Beta-thalassemia | PHASE1 |
| NCT05356195 | Beta-thalassemia I hemoglobinopathies | PHASE3 |
| NCT03655678 | Beta-thalassemia I hemoglobinopathies | PHASE2 I PHASE3 |
| NCT05477563 | Beta-thalassemia I sickle cell disease | PHASE3 |
| NCT04208529 | Beta-thalassemia I sickle cell anemia | PHASE3 |

CRISPR-Cas9 has been instrumental in creating accurate cancer models, which are crucial for understanding tumor biology and testing new therapies. For instance, CRISPR-Cas9 has been used to generate organoid models of various cancers, including colorectal and gastric cancers, by introducing mutations in key oncogenes [165]. In another example, a Cas9 KI mouse model was developed to create an adenocarcinoma model [166].

Additionally, CRISPR-Cas9 enables the systematic identification of genes involved in tumorigenesis, cancer progression, and drug resistance. By using CRISPR-Cas9 to induce targeted gene KOs, interferences, or activations across the genome, researchers can conduct high-throughput screens to pinpoint genes essential for cancer cell survival, proliferation, and metastasis [152]. This process involves CRISPR libraries, allowing for the disruption or activation of multiple genes simultaneously. The resulting phenotypic changes in cancer cells, such as reduced growth or increased sensitivity to chemotherapeutic agents, provide insights into the genetic dependencies of cancer cells and uncover potential therapeutic targets [167]. For example, CRISPR-based screens have identified genes contributing to drug resistance in cancers like melanoma and hepatocellular carcinoma [168, 169]. Additionally, a dual CRISPR interference and activation (CRISPRi/a) system has been employed in breast cancer cells, highlighting the potential of epigenetic modifications [170].

In a more direct therapeutic approach, CRISPR-Cas9 has been used to engineer CAR-T cells. CAR-T cells are genetically modified T cells that express chimeric antigen receptors (CARs) on their surface, enabling them to specifically target and kill cancer cells. While CAR-T cell therapy has shown remarkable success in treating certain blood cancers, its efficacy in solid tumors has been limited due to various immunosuppressive mechanisms employed by the tumor microenvironment. Cancer cells often express immune checkpoint proteins like *PD-1* and *CTLA-4*, which act as "brakes" on the immune system, preventing T cells from attacking cancer cells effectively. CRISPR-Cas9 has been utilized to KO these immune checkpoint genes in CAR-T cells, thereby removing these inhibitory signals and enhancing the T cells' ability to attack and destroy cancer cells. For instance, these genes have been knocked out in studies targeting non-small-cell lung cancer and B-cell lymphomas [171, 172]. Various clinical trials involve harvesting T cells from patients, genetically modifying them to disable *PD-1*, and then reintroducing them to boost the immune response against cancer cells [151]. This approach is being explored for esophageal cancer, prostate cancer, bladder cancer, and renal cell carcinoma, where tumor immune evasion mechanisms pose significant treatment challenges.

Beyond traditional CRISPR-Cas9, novel tools like base editors and prime editors are being explored for cancer therapy. Base editors enable precise single-nucleotide changes, allowing for the correction of oncogenic mutations in tumor cells [173, 174]. Prime editing offers even greater precision and flexibility in genome editing, with applications in creating and studying cancer mutations [155]. Base editors, particularly cytosine base editors (CBEs) and adenine base editors (ABEs), are useful in large-scale genome screenings, enabling the identification of gain- and loss-of-function variants critical in cancer development and progression [175]. These screenings have not only enhanced our understanding of cancer-associated variants but also revealed how specific protein modifications, such as phosphorylation, impact chemotherapy response [176]. Additionally, base editors have demonstrated therapeutic potential by modifying oncogenes or tumor suppressor genes. In triple-negative breast cancer models, CBEs were employed to modify oncogenic mutations [177], and ABEs were used to correct KRAS and TP53 mutations in patient-derived cancer organoids [178].

#### *11.4.4.3 Infectious Diseases*

CRISPR-Cas systems have also shown promise in the fight against infectious diseases, offering both therapeutic and diagnostic solutions [151].

CRISPR-Cas9 has been utilized to target and disrupt the *CCR5* gene, a crucial co-receptor for HIV-1 entry into host cells. By knocking out *CCR5*, cells become resistant to HIV infection. Advancements in CRISPR technology have improved the precision and efficiency of this approach, enabling successful *CCR5* editing in primary human CD4+ T cells and various stem cells, including HSCs and induced pluripotent stem cells (iPSCs). *In vivo* studies have demonstrated that CCR5-KO HSCs can engraft and generate HIV-resistant cells, highlighting CRISPR's potential as a curative strategy for HIV. However, further optimization is needed to enhance the proportion of edited cells and ensure long-term therapeutic efficacy [179].

Chronic hepatitis B virus (HBV) infection can lead to liver cirrhosis and hepatocellular carcinoma. CRISPR-Cas9 has been employed to target and cleave HBV DNA within infected liver cells, reducing viral replication and protein expression. In the most recent study, the CRISPR-Cas system was employed to target and cleave the covalently closed circular DNA (cccDNA) of HBV within infected hepatocytes. In a liver-humanized mouse model that closely mimics human HBV infection, this approach demonstrated successful gene editing, with significant reductions in viral replication and improvements in human hepatocyte survival.

**Table 11.6 Clinical trials using CRISPR systems to treat or investigate infectious diseases**

| Id | Conditions | Phase |
|---|---|---|
| NCT05850871 | Carbapenem-resistant Enterobacteriaceae infection | NA |
| NCT04990557 | COVID-19 respiratory infection | PHASE1 I PHASE2 |
| NCT05331976 | COVID-19 I asymptomatic COVID-19 | |
| NCT04535648 | Enterovirus infections | |
| NCT03342547 | Gastrointestinal infection | NA |
| NCT05144386 | HIV-1-infection | PHASE1 |
| NCT05143307 | HIV-1-infection | PHASE1 |
| NCT03164135 | HIV-1-infection | NA |
| NCT05107258 | SARS I COVID-19 | |
| NCT05443607 | SARS CoV2 infection I respiratory syncytial virus (RSV) | |
| NCT05034978 | SARS-CoV2 infection | |
| NCT04178382 | Severe sepsis | NA |
| NCT04560790 | Viral keratitis I Herpes simplex virus infection | NA |
| NCT04535505 | Pertussis | |
| NCT06113861 | Tuberculosis | |
| NCT04074369 | Tuberculosis, pulmonary | |

Notably, treated mice showed up to a 7.1-fold increase in human hepatocyte levels and higher serum albumin, indicating enhanced liver function [180].

CRISPR-Cas systems have rapidly advanced as powerful tools for both diagnostics and therapeutic applications, particularly highlighted during the COVID-19 pandemic. The CRISPR-Cas13 system, which targets RNA, was fundamental in developing the SHERLOCK diagnostic platform. This platform enables rapid and highly sensitive detection of SARS-CoV-2, delivering results in as little as 30 minutes, offering a significant improvement over traditional PCR-based methods, which are time-consuming and require specialized equipment. Beyond diagnostics, CRISPR-based strategies are being actively explored for antiviral therapy. Both Cas9 and Cas13 systems have shown promise in disrupting viral genomes and inhibiting replication. For example, Cas13 has been used to degrade RNA from viruses such as SARS-CoV-2 and influenza, demonstrating its potential as a therapeutic tool in preclinical models [181]. Currently, there are 16 clinical trials involving CRISPR-Cas systems for infectious diseases reported on ClinicalTrials.gov (see Table 11.6).

## 11.5 CHALLENGES AND FUTURE DIRECTIONS

### 11.5.1 Biological Barriers and Delivery Efficiency

The application of nanocarriers and CRISPR-Cas systems for therapeutic purposes has garnered significant attention, but the efficacy of these systems depends on overcoming several biological barriers. These barriers include challenges related to cellular uptake, endosomal escape, and nuclear delivery [182].

The efficiency of cellular uptake is often determined by the interaction between the delivery vehicle and the cellular membrane. Non-viral vectors, including LNPs and polymer-based carriers, are commonly used for nucleic acids delivery but often suffer from limited uptake efficiency compared to viral vectors [182].

Endosomal escape is often mediated by mechanisms such as the proton-sponge effect, where certain polymers or lipid-based vectors cause an osmotic imbalance that leads to the rupture of the endosomal membrane [182].

### 11.5.2 Immunogenicity

The immune system is responsible for distinguishing between self and non-self to defend against potential harm. Immune responses can be either non-specific, which involves the innate immune system and relies on pattern recognition through pattern recognition receptors (PRRs), or specific, which involves the adaptive immune system and is associated with humoral immunity [183].

Nanocarriers encounter several challenges based on their size. If they are too small, they can be easily drained by the lymphatic system [184]. Additionally, synthetic nanocarriers may be identified as foreign and removed by the reticuloendothelial system, and they can also exhibit some degree of immunogenicity and undesirable toxicity. To address these issues, the design and synthesis of biomimetic nanocarriers have been proposed. These biomimetic nanocarriers tend to be non-toxic and highly biocompatible [184].

Generally, the use of nanocarriers can lead to unwanted effects, such as rejection and the triggering of inflammatory and pathological responses. These effects are more pronounced when used *in vivo*, as the immune system may become hyperreactive, sometimes necessitating the use of immunosuppressants to manage immune activity [183]. Additionally, nanoparticles can induce direct or indirect modulation of oxidative enzymatic reactions, leading to the production of reactive oxygen species (ROS) [183].

Conversely, research is now exploring the use of nanocarriers to modulate immune responses, which could potentially address some of these current limitations [185].

### 11.5.3 Regulatory and Ethical Considerations

One of the primary risks associated with the nucleic acids and CRISPR-Cas systems is off-target editing, where unintended changes occur in the genome at sites other than the intended target. The alarm surrounding off-target effects is underscored by evidence from various studies, which have documented such occurrences in human, mouse, and plant cells [186]. The ethical stakes are high, as the potential for unintentional, heritable changes raises questions about the long-term safety and moral permissibility of germline modifications [187].

The ethical implications of the delivery of nucleic acids and CRISPR-Cas technology are far-reaching, particularly regarding human germline editing. The prospect of editing human embryos to prevent genetic diseases or enhance certain traits has sparked intense debate within the scientific and bioethical communities. Critics argue that germline editing could open the door to a new form of eugenics, where the potential to create "designer babies" exacerbates social inequalities and raises fundamental questions about human identity and diversity [188]. Furthermore, the lack of consensus on the moral status of the human embryo and the appropriate boundaries for genome editing complicates the regulatory landscape. For instance, while some nations strictly prohibit germline editing, others allow it under certain conditions, reflecting the diverse ethical and cultural perspectives on this issue [189].

In the United States, the regulatory landscape is complex, involving multiple federal agencies under a system known as the Coordinated Framework for the Regulation of Biotechnology. This framework, however, has been criticized for being fragmented and inconsistent, particularly in its approach to genome-edited crops versus human embryo modifications [186, 187].

The international scientific community has increasingly recognized the need for shared ethical guidelines and regulatory standards to govern the use of nucleic acids and CRISPR-Cas technology, especially in human genome editing. The World Health Organization (WHO) and other global bodies have begun to address these issues, emphasizing the importance of transparency, public engagement, and the protection of human rights in the development of genome-editing technologies.

## 11.6 CONCLUSIONS

The development of nanotechnology has significantly advanced the field of biomedical sciences, particularly in the delivery of nucleic acids for therapeutic purposes. Nanocarriers offer unique advantages, such as chemical versatility, high loading capacity, and efficient cellular internalization, making them highly effective for delivering various types of nucleic acids, including siRNA, miRNA, mRNA, DNA, and CRISPR-Cas systems. However, the efficacy of these systems is contingent upon overcoming several biological barriers, such as cellular uptake, endosomal escape, and nuclear delivery.

Despite these advancements, nanocarriers encounter several challenges based on their size and composition. If they are too small, they can be easily drained by the lymphatic system. Synthetic nanocarriers may also be identified as foreign and removed by the reticuloendothelial system, exhibiting some degree of immunogenicity and undesirable toxicity. These limitations highlight the importance of designing and synthesizing biomimetic nanocarriers, which tend to be non-toxic and highly biocompatible, minimizing immune responses and enhancing delivery efficacy.

Immunogenicity and safety concerns are also significant factors in the application of nanocarriers and CRISPR-Cas systems. The immune system's ability to distinguish between self and non-self can lead to unwanted effects, such as rejection and inflammatory responses, particularly when used *in vivo*. Nanoparticles can induce direct or indirect modulation of oxidative enzymatic reactions, leading to the production of ROS. Conversely, there is growing interest in exploring nanocarriers' potential to modulate immune responses, which could address some of these current limitations.

Regulatory and ethical considerations play a crucial role in the development and application of nucleic acids and CRISPR-Cas technologies. The risk of off-target editing, where unintended changes occur in the genome, raises significant ethical concerns, particularly regarding human germline editing. The potential for unintentional, heritable changes raises questions about the long-term safety and moral permissibility of such modifications, sparking intense debate within the scientific and bioethical communities.

The ethical implications of genome editing are far-reaching, with critics arguing that germline editing could lead to a new form of eugenics, exacerbating social inequalities and raising fundamental questions about human identity and diversity. The lack of consensus on the moral status of the human embryo and the appropriate boundaries for genome editing complicates the regulatory landscape, with varying regulations across different countries reflecting diverse ethical and cultural perspectives.

In response to these challenges, the international scientific community has increasingly recognized the need for shared ethical guidelines and regulatory standards to govern the use of nucleic acids and CRISPR-Cas technology, especially in human genome editing. Organizations such as the WHO have begun addressing these issues, emphasizing the importance of transparency, public engagement, and the protection of human rights in developing genome-editing technologies.

In conclusion, while the future of nucleic acid delivery using nanocarriers and CRISPR-Cas systems holds great promise for treating a wide range of diseases, continued research and development are essential to overcome existing challenges, improve delivery efficiency, ensure safety, and address the ethical and regulatory considerations associated with these innovative technologies.

## REFERENCES


1. P. N. Sudha, K. Sangeetha, K. Vijayalakshmi, and A. Barhoum, "Nanomaterials history, classification, unique properties, production and market," in *Emerging Applications of Nanoparticles and Architecture Nanostructures*, Elsevier, 2018, pp. 341–384. doi: 10.1016/B978-0-323-51254-1.00012-9.

2. S. Bayda, M. Adeel, T. Tuccinardi, M. Cordani, and F. Rizzolio, "The history of nanoscience and nanotechnology: from chemical–physical applications to nanomedicine," *Molecules*, vol. 25, no. 1, p. 112, Dec. 2019, doi: 10.3390/molecules25010112.

3. O. Afzal, A. S. A. Altamimi, M. S. Nadeem, S. I. Alzarea, W. H. Almalki, A. Tariq, et al., "Nanoparticles in drug delivery: from history to therapeutic applications," *Nanomaterials*, vol. 12, no. 24, p. 4494, Dec. 2022, doi: 10.3390/nano12244494.

4. M. C. Scicluna and L. Vella-Zarb, "Evolution of nanocarrier drug-delivery systems and recent advancements in covalent organic framework–drug systems," *ACS Applied Nano Materials*, vol. 3, no. 4, pp. 3097–3115, Apr. 2020, doi: 10.1021/acsanm.9b02603.

5. J. A. Kulkarni, D. Witzigmann, S. B. Thomson, S. Chen, B. R. Leavitt, P. R. Cullis, et al., "The current landscape of nucleic acid therapeutics," *Nature Nanotechnology*, vol. 16, no. 6, pp. 630–643, Jun. 2021, doi: 10.1038/s41565-021-00898-0.

6. M. Mehta, T. A. Bui, X. Yang, Y. Aksoy, E. M. Goldys, and W. Deng, "Lipid-based nanoparticles for drug/gene delivery: an overview of the production techniques and difficulties encountered in their industrial development," *ACS Materials Au*, vol. 3, no. 6, pp. 600–619, Nov. 2023, doi: 10.1021/acsmaterialsau.3c00032.

7. S. Mehraji and D. L. DeVoe, "Microfluidic synthesis of lipid-based nanoparticles for drug delivery: recent advances and opportunities," *Lab on a Chip*, vol. 24, no. 5, pp. 1154–1174, 2024, doi: 10.1039/D3LC00821E.

8. P. L. Felgner, T. R. Gadek, M. Holm, R. Roman, H. W. Chan, M. Wenz, et al., "Lipofection: a highly efficient, lipid-mediated DNA-transfection procedure," *Proceedings of the National Academy of Sciences*, vol. 84, no. 21, pp. 7413–7417, Nov. 1987, doi: 10.1073/pnas.84.21.7413.

9. X. Hou, T. Zaks, R. Langer, and Y. Dong, "Lipid nanoparticles for mRNA delivery," *Nature Reviews Materials*, vol. 6, no. 12, pp. 1078–1094, Aug. 2021, doi: 10.1038/s41578-021-00358-0.

10. J. Buck, D. Mueller, U. Mettal, M. Ackermann, H. Man Grisch-Chan, B. Thöny, et al., "Improvement of DNA vector delivery of DOTAP lipoplexes by short-chain aminolipids," *ACS Omega*, vol. 5, no. 38, pp. 24724–24732, Sep. 2020, doi: 10.1021/acsomega.0c03303.

11. X. Jiang, K. Abedi, and J. Shi, "Polymeric nanoparticles for RNA delivery," in *Encyclopedia of Nanomaterials*, Elsevier, 2023, pp. 555–573. doi: 10.1016/B978-0-12-822425-0.00017-8.

12. B. B. Mendes, J. Conniot, A. Avital, D. Yao, X. Jiang, X. Zhou, et al., "Nanodelivery of nucleic acids," *Nature Reviews Methods Primers*, vol. 2, no. 1, p. 24, Apr. 2022, doi: 10.1038/s43586-022-00104-y.

13. N. Kong, W. Tao, X. Ling, J. Wang, Y. Xiao, S. Shi, et al., "Synthetic mRNA nanoparticle-mediated restoration of p53 tumor suppressor sensitizes *p53*-deficient cancers to mTOR inhibition," *Science Translational Medicine*, vol. 11, no. 523, Dec. 2019, doi: 10.1126/scitranslmed.aaw1565.

14. X. Xu, K. Xie, X.-Q. Zhang, E. M. Pridgen, G. Y. Park, D. S. Cui, et al., "Enhancing tumor cell response to chemotherapy through nanoparticle-mediated codelivery of siRNA and cisplatin prodrug," *Proceedings of the National Academy of Sciences*, vol. 110, no. 46, pp. 18638–18643, Nov. 2013, doi: 10.1073/pnas.1303958110.

15. D. C. Luther, R. Huang, T. Jeon, X. Zhang, Y.-W. Lee, H. Nagaraj, et al., "Delivery of drugs, proteins, and nucleic acids using inorganic nanoparticles," *Advanced Drug Delivery Reviews*, vol. 156, pp. 188–213, 2020, doi: 10.1016/j.addr.2020.06.020.

16. C. Hom, J. Lu, and F. Tamanoi, "Silica nanoparticles as a delivery system for nucleic acid-based reagents," *Journal of Materials Chemistry*, vol. 19, no. 35, p. 6308, 2009, doi: 10.1039/b904197d.

17. Y. Zhang, Y. Yu, Y. Yang, Y. Wang, and C. Yu, "Engineered silica nanoparticles for nucleic acid delivery," *Small Methods*, vol. 8, no. 3, Mar. 2024, doi: 10.1002/smtd.202300812.

18. P. Guzmán-Sastoque, S. Sotelo, N. P. Esmeral, S. L. Albarracín, J. J. Sutachan, L. H. Reyes, et al., "Assessment of CRISPRa-mediated gdnf overexpression in an in vitro Parkinson's disease model," *Frontiers in Bioengineering and Biotechnology*, vol. 12, Aug. 2024, doi: 10.3389/fbioe.2024.1420183.

19. J. Zhang, T. Zhang, and J. Gao, "Biocompatible iron oxide nanoparticles for targeted cancer gene therapy: a review," *Nanomaterials*, vol. 12, no. 19, p. 3323, Sep. 2022, doi: 10.3390/nano12193323.

20. A. M. López-Estevez, R. Gref, and M. J. Alonso, "A journey through the history of PEGylated drug delivery nanocarriers," *Drug Delivery and Translational Research*, vol. 14, no. 8, pp. 2026–2031, Aug. 2024, doi: 10.1007/s13346-024-01608-8.

21. M. Yokoyama, G. S. Kwon, T. Okano, Y. Sakurai, H. Ekimoto, K. Okamoto, et al., "Composition-dependent in vivo antitumor activity of adriamycin-conjugated polymeric micelle against murine colon adenocarcinoma 26," *Drug Delivery*, vol. 1, no. 1, pp. 11–19, Jan. 1993, doi: 10.3109/10717549309031336.

22. H. Jeon, S. Han, H. Kim, and J. B. Lee, "Surface modification of RNA nanoparticles by ionic interaction for efficient cellular uptake," *Journal of Industrial and Engineering Chemistry*, vol. 70, pp. 87–93, Feb. 2019, doi: 10.1016/j.jiec.2018.10.013.

23. G. T. Hermanson, *Bioconjugate Techniques*. Elsevier, 2008. doi: 10.1016/B978-0-12-370501-3.X0001-X.

24. B. Hu, L. Zhong, Y. Weng, L. Peng, Y. Huang, Y. Zhao, et al., "Therapeutic siRNA: state of the art," *Signal Transduction and Targeted Therapy*, vol. 5, no. 1, p. 101, Jun. 2020, doi: 10.1038/s41392-020-0207-x.

25. P.-H. Tan, L.-C. Yang, and R.-R. Ji, "Therapeutic potential of RNA interference in pain medicine," *Open Pain Journal*, vol. 2, no. 1, pp. 57–63, Aug. 2009, doi: 10.2174/1876386300902010057.

26. P. Svoboda, "Key mechanistic principles and considerations concerning RNA interference," *Frontiers in Plant Science*, vol. 11, Aug. 2020, doi: 10.3389/fpls.2020.01237.

27. S. W. L. Lee, C. Paoletti, M. Campisi, T. Osaki, G. Adriani, R. D. Kamm, et al., "MicroRNA delivery through nanoparticles," *Journal of Controlled Release*, vol. 313, pp. 80–95, Nov. 2019, doi: 10.1016/j.jconrel.2019.10.007.

28. Y. Dong, D. J. Siegwart, and D. G. Anderson, "Strategies, design, and chemistry in siRNA delivery systems," *Advanced Drug Delivery Reviews*, vol. 144, pp. 133–147, Apr. 2019, doi: 10.1016/j.addr.2019.05.004.

29. M. Moazzam, M. Zhang, A. Hussain, X. Yu, J. Huang, and Y. Huang, "The landscape of nanoparticle-based siRNA delivery and therapeutic development," *Molecular Therapy*, vol. 32, no. 2, pp. 284–312, Feb. 2024, doi: 10.1016/j.ymthe.2024.01.005.

30. S. S. Ali Zaidi, F. Fatima, S. A. Ali Zaidi, D. Zhou, W. Deng, and S. Liu, "Engineering siRNA therapeutics: challenges and strategies," *Journal of Nanobiotechnology*, vol. 21, no. 1, p. 381, Oct. 2023, doi: 10.1186/s12951-023-02147-z.

31. M. A. Subhan, N. Filipczak, and V. P. Torchilin, "Advances with lipid-based nanosystems for siRNA delivery to breast cancers," *Pharmaceuticals*, vol. 16, no. 7, p. 970, Jul. 2023, doi: 10.3390/ph16070970.

32. A. Morales-Becerril, L. Aranda-Lara, K. Isaac-Olivé, B. E. Ocampo-García, and E. Morales-Ávila, "Nanocarriers for delivery of siRNA as gene silencing mediator," *EXCLI Journal*, vol. 21, pp. 1028–1052, 2022, doi: 10.17179/excli2022-4975.

33. S. Ashique, B. Almohaywi, N. Haider, S. Yasmin, A. Hussain, N. Mishra, et al., "siRNA-based nanocarriers for targeted drug delivery to control breast cancer," *Advances in Cancer Biology—Metastasis*, vol. 4, p. 100047, Jul. 2022, doi: 10.1016/j.adcanc.2022.100047.

34. J. Hong, D. Sim, B.-H. Lee, V. Sarangthem, and R.-W. Park, "Multifunctional elastin-like polypeptide nanocarriers for efficient miRNA delivery in cancer therapy," *Journal of Nanobiotechnology*, vol. 22, no. 1, p. 293, May 2024, doi: 10.1186/s12951-024-02559-5.

35. L. A. Bravo-Vázquez, A. Méndez-García, A. L. Rodríguez, P. Sahare, S. Pathak, A. Banerjee, et al., "Applications of nanotechnologies for miRNA-based cancer therapeutics: current advances and future perspectives," *Frontiers in Bioengineering and Biotechnology*, vol. 11, Jul. 2023, doi: 10.3389/fbioe.2023.1208547.

36. S. Sargazi, B. Siddiqui, M. Qindeel, A. Rahdar, M. Bilal, R. Behzadmehr, et al., "Chitosan nanocarriers for microRNA delivery and detection: a preliminary review with emphasis on cancer," *Carbohydrate Polymers*, vol. 290, p. 119489, Aug. 2022, doi: 10.1016/j.carbpol.2022.119489.

37. M. Scheideler, I. Vidakovic, and R. Prassl, "Lipid nanocarriers for microRNA delivery," *Chemistry and Physics of Lipids*, vol. 226, p. 104837, Jan. 2020, doi: 10.1016/j.chemphyslip.2019.104837.

38. S. Ghafouri-Fard, H. Shoorei, L. Noferesti, B. M. Hussen, M. H. B. Moghadam, M. Taheri, et al., "Nanoparticle-mediated delivery of microRNAs-based therapies for treatment of disorders," *Pathology—Research and Practice*, vol. 248, p. 154667, Aug. 2023, doi: 10.1016/j.prp.2023.154667.

39. A. Menon, N. Abd-Aziz, K. Khalid, C. L. Poh, and R. Naidu, "miRNA: a promising therapeutic target in cancer," *International Journal of Molecular Sciences*, vol. 23, no. 19, p. 11502, Sep. 2022, doi: 10.3390/ijms231911502.

40. G. Kara, B. Arun, G. A. Calin, and B. Ozpolat, “miRacle of microRNA-driven cancer nanotherapeutics,” *Cancers (Basel)*, vol. 14, no. 15, p. 3818, Aug. 2022, doi: 10.3390/cancers14153818.

41. B. Roy, S. Ghose, and S. Biswas, “Therapeutic strategies for miRNA delivery to reduce hepatocellular carcinoma,” *Seminars in Cell & Developmental Biology*, vol. 124, pp. 134–144, Apr. 2022, doi: 10.1016/j.semcdb.2021.04.006.

42. S. Huang, S. Duan, J. Wang, S. Bao, X. Qiu, C. Li, et al., “Folic-acid-mediated functionalized gold nanocages for targeted delivery of Anti-miR-181b in combination of gene therapy and photothermal therapy against hepatocellular carcinoma,” *Advanced Functional Materials*, vol. 26, no. 15, pp. 2532–2544, Apr. 2016, doi: 10.1002/adfm.201504912.

43. H.-Y. Xue, Y. Liu, J.-Z. Liao, J.-S. Lin, B. Li, W.-G. Yuan, et al., “Gold nanoparticles delivered miR-375 for treatment of hepatocellular carcinoma,” *Oncotarget*, vol. 7, no. 52, pp. 86675–86686, Dec. 2016, doi: 10.18632/oncotarget.13431.

44. N. A. Alden, T. J. Yeingst, H. M. Pfeiffer, N. Celik, J. H. Arrizabalaga, A. M. Helton, et al., “Near-infrared induced miR-34a delivery from nanoparticles in esophageal cancer treatment,” *Advanced Healthcare Materials*, vol. 13, no. 10, Apr. 2024, doi: 10.1002/adhm.202303593.

45. R. Pandey, F.-S. Yang, V. P. Sivasankaran, Y.-L. Lo, Y.-T. Wu, C.-Y. Chang, et al., “Comparing the variants of iron oxide nanoparticle-mediated delivery of miRNA34a for efficiency in silencing of PD-L1 genes in cancer cells,” *Pharmaceutics*, vol. 15, no. 1, p. 215, Jan. 2023, doi: 10.3390/pharmaceutics15010215.

46. H. Yang, G. Shi, C. Ge, J. Huang, L. Wan, Z. Wang, et al., “Functionalized graphene oxide as a nanocarrier for multiple suppressive miRNAs to inhibit human intrahepatic cholangiocarcinoma,” *Nano Select*, vol. 2, no. 7, pp. 1372–1384, Jul. 2021, doi: 10.1002/nano.202000130.

47. C. H. Kapadia, S. A. Ioele, and E. S. Day, “Layer-by-layer assembled PLGA nanoparticles carrying miR-34a cargo inhibit the proliferation and cell cycle progression of triple-negative breast cancer cells,” *Journal of Biomedical Materials Research Part A*, vol. 108, no. 3, pp. 601–613, Mar. 2020, doi: 10.1002/jbm.a.36840.

48. C. Cheng, Z. Zhang, S. Wang, L. Chen, and Q. Liu, “Reduction sensitive CC9-PEG-SSBPEI/miR-148b nanoparticles: synthesis, characterization, targeting delivery and application for anti-metastasis,” *Colloids and Surfaces B: Biointerfaces*, vol. 183, p. 110412, Nov. 2019, doi: 10.1016/j.colsurfb.2019.110412.

49. Q. Ning, Y.-F. Liu, P.-Y. Ye, P. Gao, Z.-P. Li, S.-Y. Tang, et al., “Delivery of liver-specific miRNA-122 using a targeted macromolecular prodrug toward synergistic therapy for hepatocellular carcinoma,” *ACS Applied Materials & Interfaces*, vol. 11, no. 11, pp. 10578–10588, Mar. 2019, doi: 10.1021/acsami.9b00634.

50. B. Soliman, M. M. Wen, E. Kandil, B. El-Agamy, A. M. Gamal-Eldeen, and M. ElHefnawi, “Preparation and optimization of MiR-375 nano-vector using two novel chitosan-coated nano-structured lipid carriers as gene therapy for hepatocellular carcinoma,” *Pharmaceutics*, vol. 16, no. 4, p. 494, Apr. 2024, doi: 10.3390/pharmaceutics16040494.

51. M. Ghaffari, S. M. Kalantar, M. Hemati, A. Dehghani Firoozabadi, A. Asri, A. Shams, et al., “Co-delivery of miRNA-15a and miRNA-16–1 using cationic PEGylated niosomes downregulates Bcl-2 and induces apoptosis in prostate cancer cells,” *Biotechnology Letters*, vol. 43, no. 5, pp. 981–994, May 2021, doi: 10.1007/s10529-021-03085-2.

52. H. Boroumand, F. Badie, S. Mazaheri, Z. S. Seyedi, J. S. Nahand, M. Nejati, et al., “Chitosan-based nanoparticles against viral infections,” *Frontiers in Cellular and Infection Microbiology*, vol. 11, Mar. 2021, doi: 10.3389/fcimb.2021.643953.

53. F. Li, D. Cao, W. Gu, D. Li, Z. Liu, and L. Cui, "Folate-targeted nanocarriers co-deliver ganciclovir and miR-34a-5p for combined anti-KSHV therapy," *International Journal of Molecular Sciences*, vol. 25, no. 5, p. 2932, Mar. 2024, doi: 10.3390/ijms25052932.

54. V. V. N. Goli, S. Tatineni, U. Hani, M. Ghazwani, S. Talath, S. B. Sridhar, et al., "Pharmacokinetics and pharmacodynamics of a nanostructured lipid carrier co-encapsulating artemether and miRNA for mitigating cerebral malaria," *Pharmaceuticals*, vol. 17, no. 4, p. 466, Apr. 2024, doi: 10.3390/ph17040466.

55. P. Anand and V. P. Stahel, "The safety of Covid-19 mRNA vaccines: a review," *Patient Safety in Surgery*, vol. 15, no. 1, p. 20, Dec. 2021, doi: 10.1186/s13037-021-00291-9.

56. M. A. Oberli, A. M. Reichmuth, J. R. Dorkin, M. J. Mitchell, O. S. Fenton, A. Jaklenec, et al., "Lipid nanoparticle assisted mRNA delivery for potent cancer immunotherapy," *Nano Letters*, vol. 17, no. 3, pp. 1326–1335, Mar. 2017, doi: 10.1021/acs.nanolett.6b03329.

57. Y. Zong, Y. Lin, T. Wei, and Q. Cheng, "Lipid nanoparticle (LNP) enables mRNA delivery for cancer therapy," *Advanced Materials*, vol. 35, no. 51, Dec. 2023, doi: 10.1002/adma.202303261.

58. P. Karmacharya, B. R. Patil, and J. O. Kim, "Recent advancements in lipid–mRNA nanoparticles as a treatment option for cancer immunotherapy," *Journal of Pharmaceutical Investigation*, vol. 52, no. 4, pp. 415–426, Jul. 2022, doi: 10.1007/s40005-022-00569-9.

59. J. Yang, B. Liu, Q. Wang, H. Yan, G. Li, X. Wang, et al., "Carboxylated mesoporous silica nanoparticle-nucleic acid chimera conjugate-assisted delivery of siRNA and doxorubicin effectively treat drug-resistant bladder cancer," *Biomedicine & Pharmacotherapy*, vol. 178, p. 117185, Sep. 2024, doi: 10.1016/j.biopha.2024.117185.

60. C. S. Diercks, D. A. Dik, and P. G. Schultz, "Adding new chemistries to the central dogma of molecular biology," *Chem*, vol. 7, no. 11, pp. 2883–2895, Nov. 2021, doi: 10.1016/j.chempr.2021.09.014.

61. X. Li, J. Qi, J. Wang, W. Hu, W. Zhou, Y. Wang, et al., "Nanoparticle technology for mRNA: delivery strategy, clinical application and developmental landscape," *Theranostics*, vol. 14, no. 2, pp. 738–760, 2024, doi: 10.7150/thno.84291.

62. F. Xu, Q. Xia, and P. Wang, "Rationally designed DNA nanostructures for drug delivery," *Frontiers in Chemistry*, vol. 8, Sep. 2020, doi: 10.3389/fchem.2020.00751.

63. M. J. Mitchell, M. M. Billingsley, R. M. Haley, M. E. Wechsler, N. A. Peppas, and R. Langer, "Engineering precision nanoparticles for drug delivery," *Nature Reviews Drug Discovery*, vol. 20, no. 2, pp. 101–124, Feb. 2021, doi: 10.1038/s41573-020-0090-8.

64. T. S. Lopes, G. G. Alves, M. R. Pereira, J. M. Granjeiro, and P. E. C. Leite, "Advances and potential application of gold nanoparticles in nanomedicine," *Journal of Cellular Biochemistry*, vol. 120, no. 10, pp. 16370–16378, Oct. 2019, doi: 10.1002/jcb.29044.

65. M. Jiao, P. Zhang, J. Meng, Y. Li, C. Liu, X. Luo, et al., "Recent advancements in biocompatible inorganic nanoparticles towards biomedical applications," *Biomaterials Science*, vol. 6, no. 4, pp. 726–745, 2018, doi: 10.1039/C7BM01020F.

66. J. Liu, M. Yu, C. Zhou, and J. Zheng, "Renal clearable inorganic nanoparticles: a new frontier of bionanotechnology," *Materials Today*, vol. 16, no. 12, pp. 477–486, Dec. 2013, doi: 10.1016/j.mattod.2013.11.003.

67. C. E. Astete and C. M. Sabliov, "Synthesis and characterization of PLGA nanoparticles," *Journal of Biomaterials Science, Polymer Edition*, vol. 17, no. 3, pp. 247–289, Jan. 2006, doi: 10.1163/156856206775997322.

68. F. Danhier, E. Ansorena, J. M. Silva, R. Coco, A. Le Breton, and V. Préat, "PLGA-based nanoparticles: an overview of biomedical applications," *Journal of Controlled Release*, vol. 161, no. 2, pp. 505–522, Jul. 2012, doi: 10.1016/j.jconrel.2012.01.043.

69. K. Divya and M. S. Jisha, "Chitosan nanoparticles preparation and applications," *Environmental Chemistry Letters*, vol. 16, no. 1, pp. 101–112, Mar. 2018, doi: 10.1007/s10311-017-0670-y.

70. C. F. Rodríguez, P. Guzmán-Sastoque, M. Gantiva-Diaz, S. C. Gómez, V. Quezada, C. Muñoz-Camargo, et al., "Low-cost inertial microfluidic device for microparticle separation: a laser-ablated PMMA lab-on-a-chip approach without a cleanroom," *HardwareX*, vol. 16, p. e00493, Dec. 2023, doi: 10.1016/j.ohx.2023.e00493.

71. S. Yousefi Adlsadabad, J. W. Hanrahan, and A. Kakkar, "mRNA delivery: challenges and advances through polymeric soft nanoparticles," *International Journal of Molecular Sciences*, vol. 25, no. 3, p. 1739, Feb. 2024, doi: 10.3390/ijms25031739.

72. A. D. Bangham, M. M. Standish, and J. C. Watkins, "Diffusion of univalent ions across the lamellae of swollen phospholipids," *Journal of Molecular Biology*, vol. 13, no. 1, pp. 238–IN27, Aug. 1965, doi: 10.1016/S0022-2836(65)80093-6.

73. M. Maeki, S. Uno, A. Niwa, Y. Okada, and M. Tokeshi, "Microfluidic technologies and devices for lipid nanoparticle-based RNA delivery," *Journal of Controlled Release*, vol. 344, pp. 80–96, Apr. 2022, doi: 10.1016/j.jconrel.2022.02.017.

74. C. Lopes, J. Cristóvão, V. Silvério, P. R. Lino, and P. Fonte, "Microfluidic production of mRNA-loaded lipid nanoparticles for vaccine applications," *Expert Opinion on Drug Delivery*, vol. 19, no. 10, pp. 1381–1395, Oct. 2022, doi: 10.1080/17425247.2022.2135502.

75. W. Li, C. Wang, Y. Zhang, and Y. Lu, "Lipid nanocarrier-based mRNA therapy: challenges and promise for clinical transformation," *Small*, vol. 20, no. 28, Jul. 2024, doi: 10.1002/smll.202310531.

76. R. Bhatia, A. Singh, S. Singh, Navneesh, and R. K. Rawal, "Emerging trends in nano-carrier based gene delivery systems for targeted cancer therapy," *Journal of Drug Delivery Science and Technology*, vol. 95, p. 105546, May 2024, doi: 10.1016/j.jddst.2024.105546.

77. M. Kozak and J. Hu, "DNA vaccines: their formulations, engineering and delivery," *Vaccines (Basel)*, vol. 12, no. 1, p. 71, Jan. 2024, doi: 10.3390/vaccines12010071.

78. S. Pagliari, B. Dema, A. Sanchez-Martinez, G. Montalvo Zurbia-Flores, and C. S. Rollier, "DNA vaccines: history, molecular mechanisms and future perspectives," *Journal of Molecular Biology*, vol. 435, no. 23, p. 168297, Dec. 2023, doi: 10.1016/j.jmb.2023.168297.

79. A. Pandya, Y. Shah, N. Kothari, H. Postwala, A. Shah, P. Parekh, et al., "The future of cancer immunotherapy: DNA vaccines leading the way," *Medical Oncology*, vol. 40, no. 7, p. 200, Jun. 2023, doi: 10.1007/s12032-023-02060-3.

80. V. Gote, P. K. Bolla, N. Kommineni, A. Butreddy, P. K. Nukala, S. S. Palakurthi, et al., "A comprehensive review of mRNA vaccines," *International Journal of Molecular Sciences*, vol. 24, no. 3, p. 2700, Jan. 2023, doi: 10.3390/ijms24032700.

81. B. Wilson and K. M. Geetha, "Lipid nanoparticles in the development of mRNA vaccines for COVID-19," *Journal of Drug Delivery Science and Technology*, vol. 74, p. 103553, Aug. 2022, doi: 10.1016/j.jddst.2022.103553.

82. T. Huang, L. Peng, Y. Han, D. Wang, X. He, J. Wang, et al., "Lipid nanoparticle-based mRNA vaccines in cancers: current advances and future prospects," *Frontiers in Immunology*, vol. 13, Aug. 2022, doi: 10.3389/fimmu.2022.922301.

83. Y. Ishino, H. Shinagawa, K. Makino, M. Amemura, and A. Nakata, "Nucleotide sequence of the iap gene, responsible for alkaline phosphatase isozyme conversion in *Escherichia coli*, and identification of the gene product," *Journal of Bacteriology*, vol. 169, no. 12, pp. 5429–5433, Dec. 1987, doi: 10.1128/jb.169.12.5429-5433.1987.

84. F. J. M. Mojica and F. Rodriguez-Valera, "The discovery of CRISPR in archaea and bacteria," *FEBS Journal*, vol. 283, no. 17, pp. 3162–3169, Sep. 2016, doi: 10.1111/febs.13766.

85. P. M. Nussenzweig and L. A. Marraffini, "Molecular mechanisms of CRISPR-Cas immunity in bacteria," *Annual Review of Genetics*, vol. 54, no. 1, pp. 93–120, Nov. 2020, doi: 10.1146/annurev-genet-022120-112523.

86. M. Jinek, K. Chylinski, I. Fonfara, M. Hauer, J. A. Doudna, and E. Charpentier, "A programmable dual-RNA–guided DNA endonuclease in adaptive bacterial immunity," *Science (1979)*, vol. 337, no. 6096, pp. 816–821, Aug. 2012, doi: 10.1126/science.1225829.

87. K. S. Makarova and E. V. Koonin, "Annotation and classification of CRISPR-Cas systems," *Methods in Molecular Biology*, 2015, pp. 47–75. doi: 10.1007/978-1-4939-2687-9_4.

88. K. Yoshimi and T. Mashimo, "Genome editing technology and applications with the type I CRISPR system," *Gene and Genome Editing*, vol. 3–4, p. 100013, Dec. 2022, doi: 10.1016/j.ggedit.2022.100013.

89. R. Molina, N. Sofos, and G. Montoya, "Structural basis of CRISPR-Cas Type III prokaryotic defence systems," *Current Opinion in Structural Biology*, vol. 65, pp. 119–129, Dec. 2020, doi: 10.1016/j.sbi.2020.06.010.

90. W. Feng, A. M. Newbigging, J. Tao, Y. Cao, H. Peng, C. Le, et al., "CRISPR technology incorporating amplification strategies: molecular assays for nucleic acids, proteins, and small molecules," *Chemical Science*, vol. 12, no. 13, pp. 4683–4698, 2021, doi: 10.1039/D0SC06973F.

91. B. Tong, H. Dong, Y. Cui, P. Jiang, Z. Jin, and D. Zhang, "The versatile type V CRISPR effectors and their application prospects," *Frontiers in Cell and Developmental Biology*, vol. 8, Feb. 2021, doi: 10.3389/fcell.2020.622103.

92. R. Nakagawa, S. Kannan, H. Altae-Tran, S. N. Takeda, A. Tomita, H. Hirano, et al., "Structure and engineering of the minimal type VI CRISPR-Cas13bt3," *Molecular Cell*, vol. 82, no. 17, pp. 3178–3192.e5, Sep. 2022, doi: 10.1016/j.molcel.2022.08.001.

93. M. Redman, A. King, C. Watson, and D. King, "What is CRISPR/Cas9?," *Archives of Disease in Childhood: Education and Practice Edition*, vol. 101, no. 4, pp. 213–215, Aug. 2016, doi: 10.1136/archdischild-2016-310459.

94. J. R. Liang and J. E. Corn, "A CRISPR view on autophagy," *Trends in Cell Biology*, vol. 32, no. 12, pp. 1008–1022, Dec. 2022, doi: 10.1016/j.tcb.2022.04.006.

95. Z. Mao, M. Bozzella, A. Seluanov, and V. Gorbunova, "DNA repair by nonhomologous end joining and homologous recombination during cell cycle in human cells," *Cell Cycle*, vol. 7, no. 18, pp. 2902–2906, Sep. 2008, doi: 10.4161/cc.7.18.6679.

96. C. K. S. Karlson, S. N. Mohd-Noor, N. Nolte, and B. C. Tan, "CRISPR/dCas9-based systems: mechanisms and applications in plant sciences," *Plants*, vol. 10, no. 10, p. 2055, Sep. 2021, doi: 10.3390/plants10102055.

97. M. Kampmann, "CRISPRi and CRISPRa screens in mammalian cells for precision biology and medicine," *ACS Chemical Biology*, vol. 13, no. 2, pp. 406–416, Feb. 2018, doi: 10.1021/acschembio.7b00657.

98. A. Kantor, M. McClements, and R. MacLaren, "CRISPR-Cas9 DNA base-editing and prime-editing," *International Journal of Molecular Sciences*, vol. 21, no. 17, p. 6240, Aug. 2020, doi: 10.3390/ijms21176240.

99. C. E. Denes, A. J. Cole, Y. A. Aksoy, G. Li, G. G. Neely, and D. Hesselson, "Approaches to enhance precise CRISPR/Cas9-mediated genome editing," *International Journal of Molecular Sciences*, vol. 22, no. 16, p. 8571, Aug. 2021, doi: 10.3390/ijms22168571.

100. F. Hejabi, M. S. Abbaszadeh, S. Taji, A. O'Neill, F. Farjadian, and M. Doroudian, "Nanocarriers: a novel strategy for the delivery of CRISPR/Cas systems," *Frontiers in Chemistry*, vol. 10, Jul. 2022, doi: 10.3389/fchem.2022.957572.

101. Y. Lin, E. Wagner, and U. Lächelt, "Non-viral delivery of the CRISPR/Cas system: DNA versus RNA versus RNP," *Biomaterials Science*, vol. 10, no. 5, pp. 1166–1192, 2022, doi: 10.1039/D1BM01658J.

102. X. Xu, C. Liu, Y. Wang, O. Koivisto, J. Zhou, Y. Shu, et al., "Nanotechnology-based delivery of CRISPR/Cas9 for cancer treatment," *Advanced Drug Delivery Reviews*, vol. 176, p. 113891, Sep. 2021, doi: 10.1016/j.addr.2021.113891.

103. Z.-Y. He, K. Men, Z. Qin, Y. Yang, T. Xu, and Y.-Q. Wei, "Non-viral and viral delivery systems for CRISPR-Cas9 technology in the biomedical field," *Science China Life Sciences*, vol. 60, no. 5, pp. 458–467, May 2017, doi: 10.1007/s11427-017-9033-0.

104. Y. Li, C. Li, J. Yan, Y. Liao, C. Qin, L. Wang, et al., "Polymeric micellar nanoparticles for effective CRISPR/Cas9 genome editing in cancer," *Biomaterials*, vol. 309, p. 122573, Sep. 2024, doi: 10.1016/j.biomaterials.2024.122573.

105. P. Kazemian, S.-Y. Yu, S. B. Thomson, A. Birkenshaw, B. R. Leavitt, and C. J. D. Ross, "Lipid-nanoparticle-based delivery of CRISPR/Cas9 genome-editing components," *Molecular Pharmaceutics*, vol. 19, no. 6, pp. 1669–1686, Jun. 2022, doi: 10.1021/acs.molpharmaceut.1c00916.

106. T. Wei, Q. Cheng, Y.-L. Min, E. N. Olson, and D. J. Siegwart, "Systemic nanoparticle delivery of CRISPR-Cas9 ribonucleoproteins for effective tissue specific genome editing," *Nature Communications*, vol. 11, no. 1, p. 3232, Jun. 2020, doi: 10.1038/s41467-020-17029-3.

107. J. S. Suk, Q. Xu, N. Kim, J. Hanes, and L. M. Ensign, "PEGylation as a strategy for improving nanoparticle-based drug and gene delivery," *Advanced Drug Delivery Reviews*, vol. 99, pp. 28–51, Apr. 2016, doi: 10.1016/j.addr.2015.09.012.

108. J. Zhang, H. Fan, D. A. Levorse, and L. S. Crocker, "Interaction of cholesterol-conjugated ionizable amino lipids with biomembranes: lipid polymorphism, structure–activity relationship, and implications for siRNA delivery," *Langmuir*, vol. 27, no. 15, pp. 9473–9483, Aug. 2011, doi: 10.1021/la201464k.

109. D. Sun and Z.-R. Lu, "Structure and function of cationic and ionizable lipids for nucleic acid delivery," *Pharmaceutical Research*, vol. 40, no. 1, pp. 27–46, Jan. 2023, doi: 10.1007/s11095-022-03460-2.

110. X. Cheng and R. J. Lee, "The role of helper lipids in lipid nanoparticles (LNPs) designed for oligonucleotide delivery," *Advanced Drug Delivery Reviews*, vol. 99, pp. 129–137, Apr. 2016, doi: 10.1016/j.addr.2016.01.022.

111. M. Schlich, R. Palomba, G. Costabile, S. Mizrahy, M. Pannuzzo, D. Peer, et al., "Cytosolic delivery of nucleic acids: the case of ionizable lipid nanoparticles," *Bioengineering & Translational Medicine*, vol. 6, no. 2, May 2021, doi: 10.1002/btm2.10213.

112. R. N. Kularatne, R. M. Crist, and S. T. Stern, "The future of tissue-targeted lipid nanoparticle-mediated nucleic acid delivery," *Pharmaceuticals*, vol. 15, no. 7, p. 897, Jul. 2022, doi: 10.3390/ph15070897.

113. C. Li, T. Yang, Y. Weng, M. Zhang, D. Zhao, S. Guo, et al., "Ionizable lipid-assisted efficient hepatic delivery of gene editing elements for oncotherapy," *Bioactive Materials*, vol. 9, pp. 590–601, Mar. 2022, doi: 10.1016/j.bioactmat.2021.05.051.

114. D. Sun, Z. Sun, H. Jiang, A. M. Vaidya, R. Xin, N. R. Ayat, et al., "Synthesis and evaluation of pH-sensitive multifunctional lipids for efficient delivery of CRISPR/Cas9 in gene editing," *Bioconjugate Chemistry*, vol. 30, no. 3, pp. 667–678, Mar. 2019, doi: 10.1021/acs.bioconjchem.8b00856.

115. F. Chen, M. Alphonse, and Q. Liu, "Strategies for nonviral nanoparticle-based delivery of CRISPR/Cas9 therapeutics," *WIREs Nanomedicine and Nanobiotechnology*, vol. 12, no. 3, May 2020, doi: 10.1002/wnan.1609.

116. R. D. Alvarez, M. W. Sill, S. A. Davidson, C. Y. Muller, D. P. Bender, R. L. DeBernardo, et al., "A phase II trial of intraperitoneal EGEN-001, an IL-12 plasmid formulated with PEG–PEI–cholesterol lipopolymer in the treatment of persistent or recurrent epithelial ovarian, fallopian tube or primary peritoneal cancer: a gynecologic oncology group study," *Gynecologic Oncology*, vol. 133, no. 3, pp. 433–438, Jun. 2014, doi: 10.1016/j.ygyno.2014.03.571.

117. X. Gao, Z. Jin, X. Tan, C. Zhang, C. Zou, W. Zhang, et al., "Hyperbranched poly(β-amino ester) based polyplex nanopaticles for delivery of CRISPR/Cas9 system and treatment of HPV infection associated cervical cancer," *Journal of Controlled Release*, vol. 321, pp. 654–668, May 2020, doi: 10.1016/j.jconrel.2020.02.045.

118. Y. Rui, D. R. Wilson, K. Sanders, and J. J. Green, "Reducible branched ester-amine quadpolymers (rBEAQs) codelivering plasmid DNA and RNA oligonucleotides enable CRISPR/Cas9 genome editing," *ACS Applied Materials & Interfaces*, vol. 11, no. 11, pp. 10472–10480, Mar. 2019, doi: 10.1021/acsami.8b20206.

119. S. Abbasi, S. Uchida, K. Toh, T. A. Tockary, A. Dirisala, K. Hayashi, et al., "Co-encapsulation of Cas9 mRNA and guide RNA in polyplex micelles enables genome editing in mouse brain," *Journal of Controlled Release*, vol. 332, pp. 260–268, Apr. 2021, doi: 10.1016/j.jconrel.2021.02.026.

120. W. Paul and C. P. Sharma, "Inorganic nanoparticles for targeted drug delivery," in *Biointegration of Medical Implant Materials*, Elsevier, 2010, pp. 204–235. doi: 10.1533/9781845699802.2.204.

121. A. D'Souza and R. Shegokar, "Polymer: lipid hybrid nanostructures in cancer drug delivery: successes and limitations," in *Nanoarchitectonics for Smart Delivery and Drug Targeting*, Elsevier, 2016, pp. 431–463. doi: 10.1016/B978-0-323-47347-7.00016-1.

122. Y. Ghosn, M. H. Kamareddine, A. Tawk, C. Elia, A. El Mahmoud, K. Terro, et al., "Inorganic nanoparticles as drug delivery systems and their potential role in the treatment of chronic myelogenous leukaemia," *Technology in Cancer Research & Treatment*, vol. 18, p. 153303381985324, Jan. 2019, doi: 10.1177/1533033819853241.

123. F. Sinclair, A. A. Begum, C. C. Dai, I. Toth, and P. M. Moyle, "Recent advances in the delivery and applications of nonviral CRISPR/Cas9 gene editing," *Drug Delivery and Translational Research*, vol. 13, no. 5, pp. 1500–1519, May 2023, doi: 10.1007/s13346-023-01320-z.

124. L. B. Truong, D. Medina-Cruz, and E. Mostafavi, "Gold nanoparticles for delivery of nucleic acid constructs for cancer treatment," in *Gold Nanoparticles for Drug Delivery*, Elsevier, 2024, pp. 141–165. doi: 10.1016/B978-0-443-19061-2.00005-5.

125. K. Lee, M. Conboy, H. M. Park, F. Jiang, H. J. Kim, M. A. Dewitt, et al., "Nanoparticle delivery of Cas9 ribonucleoprotein and donor DNA in vivo induces homology-directed DNA repair," *Nature Biomedical Engineering*, vol. 1, no. 11, pp. 889–901, Oct. 2017, doi: 10.1038/s41551-017-0137-2.

126. N. Ajinkya, X. Yu, P. Kaithal, H. Luo, P. Somani, and S. Ramakrishna, "Magnetic iron oxide nanoparticle (IONP) synthesis to applications: present and future," *Materials*, vol. 13, no. 20, p. 4644, Oct. 2020, doi: 10.3390/ma13204644.

127. M. Mohammadi Ghanbarlou, S. Abdoli, H. Omid, L. Qazizadeh, H. Bamehr, M. Raigani, et al., "Delivery of dCas9 activator system using magnetic nanoparticles technology as a vector delivery method for human skin fibroblast," *Magnetochemistry*, vol. 9, no. 3, p. 71, Feb. 2023, doi: 10.3390/magnetochemistry9030071.

128. A. M. Carvalho, R. A. Cordeiro, and H. Faneca, "Silica-based gene delivery systems: from design to therapeutic applications," *Pharmaceutics*, vol. 12, no. 7, p. 649, Jul. 2020, doi: 10.3390/pharmaceutics12070649.

129. Y. Wang, P. K. Shahi, X. Wang, R. Xie, Y. Zhao, M. Wu, et al., "In vivo targeted delivery of nucleic acids and CRISPR genome editors enabled by GSH-responsive silica nanoparticles," *Journal of Controlled Release*, vol. 336, pp. 296–309, Aug. 2021, doi: 10.1016/j.jconrel.2021.06.030.

130. A. Hendel, E. J. Fine, G. Bao, and M. H. Porteus, "Quantifying on- and off-target genome editing," *Trends in Biotechnology*, vol. 33, no. 2, pp. 132–140, Feb. 2015, doi: 10.1016/j.tibtech.2014.12.001.

131. C. Guo, X. Ma, F. Gao, and Y. Guo, "Off-target effects in CRISPR/Cas9 gene editing," *Frontiers in Bioengineering and Biotechnology*, vol. 11, Mar. 2023, doi: 10.3389/fbioe.2023.1143157.

132. M. Pacesa, C. H. Lin, A. Cléry, A. Saha, P. R. Arantes, K. Bargsten, et al., "Structural basis for Cas9 off-target activity," *Cell*, vol. 185, no. 22, pp. 4067–4081.e21, Oct. 2022, doi: 10.1016/j.cell.2022.09.026.

133. M. M. Vicente, M. Chaves-Ferreira, J. M. P. Jorge, J. T. Proença, and V. M. Barreto, "The off-targets of clustered regularly interspaced short palindromic repeats gene Editing," *Frontiers in Cell and Developmental Biology*, vol. 9, Sep. 2021, doi: 10.3389/fcell.2021.718466.

134. X. Wang, Y. Wang, X. Wu, J. Wang, Y. Wang, Z. Qiu, et al., "Unbiased detection of off-target cleavage by CRISPR-Cas9 and TALENs using integrase-defective lentiviral vectors," *Nature Biotechnology*, vol. 33, no. 2, pp. 175–178, Feb. 2015, doi: 10.1038/nbt.3127.

135. D. K. Sahel, L. K. Vora, A. Saraswat, S. Sharma, J. Monpara, A. A. D'Souza, et al., "CRISPR/Cas9 genome editing for tissue-specific in vivo targeting: nanomaterials and translational perspective," *Advanced Science*, vol. 10, no. 19, Jul. 2023, doi: 10.1002/advs.202207512.

136. A. V. Bannikov and A. V. Lavrov, "CRISPR/CAS9, the king of genome editing tools," *Molecular Biology*, vol. 51, no. 4, pp. 514–525, Jul. 2017, doi: 10.1134/S0026893317040033.

137. S.-W. Wang, C. Gao, Y.-M. Zheng, L. Yi, J.-C. Lu, X.-Y. Huang, et al., "Current applications and future perspective of CRISPR/Cas9 gene editing in cancer," *Molecular Cancer*, vol. 21, no. 1, p. 57, Feb. 2022, doi: 10.1186/s12943-022-01518-8.

138. M. Jamal, A. Ullah, M. Ahsan, R. Tyagi, Z. Habib, and K. Rehman, "Improving CRISPR-Cas9 on-target specificity," *Current Issues in Molecular Biology*, pp. 65–80, 2018, doi: 10.21775/cimb.026.065.

139. J. Kwon, M. Kim, W. Hwang, A. Jo, G.-H. Hwang, M. Jung, et al., "Extru-seq: a method for predicting genome-wide Cas9 off-target sites with advantages of both cell-based and in vitro approaches," *Genome Biology*, vol. 24, no. 1, p. 4, Jan. 2023, doi: 10.1186/s13059-022-02842-4.

140. P. Cameron, C. K. Fuller, P. D. Donohoue, B. N. Jones, M. S. Thompson, M. M. Carter, et al., "Mapping the genomic landscape of CRISPR–Cas9 cleavage," *Nature Methods*, vol. 14, no. 6, pp. 600–606, Jun. 2017, doi: 10.1038/nmeth.4284.

141. D. Kim, K. Luk, S. A. Wolfe, and J.-S. Kim, "Evaluating and enhancing target specificity of gene-editing nucleases and deaminases," *Annual Review of Biochemistry*, vol. 88, no. 1, pp. 191–220, Jun. 2019, doi: 10.1146/annurev-biochem-013118-111730.

142. S. Q. Tsai, N. T. Nguyen, J. Malagon-Lopez, V. V Topkar, M. J. Aryee, and J. K. Joung, "CIRCLE-seq: a highly sensitive in vitro screen for genome-wide CRISPR–Cas9 nuclease off-targets," *Nature Methods*, vol. 14, no. 6, pp. 607–614, Jun. 2017, doi: 10.1038/nmeth.4278.

143. V. Iyer, B. Shen, W. Zhang, A. Hodgkins, T. Keane, X. Huang, et al., "Off-target mutations are rare in Cas9-modified mice," *Nature Methods*, vol. 12, no. 6, pp. 479–479, Jun. 2015, doi: 10.1038/nmeth.3408.

144. S. Q. Tsai, Z. Zheng, N. T. Nguyen, M. Liebers, V. V. Topkar, V. Thapar, et al., "GUIDE-seq enables genome-wide profiling of off-target cleavage by CRISPR-Cas nucleases," *Nature Biotechnology*, vol. 33, no. 2, pp. 187–197, Feb. 2015, doi: 10.1038/nbt.3117.

145. B. Wienert, S. K. Wyman, C. D. Richardson, C. D. Yeh, P. Akcakaya, M. J. Porritt, et al., "Unbiased detection of CRISPR off-targets in vivo using DISCOVER-Seq," *Science (1979)*, vol. 364, no. 6437, pp. 286–289, Apr. 2019, doi: 10.1126/science.aav9023.

146. W. X. Yan, R. Mirzazadeh, S. Garnerone, D. Scott, M. W. Schneider, T. Kallas, et al., "BLISS is a versatile and quantitative method for genome-wide profiling of DNA double-strand breaks," *Nature Communications*, vol. 8, no. 1, p. 15058, May 2017, doi: 10.1038/ncomms15058.

147. R. Ballarino, B. A. M. Bouwman, and N. Crosetto, "Genome-wide CRISPR off-target DNA break detection by the BLISS method," *Methods in Molecular Biology*, 2021, pp. 261–281. doi: 10.1007/978-1-0716-0687-2_15.

148. M. K. Cromer, K. R. Majeti, G. R. Rettig, K. Murugan, G. L. Kurgan, N. M. Bode, et al., "Comparative analysis of CRISPR off-target discovery tools following ex vivo editing of CD34+ hematopoietic stem and progenitor cells," *Molecular Therapy*, vol. 31, no. 4, pp. 1074–1087, Apr. 2023, doi: 10.1016/j.ymthe.2023.02.011.

149. S.-Q. Liang, P. Liu, J. L. Smith, E. Mintzer, S. Maitland, X. Dong, et al., "Genome-wide detection of CRISPR editing in vivo using GUIDE-tag," *Nature Communications*, vol. 13, no. 1, p. 437, Jan. 2022, doi: 10.1038/s41467-022-28135-9.

150. S. Bhatia, Pooja, and S. K. Yadav, "CRISPR-Cas for genome editing: classification, mechanism, designing and applications," *International Journal of Biological Macromolecules*, vol. 238, p. 124054, May 2023, doi: 10.1016/j.ijbiomac.2023.124054.

151. M. Wang, M. Chen, X. Wu, X. Huang, and B. Yu, "CRISPR applications in cancer diagnosis and treatment," *Cellular & Molecular Biology Letters*, vol. 28, no. 1, p. 73, Sep. 2023, doi: 10.1186/s11658-023-00483-4.

152. J. B. Papizan, S. N. Porter, A. Sharma, and S. M. Pruett-Miller, "Therapeutic gene editing strategies using CRISPR-Cas9 for the β-hemoglobinopathies," *Journal of Biomedical Research*, vol. 35, no. 2, p. 115, 2021, doi: 10.7555/JBR.34.20200096.

153. A. Eid, S. Alshareef, and M. M. Mahfouz, "CRISPR base editors: genome editing without double-stranded breaks," *Biochemical Journal*, vol. 475, no. 11, pp. 1955–1964, Jun. 2018, doi: 10.1042/BCJ20170793.

154. A. V. Anzalone, P. B. Randolph, J. R. Davis, A. A. Sousa, L. W. Koblan, J. M. Levy, et al., "Search-and-replace genome editing without double-strand breaks or donor DNA," *Nature*, vol. 576, no. 7785, pp. 149–157, Dec. 2019, doi: 10.1038/s41586-019-1711-4.

155. G. F. Jacob and A. B. Raper, "Hereditary persistence of foetal haemoglobin production, and its interaction with the sickle-cell trait," *British Journal of Haematology*, vol. 4, no. 2, pp. 138–149, Apr. 1958, doi: 10.1111/j.1365-2141.1958.tb03844.x.

156. G. Lettre and D. E. Bauer, "Fetal haemoglobin in sickle-cell disease: from genetic epidemiology to new therapeutic strategies," *Lancet*, vol. 387, no. 10037, pp. 2554–2564, Jun. 2016, doi: 10.1016/S0140-6736(15)01341-0.

157. L. Ye, J. Wang, Y. Tan, A. I. Beyer, F. Xie, M. O. Muench, et al., "Genome editing using CRISPR-Cas9 to create the HPFH genotype in HSPCs: an approach for treating sickle cell disease and β-thalassemia," *Proceedings of the National Academy of Sciences*, vol. 113, no. 38, pp. 10661–10665, Sep. 2016, doi: 10.1073/pnas.1612075113.

158. D. E. Bauer, S. C. Kamran, S. Lessard, J. Xu, Y. Fujiwara, C. Lin, et al., "An erythroid enhancer of *BCL11A* subject to genetic variation determines fetal hemoglobin level," *Science (1979)*, vol. 342, no. 6155, pp. 253–257, Oct. 2013, doi: 10.1126/science.1242088.

159. Y. Shimizu-Motohashi, H. Komaki, N. Motohashi, S. Takeda, T. Yokota, and Y. Aoki, "Restoring dystrophin expression in Duchenne muscular dystrophy: current status of therapeutic approaches," *Journal of Personalized Medicine*, vol. 9, no. 1, p. 1, Jan. 2019, doi: 10.3390/jpm9010001.

160. C. S. Young, M. R. Hicks, N. V. Ermolova, H. Nakano, M. Jan, S. Younesi, et al., "A single CRISPR-Cas9 deletion strategy that targets the majority of DMD patients restores dystrophin function in hiPSC-derived muscle cells," *Cell Stem Cell*, vol. 18, no. 4, pp. 533–540, Apr. 2016, doi: 10.1016/j.stem.2016.01.021.

161. J. H. Lee and J. P. Han, "In vivo LNP-CRISPR approaches for the treatment of hemophilia," *Molecular Diagnosis & Therapy*, vol. 28, no. 3, pp. 239–248, May 2024, doi: 10.1007/s40291-024-00705-1.

162. S. Luo, Z. Li, X. Dai, R. Zhang, Z. Liang, W. Li, et al., "CRISPR/Cas9-mediated in vivo genetic correction in a mouse model of hemophilia A," *Frontiers in Cell and Developmental Biology*, vol. 9, Aug. 2021, doi: 10.3389/fcell.2021.672564.

163. C. J. Stephens, E. J. Lauron, E. Kashentseva, Z. H. Lu, W. M. Yokoyama, and D. T. Curiel, "Long-term correction of hemophilia B using adenoviral delivery of CRISPR/Cas9," *Journal of Controlled Release*, vol. 298, pp. 128–141, Mar. 2019, doi: 10.1016/j.jconrel.2019.02.009.

164. M. Matano, S. Date, M. Shimokawa, A. Takano, M. Fujii, Y. Ohta, et al., "Modeling colorectal cancer using CRISPR-Cas9–mediated engineering of human intestinal organoids," *Nature Medicine*, vol. 21, no. 3, pp. 256–262, Mar. 2015, doi: 10.1038/nm.3802.

165. R. J. Platt, S. Chen, Y. Zhou, M. J. Yim, L. Swiech, H. R. Kempton, et al., "CRISPR-Cas9 knockin mice for genome editing and cancer modeling," *Cell*, vol. 159, no. 2, pp. 440–455, Oct. 2014, doi: 10.1016/j.cell.2014.09.014.

166. H. Zhang, C. Qin, C. An, X. Zheng, S. Wen, W. Chen, et al., "Application of the CRISPR/Cas9-based gene editing technique in basic research, diagnosis, and therapy of cancer," *Molecular Cancer*, vol. 20, no. 1, p. 126, Dec. 2021, doi: 10.1186/s12943-021-01431-6.

167. O. Shalem, N. E. Sanjana, E. Hartenian, X. Shi, D. A. Scott, T. S. Mikkelsen, et al., "Genome-scale CRISPR-Cas9 knockout screening in human cells," *Science (1979)*, vol. 343, no. 6166, pp. 84–87, Jan. 2014, doi: 10.1126/science.1247005.

168. L. Wei, D. Lee, C.-T. Law, M. S. Zhang, J. Shen, D. W.-C. Chin, et al., "Genome-wide CRISPR/Cas9 library screening identified PHGDH as a critical driver for Sorafenib resistance in HCC," *Nature Communications*, vol. 10, no. 1, p. 4681, Oct. 2019, doi: 10.1038/s41467-019-12606-7.

169. X. Cui, C. Zhang, Z. Xu, S. Wang, X. Li, E. Stringer-Reasor, et al., "Dual CRISPR interference and activation for targeted reactivation of X-linked endogenous FOXP3 in human breast cancer cells," *Molecular Cancer*, vol. 21, no. 1, p. 38, Dec. 2022, doi: 10.1186/s12943-021-01472-x.

170. Y. Lu, J. Xue, T. Deng, X. Zhou, K. Yu, L. Deng, et al., "Safety and feasibility of CRISPR-edited T cells in patients with refractory non-small-cell lung cancer," *Nature Medicine*, vol. 26, no. 5, pp. 732–740, May 2020, doi: 10.1038/s41591-020-0840-5.

171. J. Zhang, Y. Hu, J. Yang, W. Li, M. Zhang, Q. Wang, et al., "Non-viral, specifically targeted CAR-T cells achieve high safety and efficacy in B-NHL," *Nature*, vol. 609, no. 7926, pp. 369–374, Sep. 2022, doi: 10.1038/s41586-022-05140-y.

172. A. C. Komor, Y. B. Kim, M. S. Packer, J. A. Zuris, and D. R. Liu, "Programmable editing of a target base in genomic DNA without double-stranded DNA cleavage," *Nature*, vol. 533, no. 7603, pp. 420–424, May 2016, doi: 10.1038/nature17946.

173. N. M. Gaudelli, A. C. Komor, H. A. Rees, M. S. Packer, A. H. Badran, D. I. Bryson et al., "Programmable base editing of A•T to G•C in genomic DNA without DNA cleavage," *Nature*, vol. 551, no. 7681, pp. 464–471, Nov. 2017, doi: 10.1038/nature24644.

174. R. Cuella-Martin, S. B. Hayward, X. Fan, X. Chen, J.-W. Huang, A. Taglialatela, et al., "Functional interrogation of DNA damage response variants with base editing screens," *Cell*, vol. 184, no. 4, pp. 1081–1097.e19, Feb. 2021, doi: 10.1016/j.cell.2021.01.041.

175. J. Li, J. Lin, S. Huang, M. Li, W. Yu, Y. Zhao, et al., "Functional phosphoproteomics in cancer chemoresistance using CRISPR-mediated base editors," *Advanced Science*, vol. 9, no. 30, Oct. 2022, doi: 10.1002/advs.202200717.

176. S. Annunziato, C. Lutz, L. Henneman, J. Bhin, K. Wong, B. Siteur, et al., "In situ CRISPR-Cas9 base editing for the development of genetically engineered mouse models of breast cancer," *EMBO Journal*, vol. 39, no. 5, Mar. 2020, doi: 10.15252/embj.2019102169.

177. Z. A. Ely, N. Mathey-Andrews, S. Naranjo, S. I. Gould, K. L. Mercer, G. A. Newby, et al., "A prime editor mouse to model a broad spectrum of somatic mutations in vivo," *Nature Biotechnology*, vol. 42, no. 3, pp. 424–436, Mar. 2024, doi: 10.1038/s41587-023-01783-y.

178. J. J. Freen-van Heeren, "Closing the door with CRISPR: genome editing of CCR5 and CXCR4 as a potential curative solution for HIV," *BioTech*, vol. 11, no. 3, p. 25, Jul. 2022, doi: 10.3390/biotech11030025.

179. D. Stone, K. R. Long, M. A. Loprieno, H. S. De S. Feelixge, E. J. Kenkel, R. M. Liley, et al., "CRISPR-Cas9 gene editing of hepatitis B virus in chronically infected humanized mice," *Molecular Therapy—Methods & Clinical Development*, vol. 20, pp. 258–275, Mar. 2021, doi: 10.1016/j.omtm.2020.11.014.

180. R. Ding, J. Long, M. Yuan, Y. Jin, H. Yang, M. Chen, et al., "CRISPR/Cas system: a potential technology for the prevention and control of COVID-19 and emerging infectious diseases," *Frontiers in Cellular and Infection Microbiology*, vol. 11, Apr. 2021, doi: 10.3389/fcimb.2021.639108.

181. I. M. S. Degors, C. Wang, Z. U. Rehman, and I. S. Zuhorn, "Carriers break barriers in drug delivery: endocytosis and endosomal escape of gene delivery vectors," *Accounts of Chemical Research*, vol. 52, no. 7, pp. 1750–1760, Jul. 2019, doi: 10.1021/acs.accounts.9b00177.

182. A. A. Aljabali, M. A. Obeid, R. M. Bashatwah, A. Serrano-Aroca, V. Mishra, Y. Mishra, et al., "Nanomaterials and their impact on the immune system," *International Journal of Molecular Sciences*, vol. 24, no. 3, p. 2008, Jan. 2023, doi: 10.3390/ijms24032008.

183. M. Qin, G. Du, and X. Sun, "Biomimetic cell-derived nanocarriers for modulating immune responses," *Biomaterials Science*, vol. 8, no. 2, pp. 530–543, 2020, doi: 10.1039/C9BM01444F.

184. C.-N. Ko, S. Zang, Y. Zhou, Z. Zhong, and C. Yang, "Nanocarriers for effective delivery: modulation of innate immunity for the management of infections and the associated complications," *Journal of Nanobiotechnology*, vol. 20, no. 1, p. 380, Aug. 2022, doi: 10.1186/s12951-022-01582-8.

185. D. Zhang, A. Hussain, H. Manghwar, K. Xie, S. Xie, S. Zhao, et al., "Genome editing with the CRISPR-Cas system: an art, ethics and global regulatory perspective," *Plant Biotechnology Journal*, vol. 18, no. 8, pp. 1651–1669, Aug. 2020, doi: 10.1111/pbi.13383.

186. R. Piergentili, A. Del Rio, F. Signore, F. Umani Ronchi, E. Marinelli, and S. Zaami, "CRISPR-Cas and its wide-ranging applications: from human genome editing to environmental implications, technical limitations, hazards and bioethical issues," *Cells*, vol. 10, no. 5, p. 969, Apr. 2021, doi: 10.3390/cells10050969.

187. T. Ishii, "The ethics of creating genetically modified children using genome editing," *Current Opinion in Endocrinology, Diabetes and Obesity*, vol. 24, no. 6, pp. 418–423, Dec. 2017, doi: 10.1097/MED.0000000000000369.

188. I. Gostimskaya, "CRISPR–Cas9: a history of its discovery and ethical considerations of its use in genome editing," *Biochemistry (Moscow)*, vol. 87, no. 8, pp. 777–788, Aug. 2022, doi: 10.1134/S0006297922080090.

189. C. Brokowski and M. Adli, "CRISPR ethics: moral considerations for applications of a powerful tool," *Journal of Molecular Biology*, vol. 431, no. 1, pp. 88–101, Jan. 2019, doi: 10.1016/j.jmb.2018.05.044.

# 12 Delivery of Proteins Using Nanocarriers

## *Enzymes, Antibodies, and Growth Factors*

*Jennifer Suurbaar, Frederick Sarfo-Antwi, and Seth Kwabena Amponsah*

### 12.1 INTRODUCTION

Protein-based therapies have utility in a wide range of diseases [1]. They are particularly beneficial in the management of diseases with complex etiologies, like cancer and autoimmune disorders [2]. Proteins as therapeutic agents are inherently efficacious and able to perform diverse functions such as binding to receptors and activating or inhibiting enzymes [3]. Given their versatility, protein-based therapies can be designed with a number of properties such as increased stability, extended half-life, and enhanced tissue penetration [4]. Due to their large size and susceptibility to degradation, protein-based therapeutic agents often require specialized delivery methods to be able to reach their target sites.

Oral route is a non-invasive and convenient route for administration of xenobiotics. For protein administration, however, challenges arise due to the likelihood of degradation by stomach acid, hydrolysis by gastrointestinal enzymes, poor permeability, and efflux mechanisms affecting absorption [5, 6]. Furthermore, nasal inhalation provides a non-invasive option for the systemic delivery of proteins, but this approach faces barriers like mucociliary clearance, enzymatic degradation, and limitations in absorption. Bioavailability via the oral and nasal routes is generally low compared to parenteral administration [3, 6]. Parenteral administration of proteins could be via the intravenous, subcutaneous, and intramuscular routes. Usually, these routes allow direct access to the bloodstream but require frequent administration due to rapid clearance [5]. Methods like implants, transdermal formulations, microneedles, iontophoresis, and penetration enhancers have been investigated as other ways of introducing proteins into circulation, but these approaches have had restricted applications so far [3, 6].

Nanocarriers are microscopic vehicles designed to encapsulate biomolecules or drugs in specific cells or tissues [1]. They present several advantages over traditional methods of biomolecules or drug delivery. Nanocarriers can enhance the bioavailability of proteins by facilitating their transport across tissue barriers. By attaching proteins to nanocarriers, they can be directed to specific tissues such as cancer cells, increasing therapeutic effect and minimizing side effects. Choosing the right polymers and particle-sized nanocarriers can improve the release process of proteins from nanocarriers. Nanocarriers can protect proteins from enzymatic degradation and immune response, increasing their stability and distribution in the body [7–9].

The use of nanocarriers for protein delivery also presents challenges, particularly regarding the immunogenicity and potential toxicity of the carriers. Further, the scalability and cost-effectiveness of large-scale production of these formulations is a major challenge [10]. Nonetheless, the delivery of proteins using nanocarriers offers a promising avenue for advancing the field of drug delivery and personalized medicine. In this review, we will highlight the benefits and current trends in this rapidly evolving field by exploring the importance of enzymes, antibodies, and growth factors in the field of protein delivery. Further, we will discuss applications of nanocarriers in protein delivery, challenges faced, and future directions.

### 12.2 DELIVERY OF PROTEINS USING NANOCARRIERS

Nanocarriers offer a promising approach for the delivery of proteins such as enzymes, growth hormones, and antibodies. They have been used to deliver enzymes in anticancer therapy. Nanocarriers have also been used for the delivery of growth hormones such as recombinant human growth hormone (rhGH). These nanoparticles (NPs) can provide sustained release of rhGH, which can be highly beneficial in improving patient compliance and reducing adverse effects. Antibodies are valuable tools in modern medicine due to their ability to target diseased cells with high specificity. Antibody–NP conjugates have been shown to have improved efficacy and reduced toxicity compared to freely delivered antibodies.

#### 12.2.1 Delivery of Enzymes Using Nanocarriers

Under modest physiological circumstances, enzymes have extraordinary catalytic activity, excellent substrate selectivity, and sensitivity [11]. Several industrial and medical fields use enzymes to diagnose and treat genetic and metabolic disorders [12, 13]. Patients with enzyme deficits can be given exogenous enzymes via the intravenous route and this approach appears to reduce patient compliance [14]. Use of natural enzymes may be hampered by proteolytic hydrolysis, oxidation, and partial denaturation which may reduce efficacy [15, 16]. The stability of nanocarriers, decreased toxicity, good bioavailability, and improved biodistribution

 DOI: 10.1201/9781003473183-12

may help address the aforementioned challenges with the use of enzymes in the field of medicine. Enzyme-responsive drug delivery systems and nanocarriers that have bioactive enzymes are currently available [17, 18]. Encapsulated medications can be delivered on demand using liposomes, polymeric, and inorganic NPs. Chemical alteration and protein or medium engineering can be employed to boost stability. Enzyme immobilization onto carriers or entrapment in a matrix, framework, or NP can delay degradation. However, it is impossible to exclude partial denaturation of proteins following attachment to solid support, aggregation, or agglomeration [19]. Increased enzyme stability, less enzymatic breakdown throughout circulation, and targeted delivery to specific sites are just a few advantages offered by NP [20, 21]. Furthermore, because of increased penetration and retention activity, nanocarriers have the benefit of preferentially accumulating in solid tumors. Moreover, prodrug treatments for cancer may employ nanostructures containing enzymes. Specifically, gold (Au) NP can be coupled with the enzyme horseradish peroxidase (HRP) to deliver it intracellularly to breast cancer cells [22, 23]. Enzymes, their nanocarriers, and their practical implications are summarized in Table 12.1.

## Table 12.1 Studies reporting the use of nanocarriers in enzyme delivery

| Enzyme | Nanocarrier Type | Description | Practical Application | Ref. |
|---|---|---|---|---|
| Antioxidant enzymes: including superoxide dismutase (SOD) and catalase (CAT) | Liposomes, polymers, conjugates, cyclodextrins, micelles, dendrimers, and NPs | Focused on antioxidant therapy in neurodegenerative diseases using NP drug delivery<br>Nanoparticle-based enzyme delivery systems were explored to enhance the effectiveness of antioxidant therapies by improving stability, targeting, and dosing | • Nanoparticle delivery enhances antioxidant therapy for neurodegenerative disease<br>• Challenges like bioavailability and stability in antioxidant therapy were addressed | [24] |
| Antioxidant enzyme: catalase | Synthetic protein nanoparticle (SPNP) was developed which incorporated the enzyme catalase as the active therapeutic agent | The catalase-based NPs show prolonged antioxidant activity and maintain higher activity levels than the enzyme in solution | • Catalase-based synthetic protein nanoparticles (CAT-SPNPs) offer a more stable and efficient delivery system for antioxidant therapy, potentially improving treatment outcomes for diseases associated with oxidative stress | [25] |
| Enzyme superoxide dismutase 1 (SOD1) was used in the form of a gene (pSOD1 plasmid) | HSA-PEI NPs: human serum albumin (HSA) NPs that were stabilized with polyethyleneimine (PEI). | The enzyme SOD1 was used in the form of a gene (pSOD1 plasmid) to be delivered by the nanocarriers for combating oxidative stress by converting free oxygen radicals into less harmful substances | • Due to their high biocompatibility and non-toxic nature, these nanocarriers could be used in gene therapy applications, minimizing side effects<br>• It might be possible to treat neuronal diseases more effectively with nanocarriers because of their ability to deliver therapeutic material precisely to the site of action | [26] |

*(continued)*

**Table 12.1 (Continued)**

| Enzyme | Nanocarrier Type | Description | Practical Application | Ref. |
|---|---|---|---|---|
| Recombinant α-galactosidase was utilized for enzyme replacement therapy (ERT) | HSA and a recombinant protein known as 30Kc19, which originates from silkworm hemolymph | Fabry disease is a genetic disorder caused by α-galactosidase deficiency, leading to the accumulation of glycosphingolipids and resulting in multi-organ dysfunction and premature death | • The 30Kc19-HSA NPs could potentially improve the efficacy of enzyme replacement therapy for Fabry disease by enhancing the delivery and stability of α-galactosidase within patient cells<br>• These NPs have shown negligible toxicity in human cells, suggesting they could be a safe carrier for drug delivery *in vivo* | [27] |
| The enzymes used in the fabrication of NP for the study included DNase I, RNase A, trypsin, chymotrypsin, catalase, HRP, and lipase, chosen for their relevance in biomedicine | Enzymatic nanoparticles (ENPs) were designed by dissolving enzymes in fluoroalcohol and mixing with 40% ethanol as an anti-solvent, followed by alcohol evaporation at high temperature and low pressure | The enzymes in the ENP retained their ability to bind substrates and catalyze reactions. The toxicity of these ENP was evaluated, revealing that only DNase NP showed significant toxicity to three cell lines, while the others were less toxic. | • ENPs can facilitate the delivery of therapeutic enzymes into cells, potentially improving treatments for diseases where enzyme function is compromised<br>• The low toxicity of most ENPs, except DNase NP, suggests they could be used safely in medical applications, reducing the risk of adverse effects on non-target cells | [16] |

### 12.2.2 Delivery of Antibodies Using Nanocarriers

Over the past two decades, the use of antibodies has shown promise in the management of several diseases. However, using antibody-based therapies requires administration via injections and antibody manufacturing costs often make the entire treatment regimen expensive. An additional drawback of antibody-based therapies is the low level of tissue penetration and poor stability. One possible way to improve antibody stability is by using nanocarriers as delivery vehicles. Furthermore, with the use of nanocarriers, there is enhanced bioavailability of the antibodies.

Low intracellular absorption and quick lysosomal degradation can also decrease the effectiveness of antibody-based therapies targeting intracellular protein-protein interactions in cancer cells [28]. Identifying and eliminating malignant cells with little impact on healthy cells is one of the main obstacles in cancer treatment. The use of antibody-based therapies with nanocarriers appears to increase entry into cells and decrease adverse effects [29]. Receptor-mediated endocytosis can be facilitated using monoclonal antibodies against cancer antigens, such as *HER2*, *TROP2*, or *Nectin-4*, which are selectively expressed on the surface of cancer cells. Antibodies, their nanocarriers and their practical implications are summarized in Table 12.2.

### 12.2.3 Delivery of Growth Factors Using Nanocarriers

Delivery of growth factors using nanotechnology represents a significant advancement in the field of tissue engineering and regenerative medicine. Growth factors are important in regulating cellular processes, including proliferation, differentiation, and migration [36]. However, their clinical application is often limited by challenges such as rapid degradation, short half-life, and non-specific distribution when administered systemically. Nanotechnology offers innovative solutions to address these challenges by enabling the controlled release, targeted delivery, and protection of growth factors from degradation [37]. In regenerative medicine, a functional replacement of damaged or diseased tissues can be achieved through delivery of growth factors [38]. These growth factors can be delivered through nanocarriers such as lipid NPs, nanofibrous structures, hydrogels, and scaffolds [39].

**Table 12.2 Studies reporting the use of nanocarriers in antibody delivery**

| Antibodies | Nanocarrier Type | Description | Practical Application | References |
|---|---|---|---|---|
| Model antibodies and therapeutic antibodies, such as herceptin, were used for assembly on NPs | Nanoparticles including gold nanorods (AuNRs), layered double hydroxide nanosheets (LDH), superparamagnetic iron oxide (SPION) NPs, and ultrasmall superparamagnetic iron oxide (USPIO) NPs were coated with a biomineralized film using zeolitic imidazole framework-8 (ZIF-8) to immobilize antibodies for enhanced cell targeting | It presents a new method for attaching antibodies to NPs using a ZIF-8 coating, which is important for targeted cell therapy | The method enables precise delivery of therapeutic agents to specific cells, potentially increasing the efficacy of treatments for diseases like cancer by ensuring that drugs are directed to the right location<br>The versatility of the coating method allows for its application across various types of NPs, suggesting a broad utility in the development of nanomedicine platforms | [30] |
| Anti-programmed cell death protein-1 (αPD1) and anti-cytotoxic T-lymphocyte antigen 4 (αCTLA4), which are immune checkpoint blockade antibodies | Fluorocarbon-modified chitosan (FCS) NPs are created by mixing therapeutic proteins with the FCS, which are then lyophilized with excipients and encapsulated for oral administration | Oral delivery of immune checkpoint blockade antibodies with FCS showed similar antitumor effects to intravenous injections and fewer side effects in tumor models | Oral delivery using FCS has the potential to reduce the immune-related adverse events often associated with intravenous injections, improving patient safety and comfort | [31, 32] |
| Anti-epidermal growth factor receptor (anti-EGFR) | Polymer-based nanocarriers, such as those made from hyaluronic acid (HA), are utilized for their ability to bind to cell surface receptors like CD44.<br>Lipid-based nanocarriers are also employed, with components like DSPE-PEG-Fab used for surface modification to target specific cells | RNA interference (RNAi) using small interfering RNA (siRNA) has the potential to treat diseases caused by gene defects or requiring gene expression inhibition, but siRNA faces challenges like poor stability and inefficient cell uptake.<br>Nanoparticulate drug delivery systems, particularly siRNA NPs modified with antibodies, can enhance the distribution and safety of siRNA therapies by targeting specific tissues | siRNA NPs can specifically target diseased cells, potentially reducing side effects and improving the efficacy of gene-silencing therapies for genetic disorders and diseases requiring gene expression inhibition. | [33] |

*(continued)*

**Table 12.2 (Continued)**

| Antibodies | Nanocarrier Type | Description | Practical Application | References |
|---|---|---|---|---|
| A specific antibody against *Staphylococcus aureus* was used | Polymeric NPs designed for encapsulation and sustained release of antibiotics | An antibody specific to *S. aureus* was attached to the NPs to direct them to the bacteria and its biofilms.<br>The targeted NPs showed better killing of *S. aureus* in both free-floating and biofilm states compared to antibiotics alone. | By specifically targeting *S. aureus*, the nanocarriers may offer a more effective treatment for stubborn biofilm infections, which are often resistant to conventional antibiotics.<br>The use of this technology could lead to improved outcomes in treating bacterial infections, especially those associated with implanted medical devices where biofilms commonly occur. | [34] |
| Recombinant human or humanized monoclonal antibodies were used. | Inorganic NPs, which include materials like gold or silica, Polymeric NPs and liposomes functionalized with maleimide chemistry for bioconjugation with monoclonal antibodies | Nanoparticle as carriers of monoclonal antibodies (mAbs) showed efficient uptake by cancer cells through receptor-mediated endocytosis in cell culture studies. | Functionalization of NPs with monoclonal antibodies can lead to more efficient uptake by cancer cells through receptor-mediated endocytosis, which could enhance the therapeutic outcomes<br>The use of NPs can improve the delivery and efficacy of monoclonal antibodies, allowing for controlled release and site-specific treatment of cancer | [35] |

Current advances in nanotechnology have aided the delivery of growth factors to tissues to facilitate wound healing and skin regeneration [40]. Growth factors that have already been used in tissue engineering and delivery by nanocarrier systems include basic fibroblast growth factor (bFGF, or FGF-2), bone morphogenetic proteins (BMPs), transforming growth factor-β (TGF-β, including TGF-β1, TGF-β2, and TGF-β3), vascular endothelial growth factor (VEGF), platelet-derived growth factor (PDGF), epidermal growth factor (EGF), hepatocyte growth factor (HGF), nerve growth factor (NGF), and insulin-like growth factor (IGF). Research in this field is likely to focus on the development of multifunctional nanocarriers that can deliver multiple growth factors or combine growth factor delivery with other therapeutic modalities. The integration of smart nanocarriers capable of responding to specific physiological stimuli for the on-demand release of growth factors is another promising area of research. Growth factors, their nanocarriers, and their practical implications are summarized in Table 12.3.

## 12.3 STRATEGIES EMPLOYED IN PROTEIN DELIVERY USING NANOCARRIERS

Protein delivery using nanocarriers is employed in several clinical scenarios, including cancer treatment. The design of these protein delivery strategies has been comprehensively summarized in this section, highlighting their mechanisms of action and potential therapeutic applications.

## Table 12.3 Studies reporting the use of nanocarriers in growth factor delivery

| Growth Factors | Nanocarrier Type | Description | Practical Application | Ref. |
|---|---|---|---|---|
| EGF | Chitosan NPs were utilized | EGF is beneficial for wound healing. Chitosan NPs were created in various ratios with EGF, and the 2:1 chitosan/EGF ratio was found to release 80% of EGF within 12 hours and maintain its biological activity | The development of chitosan NPs loaded with EGF offers a promising strategy for enhancing wound healing, as these NPs can protect EGF from degradation and ensure sustained release at the wound site | [41, 42] |
| bFGF Vascular endothelial growth factor-A (VEGF-A) | The research employed positively charged elastic nanoliposomes as the nanocarriers for delivering skin-permeable growth factors | Growth factors like bFGF and VEGF-A were modified with low-molecular-weight protamine (LMWP) to improve skin absorption and stability, aiming to enhance wound healing and skin regeneration. These modified growth factors retained their biological activity, promoting cell proliferation and wound recovery more effectively than their native forms | The development of LMWP-conjugated growth factors encapsulated in elastic nanoliposomes could lead to more effective treatments for wound healing, potentially reducing recovery time and improving the healing process. | [43] |
| Brain-derived neurotrophic factor (BDNF) | Chitosan NPs, which were prepared using a chitosan/tripolyphosphate (TPP) mixture at an optimized ratio of 1:3 for effective encapsulation of BDNF and gold nanoparticles (AuNPs) | The research aimed to create tiny particles made of chitosan, a natural substance, to carry and slowly release a special protein (BDNF) and tiny gold particles. They found the best mix for making these carrier particles and showed that they could hold onto BDNF and gold particles very well, with more than 89% and 93% efficiency, respectively | The development of chitosan NPs for delivering BDNF and AuNPs could lead to new treatments for diseases that benefit from tissue regeneration, as these particles enhance stem cell growth.<br>By controlling the release of BDNF and AuNPs, the NPs could provide a sustained therapeutic effect, potentially reducing the frequency of treatments needed for patients | [44] |
| Insulin-like growth factor 1 (IGF-1) | Amine terminal polyamidoamine (PAMAM) dendrimers as nanocarriers, were end-functionalized with variable molar ratios of PEG to control surface charge | There is poor delivery of drugs to chondrocytes in dense cartilage in osteoarthritis patients. A cationic nanocarrier, PEGylated PAMAM dendrimer, was developed to deliver growth factors like IGF-1 directly to chondrocytes, improving binding, penetration, and retention in cartilage. In a rat osteoarthritis model, a single dendrimer-IGF-1 injection was more effective than free IGF-1, reducing cartilage degeneration by 60% and osteophyte burden by 80% | The development of PEGylated PAMAM dendrimer nanocarriers offers a targeted approach to deliver growth factors directly to chondrocytes, potentially leading to more effective treatments for osteoarthritis | [40] |

*(continued)*

## Table 12.3 (Continued)

| Growth Factors | Nanocarrier Type | Description | Practical Application | Ref. |
|---|---|---|---|---|
| VEGF, specifically the VEGF165 isoform, was used | NPs made from poly(D,L-lactide-co-glycolide) (PLGA), which is a biodegradable and biocompatible polymer commonly used for drug delivery applications. | PLGA NPs were specifically designed to encapsulate and deliver the VEGF165 plasmid DNA to cardiac myocytes for gene therapy in ischemic myocardium. Nanoparticles were successful in transferring the VEGF gene into heart cells, which led to an increase in blood vessel formation in the affected area | PLGA NPs could be a promising method for delivering genes directly to heart muscles to improve function and blood supply after damage<br>The technology could pave the way for advancements in regenerative medicine, particularly in cardiac care, by enhancing the body's natural healing process through increased capillary formation. | [45] |

### 12.3.1 Encapsulation of Proteins within Nanocarriers

Nanocarriers, such as liposomes, NPs, and micelles, can provide a protective environment for proteins. The encapsulation process involves trapping proteins within the core or surface of the nanocarrier through various techniques. Encapsulating proteins within nanocarriers involves several steps. The first step encompasses selecting the protein to be encapsulated and the material for the nanocarrier. Proteins can range from small peptides to large enzymes [46, 47]. The protein and nanocarrier materials are prepared in solution. This may involve dissolving the protein in an aqueous solution and preparing the nanocarrier material in a suitable solvent that will allow for encapsulation [48].

Then the actual encapsulation of the protein within the nanocarrier may include the following common methods; dissolvation, emulsification, self-assembly or gelatinization, and inverse mini-emulsion process [48, 49]. Crosslinking agents are then used to stabilize the nanocarrier structure [49]. After encapsulation, the NPs are purified and characterized to remove unencapsulated proteins and excess materials [50]. The surface of the nanocarriers may be modified or functionalized with targeting ligands or other molecules to enhance delivery to specific cells or tissues [51, 52]. The encapsulation process aids in targeted delivery, enhances stability, and controls the release of the protein.

#### *12.3.1.1 Controlling the Release of the Protein*

Nanocarriers can be modified with targeting ligands, such as antibodies, peptides, or small molecules that specifically recognize receptors or biomarkers on the surface of target cells or tissues. This approach ensures precise delivery of proteins to the intended site, reducing off-target effects and improving therapeutic efficacy [53]. Nanocarriers can also exploit the phenomenon of passive targeting, whereby they preferentially accumulate in certain tissues or organs due to factors such as size, shape, and surface charge, and release the protein gradually over time providing sustained and prolonged delivery. For example, nanocarriers with appropriate physicochemical properties may accumulate in tumor tissues via enhanced permeability and retention (EPR) effect, allowing for selective delivery of anticancer proteins [54].

#### *12.3.1.2 Enhanced Stability*

Encapsulating proteins within NPs enhances stability by providing a stable environment for the protein molecule. The nanocarriers shield proteins from degradation by enzymes, pH changes, and other environmental factors, thus extending their half-life and improving efficacy. Moreover, as protein molecules are within a confined space, there is a low likelihood of protein–protein interactions that can lead to aggregation. The lipid bilayer of liposomes of some of the nanocarriers can act as a barrier to oxygen and other reactive species.

#### *12.3.1.3 Versatility*

Encapsulation within nanocarriers offers versatility in protein delivery. This is because a wide range of proteins, including enzymes, antibodies, growth factors, and cytokines, can be encapsulated in nanocarriers.

The nanocarriers are tailored to suit specific physicochemical properties of different proteins, including size, charge, and hydrophobicity.

### 12.3.2 Surface Modification of Nanocarriers

Surface modification of nanocarriers is a powerful tool to enhance the delivery, targeting, and stability of protein-based therapeutics. The attachment of ligands such as antibodies, peptides, or aptamers enables specific binding to receptors or biomarkers on target cells [55]. Also, coating the surface of nanocarriers with hydrophilic polymers like polyethylene glycol (PEG) creates a stealth effect, reducing recognition and clearance by the immune system [56, 57]. PEGylation prolongs the circulation time of nanocarriers in the bloodstream, enhancing their distribution.

In addition, altering the surface charge of nanocarriers by attaching molecules or polymers influences interactions with biological components and cellular uptake. For instance, cationic surface modification promotes cellular internalization via electrostatic interactions, facilitating the delivery of proteins to target cells [58, 59]. Incorporating responsive coatings onto nanocarriers allows for stimuli-triggered release of encapsulated proteins. For example, pH-sensitive coatings dissolve in acidic environments, such as those found in tumor tissues, leading to the selective release of proteins at the target site [60]. Imaging agents, such as fluorescent dyes, radiolabels, or contrast agents, can be conjugated onto the surface of nanocarriers and this aids in real-time monitoring of the biodistribution of proteins [61]. Imaging functionalized nanocarriers facilitates the non-invasive visualization of protein delivery *in vivo* [62].

Lastly, coating nanocarriers with therapeutic enhancers such as cell-penetrating peptides or cell-targeting peptides enhances their ability to deliver proteins intracellularly [63]. These surface modifications improve the efficacy of protein-based therapies by facilitating their access to the desired cellular targets.

## 12.4 CHALLENGES AND FUTURE PERSPECTIVES IN PROTEIN DELIVERY USING NANOCARRIERS

The use of nanocarriers for protein delivery presents some significant challenges. Overcoming these hurdles and harnessing the potential of this technology could lead to effective protein-based therapies. Challenges that need to be addressed to fully realize the potentials of protein delivery include structural changes that may occur during the preparation of nanocarriers which could alter the therapeutic function of proteins [64]. Second, entrapment of the protein within endosomes of liposomes, exosomes, and polymeric NPs can lead to low therapeutic efficiency [65]. Moreover, the need to overcome natural barriers within the human body, such as the blood–brain barrier or the intestinal epithelium is a challenge [66]. In the context of the blood–brain barrier, effective protein delivery to the brain during the treatment of neurodegenerative disease is impaired. However, nanocarriers including exosomes and functionalized NPs have shown potential in improving neurological disorders, but further research is needed [67].

Some nanocarriers can trigger an immune response, leading to clearance from the body before reaching the target site. Nanocarrier material can lead to adverse reactions in humans [68]. Manufacturing nanocarriers for protein delivery often involve complex and expensive processes. Developing scalable and cost-efficient production methods is essential for making these therapies accessible to a wider population [69]. Additionally, the regulatory landscape gives stringent safety and efficacy requirements such that obtaining regulatory approval for nanocarriers for protein delivery systems can be a lengthy and expensive process [70]. Another challenge associated with protein delivery using nanocarriers is the need to ensure targeted delivery to specific cells or tissues [71], which requires the modification of protein nanocarriers with ligands.

Indeed, the future of this field lies in overcoming existing challenges through innovative research and technology integration, focused on improving protein delivery's efficiency, targeting, and safety. Plant protein-based nanocarriers are being investigated for their potential to enhance the stability, bioavailability, and controlled release of bioactive substances [72]. Tailored modifications of protein-loaded nanocarriers, such as liposomes, exosomes, polymeric NPs, and nanomotors, are being explored to enhance the clinical applications of protein therapies [64]. Another perspective is the use of modular proteinaceous self-assembly delivery systems that can load diverse antigen cargos to generate high-performance nano-vaccines [67]. The development of more effective and versatile nano-systems that can trigger endocytosis and endosomal disruption, or directly deliver proteins into the cytosol, is also a future perspective in protein delivery using nanocarriers [73].

Moreover, implementing artificial intelligence tools such as advanced computer simulations can help predict and optimize the design of nanocarriers [69]. Combining protein delivery with other therapeutic modalities, such as gene therapy or small-molecule drugs, could lead to synergistic effects and more effective treatments. There is also the need to address the biocompatibility and safety concerns associated with nanocarriers used in protein delivery. There is a need to bridge the gap between the high number of nanocarriers examined in animal studies and the few that have received Food and Drug Administration (FDA) approval for clinical use in humans.

## 12.5 CONCLUSION

Nanocarriers have shown significant potential for protein delivery, particularly in cancer therapy. The nanocarriers can be made from polymers, lipids, and inorganic substances, each with their own chemical, physical, and morphological characteristics. They can be engineered to have either prolonged half-lives or enhanced cellular uptake. Nanocarriers have been used to deliver hormones, cytokines, nucleic acids, vaccines, antibodies, enzymes, and cell-based therapies for a number of diseases. They offer several advantages over traditional delivery methods and future research should focus on improving their specificity, reducing their immunogenicity, and decreasing rapid enzymatic degradation. Indeed, nanocarriers hold great promise for the future of protein delivery in healthcare.

## REFERENCES


1. Roshani, P.G., G.D. Mehetre, S.J. Gandhi, and S.P. Wadichar, *A Systematic Review of Therapeutic Proteins as a Promising Approach to Treat Various Diseases.* GSC Biological and Pharmaceutical Sciences, 2023.
2. Redman, J.M., E.M. Hill, D. AlDeghaither, and L.M. Weiner, Mechanisms of action of therapeutic antibodies for cancer. *Molecular Immunology*, 2015, 67(2): pp. 28–45.
3. Ailing, F., R. Tang, J. Hardie, and M.E. Farkas, and V.M. Rotello, Promises and pitfalls of intracellular delivery of proteins. *Bioconjugate Chemistry*, 2014, 25(9): pp. 1602–1608.
4. Dimiter, S.D., Therapeutic proteins, in *Therapeutic Proteins: Methods and Protocols*, V. Voynov and J. A. Caravella, Editors. Springer, 2012, pp. 1–26.
5. Ilaria, P. and F. Cellesi, Intracellular delivery of therapeutic proteins. New advancements and future directions. *Frontiers in Bioengineering and Biotechnology*, 2023, 11: 1211798.
6. Jeffrey, L.C., L.D. Ann, and J.M. Randall, Emerging protein delivery methods. *Current Opinion in Biotechnology*, 2001, 12(2): pp. 212–219.
7. Liu, R., L. Cong, P. Zhiqing, Z. Jinming, R. Shaobo, W. Meiying, et al., Advances of nanoparticles as drug delivery systems for disease diagnosis and treatment. *Chinese Chemical Letters*, 2023, **34**(2): p. 107518.
8. Huang, L., S. Zonghao, S. Qian, H. Zhongxi, W. Shuangxi, Y. Naidi, et al., Rational design of nanocarriers for mitochondria-targeted drug delivery. *Chinese Chemical Letters*, 2022, **33**(9): pp. 4146–4156.
9. Huang, M., Y. Sun, and C. Tan, Recent advances in emerging pectin-derived nanocarriers for controlled delivery of bioactive compounds. *Food Hydrocolloids*, 2023, p. 108682.
10. Chandrakala, V., V. Aruna, and G. Angajala, Review on metal nanoparticles as nanocarriers: current challenges and perspectives in drug delivery systems. *Emergent Materials*, 2022, **5**(6): pp. 1593–1615.
11. Nothling, M.D., Z. Xiao, A. Bhaskaran, M.T. Blyth, C.W. Bennett, M.L. Coote, et al., Synthetic catalysts inspired by hydrolytic enzymes. *ACS Catalysis*, 2018, **9**(1): pp. 168–187.
12. El-Ansary, A. and S. Al-Daihan, On the toxicity of therapeutically used nanoparticles: an overview. *Journal of Toxicology*, 2009, **2009**.
13. Wang, Y., C. Wang, K. Li, X. Song, X. Yan, L. Yu, et al., Recent advances of nanomedicine-based strategies in diabetes and complications management: diagnostics, monitoring, and therapeutics. *Journal of Controlled Release*, 2021, **330**: pp. 618–640.

14. Cioni, P., E. Gabellieri, B. Campanini, S. Bettati, and S. Raboni, Use of exogenous enzymes in human therapy: approved drugs and potential applications. *Current Medicinal Chemistry*, 2022, **29**(3): pp. 411–452.

15. Bilal, M., S.A. Qamar, D. Carballares, A. Berenguer-Murcia, and R. Fernandez-Lafuente, Proteases immobilized on nanomaterials for biocatalytic, environmental and biomedical applications: advantages and drawbacks. *Biotechnology Advances*, 2023, 70: p. 108304.

16. Morozova, O.V., N.A. Barinov, and D.V. Klinov, Stable enzymatic nanoparticles from nucleases, proteases, lipase and antioxidant proteins with substrate-binding and catalytic properties. *International Journal of Molecular Sciences*, 2023, **24**(3): p. 3043.

17. De La Rica, R., D. Aili, and M.M. Stevens, Enzyme-responsive nanoparticles for drug release and diagnostics. *Advanced Drug Delivery Reviews*, 2012, **64**(11): pp. 967–978.

18. Sobczak, M., Enzyme-responsive hydrogels as potential drug delivery systems—state of knowledge and future prospects. *International Journal of Molecular Sciences*, 2022, **23**(8): p. 4421.

19. Krakor, E., I. Gessner, M. Wilhelm, V. Brune, J. Hohnsen, et al., Selective degradation of synthetic polymers through enzymes immobilized on nanocarriers. *MRS Communications*, 2021, **11**(3): pp. 363–371.

20. Athar, M. and A.J. Das, Therapeutic nanoparticles: state-of-the-art of nanomedicine. *Advanced Materials Reviews*, 2014, **1**(1): p. 25–37.

21. Danhier, F., O. Feron, and V. Préat, To exploit the tumor microenvironment: passive and active tumor targeting of nanocarriers for anti-cancer drug delivery. *Journal of Controlled Release*, 2010, **148**(2): pp. 135–146.

22. Vivo-Llorca, G., A. Morell-Aucejo, A. Garcia-Fernandez, P. Diez, A. Llopis-Lorente, M. Orzaez, et al., Horseradish peroxidase-functionalized gold nanoconjugates for breast cancer treatment based on enzyme prodrug therapy. *International Journal of Nanomedicine*, 2022, pp. 409–422.

23. Chiu, C.-F., R.-H. Fu, S.-H. Hsu, Y.H. Yu, S.-F. Yang, T.C.-Y. Tsao, et al., Delivery capacity and anticancer ability of the berberine-loaded gold nanoparticles to promote the apoptosis effect in breast cancer. *Cancers*, 2021, **13**(21): p. 5317.

24. Ashok, A., S.S. Andrabi, S. Mansoor, Y. Kuang, B.K. Kwon, and V. Labhasetwar, Antioxidant therapy in oxidative stress-induced neurodegenerative diseases: role of nanoparticle-based drug delivery systems in clinical translation. *Antioxidants*, 2022, **11**(2): p. 408.

25. Mauser, A., D.F. Quevedo, B. Zhang, Y. Hernandez, A. Berardi, W. Brown, et al., Enzyme-based synthetic protein nanoparticles as colloidal antioxidants. *Advanced Therapeutics*, 2023, **6**(11): p. 2300007.

26. Naqvi, S., V.A. Khanadeev, B.N. Khlebtsov, N.G. Khlebtsov, M.S. Deore, and G. Packirisamy, Albumin-based nanocarriers for the simultaneous delivery of antioxidant gene and phytochemical to combat oxidative stress. *Frontiers in Cell and Developmental Biology*, 2022, **10**: p. 846175.

27. Lee, H.J., H.H. Park, Y. Sohn, J. Ryu, J.H. Park, W.J. Rhee, et al., α-Galactosidase delivery using 30Kc19-human serum albumin nanoparticles for effective treatment of Fabry disease. *Applied Microbiology and Biotechnology*, 2016, **100**: pp. 10395–10402.

28. Niamsuphap, S., C. Fercher, S. Kumble, P. Huda, S.M. Mahler, and C.B. Howard, Targeting the undruggable: emerging technologies in antibody delivery against intracellular targets. *Expert Opinion on Drug Delivery*, 2020, **17**(9): pp. 1189–1211.

29. Hossen, S., M.K. Hossain, M.K. Basher, M.N.H. Mia, M.T. Rahman, and M.J. Uddin, Smart nanocarrier-based drug delivery systems for cancer therapy and toxicity studies: a review. *Journal of Advanced Research*, 2019, **15**: pp. 1–18.

30. Zhang, Q., J. Liang, A. Bongers, J.J. Richardson, K. Liang, and Z. Gu, Site-specific antibody assembly on nanoparticles via a versatile coating method for improved cell targeting. *Advanced Science*, 2023, **10**(9): p. 2206546.

31. Zhu, W., Y. Chao, Q. Jin, L. Chen, J.-J. Shen, J. Zhu, et al., Oral delivery of therapeutic antibodies with a transmucosal polymeric carrier. *ACS Nano*, 2023, **17**(5): pp. 4373–4386.

32. Jiang, M., B. Qin, X. Li, Y. Liu, G. Guan, and J. You, New advances in pharmaceutical strategies for sensitizing anti-PD-1 immunotherapy and clinical research. *Wiley Interdisciplinary Reviews: Nanomedicine and Nanobiotechnology*, 2023; **15**(1): p. e1837.

33. Gabel, M., A. Knauss, D. Fischer, M.F. Neurath, and B. Weigmann, Surface design options in polymer- and lipid-based siRNA nanoparticles using antibodies. *International Journal of Molecular Sciences*, 2022, **23**(22): p. 13929.

34. Le, H., C. Arnoult, E. De, D. Schapman, L. Galas, D. Le Cerf, et al., *Antibody conjugated nanocarriers for targeted antibiotic delivery: Application in the treatment of bacterial biofilm infections. Biomacromolecules*, 2021, 22(4): pp. 1639–1653. Available at SSRN 3737305.

35. Tashima, T., Delivery of drugs into cancer cells using antibody–drug conjugates based on receptor-mediated endocytosis and the enhanced permeability and retention effect. *Antibodies*, 2022, **11**(4): p. 78.

36. Werner, S. and R. Grose, Regulation of wound healing by growth factors and cytokines. *Physiological Reviews*, 2003, **83**(3): pp. 835–870.

37. Sanna, V., N. Pala, and M. Sechi, Targeted therapy using nanotechnology: focus on cancer. *International Journal of Nanomedicine*, 2014, pp. 467–483.

38. Gainza, G., D.C. Bonafonte, B. Moreno, J.J. Aguirre, F.B. Gutierrez, S. Villullas, et al., The topical administration of rhEGF-loaded nanostructured lipid carriers (rhEGF-NLC) improves healing in a porcine full-thickness excisional wound model. *Journal of Controlled Release*, 2015, **197**: pp. 41–47.

39. Lee, K., E.A. Silva, and D.J. Mooney, Growth factor delivery-based tissue engineering: general approaches and a review of recent developments. *Journal of the Royal Society Interface*, 2011, **8**(55): pp. 153–170.

40. Geiger, B.C., S. Wang, R.F. Padera Jr, A.J. Grodzinsky, and P.T. Hammond, Cartilage-penetrating nanocarriers improve delivery and efficacy of growth factor treatment of osteoarthritis. *Science Translational Medicine*, 2018, **10**(469): p. eaat8800.

41. Montazeri, S., A. Rastegari, Z. Mohammadi, M. Nazari, M. Yousefi, F.Y. Samadi, et al., Chitosan nanoparticle loaded by epidermal growth factor as a potential protein carrier for wound healing: In vitro and in vivo studies. *IET Nanobiotechnology*, 2023, **17**(3): pp. 204–211.

42. Hashemi, S., E. Mihandoost, and S. Khaleghi, Preparation of chitosan-embedded recombinant human epidermal growth factor nanoparticles as accelerating compounds for skin remodelling in chronic lesions. *Micro & Nano Letters*, 2023, **18**(7): p. e12171.

43. Choi, S.W., W. Sang, M.-H. Oak, E. Yi, H.J. Kim, and J.W. Park, Preparation and in vitro evaluation of elastic nanoliposomes for topical delivery of highly skin-permeable growth factors. *Journal of Nanoscience and Nanotechnology*, 2018, **18**(2): pp. 887–892.

44. Seyedebrahimi, R., S. Razavi, and J. Varshosaz, Controlled delivery of brain derived neurotrophic factor and gold-nanoparticles from chitosan/TPP nanoparticles for tissue engineering applications. *Journal of Cluster Science*, 2020, **31**: pp. 99–108.

45. Yi, F., H. Wu, and G.L. Jia, Formulation and characterization of poly (D,L-lactide-co-glycolide) nanoparticle containing vascular endothelial growth factor for gene delivery. *Journal of Clinical Pharmacy and Therapeutics*, 2006, **31**(1): pp. 43–48.

46. Shin, J., B.D. Cole, M. Seyedmohammad, S.I. Lim, and Y. Jang, Protein nanocarriers capable of encapsulating both hydrophobic and hydrophilic drugs. *Methods in Molecular Biology*, 2024, **2720**: pp. 143–150.

47. Saif, A., L. Anjum, Z. Faisal, N. Akram, Y.A. Shah, R. Irfan, et al., Recent advances in protein-based nanoparticles and their applications in the delivery of bioactive compounds. *International Journal of Food Properties*, 2023, **26**(2): pp. 2866–2880.

48. Miao, Y., T. Yang, S. Yang, M. Yang, and C. Mao, Protein nanoparticles directed cancer imaging and therapy. *Nano Convergence*, 2022, **9**(1): p. 2.

49. Hüppe, N., J. Schunke, M. Fichter, V. Mailander, F.R. Wurm, and K. Landfester, Multicomponent encapsulation into fully degradable protein nanocarriers via interfacial azide–alkyne click reaction in miniemulsion allows the co-delivery of immunotherapeutics. *Nanoscale Horizons*, 2022, **7**(8): pp. 908–915.

50. Liu, Y. and A. Cao, Encapsulating proteins in nanoparticles: batch by batch or one by one. *Methods Enzymology*, 2017, **590**: pp. 1–31.

51. Priya, S., V.M. Desai, and G. Singhvi, Surface modification of lipid-based nanocarriers: a potential approach to enhance targeted drug delivery. *ACS Omega*, 2022, **8**(1): pp. 74–86.

52. Narwade, M., A. Shaikh, K.R. Gajbhiye, P. Kesharwani, and V. Gajbhiye, Advanced cancer targeting using aptamer functionalized nanocarriers for site-specific cargo delivery. *Biomaterials Research*, 2023, **27**(1): p. 42.

53. Shah, A.A. and A. Gupta, Encapsulation of flavonoids in nanocarriers: a novel strategy to enhance their bioefficacy and oral bioavailability, in *Innovative Approaches for Nanobiotechnology in Healthcare Systems*, T. Amna and M.S. Hassan, Editors. IGI Global: Hershey, PA, 2022, pp. 267–283.

54. Ejigah, V., O. Owoseni, P. Bataille-Backer, O.D. Ogundipe, F.A. Fisusi, and S.K. Adesina, Approaches to improve macromolecule and nanoparticle accumulation in the tumor microenvironment by the enhanced permeability and retention effect. *Polymers*, 2022, **14**(13): p. 2601.

55. Mi, P., H. Cabral, and K. Kataoka, Ligand-installed nanocarriers toward precision therapy. *Advanced Materials*, 2020, **32**(13): p. 1902604.

56. Miao, Y., L. Li, Y. Wang, J. Wang, Y. Zhou, L. Guo, et al., Regulating protein corona on nanovesicles by glycosylated polyhydroxy polymer modification for efficient drug delivery. *Nature Communications*, 2024, **15**(1): p. 1159.

57. Sharma, S. and S. Dang, Nanocarrier-based drug delivery to brain: interventions of surface modification. *Current Neuropharmacology*, 2023, **21**(3): pp. 517–535.

58. Gonzalez Solveyra, E., D.H. Thompson, and I. Szleifer, Proteins adsorbing onto surface-modified nanoparticles: effect of surface curvature, pH, and the interplay of polymers and proteins acid–base equilibrium. *Polymers*, 2022, **14**(4): p. 739.

59. Salatin, S., S.M. Dizaj, and A.Y. Khosroushahi, Effect of the surface modification, size, and shape on cellular uptake of nanoparticles. *Cell Biology International*, 2015, **39**(8): pp. 881–890.

60. Karimi, M., M. Eslami, P. Sahandi-Zangabad, F. Mirab, N. Farajisafiloo, Z. Shafaei, et al., pH-sensitive stimulus-responsive nanocarriers for targeted delivery of therapeutic agents. *Wiley Interdisciplinary Reviews: Nanomedicine and Nanobiotechnology*, 2016, **8**(5): pp. 696–716.

61. Kunjachan, S., J. Ehling, G. Storm, F. Kiessling, and T. Lammers, Noninvasive imaging of nanomedicines and nanotheranostics: principles, progress, and prospects. *Chemical Reviews*, 2015, **115**(19): pp. 10907–10937.

62. Ju, J., D. Xu, X. Mo, J. Miao, L. Xu, G. Ge, et al., Multifunctional polysaccharide nanoprobes for biological imaging. *Carbohydrate Polymers*, 2023, p. 121048.

63. Anwar, S., F. Mir, and T. Yokota, Enhancing the effectiveness of oligonucleotide therapeutics using cell-penetrating peptide conjugation, chemical modification, and carrier-based delivery strategies. *Pharmaceutics*, 2023, **15**(4): p. 1130.

64. He, X., S. Xiong, Y. Sun, M. Zhong, N. Xiao, Z. Zhou, et al., Recent progress of rational modified nanocarriers for cytosolic protein delivery. *Pharmaceutics*, 2023, **15**(6), p. 1610.

65. Mishra, N., A. Garg, and N. Upmanyu, *Therapeutic Nanocarriers in Cancer Treatment: Challenges and Future Perspective*. Bentham Science Publishers, 2023.

66. Reinholz, J., K. Landfester, and V. Mailänder, The challenges of oral drug delivery via nanocarriers. *Drug Delivery*, 2018, **25**(1): pp. 1694–1705.

67. Pan, C., J. Ye, S. Zhang, X. Li, Y. Shi, Y. Guo, et al., Production of a promising modular proteinaceous self-assembled delivery system for vaccination. *Nanoscale*, 2023, 15(25); pp. 10794–10807.

68. Yu, M., J. Wu, J. Shi, and O.C. Farokhzad, Nanotechnology for protein delivery: overview and perspectives. *Journal of Control Release*, 2016, **240**: pp. 24–37.

69. Alshawwa, S.Z., A. A. Kassem, R. M. Farid, S. K. Mostafa, and G. S. Labib, Nanocarrier drug delivery systems: characterization, limitations, future perspectives and implementation of artificial intelligence. *Pharmaceutics*, 2022. **14**(4).

70. Li, H., Q. Luo, H. Zhang, X. Ma, Z. Gu, Q. Gong, et al., Nanomedicine embraces cancer radio-immunotherapy: mechanism, design, recent advances, and clinical translation. *Chemical Society Reviews*, 2023, 52(1): pp. 47–96.

71. Cheng, Z., Y. Li, D. Zhao, W. Zhao, M. Wu, W. Zhang, et al., Nanocarriers for intracellular co-delivery of proteins and small-molecule drugs for cancer therapy. *Frontiers in Bioengineering and Biotechnology*, 2022, **10**: p. 994655.

72. Wen, C., C. Wen, J. Zhang, H. Zhang, and Y. Duan, New perspective on natural plant protein-based nanocarriers for bioactive ingredients delivery. *Foods*, 2022, **11**(12): p. 1701.

73. Porello, I. and F. Cellesi, Intracellular delivery of therapeutic proteins. New advancements and future directions. *Frontiers in Bioengineering and Biotechnology*, 2023, **11**: p. 1211798.

# 13 Nanocarrier in Gene Therapy

*Sinjini Sarkar, Swarupananda Mukherjee, Shilpa Chatterjee, Biswajit Basu, Dipanjan Karati, and Bhupendra Prajapati*

## 13.1 INTRODUCTION

Gene therapy, a groundbreaking approach in modern medicine, holds immense promise for treating a myriad of genetic disorders, cancers, and infectious diseases by manipulating the expression of genes [1]. Central to its success is the efficient delivery of therapeutic genetic material into target cells. The most widely used form of gene therapy uses DNA encoding a therapeutic gene that functions to substitute the mutated gene [1–3]. Over the years, gene therapy has become one of the most promising biomedical and pharmacological interventions with proven clinical success in various areas, owing to scientific advances in genomics that have improved our understanding of the effects of genes on disease development and treatment [4].

Several gene transfer mechanisms, also known as gene vectors, have been created to provide safe, effective, and targeted gene transfer [5]. The perfect vector would be easy to insert into the cell body and specific to the target cells. It must have the capacity for constant and abundant expression. It must also be safe, exhibiting no negative effects, and able to be produced on a big scale. Gene therapy delivery vehicles, also known as carriers, contain therapeutic genes that are intended to be delivered to specific target cells [6]. These vehicles include liposomes, which are nonviral, and viral vectors, which are viruses with genetic disabilities, like adenovirus. Numerous approaches have been studied, both with and without viruses, and each has unique benefits and drawbacks. Retrovirus (RV), lentivirus (LV), adenovirus (AD), adeno-associated virus (AAV), and other viral carriers often have a wide species and tissue spectrum and simple preparation and delivery methods [7–10]. Long-term gene expression and great transduction efficiency are also displayed by them. Nevertheless, they have some disadvantages, such as significant immunogenicity, mutagenicity, and toxicity; low target cell selectivity; high cost; and unsuitability for large-gene transfer [11].

Despite appearing less effective than viral delivery systems, nonviral delivery vehicles like nanocarriers have inherent benefits including flexibility and safety [12]. Functional molecules like integrins, lectins, proteins, and antibodies have been added to nanoparticle (NP) complexes to create nanocarriers that are very stable, highly selective for specific cells, and low in toxicity [13–15]. NPs come in many different varieties that can be used as nanocarriers. These include polymeric NPs, dendrimers, inorganic or metal NPs, quantum dots (QDs), liposomes, and micelle [16–18]. NPs typically have a broad species and tissue tropism and are easy to prepare and administer. They also exhibit long-term gene expression and great transduction efficiency. Modern genomics technology has helped uncover genes linked to several inherited disorders and other non-genetic diseases, thereby indicating the enormous potential of gene therapy in novel medical domains. It is becoming increasingly obvious that the efficiency of gene delivery—which is based on the systems, delivery vehicles, or transfer techniques used—is a prerequisite for the success of gene therapy.

This review encapsulates the evolving landscape of nanocarriers in gene therapy, showcasing their potential to revolutionize the treatment of genetic disorders and pave the way for personalized medicine. By elucidating the intricacies of viral vector alternatives, nonviral gene editing approaches, and safety-efficacy dynamics, this comprehensive exploration aims to foster a deeper understanding of the opportunities and challenges in harnessing nanocarriers for gene therapy applications.

## 13.2 VIRAL VECTOR ALTERNATIVES

While the widespread use of viral vectors in gene therapy has led to the perception that they are safe for use even if they have the ability to infect humans, their destructive nature remains a significant safety risk that must be taken into consideration while creating gene therapies. Scientists have developed a few engineering techniques for this reason, with the goal of improving the safe application of viral vectors without sacrificing their effectiveness. These include preventing viral reproduction, encouraging viral inactivation, and reducing the inherent virulence of viruses [19]. Viral diversity in terms of size, morphology, genetic material type, and natural tropism is another characteristic that lends itself to gene delivery. This diversity provides more options to select from based on the specific needs of a given gene therapy. It is possible to classify viruses using a number of different criteria. A few of these are the existence of an envelope, the symmetry of the viral capsid, the type of genetic material the virus carries—DNA or RNA—the location of the virus's replication site—the cytoplasm or nucleus—and the size of the virion [20]. When selecting a particular viral vector for gene therapy, consideration is given to the previously mentioned viral classification criteria as well as the benefits and drawbacks of viral vectors to guarantee effective gene delivery. The section that follows has a discussion of the benefits and drawbacks of using viral vectors for gene delivery.

DOI: 10.1201/9781003473183-13

### 13.2.1 Advantages and Limitations of Viral Vectors

The US Food and Drug Administration (FDA) has approved several gene products based on viral vectors, supporting the promising use of viral vectors in gene therapy. Many more items that use viruses as delivery systems are undergoing various stages of clinical testing or the approval process. These viral vectors have the following benefits [21]:

1. They offer higher effectiveness gene transfer in *in vitro* and *in vivo* settings.
2. They usually endure for extended periods of time.
3. There is a wide selection of viruses accessible.
4. They possess an innate capacity of tropism toward contamination.
5. They are able to avoid endosomes through a variety of strategies acquired through viral evolution.

**Disadvantages**

1. These may cause inflammatory and highly sensitive reactions from the immune system.
2. They have a relatively small capability for cloning.
3. Their ability to attach to certain target cells is limited.
4. By introducing their exogenous DNA into the host genome, they can induce mutagenesis.
5. More research is needed to understand the molecular mechanisms of viral infection [21].

### 13.2.2 Nonviral Approaches as Alternative

Viral vectors can express genes sustainably, but they are challenging to use because of their arduous developmental process. Moreover, the hazardous characteristics of viral vectors, such as their ability to initiate an immunological reaction and cause insertional mutagenesis, have hindered their development and sparked several safety concerns over their use. For this reason, scientists have been searching for appropriate vectors based on nonviral systems that are notably benign and able to enable sustained gene expression without inducing undesired immunological and inflammatory responses. Nonviral vectors have become more important in gene delivery research due to their high loading capacity, versatility, affordability, and lack of immunogenicity [22]. Physical and chemical methods are the two primary categories into which nonviral vector types for gene delivery are typically classified. Chemical methods include numerous inorganic materials and polymer- and lipid-based systems, whereas physical methods include techniques like electroporation, sonoporation, magnetoporation, microinjection, needle injection, and gene gun. These chemical systems are primarily cationic. Chemical vectors can be generically categorized as lipid-based, polymer-based, peptide-based, or inorganic particles (Figure 13.1) [23].

#### *13.2.2.1 Lipid-Based NPs*

Lipid nanoparticles (LNPs) are widely employed in gene therapy to treat acquired and inherited illnesses as nonviral vectors. As LNPs do not contain immunogenic viral proteins, they are safer than viral vectors. LNPs have demonstrated a strong ability to condense and transfer a variety of nucleic acid molecules, ranging in size from several million nucleotides (chromosomes) to a few nucleotides (RNA), to cells. Because of several well-established production techniques, LNPs are very simple to scale up and can be quickly modified by adding targeted ligands. Lipidic vectors can be developed in three main approaches for appropriate gene transfection. The first method involves screening lipid libraries to find the best structure and biocompatible substance for a range of uses. For instance, a number of lipids with varied structures from the lipid library were effectively chosen and created in the study by Anderson et al. to increase the therapeutic efficacy for the treatment of a variety of acute and chronic illnesses [24–26].

LNPs must have sufficient stability to withstand prolonged circulation and build up at disease sites through the increased permeability and retention (EPR) effect to function as a systemic delivery carrier. Recent research has shown that chemically modified nucleic acids can improve stability by changing their physical characteristics in addition to functioning with lipid vectors. For example, Czauderna et al. [27] demonstrated that chemical modification of small interfering RNA (siRNA) at different sites can stabilize siRNA against serum-derived nucleases and lengthen the blood's circulation period without significantly reducing RNA interference activity.

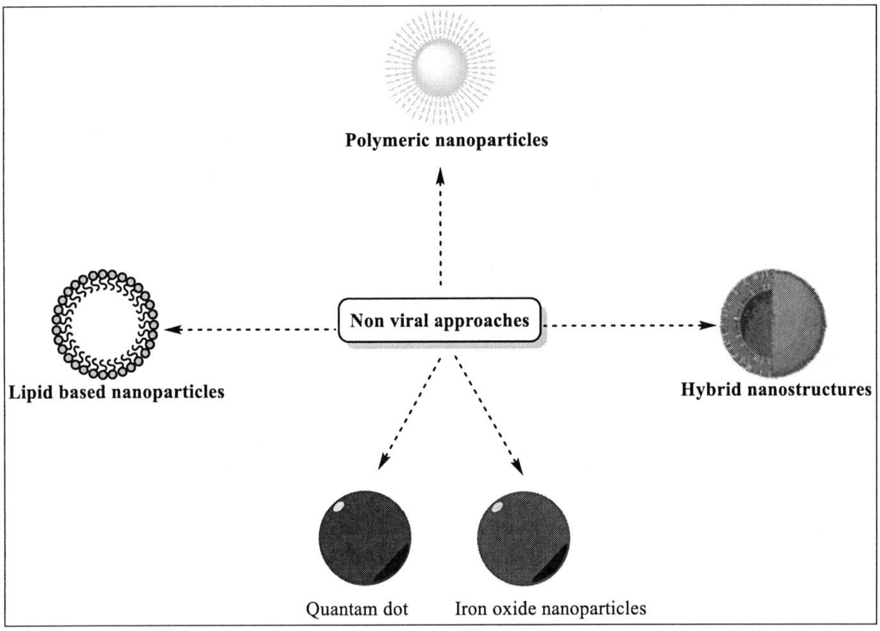


**Figure 13.1** Nonviral approaches for drug delivery.

#### *13.2.2.2 Polymeric NPs*

Biodegradable polymeric NPs have the potential to be safer alternatives to viruses for gene delivery [28]. The earliest class of nonviral methods were cationic lipid-based nonviral gene delivery systems [29]. By means of electrostatic forces, they combine with the negatively charged phosphate groups found in nucleic acids to generate lipoplexes, which are NPs. Lipoplexes can convey their genetic cargo inside of mammalian cells while shielding them from deterioration. This discovery created opportunities for nonviral vectors like polymers. Because of their easy manufacturing and malleable nature, cationic polymers are attractive options for nonviral gene transfer [30]. Gene transport can be facilitated by conjugating polymers with genetic material to produce polyplexes at physiological pH [31]. The efficacy of cationic polymers' gene transfection is influenced by a variety of parameters, including their structure, molecular weight, and surface charge. Synthetic polymers including poly(L-lysine) (PLL), poly(L-ornithine), diethylaminoethyl-dextran, poly(amidoamine) dendrimers, linear and branching polyethyleneimine, and poly(dimethylaminoethyl methacrylate) are excellent examples of polymers to be employed in gene therapy [32–36]. Research has also been done on sophisticated synthetic designs like guanidinylated bioreducible polymer and natural polymers including chitosan, dextran, and gelatin.

Effective gene transfection activity makes *PLL* one of the most researched gene carriers. PLL is a synthetic polypeptide that exhibits favorable biodegradability and biocompatibility, and that has a high charge density that is necessary for efficient pDNA condensation and complexation [37, 38].

Gene delivery technologies also primarily focus on *polyethylenimine (PEI)*. Because PEI's polyplexes (complexes of nucleic acid and polymer) have a relatively high transfection effectiveness, since its introduction in 1995 by Behr et al. [39], it has been regarded as the gold standard for polymer-based gene carriers [40–42]. The main factors contributing to PEI's success in gene transfection are its high charge density and the percentages of primary (25%) secondary (50%) and tertiary (25%) amines in it. Nitrogen is present in every third atom of PEI's branching structure, which gives it a high charge density capacity. At physiological pH 7.4, primary amines are protonated, allowing PEI to efficiently bind negatively charged genes form nanoscale complexes through electrostatic interactions. The proton-sponge effect refers to the ability of secondary and tertiary amines to trap protons in endosomes [43].

The water-soluble cationic polymer poly[2-(dimethylamino)ethyl methacrylate] (PDMAEMA) has been extensively employed as a gene transfer agent because of its intrinsic cationic charge, which can form dense complexes with negatively charged DNA via an electrostatic interaction [44–46]. With a precise, discrete molar mass and well-defined structures, PDMAEMA is easier to modify than PEI due to its monomer's ability to polymerize in distinct controlled ways. The positive charge of PDMAEMA, its capacity to form a complex with plasmid DNA, its facilitation of the proton-sponge effect, and its facile dissociation from DNA once transported into the cytosol are all factors contributing to the PDMAEMA polyplexes' effective *in vitro* transfection efficiency [44, 47].

The cationic characteristic of *chitosan* makes it a promising gene delivery vehicle. The primary amines in the chitosan backbone become positively charged at acidic pH, below the pKa. Through electrostatic contact, chitosan can bind to negatively charged DNA because of these protonated amines. Chitosan is mildly charged in neutral or alkaline environments, and the binding of chitosan to DNA may be caused by hydrogen bonding and hydrophobic interactions [48]. The first report of chitosan's potential as a gene carrier came from Mumper et al. [49]. One significant element influencing the effectiveness of gene transfer is the molecular weight of chitosan. High-molecular-weight (HMW) polymers have a greater capacity to ensnare DNA than their low-molecular-weight (LMW) counterparts, which is advantageous for safeguarding DNA within cellular endosomal/lysosomal compartments [50]. However, sometimes, this capacity limits the release of DNA within the cells, leading to either poor or delayed expression. On the other hand, complexes made of chitosans with LMW are not stable enough to offer efficient DNA protection because of their early dissociation, which results in little to no transgenic expression [51, 52]. Therefore, to produce high transfection efficiency, an intermediate degree of stability with chitosan with an acceptable molecular weight (MW) between extracellular DNA protection versus efficient intracellular DNA release must be accomplished [53, 54].

A class of synthetic, hydrolytically degradable polyamines known as *poly(amino-co-ester)s* (PAEs) were first created and thoroughly studied as cationic polymers for DNA transport [30]. The first study on poly(β-amino esters) (PBAEs) and gene transfer was carried out by Green et al. [55]. Primary or bis (secondary) aliphatic amines were added to diacrylate esters to create PBAEs. Because these biomaterials can condense plasmid DNA into stable, tiny NPs and encourage cellular uptake and endosomal escape, they hold promise for nonviral gene delivery.

Because of their decreased toxicity, water solubility, biodegradability, and biocompatibility, *poly(amidoamine)s (PAAs),* are commonly utilized for gene transfer. Lin and colleagues synthesized a series of unique polyacrylamide homo- and copolymers with distinct structures in the bisacrylamide segments and secondary and tertiary amine groups in the main chain. These PAAs were assessed as nonviral gene delivery vehicles. Polyplexes from PAAs with disulfide connections demonstrated significantly higher transfections than those from PAAs missing the disulfide linkages, according to transfection assays conducted with COS-7 cells [56].

#### *13.2.2.3 Inorganic NPs*

Since the discovery of scanning tunneling microscopy in the 1980s, which signaled the beginning of nanotechnology, human understanding of atom manipulation and the production of nanomaterials has advanced quickly [57]. Historically, nanomaterials have mostly been used as carriers of medicinal agents in the biomedical arena. To satisfy the needs of therapeutic loading capacity, biocompatibility, extended circulation time, and cellular absorption, inorganic NPs can be customized in size, shape, and elemental composition [58–61]. Sturdy inorganic nanocarriers shield therapeutic payloads against enzymatic and hydrolytic breakdown, including peptides, nucleic acids, and chemically unstable medications [62–64]. Furthermore, several inorganic NPs have special physical qualities that provide greater diversity in treatment approaches. Combinations of nucleic acids and other anticancer drug types can be loaded and delivered into nanocarriers without appreciably changing their size or shape, because of inorganic cores acting as a stiff support for polymeric coatings. Furthermore, because of their distinct magnetic and photoelectric capabilities, inorganic nanomaterials also make it possible for additional treatment modalities like hyperthermal ablation and enhanced radiation. Consequently, because of their inherent qualities that encourage the use of outside energy for therapeutic purposes as well as their capacity to transport multiple medicines at once, inorganic nanocarriers can facilitate combinatorial anticancer therapy [65]. Iron oxide ($Fe_3O_4$), silica ($SiO_2$), and noble metals (Au/Ag) are common inorganic carriers. Noble metal nanocarriers can be employed for photothermal therapy and photoacoustic imaging because they typically exhibit substantial absorption in the near-infrared (NIR) region of the electromagnetic spectrum. Iron oxide nanoparticles (IONPs) have the potential to be used in magnetic hyperthermia, contrast enhancement in magnetic resonance imaging, and magnetotransfection due to their superparamagnetism [58]. Mesoporous silica-based nanoparticles (MSNs) can be loaded with a variety of therapeutic medicines due to their pore diameters being able to be adjusted [66]. Because lanthanide ion-based nanomaterials (e.g., $Gd^{3+}$, $Yb^{3+}$, and $Er^{3+}$) can convert NIR radiation into higher-intensity visible or ultraviolet light, they have been used as dual therapeutic-diagnostic carriers [67]. Furthermore, to allow for multifunctionality, hybrid inorganic nanomaterials such as NaGdF4:$Yb^{3+}$, $Er^{3+}$/Au-MSN, and IONP/MSN were developed. These nanomaterials incorporate constituent elements from many types of nanocomposites [68, 69]. To create homogeneous, ultra-small nanomaterials (<20 nm), the previously stated nanomaterials are usually created via a bottom-up method in which precursor molecules and constituent atoms pass through growth control mechanisms in certain reaction conditions [70]. These nanomaterials are further endowed with the required biocompatibility, loading capacity, and targeting characteristics for the delivery of nucleic acids through further surface functionalization. The most researched metal NPs are gold nanoparticles (AuNPs) because of their superior physical and chemical characteristics [71], which include low cytotoxicity, high stability, and bioinertness. AuNPs are also simple to make and alter, making them ideal for delivering nucleic acids in gene therapy [72, 73]. Combining AuNPs-based therapy with currently available gene therapy has been acknowledged as a novel strategy with significant potential for treating cardiac disease [74].

#### *13.2.2.4 Hybrid Nanostructures*

In the field of gene delivery, nonviral vectors with low cytotoxicity and effective gene transfection have shown to be a double-edged sword. Modified PEI vectors, inorganic–lipid vectors, peptide–lipid vectors, and inorganic–organic hybrid vectors are a few of the current initiatives to increase transfection effectiveness while reducing cytotoxicity [75, 76].

Inorganic–organic hybrid vectors, which distribute genes to different cell types either through a platform or tailored manner, have been noted as a growing trend in the new class of non-viral vectors. Adding a mannitol group to the vector was one attempt. The usage of the multi-hydroxyl molecule mannitol as an osmolyte in clinical settings served as an inspiration for Ma et al. in 2020 [77]. They created biomimetic nonviral vectors that incorporated regulated cellular uptake and subsequent intracellular trafficking to facilitate gene delivery. Additionally, they introduced mannitol-based organic–inorganic nonviral vectors based on calcium phosphate mannitol-alendronate (CaP-MA) for the first time. The novel vectors containing mannitol groups have the potential to mimic the cellular update mediated by caveolae. They also carry the genetic payload through a non-destructive channel, so circumventing gene destruction within the lysosome. These vectors are thereby demonstrated to be more effective in transfection, biocompatibility, and toxicity compared to the unmodified CaP NPs.

The primary obstacle faced by PEI vectors is cytotoxicity, which arises from the existence of non-degradable amide bonds. These bonds cause buildup and impede metabolic function in normal cells. Furthermore, a positive correlation has been observed between the polymer molecular weight and the degree of cytotoxicity. Thus, the use of ligand-modified LMW PEIs, such as those containing folic acid or other functional peptides, offers hope for reducing toxicity [78, 79]. According to Yu et al. [79], low cytotoxicity and great transfection effectiveness were displayed by self-assembled nanoparticles (SNPs) made from cyclodextrin-grafted LMW PEI (CD-PEI800). They hypothesized that the increased gene encapsulation efficiency of SNPs was caused by the cationic hydrogel produced from CD-PEI800. In 2020, Steinman et al. [80] created the first trisaminocyclopropenium-crosslinked linear PEI (PEI/TAC) NP vector in an effort to reduce PEI toxicity. It was discovered that the DNA-containing vector particles were smaller than those made with the unaltered PEI, which had a favorable effect on cellular uptake. TAC-induced cross-linking raised the polymer's cationic charge, which made it possible for it to bind to the genetic material and encouraged endosomal escape, both of which improved transfection efficiency as a whole. Concurrently, the toxicity effects were further mitigated by reducing the amount of polymer added to the cell [80].

Another type of promising nonviral vectors for delivering nucleic acids into cells are peptide-based cationic lipids. Zhao et al. [81] created two peptide lipids with a tri-ornithine head (LOrn3) and a mono-ornithine head (LOrn1) and investigated the interaction kinetics of liposome-mediated gene transport. Hela and Hep-2 cells were exposed to green fluorescent protein (GFP) using both lipoplexes as vectors. The charge ratios of the peptide lipids had a major effect on the transfection efficiency. At charge ratios of 4:1 and 3:1, respectively, LOrn1 and LOrn3 exhibited the greatest transfection effectiveness and the least toxicity to Hela cells. LOrn3 included more amino groups than the other peptide-lipid vector, despite their comparable chemical structures, which resulted in both cellular internalization and a better proton buffering capability. Consequently, LOrn3 produced more effective transfection than LOrn1 and improved endosome escape and DNA release. The work offered theoretical guidance for future research avenues to enhance peptide–lipid vector technology.

## 13.3 NONVIRAL GENE EDITING APPROACHES

Gene editing basically involves deletion, insertion, and modifications in base pair sequences in the genome of eukaryotes and prokaryotes, by utilizing various target-specific DNA nucleases. Earliest gene editing had been performed in knock-out (KO) mice by homologous recombination (HR) [13]. However, this approach had its disadvantages like high cost to perform gene mutations/deletion, and extensive time. Thus, newer technology using programmable DNA-binding enzymes increased the efficiency for recombination and created a platform for novel gene editing techniques described below. These efficient techniques are able to model diseases, correct defective genes, and transfer traits, by regulating protein expression, and functionalize protein domains [82].

### 13.3.1 CRISPR/Cas9 and Other Genome Editing Tools

#### *13.3.1.1 CRISPR–Cas9*

##### *13.3.1.1.1 Biology of CRISPR-Cas9*

The clustered regularly interspaced short palindromic repeats (CRISPR)—CRISPR- associated protein 9 (*Cas9*) system has been repurposed to a fine genome editing technique from a bacterial defense mechanism against bacteriophage infections and plasmid transfer. This system is utilized by several bacteria and archaea which mainly involves CRISPR-associated (*Cas*) proteins, guide RNA, target DNA (infection), and protospacer adjacent motif (PAM). This RNA-guided immunization process is encoded by CRISPR loci (CRISPR array) along with *Cas* genes. Upon invasion of foreign DNA, the genetic elements are integrated into the host genome within the CRISPR repeat-spacer arrays as new spacers. This enables the organism to record an infection genetically and prepare to prevent future invasion by the same invader. The CRISPR array in the genome is processed by transcription followed by endonucleolytic cleavage of a long precursor-CRISPR transcripts to form mature, short CRISPR RNAs (crRNAs). The crRNAs contain a spacer at the 5′ end, an RNA segment complementing the foreign genetic sequence and a CRISPR repeat sequence at the 3′ end. In the advent of a second infection, hybridization between the crRNA spacer and foreign complementary target sequence/protospacer triggers the destruction of invading DNA or RNA by employing Cas nucleases forming a ribonucleoprotein (RNP) complex. PAM is a short conserved sequence motif (2–5 bp) located in close proximity to the crRNA-targeted sequence on the invading DNA [79].

CRISPR systems are categorized into six types (I–VI), with specific Cas proteins in each and crRNA for interference. The type II systems use a single DNA endonuclease, i.e., Cas9 to recognize and cleave double-stranded DNS (dsDNA). During silencing, Cas9 is directed by a paired tracrRNA and crRNA called the dual-RNA guide, to cleave the 20-nucleotide DNA matching with the target sequence beside the PAM. TracrRNA is required for crRNA maturation in type II systems. But in the case of application of this mechanism, a simplified single chimeric sgRNA works for both crRNA and tracrRNA combined, maintaining the functionality of Cas9. This system can be tailor-made for any desired DNA sequence through modifying guide RNA sequence, triggered by a double-strand break (DSB). Subsequently, these breaks are repaired either by nonhomologous end joining (NHEJ), resulting in random insertions or deletions, or through accurate homology-directed repair (HDR) using a repair template (Figure 13.2) [83].

##### *13.3.1.1.2 CRISPR-Cas9 in Drug Delivery and Treatment*

This phenomenal gene editing technique has the potential to treat a wide range of genetic disorders including cystic fibrosis, muscular dystrophy, Duchenne muscular dystrophy, sickle cell disease, thalassemia, hemoglobinopathies, hemophilia, Parkinson's disease, amyotrophic lateral sclerosis (ALS), Huntington's disease, retinitis pigmentosa, multiple sclerosis and rheumatoid arthritis, familial cardiovascular diseases, infections (HIV, hepatitis etc.), and different cancers [84]. Researchers successfully corrected the mutation at the cystic fibrosis transmembrane conductance regulator (CFTR) locus in intestinal stem cells from two cystic fibrosis patients, and restored dystrophin protein production in *DMD* in patient-derived pluripotent stem cells by CRISPR [85, 86]. *In vitro* studies revealed that modification of chemokine co-receptor type-5 (*CCR5*) via CRISPR genes in host cells prevented HIV infection with no cellular toxicity [87]. CRISPR-Cas9 was used in human studies first to engineer T-cells from lung cancer patients to delete genes like TRAC, TRBC, and *PD-1*. These T-cells showed no adverse effects and were detected even 10 months post infusion [88]. However, challenges in selecting the drug delivery systems have made it difficult to administer CRISPR systems in clinical settings and are yet restricted in the primary cells and *in vivo* applications [72].

Biomolecular therapeutic agents such as DNAs and RNAs are susceptible to enzymatic degradation in the biological fluids and thus pose a challenge in their delivery. NPs, nanocarriers, or nanoformulation shells shield the CRISPR-Cas complex/cargo from immunological clearance, renal excretion, aggregation, and premature release and also aid in targeting by coating with specific ligands. Lipid-based NPs, specifically liposomes have emerged to be a promising carrier for genetic material and editing tools. Some commercially available LNPs are lipofectamine, RNAiMAX, StemfectTM, and TurboFectTM which have been used for messenger RNA (mRNA), siRNA, gRNA, CRISPR plasmids, and Cas9-sgRNA RNP deliveries.

There are several other genome editing tools besides CRISPR/Cas9. Some of them include the following.

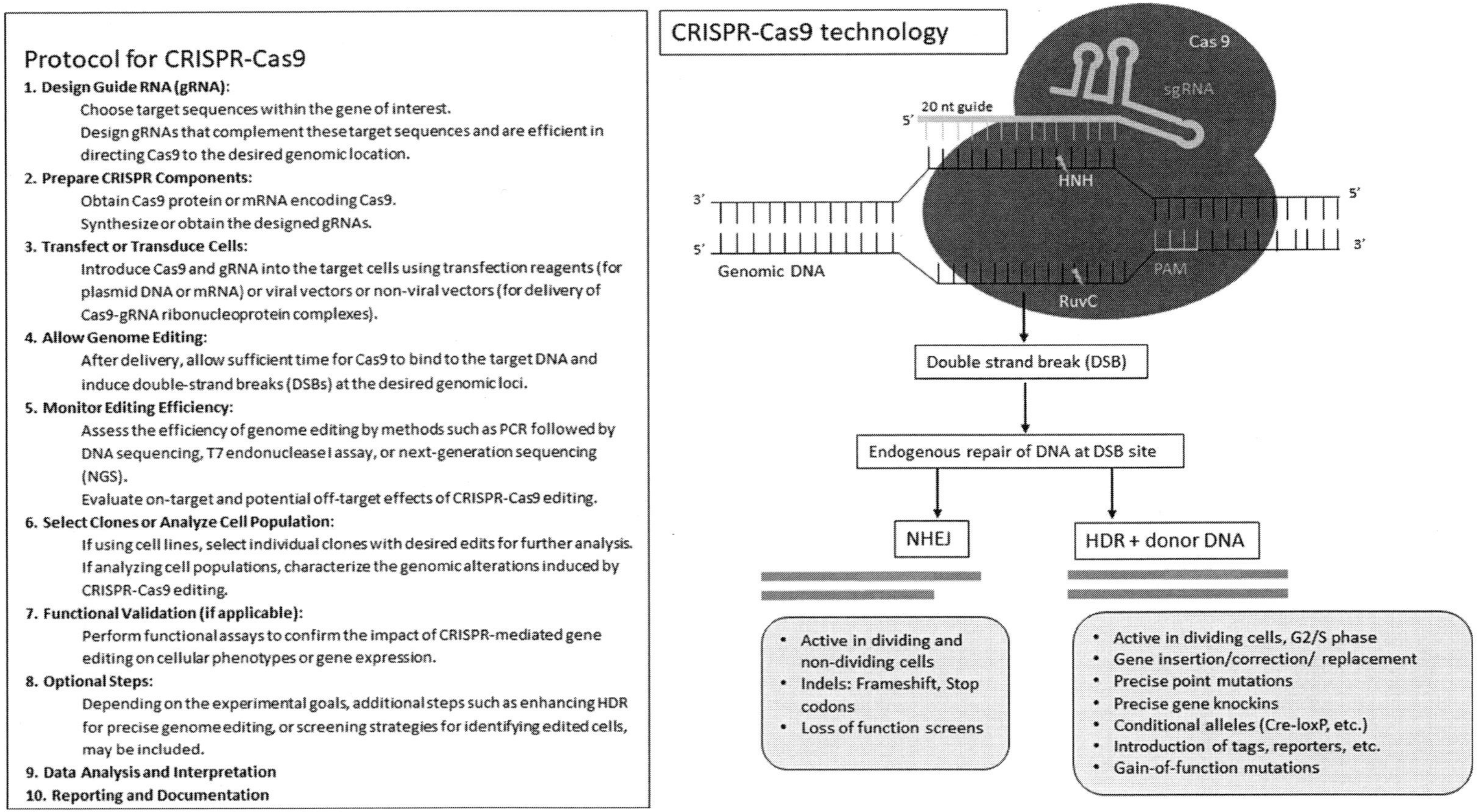


**Figure 13.2** Basic mode of CRISPR-Cas9 action for genome editing with a simple framework of the protocol.

#### *13.3.1.2 Zinc Finger Nucleases*

Zinc finger nucleases (ZFNs) are engineered chimeric proteins for targeted DNA recognition, breaks, and modifications. ZFNs use their DNA-binding domains of 3–6 zinc finger repeats to recognize 9–18 base pairs of the targeted genome and introduce DSB at the site with FoKI nuclease. Subsequently, the NHEJ and HDR perform the necessary modifications, i.e., gene corrections, KO, and insertions.

The versatility and precision have led to the widespread application of ZFN-mediated genome editing in various fields of biotechnological research, like agriculture (increased crop yield and resistance to pests) and human diseases. In biomedicine, ZFNs hold promise for the treatment of genetic disorders as scientists have successfully edited the genes in human somatic cells and pluripotent stem cells. While using a dimerized FokI, the off-target effects are reduced, achieving a high-affinity chain by assembling tandem zinc fingers is a challenge and an inherent drawback. However, advancements in ZFN engineering ensure a bright future for precision genome editing (Figure 13.3) [89–91].

#### *13.3.1.3 Transcription Activator-Like Effector Nucleases*

Transcription activator-like effector nucleases (TALENs) are also engineered proteins that use transcription activator-like effector (TALE) domains for DNA recognition. They have enhanced specificity and efficiency in targeted gene modification than its predecessor technique with ZFNs. They are regarded as molecular scissors derived from plant pathogens of *Xanthomonas* sp., engineered to induce specific gene editing. The nuclease domain, typically derived from the FokI restriction enzyme, enables the cleavage of DNA at the target site and subsequent NHEJ and HDR modifications. TALENs have the ability to recognize a single base as compared to triplets of ZFNs and thus have better flexibility. This is a versatile technique that can be modified in several ways by including Cas9 systems, site-specific recombinases and transcriptional activators. Novel laboratory techniques are developed to create an assembly of custom TALEN arrays, like commercial DNA synthesis, fast ligation-based automatable solid-phase high-throughput (FLASH), and iterative capped assembly (ICA) (Figure 13.3) [92–95].

### 13.3.2 Delivery Challenges of Gene Editing Tools

Unwanted mutagenesis is an all-pervasive risk while using gene editing tools unless delivered efficiently. There are quite a few delivery challenges to effective therapy with CRISPR-Cas9, ZFNs, and TALENs described below.

**a. Cell type specificity and off-target effects**

It is critical to deliver Cas9 RNPs in the cytosol for target specificity devoid of unintended off-target genetic modifications. The whole genome sequencing (WGS), unbiased identification of DSBs enabled

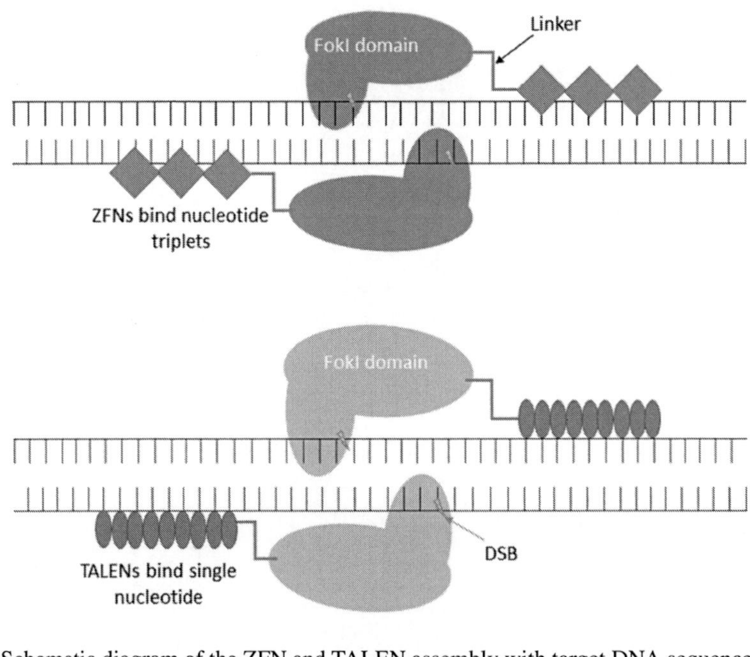


**Figure 13.3** Schematic diagram of the ZFN and TALEN assembly with target DNA sequence.

by sequencing (GUIDE-seq), discovery of in situ Cas off-targets, CHyMErA (Cas hybrid for multiplexed editing and screening applications), bioinformatics and verification by sequencing (DISCOVER-Seq), online prediction software, etc. are some of the current strategies used to predict and detect off-target effects [96].

**b. Efficiency**
The delivery method must efficiently transport the editing tool into the target cells to ensure a sufficiently high editing efficiency. Low efficiency may result in incomplete editing or require higher doses, increasing the risk of off-target effects. CRISPR-Cas9 has particularly higher efficiency than ZFNs, TALENs, and meganucleases. NHEJ-mediated DSB repair can promote high gene disruptions but can be error-prone, and hence improvement of the efficiency of HDR is important. As HDR is only active in dividing cells, this makes gene inactivation more difficult than gene insertion. This has been tried to be resolved by timed- delivery of Cas9-gRNP (guide RNA RNP) complexes that were used with high efficiency of HDR to induce site-specific double-strand breaks and new genetic information. NHEJ inhibition by small molecules and targeting DNA ligase IV has been employed for increasing efficiency [91]. Likewise, the use of 53BP1 as a genetically encoded inhibitor led to a significant enhancement in HDR-dependent genome editing efficiency, up to 5.6 times, by reducing NHEJ activity in human and mouse cells. Additionally, a substantial increase in CRISPR/Cas9- and TALEN-mediated knock-in efficiency, both *in vitro* and *in vivo*, by employing an HDR enhancer, RS-1. Interestingly, the NHEJ inhibitor SCR7 showed minimal effects in comparison [97].

**c. Immunogenicity**
Some delivery vectors or methods may trigger immune responses in the host organism, leading to clearance of the editing tool or adverse reactions. Minimizing immunogenicity is essential for successful gene editing. Cas9 proteins are mostly obtained from *Staphylococcus aureus* (SaCas9) and *Streptococcus pyogenes* (SpCas9). The human immune system recognizes these proteins as foreign because of previous infections and, therefore, develops an immune response against them upon injection, which leads to fast degradation of the Cas9 protein. This prevents the intended gene-editing function. Two ways to tackle this are: (1) implementing the CRISPR/Cas system for gene editing early in a lifetime; and (2) targeting immune-privileged organs [98].

**d. Integration vs. Non-Integration**
To avoid potential genomic instability or unwanted mutations, the editing tool may need to integrate into the host genome (as with some viral vectors) or remain episomal (non-integrating). Orip/EBNA1 is a post-synthesis function, derived from Epstein–Barr virus, and was combined with CRISPR for episomal delivery of a gene editing tool that autonomously replicated genes in eukaryotic cells without integration [99].

e. **Scale-up and Manufacturing**
Cost-effectiveness and scaling-up production of formulations incorporated with genome editing tools is a huge task especially when manufacturing for clinical applications. It has similar challenges as other ex-vivo cell-based therapies. The challenges also depend on the therapy being autologous (the patient becomes their own donor) or allogenic (the patient will be administered cellular components from a different donor). Personalization for autologous therapies is difficult because of intra-patient heterogeneity which requires a flexible cell culture platform and a scale-out approach rather than a scale-up. Allogenic therapies are similar to biologics and scaling up batch sizes is challenging due to the limited availability of equipment, reagents, and materials [100]. Contract development and manufacturing organizations (CDMO) have shown interest in this regard and are collaborating with biotech and pharma companies to improve product life cycles through developing sensors, software, and automation for better therapeutic outcomes.

LNPs are great alternatives to viral vectors like AAV due to their capacity to carry larger RNP cargos and their manufacturing does not depend on mammalian cell culture for production. With the advent of mRNA COVID-19 vaccines using LNP formulations, it is predicted that they will be preferred for gene editing tools in near future with improved versatility [101].

**f. Ethical considerations**
The challenges to the therapeutic delivery of these technologies pose immense potential in mitigating debilitating diseases but come with several ethical considerations and challenges. On one hand, the drawbacks like mosaicism (incomplete editing), and inaccurate target efficiency; on the other hand, concerns about transformed species' long-term survival, their vertical transmission through generations, and the complicated effects on the biological phenotypes [102].

#### *13.3.2.1 Encapsulation Strategies*

Encapsulation strategies involve packaging gene editing tools, such as CRISPR/Cas9 components or other editing agents, within delivery vehicles to enhance their stability, target specificity, and delivery efficiency. Here are some common encapsulation strategies:

1. *Nonviral Vectors:* Nonviral delivery systems, including liposomes, polymers, and NPs, can encapsulate gene editing components and deliver them into cells. LNPs and polymer-based NPs offer advantages such as low immunogenicity and ease of manufacturing. They can be modified with targeting ligands to enhance specificity.
2. *Exosomes:* These are naturally occurring extracellular vesicles secreted by cells, and they can be engineered to encapsulate gene editing components. Exosomes offer advantages such as biocompatibility, low immunogenicity, and the ability to cross biological barriers. They can also be modified to target specific cell types. Wan et al. (2022) precisely transported RNP complex inside exosomes derived from hepatic stellate cells [103]. Drawbacks of exosomal delivery includes variation in constitution and diameters [39, 104].
3. *Cell-Penetrating Peptides (CPPs):* CPPs, also known as protein transduction domains (PTDs) that can facilitate the delivery of cargo molecules, including gene editing tools, across cell membranes were first identified ~25 years ago. They are small, synthetic, or naturally occurring peptides, 6–30 amino acid long, have less cellular toxicity, and induce fewer off-target mutations [75, 82, 105]. Protocols can be modified/optimized to deliver to even hard-to-transfect cell lines [106]. Several hurdles exist to the clinical use of CPP mainly due to efficacy concerns, decreased cell specificity toward cationic and hydrophobic CPPs, immunogenicity, and lack of oral bioavailability [107].
4. *Hydrogels:* Hydrogels are three-dimensional networks of hydrophilic polymers that can encapsulate gene editing components and provide sustained release upon injection into target tissues. Hydrogels offer advantages such as biocompatibility, tunable properties, and the ability to localize the delivery of gene editing tools.
5. *Electroporation and Microinjection:* These physical methods involve the direct delivery of gene editing components into cells using electric pulses (electroporation) or microinjection techniques. While these methods offer high delivery efficiency, they are typically limited to *ex vivo* applications or specific cell types.
6. *Chemical Modification:* Gene editing components can be chemically modified to enhance their stability and cellular uptake. For example, chemical modifications of gRNA, mRNA, for base and prime editors, changes in the nucleic acid chemistry adjusts hydrophobicity, alter target binding affinity, improves nuclease resistance and reduce the immune response [100, 108].

These encapsulation strategies can be tailored to specific applications and target cell types to optimize the delivery of gene editing tools while minimizing off-target effects and immunogenicity.

##### *13.3.2.2 Targeting Specific Cells and Tissues*

Targeting specific tissues or organs presents additional challenges, especially if they are protected by barriers such as the blood–brain barrier or the placental barrier. Addressing these challenges often involves a combination of optimizing existing delivery methods (e.g., viral vectors, NPs, liposomes) and developing novel approaches tailored to specific applications and target cell types/barriers. Some delivery methods may have limitations in achieving cell type specificity. Controlling the size, uniformity, and stability of LNPs poses significant challenges, particularly for local administration *in vivo* to areas like the inner ear, muscle, brain, and tumors. Currently, creating stable NPs for the systemic delivery of RNPs to target organs remains a formidable task [100].

Here are some approaches to achieving cell and tissue specificity:

*Cell Surface Receptor Targeting and Tissue-Specific Promoters:* Modify the delivery vehicle (e.g., viral vectors, NPs) coated with ligands that bind to specific cell surface receptors to facilitate receptor-mediated endocytosis, and enhance uptake into the desired cells.

Use of tissue-specific promoters to drive the expression of gene editing tools within target cells is also achievable. An innovative approach is the selective organ targeting (SORT) system which is LNP-based and adsorbs proteins from the plasma which mediates the uptake in specific cells or tissues [101].

*Biomaterial-Based Delivery:* Hydrogels, smart hydrogels, and other biomaterial-based delivery systems have the potential to deliver gene editing tools to targeted tissues with the advantage of being non-invasive and can be remote-controlled [109].

*Extracellular Matrix Targeting:* Modified hydrogels that interact with components of the extracellular matrix (ECM) as cell-binding ligands can facilitate cell-targeted delivery and improve therapeutic efficiency [110].

*Cellular Targeting and In Vivo Selection:* Employ *in vivo* selection strategies like the addition of antibodies, peptides, or aptamer and synthetic nucleocapsids to enrich for cells that have taken up the gene editing tools can prove to be beneficial in a complex pathological microenvironment [111].

*Physical Targeting:* These methods include the use of ultrasound, magnetic guidance, hydrodynamic gene transfer, Elevated temperature, laser, microinjection, bolistic and electroporation, and innovative microfluidic devices to facilitate the targeted delivery of gene editing tools to specific tissues or organs [112].

*Local Administration:* Formulated gene editing tools can be administered via intratumoral injection, intravitreal injection, or intrathecal injection to reduce systemic exposure and target organ-specific delivery.

## 13.4 SAFETY CONSIDERATIONS

Gene therapy offers great potential for the treatment of several inherited medical conditions by introducing therapeutic genes into the targeted cells. The effective delivery of these genes to the targeted cells is still a major obstacle, though. Because they can prevent genetic material from degrading, increase cellular uptake, and enable targeted distribution, nanocarriers such as liposomes, polymeric NPs, and viral vectors have been thoroughly investigated as delivery vehicles. Despite their potential, safety concerns surrounding nanocarrier-mediated gene therapy persist, warranting a comprehensive safety assessment. Safety assessment of nanocarriers in gene therapy is crucial due to the potential risks associated with their use. Here are some key aspects typically considered in safety assessments (Figure 13.4).

### 13.4.1 Biocompatibility

One of the primary considerations in the safety assessment of nanocarriers is their biocompatibility. Biocompatible nanocarriers should exhibit minimal adverse effects on biological systems, including cells, tissues, and organs. Various factors influence the biocompatibility of nanocarriers, including their physicochemical properties, surface modifications, and interactions with biological components [113, 114]. Assessing biocompatibility involves evaluating cytotoxicity, hemocompatibility, and interactions with the immune system [115]. Strategies to enhance the biocompatibility of nanocarriers include surface functionalization with biocompatible polymers, such as polyethylene glycol (PEG), and optimization of particle size and surface charge [115, 116].

### 13.4.2 Stability

Nanocarriers should remain stable in biological environments to ensure the effective delivery of genetic material without degradation or premature release. Recent advances in nanotechnology have led to the

**Figure 13.4** Comprehensive safety assessment parameters of nanocarriers.

development of nanocarriers with enhanced stability profiles, overcoming previous limitations such as premature cargo release, aggregation, and degradation [116]. The improvement of nanocarrier stability includes surface modification with biocompatible polymers (e.g., PEGylation), incorporation of stabilizing agents (e.g., lipids, surfactants), and engineering of stimuli-responsive systems for controlled release [115, 117]. These approaches have demonstrated promising results in preclinical studies, showcasing prolonged circulation times, reduced immunogenicity, and enhanced therapeutic efficacy.

##### *13.4.2.1 Challenges and Limitations*

Despite significant progress, several challenges persist in ensuring the stability of nanocarriers for gene therapy. One major concern is the potential for off-target effects and cytotoxicity associated with non-specific interactions between nanocarriers and biological components. Additionally, the complex physiological environment, including enzymatic degradation and immune responses, poses challenges to maintaining nanocarrier stability *in vivo*. Furthermore, the scalability and reproducibility of manufacturing processes remain key hurdles in translating nanocarrier-based gene therapies from bench to bedside.

Addressing the challenges in nanocarrier stability requires interdisciplinary efforts combining expertise in nanotechnology, molecular biology, pharmacology, and immunology. Future research directions may focus on developing advanced nanocarrier formulations with tunable properties for improved stability, biocompatibility, and targeting specificity. Furthermore, innovative strategies for real-time monitoring of nanocarrier behavior *in vivo* could provide valuable insights into their stability and performance [118]. Moreover, the integration of emerging technologies such as CRISPR-based genome editing and mRNA therapeutics holds promise for advancing the field of gene therapy with stable nanocarriers [118].

### 13.4.3 Targeting and Specificity

Nanocarriers should possess targeting capabilities to deliver genes specifically to the desired cells or tissues, minimizing off-target effects.

Nanocarriers can be engineered to achieve targeting and specificity through various strategies, including ligand-based targeting, stimuli-responsive systems, and tissue-specific delivery [119]. Ligand-mediated targeting involves the conjugation of specific ligands (e.g., antibodies, peptides, aptamers) to the surface of nanocarriers, enabling recognition and binding to target receptors expressed on the surface of target cells [119]. Stimuli-responsive nanocarriers exploit endogenous or exogenous stimuli (e.g., pH, temperature, enzymes) to trigger the controlled release of therapeutic cargo within the target site. Tissue-specific targeting aims to exploit unique physiological characteristics of diseased tissues (e.g., tumor microenvironment) for selective accumulation of nanocarriers.

### 13.4.4 Advancements and Applications

Recent advancements in nanotechnology have enabled the development of highly sophisticated nanocarriers with enhanced targeting capabilities for gene therapy applications. For instance, the use of targeted liposomes decorated with tumor-specific ligands has demonstrated improved accumulation and uptake in cancer cells, leading to enhanced therapeutic efficacy and reduced systemic toxicity [120]. Similarly, the development of smart nanocarriers responsive to tumor-specific microenvironmental cues (e.g., acidic pH, elevated enzyme levels) has facilitated site-specific release of therapeutic genes, enhanced treatment precision, and minimizing off-target effects [120]. Furthermore, innovative strategies such as cell-specific targeting and homing approaches hold promise for further enhancing the specificity of nanocarriers in gene therapy.

### 13.4.5 Delivery Efficiency

Nanocarriers play a pivotal role in gene therapy by facilitating the efficient delivery of nucleic acid payloads to desired cellular targets [71, 87].

The design of nanocarriers significantly influences their gene delivery efficiency. Factors such as NP size, shape, surface charge, and composition play crucial roles in determining cellular uptake, intracellular trafficking, and gene expression levels. Engineered NPs can be tailored to overcome biological barriers, enhance stability, and improve interactions with target cells, thereby enhancing gene delivery efficiency [121].

Moreover, surface modification of nanocarriers offers a versatile approach to improve gene delivery efficiency. Functionalization with targeting ligands, such as antibodies, peptides, or aptamers, enables specific recognition and binding to cell surface receptors, enhancing cellular uptake and intracellular trafficking of nanocarriers [122]. Additionally, surface PEGylation can confer stealth properties, prolonging circulation time and enhancing systemic biodistribution while reducing nonspecific interactions and immune responses.

In addition, targeted delivery of nanocarriers to specific cell types or tissues is essential for enhancing gene delivery efficiency and minimizing off-target effects. Active targeting strategies exploit overexpressed receptors or biomarkers on target cells, enabling selective accumulation and internalization of nanocarriers.

Passive targeting approaches leverage physiological characteristics, such as the enhanced permeability and retention effect, to achieve preferential accumulation in diseased tissues, further enhancing gene delivery efficiency [77, 108].

While significant progress has been made in improving nanocarrier-mediated gene delivery efficiency in preclinical studies, several challenges remain for clinical translation. Addressing concerns related to safety, scalability, and regulatory compliance is crucial for advancing nanocarrier-based gene therapies toward clinical applications. Future research directions may focus on integrating multimodal imaging and therapeutic functionalities into nanocarrier designs, enabling real-time monitoring of gene delivery and therapeutic responses. Assessment of the efficiency of gene delivery is essential to ensure that an adequate amount of genetic material reaches the target cells to achieve the desired therapeutic effect [123].

### 13.4.6 Immunogenicity

In the rapidly evolving landscape of gene therapy, nanocarriers have emerged as promising vehicles for delivering therapeutic genes to target cells. However, the immunogenicity of these nanocarriers remains a significant concern, potentially impacting their efficacy and safety. Nanocarriers, such as liposomes, polymeric NPs, and viral vectors, possess inherent properties that can trigger immune responses upon administration. These responses may range from mild inflammation to severe immune reactions, limiting the therapeutic efficacy of gene delivery systems [124].

Several factors influence the immunogenicity of nanocarriers, including their physicochemical properties, surface characteristics, and route of administration. Additionally, the presence of foreign antigens, such as viral proteins in viral vectors, can stimulate immune recognition and activation [116]. Furthermore, the choice of biomaterials and coating strategies can modulate the interaction of nanocarriers with the immune system, affecting their immunogenicity profile.

The immunogenicity of nanocarriers poses challenges for their clinical translation in gene therapy. Immune responses may lead to clearance of the therapeutic payload, reduced transfection efficiency, or adverse effects such as cytokine release syndrome. Therefore, strategies to mitigate immunogenicity are essential to enhance the safety and efficacy of nanocarrier-based gene delivery systems in clinical settings [125].

Several approaches can be employed to reduce the immunogenicity of nanocarriers, including surface modification with stealth polymers, immunomodulatory agents, or masking antigens. Rational design of nanocarriers can also minimize immune recognition while maximizing target specificity and payload delivery. Furthermore, optimizing the route of administration and dosing regimen can help minimize immune responses and improve therapeutic outcomes [125].

Multidisciplinary approaches integrating immunology, nanotechnology, and molecular biology will be instrumental in advancing the field toward safer and more efficacious gene delivery platforms. By addressing immunogenicity concerns, nanocarriers hold great potential to revolutionize gene therapy and address unmet medical needs in various disease settings.

### 13.4.7 Off-Target Effects

Evaluation of potential off-target effects is necessary to minimize unintended consequences of gene delivery, such as unintended gene expression or interference with normal cellular functions. Off-target effects in gene therapy refer to unintended interactions or consequences resulting from the delivery of genetic material to cells, tissues, or organs. When nanocarriers are utilized to deliver therapeutic genes, they can introduce several off-target effects such as, non-specific binding, immune response, insertional mutagenesis, off-target gene expression, systemic distribution, cytotoxicity, and dose-dependent effects [126].

The occurrence of off-target effects has significant implications for the clinical translation of nanocarrier-based therapies. Inefficacious targeting and unintended toxicity can undermine the therapeutic benefits of nanocarriers, limiting their utility in disease treatment. Moreover, off-target effects may raise safety concerns, necessitating thorough preclinical evaluation and stringent regulatory oversight. Addressing off-target effects is thus imperative to harness the full potential of nanocarriers in precision medicine.

Mitigating off-target effects presents several challenges, including optimizing nanocarrier design, improving targeting strategies, and enhancing biocompatibility [13]. Rational design approaches, such as surface modifications and functionalization, can enhance the specificity of nanocarriers toward target tissues while minimizing non-specific interactions. Moreover, leveraging advanced targeting ligands, such as antibodies or peptides, enables precise localization of nanocarriers, reducing off-target deposition. Additionally, incorporating biocompatible materials and formulations can mitigate immunogenicity and cytotoxicity, enhancing the safety profile of nanocarriers. Additionally, close monitoring and surveillance during clinical trials and post-marketing surveillance are essential for detecting and managing any off-target effects that may arise in patients receiving gene therapy treatments [123].

### 13.4.8 Long-Term Effects

Nanocarriers have revolutionized drug delivery by enhancing therapeutic efficacy and reducing systemic toxicity. However, concerns regarding their long-term effects have garnered increasing attention. The long-term effects of nanocarriers encompass a broad spectrum of biological responses, including biodistribution, biocompatibility, and biodegradability. Understanding the fate of nanocarriers within the body over extended periods is crucial for assessing their safety and efficacy. Biodistribution studies reveal the accumulation of nanocarriers in various organs and tissues over time, influencing their pharmacokinetics and potential toxicity profiles. Moreover, the biocompatibility of nanocarriers plays a pivotal role in determining their long-term impact on physiological functions and host immune responses. Additionally, the biodegradability of nanocarriers influences their persistence *in vivo* and the eventual clearance of degradation products, which may exert unforeseen effects on biological systems [127].

The long-term effects of nanocarriers pose critical implications for their clinical translation and regulatory approval. Uncertainty regarding their safety profiles, especially over prolonged exposure durations, necessitates comprehensive preclinical evaluation and rigorous risk assessment. Moreover, the potential for accumulation and persistence of nanocarriers in vital organs raises concerns regarding chronic toxicity and adverse effects. Addressing these concerns is essential to ensure the safe and effective utilization of nanocarrier-based therapies in clinical practice.

Several challenges hinder our understanding of the long-term effects of nanocarriers and their translation into clinical applications. These include the need for standardized methodologies for assessing long-term toxicity, elucidating mechanisms underlying nanocarrier interactions with biological systems, and developing predictive models for evaluating their long-term safety profiles. Furthermore, advancing nanocarrier design to enhance biocompatibility, biodegradability, and targeted delivery holds promise for mitigating long-term adverse effects and improving therapeutic outcomes [128].

### 13.4.9 Biodistribution

Understanding the biodistribution and pharmacokinetics of nanocarriers is essential for predicting the *in vivo* behavior and potential toxicity. Nanocarriers should exhibit favorable pharmacokinetic profiles, characterized by prolonged circulation times, efficient accumulation at target sites, and minimal off-target distribution. Techniques such as fluorescence imaging, positron emission tomography (PET), and magnetic resonance imaging (MRI) enable non-invasive monitoring of nanocarrier biodistribution in preclinical studies. Optimizing particle size, surface properties, and targeting ligands can enhance the biodistribution and targeting efficiency of nanocarriers [129].

### 13.4.10 Clearance

The clearance of the nanocarriers from the body remains a critical consideration, as it influences their efficacy, safety, and clinical translation. Evaluation of the clearance pathways of nanocarriers helps determine their persistence in the body and potential for accumulation in organs or tissues. The clearance of nanocarriers from systemic circulation involves intricate interactions with various biological components, including the mononuclear phagocyte system (MPS), renal clearance, and other clearance pathways. MPS-mediated clearance predominantly occurs in the liver and spleen, where phagocytic cells recognize and engulf nanocarriers, leading to their sequestration and subsequent degradation. Renal clearance involves the filtration of smaller nanocarriers by the kidneys, followed by excretion in urine. Additionally, interactions with serum proteins, opsonization, and recognition by immune cells influence nanocarrier clearance kinetics [129].

Several factors influence the clearance of nanocarriers in gene therapy, including size, surface charge, shape, hydrophobicity, and surface modifications. Nanocarriers with larger sizes tend to undergo rapid clearance by MPS-mediated phagocytosis, while smaller nanocarriers may be cleared through renal filtration. Surface charge and hydrophobicity affect protein adsorption and opsonization, thereby influencing clearance rates. Moreover, surface modifications such as PEGylation can impart stealth properties, prolonging circulation time and reducing clearance rates [130].

To enhance the clearance of nanocarriers in gene therapy, several strategies can be employed. Rational design approaches, such as size optimization to evade MPS clearance or surface modifications to reduce opsonization, can prolong circulation time and enhance biodistribution to target tissues. Engineering nanocarriers with stimuli-responsive or biodegradable properties can facilitate controlled clearance post-delivery, minimizing systemic exposure and toxicity. Additionally, combinatorial strategies incorporating targeting ligands or stimuli-responsive release mechanisms can further enhance clearance efficiency while ensuring targeted delivery [130].

### 13.4.11 Regulatory Compliance

Compliance with regulatory guidelines and standards for gene therapy products is critical to ensure patient safety and facilitate approval for clinical use. Gene therapy, particularly when utilizing nanocarriers as delivery vehicles for genetic material, poses unique regulatory challenges due to the complex nature of these systems and the potential risks involved [131, 132]. The regulatory considerations are:

1. *Safety Assessments:* Regulatory agencies require extensive safety assessments for nanocarriers intended for gene therapy. This includes evaluating potential toxicities, immunogenicity, and off-target effects of the nanocarriers themselves, as well as any payloads they carry.
2. *Characterization of Nanocarriers:* Comprehensive characterization of nanocarriers is essential for regulatory approval. This involves detailing their physicochemical properties such as size, surface charge, morphology, stability, and drug release kinetics.
3. *Quality Control Standards:* Establishing quality control standards is crucial to ensure the consistency and reproducibility of nanocarrier production. Compliance with Good Manufacturing Practices (GMP) is necessary to meet regulatory requirements for manufacturing processes, raw materials, and final product quality.
4. *Preclinical Studies:* Robust preclinical studies are needed to assess the safety and efficacy of nanocarriers in relevant animal models. These studies should evaluate biodistribution, pharmacokinetics, and tissue-specific targeting of the nanocarriers, as well as their ability to deliver genetic material and induce therapeutic effects.
5. *Risk Assessment and Mitigation:* Regulatory agencies require thorough risk assessments to identify and mitigate potential hazards associated with nanocarrier-mediated gene therapy. This includes assessing risks related to immunogenicity, insertional mutagenesis, and unintended gene expression.
6. *Clinical Trial Design:* Clinical trials involving nanocarrier-based gene therapies must adhere to regulatory guidelines for protocol design, patient eligibility criteria, dosing regimens, and safety monitoring. Close collaboration with regulatory authorities is necessary to obtain approval for clinical trials and ensure compliance with regulatory standards.
7. *Regulatory Submissions:* Developers of nanocarrier-based gene therapies are required to submit comprehensive regulatory dossiers containing detailed data on preclinical and clinical studies, manufacturing processes, quality control measures, and risk assessments. These submissions undergo rigorous review by regulatory agencies before approval is granted for marketing authorization [132].
8. *Post-Marketing Surveillance:* After approval, ongoing post-marketing surveillance is essential to monitor the long-term safety and effectiveness of nanocarrier-based gene therapies in real-world clinical settings. Adverse event reporting systems help regulatory agencies detect and respond to any safety concerns that may arise post-approval.

Regulatory compliance of nanocarriers in gene therapy requires adherence to stringent safety, quality, and efficacy standards throughout the product development lifecycle. Close collaboration between developers, regulatory agencies, and healthcare stakeholders is essential to ensure that these innovative therapies meet the highest regulatory standards while advancing patient care [132].

## 13.5 CHALLENGES AND FUTURE DIRECTIONS

Despite significant progress, several challenges remain in achieving optimal targeting and specificity of nanocarriers for gene therapy. These include the need for improved understanding of target cell biology, optimization of ligand–receptor interactions, and mitigation of potential immunogenicity and cytotoxicity associated with targeting moieties [120]. Additionally, challenges related to systemic delivery, pharmacokinetics, and scalability of manufacturing processes need to be addressed to facilitate the clinical translation of targeted nanocarrier-based gene therapies [120]. Future research directions may focus on integrating advanced nanomaterials, bioinformatics, and imaging techniques to design next-generation nanocarriers with enhanced targeting specificity and therapeutic efficacy.

## 13.6 CONCLUSIONS

Novel bio-products with a range of uses can be created using nanostructured systems. The use of such products, meanwhile, may be hazardous to one's health. Thus, it is necessary to learn about the possible hazards associated with their use in biotechnology, electronics, medicine, and nanotechnology. A deeper comprehension of the hazards to health and safety posed by using these materials will be crucial to future advances.

The field of marketing nanomaterials is expanding dramatically. To meet this need, relevant educational programs must be created, and advisory services on a range of topics—such as risk management, commercialization lifecycles, and strategic decision-making—must be offered. To upskill their workforce, the government, manufacturers, nano-research labs, investors, solicitors, and professional media should take use of these opportunities. Private nano businesses must offer training, courses, and advisory services to nano-research labs with a focus on nano-toxicological problems. By proactively addressing potential safety concerns, standard services can help reduce risk and achieve significant cost savings.

Nanocarriers hold immense potential as delivery vehicles for gene therapy, offering the ability to overcome many of the challenges associated with nucleic acid delivery. However, ensuring their safety is paramount for their successful clinical translation. Comprehensive safety assessment strategies encompassing biocompatibility, immunogenicity, biodistribution, and toxicity are essential for the development of safe and effective nanocarriers. Continued research efforts aimed at understanding the complex interactions between nanocarriers and biological systems will pave the way for the development of next-generation gene delivery platforms with improved safety profiles and clinical outcomes.

## REFERENCES


1. N. He, X. Zeng, W. Wang, K. Deng, Y. Pan, L. Xiao, et al., "Challenges and future expectations of reversed gene therapy," *Journal of Nanoscience and Nanotechnology*, vol. 11, no. 10, pp. 8634–8638, 2011.

2. K. G. Lee, J. C. Kim, R. Wi, J. S. Min, J. K. Ahn, and D. H. Kim, "Synthesis of stable silica-dye hybrid nanomaterial as DNA carrier," *Journal of Nanoscience and Nanotechnology*, vol. 11, no. 1, pp. 686–690, 2011.

3. T. Liu, X. Yu, B. Kan, Q. Guo, X. Wang, S. Shi, et al., "Enhanced gene delivery using biodegradable poly(ester amine)s (PEAs) based on low-molecular-weight polyethylenimine and poly(ε-caprolactone)-pluronic-poly (ε-caprolactone)," *Journal of Biomedical Nanotechnology*, vol. 6, no. 4, pp. 351–359, 2010.

4. M. E. Davis, J. E. Zuckerman, C. H. J. Choi, D. Seligson, A. Tolcher, C. A. Alabi, et al., "Evidence of RNAi in humans from systemically administered siRNA via targeted nanoparticles," *Nature*, vol. 464, no. 7291, pp. 1067–1070, 2010.

5. B. Shah, S. Kona, T. A. Gilbertson, and K. T. Nguyen, "Effects of poly-(lactide-co-glycolide) nanoparticles on electrophysiological properties of enteroendocrine cells," *Journal of Nanoscience and Nanotechnology*, vol. 11, no. 4, pp. 3533–3542, 2011.

6. A. M. Ribeiro, A. C. O. Souza, A. C. Amaral, N. M. Vasconcelos, M. S. Jeronimo, F. P. Carneiro, et al., "Nanobiotechnological approaches to delivery of DNA vaccine against fungal infection," *Journal of Biomedical Nanotechnology*, vol. 9, no. 2, pp. 221–230, 2013.

7. Y. Kawasaki, A. Tamamoto, M. Takagi-Kimura, Y. Maeyama, N. Yamaoka, N. Terada, et al., "Replication-competent retrovirus vector-mediated prodrug activator gene therapy in experimental models of human malignant mesothelioma," *Cancer Gene Therapy*, vol. 18, no. 8, pp. 571–578, 2011.

8. Z. Liu, F. Ke, C. Duan, H. Lan, J. Li, C. Gao, et al., "Mannan-conjugated adenovirus enhanced gene therapy effects on murine hepatocellular carcinoma cells in vitro and in vivo," *Bioconjugate Chemistry*, vol. 24, no. 8, pp. 1387–1397, 2013.

9. X. Zhang, L. Zhao, C. Chen, J. Yan, C. Zhou, G. Yue, et al., "The effect of lentivirus-mediated expression of tumor necrosis factor related apoptosis-inducing ligand and shRNA against Bcl-2 on the growth of lymphoma cells," *Leukemia & Lymphoma*, vol. 53, no. 4, pp. 710–717, 2012.

10. C. D. Heldermon, E. Y. Qin, K. K. Ohlemiller, E. D. Herzog, J. R. Brown, C. Vogler, et al., "Disease correction by combined neonatal intracranial AAV and systemic lentiviral gene therapy in Sanfilippo syndrome type B mice," *Gene Therapy*, vol. 20, no. 9, pp. 913–921, 2013.

11. T. Xiong, Y. Li, F. Ni, and F. Zhang, "Monitoring of bystander effect of herpes simplex virus thymidine kinase/acyclovir system using fluorescence resonance energy transfer technique," *Journal of Biomedical Nanotechnology*, vol. 8, no. 1, pp. 74–79, 2012.

12. M. A. Islam, J. Firdous, Y.-J. Choi, C.-H. Yun, and C.-S. Cho, "Regulation of endocytosis by non-viral vectors for efficient gene activity," *Journal of Biomedical Nanotechnology*, vol. 10, no. 1, pp. 67–80, 2014.

13. L. Qu, M. Jiang, Z. Li, F. Pu, L. Gong, L. Sun, et al., "Inhibitory effect of biosynthetic nanoscale peptide Melittin on hepatocellular carcinoma, driven by survivin promoter," *Journal of Biomedical Nanotechnology*, vol. 10, no. 4, pp. 695–706, 2014.

14. Z. He, Y. Yu, Y. Zhang, Y. Yan, Y. Zheng, J. He, et al., "Gene delivery with active targeting to ovarian cancer cells mediated by folate receptor α," *Journal of Biomedical Nanotechnology*, vol. 9, no. 5, pp. 833–844, 2013.

15. D. K. Kim, J. H. Chang, and Y. J. Kang, "Efficient internalization of peptide-conjugated SPIONs in dendritic cells for tumor targeting," *Journal of Nanoscience and Nanotechnology*, vol. 12, no. 7, pp. 5191–5198, 2012.

16. P. Ojer, L. Neutsch, F. Gabor, J. M. Irache, and A. L. de Cerain, "Cytotoxicity and cell interaction studies of bioadhesive poly (anhydride) nanoparticles for oral antigen/drug delivery," *Journal of Biomedical Nanotechnology*, vol. 9, no. 11, pp. 1891–1903, 2013.

17. N. Ma, C. Ma, Y. Deng, T. Wang, and N. He, "Advances in applications of dendritic compounds," *Journal of Nanoscience and Nanotechnology*, vol. 13, no. 1, pp. 33–39, 2013.

18. X. Wang, B. Chen, X. Yang, J. Zhang, L. Zhao, and J. Tang, "Functionalized superparamagnetic nanoparticles for highly-efficient gene delivery," *Journal of Nanoscience and Nanotechnology*, vol. 13, no. 2, pp. 746–750, 2013.

19. K. Lundstrom, "Viral vectors in gene therapy," *Diseases*, vol. 6, no. 2, p. 42, 2018.

20. M. A. Kotterman, T. W. Chalberg, and D. V. Schaffer, "Viral vectors for gene therapy: translational and clinical outlook," *Annual Review of Biomedical Engineering*, vol. 17, no. 1, pp. 63–89, 2015.

21. M. H. Butt, M. Zaman, A. Ahmad, R. Khan, T. H. Mallhi, M. M. Hasan, et al., "Appraisal for the potential of viral and nonviral vectors in gene therapy: a review," *Genes*, vol. 13, no. 8, p. 1370, 2022.

22. L. Liu, J. Yang, K. Men, Z. He, M. Luo, Z. Qian, et al., "Current status of nonviral vectors for gene therapy in China," *Human Gene Therapy*, vol. 29, no. 2, pp. 110–120, 2018.

23. M. Ramamoorth and A. Narvekar, "Non viral vectors in gene therapy—an overview," *Journal of Clinical and Diagnostic Research: JCDR*, vol. 9, no. 1, p. GE01, 2015.

24. D. Chen, K. T. Love, Y. Chen, A. A. Eltoukhy, C. Kastrup, G. Sahay, et al., "Rapid discovery of potent siRNA-containing lipid nanoparticles enabled by controlled microfluidic formulation," *Journal of the American Chemical Society*, vol. 134, no. 16, pp. 6948–6951, 2012.

25. Y. Dong, K. T. Love, J. R. Dorkin, S. Sirirungruang, Y. Zhang, D. Chen, et al., "Lipopeptide nanoparticles for potent and selective siRNA delivery in rodents and nonhuman primates," *Proceedings of the National Academy of Sciences*, vol. 111, no. 11, pp. 3955–3960, 2014.

26. K. A. Whitehead, J. R. Dorkin, A. J. Vegas, P. H. Chang, O. Veiseh, J. Matthews, et al., "Degradable lipid nanoparticles with predictable in vivo siRNA delivery activity," *Nature Communications*, vol. 5, no. 1, p. 4277, 2014.

27. F. Czauderna, M. Fechtner, S. Dames, H. Aygün, A. Klippel, G. J. Pronk, et al., "Structural variations and stabilising modifications of synthetic siRNAs in mammalian cells," *Nucleic Acids Research*, vol. 31, no. 11, pp. 2705–2716, 2003.

28. L. Li, Y. Wei, and C. Gong, "Polymeric nanocarriers for non-viral gene delivery," *Journal of Biomedical Nanotechnology*, vol. 11, no. 5, pp. 739–770, 2015.

29. P. L. Felgner, T. R. Gadek, M. Holm, R. Roman, H. W. Chan, M. Wenz, et al., "Lipofection: a highly efficient, lipid-mediated DNA-transfection procedure," *Proceedings of the National Academy of Sciences*, vol. 84, no. 21, pp. 7413–7417, 1987.

30. S. K. Samal, M. Dash, S. Van Vlierberghe, D. L. Kaplan, E. Chiellini, C. Van Blitterswijk, et al., "Cationic polymers and their therapeutic potential," *Chemical Society Reviews*, vol. 41, no. 21, pp. 7147–7194, 2012.

31. D. Lechardeur, A. S. Verkman, and G. L. Lukacs, "Intracellular routing of plasmid DNA during non-viral gene transfer," *Advanced Drug Delivery Reviews*, vol. 57, no. 5, pp. 755–767, 2005.

32. A. Bielinska, R. A. Shivdasani, L. Zhang, and G. J. Nabel, "Regulation of gene expression with double-stranded phosphorothioate oligonucleotides," *Science*, vol. 250, no. 4983, pp. 997–1000, 1990.

33. O. Boussif, M. A. Zanta, and J. P. Behr, "Optimized galenics improve in vitro gene transfer with cationic molecules up to 1000-fold," *Gene Therapy*, vol. 3, no. 12, pp. 1074–1080, 1996.

34. P. Erbacher, S. Zou, T. Bettinger, A.-M. Steffan, and J.-S. Remy, "Chitosan-based vector/DNA complexes for gene delivery: biophysical characteristics and transfection ability," *Pharmaceutical Research*, vol. 15, pp. 1332–1339, 1998.

35. S. Ferrari, E. Moro, A. Pettenazzo, J. P. Behr, F. Zacchello, and M. Scarpa, "ExGen 500 is an efficient vector for gene delivery to lung epithelial cells in vitro and in vivo," *Gene Therapy*, vol. 4, no. 10, pp. 1100–1106, 1997.

36. B. F. Smith, H. J. Baker, D. T. Curiel, W. Jiang, and R. M. Conry, "Humoral and cellular immune responses of dogs immunized with a nucleic acid vaccine encoding human carcinoembryonic antigen," *Gene Therapy*, vol. 5, no. 7, pp. 865–868, 1998.

37. Z. Kadlecova, Y. Rajendra, M. Matasci, D. Hacker, L. Baldi, F. M. Wurm, et al., "Hyperbranched polylysine: a versatile, biodegradable transfection agent for the production of recombinant proteins by transient gene expression and the transfection of primary cells," *Macromolecular Bioscience*, vol. 12, no. 6, pp. 794–804, 2012.

38. H. He, E. Luedke, X. Zhang, B. Yu, A. Schmitt, B. McClarren, et al., "A nanoporous cell-therapy device with controllable biodegradation for long-term drug release," *Journal of Controlled Release*, vol. 165, no. 3, pp. 226–233, 2013.

39. J.-P. Behr, "The proton sponge: a trick to enter cells the viruses did not exploit," *Chimia*, vol. 51, no. 1–2, pp. 34–34, 1997.

40. M. Jäger, S. Schubert, S. Ochrimenko, D. Fischer, and U. S. Schubert, "Branched and linear poly (ethylene imine)-based conjugates: synthetic modification, characterization, and application," *Chemical Society Reviews*, vol. 41, no. 13, pp. 4755–4767, 2012.

41. U. Lungwitz, M. Breunig, T. Blunk, and A. Göpferich, "Polyethylenimine-based non-viral gene delivery systems," *European Journal of Pharmaceutics and Biopharmaceutics*, vol. 60, no. 2, pp. 247–266, 2005.

42. A. Kichler, C. Leborgne, E. Coeytaux, and O. Danos, "Polyethylenimine-mediated gene delivery: a mechanistic study," *Journal of Gene Medicine*, vol. 3, no. 2, pp. 135–144, 2001.

43. M. C. Garnett, "Gene-delivery systems using cationic polymers," *Critical Reviews™ in Therapeutic Drug Carrier Systems*, vol. 16, no. 2, 1999.

44. P. van de Wetering, E. E. Moret, N. M. E. Schuurmans-Nieuwenbroek, M. J. van Steenbergen, and W. E. Hennink, "Structure–activity relationships of water-soluble cationic methacrylate/methacrylamide polymers for nonviral gene delivery," *Bioconjugate Chemistry*, vol. 10, no. 4, pp. 589–597, 1999.

45. P. van de Wetering, N. M. E. Schuurmans-Nieuwenbroek, W. E. Hennink, and G. Storm, "Comparative transfection studies of human ovarian carcinoma cells in vitro, ex vivo and in vivo with poly (2-(dimethylamino) ethyl methacrylate)-based polyplexes," *Journal of Gene Medicine*, vol. 1, no. 3, pp. 156–165, 1999.

46. P. van de Wetering, J.-Y. Cherng, H. Talsma, and W. E. Hennink, "Relation between transfection efficiency and cytotoxicity of poly (2-(dimethylamino) ethyl methacrylate)/plasmid complexes," *Journal of Controlled Release*, vol. 49, no. 1, pp. 59–69, 1997.

47. J. M. Layman, S. M. Ramirez, M. D. Green, and T. E. Long, "Influence of polycation molecular weight on poly (2-dimethylaminoethyl methacrylate)-mediated DNA delivery in vitro," *Biomacromolecules*, vol. 10, no. 5, pp. 1244–1252, 2009.

48. I. Messai, D. Lamalle, S. Munier, B. Verrier, Y. Ataman-Önal, and T. Delair, "Poly (D,L-lactic acid) and chitosan complexes: interactions with plasmid DNA," *Colloids and Surfaces A: Physicochemical and Engineering Aspects*, vol. 255, no. 1–3, pp. 65–72, 2005.

49. R. J. Mumper, "Novel polymeric condensing carriers for gene transfer," *Proceedings of the International Symposium on Controlled Release of Bioactive Materials*, vol. 22, pp. 178–179, 1995.

50. T. Kiang, J. Wen, H. W. Lim, and K. W. Leong, "The effect of the degree of chitosan deacetylation on the efficiency of gene transfection," *Biomaterials*, vol. 25, no. 22, pp. 5293–5301, 2004.

51. M. Köping-Höggård, I. Tubulekas, H. Guan, K. Edwards, M. Nilsson, K. M. Vårum, et al., "Chitosan as a nonviral gene delivery system. Structure–property relationships and characteristics compared with polyethylenimine in vitro and after lung administration in vivo," *Gene Therapy*, vol. 8, no. 14, pp. 1108–1121, 2001.

52. F. C. MacLaughlin, R. J. Mumper, J. Wang, J. M. Tagliaferri, I. Gill, M. Hinchcliffe, et al., "Chitosan and depolymerized chitosan oligomers as condensing carriers for in vivo plasmid delivery," *Journal of Controlled Release*, vol. 56, no. 1–3, pp. 259–272, 1998.

53. M. Huang, C.-W. Fong, E. Khor, and L.-Y. Lim, "Transfection efficiency of chitosan vectors: effect of polymer molecular weight and degree of deacetylation," *Journal of Controlled Release*, vol. 106, no. 3, pp. 391–406, 2005.

54. M. Köping-Höggård, K. M. Vårum, M. Issa, S. Danielsen, B. E. Christensen, B. T. Stokke, et al., "Improved chitosan-mediated gene delivery based on easily dissociated chitosan polyplexes of highly defined chitosan oligomers," *Gene Therapy*, vol. 11, no. 19, pp. 1441–1452, 2004.

55. J. J. Green, R. Langer, and D. G. Anderson, "A combinatorial polymer library approach yields insight into nonviral gene delivery," *Accounts of Chemical Research*, vol. 41, no. 6, pp. 749–759, 2008.

56. C. Lin, Z. Zhong, M. C. Lok, X. Jiang, W. E. Hennink, J. Feijen, et al., "Linear poly (amido amine) s with secondary and tertiary amino groups and variable amounts of disulfide linkages: synthesis and in vitro gene transfer properties," *Journal of Controlled Release*, vol. 116, no. 2, pp. 130–137, 2006.

57. C. Toumey, "Probing the history of nanotechnology," *Nature Nanotechnology*, vol. 7, no. 4, pp. 205–206, 2012.

58. Q. A. Pankhurst, N. T. K. Thanh, S. K. Jones, and J. Dobson, "Progress in applications of magnetic nanoparticles in biomedicine," *Journal of Physics D: Applied Physics*, vol. 42, no. 22, p. 224001, 2009.

59. M. A. Wall, S. Harmsen, S. Pal, L. Zhang, G. Arianna, J. R. Lombardi, et al., "Surfactant-free shape control of gold nanoparticles enabled by unified theoretical framework of nanocrystal synthesis," *Advanced Materials*, vol. 29, no. 21, p. 1605622, 2017.

60. X. Sun, C. Zheng, F. Zhang, Y. Yang, G. Wu, A. Yu, et al., "Size-controlled synthesis of magnetite ($Fe_3O_4$) nanoparticles coated with glucose and gluconic acid from a single Fe (III) precursor by a sucrose bifunctional hydrothermal method," *Journal of Physical Chemistry C*, vol. 113, no. 36, pp. 16002–16008, 2009.

61. P.-C. Chen, G. Liu, Y. Zhou, K. A. Brown, N. Chernyak, J. L. Hedrick, et al., "Tip-directed synthesis of multimetallic nanoparticles," *Journal of the American Chemical Society*, vol. 137, no. 28, pp. 9167–9173, 2015.

62. F.-H. Du, K.-X. Wang, and J.-S. Chen, "Strategies to succeed in improving the lithium-ion storage properties of silicon nanomaterials," *Journal of Materials Chemistry A*, vol. 4, no. 1, pp. 32–50, 2016.

63. K. Wang, F. M. Kievit, J. G. Sham, M. Jeon, Z. R. Stephen, A. Bakthavatsalam, et al., "Iron-oxide-based nanovector for tumor targeted siRNA delivery in an orthotopic hepatocellular carcinoma xenograft mouse model," *Small*, vol. 12, no. 4, pp. 477–487, 2016.

64. Z. Wang, G. Liu, H. Zheng, and X. Chen, "Rigid nanoparticle-based delivery of anti-cancer siRNA: challenges and opportunities," *Biotechnology Advances*, vol. 32, no. 4, pp. 831–843, 2014.

65. G. Lin, R. A. Revia, and M. Zhang, "Inorganic nanomaterial-mediated gene therapy in combination with other antitumor treatment modalities," *Advanced Functional Materials*, vol. 31, no. 5, p. 2007096, 2021.

66. P. Yang, S. Gai, and J. Lin, "Functionalized mesoporous silica materials for controlled drug delivery," *Chemical Society Reviews*, vol. 41, no. 9, pp. 3679–3698, 2012.

67. G. Chen, H. Ågren, T. Y. Ohulchanskyy, and P. N. Prasad, "Light upconverting core–shell nanostructures: nanophotonic control for emerging applications," *Chemical Society Reviews*, vol. 44, no. 6, pp. 1680–1713, 2015.

68. S. Santra, R. Tapec, N. Theodoropoulou, J. Dobson, A. Hebard, and W. Tan, "Synthesis and characterization of silica-coated iron oxide nanoparticles in microemulsion: the effect of nonionic surfactants," *Langmuir*, vol. 17, no. 10, pp. 2900–2906, 2001.

69. S. Liu, G. Chen, T. Y. Ohulchanskyy, M. T. Swihart, and P. N. Prasad, "Facile synthesis and potential bioimaging applications of hybrid upconverting and plasmonic $NaGdF_4$: $Yb^{3+}$, $Er^{3+}$/silica/gold nanoparticles," *Theranostics*, vol. 3, no. 4, p. 275, 2013.

70. G. Chen, I. Roy, C. Yang, and P. N. Prasad, "Nanochemistry and nanomedicine for nanoparticle-based diagnostics and therapy," *Chemical Reviews*, vol. 116, no. 5, pp. 2826–2885, 2016.

71. X. Zhang, H. Jin, X. Huang, B. Chaurasiya, D. Dong, T. P. Shanley, et al., "Robust genome editing in adult vascular endothelium by nanoparticle delivery of CRISPR-Cas9 plasmid DNA," *Cell Reports*, vol. 38, no. 1, 2022.

72. L. Tang, A. Zhang, Z. Zhang, Q. Zhao, J. Li, Y. Mei, et al., "Multifunctional inorganic nanomaterials for cancer photoimmunotherapy," *Cancer Communications*, vol. 42, no. 2, pp. 141–163, 2022.

73. M. Lee, S. Kang, S. Kim, and N. Park, "Advances and trends in miRNA analysis using DNAzyme-based biosensors," *Biosensors*, vol. 13, no. 9, p. 856, 2023.

74. R. Pala, V. T. Anju, M. Dyavaiah, S. Busi, and S. M. Nauli, "Nanoparticle-mediated drug delivery for the treatment of cardiovascular diseases," *International Journal of Nanomedicine*, pp. 3741–3769, 2020.

75. H.-X. Wang, Z. Song, Y.-H. Lao, X. Xu, J. Gong, D. Cheng, et al., "Nonviral gene editing via CRISPR/Cas9 delivery by membrane-disruptive and endosomolytic helical polypeptide," *Proceedings of the National Academy of Sciences*, vol. 115, no. 19, pp. 4903–4908, 2018.

76. R. Mohammadinejad, A. Dehshahri, V. S. Madamsetty, M. Zahmatkeshan, S. Tavakol, P. Makvandi, et al., "In vivo gene delivery mediated by non-viral vectors for cancer therapy," *Journal of Controlled Release*, vol. 325, pp. 249–275, 2020.

77 X.-X. Ma, J.-L. Xu, Y.-Y. Jia, Y.-X. Zhang, W. Wang, C. Li, et al., "Enhance transgene responses through improving cellular uptake and intracellular trafficking by bio-inspired non-viral vectors," *Journal of Nanobiotechnology.*, vol. 18, no. 1, p. 26, 2020.

78. Z. Chen, Z. Lv, Y. Sun, Z. Chi, and G. Qing, "Recent advancements in polyethyleneimine-based materials and their biomedical, biotechnology, and biomaterial applications," *Journal of Materials Chemistry B*, vol. 8, no. 15, pp. 2951–2973, 2020.

79. Q. Yu, M. Zhang, Y. Chen, X. Chen, S. Shi, K. Sun, et al., "Self-assembled nanoparticles prepared from low-molecular-weight PEI and low-generation PAMAM for EGFRvIII-chimeric antigen receptor gene loading and T-cell transient modification," *International Journal of Nanomedicine*, 15: pp. 483–495, 2020.

80. N. Y. Steinman, L. M. Campos, Y. Feng, A. J. Domb, and H. Hosseinkhani, "Cyclopropenium nanoparticles and gene transfection in cells," *Pharmaceutics*, vol. 12, no. 8, p. 768, 2020.

81. Y. Zhao, T. Zhao, Y. Du, Y. Cao, Y. Xuan, H. Chen, et al., "Interaction kinetics of peptide lipids-mediated gene delivery," *Journal of Nanobiotechnology*, vol. 18, pp. 1–14, 2020.

82. I. Lostalé-Seijo, I. Louzao, M. Juanes, and J. Montenegro, "Peptide/Cas9 nanostructures for ribonucleoprotein cell membrane transport and gene edition," *Chemical Science*, vol. 8, no. 12, pp. 7923–7931, 2017.

83. F. Jiang and J. A. Doudna, "CRISPR–Cas9 structures and mechanisms," *Annual Review of Biophysics*, vol. 46, no. 1, pp. 505–529, 2017.

84. S. K. Saha, F. K. Saikot, M. S. Rahman, A. H. M. Jamal, S. M. K. Rahman, S. M. R. Islam, et al., "Programmable molecular scissors: applications of a new tool for genome editing in biotech," *Molecular Therapy-Nucleic Acids*, vol. 14, pp. 212–238, 2019.

85. G. Schwank, B.-K. Koo, V. Sasselli, J. F. Dekkers, I. Heo, T. Demircan, et al., "Functional repair of CFTR by CRISPR/Cas9 in intestinal stem cell organoids of cystic fibrosis patients," *Cell Stem Cell*, vol. 13, no. 6, pp. 653–658, 2013.

86. H. L. Li, N. Fujimoto, N. Sasakawa, S. Shirai, T. Ohkame, T. Sakuma, et al., "Precise correction of the dystrophin gene in Duchenne muscular dystrophy patient induced pluripotent stem cells by TALEN and CRISPR-Cas9," *Stem Cell Reports*, vol. 4, no. 1, pp. 143–154, 2015.

87. Z. Liu, S. Chen, X. Jin, Q. Wang, K. Yang, C. Li, et al., "Genome editing of the HIV co-receptors CCR5 and CXCR4 by CRISPR-Cas9 protects CD4+ T cells from HIV-1 infection," *Cell & Bioscience*, vol. 7, pp. 1–15, 2017.

88. E. A. Stadtmauer, J. A. Fraietta, M. M. Davis, A. D. Cohen, K. L. Weber, E. Lancaster, et al., "CRISPR-engineered T cells in patients with refractory cancer," *Science*, vol. 367, no. 6481, p. eaba7365, 2020.

89. V. Sebastiano, M. L. Maeder, J. F. Angstman, B. Haddad, C. Khayter, D. T. Yeo, et al., "In situ genetic correction of the sickle cell anemia mutation in human induced pluripotent stem cells using engineered zinc finger nucleases," *Stem Cells*, vol. 29, no. 11, pp. 1717–1726, 2011.

90. M. H. Porteus and D. Baltimore, "Chimeric nucleases stimulate gene targeting in human cells," *Science*, vol. 300, no. 5620, pp. 763–763, 2003.

91. F. D. Urnov, J. C. Miller, Y.-L. Lee, C. M. Beausejour, J. M. Rock, S. Augustus, et al., "Highly efficient endogenous human gene correction using designed zinc-finger nucleases," *Nature*, vol. 435, no. 7042, pp. 646–651, 2005.

92. D. R. Ede, N. Farhang, J. D. Stover, and R. D. Bowles, "4.32 Gene editing tools," in *Comprehensive Biomaterials II*, Elsevier, 2017, pp. 589–599.

93. D. Reyon, S. Q. Tsai, C. Khayter, J. A. Foden, J. D. Sander, and J. K. Joung, "FLASH assembly of TALENs for high-throughput genome editing," *Nature Biotechnology*, vol. 30, no. 5, pp. 460–465, 2012.

94. Z. Wang, J. Li, H. Huang, G. Wang, M. Jiang, S. Yin, C. Sun, et al., "An integrated chip for the high-throughput synthesis of transcription activator-like effectors," *Angewandte Chemie (International Edition in English)*, vol. 51, no. 34, pp. 8505–8508, 2012.

95. A. W. Briggs, X. Rios, R. Chari, L. Yang, F. Zhang, P. Mali, et al., "Iterative capped assembly: rapid and scalable synthesis of repeat-module DNA such as TAL effectors from individual monomers," *Nucleic Acids Research*, vol. 40, no. 15, pp. e117–e117, 2012.

96. J. Li, S. Hong, W. Chen, E. Zuo, and H. Yang, "Advances in detecting and reducing off-target effects generated by CRISPR-mediated genome editing," *Journal of Genetics and Genomics*, vol. 46, no. 11, pp. 513–521, 2019.

97. M. D. Canny, N. Moatti, L. C. K. Wan, A. Fradet-Turcotte, D. Krasner, P. A. Mateos-Gomez, et al., "Inhibition of 53BP1 favors homology-dependent DNA repair and increases CRISPR–Cas9 genome-editing efficiency," *Nature Biotechnology*, vol. 36, no. 1, pp. 95–102, 2018.

98. M. F. Rasul, B. M. Hussen, A. Salihi, B. S. Ismael, P. J. Jalal, A. Zanichelli, et al., "Strategies to overcome the main challenges of the use of CRISPR/Cas9 as a replacement for cancer therapy," *Molecular Cancer*, vol. 21, no. 1, pp. 1–30, 2022.

99. N. Duan, S. Tang, B. Zeng, Z. Hu, Q. Hu, L. Wu, et al., "An episomal CRISPR/Cas12a system for mediating efficient gene editing," *Life*, vol. 11, no. 11, p. 1262, 2021.

100. Q. Chen, Y. Zhang, and H. Yin, "Recent advances in chemical modifications of guide RNA, mRNA and donor template for CRISPR-mediated genome editing," *Advanced Drug Delivery Reviews*, vol. 168, pp. 246–258, 2021.

101. Q. Cheng, T. Wei, L. Farbiak, L. T. Johnson, S. A. Dilliard, and D. J. Siegwart, "Selective organ targeting (SORT) nanoparticles for tissue-specific mRNA delivery and CRISPR–Cas gene editing," *Nature Nanotechnology*, vol. 15, no. 4, pp. 313–320, 2020.

102. C. Brokowski and M. Adli, "CRISPR ethics: moral considerations for applications of a powerful tool," *Journal of Molecular Biology*, vol. 431, no. 1, pp. 88–101, 2019.

103. T. Wan, J. Zhong, Q. Pan, T. Zhou, Y. Ping, and X. Liu, "Exosome-mediated delivery of Cas9 ribonucleoprotein complexes for tissue-specific gene therapy of liver diseases," *Science Advances*, vol. 8, no. 37, eabp9435.

104. M. Behr, J. Zhou, B. Xu, and H. Zhang, "In vivo delivery of CRISPR-Cas9 therapeutics: progress and challenges," *Acta Pharmaceutica Sinica B*, vol. 11, no. 8, pp. 2150–2171, 2021.

105. S. Ramakrishna, A.-B. K. Dad, J. Beloor, R. Gopalappa, S.-K. Lee, and H. Kim, "Gene disruption by cell-penetrating peptide-mediated delivery of Cas9 protein and guide RNA," *Genome Research*, vol. 24, no. 6, pp. 1020–1027, 2014.

106. T. Del'Guidice, J. P. Lepetit-Stoffaes, L.-J. Bordeleau, J. Roberge, V. Theberge, C. Lauvaux, et al., "Membrane permeabilizing amphiphilic peptide delivers recombinant transcription factor and CRISPR-Cas9/Cpf1 ribonucleoproteins in hard-to-modify cells," *PLOS One*, vol. 13, no. 4, p. e0195558, 2018.

107. R. E. Taylor and M. Zahid, "Cell penetrating peptides, novel vectors for gene therapy," *Pharmaceutics*, vol. 12, no. 3, p. 225, 2020.

108. H. Zu and D. Gao, "Non-viral vectors in gene therapy: recent development, challenges, and prospects," *AAPS Journal*, vol. 23, no. 4, p. 78, 2021.

109. S. Mantha, S. Pillai, P. Khayambashi, A. Upadhyay, Y. Zhang, O. Tao, et al., "Smart hydrogels in tissue engineering and regenerative medicine," *Materials*, vol. 12, no. 20, p. 3323, 2019.

110. J. Hwang, M. O. Sullivan, and K. L. Kiick, "Targeted drug delivery via the use of ECM-mimetic materials," *Frontiers in Bioengineering and Biotechnology*, vol. 8, p. 69, 2020.

111. A. Olshefsky, H. Benasutti, M. Sylvestre, G. L. Butterfield, G. J. Rocklin, C. Richardson, et al., "In vivo selection of synthetic nucleocapsids for tissue targeting," *Proceedings of the National Academy of Sciences*, vol. 120, no. 46, p. e2306129120, 2023.

112. I. Ates, T. Rathbone, C. Stuart, P. H. Bridges, and R. N. Cottle, "Delivery approaches for therapeutic genome editing and challenges," *Genes*, vol. 11, no. 10, p. 1113, 2020.

113. B. Yameen, W. I. Choi, C. Vilos, A. Swami, J. Shi, and O. C. Farokhzad, "Insight into nanoparticle cellular uptake and intracellular targeting," *Journal of Controlled Release*, vol. 190, pp. 485–499, 2014.

114. X.-Q. Zhang, X. Xu, N. Bertrand, E. Pridgen, A. Swami, and O. C. Farokhzad, "Interactions of nanomaterials and biological systems: implications to personalized nanomedicine," *Advanced Drug Delivery Reviews*, vol. 64, no. 13, pp. 1363–1384, 2012.

115. C. D. Walkey, J. B. Olsen, H. Guo, A. Emili, and W. C. W. Chan, "Nanoparticle size and surface chemistry determine serum protein adsorption and macrophage uptake," *Journal of the American Chemical Society*, vol. 134, no. 4, pp. 2139–2147, 2012.

116. I. M. Verma and N. Somia, "Gene therapy-promises, problems and prospects," *Nature*, vol. 389, no. 6648, pp. 239–242, 1997.

117. L. Zhang, F. X. Gu, J. M. Chan, A. Z. Wang, R. S. Langer, and O. C. Farokhzad, "Nanoparticles in medicine: therapeutic applications and developments," *Clinical Pharmacology & Therapeutics*, vol. 83, no. 5, pp. 761–769, 2008.

118. H. Lee, J. H. Jeong, and T. G. Park, "A new gene delivery formulation of polyethylenimine/DNA complexes coated with PEG conjugated fusogenic peptide," *Journal of Controlled Release*, vol. 76, no. 1–2, pp. 183–192, 2001.

119. J. Shi, P. W. Kantoff, R. Wooster, and O. C. Farokhzad, "Cancer nanomedicine: progress, challenges and opportunities," *Nature Reviews Cancer*, vol. 17, no. 1, pp. 20–37, 2017.

120. S. Wilhelm, A. J. Tavares, Q. Dai, S. Ohta, J. Audet, H. F. Dvorak, et al., "Analysis of nanoparticle delivery to tumours," *Nature Reviews Materials*, vol. 1, no. 5, pp. 1–12, 2016.

121. D. W. Pack, A. S. Hoffman, S. Pun, and P. S. Stayton, "Design and development of polymers for gene delivery," *Nature Reviews Drug Discovery*, vol. 4, no. 7, pp. 581–593, 2005.

122. W. Sun, W. Ji, J. M. Hall, Q. Hu, C. Wang, C. L. Beisel, et al., "Self-assembled DNA nanoclews for the efficient delivery of CRISPR–Cas9 for genome editing," *Angewandte Chemie*, vol. 127, no. 41, pp. 12197–12201, 2015.

123. E. Blanco, H. Shen, and M. Ferrari, "Principles of nanoparticle design for overcoming biological barriers to drug delivery," *Nature Biotechnology*, vol. 33, no. 9, pp. 941–951, 2015.

124. S. L. Ginn, A. K. Amaya, I. E. Alexander, M. Edelstein, and M. R. Abedi, "Gene therapy clinical trials worldwide to 2017: an update," *Journal of Gene Medicine*, vol. 20, no. 5, p. e3015, 2018.

125. P. M. Kharkar, A. M. Kloxin, and K. L. Kiick, "Dually degradable click hydrogels for controlled degradation and protein release," *Journal of Materials Chemistry B*, vol. 2, no. 34, pp. 5511–5521, 2014.

126. D. Peer, J. M. Karp, S. Hong, O. C. Farokhzad, R. Margalit, and R. Langer, "Nanocarriers as an emerging platform for cancer therapy," *Nano-Enabled Medical Applications*, pp. 61–91, 2020.

127. B. M. Geilich, A. L. van de Ven, G. L. Singleton, L. J. Sepúlveda, S. Sridhar, and T. J. Webster, "Silver nanoparticle-embedded polymersome nanocarriers for the treatment of antibiotic-resistant infections," *Nanoscale*, vol. 7, no. 8, pp. 3511–3519, 2015.

128. J. Wang, J. D. Byrne, M. E. Napier, and J. M. DeSimone, "More effective nanomedicines through particle design," *Small*, vol. 7, no. 14, pp. 1919–1931, 2011.

129. K. Xiao, Y. Li, J. Luo, J. S. Lee, W. Xiao, A. M.Gonik, et al., "The effect of surface charge on in vivo biodistribution of PEG-oligocholic acid based micellar nanoparticles," *Biomaterials*, vol. 32, no. 13, pp. 3435–3446, 2011.

130. F. Alexis, E. Pridgen, L. K. Molnar, and O. C. Farokhzad, "Factors affecting the clearance and biodistribution of polymeric nanoparticles," *Molecular Pharmaceutics*, vol. 5, no. 4, pp. 505–515, 2008.

131. C. European, "Guidelines on good manufacturing practice specific to advanced therapy medicinal products," *European Commission Journal*, vol. 4, pp. 1–32, 2017.

132. European Medicines Agency, "Guideline on the quality, non-clinical and clinical aspects of gene therapy medicinal products," *European Medicines Agency Guide*, vol. 44, pp. 1–41, 2018.

# 14 Nanocarriers in Cancer Immunotherapy

## *Personalized Cancer Vaccines, Immune Checkpoint Modulation, Adoptive T-Cell Therapy*

*Vineet Mahajan and Yashwant V. Pathak*

## 14.1 INTRODUCTION

### 14.1.1 Background on Cancer Immunotherapy

#### *14.1.1.1 Evolution of Cancer Treatment*

Cancer treatment has undergone a transformative journey, progressing from conventional approaches such as surgery, chemotherapy, and radiation therapy [1]. The emergence of cancer immunotherapy represents a paradigm shift as it leverages the body's own immune system to combat malignancies [2, 3].

#### *14.1.1.2 Immunotherapy Overview*

Immunotherapy has emerged as a powerful approach to fight cancer. Unlike traditional treatments that directly target cancer cells, immunotherapy enhances the immune response of the body to recognize and eliminate cancerous cells [2]. Notable successes, such as immune checkpoint inhibitors and adoptive T-cell therapies, underscore the potential of immunotherapy to achieve durable responses in various types of cancer [4, 5].

#### *14.1.1.3 Success Stories and Limitations*

Although immunotherapy has demonstrated remarkable success in certain cases, several challenges persist. Response rates can vary, and resistance can develop. In addition, not all patients benefit equally from these treatments. Understanding limitations is crucial to refine existing approaches and explore innovative strategies to improve therapeutic results [1].

### 14.1.2 Rationale for Nanocarriers in Cancer Immunotherapy

#### *14.1.2.1 Enhancing Drug Delivery*

Traditional immunotherapy faces challenges related to the delivery of therapeutic agents [6]. Nanocarriers offer a solution by providing a platform for precise drug delivery to tumor sites [7]. Their small size and tunable properties enable efficient transport of nanocarriers through biological barriers, ensuring targeted delivery [8].

#### *14.1.2.2 Benefits of Nanocarriers*

Nanocarriers offer several advantages over immunotherapy for cancer. They improve the bioavailability of therapeutic agents, prolong the circulation time in the bloodstream, and protect fragile payloads from degradation [9]. This improved pharmacokinetic profile contributes to maximizing therapeutic impact while minimizing systemic side effects [6].

#### *14.1.2.3 Tailored Therapeutic Strategies*

Use of nanocarriers allows for the customization of therapeutic strategies. By encapsulating immunotherapeutic agents within these carriers, it becomes possible to design personalized treatment approaches. This tailoring ensures that the delivered payload reaches its intended destination, thus optimizing treatment efficacy [10].

### 14.1.3 Scope of the Chapter

#### *14.1.3.1 Overview of Nanocarriers*

This chapter provides an in-depth exploration of nanocarriers, including nanoparticles, liposomes, and other advanced delivery systems. These carriers serve as essential tools to overcome the challenges associated with traditional cancer immunotherapy, offering precise and controlled delivery of therapeutic payloads.

#### *14.1.3.2 Focus on Three Key Areas*

The chapter focuses on three critical areas within cancer immunotherapy: personalized cancer vaccines, immune checkpoint modulation, and adoptive T-cell therapy. Each section delves into the specific role of nanocarriers in enhancing the effectiveness of these therapeutic modalities.

DOI: 10.1201/9781003473183-14

#### *14.1.3.3 Integration of Research and Clinical Perspectives*

By integrating insights from both preclinical and clinical studies, this chapter provides a comprehensive understanding of the current state of nano-carrier-based cancer immunotherapy. Bridging the gap between research and clinical application is vital to translating promising findings into tangible advances in patient care.

#### *14.1.3.4 Significance for Future Therapeutics*

This chapter not only reviews existing knowledge but also highlights the significance of ongoing research and emerging technologies. The exploration of nanocarriers in cancer immunotherapy is positioned as a cornerstone in shaping the future landscape of cancer therapeutics, offering hope for more targeted, efficient, and personalized treatment strategies.

## 14.2 PERSONALIZED CANCER VACCINES

This detailed section aims to provide a comprehensive exploration of personalized cancer vaccines, covering their definition, types, mechanisms, the importance of personalization, the role of nanocarriers, and specific case studies involving mRNA and peptide-based vaccines.

### 14.2.1 An Overview of Personalized Cancer Vaccines: A Paradigm Shift in Cancer Immunotherapy

#### *14.2.1.1 Definition and Principles*

Personalized cancer vaccines represent a transformative approach to cancer immunotherapy, ushering in an era of precision oncology. These vaccines are tailored to the unique genetic and molecular landscape of an individual patient's tumor, marking a departure from the traditional one-size-fits-all approach [11]. Unlike conventional vaccines that target generic tumor antigens shared across different malignancies, personalized vaccines harness the patient's own immune system to recognize and eliminate cancer cells harboring distinct tumor-specific antigens [12].

#### *14.2.1.2 Types of Personalized Vaccines*

Personalized cancer vaccines encompass a diverse array of platforms, each with its own unique characteristics and mechanisms of action. Among the most prominent types are:

- *mRNA Vaccines:* These vaccines deliver messenger RNA (mRNA) encoding tumor-associated antigens directly into the patient's cells [12]. Once internalized, the mRNA vaccine instructs the cells to synthesize the target antigens, effectively transforming the patient's cells into antigen-producing factories [13].
- *Peptide-Based Vaccines:* These vaccines comprise synthetic peptides derived from specific tumor-associated antigens. Upon administration, these peptides are processed and presented by antigen-presenting cells (APCs), triggering the activation of tumor-specific T cells [14].
- *Dendritic Cell Vaccines:* These vaccines harness the power of dendritic cells, specialized immune cells that play a pivotal role in T-cell activation. Dendritic cells are isolated from the patient, incubated with tumor-specific antigens, and reintroduced into the patient, where they efficiently prime T cells against the tumor antigens [15].

#### *14.2.1.3 Mechanism of Action*

The essence of personalized cancer vaccines lies in their ability to elicit an immune response directed against the patient's unique tumor antigens. These antigens may be tumor-specific antigens, neoantigens (antigens that arise from tumor-specific mutations), or antigens derived from the patient's own cancer cells. By targeting these unique antigens, personalized vaccines bypass the tolerance mechanisms that often hinder conventional immunotherapy approaches [11].

### 14.2.2 Unleashing the Power of Personalization in Cancer Treatment

#### *14.2.2.1 Tackling Heterogeneity: A Hallmark of Cancer*

Cancer is characterized by remarkable heterogeneity, both between different types of cancer and within individual tumors [16]. This heterogeneity poses a significant challenge to conventional cancer therapies, which often fail to account for the vast molecular diversity within tumors. Personalized cancer vaccines address this heterogeneity directly by tailoring treatments to the specific genetic and molecular makeup of each patient's tumor. This precision approach enhances the likelihood of treatment success, as it targets the unique vulnerabilities of each patient's cancer.

#### *14.2.2.2 Precision Medicine: A New Era in Cancer Treatment*

Personalized cancer vaccines align seamlessly with the principles of precision medicine, an emerging paradigm that emphasizes the importance of treating patients based on their individual characteristics. By leveraging the patient's own genetic and molecular information, personalized vaccines offer the potential for more effective and better-tolerated treatments compared to standard therapies.

#### *14.2.2.3 Enhancing Patient Outcomes: The Promise of Personalized Vaccines*

Personalized cancer vaccines have demonstrated promising results in improving patient outcomes, including increased response rates and extended survival. Clinical trials have shown that these vaccines can induce durable antitumor responses in patients with a range of malignancies, particularly those that have proven resistant to conventional therapies. The personalized approach also minimizes the risk of adverse effects associated with non-specific treatments, contributing to a more favorable risk–benefit profile. This is particularly important for patients who have already undergone conventional therapies, as they may be more susceptible to side effects.

### 14.2.3 Role of Nanocarriers in Delivering Personalized Cancer Vaccines

In the realm of personalized cancer immunotherapy, nanocarriers have emerged as transformative tools for targeted antigen delivery and enhanced antigen presentation. These nanoscale constructs offer a unique potential to revolutionize cancer treatment by precisely delivering vaccine components to tumor sites and augmenting antitumor immunity [17].

#### *14.2.3.1 Targeted Antigen Delivery*

Nanocarriers, such as liposomes or nanoparticles, serve as versatile vehicles for transporting antigens directly to cancer cells [17]. Their tunable surface properties allow for the conjugation of specific targeting moieties, enabling them to navigate the complex biological milieu and reach their desired destinations [18, 19]. One compelling example of targeted antigen delivery is the use of gold nanoparticles conjugated with antibodies against tumor-associated antigens [20, 21]. These nanoparticles selectively bind to cancer cells, thereby facilitating the localized release of antigens and minimizing systemic exposure. This targeted approach enhances the efficacy of personalized cancer vaccines while reducing potential side effects [22].

#### *14.2.3.2 Improved Antigen Presentation*

Nanocarriers not only facilitate the delivery of antigens but also enhance their presentation to immune cells, particularly APCs. APCs, such as dendritic cells, play a crucial role in initiating immune responses by engulfing and processing antigens, ultimately presenting them to T cells for activation [23, 22]. Nanocarriers can modulate the antigen presentation process by encapsulating antigens within a controlled release matrix, ensuring sustained antigen release and prolonging the interaction between antigens and APCs [23]. Additionally, nanocarriers can be engineered to co-deliver antigens along with immunostimulatory molecules, further enhancing T-cell activation and amplifying the overall immune response [23].

### 14.2.4 Case Studies and Clinical Applications

#### *14.2.4.1 mRNA Vaccines*

mRNA vaccines represent a promising avenue for personalized cancer immunotherapy. These vaccines deliver mRNA-encoding tumor-associated antigens directly into cells, enabling the production of antigens within the patient's own cells. This approach offers the advantage of tailoring the vaccine to each patient's unique tumor antigen profile. A notable example of mRNA vaccines in personalized cancer treatment is the clinical trial of an mRNA vaccine targeting neoantigens, which are antigens unique to individual tumors. This trial demonstrated promising results, with patients exhibiting durable antitumor responses and improved overall survival [12].

#### *14.2.4.2 Peptide-Based Vaccines*

Peptide-based vaccines offer a more direct approach to personalized cancer immunotherapy [24]. These vaccines consist of synthetic peptides derived from specific tumor-associated antigens, allowing for precise activation of T cells against these antigens [24]. One compelling example of peptide-based personalized cancer vaccines is the use of peptides derived from the *MAGE-A3* antigen, which is expressed in a variety of cancers. Clinical studies have shown that *MAGE-A3* peptide vaccines can elicit antitumor responses in patients with melanoma, lung cancer, and other malignancies [25].

#### *14.2.4.3 Challenges and Future Directions*

While nanocarriers and personalized cancer vaccines have shown promising potential, challenges remain in translating these approaches into routine clinical practice. One challenge lies in optimizing the design and formulation of nanocarriers to ensure their stability, targeting efficiency, and biocompatibility [26]. Additionally, further research is needed to identify and validate tumor-specific antigens that can serve as effective targets for personalized immunotherapy. Despite these challenges, the field of personalized cancer immunotherapy holds immense promise for revolutionizing cancer treatment. With continued research and development, nanocarriers and personalized cancer vaccines have the potential to transform the way we combat cancer, offering patients more effective and individualized treatment options [26].

### 14.2.5 Immune Checkpoints in Cancer Immunotherapy: A Balancing Act

- *Introduction to Immune Checkpoints:* Immune checkpoints are crucial regulatory mechanisms that maintain immune system homeostasis by preventing excessive activation and autoimmunity. These checkpoints act as molecular brakes, preventing T cells from indiscriminately attacking both healthy and cancerous cells. Some examples of immune checkpoints include *PD-1*, *PD-L1*, and *CTLA-4* [27, 28].
- *Role of Immune Checkpoints in Cancer Immunotherapy:* In cancer, tumors often exploit these checkpoints to evade immune detection and destruction. Cancer cells express checkpoint ligands, such as *PD-L1*, which bind to checkpoint receptors, such as *PD-1*, on T cells, sending an inhibitory signal that suppresses T cell activation and prevents them from effectively mounting an antitumor response [29].

### 14.2.6 Harnessing Immune Checkpoint Blockade: A Paradigm Shift in Cancer Treatment

- *Immune Checkpoint Molecules as Key Targets:* Immune checkpoint molecules, such as *PD-1*, *PD-L1*, and *CTLA-4*, have emerged as key targets in cancer immunotherapy. Inhibiting these checkpoints has revolutionized cancer treatment, aiming to unleash the immune system's inherent ability to recognize and eliminate cancer cells [30–32].
- *Advent of Immune Checkpoint Inhibitors:* The advent of immune checkpoint inhibitors (ICIs) marks a pivotal moment in cancer therapy. These agents, such as ipilimumab (anti-CTLA-4) and nivolumab (anti-PD-1), have demonstrated remarkable efficacy in treating a range of previously intractable cancers, leading to durable remissions and improved survival rates [31, 32].

### 14.2.7 Challenges in Immune Checkpoint Blockade: Navigating the Complexities of Cancer Immunotherapy

- *Limited Response Rates:* Despite significant breakthroughs, not all patients respond to ICIs, and response rates vary widely across different cancer types. Understanding the factors contributing to limited response rates is crucial for improving the overall efficacy of these therapies [33].
- *Autoimmune Side Effects:* ICIs can lead to autoimmune side effects, as the removal of inhibitory signals may result in the immune system attacking normal tissues. These side effects, while manageable in most cases, can be severe and require careful monitoring and management [34].

### 14.2.8 Nanocarriers: Enhancing Precision and Overcoming Challenges

- *Targeted Delivery:* Encapsulating ICIs within nanoparticles or liposomes enables targeted delivery to tumor sites, enhancing drug delivery precision, reducing off-target effects, and minimizing systemic toxicity [35].
- *Combination Therapies:* Nanocarriers facilitate the combination of ICIs with other therapeutic agents, such as chemotherapy or other immunomodulators, enabling synergistic effects and addressing multiple aspects of tumor evasion [36].
- *Controlled Release Strategies:* Nanocarriers enable controlled release strategies, allowing for sustained and localized delivery of ICIs, potentially overcoming resistance mechanisms and maximizing therapeutic efficacy [37].
- *Co-delivery with Immunomodulatory Agents:* Nanocarriers can be employed for the co-delivery of ICIs with immunomodulatory agents to fine-tune the immune response, mitigating autoimmune side effects while maximizing antitumor effects [38].

### 14.2.9 A Promising Future for Immune Checkpoint Modulation with Nanocarriers

- *Transformative Approach:* Nanocarriers offer a transformative approach to immune checkpoint modulation, thereby addressing the limitations of current therapies and paving the way for more effective and personalized cancer treatment [39].
- *Increasing Pivotal Role:* As research continues to advance, nanocarriers are poised to play an increasingly pivotal role in revolutionizing cancer immunotherapy and improving patient outcomes [40].

## 14.3 ADOPTIVE T-CELL THERAPY: UNLEASHING THE POWER OF THE IMMUNE SYSTEM

### 14.3.1 Overview of Adoptive T-Cell Therapy: A Personalized Approach to Cancer Treatment

#### *14.3.1.1 Definition and Principles*

Adoptive T-cell therapy, a form of immunotherapy, involves harnessing the patient's own immune system to fight cancer. This approach involves the isolation, expansion, and reinfusion of a patient's T cells, which are a type of white blood cell that plays a crucial role in the immune response [41, 42].

After isolating T cells from the patient's blood, they undergo a process called activation, either through genetic modification or *ex vivo* stimulation. This activation process enhances the T cells' ability to recognize and attack cancer cells. Once activated, the T cells are reinfused back into the patient, where they can specifically target and eliminate tumor cells [43].

#### *14.3.1.2 Types of Adoptive T-Cell Therapy*

Two primary types of adoptive T-cell therapy are:

- *Chimeric Antigen Receptor T-cell (CAR-T) Therapy:* This approach involves genetically engineering T cells to express a synthetic receptor, called a chimeric antigen receptor (CAR), on their surface. CARs are designed to recognize specific tumor antigens, allowing the T cells to specifically target and attack cancer cells that express these antigens [41].
- *Tumor-Infiltrating Lymphocyte (TIL) Therapy:* This approach involves isolating and expanding naturally occurring tumor-specific T cells, known as tumor-infiltrating lymphocytes (TILs), from the patient's tumor. TILs are T cells that have already infiltrated the tumor and have developed some level of reactivity against the tumor cells [44].

### 14.3.2 Limitations and Challenges: Navigating the Complexities of Adoptive T-Cell Therapy

Despite having immense promise, adoptive T-cell therapy faces several limitations and challenges:

- *Limited Persistence:* Adoptively transferred T cells may exhibit limited persistence in the body, meaning they may not survive for a long period. This can affect their long-term anti-tumor activity and the durability of the treatment response [43].
- *Off-Target Toxicities:* In some cases, activated T cells may inadvertently target normal tissues, leading to severe off-target toxicities. Enhancing precision in T-cell targeting and minimizing these adverse effects are crucial for improving the safety profile of adoptive T-cell therapies [45].
- *Tumor Microenvironment Suppression:* The tumor microenvironment, the surrounding tissue, and cells that support tumor growth can be immunosuppressive. It implies that such microenvironment can hinder T-cell function and limit their effectiveness in infiltrating and eradicating tumors. Overcoming these inhibitory signals is a significant challenge in optimizing adoptive T-cell therapy outcomes [41, 43].

### 14.3.3 Enhancing Adoptive T-Cell Therapy with Nanocarriers: A Promising Approach

Adoptive T-cell therapy has emerged as a promising treatment modality for cancer, harnessing the patient's own immune system to attack cancer cells. However, several limitations hinder the efficacy and safety of this approach, including limited T-cell persistence, off-target toxicities, and the immunosuppressive tumor microenvironment. Nanocarriers, nanoscale constructs such as nanoparticles or liposomes, hold immense potential to address these limitations and revolutionize adoptive T-cell therapy [46].

#### *14.3.3.1 Improving T-Cell Trafficking*

Nanocarriers can facilitate an efficient delivery of T cells directly to tumor sites, enhancing their ability to overcome biological barriers and reach the tumor microenvironment [47, 48]. This is crucial for effective antitumor activity, as T cells often encounter obstacles in reaching the tumor due to the complex anatomy of the tumor microenvironment and the presence of tight junctions. Nanocarriers can potentially overcome these barriers by modifying their surface properties to enhance interaction with specific tissues or directly penetrating through cellular barriers [39]. Moreover, nanocarriers can be used to deliver chemokines, which are molecules that attract and guide T cells toward tumor sites, further promoting T-cell recruitment to the tumor microenvironment [47–49].

##### *14.3.3.2 Targeted Delivery of T-Cell Activating Agents*

Nanocarriers can serve as platforms for the targeted delivery of cytokines or other immune-stimulating factors to the tumor microenvironment. By delivering these activating agents directly to the tumor site, nanocarriers can create a supportive environment for T-cell activation and persistence [39]. This is particularly important for overcoming the immunosuppressive tumor microenvironment, where tumor cells and stromal cells secrete inhibitory factors that can dampen T-cell activity [49]. Nanocarriers can encapsulate cytokines or other immune-modulating agents and release them at controlled rates, ensuring sustained activation of T cells in the tumor microenvironment [47]. This targeted delivery approach can minimize systemic exposure to these agents, reducing the risk of off-target toxicities.

##### *14.3.3.3 Reducing Off-Target Toxicities*

Nanocarriers can be used to encapsulate T-cell activating agents and control their release, thereby minimizing systemic exposure to these agents and reducing the risk of off-target toxicities [48]. This is particularly important for cytokines, which can have pleiotropic effects on various cells, leading to adverse events such as cytokine release syndrome (CRS). By controlling the release of cytokines from nanocarriers, their concentration in the body can be carefully regulated, preventing excessive activation of immune cells that could cause harm to normal tissues [50].

Nanocarriers offer a promising solution to address the limitations of adoptive T-cell therapy, enhancing efficacy and safety through improved T-cell trafficking, targeted delivery of activating agents, and reduced off-target toxicities. As research continues to evolve, nanocarriers are poised to play an increasingly pivotal role in revolutionizing adoptive T-cell therapy and improving patient outcomes in the fight against cancer.

### 14.3.4 Case Studies and Clinical Progress: Real-World Applications of Adoptive T-Cell Therapy

Clinical trials and case studies have demonstrated promising results with adoptive T-cell therapy, particularly in the treatment of hematological malignancies and solid tumors:

- *CAR-T Cell Therapies:* CAR-T cell therapy has shown remarkable success in treating B-cell lymphomas, demonstrating durable responses and improved patient outcomes. For instance, CD19-targeted CAR-T cells have been successfully used to treat patients with refractory or relapsed B-cell acute lymphoblastic leukemia (B-ALL) [51].
- *Tumor-Infiltrating Lymphocytes:* TIL therapy has shown promise in treating solid tumors, particularly melanoma. Clinical studies have demonstrated that TILs can induce tumor regression and improve patient outcomes in patients with advanced melanoma [52].

Adoptive T-cell therapy holds immense promise as a personalized and effective cancer treatment modality. The integration of nanocarriers into adoptive T-cell therapy has the potential to address the limitations of this approach and further enhance its efficacy and safety.

## 14.4 COMBINATION STRATEGIES: HARNESSING SYNERGISTIC EFFECTS FOR ENHANCED CANCER IMMUNOTHERAPY

### 14.4.1 Rationale for Combination Approaches: A Multifaceted Approach to Cancer Treatment

In the realm of cancer immunotherapy, combination strategies have emerged as a promising approach to overcome the limitations of monotherapies and achieve more robust and durable anti-tumor responses. These strategies involve the integration of different therapeutic modalities, such as immune checkpoint blockade, adoptive T-cell therapy, and personalized vaccines, to synergistically target multiple aspects of tumor evasion and resistance [6, 11, 38, 53].

##### *14.4.1.1 Enhancing Therapeutic Synergy: Unleashing the Power of Combined Modalities*

Combination approaches capitalize on the complementary mechanisms of action of different therapeutic modalities, creating a synergistic effect that surpasses the efficacy of individual treatments. For instance, combining immune checkpoint blockade with adoptive T-cell therapy can simultaneously enhance T-cell activation and overcome the immunosuppressive tumor microenvironment [36, 54].

#### *14.4.1.2 Overcoming Limitations of Monotherapy: Addressing Resistance and Variability*

While monotherapies have demonstrated success in certain cancer types, they often face challenges such as tumor resistance and variable response rates. Combination strategies aim to address these limitations by utilizing the strengths of different therapeutic modalities to achieve more comprehensive and sustained anti-tumor responses [37, 54].

### 14.4.2 Nanocarriers as Platforms for Combinatorial Therapies: A Versatile Delivery System

Nanocarriers, nanoscale constructs such as nanoparticles, liposomes, and polymers, offer a versatile platform for implementing combination strategies in cancer immunotherapy. Their unique properties enable the co-delivery of multiple therapeutic agents, facilitating synergistic interactions and enhancing the overall efficacy of the treatment [18, 46].

#### *14.4.2.1 Personalized Vaccines with Immune Checkpoint Blockade: A Synergistic Combination*

Nanocarriers can serve as powerful tools for combining personalized cancer vaccines with immune checkpoint blockade, by encapsulating tumor-specific antigens and checkpoint inhibitors within the same carrier, Nanocarriers can deliver these agents in a synchronized manner, promoting a robust and targeted immune response against cancer cells [22, 38, 50, 55].

#### *14.4.2.2 Integrating Adoptive T-Cell Therapy with Other Modalities: Enhancing T-Cell Efficacy*

Nanocarriers play a pivotal role in integrating adoptive T-cell therapy with other treatment modalities. By co-delivering engineered T cells and additional therapeutic agents, such as cytokines or chemokines, nanocarriers can enhance T-cell trafficking, persistence, and efficacy, contributing to a more potent and sustained anti-tumor response [37].

### 14.4.3 Overcoming Immunotherapy Resistance: Tackling Tumor Evasion Mechanisms

Combination strategies offer the potential to overcome immunotherapy resistance by addressing the diverse mechanisms employed by tumors to evade immune recognition and attack. Nanocarriers can be employed to deliver agents that disrupt immune evasion pathways, such as antibodies that block inhibitory molecules or agents that modulate tumor metabolism, making tumors more susceptible to immune attack [22].

#### *14.4.3.1 Targeting Immune-Suppressive Microenvironments: Transforming the Tumor Battleground*

Nanocarriers provide a platform for delivering therapeutics that target the immune-suppressive microenvironments within tumors [47]. These microenvironments, characterized by the presence of inhibitory cells and factors, can hinder T-cell infiltration and function. Combinatorial approaches, incorporating agents that modulate the tumor microenvironment, such as chemokines that attract T cells or agents that deplete inhibitory cells, can enhance the infiltration and function of immune cells, overcoming the inhibitory signals that promote tumor immune escape [47, 48].

Nanocarriers have emerged as transformative tools in the field of cancer immunotherapy, thus enabling the implementation of combination strategies that address the multifaceted challenges of tumor evasion and resistance. By facilitating the co-delivery of multiple therapeutic agents and targeting immune-suppressive microenvironments, nanocarriers pave the way for more effective and durable cancer treatment approaches.

## 14.5 CHALLENGES AND FUTURE DIRECTIONS

Nanocarrier-based cancer immunotherapy has emerged as a promising therapeutic approach, offering the potential for enhanced precision, efficacy, and personalization of cancer treatment. However, despite significant advancements, several challenges and unmet needs need to be addressed to fully realize the potential of this approach.

### 14.5.1 Current Challenges in Nanocarrier-Based Cancer Immunotherapy

Despite the promise of nanocarriers in cancer immunotherapy, several challenges hinder their widespread clinical application:

#### *14.5.1.1 Biocompatibility and Toxicity*

- Ensuring the biocompatibility and safety of nanocarriers is crucial, as their potential for inducing adverse effects remains a concern. Thorough preclinical and clinical evaluation is essential to assess the potential toxicity of nanocarriers and their biological impact on the body [50].

#### *14.5.1.2 Targeting Precision*

- Achieving precise and selective targeting of tumor sites while minimizing off-target effects is a persistent challenge [56]. The development of nanocarriers with enhanced tumor-homing capabilities and improved specificity is crucial for optimizing therapeutic efficacy and reducing systemic exposure to therapeutic agents [46, 47].

#### *14.5.1.3 Immune System Interactions*

- The interaction of nanocarriers with the immune system can significantly impact treatment outcomes. Understanding and modulating the immunogenicity of nanocarriers is essential to prevent immune clearance and maximize their therapeutic potential. Strategies such as surface modification of nanocarriers with stealth coatings or co-delivery of immunosuppressive agents can help mitigate immune recognition and clearance [23, 50, 56].

### 14.5.2 Unmet Needs and Opportunities for Improvement

#### *14.5.2.1 Patient-Specific Approaches*

Tailoring nanocarrier-based therapies to the unique characteristics of each patient's tumor is an unmet need [39]. Advances in personalized medicine, including the development of patient-specific nanocarriers, hold the potential to significantly improve treatment outcomes [57]. By incorporating patient-specific tumor profiling and biomarker analysis, researchers can design nanocarriers that target specific tumor antigens and modulate the immune microenvironment to achieve personalized treatment plans [22, 47].

#### *14.5.2.2 Combination Therapies Optimization*

- Optimizing the design of combination therapies using nanocarriers is an area for further exploration [23].
- Identifying the most effective combinations of therapeutic agents, determining optimal dosing regimens, and establishing the ideal sequence of treatments are critical steps in achieving synergistic effects and overcoming resistance mechanisms [56].

#### *14.5.2.3 Biomarker Discovery*

- The identification of reliable biomarkers for predicting treatment response and monitoring therapeutic efficacy is a pressing need. Biomarkers can serve as valuable tools for patient stratification, thus enabling the selection of patients who are most likely to benefit from specific nanocarrier-based immunotherapies [58].

### 14.5.3 Emerging Technologies and Innovations

#### *14.5.3.1 Advanced Nanocarrier Designs*

- Emerging technologies are paving the way for the development of more sophisticated nanocarrier designs. Innovations such as smart nanocarriers with stimuli-responsive properties, improved drug loading capacities, and enhanced stability hold promise for addressing current limitations. These advanced nanocarriers can release their therapeutic payloads in a controlled manner, respond to specific triggers in the tumor microenvironment, and maximize their accumulation at tumor sites [22].

#### *14.5.3.2 Immunomodulatory Nanocarriers*

- The integration of immunomodulatory properties into nanocarriers represents a promising innovation. Nanocarriers capable of modulating the immune response locally, such as those that can deliver cytokines or chemokines directly to tumor sites, may enhance the overall immunotherapeutic impact while minimizing systemic side effects [22, 46, 48].

#### 14.5.3.4 Future Prospects for Personalized Cancer Immunotherapy

##### *14.5.3.1 Integration of Multi-Omics Data*

The future of personalized cancer immunotherapy lies in the integration of multi-omics data. Incorporating genomics, transcriptomics, and proteomics data into treatment decisions can refine patient stratification, inform nanocarrier design, and enable the development of personalized treatment plans [59].

##### *14.5.3.2 Advancements in Target Identification*

Advances in target identification technologies will play a pivotal role in the development of personalized therapies. Precise identification of patient-specific antigens, immune checkpoints, and other therapeutic targets will guide the design of nanocarriers tailored to the unique features of each individual's cancer [60].

##### *14.5.3.3 In-vivo Imaging and Monitoring*

The incorporation of advanced *in vivo* imaging and monitoring technologies will enhance treatment assessment and refinement. Real-time imaging of nanocarrier distribution, therapeutic payload release, and immune response activation can provide valuable insights for adjusting treatment strategies on an individualized basis. This real-time monitoring will enable clinicians to personalize treatment parameters, optimize dosing regimens, and identify potential adverse effects early on [22].

Nanocarrier-based cancer immunotherapy holds immense potential to revolutionize cancer treatment by enabling personalized, targeted, and synergistic therapies. Addressing the current challenges, harnessing emerging technologies, and embracing personalized medicine will shape the next generation of nanocarrier-based approaches, thereby contributing to more effective and individualized cancer treatments [61].

## 14.6 CONCLUSION

### 14.6.1 Recapitulation of Key Findings

Throughout this exploration, it becomes evident that nanocarriers play a pivotal role in advancing cancer immunotherapy. Their unique properties and versatility enable them to address critical challenges and enhance the efficacy of immunotherapeutic agents as shown in Figure 14.1.

Key findings include.

#### *14.6.1.1 Enhanced Targeted Drug Delivery*

Nanocarriers overcome the limitations of traditional delivery methods, delivering therapeutic agents directly to tumor sites with minimal off-target effects. This precise targeting minimizes systemic toxicity and maximizes therapeutic impact.

#### *14.6.1.2 Synergy Achieved Through Combination Therapies*

Nanocarriers facilitate the co-delivery of multiple therapeutic agents, enabling synergistic interactions that amplify the anti-tumor response. This approach holds promise for overcoming resistance mechanisms and achieving more durable treatment outcomes.

#### *14.6.1.3 Overcoming Challenges in Immune Checkpoint Modulation*

Nanocarriers can effectively deliver immune checkpoint inhibitors to tumor microenvironments, modulating the immune response and enabling tumor recognition and attack. This strategy has demonstrated promising results in treating various cancer types.

### 14.6.2 Implications for Clinical Practice

#### *14.6.2.1 Enhanced Treatment Efficacy*

- The use of nanocarriers in cancer immunotherapy has significant implications for improving treatment efficacy.
- By overcoming challenges such as off-target effects and improving drug delivery precision, nanocarriers have the potential to enhance the therapeutic impact of immunotherapeutic agents.

#### *14.6.2.2 Tailored Treatment Approaches*

- The incorporation of nanocarriers facilitates the development of tailored treatment approaches.
- Clinicians can leverage nanocarriers to customize therapies based on individual patient profiles, contributing to more effective and personalized cancer care.

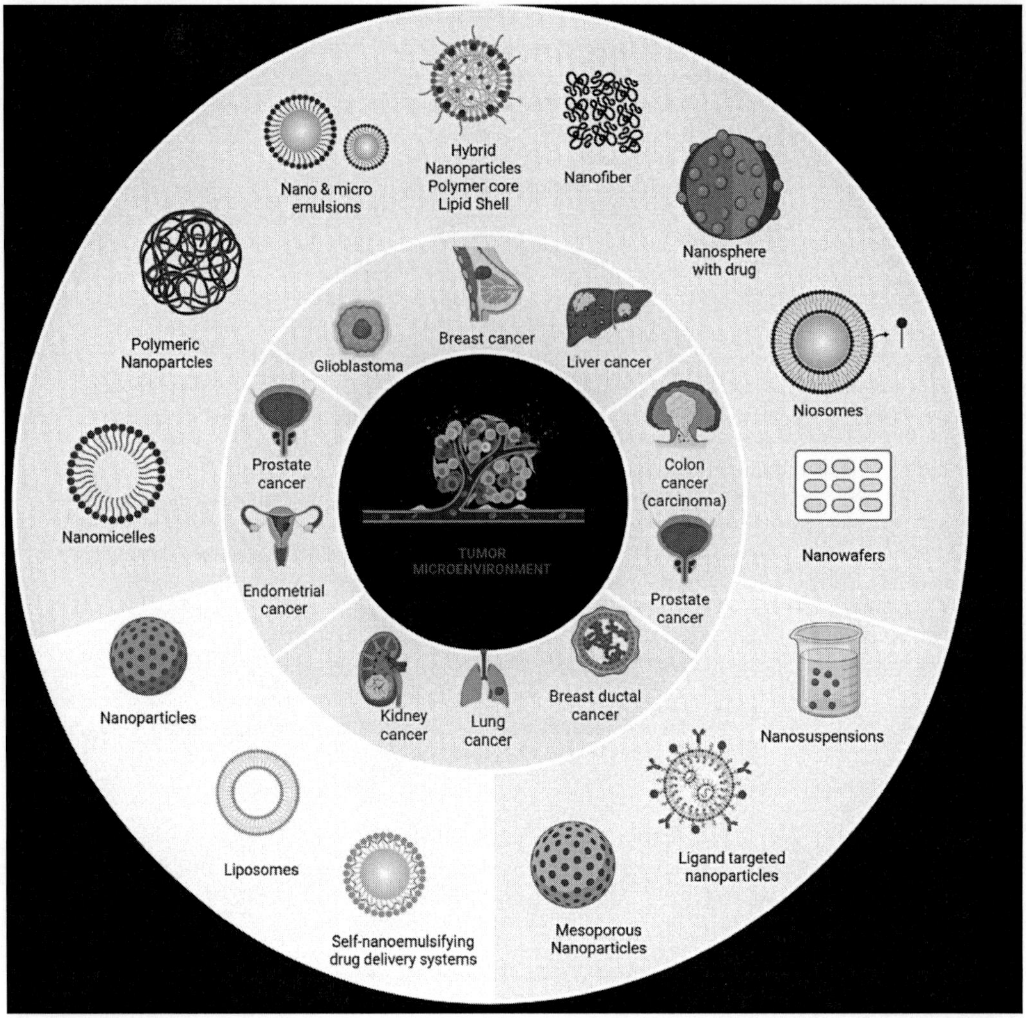


**Figure 14.1** Nanocarrier-based drug delivery systems offer a promising approach for targeting the complex tumor microenvironment (TME). This schematic illustrates the potential of diverse nanoparticle types (including liposomes, polymeric nanoparticles, micelles, niosomes, nanowafers, and nanosuspensions) to deliver therapeutic agents directly to the TME. These systems can be tailored for specific targeting in a range of cancers, including liver, lung, colon, prostate, and endometrial cancers. Additionally, specific drug delivery systems like self-nanoemulsifying drug delivery systems (SNEDDS) and mesoporous nanoparticles provide further specialized advantages within this therapeutic strategy.

##### *14.6.2.3 Optimizing Combination Therapies*

- Combination therapies, made possible by nanocarriers, offer new avenues for optimizing cancer treatment.
- Clinicians can explore novel combinations, adjust dosing regimens, and fine-tune treatment strategies to achieve synergistic effects and overcome resistance.

#### 14.6.3 Call to Action for Future Research

##### *14.6.3.1 Further Understanding of Resistance Mechanisms*

- Future research should focus on gaining a deeper understanding of resistance mechanisms in nanocarrier-based immunotherapies.

- Investigating how tumors evade immune recognition and nano carrier-mediated interventions will guide the development of more resilient and effective treatments.

#### *14.6.3.2 Clinical Translation and Validation:*

- The translation of promising preclinical findings to clinical practice requires rigorous validation.
- Future research efforts should prioritize large-scale clinical trials to validate the safety, efficacy, and long-term outcomes of nanocarrier-based cancer immunotherapies.

#### *14.6.3.3 Exploration of Emerging Technologies*

- Continued exploration of emerging technologies is essential for pushing the boundaries of nanocarrier-based immunotherapy.
- Researchers should delve into advanced nanocarrier designs, immunomodulatory properties, and innovative combination strategies to stay at the forefront of cancer treatment innovation.

#### *14.6.3.4 Integration of Multi-Omics Approaches*

- Future research should emphasize the integration of multi-omics approaches for a more comprehensive understanding of individual tumors.
- Integrating genomics, transcriptomics, and proteomics data will refine the identification of targetable antigens and immune checkpoints, thus optimizing the design of nanocarriers for personalized immunotherapies.

In conclusion, the integration of nanocarriers in cancer immunotherapy represents a transformative approach with implications for clinical practice and the future of cancer care. By addressing challenges, refining treatment strategies, and embracing emerging technologies, researchers and clinicians can usher in a new era of precision and personalized cancer immunotherapy.

## REFERENCES

1. Topalian, S.L., C.G. Drake, and D.M. Pardoll, *Immune checkpoint blockade: a common denominator approach to cancer therapy. Cancer Cell*, 2015. **27**(4): p. 450–61.

2. Mellman, I., G. Coukos, and G. Dranoff, *Cancer immunotherapy comes of age. Nature*, 2011. **480**(7378): p. 480–9.

3. Finck, A., S.I. Gill, and C.H. June, *Cancer immunotherapy comes of age and looks for maturity. Nat Commun*, 2020. **11**(1): p. 3325.

4. Pardoll, D.M., *The blockade of immune checkpoints in cancer immunotherapy. Nat Rev Cancer*, 2012. **12**(4): p. 252–64.

5. Rosenberg, S.A., N. P. Restifo, J. C. Yang, R. A. Morgan, and M. E. Dudley, *Adoptive cell transfer: a clinical path to effective cancer immunotherapy. Nat Rev Cancer*, 2008. **8**(4): p. 299–308.

6. Anselmo, A.C., and S. Mitragotri, *Nanoparticles in the clinic: an update. Bioeng Transl Med*, 2019. **4**(3): p. e10143.

7. Wilhelm, S., A.J. Tavares, Q. Dai, S. Ohta, J. Audet, H.F. Dvorak, et al., Analysis of nanoparticle delivery to tumours. *Nature Reviews Materials*, 2016. 1(5): p. 1–12.

8. Popescu, M.T., and C. Tsitsilianis, *Gold/pentablock terpolymer hybrid multifunctional nanocarriers for controlled delivery of tamoxifen: effect of nanostructure on release kinetics. Molecules*, 2022. **27**(12): p. 3764.

9. Torchilin, V.P., *Multifunctional nanocarriers. Adv Drug Deliv Rev*, 2006. **58**(14): p. 1532–55.

10. Bae, Y.H., and K. Park, *Targeted drug delivery to tumors: myths, reality and possibility. J Control Release*, 2011. **153**(3): p. 198–205.

11. Sahin, U., and O. Tureci, *Personalized vaccines for cancer immunotherapy. Science*, 2018. **359**(6382): p. 1355–60.

12. Yuan, Y., F. Gao, Y. Chang, Q. Zhao, and X. He, *Advances of mRNA vaccine in tumor: a maze of opportunities and challenges. Biomark Res*, 2023. **11**(1): p. 6.

13. Yarchoan, M., B.A Johnson III, E. R. Lutz, D.A. Laheru, and E.M. Jaffee, *Targeting neoantigens to augment antitumour immunity. Nat Rev Cancer*, 2017. **17**(4): p. 209–222.

14. Parmiani, G., V. Russo, C. Maccalli, D. Parolini, N. Rizzo, and M. Maio, *Peptide-based vaccines for cancer therapy. Hum Vaccin Immunother*, 2014. **10**(11): p. 3175–8.

15. Benteyn, D., C. Heirman, A. Bonehill, K. Thielemans, and K. Breckpot, *mRNA-based dendritic cell vaccines. Expert Rev Vaccines*, 2015. **14**(2): p. 161–76.

16. Dagogo-Jack, I., and A.T. Shaw, *Tumour heterogeneity and resistance to cancer therapies. Nat Rev Clin Oncol*, 2018. **15**(2): p. 81–94.

17. Sunoqrot, S., S.A. Abdel Gaber, R. Abujaber, M. Al-Majawleh, and S. Talhouni, Lipid- and polymer-based nanocarrier platforms for cancer vaccine delivery. *ACS Appl Bio Mater*, 2024. 7(8): p. 4998–5019.

18. Bai, X., Z.L. Smith, Y. Wang, S. Butterworth, and A. Tirella, *Sustained drug release from smart nanoparticles in cancer therapy: a comprehensive review. Micromachines (Basel)*, 2022. **13**(10): p. 1623.

19. Kong, X., Y. Liu, X. Huang, S. Huang, F. Gao, P. Rong et al., *Cancer therapy based on smart drug delivery with advanced nanoparticles. Anticancer Agents Med Chem*, 2019. **19**(6): p. 720–730.

20. Kumar, A., X. Zhang, and X.J. Liang, *Gold nanoparticles: emerging paradigm for targeted drug delivery system. Biotechnol Adv*, 2013. **31**(5): p. 593–606.

21. Sui, C., R. Tan, Z. Liu, X. Li, and W. Xu, *Smart chemical oxidative polymerization strategy to construct Au@PPy core-shell nanoparticles for cancer diagnosis and imaging-guided photothermal therapy. Bioconjug Chem*, 2023. **34**(1): p. 257–268.

22. Sun, L., H. Liu, Y. Ye, Y. Lei, R. Islam, and S. Tan et al., *Smart nanoparticles for cancer therapy. Signal Transduct Target Ther*, 2023. **8**(1): p. 418.

23. Zhao, B., X. Li, Y. Kong, W. Wang, T. Wen, Y. Zhang et al., *Recent advances in nano-drug delivery systems for synergistic antitumor immunotherapy. Front Bioeng Biotechnol*, 2022. **10**: p. 1010724.

24. Liu, W., H. Tang, L. Li, X. Wang, Z. Yu, and J. Li, *Peptide-based therapeutic cancer vaccine: current trends in clinical application. Cell Prolif*, 2021. **54**(5): p. e13025.

25. Neek, M., J.A Tucker, T.I. Kim, N.M. Molino, E.L. Nelson, and S.-W. Wang, *Co-delivery of human cancer-testis antigens with adjuvant in protein nanoparticles induces higher cell-mediated immune responses. Biomaterials*, 2018. **156**: p. 194–203.

26. Tay, B.Q., Q. Wright, R. Ladwa, C. Perry, G. Leggatt, F. Simpson et al., *Evolution of cancer vaccines-challenges, achievements, and future directions. Vaccines (Basel)*, 2021. **9**(5): p. 535.

27. Guo, X., Y. Yang, Z. Qian, M. Chang, Y. Zhao, W. Ma et al., Immune landscape and progress in immunotherapy for pituitary neuroendocrine tumors. *Cancer Lett*, 2024. **592**: p. 216908.

28. Sharma, S., N. Singh, A.A. Turk, I. Wan, A. Guttikonda, J.L. Dong et al., *Molecular insights into clinical trials for immune checkpoint inhibitors in colorectal cancer: unravelling challenges and future directions. World J Gastroenterol*, 2024. **30**(13): p. 1815–1835.

29. Benson, D.M., Jr., *Checkpoint inhibition in myeloma. Hematology Am Soc Hematol Educ Program*, 2016. **2016**(1): p. 528–533.

30. Collin, M., *Immune checkpoint inhibitors: a patent review (2010–2015). Expert Opin Ther Pat,* 2016. **26**(5): p. 555–64.

31. Penter, L., Y. Zhang, A. Savell, T. Huang, N. Cieri, E.M. Thrash et al., *Molecular and cellular features of CTLA-4 blockade for relapsed myeloid malignancies after transplantation. Blood*, 2021. **137**(23): p. 3212–3217.

32. Xu, J., Y. Kong, P. Zhu, M. Du, X. Liang, Y. Tong et al., *Progress in small-molecule inhibitors targeting PD-L1. RSC Med Chem*, 2024. **15**(4): p. 1161–1175.

33. Li, J., J. Xie, Z. Jintao, and J. Tao, *Multifunctional nanoparticle approach for targeting melanoma. J Investig Dermatol Symp Proc*, 2018. **19**(2): p. S89–S90.

34. Postow, M.A., R. Sidlow, and M.D. Hellmann, *Immune-related adverse events associated with immune checkpoint blockade. N Engl J Med*, 2018. **378**(2): p. 158–168.

35. Yoo, J., C. Park, G. Yi, D. Lee, and H. Koo, *Active targeting strategies using biological ligands for nanoparticle drug delivery systems. Cancers (Basel*), 2019. **11**(5): p. 640.

36. Kiaie, S.H., H. Salehi-Shadkami, M.J. Sanaei, M. Azizi, M. Shokrollahi Barough, M.S. Nasr et al., *Correction: nano-immunotherapy: overcoming delivery challenge of immune checkpoint therapy. J Nanobiotechnology*, 2023. **21**(1): p. 433.

37. Bozzer, S., M. Dal Bo, M.C. Grimaldi, G. Toffoli, and P. Macor, *Nanocarriers as a delivery platform for anticancer treatment: biological limits and perspectives in B-cell malignancies. Pharmaceutics*, 2022. **14**(9): p. 1965.

38. Zhang, J., S. Wang, D. Zhang, X. He, X. Wang, H. Han and Y. Qin, *Nanoparticle-based drug delivery systems to enhance cancer immunotherapy in solid tumors. Front Immunol*, 2023. **14**: p. 1230893.

39. Chehelgerdi, M., M. Chehelgerdi, O.Q.B. Allela, R.D.C. Pecho, N. Jayasankar, D.P. Rao et al., *Progressing nanotechnology to improve targeted cancer treatment: overcoming hurdles in its clinical implementation. Mol Cancer*, 2023. **22**(1): p. 169.

40. Das, S.S., P. Bharadwaj, M. Bilal, M. Barani, A. Rahdar, P. Taboada et al., *Stimuli-responsive polymeric nanocarriers for drug delivery, imaging, and theragnosis. Polymers (Basel*), 2020. **12**(6): p. 1397.

41. Zhang, T., Z. Tai, F. Miao, X. Zhang, J. Li, Q. Zhu et al., *Adoptive cell therapy for solid tumors beyond CAR-T: current challenges and emerging therapeutic advances. J Control Release*, 2024. **368**: p. 372–396.

42. Gitto, S.B., C.J.N. Ihewulezi, and D.J. Powell, Jr., *Adoptive T cell therapy for ovarian cancer. Gynecol Oncol*, 2024. **186**: p. 77–84.

43. Albarran, V., M. San Roman, J. Pozas, J. Chamorro, D.I. Rosero, P. Guerrero et al., *Adoptive T cell therapy for solid tumors: current landscape and future challenges. Front Immunol*, 2024. **15**: p. 1352805.

44. Zhao, Y., J. Deng, S. Rao, S. Guo, J. Shen, F. Du et al., *Tumor infiltrating lymphocyte (TIL) therapy for solid tumor treatment: progressions and challenges. Cancers (Basel*), 2022. **14**(17): p. 4160.

45. Bagley, E.M., and N.A. Wages, *Dose-finding based on feasibility and late-onset toxicity in adoptive cell therapy trials. J Biopharm Stat*, 2024. **34**(2): p. 151–163.

46. Rana, I., J. Oh, J. Baig, J.H. Moon, S. Son, and J. Nam, *Nanocarriers for cancer nano-immunotherapy. Drug Deliv Transl Res*, 2023. **13**(7): p. 1936–1954.

47. Huang, Y., H. Fan, and H. Ti, *Tumor microenvironment reprogramming by nanomedicine to enhance the effect of tumor immunotherapy. Asian J Pharm Sci*, 2024. **19**(2): p. 100902.

48. Raker, V.K., C. Becker, K. Landfester, and K. Steinbrink, *Targeted activation of T cells with IL-2-coupled nanoparticles. Cells*, 2020. **9**(9): p. 2063.

49. Parayath, N.N., S.B. Stephan, A.L. Koehne, P.S. Nelson, and M.T. Stephan, *In vitro-transcribed antigen receptor mRNA nanocarriers for transient expression in circulating T cells in vivo. Nat Commun*, 2020. **11**(1): p. 6080.

50. Su, S., and M. Kang P, *Recent advances in nanocarrier-assisted therapeutics delivery systems. Pharmaceutics*, 2020. **12**(9): p. 837.

51. Aldoss, I., M. C. Clark, X. Wang, and S.J. Forman, *Leveraging CD19CAR T cells early in the treatment of older patients with B-ALL: are we there yet? Leuk Lymphoma*, 2024. **65**(4): p. 440–448.

52. Topalian, S.L., D. Solomon, F.P. Avis, A.E. Chang, D.L. Freerksen, W.M. Linehan et al., *Immunotherapy of patients with advanced cancer using tumor-infiltrating lymphocytes and recombinant interleukin-2: a pilot study. J Clin Oncol*, 1988. **6**(5): p. 839–53.

53. Bezbaruah, R., V.P. Chavda, L. Nongrang, S. Alom, K. Deka, T. Kalita et al., *Nanoparticle-based delivery systems for vaccines. Vaccines (Basel)*, 2022. **10**(11): p. 1946.

54. Wu, F., F. Qiu, S.A. Wai-Keong, and Y. Diao, *The smart dual-stimuli responsive nanoparticles for controlled anti-tumor drug release and cancer therapy. Anticancer Agents Med Chem*, 2021. **21**(10): p. 1202–1215.

55. Nel, A.E., K.-C. Mei, Y.-P. Liao, and X. Liu, *Multifunctional lipid bilayer nanocarriers for cancer immunotherapy in heterogeneous tumor microenvironments, combining immunogenic cell death stimuli with immune modulatory drugs. ACS Nano*, 2022. **16**(4): p. 5184–5232.

56. Dubey, S.K., M. Kali, S. Hejmady, R.N. Saha, A. Alexander, and P. Kesharwani, *Recent advances of dendrimers as multifunctional nano-carriers to combat breast cancer. Eur J Pharm Sci*, 2021. **164**: p. 105890.

57. Shams, F., A. Golchin, A. Azari, L. Mohammadi Amirabad, F. Zarein, A. Khosravi et al., *Nanotechnology-based products for cancer immunotherapy. Mol Biol Rep*, 2022. **49**(2): p. 1389–1412.

58. Ahmad, A., M. Imran, and H. Ahsan, *Biomarkers as biomedical bioindicators: approaches and techniques for the detection, analysis, and validation of novel biomarkers of diseases. Pharmaceutics*, 2023. **15**(6): p. 1630.

59. Donisi, C., A. Pretta, V. Pusceddu, P. Ziranu, E. Lai, M. Puzzoni, S. Mariani et al., *Immunotherapy and cancer: the multi-omics perspective. Int J Mol Sci*, 2024. **25**(6): p. 3563.

60. Li, J., Q. Wang, G. Xia, N. Adilijiang, Y. Li, Z. Hou, Z. Fan et al., *Recent advances in targeted drug delivery strategy for enhancing oncotherapy. Pharmaceutics*, 2023. **15**(9): p. 2233.

61. Garcia-Dominguez, D.J., S. Lopez-Enriquez, G. Alba, C. Garnacho, C. Jimenez-Cortegana, R. Flores-Campos et al., *Cancer nano-immunotherapy: the novel and promising weapon to fight cancer. Int J Mol Sci*, 2024. **25**(2): p. 1195.

# 15 Future Perspectives and Challenges

## *Clinical Translation and Regulatory Aspects, Scale-Up and Manufacturing, Ethical Considerations*

*Aparoop Das, Kalyani Pathak, Riya Saikia, Manash Pratim Pathak, Urvashee Gogoi, Dibyajyoti Das, Jon Jyoti Sahariah, Saptasikha Gogoi, and Sukanya Sonowal*

## 15.1 INTRODUCTION

The advent of nanotechnology has revolutionized the biomedical field, especially in delivering nucleic acids and proteins. Nanocarriers, with their unique properties, show great potential in enhancing therapeutic interventions' efficacy, specificity, and safety. However, transitioning from lab research to clinical applications is complex and multifaceted. Nanocarriers must offer superior therapeutic benefits and comply with stringent regulatory standards [1]. Despite high expectations for nanomedicine, regulatory guidance remains sparse, creating uncertainty that could negatively affect funding, research, and public perception. The lack of a unified definition for nanomaterials across organizations, such as the National Institutes of Health (NIH), the European Science Foundation, and the Food and Drug Administration (FDA), exacerbates this issue. Additionally, the clinical development of nanomedicines is more expensive and time-consuming than conventional drugs due to biological challenges, large-scale manufacturing, biocompatibility and safety issues, intellectual property concerns, and stringent regulations [2]. For successful clinical implementation, nanomedicines must demonstrate a positive benefit-to-risk ratio and overall healthcare cost reduction. However, the necessary infrastructure and expertise for their commercial development are not yet widespread in most pharmaceutical companies. Addressing these gaps is essential for the successful translation of nanomedicines from the lab to the clinic [1].

Despite the lack of specific regulatory guidance, over 50 nanomedicines have reached the market, primarily in cancer therapy, due to the challenging nature of tumor treatment. Notable examples include Doxil®, AmBisome®, Abraxane®, and Eligard®. However, the global inconsistency in the regulation of nanomedicines, with some classified as medical devices and others as medicines, creates significant hurdles. This inconsistency hampers the global licensing of nanomedicines and necessitates urgent coherence across regulatory bodies to meet development expectations. Efforts to establish National Characterisation Laboratories have been made, but more explicit and stringent guidance is needed from main governing bodies like the FDA and Medicines and Healthcare products Regulatory Authority (MHRA). A global consortium for nanomaterial regulation is essential to provide formal guidance to research communities. Without clear leadership and guidance from regulatory bodies, the billions of dollars invested in nanomedicine development over the past two decades may not lead to marketable products, and future investments may decline [2].

Nanotechnology is broadly defined by the use and application of materials within the nanometer range. This extremely small-scale offers numerous benefits across various applications, including electronics, sunscreens, cosmetics, energy storage, and drug delivery [3–7]. The unique chemical, physical, and biological properties of nanoparticles often provide advantages over their larger counterparts [8]. Nanoscale medicines are particularly promising, as many biologically significant molecules such as water, antibodies, proteins, glucose, enzymes, hemoglobin, and receptors fall within this size range [9]. Although nanotechnology was expected to revolutionize healthcare, the impact has not yet met initial expectations. Many experts attribute this to nanomedicine's infancy and the lack of clear regulations for clinical use [10, 11]. Despite limited knowledge and data on the pharmacokinetics, pharmacodynamics, and toxicity of many nanomaterials in humans, the potential benefits of these technologies are significant. There remains strong anticipation about nanomedicine's impact on the pharmaceutical industry, but urgent regulatory guidance is needed to provide legal certainty for manufacturers, policymakers, healthcare providers, and the general public.

### 15.1.1 The Need for Nanomedicine Regulation

Despite the excitement surrounding nanomedicine, the field currently lacks clear regulatory guidance. Many nanomedicines interact directly with genetic materials or biomolecules essential for normal genome function and cell division, potentially causing genotoxicity and mutagenicity [12]. Such toxicity arises from the inflammatory response of neutrophils and macrophages, which produce reactive oxygen and nitrogen species, leading to oxidative and nitrosative stress [13]. The accumulation of free radicals can cause extensive bodily damage, including oxidative DNA damage, protein denaturation, and lipid peroxidation, which can lead to cancer, mitochondrial membrane damage, cell death, necrosis, and transcription of genes responsible for carcinogenesis and fibrosis [14].

DOI: 10.1201/9781003473183-15

Intravenous administration of nanomedicines often results in particle accumulation within the liver and translocation to the central nervous, cardiovascular, and renal systems [15]. For particles that cannot be traced post-administration, there are numerous unknowns posing potential safety threats. The precise interactions of many nanomedicines with biological systems are not fully understood, making it challenging to comprehend their physicochemical and toxicological properties.

Standardized regulatory guidance is essential for progress. However, it is crucial to recognize that a "one-size-fits-all" approach is impractical, as the unique properties at the nanoscale depend significantly on nanoparticle type, surface properties, administration route, and morphology—factors that complicate the regulatory process.

### 15.1.2 The Caution Approach of Regulatory Agencies

Regulatory agencies are justifiably cautious, as past market approvals for nanoparticles used in medical imaging have led to their withdrawal due to unforeseen patient events post-administration [16]. For instance, Sinerem®, an ultra-small superparamagnetic iron oxide (USPIO) contrast agent for MRI, was denied marketing authorization and withdrawn by the European Medicines Agency (EMA) in 2008 due to severe adverse reactions observed in clinical trials. These included muscle pains, especially in the lower back, and severe allergic reactions resulting in one death. The risks associated with this nanomolecule were deemed to outweigh any potential benefits, leading to its market withdrawal.

Advances in regenerative medicine are rapidly transforming healthcare, with a key component being the introduction of *in vitro* grown or manipulated cells. Crucial safety questions arise post-reintroduction of these cells, including their localization, survival, and migration, which can affect both safety (e.g., tumorigenesis) and efficacy (e.g., injection site accuracy). The EMA suggests applying suitable stem cell tracking methods when available. This discussion addresses the European regulatory framework and the scientific evidence needed to facilitate the clinical implementation of MRI contrast media for tracking implanted/injected cells in human studies [17].

However, this cautious approach has created inertia within the field. The approval benchmarks often remain unclear and are based on regulations for small drug molecules, which do not accurately reflect the potential of nanomaterials. Clear guidance is essential to provide manufacturers, healthcare providers, the public, and policymakers with the necessary clarity and legal certainty. The U.S. FDA, Environmental Protection Agency (EPA), and the Health and Consumer Protection Directorate of the European Commission have initiated measures to address the potential risks posed by nanoparticles [18]. Additionally, initiatives like the REFINE project aim to define regulatory criteria for nanomedicines and nanomaterials for clinical use [19].

### 15.1.3 Need for Consistent Nanomedicine Regulation

Many feel that no firm and consistent lines have been drawn to ensure uniform regulation worldwide or to guarantee that regulatory agencies will act upon such guidance. The REFINE project's 2019 white paper outlines their objectives, including the "development and validation of new analytical or experimental methods" [20]. This need is echoed throughout the community, as nano-based interventions often fail in clinical stages due to inconsistent or inappropriate preclinical testing models [21].

Another factor to consider is the potential environmental impact of nanomedicines during production, use, and disposal [22]. Conventional pharmaceuticals eventually enter the environment, and nanomedicines are expected to do the same, potentially causing environmental harm [23]. The FDA cites a lack of data to determine safety for humans and the environment, making it difficult to formulate criteria for the safe and effective development of nano-products, whether drugs, devices, or biologics. In response to criticism for their lack of nanoparticle regulation, the FDA released a draft guidance in June 2011, but a final document has yet to be produced [24].

Despite the need for formal regulation, the FDA continues to overlook existing data on toxicity profiles, instead taking a precautionary approach to avoid future negative public opinion by treating nanoparticles as equivalent to their bulk counterparts. This approach negatively impacts nanomedicine development and future use, as uncertainty affects funding, research, and public acceptance, potentially delaying the commercialization of nano-products [25].

In both the USA and the EU, medical product assessments include criteria based on estimated environmental effects. In the EU, all marketing authorization applications must undergo an environmental risk assessment, including a pre-screening stage to estimate the predicted environmental concentration (PEC) for surface water, with an acceptable limit of 0.01 ppb. If the PEC is below this limit and no other environmental concerns arise, no further action is taken regarding environmental risk. In the USA, the FDA requires an environmental assessment for new drug applications (NDAs) unless exempt, but if the expected environmental concentration exceeds 1 ppb, an exemption cannot be granted.

### 15.1.4 Regulatory Challenges for Nanomedicines

One of the most significant issues is that regulatory bodies like the FDA rely on safety data based on bulk materials, which do not exhibit the same pharmacodynamic and pharmacokinetic behaviors as nanomedicines [26]. Consequently, the safety and efficacy data collected may not accurately reflect what happens when nanomedicines are used clinically after achieving marketing authorization. This discrepancy creates difficulties in establishing safety and efficacy regulations, as non-nano versions might meet regulatory standards while nanomedicines might not [25].

Another major challenge is the classification of nanomedicines, which can be categorized as either medicines or medical devices, with no consistency across global regulators. This inconsistency means a nanomedicine could be considered a medicine in one country and a medical device in another, leading to varying regulations and safety and efficacy standards. As a result, a nanomedicine approved in one country might not meet the regulatory standards in another [27].

In 2009, Rannard and Owen warned that a one-size-fits-all approach is inappropriate for nanomedicines due to their clinical needs, application routes, and physiology [28]. However, this advice has largely been ignored by current regulatory frameworks. The complex structure, form, size, and clinical application of nanomedicines pose challenges for regulatory bodies in characterizing and categorizing them. For example, dynamic light scattering estimates hydrodynamic size by assuming spherical shapes, which is inaccurate for rod-shaped materials [29]. Additionally, other size measurement techniques may alter the nanomaterial's form compared to its state in the human body. Transmission electron microscopy, for instance, involves drying samples, which can change their shape or size compared to their solution phase. Protein corona formation upon injection into the bloodstream can also significantly affect size estimates [30]. There is no consensus in the literature on the best nanometrology or characterization standards [31]. Until rigorous clinical regulatory guidance or intervention is established, preclinical nanomedicine characterization will remain inconsistent.

Scaling up and ensuring stability during manufacturing is often challenging for nanomaterials and nanomedicines [32]. Therefore, stringent protocols and assurances are necessary for approval. Identifying and controlling manufacturing processes at critical points is essential, as is defining critical quality attributes (CQA) to enhance the understanding of nanomedicine manufacturing [33]. Due to their complex structures and properties, establishing a consistent manufacturing process that ensures quality, efficacy, stability, and safety is difficult [34]. Detailed clarification of CQA involves identifying and analyzing nanomedicine properties in small-scale manufacturing, facilitating the understanding of large-scale manufacturing processes [35].

To address these concerns, consortiums and government agencies worldwide have established semi-regulatory testing facilities, such as the US and EU Nanotechnology Characterization Laboratories. However, researchers often do not access these resources until much later in their development pipeline, typically at the end of preclinical testing.

### 15.1.5 Addressing Regulatory Challenges in Nanotoxicology and Cellular Response

Nanotoxicology and cellular response present significant challenges for regulators [36]. Various methods have been proposed to measure nanotoxicity. Traditional toxicity assessments, such as extensive animal testing used for small drug molecules, are now considered unethical, costly, and impractical for nanotoxicity measurement [37]. *In vitro* toxicity methods are typically used as an initial approach for assessing nanoparticles. These methods are cost-effective, time-efficient, and allow for better control over experimental conditions compared to animal testing. However, common 2-dimensional assays often overlook the complexity of the human body, which employs compensation mechanisms, pathological responses to toxins, and complex metabolic activities [38]. Additionally, evidence suggests that traditional assays for *in vitro* testing of small compounds are not suitable for nanomaterials [39]. Nanoparticles frequently interact with *in vitro* assay reagents or interfere with detection mechanisms, leading to false positives or invalid data. Nanomaterial properties such as high adsorption capacity, optical properties, catalytic activity, acidity or alkalinity, magnetic properties, and dissolution can interfere with *in vitro* testing reagents or measurements [40, 41]. Consequently, new assays are needed to accurately measure the toxicity of nanomaterials and nanomedicines before proper regulatory guidance can be developed, which is a significant obstacle in this field [38].

When developing nanomedicines for clinical use, the drug delivery mechanism and action require extensive preclinical safety data, including information on adverse effects, before approval [42]. The toxic effects of high drug doses in nano form may impact specific cells or organs, potentially being lethal for patients with chronic kidney disease or diabetes, or leading to antibiotic resistance. Additionally, the mobility of smaller particles poses a threat as they can cross the blood-brain barrier, potentially compromising brain function or causing edema [43]. This is a crucial factor, and without sufficient data to demonstrate the new form of medicine's safety, approvals may be later revoked.

Defining the pharmacokinetics of nanomedicines is a major regulatory challenge [44, 45]. Nanomedicines deviate from the normal and expected course experienced by small drug molecules, resulting in sustained bioavailability. This prolonged presence raises concerns about health hazards if nanomedicine products were to be available over the counter. Regulatory bodies must determine whether a given nanomedicine should be marketed under strict supervision or be available as over-the-counter products. However, the lack of toxicity information and data makes it difficult to reach a conclusive decision on this matter.

### 15.1.6 Regulatory Challenges and Considerations for Nanomedicines and Nanosimilars

In the past decade, nanomedicines and their subsequent products, nanosimilars, have entered the market [34]. The challenges associated with the pharmaceutical development and manufacturing processes apply to both nanomedicines and their follow-on products. The increasing awareness of these products has prompted regulatory bodies to reassess their existing regulations. In the EU, follow-on products can be approved by the EMA under the categories of generic or hybrid applications. However, due to the unknown CQA of nanomedicines, the generic approach is considered invalid. The term "nanosimilar" is deemed more appropriate due to the complexity of these products. The regulation of generic applications is based on the equation PE + BE = TE, where PE stands for pharmaceutical equivalence, BE for bioequivalence, and TE for therapeutic equivalence. Nanomedicines and nanosimilars heavily challenge all these parameters. PE implies identical active ingredients in the same composition, but it is difficult to isolate or identify the "active ingredient" in many nanomedicines because they exist as complex nanoparticulate structures rather than homo-molecular structures [40]. Furthermore, without established CQA, identifying which parameters are identical or novel remains uncertain.

Another issue in the regulation of nanomedicines is determining who should be responsible for formulating the guidelines. This decision requires consultation with many stakeholders, including academics and clinicians. Additionally, there is an urgent need to establish regulatory, high-caliber laboratories at the federal level, along with developing risk assessment guidelines and technical standards for personnel. Key bodies often lack scientific expertise due to the novelty and diversity of nanomedicines. This limited knowledge makes it challenging to create adequate regulations that ensure patient safety and effectively regulate the clinical use of nanomedicines [45]. While strong national consortiums and National Characterization Laboratories already exist, the integration of their information and guidance into regulatory frameworks is still lacking.

The challenges mentioned significantly hinder the future of nanomedicine, obstructing the production of safe, high-quality nanomedicine products and potentially leading to ineffective control of nanoparticles due to the absence of nanomedicine-specific safety protocols. Without clear and consistent guidance from governing agencies, it is unlikely that the full potential of nanomedicine breakthroughs will be realized.

## 15.2 CLINICAL TRANSLATION

Translational clinical research is a specialized field that aims to convert laboratory-based research into clinical interventions by bridging the gap between basic scientific discoveries and their practical use in patient care [46]. Translational research serves as a catalyst for the comprehensive and interdisciplinary convergence of fundamental, patient-centered, and population-based research, with the ultimate objective of enhancing public health outcomes [47]. There are two perspectives of clinical translation. For basic scientists, translational research encompasses the process of converting fundamental scientific knowledge into tangible clinical applications such as novel drugs, devices, or treatment modalities. This research domain, situated at the crossroads of basic science and clinical practice, ultimately aims to develop promising therapeutic interventions suitable for clinical deployment or commercialization. The effective transformation of foundational scientific discoveries into innovative prevention, diagnostic, and treatment strategies is imperative for advancing human health. And for health services researchers and public health investigators, translational research focuses on the effective dissemination and implementation of research findings into real-world healthcare settings and populations. In contrast to the bench-to-bedside model, which culminates in the development of a novel therapeutic, this phase of translational research commences at the point of a promising new treatment and seeks to optimize its delivery and impact on patient outcomes and population health [48]. The first phase focuses on transforming basic scientific knowledge into potential clinical applications, such as new drugs or medical devices. The second phase concentrates on effectively delivering and implementing these innovations within healthcare systems and populations, ensuring maximum benefit to patients.

Because translational research has the potential to greatly improve healthcare delivery and results, the efficient translation of research into clinical practice is a priority acknowledged on a global scale. The integration of relevant research information into typical clinical care is an ongoing difficulty in many places globally, despite the significant increase and complexity of health research. A recent study found that the main obstacles to putting new research evidence into practice were individual-level factors, with organizational challenges having a less significant effect. Specifically, the most significant individual-level barriers were deficiencies in the knowledge and skills needed to conduct, organize, utilize, and appraise research literature. The main organizational challenges were limited access to research evidence and inadequate resources, such as equipment [49]. The study opined that building cooperative relationships between policymakers and healthcare providers at every stage of the research project is essential to overcoming these obstacles.

In recent decades, there has been a growing chorus of criticism alleging substantial waste and inefficiency within the clinical translation process. Patient advocacy groups contend that overly stringent regulations impede timely access to life-saving treatments. Research funders express concern over the underutilization of basic scientific discoveries within academic settings. Journal editors claim that biased reporting undermines the public return on investment in biomedical research. Furthermore, both pharmaceutical companies and their critics assert that excessive expenditures characterize the drug development process [50]. A recent study brought to light the high rates of attrition in the drug development pipeline as well as the ongoing challenges in reproducing and applying preclinical research findings to human subjects. The study also highlighted a minimal clinical impact in comparison to the funds invested in this area of study. Improving translational research requires a multifaceted approach. Rigorous hypothesis development and testing can optimize research efforts, while advancements in drug discovery methodologies hold promise for faster and more efficient development of new therapies [51]. Shifting from traditional cell-based models to simpler biological systems, coupled with innovative technologies like Clinical Trials in a Dish (CTiD), has the potential to significantly reduce drug development costs and time, ultimately leading to improved patient outcomes [52–54].

For a successful clinical translation, the procedure is intricate and needs smooth cooperation between physicians, engineers, basic scientists, and other relevant stakeholders. When it functions, it can completely transform healthcare by offering cutting-edge cures, early detection systems, and diagnostic tools. Below, we have reported several successful examples of clinical translation research that have saved countless lives from devastating diseases.

Gene treatments for neurological illnesses have long been the subject of research and development funded by the National Institute of Neurological illnesses and Stroke (NINDS). The development of gene therapy for Spinal Muscular Atrophy (SMA) stands as a prominent example of successful clinical translation. Researchers at Nationwide Children's Hospital in Ohio engineered an intrathecally administered AAV9-SMN vector, designated AVXS-101, to introduce a fully functional SMN gene. This therapeutic approach aimed to augment SMN protein production within patients to levels sufficient to reverse disease progression. In 2018, the pharmaceutical conglomerate Novartis acquired AveXis, the originator of AVXS-101. Subsequently, Novartis rebranded AVXS-101 as Zolgensma (onasemnogeneabeparvovec-xioi).

A novel therapeutic approach for glioblastoma involves oncolytic virotherapy, a strategy whereby cancer-eliminating genes are delivered to patients via harmless viral vectors. Four such therapies, all of which have progressed to clinical trials, offering promising prospects for the treatment of glioblastoma [55]. Researchers have long endeavored to develop a cellular-based cure for type 1 diabetes, a goal that has become increasingly feasible in recent years [56]. As the condition arises from the immune system's destruction of insulin-producing beta cells within the pancreatic islets, research has centered on replacing these dysfunctional cells with functional counterparts. Recent advancements have enabled the generation of beta cells from stem cells, undifferentiated cells capable of developing into various specialized cell types [57]. In a significant breakthrough, Vertex conducted the first successful transplantation of stem cell-derived beta cells into a type 1 diabetes patient in 2021 [58]. Concurrently, ViaCyte published the initial peer-reviewed studies demonstrating the insulin-producing capabilities of its own stem cell therapy in individuals with type 1 diabetes [59].

Coronary artery disease, a prevalent condition characterized by the narrowing of coronary arteries, has long been a primary focus of interventional cardiology. Percutaneous coronary intervention (PCI) using bare metal stents (BMS) emerged as a cornerstone treatment, though it was often hampered by in-stent restenosis—a process where the artery re-narrows due to excessive tissue growth. Recognizing this limitation, the development of drug-eluting stents (DES) represented a significant translational research triumph. These specialized stents are engineered to deliver therapeutic agents directly to the arterial wall, inhibiting excessive tissue growth and preserving vessel patency. By mitigating the inflammatory response and preventing neointimal hyperplasia, DES have markedly reduced the incidence of restenosis, thereby improving patient outcomes and decreasing the need for repeat interventions. This technological advancement underscores the pivotal role of translational research in transforming scientific discoveries into clinically effective therapies [60].

The invention of chimeric antigen receptor T-cell (CAR-T) therapy for specific cancer types is one of the most notable cases of successful clinical translation research. This novel method targets and kills cancer cells by genetically altering a patient's T cells. The ability of T cells to specifically target cancer cells was discovered, and techniques for modifying them using CARs were developed. Patients with resistant hematological malignancies showed exceptional response to CAR-T cell treatment in early clinical trials. Patients with limited options for treatment now have hope because of the advancements in CAR-T cell therapy for these tumors [61, 62].

## 15.3 REGULATORY ASPECTS

### 15.3.1 Overview of the Current Regulatory Landscape

Nanomedicine is a rapidly growing field with full of potential, yet it currently lacks clear regulatory guidelines. Many nanomedicines exert their therapeutic effects by directly interacting with genetic material or by influencing biomolecules essential for normal cellular processes [63]. This mode of action can inadvertently trigger genotoxicity and mutagenicity [64].

Such toxic effects often stem from an inflammatory response initiated by neutrophils and macrophages, which produce reactive oxygen and nitrogen species [12]. The accumulation of these free radicals can wreak havoc on the body, causing oxidative DNA damage, protein degradation, lipid peroxidation, and mitochondrial dysfunction [13, 14]. These cellular insults can ultimately lead to cancer, cell death, and tissue fibrosis. Intravenously administered nanoparticles have been shown to accumulate in the liver and subsequently migrate to vital organs like the brain, heart, and kidneys [15]. The lack of tracking capabilities for many nanoparticles exacerbates the uncertainty surrounding their potential safety risks. Current understanding of how nanomedicines interact with biological systems is limited, making it challenging to fully characterize their physicochemical and toxicological properties. Establishing standardized regulatory frameworks is crucial to address these challenges. It is essential to recognize that a one-size-fits-all approach is insufficient due to the diverse properties of nanoparticles, including size, shape, surface characteristics, and administration route. These factors significantly influence nanoparticle behavior and complicate the regulatory process.

Regulatory bodies have adopted a prudent approach to nanomedicine approval. History underscores the importance of such caution. For instance, some nanoparticles initially approved for medical imaging were later withdrawn due to unforeseen adverse events in patients [16]. Sinerem®, a USPIO contrast agent for MRI, serves as a notable example. In 2008, the EMA rejected its marketing authorization following concerns raised during clinical trials. These concerns centered on severe adverse effects, including lower back muscle pain and, more critically, fatal allergic reactions. Consequently, the EMA determined that the risks associated with Sinerem® outweighed its potential benefits [65]. Despite the understandable caution, the regulatory landscape has become overly cautious, hindering progress in the field. Current approval benchmarks often remain unclear and are inappropriately adapted from regulations governing traditional small-molecule drugs, failing to account for the unique characteristics of nanomaterials.

Clear regulatory guidance is essential to provide clarity and legal certainty for manufacturers, healthcare providers, the public, and policymakers. Recognizing this need, regulatory bodies such as the U.S. FDA, EPA, and the European Commission's Health and Consumer Protection Directorate have initiated efforts to address the potential risks posed by nanoparticles [66]. Additionally, local and regional initiatives, like the REFINE project, are working to establish specific regulatory criteria for nanomedicines and nanomaterials intended for clinical use [67]. Despite these efforts, a standardized global regulatory framework for nanomedicines remains elusive. There is a lack of consensus on specific guidelines, and there is no guarantee that regulatory agencies will consistently implement existing recommendations. The REFINE project's 2019 white paper highlights the need for developing and validating new analytical and experimental methods [20]. This need is widely acknowledged within the field, as numerous nano-based interventions have failed in clinical trials due to inadequate or inconsistent preclinical testing [21].

The environmental impact of nanomedicines throughout their lifecycle, from production to disposal, is another critical concern [22]. Given the known environmental persistence of traditional pharmaceuticals, it is reasonable to anticipate a similar fate for nanomedicines, potentially leading to adverse ecological consequences [23]. The FDA has acknowledged the dearth of data on the safety of nanomaterials for both humans and the environment, hindering the development of comprehensive regulatory standards for nanomedicines, devices, and biologics [24]. Although the FDA released a preliminary guidance document in June 2011 in response to public pressure, a definitive regulatory framework for nanomedicines remains absent. Despite a pressing need for a comprehensive regulatory framework, the FDA has been criticized for its cautious approach to nanomedicine regulation. Rather than leveraging existing toxicity data, the agency has opted for a precautionary stance, potentially motivated by concerns about public perception. By treating nanomaterials as equivalent to their bulk counterparts, the FDA may be inadvertently hindering nanomedicine development. This regulatory uncertainty undermines research funding, stifles innovation, and erodes public trust, potentially delaying the commercialization of nanotechnology-based products [25]. Both the U.S. and the EU employ inclusion and exclusion criteria for medical products based on their estimated environmental impact. In the EU, all marketing authorization applications undergo an environmental risk assessment, including a pre-screening stage that evaluates PECs in surface water against a threshold of 0.01 ppb. If this threshold is not exceeded and no other environmental concerns arise, further assessment is not required. The FDA also mandates environmental assessments for NDAs, unless exempt, but imposes a stricter threshold of 1 ppb for environmental concentration to avoid exemptions.

### 15.3.2 Challenges in Regulatory Approval for New Therapies

To actualize the different benefits nanotechnology has to offer to us humans at different spheres of our lives, regulatory bodies globally are in continuous efforts to develop strategies to regulate nanomedicines falling within their jurisdiction. Despite these advancements, it is a known fact that the current regulations are not relevant and are not efficient enough to deal with the aspects of nanotechnology-based products, especially their toxicity aspect [68]. This section will highlight the regulations set by different global regulatory bodies.

#### *15.3.2.1 USA*

The U.S. FDA, a leading global regulatory authority, has a long history of overseeing therapeutic interventions, including nanomedicines (nanomedicines). A primary goal of the FDA is to establish product-specific review standards for nanotechnology-based products and to deepen understanding of nanomedicine properties [68]. In line with these objectives, the FDA published draft guidance documents in 2017 and 2018 to assist industry in the registration and approval of nanomedicines containing nanomaterials [69, 70]. These initiatives align with the FDA's Good Guidance Practice and recommendations from the 2007 Nanotechnology Task Force Report. Rather than providing a definitive classification of safe and harmful nanomedicines, the FDA adopts a case-by-case approach to evaluating nano-products [71]. The agency emphasizes the importance of considering the unique properties and characteristics imparted by nanotechnology when assessing the safety, efficacy, global health implications, and regulatory status of engineered nanomaterials (ENM).

It is advised to the product developers to have proper consultation with the USFDA during the development process to come to terms with mutual agreements regarding the regulations and to avoid subjection to premarket review [72]. Safety information review and post-marketing safety designs are provided by the USFDA to the manufacturers. For example: Data obtained from comparison studies should be available to characterize and measure variances in Absorption, Distribution, Metabolism and Excretion (ADME) of the active component among liposome and non-liposome medication formulations. The CQAs of the product are defined by the liposome's physiochemical characteristics (e.g., zeta potential, mean particle size, size distribution of liposomes, physical stability, and osmolality). The role of the USFDA is not limited to this; this regulatory body also scrutinizes the products after reaching the market in the best interest of consumers. The Nanotechnology Task Force and Nanotechnology Interest Group, comprising several representatives from many regulatory centers, was formed by the USFDA with the objective of handling the regulation issues of nanotechnology worldwide. Despite all these efforts, the USFDA is yet to produce a clear set of guidelines that gives transparent information for the interests of manufacturers and consumers as well. The USFDA regulates product, science-based policy. Product-specific technical analyses consider the impacts of nanomedicines in the biological and mechanical context of each product and its intended application. [69, 70].

The FDA defines regulatory science as the development of new tools and methods to assess product safety, efficacy, quality, and performance [73]. Recognizing the inherent uncertainties of emerging technologies, the FDA has opted against creating specific nanomaterial regulations. Instead, the agency relies on its existing framework, coupled with proactive measures like horizon scanning and product submission reviews, to oversee nanomaterials and nanomaterial-containing products. The FDA has established nanotechnology core facilities to develop testing methods and standards for assessing the safety of nanomaterials and nanomaterial-containing products. The FDA's CORES program promotes collaborative research on product characterization and safety. To enhance regulatory expertise, the FDA provides specific training and conducts gap analyses, resulting in guidance documents for industry [74]. The FDA collaborates with other U.S. government agencies through the National Nanotechnology Initiative (NNI) to address nanotechnology challenges. The NNI focuses on measurement infrastructure, human exposure and health, environmental impact, risk assessment, and informatics [75]. This collaboration has led to guidance documents, webinars, and worker safety protocols. To foster international cooperation, the FDA participates in the EU-US Nanotechnology Communities of Research. The collaboration between the EU Nanomedicine Characterization Laboratory

(EU-NCL) and the US National Cancer Institute's Nanotechnology Characterization Laboratory (NCI-NCL) is crucial for accelerating the development of innovative nanotechnology-based therapies and diagnostics [73]. FDA fosters global collaboration, particularly in nanomaterial characterization, through partnerships with countries such as Canada and India, as well as participation in international standardization organizations. While significant progress has been made, the number of approved nanotechnology-based products remains limited. Despite promising scientific advancements in cancer research, greater efforts are needed to translate these findings into marketable products. The FDA is witnessing a surge in product submissions containing nanomaterials, with an increasing number entering clinical trials and gaining approval for future drugs and medical devices [73].

An informal U.S. Nanoplastics Interest Group, comprising over 20 agencies, is studying the implications of nanoplastics. The group aims to leverage nanotechnology advancements, foster collaboration, and mitigate environmental and food chain contamination risks. The FDA Europe Office promotes collaboration and knowledge sharing with European regulatory counterparts, including those in Russia and Central Asia [76]. The office focuses on policy alignment, regulatory dialogue, and strategic partnerships, as outlined in the EU-US Mutual Recognition Agreement (MRA). This agreement, a result of 5-year collaboration, allows the U.S. and EU to rely on each other's pharmaceutical Good Manufacturing Practice (GMP) inspections, streamlining processes and redirecting resources toward global health priorities. Emerging technologies, including nanotechnology, are a key focus of this cooperation [73].

#### *15.3.2.2 EU*

The EU prioritizes a "safe-by-design" approach to ensure the safety of nanomaterials throughout their lifecycle, from production to disposal. This concept, rooted in industrial innovation, was initially defined for nanomaterials within the EU's NANoREG flagship project [77]. By integrating safety considerations into the development process, safe-by-design aims to accelerate innovation. To support this approach, the EU has established digital innovation hubs and open innovation test beds. Digital innovation hubs serve as one-stop resources for businesses, particularly SMEs (small and medium-sized enterprise), startups, and mid-sized companies, providing access to technology, financial guidance, and market insights [78]. Open innovation test beds offer physical facilities and expertise for developing, testing, and scaling advanced materials in industrial settings.

The European Chemicals Agency (ECHA) oversees the safety assessment of chemicals, including manufactured nanomaterials, under the REACH regulation (EC 1907/2006) [79]. REACH mandates chemical registration, evaluation, authorization, and restriction within the EU. While REACH is a cornerstone, nanomaterials are also subject to regulations governing occupational safety, cosmetics, food packaging, biocides, and food [80]. In 2018, REACH was amended to include specific information requirements and safety assessment provisions for nanomaterials. To date, 37 substance registrations include nanoform data. The absence of a transition period and standardized test methods emphasizes the urgent need for developing new testing protocols. Beyond its regulatory role, ECHA operates the European Union Observatory for Nanomaterials, providing public access to information on nanomaterial innovation and safety [80, 81].

The European Food Safety Authority (EFSA) is responsible for assessing food and feed safety risks, encompassing areas such as nutrition, biological hazards, animal welfare, contaminants, and additives. EFSA provides scientific advice to inform European policies and supports regulatory implementation for human, animal, and plant health. The agency also evaluates the environmental impact of the food chain on biodiversity. To address the challenges posed by nanotechnologies and nanomaterials in the food and feed sector, EFSA established a scientific network to facilitate information sharing among EU member states and prioritize risk assessment activities [82]. A dedicated working group on nanotechnologies has been formed, and the agency has published guidance on risk assessment in this area. A comprehensive review of nanotechnology regulations in the agriculture, feed, and food sectors within and outside the EU was conducted by Amenta et al. in 2015 [82]. The EMA is another agency of the European Union. It operates a working definition of nanomedicines as medicinal products that are purposely designed systems for clinical applications and have at least one component at the nanoscale, leading to definable specific properties and characteristics providing clinical advantages, the nanotechnology offering specific benefits in, for example, dosage, drug targeting, or reduced toxicity [83].

The European Green Deal [84] outlines a comprehensive strategy for Europe to become climate-neutral, pollution-free, and sustainable. This ambitious initiative aligns with the UN's 2030 Agenda for Sustainable Development and drives the EU's New Industrial Strategy [85, 86]. A key focus of the Green Deal is addressing plastic pollution. Building upon its 2018 Plastics Strategy [84], the EU is implementing measures to reduce microplastics, both intentionally added and unintentionally released. This includes developing regulations for biodegradable and bio-based plastics. Furthermore, the EU is tackling plastic pollution from sources like urban runoff and exploring additional measures to combat pollution from harmful substances.

The EU's Circular Economy Action Plan recognized plastics as a priority area, aiming to promote sustainable resource use [85]. The subsequent Plastics Strategy of 2018 sought to revolutionize plastic product lifecycle management, emphasizing design, production, use, and recycling improvements. This includes increasing recycling rates and enhancing the quality of recycled plastics to stimulate market demand.

The EU has taken steps to restrict intentionally added microplastics, with ECHA proposing a ban on their use in consumer and professional products. This measure could significantly reduce microplastic pollution. While addressing the primary issue of plastic waste generation and leakage, the EU's Single-Use Plastics Directive indirectly contributes to curbing microplastic formation from plastic waste fragmentation [86]. Recognizing the growing concern over microplastics from various sources, the EU plans to introduce a regulatory framework for biodegradable plastics. This initiative aims to decelerate the accumulation of micro- and nanoplastics in the environment.

#### *15.3.2.3 Australia*

Therapeutic Goods Administration (TGA) is the regulatory body governing the legislations of cosmetics, pharmaceutical, and medical products in Australia, which have been jotted in the Therapeutic Goods Act 1989. In Australia, nanomedicines are considered a type of medicine and nano health products as a type of therapeutic goods; hence, it is stipulated that Australia does not require special provisions for nanomedicines as the Laws of Therapeutic Goods Act 1989 are capable of addressing the potential benefits and risks associated with nanotechnology-based products. This stipulation is backed by a detailed study conducted in 2007 on the need for nanomedicine-specific legislation [87]. The National Industrial Chemicals Notification and Assessment Schemes (NICNAS) is the primary framework for regulating the industrial use of nanomedicines in Australia. As of now, Australia does not have a working definition of nanomedicines, but there is a definition for nanomedicines as the country does not consider nanomedicine production to constitute an entire industry [87]. However, the explosive potential of nanotechnology-based products has been examined by the chemical industries of Australia; their safe working environment provisions have been made, but little importance has been shown toward the implications of their exposure to the environment [88].

#### *15.3.2.4 Asia*

Asia, despite widespread use of nanomaterials (nanomedicines) and nanomedicines (nanomedicines), has a relatively underdeveloped regulatory framework for nanotechnology (nanotechnology) [89]. Japan, a notable exception, prioritizes translating research into practical applications that benefit public health. Regulatory oversight is shared among the Ministry of Health, Labour, and Welfare (MHLW), the Pharmaceutical and Medical Devices Agency (PMDA), the National Institute of Health Sciences (NIHS), and the Agency for Medical Research and Development (AMED).

While Japan lacks specific nanotechnology legislation, existing regulations for handling small particles, such as the Chemical Substances Control Act and the Industrial Safety and Health Act, are applied [90]. The MHLW has issued guidelines to prevent workplace exposure to nanomedicines during manufacturing and handling [91]. Japan's regulatory focus on exposure prevention differs significantly from the registration and reporting-centric approach of regulatory bodies in regions like the EU and the U.S. [73]. Japan has taken a proactive approach to regulating nanomedicine. The establishment of the PMDA's Regulatory Science Center in 2018 signifies a commitment to fostering innovation and addressing emerging technologies like nanotechnology. This center's focus on horizon mapping ensures the regulatory framework stays aligned with rapid advancements in the field. Leveraging clinical data and electronic health records, Japan is enhancing its ability to conduct large-scale research and assess product safety and efficacy. The Act on Promoting Health Industries and Development of Healthcare Technologies has further strengthened the regulatory and research landscape for cutting-edge pharmaceuticals. In 2019, the Japanese government issued guidance on nanopharmaceutical assessment, emphasizing the need for robust scientific evidence to support product development. Comparative studies with existing drugs are encouraged to demonstrate improvements in safety, efficacy, and cost-effectiveness [73]. Research institutions like the National University of Singapore are collaborating with industry and regulators to address critical challenges in nanomedicine development. By analyzing large clinical datasets, researchers are identifying key quality parameters and optimizing preclinical testing strategies. These initiatives highlight Japan's commitment to creating a supportive regulatory environment for nanomedicine while ensuring patient safety and product efficacy [92].

#### *12.3.2.5 Canada*

Various Canadian ministries and organizations, including Health Canada, Environmental & Climate Science Canada, Agricultural and Agri-Food Canada, and the Canadian Food Inspection Agency, are working together on initiatives to ensure the safe use of nanomaterials (nanomedicines) and to evaluate the risks associated with ENM (nanomedicines). These organizations are concentrating their regulatory efforts on minimizing risks, creating a database of market items containing nanomedicines, and assessing the biological impacts of specific nanomedicines. This work involves understanding how nanomaterial items are used or consumed, as well as considering their lifecycle and exposure potential. Under the Chemicals Management Program (CMP) of the Government of Canada, there is a mandate to address potentially hazardous new and existing substances, including nanomedicines [93]. The regulatory framework in Canada follows the Organisation for Economic Co-operation and Development (OECD) Council guideline for the safety testing and assessment of manufactured nanomedicines as outlined under the Canada Environment Act [94]. For nanoparticles, regulations such as the Canadian Food and Drugs Act, the Consumer Product Safety and Hazardous Products Act, the Pest Control Products Act, the Feeds Act, the Fertilizers Act, and the Animal Health Act may apply. Health Canada's definition of nanoparticles is one of the tools used by the CMP [73]. Given the similarities between U.S. and Canadian definitions of nanomaterials (nanomedicines), the Canada–U.S. Regulatory Cooperation Council facilitates close collaboration between relevant agencies. They work together to develop a unified strategy for prioritizing activities and conducting comprehensive risk assessments for nanomaterials. This includes a standardized classification system for nanomedicines based on their chemical composition and unique characteristics, helping to distinguish nanomedicines from non-nanoscale materials. The Canadian National Centre of Excellence in Engineered Nanomaterials (nanomedicines) focuses on drug delivery, new treatments, and diagnostics, with a particular emphasis on cancer and gene therapy. Non-viral nanoparticle systems in Canada have shown promise in efficiently delivering genetic material, simplifying production processes, and reducing costs. Despite these promising research results, several regulatory challenges remain, especially concerning personalized medicine tailored to individuals or small patient groups. Canadian agencies also collaborate closely with European Union regulatory authorities through the OECD's initiatives [95].

##### *15.3.2.6 International (International Council for Harmonization of Technical Requirements for Pharmaceuticals for Human Use, International Organization for Standardization, and World Health Organization)*

The International Council for Harmonization of Technical Requirements for Pharmaceuticals for Human Use (ICH) oversees the International Pharmaceutical Regulators Programme (IPRP), which addresses emerging issues in pharmaceutical regulation. This includes ENM (nanomedicines), nanomaterials (nanomedicines) in pharmaceutical products, borderline and combination medicines, and innovative assessment methodologies [73]. The IPRP nanomedicines steering committee includes representatives from government and agencies across the Americas, Asia, Europe, and Oceania. The committee's objectives include sharing non-confidential information, harmonizing regulations, coordinating training providers, and engaging with groups of nanomedicine developers and other stakeholders [96].

Regulations specify the required disclosures about nanomedicines, including the metrics to be used, such as particle size or weight. For instance, under the EU's REACH regulations, the particle size distribution must be provided based on the number of particles for nanoform characterization. In contrast, the USEPA's Toxic Substances Control Act requires reporting based on the weight percentage of nanostructures [97]. In some cases, both measurements are necessary, such as when reporting a nanomaterial to the European Commission's Cosmetics Products Notification Portal (CPNP).

### 15.3.3 Impact of Regulatory Frameworks on Innovation

Nanotechnology products have entered the global marketplace, yet our understanding of their environmental and health impacts remains incomplete. While significant efforts by the U.S., EU, and international organizations like the OECD are underway to assess the safety of nanomaterials, substantial challenges persist [73]. This section examines global initiatives aimed at standardizing nanomaterial assessment through test guidelines and risk frameworks. The US EPA is particularly focused on understanding the toxicity and accumulation of nanomaterials in marine environments. Research indicates that salinity significantly influences nanomaterial stability and bioavailability. While large nanoparticle clusters tend to settle in sediments, some nanomaterials release toxic ions into water, causing harm [98].

However, accurately detecting and identifying nanomaterials in marine ecosystems is complex due to their often minute quantities and potential mixing with natural materials. Current methods primarily assess acute toxicity, with a need for additional tools to measure sub-lethal effects [98]. The EU Cosmetics Regulation (EC) No. 1223/2009 encompasses the safe utilization of nanomaterials in cosmetic products [73]. Sunscreens incorporating nanoform titanium dioxide exemplify this, offering enhanced human health protection. Specific regulatory stipulations for nanomaterials mandate a six-month premarket notification. The EU Scientific Committee on Consumer Safety (SCCS) investigates potential safety concerns, such as membrane penetration, accumulation, and persistence. While the SCCS has provided guidance and opinions on nanomaterials in cosmetics [90], comprehensive safety dossiers remain absent [99].

Challenges in evaluating nanomaterial-containing cosmetic dossiers persist due to uncertainties in physicochemical properties, environmental impact, and toxicological effects. Early-stage testing often diverges from the final product's behavior. Key challenges include the need for precise material characterization, verification of test material consistency with the dossier, accounting for potential internal exposure, incorporating non-animal testing methods, and developing a robust weight-of-evidence approach to safety assessment [73]. A recent in-depth analysis of titanium dioxide as a food additive in confectionery highlighted the critical role of precise analytical methods for detection and quantification. Furthermore, the study underscored the importance of effective sample preparation techniques [100]. These findings emphasize the general need for rigorous characterization and safety assessment of all nanomaterials. The EMA assesses the safety of diverse nanomedicine types, including liposomes, iron nanoparticles, and lipid nanoparticles (LNPs), with numerous other formulations under development [101]. Given the broad spectrum of nanomedicine materials and structures, a standardized evaluation approach is impractical. The EMA aids drug developers in producing safe and effective products through guidance documents [73]. Early emphasis on CQA influencing efficacy and safety, the establishment of suitable analytical methods, and the feasibility of scalable, consistent manufacturing is crucial.

The U.S. FDA's Center for Devices and Radiological Health (CDRH) oversees the safety evaluation of nanotechnology-based medical devices. Its Nanotechnology Regulatory Science Research Program focuses on physicochemical characterization, *in vitro* and *in vivo* models, and risk assessment. These devices, spanning antimicrobial, dental, and orthopedic applications to *in vitro* diagnostics, utilize a variety of nanomaterials including silver, titanium, and graphene. Safety assessments must quantify nanomaterial release into the body and employ suitable *in vitro* tests. Advanced risk assessment methodologies are crucial for understanding potential adverse health outcomes from exposure. Moreover, the impact of nanoparticles on standard biocompatibility and toxicity assays needs investigation. Due to the vast diversity of nanomaterials, the development of *in vitro* (2D, 3D, organ-on-chip) and in silico models is essential for predicting human responses and refining *in vivo* extrapolations [73]. The safety evaluation of nanomedicines is equally critical. Pharmaceutical companies utilize nanoparticles, such as drug-encapsulating polymeric nanoparticles and polymer-conjugated drugs (dendrimers), in their formulations. Comprehending their behavior in physiological environments necessitates investigating nanoparticle aggregation and solution properties. This is especially crucial for dendrimer drugs where the active pharmaceutical ingredient is covalently linked to the nanoparticle, and drug release is influenced by dendrimer composition and time. Encapsulated drugs, like accurrins, exhibit more complex release profiles. Drug molecules at the nanoparticle core release more slowly due to their greater distance from the surface. Additionally, nanoparticle size impacts drug release kinetics. In silico simulations are employed to optimize release patterns. Controlled release transitions from initial uniform drug distribution to later stages with varying drug concentrations and distribution [95].

Despite extensive national and global efforts to assess nanomaterial safety, significant gaps persist in risk assessment methodologies due to the vast array of nanoparticle characteristics. While recent studies have facilitated the development of tolerable intake levels for primary exposure routes (oral, inhalation, dermal), robust exposure and dose-response assessments remain limited. Establishing safety margins is challenging given the trend towards reduced animal testing. Alternative methods must be developed and integrated into safety evaluations. Regulatory science must prioritize nanomaterial dosimetry (mass, surface area, number) and ensure clinically relevant dosing in risk assessments, yet accurate measurement methods for these parameters in products are lacking. Enhanced collaboration between academia and regulatory bodies is essential for harmonizing risk assessment approaches and promoting transparency. Expanding worldwide proficiency testing and method validation is crucial, although hindered by a dearth of suitable reference materials.

### 15.3.4 Future Directions in Regulatory Science

Despite the absence of specific regulatory guidelines, over 50 nanomedicines have successfully entered the market, with this number steadily increasing. The cancer therapy domain dominates this field due to the challenges posed by toxic compounds and complex tumor environments. Notable examples include liposomal formulations like Doxil® and AmBisome®, as well as more recent advancements in albumin-drug nanoparticles such as Abraxane® and polymeric micelles like Eligard® [102]. A lack of standardized regulations governing nanomedicine and nanomaterial production for healthcare applications is a global concern. Inconsistencies between government agencies often lead to the classification of nanomedicines as either medical devices or drugs. This regulatory disparity hinders global product approval, unlike the situation with small molecules. The nanomedicine community urgently requires a harmonized regulatory framework to facilitate continued development and meet stakeholder expectations.

Despite the formation of various clusters and working groups, tangible progress in nanomaterial regulation remains limited. Given the increasing urgency for treatments of numerous diseases, the current regulatory framework proves inadequate. While some academic and governmental initiatives have established National Characterization Laboratories, more specific and rigorous guidelines from key regulatory authorities like the FDA and MHRA are imperative. The global nature of many diseases necessitates a unified international consortium to develop and disseminate standardized regulations for nanomaterials. Substantial investments in nanomedicine over the past two decades risk being wasted without clear leadership and guidance from regulatory bodies. To ensure product commercialization and attract continued investment, a robust regulatory framework is essential.

As nanotechnology (nanotechnology) becomes increasingly ubiquitous, consumer concerns about exposure to nanomaterials (nanomedicines) grow. Despite a surge in nanotechnology-based products, a universally accepted regulatory framework for nanomaterials manufacturing, use, and disposal is absent. While regulatory agencies such as the USFDA, EMA, TGA, and MHRA strive for harmonization, existing regulations primarily address bulk chemicals rather than nanomedicines. Urgent revisions and new regulations aligned with global legislation are necessary. Collaborative efforts among stakeholders are crucial to address the unique challenges posed by nanomedicines, including their environmental impact. A comprehensive approach considering both nanomaterials applications and their environmental implications will be essential for developing effective regulatory frameworks.

## 15.4 SCALE-UP AND MANUFACTURING

Scale-up and manufacturing of nanocarriers for nucleic acids and proteins are crucial for their widespread application in biomedical and pharmaceutical fields. These nanocarriers, such as LNPs, polymeric nanoparticles, and inorganic nanoparticles, require efficient and scalable manufacturing methods to ensure their quality, consistency, and cost-effectiveness. Scale-up and manufacturing in nanocarriers for nucleic acids and proteins present significant challenges in the pharmaceutical industry, including aggregation, contamination, and degradation of nanoparticles along with low yield, questioning the economic aspect of scalability. These lead to hindrances in the industry acceptability of nanoparticle production [103]. A significant challenge in the manufacturing of nanoformulations is the difficulty in transitioning from small-scale preparation, which is often tailored to the physicochemical properties of a specific drug, to industrial-scale production. This transition requires the simplification of the numerous complex steps involved in the manufacturing process and the development of streamlined, one-step industrial scale-up solutions. These solutions should be supported by advanced characterization techniques to ensure stability, quality control, and therapeutic efficacy [104]. The production of nanoparticles, crucial for efficient delivery, involves top-down and bottom-up approaches, with issues such as aggregation, contamination, and low yield hindering industrial scalability [103]. Delivery systems for nucleic acids and proteins, particularly non-viral systems, particularly nanoparticles, have shown promise in delivering therapeutic cargoes effectively, with a focus on overcoming biological barriers and enhancing cellular internalization. Engineering these vectors to overcome biological barriers without causing significant perturbations requires precise control over their physicochemical properties. The need for consistent and scalable production methods to ensure these sophisticated delivery systems retain their efficacy and safety [105]. Nucleic acid-based therapeutics hold significant promise for treating genetic disorders and developing next-generation vaccines. However, the efficient delivery of these therapeutics remains challenging due to the inherent properties of nucleic acids. Nanoparticles play a critical role as carriers, enhancing the delivery to target cells and addressing these delivery challenges. Understanding the global patent landscape in this field is crucial for fostering innovation and informing decisions for researchers, the pharmaceutical industry, and regulatory agencies. To successfully translate nucleic acid-based therapeutics into clinical practice, there is a pressing need for streamlined manufacturing processes and robust clinical evidence [106].

### 15.4.1 Importance of Scale-Up in Medical Innovation

Scale-up in medical innovation plays a crucial role in advancing nanocarriers for nucleic acids and proteins, ensuring reproducibility and scalability for effective biomedical applications and enables the pharmaceutical industry to leverage the versatile uses of nanoparticles, such as drug delivery, targeting, enhancing bioavailability, imaging, biosensing, and controlled release. Overcoming the challenges of scaling up nanoparticles with uniform size, distribution, and shape is essential to ensure reproducibility, economic viability, and industry acceptance, thereby facilitating the translation of nanoparticle-based innovations into clinical practice [103]. Vital for the effective biomedical application of nanoparticles, as it ensures the reproducibility of key parameters such as size, shape, morphology, size distribution, targetability, and functionality. Achieving reproducible drug release profiles and batch-to-batch uniformity is essential for maintaining quality performance. Emerging production methods like membrane extrusion, supercritical fluid technology, and microfluidizer technology offer scale-up capabilities, which are crucial for the industrial applicability and successful clinical use of nanoparticle-based formulations. Ensuring purity and safety according to regulatory guidelines further emphasizes the importance of scalable production methods in translating nanoparticle innovations into clinical nanomedicine [107]. Furthermore, it is also essential for the development and translation of nanotherapeutics into clinical practice. It ensures the reproducible production of safe and efficacious nanomedicines that meet NDA specifications. Key aspects include raw materials, synthesis routes, batch sizes, stability checks, analytical methods, and documentation. Successful scale-up enables robust preclinical studies and clinical trials, and advances such as microfluidics, iterative CMC (chemistry, manufacturing, and controls) optimization, and computer software enhance the preparation of large-batch nanoparticles. This facilitates personalized and timely nanomedicine [108]. Therapeutically, Biomedical Nanotechnology (BNT) has generated many novel nanocarriers that each express specifically designed physicochemical properties that optimize their desired pharmacokinetic profile. Nanocarriers are also being integrated into nanoscale platforms that further enhance their delivery by controlling and prolonging their release profile. Regarding diagnostics, nanocarriers are being designed to perform targeted delivery of luminescent tags and contrast agents that enhance the nanocarrier-aided imaging capabilities and resulting diagnostic accuracy of the presence of diseased cells. Nanomedicine involves the use of nanosized cargos of drugs/vaccines/DNA therapeutics, including nanoparticles, nanoclusters, and nanospheres, for controlled delivery of bioactives or diagnosis of diseases at early stages. The formulation scientist faces numerous challenges related to their development, scale-up feasibilities, regulatory aspects, and commercialization. By addressing these scale-up challenges and optimizing production methods, nanocarriers for nucleic acids and proteins can revolutionize medical innovation with improved therapeutic delivery systems and diagnostic techniques [109, 110].

### 15.4.2 Challenges in Manufacturing New Therapies

The manufacturing of new therapies using nanocarriers for nucleic acids and proteins faces significant challenges, particularly in ensuring safety and minimizing potential harmful effects due to insufficient data on their safe use. Understanding the interactions between nanostructures, cells, and embryos is crucial, as is achieving consistent reproducibility at large scales. Regulatory compliance and thorough toxicity research are essential for advancing nanotechnology in genetic modifications and therapeutic applications [111]. Manufacturing new therapies using nanocarriers for nucleic acids and proteins faces challenges despite their potential in gene therapy and drug delivery. The development of stimuli-responsive delivery systems has shown promise in overcoming barriers to efficient nucleic acid delivery [112]. Non-viral vectors, such as inorganic nanoparticles, carbon nanotubes, liposomes, and polymeric materials, offer several advantages over viral vectors, including decreased immune response and greater design flexibility. These non-viral vectors can be functionalized and targeted to specific sites within a biological system, often with lower cytotoxicity. Despite these benefits, challenges remain, particularly in achieving satisfactory transfection efficiency and addressing safety concerns. These issues hinder the clinical translation of non-viral vectors for gene delivery. Therefore, while non-viral vectors, such as inorganic nanoparticles and liposomes, offer significant advantages like reduced immune response and design flexibility, the hurdles of unsatisfactory transfection efficiency and safety concerns must be overcome to realize their full potential in clinical applications [113]. Understanding the physiochemical properties and behavior of nanocarriers is crucial for their clinical application, with a focus on improving characteristics, delivery, and drug release. Recent research on nanoscale nucleic acid carriers has highlighted their distinctive and controllable properties, but there is a need for deeper knowledge on how to efficiently utilize these systems in clinical settings. Significant innovations in the synthesis, physical properties, and utilization of nucleic acid nanocarriers have been made in the past five years, emphasizing the importance of these properties for enhancing nanocarrier characteristics. Addressing challenges related to the transport and therapeutic release of these nanocarriers, as well as overcoming barriers to effective tissue transport, is essential for translating these tools into clinical applications [114]. Advanced nanoscale biotechnology strategies, including controlling particle properties and response to environmental signals, are essential for effective *in vivo* and *in vitro* delivery of nucleic acid therapeutics [115]. Addressing these challenges through innovative manufacturing processes is key to realizing the full potential of nanocarriers for nucleic acids and proteins in therapeutic applications.

### 15.4.3 Strategies for Efficient Scale-Up and Production

Efficient scale-up and production strategies in nanocarriers for nucleic acids and proteins involve various approaches outlined in the provided research contexts. For instance, lipidic nanocarriers have shown potential for delivering nucleic acid therapeutics, with computational models aiding in predicting physiochemical properties and payload delivery efficiency [115]. Additionally, the use of microfluidic processes has been proposed to develop a scale-independent platform for manufacturing and monitoring nanoparticles, enhancing control over particle size and characteristics, especially in poly(lactic-co-glycolic acid) (PLGA) particles [116]. Furthermore, the development of nanohydrogels (NHGs) complexed with DNA has demonstrated sustained protein expression without cellular toxicity, showcasing the potential for efficient gene delivery and foreign protein expression over extended periods [117]. Efficient scale-up strategies for mRNA delivery using crosslinked nucleic acid nanogels show promise for therapeutic protein expression, enhancing production potential in nanocarriers for nucleic acids and proteins [118]. Strategies for efficient scale-up in electrospun nanofiber production include increasing needle numbers, using auxiliary electrodes, new nozzles, automatic wiping, and controlling feed rate and solution concentration [119]. Continuous LNP formation through microfluidic mixing enables efficient scale-up for mRNA production, ensuring high quality and consistency in nucleic acid encapsulation. The study demonstrates that continuous microfluidic mixing significantly improves product uniformity and particle size compared to traditional pipetting methods, achieving a particle size of 71 nm and an encapsulation efficiency (EE) of 88%, closely aligning with the specifications required for mRNA vaccines such as Pfizer's. This approach, supported by predictive process models and digital twins, enhances productivity by 15–20% and reduces personnel and chemical use by approximately 30%, establishing a robust foundation for the continuous, bottleneck-free production of mRNA and other therapeutic nucleic acids in a GMP- and QbD-compliant Biopharma 4.0 facility [120]. These diverse strategies highlight the importance of technological advancements and innovative approaches in optimizing the scale-up and production of nanocarriers for nucleic acids and proteins.

### 15.4.4 Case Studies of Successful Scale-Up Initiatives

Successful scale-up initiatives for nanocarriers, including solid lipid nanoparticles (SLNs), have been driven by the need for large-scale production, physicochemical stability, and cost-effective design. SLNs have gained attention for their ability to deliver a range of substances, such as drugs, nucleic acids, and proteins, due to their ease of preparation and stability. The need for large-scale manufacturing units is crucial for producing sufficient formulations for clinical trials and commercial applications. Continuous processing methods are becoming increasingly popular to minimize batch-to-batch variability. Recent advancements in scale-up technologies, including high-pressure homogenization, hot melt extrusion, microchannels, nanoprecipitation, and microemulsion-based methods, highlight the progress toward the commercialization of SLNs. These technologies not only support large-scale production but also ensure the physicochemical stability and cost-effectiveness required for successful market entry. The rapid development and scale-up of LNP-encapsulated mRNA vaccines for COVID-19 was a major success story. Moderna and Pfizer/BioNTech were able to leverage their existing LNP platforms to rapidly develop and manufacture mRNA vaccines at unprecedented scale. This demonstrated the potential of LNPs for scalable delivery of nucleic acids [121, 122]. Patisiran, an RNAi therapeutic delivered via LNPs, was approved by the FDA in 2018 for the treatment of hereditary transthyretin-mediated amyloidosis. This was the first approved RNAi therapeutic and the first systemically administered LNP product, validating the potential of LNPs for delivering nucleic acid drugs. Conjugating siRNAs to N-acetylgalactosamine (GalNAc) ligands enables targeted delivery to hepatocytes via the asialoglycoprotein receptor. This approach has enabled multiple GalNAc-siRNA conjugates to advance to late-stage clinical trials for liver diseases, demonstrating the potential for scalable chemical conjugation to deliver nucleic acids [123]. Another example involves polymeric micelles, which have emerged as versatile delivery vehicles capable of overcoming extracellular and intracellular barriers. These micelles can be engineered with stimuli-sensitive groups or specific ligands to enhance targeting and accumulation at the desired site of action, protecting the nucleic acids from degradation and promoting cellular uptake. The design and optimization of these polymeric micelles have been pivotal in advancing their clinical translation, demonstrating their potential in gene therapy and combined therapeutic strategies [124]. Microfluidics technology has demonstrated successful scale-up of protein-loaded liposomes from bench scale to GMP production, maintaining consistent physicochemical attributes. This study investigates the use of microfluidic manufacture for nanomedicines, specifically focusing on protein-loaded liposomes. By employing two different laminar flow microfluidic cartridge designs, including a novel toroidal mixer design, the research demonstrates seamless scale-up from small-scale (1 mL) to GMP volume production (200 mL/min) with consistent size, polydispersity index (PDI), protein loading, and release properties. The application of tangential flow filtration for downstream processing further ensures high product yield. This work confirms the feasibility of a scale-independent production process, reducing risks associated with transitioning from bench scale to approved product [125]. Protein nanocages derived from natural proteins like ferritin provide a biocompatible platform for delivering nucleic acids. Their inherent stability, ability to be engineered, and capacity for targeted delivery make them a promising scalable approach. Protein nanocages have shown promise in preclinical studies for delivering DNA, RNA, and proteins [126].

## 15.5 ETHICAL CONSIDERATIONS

### 15.5.1 Ethical Issues in Clinical Translation and Medical Innovation

Nanoscience has transformed the pharmaceutical industry by facilitating the development of advanced therapeutic drugs that are more effective and less toxic. As nanomedical technology continues to evolve, it is essential to address the legal, social, and ethical implications of their widespread use in therapy while being mindful of potential risks. Several ethical and legal challenges have been recognized in the clinical application and medical innovation of nanocarriers [127].

#### *15.5.1.1 Exposure to Harm and Health Risks*

An important concern highlighted is the possibility of harm resulting from nanomedical technology. Due to their heightened reactivity compared to their bulk counterparts, nanocarriers have the potential to cause health damage [25]. Clinical studies have demonstrated that nanocarriers like hematite and magnetite can lead to significant DNA damage [128], and inhaling these particles can result in pulmonary fibrosis [129]. Additionally, nanocarriers can elevate toxicity levels in the body [130], which may cause blood clotting in vessels [131] and increase the risk of breast cancer in patients [132].

#### *15.5.1.2 Issues of Consent in Nanomedical Research*

Research and thorough investigations are crucial for the advancement of safe and effective nanomedical technology. However, clinical researchers face unique challenges in this field. One major issue is obtaining truly informed consent from participants in nanomedical device and nano-chemical drug trials, given the novel and evolving nature of the research. Moreover, participants might downplay the potential hazards of nanocarrier drugs [133], complicating the process of securing fully informed consent.

#### *15.5.1.3 Patient Privacy*

As nanomedical devices advance and increasingly integrate with technologies like artificial intelligence (AI) and machine learning, concerns arise regarding their capacity to generate patient data [134]. When this data is logged for purposes such as comparing groups or training AI systems, it may be stored in large digital databases. This situation can potentially leave patients vulnerable, as their health data could be used in ways that might not align with their interests [135].

#### *15.5.1.4 Access to Nanomedical Technology*

The substantial costs linked to nanomedical technology will inevitably be transferred to patients, impacting the expense of their healthcare, particularly in insurance-based health systems. These financial burdens make nanocarriers prohibitively expensive for many patients who might otherwise benefit from such technologies [135]. Geographical accessibility poses another challenge. As noted in the UNESCO Science Report [136], countries like the Republic of Korea, Canada, and Germany are at the forefront of nanomedicine innovation. While nanomedical technology has the potential to enhance the quality of life for many, patients in developing countries often face significantly limited access to nanochemical drugs and nanomedical devices [137].

### 15.5.2 Ethical Frameworks and Guidelines

Nanocarrier products must undergo rigorous preclinical and clinical testing before they can be used for disease diagnosis, prevention, or treatment. Researchers are still in the early stages of examining the toxicological, pharmacological, and immunological properties of various nanocarriers. Additionally, there are no standardized international guidelines for nanopharmaceuticals. The standard approach for assessing the quality, safety, and efficacy of nanotechnology-based products involves a "case-by-case" evaluation. This method considers the physical, chemical, and biological properties of the nanoparticles, the product itself, the route of administration, the intended use of the product, and other relevant factors.

The U.S. EPA, the National Institute of Environmental Health Sciences, the National Science Foundation, and the National Institute of Occupational Safety and Health have initiated several programs to study the risks associated with nanomaterials. Furthermore, the National Cancer Institute has set up a laboratory dedicated to characterizing the *in vitro* response to ENM that could be used in cancer diagnosis or treatment [138].

Evaluating the safety of nanomaterials is challenging due to their only common characteristic being their size (1–100 nm). Because nanomaterials do not constitute a uniform class of compounds, each type must be individually assessed based on its specific properties.

#### *15.5.2.1 USA*

Over the course of time, the FDA has issued guidelines for the use of nanomaterials in food, cosmetics, and animal food, but specific guidance for nanomedicine has yet to be published. In 2017, the FDA released draft guidance on drug products, including biological products, containing nanomaterials. The FDA evaluates each nanotechnology individually rather than categorizing nanotechnology as inherently safe or harmful [139]. Manufacturers are encouraged to engage with the FDA early in the development of their nanotechnology products to ensure a shared understanding of regulatory concerns. The Nanotechnology Characterization Laboratory of the National Cancer Institute (NCL-NCI) has also played a significant role in regulating nanomedicines for over a decade. Additionally, the FDA established the Nanotechnology Task Force and the Nanotechnology Interest Group, which include representatives from various regulatory centers, to address the global challenges of regulating nanotechnology effectively.

#### *15.5.2.2 UK*

In the UK, the MHRA oversees the regulation of medicines. Similar to the FDA, there is no specific guidance for the approval of nanomedicines, and these products are evaluated on an individual basis. Similar to the United States, organizations like the EU-NCL operate across the UK and EU, continually advancing knowledge on preclinical characterization assays for nanomedicine [140].

#### *15.5.2.3 India*

The Central Drugs Standard Control Organization (CDSCO), India's Central Drug Authority, oversees the regulation of nanomedicines under the Drugs and Cosmetics Act. According to CDSCO guidelines, all nanopharmaceutical preparations are permanently classified as Investigational New Drugs (IND). Safety studies should adhere to the broad principles outlined in Schedule Y of the Drugs and Cosmetics Rules, 1945, with additional specific criteria tailored for nanopharmaceuticals as needed. Given the inherent intricacies of nanotechnology-based products, an individualized assessment is advised to evaluate their quality, safety, and efficacy. Furthermore, continuous pharmacovigilance is essential throughout the life cycle of nanopharmaceuticals to ensure sustained safety and effectiveness.

#### *15.5.2.4 EU*

The European Union Nanomedicine Characterisation Laboratory (EU-NCL) and the REFINE (Regulatory Science Framework for the risk-benefit assessment of medical products and medical devices that are based on nanomedicines) project have received government funding to contribute to the progress in the field of nanomedicines. Unlike the UK, the European regulatory body, the EMA, has issued a set of preliminary guidelines for the preparation standards of nanomedicines. The EU-NCL collaborates with these regulatory bodies to provide insights and influence decisions regarding the regulation and safety of nanocarriers.

#### *15.5.2.5 Canada*

In Canada, the approval of nanotechnology products is managed using existing regulatory frameworks. Health Canada recommends that manufacturers engage with the relevant regulatory authority early in the development process to identify and evaluate the risks and properties of the product (Health Canada). Canada has established the Health Portfolio Nanotechnology Working Group, which comprises delegates from regulatory bodies such as Health Canada and the Canadian Institutes of Health Research (CIHR). This group addresses issues related to nanotechnology. Moreover, Health Canada has provided comprehensive guidance concerning health products and food based on nanotechnology.

#### *15.5.2.6 Japan*

In Japan, medicines are regulated by the MHLW and the PMDA. While these governing bodies have not yet set forth a precise definition or regulations for nanomedicines, they published a guideline for the manufacture of liposomes in 2016. Nanomedicines are presently governed under the Pharmaceutical Affairs Law framework, which employs broad medicinal product legislation on an individual basis [2].

### 15.5.3 Future Ethical Challenges and Considerations

Despite the absence of specific regulatory guidance, more than 50 nanomedicines have successfully entered the market, with the number steadily increasing. These products are predominantly focused on cancer therapy, addressing the challenges posed by toxic compounds and the complex tumor microenvironment, which often hinders effective drug treatment. The global absence of formal regulations for nanomedicines and nanomaterial production in health applications presents a significant challenge. Varying standards among different government agencies categorize some nanomedicines as medical devices and others as pharmaceuticals. What meets regulatory criteria in one region may not be accepted elsewhere, particularly hindering global licensing for small molecules. Urgent alignment across regulatory frameworks is crucial for the nanomedicine field to progress in line with expectations. Despite the formation of clusters and working groups, actionable outcomes have been limited. Nanomaterials, despite their established presence, continue to face regulatory ambiguities that hamper the timely development of treatments for various diseases and conditions.

Efforts within academic and government circles to establish National Characterization Laboratories are notable, yet clearer and more stringent guidance is imperative from leading regulatory bodies like the FDA and MHRA. Given that diseases affect populations worldwide indiscriminately, there is a pressing need for a global consortium to harmonize regulations on nanocarriers and issue comprehensive guidance to researchers. Despite significant financial investments in nanocarrier development over the past 20 years, without decisive leadership and regulatory direction, the potential for bringing products to market remains uncertain, potentially diverting future investments to other areas.

## 15.6 INTERDISCIPLINARY COLLABORATION

### 15.6.1 Importance of Collaboration Across Disciplines

The importance of interdisciplinary collaboration in the clinical translation and regulatory aspects of nanomedicine cannot be overstated. This collaboration is crucial for addressing the complex challenges involved in developing safe and effective nanomedicine products. Nanomedicines often involve novel materials with unique properties that require expertise from multiple disciplines like Nanotechnology, Biomedical Engineering, Pharmacology and Toxicology. Researchers study the interaction of nanomaterials with biological systems and assess safety and efficacy [141]. Translating nanomedicine from the lab to clinical use involves multiple stages, including preclinical testing and clinical trials. Collaboration ensures that each phase adheres to rigorous standards [142]. Regulatory bodies need to develop guidelines that account for the unique properties of nanomedicines. Regulatory Scientists and Policy Makers Develop and update regulations based on new scientific findings. Industry Experts and Regulatory Affairs Specialists Ensure compliance with evolving standards and facilitate the approval process [143].

### 15.6.2 Case Studies of Successful Interdisciplinary Efforts

Successful interdisciplinary efforts in nanomedicine often involve collaboration across various fields to address complex challenges in clinical translation and regulatory aspects.

Doxil® is an FDA-approved liposomal formulation of the chemotherapy drug doxorubicin. It significantly improves the drug's pharmacokinetics and reduces side effects compared to conventional doxorubicin. Development of liposomal carriers to encapsulate doxorubicin in nanotechnology. Optimization of the liposome formulation for enhanced drug delivery and reduced toxicity in pharmaceutical sciences. Conducted clinical trials to demonstrate the efficacy and safety of Doxil®. Regulatory Affairs in coordination with the FDA for approval, involving extensive documentation and compliance with guidelines for liposomal drug products [144].

Abraxane® is a formulation of the chemotherapy drug paclitaxel bound to albumin nanoparticles. This formulation enhances drug solubility and reduces the risk of hypersensitivity reactions. Development of albumin-bound nanoparticles to improve paclitaxel delivery in nanotechnology. Trials demonstrated the effectiveness of Abraxane® in treating metastatic breast cancer, non-small cell lung cancer, and pancreatic cancer [145].

Cytalux® is a novel imaging agent for detecting ovarian cancer. It is a fluorescent imaging agent that binds to cancer cells, allowing surgeons to visualize and remove tumors more effectively during surgery. Development of the fluorescent agent based on a nanoparticle platform. Integration of the imaging agent into surgical procedures to improve tumor detection. Cytalux® has improved the accuracy of ovarian cancer surgeries, demonstrating the value of interdisciplinary collaboration in developing and implementing novel imaging technologies [146].

NanoTherm® therapy involves the use of iron oxide nanoparticles that are activated by an external magnetic field to generate localized hyperthermia and treat tumors, highlighting the effectiveness of interdisciplinary efforts in developing advanced therapeutic modalities [147].

### 15.6.3 Future Trends in Collaborative Research and Development

AI and machine learning are increasingly being used to analyze large datasets, optimize nanomaterial design, predict drug interactions, and personalize treatments. These technologies can enhance both preclinical and clinical stages by providing insights that improve decision-making processes [148]. Nanomedicine is increasingly being integrated with precision medicine approaches, where treatments are tailored to individual genetic, environmental, and lifestyle factors. This trend involves the development of nanocarriers that can deliver therapeutics based on a patient's specific molecular profile [149]. As nanomedicine technologies evolve, there is a growing need for updated and harmonized regulatory frameworks. Future trends include the development of more flexible and dynamic regulatory pathways that can accommodate the rapid pace of nanotechnology advancements [150, 151]. Future research is likely to focus on developing multifunctional nanoparticles that combine imaging, diagnostics, and therapeutic capabilities into a single platform [152]. Increased global collaboration will be essential to address the challenges of nanomedicine research and development. International partnerships will facilitate knowledge exchange, harmonize regulatory practices, and accelerate the development and adoption of nanomedicine technologies [153, 154].

## 15.7 CONCLUSION

The integration of nanotechnology into the biomedical field marks a significant advancement in the delivery of nucleic acids and proteins, offering new avenues for enhancing therapeutic interventions. Nanocarriers, with their unique physicochemical properties, have demonstrated the potential to improve drug efficacy, specificity, and safety. However, the journey from laboratory research to clinical application is fraught with challenges that necessitate meticulous attention to regulatory, ethical, and manufacturing aspects. Regulatory frameworks for nanomedicines are still evolving, with global inconsistencies posing significant hurdles. The lack of unified guidelines complicates the development and approval process, underscoring the need for a harmonized regulatory approach. Ethical considerations, particularly regarding patient safety and informed consent, must be meticulously addressed to maintain public trust and support for nanomedicine advancements. Manufacturing nanocarriers at scale while ensuring their stability and reproducibility remains a critical challenge. Advances in production technologies and innovative strategies are essential to overcome these obstacles and make nanomedicine commercially viable. Successful case studies highlight the importance of interdisciplinary collaboration in achieving these goals, demonstrating that concerted efforts across various scientific and regulatory domains can lead to significant breakthroughs. Looking ahead, the future of nanomedicine relies heavily on continued innovation and robust regulatory support. Interdisciplinary collaboration will be pivotal in navigating the complexities of translating nanotechnology from the bench to the bedside. By addressing the existing gaps and fostering a collaborative environment, the full potential of nanocarriers in revolutionizing therapeutic delivery can be realized, ultimately improving patient outcomes and transforming the landscape of modern medicine.

## REFERENCES

1. Hua, S., de Matos, M. B. C., Metselaar, J. M., & Storm, G. (2018). Current trends and challenges in the clinical translation of nanoparticulate nanomedicines: Pathways for translational development and commercialization. *Frontiers in Pharmacology*, 9, 790. https://doi.org/10.3389/fphar.2018.00790

2. Foulkes, R., Man, E., Thind, J., Yeung, S., Joy, A., & Hoskins, C. (2020). The regulation of nanomaterials and nanomedicines for clinical application: Current and future perspectives. *Biomaterials Science*, 8(17), 4653–4664. https://doi.org/10.1039/D0BM00558D

3. Yeh, N.-C. (2013). Nanotechnology for electronics & photonics. *Technovation*, 33(4), 108. https://doi.org/10.1016/j.technovation.2013.03.008

4. Hanigan, D., Truong, L., Schoepf, J., Nosaka, T., Mulchandani, A., Tanguay, R. L., & Westerhoff, P. (2018). Trade-offs in ecosystem impacts from nanomaterial versus organic chemical ultraviolet filters in sunscreens. *Water Research*, 139, 281–290. https://doi.org/10.1016/j.watres.2018.03.062

5. Mihranyan, A., Ferraz, N., & Strømme, M. (2012). Current status and future prospects of nanotechnology in cosmetics. *Progress in Materials Science*, 57(5), 875–910. https://doi.org/10.1016/j.pmatsci.2011.10.001

6. Pomerantseva, E., Bonaccorso, F., Feng, X., Cui, Y., & Gogotsi, Y. (2019). Energy storage: The future enabled by nanomaterials. *Science (New York, N.Y.)*, 366(6468), eaan8285. https://doi.org/10.1126/science.aan8285

7. Min, Y., Caster, J. M., Eblan, M. J., & Wang, A. Z. (2015). Clinical translation of nanomedicine. *Chemical Reviews*, 115(19), 11147–11190. https://doi.org/10.1021/acs.chemrev.5b00116

8. Ventola, C. L. (2012). The nanomedicine revolution: Part 1: Emerging concepts. *P & T: A Peer-Reviewed Journal for Formulary Management*, 37(9), 512–525.

9. Seigneuric, R., Markey, L., Nuyten, D. S. A., Dubernet, C., Evelo, C. T. A., Finot, E., & Garrido, C. (2010). From nanotechnology to nanomedicine: Applications to cancer research. *Current Molecular Medicine,* 10(7), 640–652. https://doi.org/10.2174/156652410792630634

10. He, H., Liu, L., Morin, E. E., Liu, M., & Schwendeman, A. (2019). Survey of clinical translation of cancer nanomedicines—lessons learned from successes and failures. *Accounts of Chemical Research*, 52(9), 2445–2461. https://doi.org/10.1021/acs.accounts.9b00228

11. Martins, J. P., das Neves, J., de la Fuente, M., Celia, C., Florindo, H., Günday-Türeli, N., Popat, A., Santos, J. L., Sousa, F., Schmid, R., Wolfram, J., Sarmento, B., & Santos, H. A. (2020). The solid progress of nanomedicine. *Drug Delivery and Translational Research*, 10(3), 726–729. https://doi.org/10.1007/s13346-020-00743-2

12. Smolkova, B., Dusinska, M., & Gabelova, A. (2017). Nanomedicine and epigenome. Possible health risks. *Food and Chemical Toxicology: An International Journal Published for the British Industrial Biological Research Association*, 109(Pt 1), 780–796. https://doi.org/10.1016/j.fct.2017.07.020

13. Lobo, V., Patil, A., Phatak, A., & Chandra, N. (2010). Free radicals, antioxidants and functional foods: Impact on human health. *Pharmacognosy Reviews*, 4(8), 118–126. https://doi.org/10.4103/0973-7847.70902

14. Manke, A., Wang, L., & Rojanasakul, Y. (2013). Mechanisms of nanoparticle-induced oxidative stress and toxicity. *BioMed Research International*, 2013(1), 942916. https://doi.org/10.1155/2013/942916

15. Kermanizadeh, A., Balharry, D., Wallin, H., Loft, S., & Møller, P. (2015). Nanomaterial translocation—the biokinetics, tissue accumulation, toxicity and fate of materials in secondary organs—a review. *Critical Reviews in Toxicology*, 45(10), 837–872. https://doi.org/10.3109/10408444.2015.1058747

16. Modo, M., Kolosnjaj-Tabi, J., Nicholls, F., Ling, W., Wilhelm, C., Debarge, O., Gazeau, F., & Clement, O. (2013). Considerations for the clinical use of contrast agents for cellular MRI in regenerative medicine. *Contrast Media & Molecular Imaging*, 8(6), 439–455. https://doi.org/10.1002/cmmi.1547

17. European Medicines Agency. *Withdrawal Assessment Report for Sinerem*. www.ema.europa.eu/documents/withdrawal-report/withdrawal-assessment-report-sinerem_en.pdf [Accessed July 12, 2024].

18. Falkner, R., & Jaspers, N. (2012). Regulating nanotechnologies: risk, uncertainty and the global governance gap. *Global Environmental Politics*, 12(1), 30–55. https://doi.org/10.1162/GLEP_a_00096

19. Regulatory Science Framework for Nano(bio)material-based Medical Products and Devices. http://refine-nanomed.eu/ [Accessed July 12, 2024].

20. Halamoda-Kenzaoui, B., Box, H., van Elk, M., Gaitan, S., Geertsma, R. E., Lafuente, E. G., & Bremer-Hoffmann, S. (2019). *Anticipation of Regulatory Needs for Nanotechnology Enabled Health Products. The REFINE White Paper*. Publications office of the European Union, Luxemburg.

21. Landesman-Milo, D., & Peer, D. (2016). Transforming nanomedicinesfrom lab scale production to novel clinical modality. *Bioconjugate Chemistry*, 27(4), 855–862. https://doi.org/10.1021/acs.bioconjchem.5b00607

22. Mahapatra, I., Clark, J. R. A., Dobson, P. J., Owen, R., Lynch, I., & Lead, J. R. (2018). Expert perspectives on potential environmental risks from nanomedicines and adequacy of the current guideline on environmental risk assessment. *Environmental Science: Nano*, 5(8), 1873–1889. https://doi.org/10.1039/C8EN00053K

23. Baun, A., & Hansen, S. F. (2008). Environmental challenges for nanomedicine. *Nanomedicine (London, England)*, 3(5), 605–608. https://doi.org/10.2217/17435889.3.5.605

24. Food and Drug Administration. (2017). *Drug Products, Including Logical Products, That Contain Nanomaterials Guidance for Industry*. Available from www.fda.gov/regulatory-information/search-fda-guidance-documents/drug-products-including-biological-products-contain-nanomaterials-guidance-industry [Accessed July 12, 2024].

25. Bawa, R. (2011). Regulating nanomedicine—can the FDA handle it? *Current Drug Delivery*, 8(3), 227–234. https://doi.org/10.2174/156720111795256156

26. Bawa, R., Audette, G. F., & Reese, B. (2016). *Handbook of Clinical Nanomedicine: Law, Business, Regulation, Safety, and Risk*. CRC Press.

27. Kelly. (2010). *Nanomedicines: Regulatory Challenges and Risks Ahead. Regulatory Affairs Pharma, 10, 14–17 [online]*. Available at: www.cov.com/-/media/files/corporate/publications/2010/10/nanomedicines regulatory.pdf [Accessed July 12, 2024].

28. Rannard, S., & Owen, A. (2009). Nanomedicine: Not a case of "One size fits all." *Nano Today*, 4(5), 382–384. https://doi.org/10.1016/j.nantod.2009.07.002

29. Cleland, A. N., Fraikin, J. L., Meinhold, P., & Monzon, F. (2016). Nanoparticle characterization—one size does not fit all: Nanoparticle size analysis for nanomedicine applications. *Drug Development & Delivery*, 6, 20.

30. Oh, J. Y., Kim, H. S., Palanikumar, L., Go, E. M., Jana, B., Park, S. A., Kim, H. Y., Kim, K., Seo, J. K., Kwak, S. K., Kim, C., Kang, S., & Ryu, J.-H. (2018). Cloaking nanoparticles with protein corona shield for targeted drug delivery. *Nature Communications*, 9(1), 4548. https://doi.org/10.1038/s41467-018-06979-4

31. Leong, H. S., Butler, K. S., Brinker, C. J., Azzawi, M., Conlan, S., Dufés, C., Owen, A., Rannard, S., Scott, C., Chen, C., Dobrovolskaia, M. A., Kozlov, S. V., Prina-Mello, A., Schmid, R., Wick, P., Caputo, F., Boisseau, P., Crist, R. M., McNeil, S. E., … Pastore, C. (2019). On the issue of transparency and reproducibility in nanomedicine. *Nature Nanotechnology*, 14(7), 629–635. https://doi.org/10.1038/s41565-019-0496-9

32. Muthu, M. S., & Wilson, B. (2012). Challenges posed by the scale-up of nanomedicines. *Nanomedicine (London, England)*, 7(3), 307–309. https://doi.org/10.2217/nnm.12.3

33. Soares, S., Sousa, J., Pais, A., & Vitorino, C. (2018). Nanomedicine: Principles, properties, and regulatory issues. *Frontiers in Chemistry*, 6, 360. https://doi.org/10.3389/fchem.2018.00360

34. Mühlebach, S. (2018). Regulatory challenges of nanomedicines and their follow-on versions: A generic or similar approach? *Advanced Drug Delivery Reviews*, 131, 122–131. https://doi.org/10.1016/j.addr.2018.06.024

35. Sainz, V., Conniot, J., Matos, A. I., Peres, C., Zupancic, E., Moura, L., Silva, L. C., Florindo, H. F., & Gaspar, R. S. (2015). Regulatory aspects on nanomedicines. *Biochemical and Biophysical Research Communications*, 468(3), 504–510. https://doi.org/10.1016/j.bbrc.2015.08.023

36. Paradise, J. (2019). Regulating nanomedicine at the Food and Drug Administration. *AMA Journal of Ethics*, 21(4), E347–E355. https://doi.org/10.1001/amajethics.2019.347

37. Kroll, A., Pillukat, M. H., Hahn, D., & Schnekenburger, J. (2009). Current in vitro methods in nanoparticle risk assessment: Limitations and challenges. *European Journal of Pharmaceutics and Biopharmaceutics*, 72(2), 370–377. https://doi.org/10.1016/j.ejpb.2008.08.009

38. Edmondson, R., Broglie, J. J., Adcock, A. F., & Yang, L. (2014). Three-dimensional cell culture systems and their applications in drug discovery and cell-based biosensors. *Assay and Drug Development Technologies*, 12(4), 207–218. https://doi.org/10.1089/adt.2014.573

39. Dickinson, A. M., Godden, J. M., Lanovyk, K., & Ahmed, S. S. (2019). Assessing the safety of nanomedicines: A mini review. *Applied In Vitro Toxicology*, 5(3), 114–122. https://doi.org/10.1089/aivt.2019.0009

40. Flühmann, B., Ntai, I., Borchard, G., Simoens, S., & Mühlebach, S. (2019). Nanomedicines: The magic bullets reaching their target? *European Journal of Pharmaceutical Sciences: Official Journal of the European Federation for Pharmaceutical Sciences*, 128, 73–80. https://doi.org/10.1016/j.ejps.2018.11.019

41. Hoskins, C., Wang, L., Cheng, W. P., & Cuschieri, A. (2012). Dilemmas in the reliable estimation of the in-vitro cell viability in magnetic nanoparticle engineering: Which tests and what protocols? *Nanoscale Research Letters*, 7(1), 77. https://doi.org/10.1186/1556-276X-7-77

42. Siegrist, S., Cörek, E., Detampel, P., Sandström, J., Wick, P., & Huwyler, J. (2019). Preclinical hazard evaluation strategy for nanomedicines. *Nanotoxicology*, 13(1), 73–99. https://doi.org/10.1080/17435390.2018.1505000

43. Sharma, H. S., Hussain, S., Schlager, J., Ali, S. F., & Sharma, A. (2010). Influence of nanoparticles on blood-brain barrier permeability and brain edema formation in rats. *ActaNeurochirurgica. Supplement*, 106, 359–364. https://doi.org/10.1007/978-3-211-98811-4_65

44. Agrahari, V., & Hiremath, P. (2017). Challenges associated and approaches for successful translation of nanomedicines into commercial products. *Nanomedicine (London, England)*, 12(8), 819–823. https://doi.org/10.2217/nnm-2017-0039

45. Limaye, V., Fortwengel, G., & Limaye, D. (2014). Regulatory roadmap for nanotechnology based medicines. *International Journal of Drug Regulatory Affairs*, 2(4),33–41. https://doi.org/10.22270/ijdra.v2i4.151

46. Patel, V. L., Zhang, J., Yoskowitz, N. A., Green, R. and Sayan, O. R. (2008). Translational cognition for decision support in critical care environments: A review. *Journal of Biomedical Informatics,* 41(3), 413–431.

47. Rubio, D. M., Schoenbaum, E. E., Lee, L. S., Schteingart, D. E., Marantz, P. R., Anderson, K. E., Platt, L. D., Baez, A., & Esposito, K. (2010). Defining translational research: Implications for training. *Academic Medicine*, 85(3), 470–475.

48. Woolf, S. H. (2008). The meaning of translational research and why it matters. *JAMA*, 299(2), 211–213.

49. Abu-Odah, H., Said, N. B., Nair, S. C., Allsop, M. J., Currow, D. C., Salah, M. S., Hamad, B. A., Elessi, K., Alkhatib, A., ElMokhallalati, Y., & Bayuo, J. (2022). Identifying barriers and facilitators of translating research evidence into clinical practice: A systematic review of reviews. *Health & Social Care in the Community*, 30(6), e3265–e3276.

50. Kimmelman, J., & London, A. J. (2015). The structure of clinical translation: Efficiency, information, and ethics. *Hastings Center Report*, 45(2), 27–39.

51. Seyhan, A. A. (2019). Lost in translation: The valley of death across preclinical and clinical divide—identification of problems and overcoming obstacles. *Translational Medicine Communications*, 4(1), 1–19.

52. Chung, K. (2018). Rapid drug screen using 3D tumororganoids. *Science Translational Medicine*, 10(423), eaar7507.

53. Fermini, B., Coyne, S. T., & Coyne, K. P. (2018). Clinical trials in a dish: A perspective on the coming revolution in drug development. *Slas Discovery: Advancing Life Sciences R&D*, 23(8), 765–776.

54. Kwok, T. C., Ricker, N., Fraser, R., Chan, A. W., Burns, A., Stanley, E. F., McCourt, P., Cutler, S. R., & Roy, P. J. (2006). A small-molecule screen in *C. elegans* yields a new calcium channel antagonist. *Nature*, 441(7089), 91–95.

55. Translational Research Success Stories. (2023). *National Institute of Neurological Illnesses and Stroke (NINDS)*. www.ninds.nih.gov/about-ninds/what-we-do/impact/translational-research-success-stories [Accessed on July 20, 2024].

56. Cona, L. A. (2024). *Type 1 Diabetes Cure: Current Research*. www.dvcstem.com/post/type-1-diabetes-cure [Accessed on July 20, 2024].

57. Wang, X., Gao, M., Wang, Y., & Zhang, Y. (2022). The progress of pluripotent stem cell-derived pancreatic β-cells regeneration for diabetic therapy. *Frontiers in Endocrinology,* 13, 927324.

58. Tucker, M. E. (2021). *Stem Cell Transplant Seen as Major Type 1 Diabetes Advance*. www.medscape.com/viewarticle/961696?form=fpf [Accessed on July 20, 2024].

59. Briskin, A. (2022). *ViaCyte and CRISPR Introduce New Stem Cell Therapy for Type 1 Diabetes*. https://diatribe.org/diabetes-research/viacyte-and-crispr-introduce-new-stem-cell-therapy-type-1-diabetes [Accessed on July 20, 2024].

60. Burt, H. M., & Hunter, W. L. (2006). Drug-eluting stents: A multidisciplinary success story. *Advanced Drug Delivery Reviews*, 58(3), 350–357.

61. Acharjee, A. (2023). Translational research and key aspects to make it successful. *Translational Medicine Communications*, 8(1), 19.

62. National Cancer Institute (NCI). (2022). *CAR T Cells: Engineering Patients' Immune Cells to Treat Their Cancers*. www.cancer.gov/about-cancer/treatment/research/car-t-cells#:~:text=They%20are%20made%20by%20collecting,their%20surface%20called%20chimeric%20antigen

63. Zhang, X. Q., Xu, X., Bertrand, N., Pridgen, E., Swami, A., Farokhzad, O. C. (2012). Interactions of nanomaterials and biological systems: Implications to personalized nanomedicine. *Advanced Drug Delivery Reviews*, 64(13), 1363–1384.

64. Singh, N., Manshian, B., Jenkins, G. J., Griffiths, S. M., Williams, P. M., Maffeis, T. G., Wright, C. J., & Doak, S. H. (2009). NanoGenotoxicology: The DNA damaging potential of engineered nanomaterials. *Biomaterials,* 30(23–24), 3891–3914.

65. European Medicines Agency: Withdrawn applications for human and veterinary medicines, December 2023. www.ema.europa.eu/en/documents/other/withdrawn-applications-human-and-veterinary-medicines-download-table-4-december-2023_en.xlsx [Accessed August 24, 2024].

66. Faulkner, R., & Jasper, N. (2012). Regulating nanotechnologies: risk, uncertainty and the global governance gap. *Global Environmental Pollution*, 12(1), 30–35.

67. REFINE Consortium. Regulatory Science Framework for Nano(bio)material-based Medical Products and Devices. http://refine-nanomed.eu/ [Accessed July 12, 2024].

68. Hariyadi, D. M., Athiyah, U., & Pathak, Y. V. (2020). Nanomedicine: Risk, safety, regulation, and public health. In: *Nanomedicine and Nanosafety*. Springer Singapore, pp. 561–578.

69. Liposome Drug Products. *Chemistry, Manufacturing, and Controls; Human Pharmacokinetics and Bioavailability; and Labeling Documentation*. FDA. www.fda.gov/regulatory-information/search-fda-guidance-documents/liposome-drug-products-chemistry-manufacturing-and-controls-human-pharmacokinetics-and [Accessed July 24, 2024].

70. U.S. Food and Drug Administration. (2022). *Drug Products, Including Biological Products, That Contain Nanomaterials—Guidance for Industry*. FDA. Docket Number: FDA-2017-D-0759. www.fda.gov/regulatory-information/search-fda-guidance-documents/drug-products-including-biological-products-contain-nanomaterials-guidance-industry [Accessed July 24, 2024].

71. U.S. Food and Drug Administration. (2018). Nanotechnology. fda.gov/ScienceResearch/SpecialTopics/Nanotechnology/ucm402230.htm [Accessed March 19, 2025].

72. U.S. Food and Drug Administration. (2013). *Ingredients, Packaging, Labeling*. www.fda.gov/Food/IngredientsPackagingLabeling/default.htm [Accessed July 24, 2024].

73. Allan, J., Sokull-Kluettgen, B., & Patri A. K. (2020). *Gobal Summit on Regulatory Science 2019 Nanpotechnology and Nanoplastics, EUR 30195 EN*. Publications Office of the European Union, Luxembourg. ISBN 978-92-76-18435-5, doi:10.2760/517689, JRC120318.

74. US National Institute for Standards and Technology (NIST). (2020). *Standard Reference Materials*. www.nist.gov/srm

75. US Food & Drug Administration (FDA). (2020). *Nanotechnology Guidance Documents*. www.fda.gov/science-research/nanotechnology-programs-fda/nanotechnology-guidance-documents

76. US Food & Drug Administration (FDA). (2020). *Europe Office*. www.fda.gov/about-fda/office-global-operations/europe-office

77. Gottardo, S., Alessandrelli, M., Valeria, A., Atluri, R., Barberio, G., Bekker, C., Bergonzo, P., Bleeker, E., Booth, A., Borges, T., Buttol, P., Carlander, D., Castelli, S., Chevillard, S., Clavaguera, S., Dekkers, S., Delpivo, C., Di Prospero Fanghella, P., Dusinska, M., … Crutzen, H. (2017). *NANoREG Framework for the Safety Assessment of Nanomaterials, EUR 28550 ENEuropean Commission H2020 Programme—Explanatory Notes on Open Innovation Test Beds*. https://ec.europa.eu/research/participants/data/ref/h2020/other/guides_for_applicants/h2020-supp-info-innotestbeds-18-20_en.pdf. https://publications.jrc.ec.europa.eu/repository/handle/JRC105651

78. European Parliament and Council of the European Union. (2009). Regulation (EC) No 1223/2009 of the European Parliament and of the Council of 30 November 2009 on cosmetic products (recast). Official Journal of the European Union, L 342/59-342/209. https://health.ec.europa.eu/system/files/2016-11/cosmetic_1223_2009_regulation_en_0.pdf [Accessed March 19, 2025].

79. European ChemicalsAgency (ECHA). (2020). *Nanomaterials*. https://echa.europa.eu/regulations/nanomaterials

80. European Commission: Joint Research Centre, Allan, J., Sokull-Kluettgen, B. and Patri, A. (2020). *Global Summit on Regulatory Science 2019. Nanotechnology and Nanoplastics – Perspectives on nanotechnology and nanoplastics, Publications Office, 2020*. https://data.europa.eu/doi/10.2760/517689.

81. European Food Safety Authority (EFSA). (2020). *EFSA Scientific Network for Risk Assessment of the Use of Nanotechnologies in Food and Feed*. www.efsa.europa.eu/sites/default/files/Nanonetwork.pdf

82. Amenta, V., Aschberger, K., Arena, M., Bouwmeester, H., Moniz, F. B., Brandhoff, P., Gottardo, S., Marvin, H. J. P., Mech, A., Pesudo, L. Q., Rauscher, H., Schoonjans, R., Vettori, M. V., Weigel, S., Peters, R. J., & Amenta, V. (2015). Regulatory aspects of nanotechnology in the agri/food/food sector in EU and non-EU countries. *Regulatory Toxicology and Pharmacology*, 73(1), 463–476.

83. European Commission. (2019). Communication from the Commission to the European Parliament, the European Council, the Council, the European Economic and Social Committee and the Committee of the Regions: The European Green Deal. COM(2019)640 final. Publications Office of the European Union. https://op.europa.eu/en/publication-detail/-/publication/b828d165-1c22-11ea-8c1f-01aa75ed71a1 [Accessed March 19, 2025].

84. European Commission. (2020). *Communication from the Commission to the European Parliament, the European Council, the Council, the European Economic and Social Committee and the Committee of the Regions: A New Industrial Strategy for Europe. COM(2020) 102 final. Publications Office of the European Union.* https://eur-lex.europa.eu/legal-content/EN/TXT/?uri=CELEX:5202 0DC0102 [Accessed March 19, 2025].

85. United Nations. (2015). *Transforming Our World: The 2030 Agenda for Sustainable Development.* Resolution Adopted by the General Assembly on 25 September 2015, A/RES/70/1. https://sdgs. un.org/2030agenda [Accessed March 19, 2025].

86. European Commission. (2018). Communication from the Commission to the European Parliament, the European Council, the Council, the European Economic and Social Committee and the Committee of the Regions: A European Strategy for Plastics in a Circular Economy. COM(2018) 28 final. Publications Office of the European Union. https://eur-lex.europa.eu/legal-content/EN/TXT/ ?uri=COM%3A2018%3A28%3AFIN [Accessed March 19, 2025].

87. Rahim, M. M. (2018). Nanomedicine regulation in Australia. *Alternative Law Journal*, 44(2), 1037969X1881573.

88. Oomen, A. G., Bennink, M., Engelen, J. G. M., & Sips, A. J. A. M. (2011). *Nanomaterial in Consumer Products: Detection, Characterisation and Interpretation.* National Institute for Public Health and the Environment.

89. Karim, M. E., & Munir, A. B. (2014). Nanotechnology in Asia: A preliminary assessment of the existing legal framework. *KLRI Journal of Law and Legislation*, 4(2), 171–223.

90. Nasu, H., & Faunce, T. (2013). Nanotechnology in Japan: A route to energy security after Fukushima? *Bulletin of the Atomic Scientists*, 69(5), 68–74.

91. Bowman, D. M, & Hodge, G. A. (2007). A small matter of regulation: An international review of nanotechnology regulation. *Columbia Science and Technology Law Review*, 8, 1–36.

92. Nasu, H., & Faunce, T. (2011). The proposed ban on certain nanomaterials for electrical and electronic equipment in Europe and its global security implications: A search for an alternative regulatory approach. *European Journal of Law and Technology*, 2(3). https://ejlt.org/index.php/ejlt/ article/view/96 [Accessed March 19, 2025].

93. Organisation for Economic Co-operation and Development. www.oecd.org/en/topics/sub-issues/ emerging-technologies.html [Accessed March 19, 2025].

94. Nanotechnology Industries Association. Voice of the Nanotechnology Industries. https://nanotechia. org [Accessed March 19, 2025].

95. Health Canada. *Policy Statement on Health Canada's Working Definition for Nanomaterial.* /www. canada.ca/en/health-canada/services/science-research/reports-publications/nanomaterial/policy-statement-health-canada-working-definition.html [Accessed March 19, 2025].

96. International Pharmaceutical Regulators Programme. www.iprp.global/home [Accessed August 21, 2024].

97. US Environmental Protection Agency (USEPA). (2017). Chemical substances when manufactured or processed as nanoscale materials; TSCA reporting and record keeping requirements. *Federal Register*, 82(8), 3641.

98. Organisation for Economic Co-operation and Development. www.oecd.org/en/topics/sub-issues/nanomaterials-and-advanced-materials.html

99. Scientific Committee on Consumer Safety (SCCS). (2019). *Guidance on the Safety Assessment of Nanomaterials in Cosmetics*. European Union. ISSN 1831-4767 ISBN 978-92-76-15271-2.

100. Geiss, O., Ponti, J., Senaldia, C., Bianchia, I., Mehna, D., Barrero, J., Gillilanda, D., Matissek, R., & Anklam, E. (2020). Characterisation of food grade titania with respect to nanoparticle content in pristine additives and in their related food products. *Food Additives and Contaminants*, 37(2), 239–253.

101. Genchi, G. G., Marino, A., Tapeinos, C., & Ciofani, G. (2017). Smart materials meet multifunctional biomedical device: Current and prospective implications for nanomedicine. *Frontiers in Bioengineering and Biotechnology*, 5, 80.

102. European Medicines Agency (EMA). (2020). *Multidisciplinary: Nanomedicines*. www.ema.europa.eu/en/human-regulatory/research-development/scientific-guidelines/multidisciplinary/multidisciplinary-nanomedicines

103. Fernandes, C., Jathar, M., Sawant, B. K. S., & Warde, T. (2023). Scale-up of nanoparticle manufacturing process. In: Jindal, A. B. (ed.), *Pharmaceutical Process Engineering and Scale-up Principles. AAPS Introductions in the Pharmaceutical Sciences*, vol. 13. Springer, pp. 173–203. https://doi.org/10.1007/978-3-031-31380-6_12

104. Goel, H., Siddiqui, L., Mahtab, A., & Talegaonkar, S. (2022). Fabrication design, process technologies, and convolutions in the scale-up of nanotherapeutic delivery systems. In: Kesharwani, P., & Singh, K. K. (eds.), *Nanoparticle Therapeutics*. Academic Press, pp. 47–131. https://doi.org/10.1016/B978-0-12-820757-4.00017-X

105. Torres-Vanegas, J. D., Cruz, J. C., & Reyes, L. H. (2021). Delivery systems for nucleic acids and proteins: Barriers, cell capture pathways and nanocarriers. *Pharmaceutics*, 13(3), 428. https://doi.org/10.3390/PHARMACEUTICS13030428

106. dos Santos Fonseca, L. M., Machado, B. A. S., Oliveira, F. O., de Jesus Santos, J. R., da Silva, J. W., Hodel, K. V. S., Rosatti, B. G., Pinto, C. D., & Soares, M. B. P. (2024). An overview on recent patents and technologies on nanoparticles for nucleic acid delivery. *Expert Opinion on Therapeutic Patents*, 34(3), 171–186. https://doi.org/10.1080/13543776.2024.2338097

107. Patel, D. M., Patel, N. N., & Patel, J. K. (2021). Nanomedicine scale-up technologies: Feasibilities and challenges. In: Patel, J. K., & Pathak, Y. V. (eds.), *Emerging Technologies for Nanoparticle Manufacturing*. Springer, pp. 511–539. https://doi.org/10.1007/978-3-030-50703-9_24

108. Liu, X., & Meng, H. (2021). Consideration for the scale-up manufacture of nanotherapeutics—a critical step for technology transfer. *View*, 2(5), 20200190. https://doi.org/10.1002/VIW.20200190

109. Singer, A., Markoutsa, E., Limayem, A., Mohapatra, S., & Mohapatra, S. S. (2018). Nanobiotechnology medical applications: Overcoming challenges through innovation. *The EuroBiotech Journal*, 2(3), 146–160. https://doi.org/10.2478/EBTJ-2018-0019

110. Paliwal, R., Babu, R. J., & Palakurthi, S. (2014). Nanomedicine scale-up technologies: Feasibilities and challenges. *AAPS PharmSciTech*, 15(6), 1527–1534. https://doi.org/10.1208/s12249-014-0177-9

111. Quintão, C. C. R., Camargo, L. S. de A., Brandão, H. de M., Saraiva, N. Z., & Munk, M. (2022). Challenges in the use of nanostructures as carriers of nucleic acids in clinical practice. *Einstein (Sao Paulo, Brazil)*, 20, 1. https://doi.org/10.31744/EINSTEIN_JOURNAL/2022RB5898

112. Lin, M., & Qi, X. (2023). Advances and challenges of stimuli-responsive nucleic acids delivery system in gene therapy. *Pharmaceutics*, 15(5), 1450. https://doi.org/10.3390/PHARMACEUTICS15051450

113. Alkhawaldeh, A. K., Rheima, A. M., Kadhim, M. M., sabri Abbas, Z., dhari jawad al-bayati, A., Abed, Z. T, mohamed dashoor Al-Jaafari, F., Jaber, A. S., Hachim, S. K., Ali, F. K., Mahmoud, Z. H., Behzadi pour, G., & Kianfar, E. (2023). Nanomaterials as transmitters of non-viral gene vectors: A review. *Case Studies in Chemical and Environmental Engineering*, 8, 100372. https://doi.org/10.1016/J.CSCEE.2023.100372

114. Mosley, R. J., Rucci, B., & Byrne, M. E. (2023). Recent advancements in design of nucleic acid nanocarriers for controlled drug delivery. *Journal of Materials Chemistry B*, 11(10), 2078–2094. https://doi.org/10.1039/D2TB02325C

115. Weng, Y., Huang, Q., Li, C., Yang, Y., Wang, X., Yu, J., Huang, Y., & Liang, X. J. (2020). Improved nucleic acid therapy with advanced nanoscale biotechnology. *Molecular Therapy—Nucleic Acids*, 19, 581–601. https://doi.org/10.1016/J.OMTN.2019.12.004

116. Roces, C. B., Christensen, D., & Perrie, Y. (2020). Translating the fabrication of protein-loaded poly(lactic-co-glycolic acid) nanoparticles from bench to scale-independent production using microfluidics. *Drug Delivery and Translational Research*, 10(3), 582–593. https://doi.org/10.1007/s13346-019-00699-y

117. Eswaran, L., Kazimirsky, G., Yehuda, R., & Byk, G. (2023). A new strategy for nucleic acid delivery and protein expression using biocompatible nanohydrogels of predefined sizes. *Pharmaceutics*, 15(3), 961. https://doi.org/10.3390/PHARMACEUTICS15030961

118. Huang, X., Zheng, R., Ding, F., Yang, J., Xie, M., Liu, X., Li, J., Feng, J., Zhu, X., & Zhang, C. (2020). Efficient delivery of mRNA using crosslinked nucleic acid nanogel as a carrier. *ACS Materials Letters*, 2(11), 1509–1515. https://doi.org/10.1021/ACSMATERIALSLETT.0C00375

119. Liu, Y., Hao, M., Zhou, C., Yang, B., Jiang, S., Huang, J., Chen, Z., Liu, Y., & Ramakrishna, S. (2023). Scale-up strategies for electrospun nanofiber production. In: Kargari, A., Matsuura, T., & Shirazi, M. M. A. (eds.), *Electrospun and Nanofibrous Membranes: Principles and Applications.* Elsevier, pp. 205–266. https://doi.org/10.1016/B978-0-12-823032-9.00020-9

120. Hengelbrock, A., Schmidt, A., & Strube, J. (2023). Formulation of nucleic acids by encapsulation in lipid nanoparticles for continuous production of mRNA. *Processes*, 11(6), 1718. https://doi.org/10.3390/PR11061718

121. Khairnar, S. V., Pagare, P., Thakre, A., Nambiar, A. R., Junnuthula, V., Abraham, M. C., Kolimi, P., Nyavanandi, D., & Dyawanapelly, S. (2022). Review on the scale-up methods for the preparation of solid lipid nanoparticles. *Pharmaceutics*, 14(9), 1886. https://doi.org/10.3390/PHARMACEUTICS14091886

122. Baker, A., Lorch, J., VanderWeele, D., & Zhang, B. (2023). Smart nanocarriers for the targeted delivery of therapeutic nucleic acid for cancer immunotherapy. *Pharmaceutics*, 15(6), 1743. https://doi.org/10.3390/PHARMACEUTICS15061743

123. Freitag, F., & Wagner, E. (2021). Optimizing synthetic nucleic acid and protein nanocarriers: The chemical evolution approach. *Advanced Drug Delivery Reviews*, 168, 30–54. https://doi.org/10.1016/J.ADDR.2020.03.005

124. Rangelov, S., Dai, J., Haladjova, E., Sinani, G., Durgun, M. E., Cevher, E., & Özsoy, Y. (2023). Polymeric-micelle-based delivery systems for nucleic acids. *Pharmaceutics*, 15(8), 2021. https://doi.org/10.3390/PHARMACEUTICS15082021

125. Webb, C., Forbes, N., Roces, C. B., Anderluzzi, G., Lou, G., Abraham, S., Ingalls, L., Marshall, K., Leaver, T. J., Watts, J. A., Aylott, J. W., & Perrie, Y. (2020). Using microfluidics for scalable manufacturing of nanomedicines from bench to GMP: A case study using protein-loaded liposomes. *International Journal of Pharmaceutics*, 582, 119266. https://doi.org/10.1016/J.IJPHARM.2020.119266

126. Zhang, Y., Luo, J., Gui, X., Zheng, Y., Schaar, E., Liu, G., & Shi, J. (2023). Bioengineered nanotechnology for nucleic acid delivery. *Journal of Controlled Release: Official Journal of the Controlled Release Society*, 364, 124. https://doi.org/10.1016/J.JCONREL.2023.10.034

127. Wasti, S., Lee, I. H., Kim, S., Lee, J.-H., & Kim, H. (2023). Ethical and legal challenges in nanomedical innovations: A scoping review. *Frontiers in Genetics,* 14, 1163392. https://doi.org/10.3389/fgene.2023.1163392

128. Karlsson, H. L., Cronholm, P., Gustafsson, J., & Möller, L. (2008). Copper oxide nanoparticles are highly toxic: A comparison between metal oxide nanoparticles and carbon nanotubes. *Chemical Research in Toxicology*, 21, 1726–1732. https://doi.org/10.1021/tx800064j

129. Lu, X., Zhu, T., Chen, C., & Liu, Y. (2014). Right or left: The role of nanoparticles in pulmonary diseases. *International Journal of Molecular Sciences*, 15, 17577–17600. https://doi.org/10.3390/ijms151017577

130. Trisolino, A. (2014). Nanomedicine: Building a bridge between science and law. *Nanoethics*, 8, 141–163. https://doi.org/10.1007/s11569-014-0193-y

131. Evans, D. (2007). *Ethics, Nanotechnology and Health.* 1st ed. UNESCO Publishing.

132. Peng, F., Setyawati, M. I., Tee, J. K., Ding, X., Wang, J., Nga, M. E., Ho, H. K., & Leong, D. T. (2019). Nanoparticles promote in vivo breast cancer cell intravasation and extravasation by inducing endothelial leakiness. *Nature Nanotechnology*, 14, 279–286. https://doi.org/10.1038/s41565-018-0356-z

133. Resnik, D. B., & Tinkle, S. S. (2007). Ethics in nanomedicine. *Nanomedicine*, 2, 345–350. https://doi.org/10.2217/17435889.2.3.345

134. Toumey, C. (2007). Privacy in the shadow of nanotechnology. *Nanoethics*, 1, 211–222. https://doi.org/10.1007/s11569-007-0023-6

135. Fisher, E., Boenink, M., van der Burg, S., & Woodbury, N. (2012). Responsible healthcare innovation: Anticipatory governance of nanodiagnostics for theranostics medicine. *Expert Review of Molecular Diagnostics*, 12, 857–870. https://doi.org/10.1586/erm.12.125

136. Brownsword, R. (2008). Regulating nanomedicine—the smallest of our concerns? *Nanoethics*, 2, 73–86. https://doi.org/10.1007/s11569-008-0030-2

137. Meetoo, D. (2009). Nanotechnology: Is there a need for ethical principles? *British Journal of Nursing*, 18, 1264–1268. https://doi.org/10.12968/bjon.2009.18.20.45125

138. Resnik, D. B., & Tinkle, S. S. (2007). Ethical issues in clinical trials involving nanomedicine. *Contemporary Clinical Trials*, 28, 433–441. https://doi.org/10.1016/j.cct.2006.11.001

139. Hamburg, M. A. (2012). FDA's approach to regulation of products of nanotechnology. *Science*, 336(6079), 299–300.

140. Bremer-Hoffmann, S., Halamoda-Kenzaoui, B., & Borgos, S. E. (2018). Identification of regulatory needs for nanomedicines. *Journal of Interdisciplinary Nanomedicine*, 3, 4–15. https://doi.org/10.1002/jin2.34.

141. Moghimi, S. M., Hunter, A. C., & Murray, J. C. (2014). Nanomedicine: A review of clinical developments and further perspectives. *Biomaterials*, 35(23), 8534–8545.

142. Đorđević, S., Gonzalez, M. M., Conejos-Sánchez, I., Carreira, B., Pozzi, S., Acúrcio, R. C. et al. (2022). Current hurdles to the translation of nanomedicines from bench to the clinic. *Drug Delivery and Translational Research*, 12, 500–525. https://doi.org/10.1007/s13346-021-01024-2.

143. Sainz, V., Conniot, J., Matos, A. I., Peres, C., Zupancic, E., Moura, L., et al. (2015). Regulatory aspects on nanomedicines. *Biochemical and Biophysical Research Communications*, 468(3), 504–510. doi: 10.1016/j.bbrc.2015.08.023. Epub 2015 Aug 8. PMID: 26260323.

144. Barenholz, Y. (2012). Doxil®—The first FDA-approved nano-drug: Lessons learned. *Molecular Pharmaceutics*, 9(6), 1595–1610.

145. Gabizon, A., Shmeeda, H., & Barenholz, Y. (2003). Pharmacokinetics of pegylated liposomal doxorubicin: Review of animal and human studies. *Clinical Pharmacokinetics*, 42(5), 419–436.

146. Cytalux, a novel imaging agent, FDA approved for diagnosing inoperable ovarian cancer lesions. (n.d.). *Journal of Hematology Oncology Pharmacy*. https://jhoponline.com/issue-archive/2021-issues/december-2021-vol-11-no-6/19104:cytalux-a-novel-imaging-agent-fda-approved-for-diagnosing-inoperable-ovarian-cancer-lesions.

147. Soetaert, F., Korangath, P., Serantes, D., Fiering, S., & Ivkov, R. (2020). Cancer therapy with iron oxide nanoparticles: Agents of thermal and immune therapies. *Advanced Drug Delivery Reviews*, 163–164, 65–83. doi: 10.1016/j.addr.2020.06.025. Epub 2020 Jun 27. PMID: 32603814; PMCID: PMC7736167.

148. Mendes, B. B., Zhang, Z., Conniot, J., Sousa D. P., Ravasco, J. M. J. M., Onweller, L. A. et al. (2024) A large-scale machine learning analysis of inorganic nanoparticles in preclinical cancer research. *Nature Nanotechnology* 19, 867–878. https://doi.org/10.1038/s41565-024-01673-7.

149. Mitchell, M. J., Billingsley, M. M., Haley, R. M., Wechsler M. E., Peppas, N. A. & Langer, R. (2021). Engineering precision nanoparticles for drug delivery. *Nature Reviews Drug Discovery*, 20, 101–124. https://doi.org/10.1038/s41573-020-0090-8.

150. Mühlebach S. (2018). Regulatory challenges of nanomedicines and their follow-on versions: A generic or similar approach? *Advanced Drug Delivery Reviews*,131, 122–131. doi: 10.1016/j.addr.2018.06.024. Epub 2018 Jun 30. PMID: 29966685.

151. Caputo, F., G. Favre, G. Borchard, L. Calzolai, P. Fisicaro, E. Frejafon et al. (2024). Toward an international standardisation roadmap for nanomedicine. *Drug Delivery and Translational Research*, 14(9), 2578–2588 124.

152. Raheem, M. A., Rahim, M. A., Gul, I., Zhong, X., Xiao, C., Zhang, H., et al. (2023). Advances in nanoparticles-based approaches in cancer theranostics. *OpenNano*, 12, 100152. https://doi.org/10.1016/j.onano.2023.100152.

153. Chen, C., Yuan, P., & Zhang, Z. Y. (2024). Nanomedicine-based cancer immunotherapy: A bibliometric analysis of research progress and prospects. *Frontiers in Immunology*, 14, 1446532. doi:10.3389/fimmu.2024.1446532.

154. Prasad, R., A. Ghosh, V. Patel, B. Peng, B. Mendes, E. H. A. Win, L. et al. (2025). Voices of nanomedicine: Blueprint guidelines for collaboration in addressing global unmet medical needs. *ACS Nano*, 19, 2979–2991.

# 16 Regulatory Toxicology of Biologics

## *Safeguarding Health in Biotherapeutic Development*

*Kinjal Parikh, Nishant Talati, Shahin Vhora, Nidhi Kotadiya, Mitul Talala, Ruby Christian, Vikas Jain, Tosha Pandya, Disha Suthar, and Satish Rojekar*

### 16.1 INTRODUCTION

Biotherapeutic products, also known as biologics, encompass diverse therapeutic agents derived from living organisms or their components. Unlike conventional pharmaceuticals synthesized through chemical processes, biologics are complex structured molecules, including proteins, nucleic acids, and cellular components (1). This category includes vaccines, blood components, gene therapies, and monoclonal antibodies.

Biological products have revolutionized modern medicine by offering innovative treatments for diseases ranging from cancer to autoimmune disorders. The development of biologics that are derived from living organisms or their products has opened new avenues in medicine, offering targeted therapies for a range of diseases (2). However, the unique nature of biologics demands rigorous regulatory oversight to ensure their safety and efficacy. Regulatory toxicology is a vital component in the development of biologics, providing crucial insights into potential risks and guiding safety assessments from preclinical stages to clinical trials and beyond.

Biologics, as defined by the U.S. Food and Drug Administration (FDA), are products derived from living organisms or their cellular components intended to prevent, treat, or cure diseases. They are categorized based on their source, mode of action (MoA), and therapeutic purpose (3). For instance, recombinant proteins, such as insulin or growth factors, are produced through genetic engineering techniques. Biologics include therapeutic proteins, enzymes, and monoclonal antibodies engineered to target specific molecules or pathways in the body. Monoclonal antibodies are designed to target specific antigens implicated in diseases like cancer or autoimmune disorders. Vaccines stimulate the immune system to provide defense against infectious diseases. Gene therapy is used by introducing genetic material into cells to correct genetic defects or modulate cellular functions. Cell-based therapies utilize living cells, such as stem cells or immune cells, for therapeutic purposes, often in regenerative medicine or cancer treatment.

#### 16.1.1 Unique Characteristics of Regulatory Toxicology

Toxicology, as a distinct scientific discipline, is relatively modern, but the understanding and awareness of poisons trace back to ancient civilizations. Ancient texts from around 450 BCE detail the toxicity of venom from snakebites and their treatments, showcasing early human encounters with toxic substances (4). This historical context highlights humanity's long-standing quest to understand and mitigate the harmful effects of toxic substances. In an era where drug safety remains a paramount concern in public health, the significance of predictivity in modern predictive toxicology cannot be understated. This evolving discipline stratifies the intersection between state-of-the-art scientific methodologies and the urgent need to accurately predict the potential toxicological impacts of new chemical entities. By harnessing the predictive power of various *in vitro* and silico models alongside the advancements in biomarkers and computational toxicology, the field aims to mitigate risks associated with hepatotoxicity, nephrotoxicity, and other adverse drug reactions (5). Furthermore, these approaches serve as viable animal testing alternatives, highlighting their importance in both ethical and scientific domains of toxicity testing.

Biologics exhibit distinct characteristics that present unique challenges in regulatory toxicology assessments due to their complex structure. Their large molecular size, structural complexity, and potential immunogenicity require specialized evaluation methods. Traditional toxicological studies may not adequately capture the nuances of biological products, requiring tailored approaches to assess their safety profiles. Moreover, the potential for variability in manufacturing processes and biological activity adds another layer of complexity to regulatory evaluations. Complex structured biologics are large, complex molecules with complicated three-dimensional structures. It can influence pharmacokinetics, immunogenicity, and interactions with biological systems. Immunogenicity biologics may bring potential immune responses in recipients, leading to adverse effects such as hypersensitivity reactions or neutralizing antibody formation. Batch-to-batch variability for biological products may show variability in composition and activity due to various factors like manufacturing processes, cell lines, and culture conditions, requiring thorough characterization and quality control measures (6).

DOI: 10.1201/9781003473183-16

### 16.1.2 Importance of Regulatory Toxicology in the Development of Biologics

Regulatory toxicology plays a critical role in ensuring the safety and efficacy of biologics throughout their development and commercialization.

#### *16.1.2.1 Ensuring the Safety of Human Subjects*

Preclinical toxicology studies are conducted to assess the safety profile of biologics before advancing to clinical trials. Preclinical studies, including *in vitro* assays and animal models, evaluate potential toxicities, such as immunogenicity, cytokine release, and organ toxicity (7). These studies provide crucial insights into the potential risks associated with biological therapies, guiding dose selection, and risk mitigation strategies in clinical trials. By identifying potential safety concerns early in the development process, regulatory toxicology helps minimize the risks to patients and ensures the ethical conduct of clinical research.

A few of the key considerations include "Repeat-Dose toxicity," which evaluates the effects of prolonged exposure to biologics on various organ systems, helping to establish safe dosage regimens for human use. "Immunotoxicity" is assessing the impact of biologics on the immune system, including the potential for immunosuppression, cytokine release syndrome, or autoimmune reactions. "Reproductive and developmental toxicity" investigates the potential risks to fertility, pregnancy, and fetal development, ensuring the safety of biologics for women of childbearing age and pregnant individuals (8).

#### *16.1.2.2 Understanding Regulatory Compliance and Risk Mitigation*

Compliance with regulatory guidelines is essential for biologics' successful development and approval. Regulatory agencies, such as the FDA and the European Medicines Agency (EMA), have established rigorous guidelines for the preclinical evaluation of biologics, encompassing safety, efficacy, and quality considerations. Regulatory toxicology studies are designed to adhere to these guidelines, providing comprehensive data to support regulatory submissions. Risk mitigation strategies, such as post-marketing surveillance and pharmacovigilance, play an important role in monitoring the long-term safety of biological therapies. By proactively identifying and addressing potential risks, regulatory toxicology contributes to the ongoing safety assessment of biological products throughout their lifecycles. Regulatory toxicology studies aid in demonstrating comparability by conducting comparative toxicology studies to establish the similarity of biosimilar products to reference biologics in terms of safety and efficacy. Risk mitigation by identifying and mitigating potential safety concerns early in development through comprehensive toxicological assessments, minimizing risks to patients during clinical trials and post-marketing surveillance (9).

## 16.2 FUNDAMENTALS OF REGULATORY TOXICOLOGY

### 16.2.1 Dose–Response Relationships

It is essential to understand how dose, drug concentration in blood, and clinical responses are interrelated. This relationship is determined using the dose–response curve. This forms the basis of determining the starting dose of the drug and the limit to which dose can be increased before it causes toxicity (10).

A typical dose–response curve is shown in Figure 16.1A. The curve reaches an asymptote when all the receptor sites are occupied. The response plotted in this manner is too wide for the linear format. Hence, the log concentration of the drug against the response measured is plotted, resulting in a sigmoid shape (Figure 16.1B). Four main parameters obtained from a graded dose-response curve are as follows:

- *Potency ($EC_{50}$):* Indicates a drug's affinity for its receptor.
- *Slope:* A shallow slope suggests a higher risk of overlapping desired effects and side effects, while a steep slope indicates difficulty in precisely controlling the effects.
- *Maximum Effect ($E_{max}$):* The highest achievable effect may not always be available, even from *in vitro* data.
- *Threshold Dose:* The lowest dose at which a drug effect is observed may not always be the zero point due to pre-existing baseline effects.

#### *16.2.1.1 Cumulative Dose-Response Curves*

Sum frequency-response curves are plotted against the logarithm of the dose (Figure 16.2). The curve typically appears linear between 0% and 100% response, thus facilitating the identification of key dose–response relationship features as described above. Figure 16.2 illustrates a cumulative dose–response curve where the highest dose produces no response in test animals, and all lower doses are called no-observable-effect levels (NOELs). These NOELs are important as they represent safe doses. The highest NOEL is the "threshold" dose, defined as the dose below which no toxicity occurs. Above the threshold dose, responses increase with dose until a 100% response rate is achieved, where the curve flattens as higher doses no longer

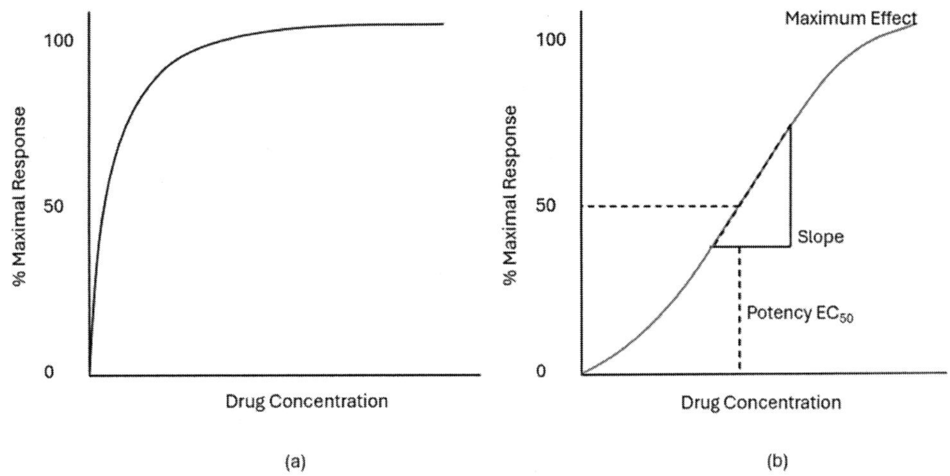


**Figure 16.1** (A) Dose–response curve. (B) Log dose–response curve.

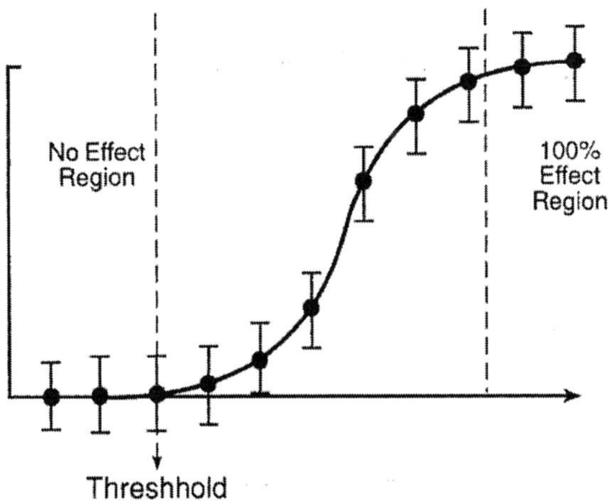


**Figure 16.2** No effect and 100% effect region in cumulative dose–response curve.

change the response rate. For therapeutic effects, physicians aim for doses in the flat region of the curve to ensure a 100% response, ensuring drug efficacy despite individual variations. Conversely, toxicologists focus on identifying the threshold dose and the no-effect region to avoid undesirable effects induced by chemicals.

$ED_{50}$, i.e., dose causing 50% of maximum effect, is often used as the starting dose due to its position on the steep part of the curve. However, individual $ED_{50}$ can vary due to body composition, pharmacokinetics, pharmacodynamics, and drug interactions. Dosage guidelines for anticancer drugs are often based on surrogate measures like tumor size reduction rather than long-term outcomes like mortality or quality of life. Animal studies data and *in vitro* studies guide determining $ED_{50}$; however, transitioning from homogeneous cell studies to heterogeneous tumors add complexity due to changing tumor biology, drug resistance, and varied tumor types and locations. The science of posology encompasses all these factors to determine optimal dosing and regimen.

### 16.2.2 Species Differences and Extrapolation to Humans

The FDA, Center for Drug Evaluation and Research (CDER), issued guidance in 2005 on estimating the maximum safe starting dose in initial clinical trials for humans. The steps include:

1. Determine the NOAEL in at least two pre-clinical animal species.

2. Convert NOAEL to a human equivalent dose (HED), normalized by body surface area or body weight.
3. Select HED from the most appropriate species, typically the most sensitive one.
4. Apply a safety factor, generally at least 10-fold, to determine the maximum recommended starting dose (MRSD).
5. Adjust MRSD based on the pharmacologically active dose if necessary.

This approach is simple and historically supported but focuses primarily on NOAEL rather than pharmacologically active doses and oversimplifies scaling to humans.

After the determination of NOAEL, it is then converted to HEDs. Toxic end points such as MTD for the systemically administered drug in animals scale well between species when body surface area (i.e., mg/m$^2$) normalization is considered. Literature has reported, that for anticancer drugs, dose toxic to 10% rodents ($LD_{10}$) and MTD in non-rodents correlate well with humans when the doses were normalized to the same administration schedule and expressed as mg/m$^2$. Also, extrapolation is best between species for such drugs when doses are normalized to $W^{0.75}$ rather than $W^{0.67}$.

Factors for converting animal NOAEL to HEDs (Table 16.1) assume a 1:1 dose ratio between species based on body surface area. However, scaling based on body weight (HED (mg/kg) = NOAEL (mg/kg)) may sometimes be more appropriate. This requires data showing similar NOAEL doses (mg/kg) across species. Before using mg/kg scaling, the following conditions must be met:

1. NOAELs occur at a similar mg/kg dose across test species.
2. If only two NOAELs from different species are available, one of these must also be true:

- The therapeutic is orally administered and limited by local toxicities, with GI compartment weight scaling by $W^{0.94}$
- Human toxicity (for a particular class) is highly correlated with dose on a mg/kg basis across species
- Other pharmacologic and toxicological endpoints scaled by mg/kg, such as MTD, lowest lethal dose, and pharmacologically active dose
- There is a strong correlation between plasma drug levels ($C_{max}$ and AUC) and dose in mg/kg.

## Table 16.1 Conversion of animal doses to human equivalent doses based on body surface area

| | | To Convert Animal Dose in mg/kg to HED in mg/kg, Either: | |
|---|---|---|---|
| **Species** | **To Convert Animal Dose in mg/kg to Dose in mg/m², Multiply by km** | **Divide Animal Dose By** | **Multiply Animal Dose By** |
| **Human** | 37 | — | — |
| Child (20 kg) | 25 | — | — |
| Mouse | 3 | 12.3 | 0.08 |
| Hamster | 5 | 7.4 | 0.13 |
| Rat | 6 | 6.2 | 0.16 |
| Ferret | 7 | 5.3 | 0.19 |
| Guinea pig | 8 | 4.6 | 0.22 |
| Rabbit | 12 | 3.1 | 0.32 |
| Dog | 20 | 1.8 | 0.54 |
| **Primates** | | | |
| Monkeys | 12 | 3.1 | 0.32 |
| Marmoset | 6 | 6.2 | 0.16 |
| Squirrel monkey | 7 | 5.3 | 0.19 |
| Baboon | 20 | 1.8 | 0.54 |
| Micro-pig | 27 | 1.4 | 0.73 |
| Mini-pig | 35 | 1.1 | 0.95 |

*Source:* Data adapted and modified from FDA draft guidelines.

If these conditions are not met, use the mg/m² method for a safer MRSD. Note that mg/kg scaling results in HEDs 12, 6, and 2 times higher than the mg/m² method for mice, rats, and dogs, respectively.

### 16.2.3 Key Toxicological Endpoints

Biotherapeutics derived from biological sources require rigorous toxicological assessment to ensure their safety and efficacy in humans. Acute toxicity studies, conducted within 24 hours, determine immediate toxic effects and lethal dose ($LD_{50}$). Chronic toxicity studies, lasting 6–24 months, evaluate long-term effects and potential organ damage. Regulatory agencies provide guidelines for standardized toxicity studies. Dose selection, animal model choice, and immunogenicity evaluation are crucial considerations for accurately assessing biotherapeutics (11).

To determine the immediate toxic effects of biologics, acute toxicity studies are conducted within 24 hours. These studies aim to determine the $LD_{50}$ and assess any signs of morbidity, mortality, toxicity, changes in body weight, and physical abnormalities. Chronic toxicity studies, lasting 6–24 months, evaluate the long-term effects of prolonged exposure to biologics on animal lifespan. These studies involve periodic blood samples, monitoring for toxicity signs, and tissue biopsy to assess organ damage. Regulatory agencies like FDA, EMA, and International Council for Harmonisation (ICH) provide guidelines for standardized toxicity studies. Dose selection, necropsy, and histopathological examinations are crucial components of toxicity studies. Data analysis involves statistical comparisons, identification of dose–response relationships, and determination of NOAEL and LOAEL.

Immunotoxicity, the adverse effects of drugs on the immune system, can involve suppression or enhancement of immune responses. Drugs may trigger allergic reactions due to foreign recognition. Antibody measurement in repeated dose toxicity studies is crucial for assessing immunogenicity in biotechnology-derived pharmaceuticals. Antibody responses should be characterized and correlated with pharmacological and toxicological changes. However, the detection of antibodies alone is not grounds for early study termination unless they significantly neutralize drug effects (12).

Immunotoxicological testing strategies may require screening and mechanistic studies. Routine testing approaches are not recommended for biotechnology-derived pharmaceuticals. Assessing immunogenicity/allergenicity involves careful selection of species and consideration of the limits of predictivity. Many biotechnology-derived pharmaceuticals' biological activity and species specificity often preclude standard toxicity testing designs. Immunotoxicity testing programs should use relevant species, which may require nonhuman primates or transgenic mice. *In vitro* assessment using human materials may be advisable when no relevant species exists (13).

## 16.3 PRECLINICAL SAFETY ASSESSMENT

### 16.3.1 Nonclinical Study Design

#### *16.3.1.1 Selection of Relevant Animal Models*

Choosing appropriate animal models for biotherapeutics is crucial due to their complex nature (14). Factors like mechanism of action, pharmacokinetics, species specificity, ethics, and regulations must be considered. Suitable models should mimic human physiology, have genetic similarities, and reflect the immunogenicity of biologics. Additionally, animal species' pharmacokinetic profile, availability, and size should be considered. Drug characteristics, safety considerations, regulatory requirements, and special considerations for specific biologics must also be addressed. Transgenic models and combination models can enhance prediction value. Collaborating with experts is essential for selecting and utilizing suitable animal models effectively.

When selecting animal models for nonclinical studies, human physiology, species specificity, and genetic similarity are key considerations. Animal models should closely mimic human physiology for accurate assessment of target organ or system effects. Common species include rodents, rabbits, dogs, and non-human primates (NHPs). For species-specific effects, like some cytokines, appropriate models are essential. Genetic homology between animal models and humans, especially for targeted proteins, is another key consideration. Transgenic animals expressing human genes can provide more relevant data.

Animal models should reflect the pharmacokinetic profile of biologics, including absorption, distribution, metabolism, and excretion (ADME), for accurate toxicity estimation. Species longevity, population size, and availability influence model selection. Ethical considerations and animal welfare must be prioritized, with alternatives like *in vitro* tests or computer modeling considered when possible. Specific biologics may require tailored animal models based on their mechanism of action and target tissues. Transgenic models can aid in studying biological effects on specific pathways or disorders. Combining multiple animal models can enhance the predictive value of toxicity studies (15).

Animal models that mimic human immune systems should be used to elucidate the immune responses generated by biologics. Non-human primates have been found to have immune functions similar to humans, and thus, they are preferred for evaluating immunogenicity. When choosing an appropriate animal model, it is important to consider the physical characteristics of the medication, its intended mechanism of action, and the dose–response relationship for desirable outcomes against unwanted effects.

Adherence to regulatory guidelines for toxicological testing, which may vary by jurisdiction, is most crucial. Ensure that study designs and animal models comply with legal safety standards. Collaborate closely with veterinarians, pharmacologists, toxicologists, and other experts to plan and conduct research using suitable animal models.

Rodents (mice, rats), rabbits, and NHPs are commonly used animal models. Pharmacological relevance is evaluated using a variety of techniques, including target sequence homology, binding affinity, and target expression comparison (16). Preclinical toxicity studies in drug development seek to forecast therapeutic medicines' human characteristics. Preclinical toxicity studies require new methods because there is not always a substantial correlation between animal and human toxicity. The extrapolation of animal toxicity to people is impacted by issues such as species variances and dose scale, even though a combination of rodent and non-rodent research demonstrates positive concordance with human values.

#### *16.3.1.2 Duration and Dosing Regimens*

These studies can have different lengths of time and dosage schedules based on things like the type of biotherapeutic, how it will be used, and legal and regulatory constraints. Nonetheless, the following broad rules and criteria can be adhered to:

The duration of the study is decided based on the intended duration of treatment in humans. Based on this, the non-clinical toxicity studies often vary in time span from several weeks to months, with chronic studies requiring longer durations.

*ICH M3 (R2) Guideline:* The ICH guideline M3 (R2) is the primary reference for designing non-clinical safety studies for pharmaceuticals (17). This guideline recommends study durations based on the intended clinical use of the biotherapeutics, with some examples given in Table 16.2.

Dosing regimens include single-dose, repeated, and dose-escalation studies. Single-dose studies observe responses after a single dose. Repeated dose studies involve multiple doses over extended periods, with typical daily, weekly, or biweekly dosing schedules for weeks to months. Dose escalation studies assess the effects of toxicity with increasing doses.

The route of administration (e.g., intravenous, subcutaneous, intramuscular) should closely mimic the intended clinical route. Dose selection should include a variety of doses, including therapeutic and higher levels, to assess safety margins. Consider the expected therapeutic dose in humans, the maximum feasible dose (MFD), and the limit dose. The dosing frequency should match the intended clinical use. Non-clinical toxicity studies should be well-designed with appropriate controls, randomization, blinding, and sample sizes. Consult regulatory agencies and toxicologists to ensure the study design meets regulatory standards and provides sufficient safety data.

### 16.3.2 *In Vitro* Studies

#### *16.3.2.1 Cell-Based Assays*

*In vitro*, cell-based assays are valuable tools for assessing biotherapeutic toxicity prior to *in vivo* studies. These assays provide cost-effective and efficient methods to evaluate cellular functions, viability, and proliferation. They reduce animal use and can offer greater sensitivity for detecting certain types of toxicity (18).

*Cell Line Selection:* Researchers should select relevant cell lines for *in vitro* toxicity testing, considering factors such as the therapeutic target, sensitivity, ease of culture, and availability. Common cell lines include

**Table 16.2 Duration of toxicity study of biologics as per ICH M3 (R2) guideline**

| Intended Clinical Use | Route of Administration | Biotherapeutic Class | Recommended Minimum Duration |
|---|---|---|---|
| Short-term (e.g., vaccine) | Intramuscular injection | Viral vector vaccine | 14–28 days |
| Long-term (e.g., chronic inflammatory disease) | Subcutaneous injection | Monoclonal antibody | 28 days–3 months |
| Chronic (e.g., cancer treatment) | Intravenous infusion | Bispecific antibody | 6–12 months |

MDCK, Caco, V79, HUH7, HepG2, NRK-52E, 3T3, and HEK293. These cell lines should be cultured and exposed to biotherapeutics to assess effects on cell viability, cytotoxicity, and inflammatory markers. Concentration curves should be established using multiple assays to determine toxicity levels accurately.
*Assay Format:*

- *Viability Assay:* Researchers can assess cell viability using various methods, including MTT, MTS, resazurin, and ATP-based assays. These assays measure metabolic activity or ATP levels to indicate cell viability. Common assays include the MTT assay, which measures mitochondrial dehydrogenase enzyme activity; the XTT assay, which is similar to the MTT assay but uses a different tetrazolium salt; and the LDH assay, which measures lactate dehydrogenase release.
- *Proliferation Assay:* One can measure cell proliferation using various methods, including cell counting, BrdU incorporation, or EdU incorporation assays.
- *Apoptosis Assays:* One can measure apoptosis using various assays, including annexin V staining, TUNEL assay, and caspase activity assays. These assays quantify the number of dead cells in a culture. The propidium iodide (PI) exclusion assay stains dead cells with PI for flow cytometry analysis. The annexin V-FITC assay stains apoptotic cells with annexin V-FITC for flow cytometry analysis.
- *Cellular Function Assays:* Assess specific cellular functions affected by the biotherapeutic, such as cytokine secretion, receptor activation, or signaling pathway activation.

Establish a concentration range for biotherapeutic testing, including therapeutic and higher concentrations. Include appropriate controls in the assay. Optimize assay conditions for robust results, considering factors like cell seeding density, incubation time, culture medium composition, pH, osmolarity, and temperature stability (19). Determine relevant endpoints and analyze assay data to quantify the effects of the biotherapeutic on cellular parameters. Ensure assay validation for reliability and reproducibility. Monitor and control for variability sources, such as cell culture conditions, passage number, and assay reagents. Ensure regulatory compliance for the safety assessment of biotherapeutics, adhering to relevant guidelines from agencies like the FDA, EMA, or ICH. Before moving on to *in vivo* investigations, researchers can obtain important insights into the possible toxicity of biotherapeutics by carefully planning and carrying out *in vitro* cell-based experiments. It is crucial to remember that these assays are not flawless and cannot always forecast how a biotherapeutic will act *in vivo*. A biotherapeutic cannot be used on humans unless its safety has been verified through *in vivo* testing.

#### *16.3.2.2 Evaluation of Immunogenicity*

Evaluating the immunogenicity of biologics is essential in guaranteeing the security and effectiveness of these medications. Large, complex chemicals derived from living things called biologics can cause immunological reactions in individuals. Anti-drug antibodies (ADAs) may develop as a result of these immune reactions, which may impair the medication's effectiveness and perhaps result in negative side effects (20).

Immunogenicity assays are crucial for assessing the risk of adverse events associated with biologics. ADAs can bind to biologics, reducing their efficacy by preventing them from reaching their targets. Additionally, ADAs can trigger immune reactions, leading to side effects like infusion reactions, allergic reactions, and autoimmune reactions.

Immunogenicity evaluation of biologics is a multi-step process involving preclinical and clinical studies.

I. *Preclinical Studies:* Preclinical studies assess the immunogenicity of biologics using *in vitro* and *in vivo* methods. In silico analysis predicts potential immunogenic epitopes. *In vitro* studies measure T-cell proliferation, dendritic cell maturation, and B-cell activation to assess immunogenicity. Animal studies, including rodent models and NHPs, provide insights into immune responses and potential immunogenicity.

II. *Clinical Studies:* Clinical trials are designed to assess the incidence and clinical impact of ADA formation in patients receiving the biologic. Blood samples are collected from patients at regular intervals to measure ADA levels. Other clinical assessments are also performed to monitor for any adverse events related to ADA formation.

- *Phase I Trials:* These trials typically evaluate the safety and pharmacokinetics of the biotherapeutic in a small group of healthy volunteers. Immunogenicity assessments may include monitoring for ADAs development.
- *Phase II/III Trials:* Larger trials involve patients with the target disease. Immunogenicity assessments continue, including monitoring for ADAs and their potential impact on efficacy and safety.

## 16.4 IMMUNOGENICITY ASSESSMENT

### 16.4.1 Immunogenicity Challenges in Biologics

Immunogenicity is a critical concern in the development and application of biologic medicines. Vaccines, monoclonal antibodies, and therapeutic proteins can interact with the immune system, but their complex structures and biological origins may also elicit immune responses in some individuals. This can affect treatment safety and efficacy.

Managing and understanding immunogenicity is a significant challenge for biopharmaceutical developers and regulators. This book chapter will explore the factors that contribute to immunogenicity, its impact on safety and efficacy, and strategies for risk management. To guarantee the health of patients and the success of their treatment, it is essential to evaluate the influence that immunogenicity has on the relative safety and effectiveness of biologic medicines (21).

To examine these impacts, several different assays and methodologies are utilized, making it possible to get significant insights into the immunological responses that are generated by these medicines (22).

**Immune Response Assays** allow for the measurement of the immunological response triggered by biologics, including the generation of ADAs. The quantification of ADAs in patient samples is often accomplished by utilizing enzyme-linked immunosorbent assays (ELISAs) and radioimmunoassays (RIAs). **Neutralizing antibody assays** determine whether ADAs can negate the biological activity of certain biologic therapies. To determine the neutralizing antibody activity, cell-based assays or bioassays are frequently utilized.

**Clinical Outcome Assessment** is of the utmost importance in monitoring clinical outcomes in patients with biologic treatments. To better understand the influence of immunogenicity on the safety and effectiveness of a treatment, changes in disease activity, treatment response, and adverse events are examined. **Biomarkers** have the potential to offer insights into the mechanisms that underlie immunogenicity as well as the influence that it has on both protection and effectiveness. The levels of cytokines and the profiles of gene expression, for instance, are examples of biomarkers that can be utilized to evaluate immunological activity. To visualize the effects of immunogenicity, such as inflammation or tissue damage, in patients receiving biologic therapy, **imaging techniques** such as positron emission tomography (PET) and magnetic resonance imaging (MRI) can be utilized. **Pharmacokinetic and pharmacodynamic studies** are conducted to understand better the connection between drug exposure, immune response, and treatment outcomes. It is important to note that these studies offer vital information regarding treatment techniques and dosing regimens. Before clinical trials are conducted, it is possible to conduct ***in vitro* studies** utilizing cell cultures or animal models to examine the mechanisms of immunogenicity and evaluate the potential influence on safety and efficacy.

#### *16.4.1.1 Methods for Detection and Characterization*

Identifying and characterizing biologics are crucial in their development, production, and quality control. These techniques ensure the safety, efficacy, and quality of biologic products and are essential for post-market surveillance and regulatory filings (23). Various techniques with advantages and limitations are used to detect and characterize biologics.

*Analytical Ultracentrifugation (AUC):* It is a potent method that may be used to analyze biologics and yield details on their molecular weight, size, and structure. It is beneficial for researching the creation of aggregates and interactions between proteins.

*Chromatography:* Based on their physicochemical characteristics, biologics are separated and quantified using a variety of chromatographic techniques, including size-exclusion chromatography (SEC), ion-exchange chromatography (IEC), and high-performance liquid chromatography (HPLC).

*Mass Spectrometry (MS):* MS provides details on molecular weight, structure, and post-translational changes and is utilized in the analysis of biologic products. It is beneficial for the characterization of protein therapies.

*Electrophoresis:* Proteins are separated and analyzed according to their charge and size using electrophoretic techniques, including capillary electrophoresis (CE) and sodium dodecyl sulfate–polyacrylamide gel electrophoresis (SDS–PAGE).

*Immunoassays:* Based on the interactions between antigen and antibody, immunoassays, such as Western blotting and ELISAs are used to detect and quantify biologics.

*Assays Biological:* Biologics' biological activity is evaluated, and their potency and efficacy are confirmed using biological tests such as *in vivo* bioassays and cell-based assays.

*Nuclear Magnetic Resonance (NMR):* NMR spectroscopy provides details on the conformation, dynamics, and interactions of biologics and is used to characterize their structure.

*X-ray Crystallography:* This technique offers comprehensive details on the three-dimensional structure of biologics through high-resolution structural analysis.

*Differential Scanning Calorimetry (DSC):* DSC provides information on biologics' folding and aggregation behavior and is used to investigate their thermal stability.

### 16.4.2 Risk Mitigation Strategies

Risk mitigation techniques are crucial for biologics to be safe, effective, and high-quality treatments. These approaches cover several facets of research and development, production, and post-marketing monitoring to reduce possible risks related to immunogenicity, contamination, and other variables.

#### *16.4.2.1 Engineering Approaches to Reduce Immunogenicity*

One of the main objectives of creating biologic medicines is to reduce immunogenicity. By modifying biologic molecules, engineering approaches seek to maximize their therapeutic efficacy while reducing their potential to trigger immunological responses.

**Humanization** is a process of modifying biologic molecules by replacing non-human sequences with human sequences to reduce immunogenicity. Chimerization and humanization are techniques used for this purpose. Chimerization replaces the entire non-human part of the molecule, while humanization specifically swaps out non-human complementarity-determining regions (CDRs) in monoclonal antibodies (24). **Deimmunization** is a process of identifying and modifying T-cell epitopes in biologic molecules to minimize the risk of triggering an immune response. Computational methods or experimental screening can be used to detect and alter immunogenic epitopes while preserving the therapeutic efficacy of the molecule. **Site-directed mutagenesis** is a technique that modifies specific amino acid residues in biologic molecules to reduce immunogenicity. Researchers can potentially minimize the risk of an immune response by targeting residues involved in T-cell recognition or antibody binding. **PEGylation** is a process that involves attaching polyethylene glycol (PEG) chains to biologic molecules. This modification increases their size and reduces immunogenicity. PEGylation can also enhance the pharmacokinetic properties of biologics, improving their stability and reducing clearance. **Fc engineering**, a biophysical modification of antibodies, can modulate their interaction with the immune system. Altering the Fc region may reduce affinity for Fc receptors, decreasing the likelihood of immune complex formation and antibody-dependent cellular cytotoxicity (ADCC). Fragmenting antibodies into smaller fragments, such as Fab or F(ab')2, can reduce immunogenicity while preserving antigen-binding specificity. T cells may be less likely to detect these fragments and are often rapidly eliminated from the body. **Fusion proteins** are created by combining a biologic molecule with another protein or peptide to reduce immunogenicity. **Tolerance induction strategies** promote immune tolerance to biologic molecules, preventing the immune system from recognizing them as foreign. This can be achieved through various techniques, such as oral or nasal administration of the biologic or tolerogenic adjuvants (25).

#### *16.4.2.2 Patient Monitoring and Management*

Patient monitoring and management for biologics require careful consideration due to their unique characteristics and potential side effects. Safety monitoring is essential to detect immune responses and infusion reactions. Efficacy monitoring involves assessing biomarkers associated with the target disease. Adverse events, such as cytokine release syndrome or infusion-related reactions, must be closely monitored and managed. Regular laboratory tests may be necessary to evaluate potential toxicities. Patient education regarding side effects and seeking medical attention is crucial. Monitoring patient compliance ensures optimal treatment effectiveness. Remote monitoring tools can efficiently track vital signs and symptoms, enabling timely intervention. Effective care coordination among healthcare providers is essential for providing the best possible patient care.

## 16.5 REGULATORY GUIDELINES AND COMPLIANCE

### 16.5.1 FDA and International Regulatory Agencies

Various products fall under biological products, including tissues, allergenics, vaccines, blood and blood components, somatic cells, gene therapy, and recombinant therapeutic proteins. Biologics can be living things like cells and tissues or consist of complex combinations of carbohydrates, proteins, and nucleic acids. Biosimilars are biologic medicines that do not differ clinically from the original reference product. Patients with autoimmune diseases, malignancies, diabetes, and other chronic, frequently incapacitating disorders have therapeutic choices using biological medications.

#### *16.5.1.1 European Medical Agency*

The EMA was the first to create biosimilar guidelines in 2005, with the first approval of Omnitrope in 2006. The EU is a pioneer in the biosimilar market, with a well-documented legal and regulatory pathway for approval. The guidelines outline detailed analytical, preclinical, and clinical data requirements, requiring comprehensive comparability studies between biosimilars and reference products. The EMA initially issued general guidelines for quality and nonclinical issues, followed by detailed product-specific guidelines by the EMA/Committee for Medicinal Products for Human Use (CHMP), adopted in countries like Australia, Canada, Japan, Korea, and South Africa.

#### *16.5.1.2 WHO Regulatory Framework*

To assure the safety, efficacy, and quality of biotherapeutic products, World Health Organization (WHO) guidelines draw certain basic principles mandatory for the licensure of a biosimilar product. This ensures successful regulatory submissions in most major pharmaceutical geographies around the world.

- *Reference Product:* The chosen reference biotherapeutic product (RBP) should be licensed based on full quality, safety, and efficacy data.
- *Quality:* All aspects of quality and heterogeneity should be assessed, including head-to-head comparisons with the reference product.
- *Nonclinical Data:* Data should include pharmacodynamic, pharmacokinetic, and comparative repeat-dose toxicity studies in a relevant species.
- *Clinical Studies:* Similarity of the similar biotherapeutic products (SBPs) and RBPs should be demonstrated in adequately powered, randomized, and controlled clinical trial(s).
- *Pharmacovigilance:* Monitoring or a pharmacovigilance plan should be included in the post-marketing phase to supplement the limited clinical data during authorization.

Based on the EMA guidelines, the WHO guidelines were not geography-specific and were used as a template for other national regulatory agencies. The US FDA, a late entrant to the biosimilars regulatory pathway, signed the Biologics Price Competition and Innovation Act (BPCIA) in 2010, creating a new biosimilars license pathway within the FDA's jurisdiction. The Act defined the basic requirement of demonstrating biosimilarity between reference and biosimilars, proposed that the sponsor's product be shown as interchangeable with the originator's product, and provided exclusivity periods for reference biologics and some approved follow-on biologics. However, specific criteria for biosimilarity or interchangeability were not defined. Currently, the BPCIA provides case-specific determinations to the sponsor's individual molecules.

#### *16.5.1.3 US FDA Draft Guidance on Biosimilarity*

The US FDA issued three draft guidance documents in 2012, emphasizing the importance of "totality of evidence" in demonstrating biosimilarity. The Scientific Considerations guidance calls for safety, potency, and purity demonstrations between the reference and biosimilar, with no clinically meaningful differences. The Quality Considerations guidance focuses on physicochemical characterization needs, considering factors like manufacturing process consistency, structural similarity, potency assays, and comparative pharmacokinetic. The Questions and Answers Guidance provides answers to biosimilar sponsors, including formulation differences, exclusivity issues, and meeting schedules. The guidance also provides definitions for terms like "chemically synthesized polypeptide," "biological product," and "protein." Sponsors are advised to establish early meetings with FDA officials and understand submission requirements (26, 27).

#### *16.5.1.4 Japan (Pharmaceuticals and Medical Devices Agency)*

Japan's Pharmaceuticals and Medical Devices Agency (PMDA) has established a regulatory pathway for biosimilars since 2009, following guidelines from the ICH. The pathway includes recombinant plasma proteins, recombinant vaccines, PEGylated recombinant proteins, and highly purified nonrecombinant proteins. However, some EU inclusions, like polyglycans and synthetic peptides, have been excluded. Comparative stability studies are not mandatory in Japan (28, 29).

The detailed guidelines are outlined in:

- "Guidelines for the Quality, Safety, and Efficacy Assurance of Follow-on Biologics" (Yakushokushinsahatu 0304007 by MHLW/March 4, 2009)
- "Nonproprietary Name and Brand Name of Follow-on Biologics" (Yakushokushinsahatu 0304011 by MHLW/March 4, 2009)

- "Revision of Marketing Approval Application" (Yakushokushinsahatu 0331015 by MHLW/March 4, 2009).

The EMA guidelines are used to develop biosimilars, but the need for comparative stability in Japan can be a differentiator. The Japanese Pharmaceutical Development Authority (PDA) has supported biosimilar development by consulting with pharmaceutical companies and reducing approval time. A survey of key leaders from various companies revealed that incentives for innovator biopharmaceuticals should not suffer in the competitive race with biosimilars. Only one to two companies have a biosimilar product under development, while others focus on new or second-generation biopharmaceuticals. Offshore manufacturing contractors in India or China could be considered.

#### *16.5.1.5 Korea (KFDA)*

Samsung, a Korean electronics company, invested $389 million in biosimilars in 2009, marking a significant entry into the market. The regulatory framework for biosimilar products in Korea is governed by the Pharmaceutical Affairs Act, Notification of the Regulation on Review and Authorization of Biological Products, and the Guideline on Evaluation of Biosimilar Products. The Korean guidelines align with WHO guidelines, except for clinical evaluation and equivalence testing, which allows for extrapolation of efficacy data (30).

#### *16.5.1.6 Canada (Health Canada)*

Canada's Health Canada issued guidelines in 2010 to evaluate the safety, efficacy, and quality of subsequent entry biologics developed after patent expiry. The guidelines require a complete clinical data package with sufficient safety and efficacy data, advanced analytical testing, and meeting defined criteria. The Canadian guidelines also outline post-marketing requirements, such as adverse drug reaction reports and safety update reports. The recent authorization of monoclonal antibodies, remsima and inflectra, indicates acceptance of affordable biosimilars, potentially causing the first 351k biosimilar filing with the US FDA (31).

#### *16.5.1.7 The People's Republic of China (State Food and Drug Administration)*

The People's Republic of China and India are often compared due to their large economies, with China focusing on manufacturing chemicals and active substances, and India on sourcing active substances and finished dose products. However, both countries face challenges like devaluation of the rupee and talent retention. India has an edge in language skills and regulatory authorities, and has over 40 approved biopharmaceutical products, including biosimilars, based on EMA guidelines.

Guidelines for Registration of Drugs (2010)

- Guidance for Review and Approval of Recombinant Protein Drugs (2002)
- Guidance for Review and Approval of Biosimilar Product (2008)
- Points to Consider for Common Technical Documents (CTD) in Review and Approval of Biosimilar Product (2010).

#### *16.5.1.8 India (Central Drugs Standard Control Organization)*

Mohal Sarabhai highlighted India's strengths, including regulatory compliance, large talent pool, technology development, and low manufacturing costs, making it the "Pharmacy of the World."

The regulatory agencies under the government of India that are involved in the approval process of SBPs are:

- Central Drugs Standard Control Organization
- The Office of the Drug Controller General of India
- Review Committee on Genetic Manipulation
- Genetic Engineering Approval Committee.

The Indian biosimilar industry, valued at $338 million, has grown at a 30% CAGR since 2008. Currently, 25 Indian companies market around 50 biosimilar products, including erythropoietin, human growth hormone, recombinant human insulin, granulocyte colony-stimulating factor, and interferon. These products dominate the market, with erythropoietin alfa holding over 40% market share in 2010.

#### *16.5.1.9 Islamic Republic of Iran (National Regulatory Authority)*

The Iranian National Regulatory Authority (NRA) is the Ministry of Health and Medical Education responsible for establishing quality standards for medicinal product regulation. The divisions include the Pharmaceuticals and Narcotic Affairs Division, Drug Control Laboratory (FDCL), Adverse Drug Reaction Centre, Clinical Trials Evaluation Committee, and Centre for Diseases Control. The Office of Biologics works with FDCL and the Adverse Drug Reaction Centre for marketing authorization, licensing, lot release, and regulatory inspections of biologics products. The Iranian guidelines for biosimilar registration are similar to WHO guidelines but with some differences. Iran does not require comprehensive clinical trials but relies on national post-marketing surveillance data for drug safety. Iran imposes an import tariff of up to 65% on imported drugs to promote locally manufactured drugs, which may not be considered "highly similar" under EMA/US FDA/WHO regulations.

#### *16.5.1.10 Jordan (Jordan Food and Drug Administration)*

The Jordan Food and Drug Administration (JFDA) has two main directorates, the Drug Directorate and the Food Directorate, which are responsible for drug and food safety. The Technical Committee for the Registration of New Drugs is responsible for registering originators, new drugs, and biological and biosimilar products. The committee decides on registration within 180 days and ensures information protection. Since 2008, the JFDA has reviewed biosimilar applications based on EMA guidelines, emphasizing post-marketing surveillance. The choice of RBP is the first product registered internationally with a particular active ingredient.

#### *16.5.1.11 Middle East*

A group of Middle Eastern physicians and neurologists discussed biosimilar approval procedures, recommending implementing EMA guidelines for safety, efficacy, and pharmacovigilance. They recommended consistent quality control, comparable pharmacokinetic and pharmacodynamic, double-arm direct clinical trials, and post-marketing surveillance (32).

### 16.5.2 Interactions with Regulatory Authorities

The FDA provides guidance and advice to Investigational New Drugs (IND) sponsors at any stage of the development and submission process. Sponsors may request meetings with the appropriate review division before submission. The FDA has a regulatory guidance document titled "Guidance for Industry: IND Meetings for Human Drugs and Biologics- Chemistry, Manufacturing, and Controls Information," which provides guidelines for formal meetings between sponsors and the Centre for Drug Evaluation and Research (Table 16.3).

## Table 16.3 Meeting with FDA

**FDA Meeting Timelines**

| Type of Meeting | **FDA Response to Request** | **FDA Receipt of Meeting Package** | **FDA Preliminary Responses to Requester** | **Requester Responses to FDA Preliminary Responses** | **FDA Scheduled Meeting Date (Days from Receipt of Request)** | **FDA Meeting Minutes** |
|---|---|---|---|---|---|---|
| Interact | 21 days | 90 days before meeting (with request) | 1 day before meeting | — | Within 90 days | — |
| Pre-IND (Type B) | 21 days | 30 days before meeting | 2 days before meeting | — | Within 60 days | 30 days after meeting |
| Type B (EOP II, Pre-NDA, Pre-BLA) | 14 days | 50 days before meeting | 5 days before meeting | 3 days before responses | Within 70 days | 30 days after meeting |
| Type C | 21 days | 47 days before meeting | 5 days before meeting | 3 days before responses | Within 75 days | 30 days after meeting |

#### *16.5.2.1 General Aspects of IND Meetings*

IND meetings aim to seek FDA support in addressing scientific or outstanding concerns during clinical investigation periods for drug/biologics product development and regulatory processes. They help prevent clinical hold issues and aid sponsors in developing a complete IND. Sponsors should forward a request for a meeting with a draft list of CMC-related questions, clear objectives, and desired outcomes. The package should summarize relevant CMC information, developmental status, and future drug development plans. Once approved, a multidisciplinary meeting between technical experts from the sponsor and the Agency will be scheduled. The meeting should focus on specific CMC-related issues and may include a separate CMC-specific meeting if necessary. Depending on the application stage, sponsors may request different types of meetings, such as pre-IND, end-of-phase 1, end-of-phase 2, and pre-NDA or pre-BLA.

*Pre-IND Meeting:* Pre-IND meetings (Figure 16.3) are formal PDUFA Type B meetings between sponsors and CDER or Center for Biologics Evaluation and Research (CBER), crucial for pharmacology, toxicology, drug activity studies, initial drug development plans, regulatory requirements, CMC issues, novel excipients, and viral clearance studies. The FDA advises sponsors to request a pre-IND meeting for certain drugs, such as those not previously approved or licensed, new molecular entities, planned marketing applications, drugs critical to public health, and drugs with substantial early development outside the US. To request a pre-IND meeting, a PIND number must be requested 30 days before the requested date. A briefing document must be prepared and submitted 30 days before the scheduled meeting date.

*End-of-Phase 1 (EOP1) Meetings:* EOP1 meetings (Figure 16.4) are formal PDUFA Type B meetings between IND sponsors and CDER or CBER, crucial for evaluating phase 2 controlled clinical trial designs and discussing drug development program issues, especially in pediatric research. The FDA encourages sponsors to request EOP1 meetings for drugs that treat significant life-threatening illnesses, particularly where phase 2 trials or accelerated approval are appropriate. EOP1 meetings cover pharmacokinetics/pharmacodynamics, proposed protocol, population identification, pediatric studies, adaptive trial designs, and nonclinical data adequacy. The FDA provides preliminary written responses to sponsors, with responses usually sent within 5 days. The FDA responds with the date and time if a written response is requested.

*End-of-Phase 2A (EOP2A) Meetings:* EOP2A meetings (Figure 16.4) are formal PDUFA Type C meetings held between IND sponsors and CDER or CBER to optimize drug development. They discuss trial designs,

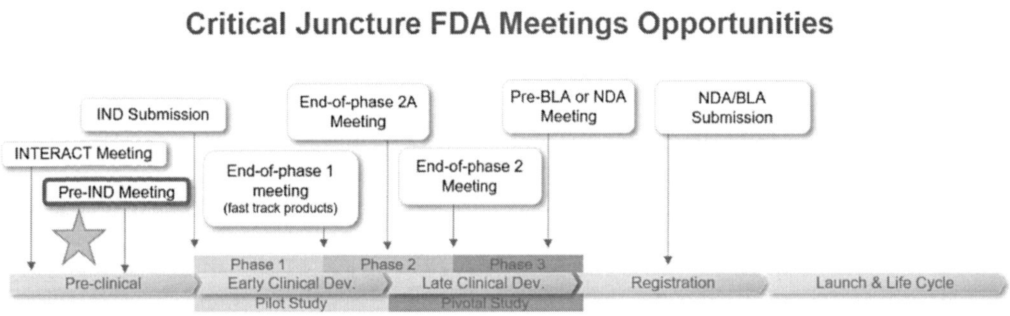


**Figure 16.3** Critical juncture FDA meetings opportunities—pre IND meeting.

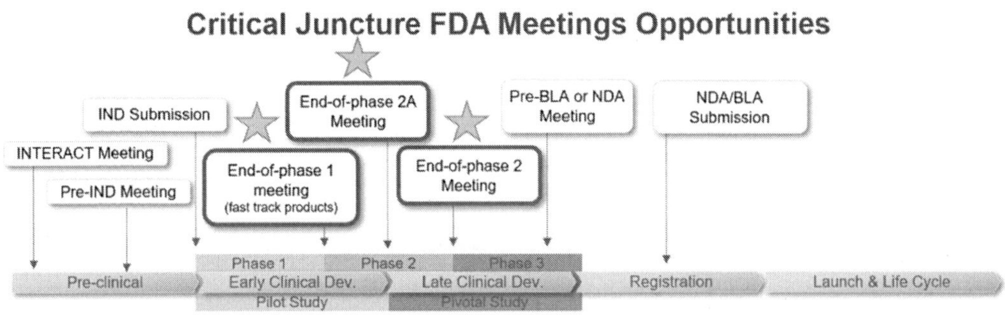


**Figure 16.4** Critical juncture FDA meetings opportunities—end-of-phase meeting.

modeling strategies, and clinical trial simulation scenarios to improve the quantification of exposure-response information from early drug development. The objectives are to select dosing regimens for the next phase of drug development and design well-planned dose-response trials that inform later-phase clinical trials by best incorporating prior quantitative knowledge. EOP2A meetings are held once phase 2A clinical data is available and before phase 2B and phase 3 clinical studies are initiated. Sponsors are encouraged to submit all relevant information with their meeting request, including data, models, simulations of trial design, or disease or outcome models. The FDA responds within 21 days of request receipt, and if granted, the meeting will be scheduled within 75 days.

*EOP2 Meetings:* EOP2 meetings (Figure 16.4) are formal PDUFA Type B meetings between IND sponsors and CDER or CBER to evaluate and determine phase 3 plan, protocols, and study accuracy. These meetings are typically held before major commitments to specific phase 3 testing have been made. The FDA answers requests for EOP2 meetings within 14 calendar days, and meetings are usually arranged within 70 days. Meeting packets must be submitted at least 50 days before the appointed date. If a face-to-face, teleconference, or videoconference meeting is approved, preliminary written replies are sent no later than 5 days before the planned meeting. If a meeting is rejected, the FDA will contact the sponsor and offer a rationale within 14 calendar days.

*Pre-NDA or Pre-BLA Meeting:* The sponsor requests a pre-NDA or pre-BLA meeting with the review team to critically evaluate the filing and formatting issues for submitting a complete and well-organized NDA or BLA application. Ideally, a sponsor should plan this meeting 6 months before the submission date for the NDA or BLA. The meeting discussion should be focused on the necessary formatting details required for the smooth NDA or BLA filing. It can also be extended to discuss the address of the EOP2 meeting outstanding problems and other forecasted issues, ensuring no further delay.

## 16.6 CASE STUDIES IN REGULATORY TOXICOLOGY

### 16.6.1 Success Stories in Biologics Development

In the current pharmaceuticals and biotech industries, the development and approval of biologics are at the topmost positions in innovations and regulatory requirements (33). Biologics are derived from living organisms and are available for a range of various diseases. The process from drug discovery to the approval of the product is apprehensive, with challenges requiring toxicologic evaluations to confirm the drug's safety and efficacy. Case studies in regulatory toxicology ease the complex journey of biologics development, foreseeing the challenges, product approval, and, ultimately, toxicological assessment, which helps protect public health. Here we will discuss the success stories during biologics development, focusing on demonstrating safety during the preclinical phase, various regulatory requirements, and valuable lessons from setbacks.

#### *16.6.1.1 Demonstrating Safety in the Preclinical Phase*

Confirming the drug's safety in the preclinical phase is very important. Take an example of immunogenicity during the preclinical development of biological therapeutics (34). Immunogenicity refers to the propensity of a therapeutic agent to induce an immune response in the body, which may lead to adverse effects such as infusion reactions, decreased efficacy, or the development of ADAs. Researchers engage with various strategies to evaluate and mitigate immunogenicity risks to establish safety in the preclinical phase. These strategies typically involve conducting comprehensive preclinical toxicology assessments using *in vitro* and *in vivo* assays. For example, let us consider the development of a novel monoclonal antibody (mAb) targeting a key cytokine implicated in autoimmune diseases. During the preclinical phase, researchers would conduct a series of studies to evaluate the safety profile of the mAb, with a particular focus on immunogenicity.

In -vtro assays may include studies to assess the reaction between the mAb and the target antigen and to measure the probability of producing an immune response in human cells. This will provide valuable insights into the specificity and potential immunogenicity of mAb (35).

*In vivo*, assay is performed conducting in animal models and is important for assessing the systemic safety of the mAb. Researchers administer the mAb to animals and then closely observe for any signs of adverse reactions, such as infusion-related reactions or changes in immune parameters. Also, these immunogenicity studies in animals may help identify the potential for developing ADAs and evaluate their impact on the pharmacokinetics and pharmacodynamics of the mAb.

By rigorous brainstorming and evaluation of the immunogenicity in the preclinical phase, researchers may be able to identify potential risks at the early stage of development and implement strategies to mitigate them. For example, modifying the mAb's structure or formulation, updating the dosing regimens, or adding immunomodulatory agents may help reduce immunogenicity and enhance the safety profile of the therapeutic agent.

Overall, demonstrating safety in the preclinical phase through thorough immunogenicity assessment is essential for advancing large molecule-based biologics toward clinical trials and regulatory approval, including cutting-edge therapies such as CAR T-cell and gene therapies.

##### *16.6.1.2 Overcoming Challenges and Gaining Regulatory Approval*

Developing biosimilar products that are highly similar to approved reference biological drug products presents unique challenges in regulatory toxicology. Below will be a discussion on various challenges such as immunogenicity, complexity and conducting clinical trials, and important parameters for gaining regulatory approvals such as analytical similarity, clinical comparability, and post-marketing surveillance.

A few of the challenges pertaining to **biological product development** are mentioned below:

*Immunogenicity:* Biosimilars can elicit immune responses in patients, leading to potential safety concerns. Assessing immunogenicity and ensuring the biosimilar's safety profile is comparable to the reference product is an essential key point for regulatory approval (36).

*Complexity:* Biosimilars are highly complex biological drugs manufactured using living organisms. Developing biosimilars requires an extensive understanding of the structure and function of the reference product, which can be challenging due to the complexity of these large molecules.

*Clinical Trials:* Conducting clinical trials for biosimilars in oncology comes with unique challenges. Demonstrating similarity in terms of efficacy and safety against the reference product requires large-scale clinical studies, which can be expensive and time-consuming.

A few of the important parameters for gaining **regulatory approvals** are as mentioned below:

*Analytical:* The regulatory approval process for the biosimilar drug products involves demonstrating analytical similarity to the reference product through a wide range of structural and functional characterization studies.

*Clinical Comparability:* The clinical studies are performed to establish comparability in efficacy and safety between the biosimilar test product and the reference biological product. These studies generally include pharmacokinetic, pharmacodynamic, and clinical outcome evaluations.

*Post-Marketing Surveillance:* After regulatory approval of the drug product, post-marketing surveillance is conducted to monitor the safety and effectiveness of biosimilars. This ongoing monitoring helps to identify if there are any potential safety concerns and ensures patient safety.

### 16.6.2 Lessons Learned from Setbacks

Despite meticulous planning and rigorous testing, unexpected toxicological findings may arise, leading to setbacks in regulatory approvals.

#### *16.6.2.1 Addressing Unexpected Toxicological Findings*

Addressing unexpected toxicological findings and setbacks in drug product development involves a stepwise approach to hazard identification, hazard characterization, risk evaluation, and risk management. It requires careful scientific evaluations, investigations, and collaboration with internal and external subject matter experts. The steps include recognizing and assessing the unexpected adverse preclinical findings (APFs), characterizing and measuring the observed APF, determining the potential relevance of the APF to humans, calculating risk–benefit ratios, implementing precautions, and securing support from management and regulatory authorities. The process also involves understanding the drug candidate's MoA, calculating safety ratios, and conducting a weight of evidence (WoE) analysis. The goal is to optimize the chances of successful drug development while ensuring the safety of patients and the company.

The process and steps involved in dealing with unexpected APFs during drug development can be divided into various steps. It may guide addressing these findings and optimize the chances of successful drug development. Also hazard identification and hazard characterization in understanding the nature and significance of the APFs is critical. We can discuss below a few of the lessons learned are as follows:

1. APFs can arise unexpectedly during drug product development and fall into categories such as functional disturbances, genotoxicity, morphological toxicity, carcinogenicity, and reproductive toxicity.
2. Hazard identification is the first step in recognizing and qualitatively assessing APFs. It involves determining if there is any adverse effect, reviewing what is known about the drug in question, considering other relevant findings, evaluating the technical validity of the study, assessing the validity of the animal model, and considering other modifying factors.

3. Hazard characterization in the next step involves characterizing and quantifying the observed APF in more detail. This step includes obtaining additional data on dose–response, early markers for the APF, and the drug candidate's MoA.
4. Tailor-made mechanistic studies may be conducted to better understand APF and its potential implications. These studies can involve additional investigations of available samples, such as hormone measurements, gene expression analysis, and morphological investigations. *In vitro*, studies with cell or tissue cultures can also help investigate metabolism, subcellular effects, and gene expression.
5. Increased cell proliferation is often associated with tumors in lifetime bioassays. It can be caused by subacute to chronic cytotoxicity, chronic tissue irritation, age-related "spontaneous" lesions, or increased physiological stimulation of cells.

Overall, it is important to thoroughly evaluate and characterize APFs during drug development to ensure the safety and efficacy of drug candidates.

#### *16.6.2.2 Strategies for Recovery and Resubmission*

If an applicant receives a refusal or query from the agency, it is important to carefully review the reason provided. Identify the issues and develop a plan to address them.

*Additional Data:* If the query/refusal is due to insufficient data or evidence, gather additional data to support the safety, efficacy, and quality to support the drug product. This may involve conducting further preclinical or clinical studies, collecting additional data on manufacturing processes, or providing more comprehensive information on the product's characteristics.

*Get Expert Advice:* Consult with appropriate regulatory experts or seek guidance from regulatory agencies via an appropriate route to understand the specific requirements for recovery and resubmission. They can provide valuable insights and recommendations on addressing the issues raised during the review. While communicating with the regulatory agencies, maintain open and transparent communication throughout the recovery and resubmission process. Seek clarification on any issues or concerns and promptly respond to their requests for additional information.

*Revise Documentation:* Ensure all documentation in the dossier is complete, accurate, and well-organized. Present the data and evidence supporting the drug product's safety, efficacy, and quality.

*Experiences:* Analyze the reasons for query/refusal and learn from any mistakes or shortcomings in the application. Use this knowledge to improve the resubmission strategy and address any weaknesses in the drug product development, manufacturing, or clinical data.

It is important to note that these strategies are general recommendations and may need to be tailored to the drug product's specific regulatory requirements and guidelines submitted for approval. Consulting with regulatory experts and seeking guidance from the relevant regulatory agencies is crucial for a successful recovery and resubmission process.

## 16.7 EMERGING TRENDS AND TECHNOLOGIES

### 16.7.1 Advances in Predictive Toxicology

Toxicology has evolved significantly, driven by societal and technological changes (37). The publication of Silent Spring marked a pivotal moment, raising awareness of chemical toxicity and prompting more rigorous investigations. This era transitioned from limited facilities to sophisticated methods and comprehensive toxicological studies.

Technological advancements have been instrumental in this evolution. High-throughput screening (HTS) enables rapid and efficient testing of numerous compounds. Organ-on-a-chip (OOC) technology offers a more realistic simulation of human organ responses, providing a cost-effective alternative to animal models. Systems toxicology, genomics, transcriptomics, and metabolomics have introduced new approaches to studying molecular interactions, identifying biomarkers, and developing accurate toxicity prediction models.

Predictive toxicology is a cornerstone of modern medical and chemical research, enabling the prediction of substance toxicity before widespread use (38). By combining traditional toxicity testing with advanced computational models and *in vitro* systems, predictive toxicology reduces reliance on animal testing and improves safety assessment accuracy. Its focus on safeguarding human health and the environment and streamlining drug development makes personalized medicines more accessible. These advancements highlight the pivotal role of predictive toxicology in shaping the future of therapeutics and regulatory practices.

Recent technological advancements are poised to revolutionize predictive toxicology. Breakthroughs in systems biology, stem cell research, engineered tissues, and mathematical modeling offer unprecedented opportunities to improve the accuracy of toxicological assessments. These advancements are expected to accelerate medical product development and prevent hazardous products from reaching the market.

The field explores various techniques, including computational methods, sophisticated *in vitro* systems, and innovative animal models like humanized animals or genetically altered organisms. While predicting the most successful technologies is challenging, the FDA's active involvement highlights the potential for a significant impact on toxicology-related areas.

#### *16.7.1.1 High-Content Screening in Predictive Toxicology*

High-content screening (HCS) technology merges automated microscopy with quantitative image analysis, providing a robust method for addressing complex biological questions in predictive toxicology. This technology has been pivotal in academia and the pharmaceutical industry, particularly in the research field of *in vitro* toxicology.

*Mechanism of Action:* HCS is instrumental in identifying genes essential for specific biological processes through genome-wide RNA interference (RNAi) screens. This capability is crucial for understanding the genetic bases of diseases and the effects of toxic substances at a molecular level. Moreover, HCS is a vital *in vitro* tool for testing compound toxicity and elucidating MOAs, significantly reducing the reliance on animal testing. By providing detailed information at the cellular level, HCS offers insights into the spatial distribution and dynamics of biological responses, making it an invaluable tool in addressing the challenges posed by the "Toxicity testing in the 21st century" framework.

*Applications in Drug Safety Assessment:* In drug safety, HCS is extensively used to explore the cardiotoxic effects of compounds. Techniques such as kinetic imaging cytometry and HCS assays for $Ca^{2+}$ dynamics assess cardiotoxicity using human pluripotent stem cell-derived cardiomyocytes. These assays evaluate critical parameters like nuclear remodeling, mitochondrial status, apoptosis, necrosis, and alterations in ion channel permeability. Additionally, combined phenotypic and HCS assays provide a comprehensive analysis of cell viability, mitochondrial integrity, and reactive oxygen species, facilitating the identification of potential cardiotoxic compounds and their MOAs.

In the pharmaceutical industry, HCS assays have become a cornerstone for identifying mechanisms implicated in toxicity. Multi-parametric HCS assays, for instance, are utilized to screen and classify hepatotoxic compounds. These assays can detect early and late events in the hepatotoxic process, allowing for the stratification of compounds according to their specific degree of injury. This multi-parametric approach underscores the versatility and effectiveness of HCS in predictive toxicology, offering a more refined understanding of compound toxicity and enhancing the safety profiles of pharmaceuticals (39).

#### *16.7.1.2 Advancements in In Vitro Models*

*Development and Validation of New Models:* Recent advancements *in vitro* models have significantly contributed to predictive toxicology by enhancing the ability to mimic *in vivo* systems through the utilization of stem cells in both two-dimensional (2D) and three-dimensional (3D) cultures (40, 41). Spheroids, which are self-aggregating clusters of cells, and organoids, a 3D heterogeneous collection of cells that self-organize to recreate organ microanatomy, represent two major developments in this area. Spheroids enable studying cell interactions and drug responses in a more tissue-like environment. At the same time, organoids provide a more complete model by allowing drug penetration and the establishment of a toxicological gradient.

However, traditional 2D cultures have limitations, as they cannot fully replicate the complex physiology of *in vivo* systems, such as the presence of connective tissue or extracellular matrix, leading to a poor standard of pre-clinical *in vitro* hepatotoxicity assays. In contrast, HepG2 cells cultured in a rigid 3D environment show improved hepatic functions, including glycogen storage and bile salt transportation, demonstrating the potential of 3D cultures to overcome these limitations.

Organoid models, particularly those developed from human fatal liver progenitor cell-derived hepatocytes, have shown significant improvements in urea and albumin production compared to current 3D models, indicating their potential to enhance predictive toxicology practices. Furthermore, cardiomyocyte spheroids treated with various drugs have accurately identified cardiotoxic compounds *in vitro*, aligning with clinical data and highlighting the efficacy of these models in predicting cardiotoxicity. Gastrointestinal organoids, generated from stem cell isolation, have established metabolic profiles that are comparable to donor-matched tissue, further demonstrating the applicability of these models in toxicological studies (42).

#### *16.7.1.3 Computational Models in Predictive Toxicology*

*In Silico Approaches and Their Impact:* In silico toxicology, a field gaining momentum, focuses on reducing animal testing in line with the 3R principles (Replacement, Reduction, and Refinement). The primary goal of in silico toxicology is to predict hazards like mutagenicity or organ toxicities using computational models, which may be expert-driven (e.g., structural alerts or read-across) or automated (e.g., machine learning techniques). The foundational concept in this field is the relationship between a chemical's structure and its biological activity, suggesting that physiological actions are dictated by chemical constitution, a theory dating back to 1868 by Brown and Fraser.

Expert methods in in silico toxicology utilize specialists' accumulated knowledge and experience to deduce or elucidate toxicity mechanisms for both individual compounds and entire classes of compounds. Machine learning, encompassing supervised and unsupervised techniques, is crucial in enhancing predictive accuracy for toxicity. Neural networks, for instance, have evolved since the mid-20th century and have seen significant advancements since 2006, offering flexible structural representations and multi-task predictions, which allow for the simultaneous prediction of multiple toxicities (43).

*Integration with Experimental Data:* Integrating experimental data significantly enhances the performance of in silico models. For example, combining cellular morphological descriptors with molecular fingerprints has improved the prediction accuracy for cytotoxicity. Similarly, including *in vivo* data in silico frameworks has proven beneficial for assessing conditions such as hepatic steatosis.

Furthermore, combining classic mixture models with in silico methods has been explored to predict the joint toxicity of nanoparticles, demonstrating the potential for broader applications in toxicological assessments. The choice of molecular descriptors is critical in these models, as they must accurately capture the molecular features responsible for the target effects.

A notable case study illustrating the practical application of in silico models involved using the collaborative acute toxicity modeling suite to predict acute oral systemic toxicity. This model achieved 72–76% concordance with *in vivo* data, showcasing the reliability of in silico predictions in regulatory settings. This example underscores the importance of data curation, access, and the mechanistic interpretation of results to facilitate the acceptance of these models in regulatory frameworks.

Integrating in silico models with experimental and real-world data improves the accuracy of predictions and enhances the understanding of complex toxicological *phenomena*, paving the way for more informed safety assessments and regulatory decisions.

#### *16.7.1.4 Use of Biomarkers in Toxicity Assessment*

Biomarkers play an essential role in toxicity assessment as they provide measurable indicators of the biological effects caused by exposure to toxic substances. These indicators can be molecular, cellular, or physiological. In toxicity assessments, biomarkers are used to evaluate toxic effects' presence, magnitude, and duration. They can provide valuable information about the mechanism of toxicity, target organ(s) affected, and the overall health status of an organism. By measuring specific biomarkers, toxicologists can assess the extent of exposure, identify early warning signs of toxicity, monitor the progression of toxicity over time, and determine the effectiveness of intervention or treatment strategies.

Biomarkers can be detected through various techniques such as blood tests, imaging techniques, genetic analysis, and molecular assays. Examples of biomarkers commonly used in toxicity assessment include liver enzymes, oxidative stress markers, DNA damage markers, inflammatory markers, and specific protein or gene expression patterns. Overall, using biomarkers in toxicity assessment helps identify and understand the adverse effects of toxic substances, facilitating more effective risk assessment, and enabling timely intervention to minimize the detrimental impacts on human and environmental health.

There have been several exciting advancements in the use of biomarkers in toxicology. Here are a few of the latest developments:

- *Omics Technologies:* Omics technologies such as genomics, transcriptomics, proteomics, and metabolomics have revolutionized toxicology research. These approaches enable the simultaneous analysis of thousands of genes, proteins, or metabolites, providing a comprehensive view of toxicity-related molecular changes.
- *Non-Invasive Biomarkers:* Non-invasive sampling methods, such as urine, saliva, or exhaled breath, are gaining popularity in toxicity assessment. These approaches offer easier and less invasive sample collection, making them more convenient for repeated measurements in human and animal studies.
- *High-Throughput Screening:* HTS allows for rapidly evaluating thousands of chemicals for potential toxicity. By incorporating biomarker-based assays into HTS platforms, researchers can efficiently identify compounds with specific molecular effects or targeting particular pathways, enabling faster and more effective toxicity screening.

- *Microfluidics and Lab-on-a-Chip Devices:* These technologies enable the miniaturization of assays, allowing for the simultaneous analysis of multiple biomarkers on a single chip. Microfluidics-based platforms offer improved sensitivity, reduced sample volume requirements, and faster analysis times, making them promising tools for toxicology research and diagnostics.
- *Integration of Computational Approaches:* Advanced computational models like machine learning and systems toxicology can analyze large datasets and predict toxicity outcomes based on biomarker profiles. These models offer a more comprehensive understanding of complex toxicological processes and aid in identifying new biomarkers or patterns that may be missed using traditional approaches.

These biomarker research advancements enhance our ability to detect early signs of toxicity, understand mechanisms of action, and improve risk assessment strategies, ultimately leading to more effective toxicology studies and safer chemical management practices.

#### *16.7.1.5 Micro-Physiological Systems*

Micro-physiological systems (MPS), advanced cellular models, are among the most notable contributions to predictive toxicology. These systems, along with advancements in data science, including omics data analysis and systems toxicology, as well as physiologically based pharmacokinetic/toxicokinetic modeling and simulation, are at the forefront of the field's evolution. These technologies complement and, in some instances, are on the path to replacing traditional animal testing, marking a scientific revolution in toxicology.

### 16.7.2 Future Directions in Regulatory Toxicology

Future directions in regulatory toxicology are increasingly focused on personalized medicine and tailored safety assessments. These approaches recognize that individuals can respond differently to toxic substances based on their unique genetic makeup, lifestyle factors, and environmental exposures.

#### *16.7.2.1 Personalized Medicine and Tailored Safety Assessments*

Personalized medicine and tailored safety assessments aim to optimize healthcare outcomes by considering an individual's unique characteristics and needs. A brief explanation of these concepts is given:

*Personalized Medicine:* Personalized medicine involves tailoring medical treatments and interventions to an individual's specific characteristics. This approach takes into account factors such as genetics, biomarkers, lifestyle, and environmental factors to design treatment plans that are more targeted and effective. By analyzing an individual's genetic profile and considering other relevant factors, healthcare providers can make more informed decisions regarding medication selection, dosage, and treatment strategies. Personalized medicine utilizes genetic testing as a crucial tool to identify genetic variations that can impact an individual's response to medications and treatment options. Genetic testing contributes to determining suitable treatment options.

- *Pharmacogenomics:* Pharmacogenomic testing focuses on how an individual's genetic makeup affects their drug response. It involves analyzing specific genes involved in drug metabolism, transport, and receptor interactions. By identifying genetic variations in these genes, healthcare providers can predict how an individual is likely to respond to a particular medication. This information helps tailor the choice of medication, dosage, and treatment plan to maximize effectiveness and minimize adverse reactions.
- *Drug Efficacy and Safety:* Genetic testing can provide insights into an individual's likelihood of responding to a particular therapy or experiencing adverse drug reactions. For example, certain genetic variations may affect drug metabolism enzymes, influencing how quickly or slowly a person metabolizes a medication. This information helps determine the most suitable treatment option by selecting more effective drugs or adjusting dosages for optimal therapeutic outcomes.
- *Disease Risk Assessment:* Genetic testing can also contribute to disease risk assessment, enabling early intervention, prevention, and personalized screening strategies. By identifying genetic markers associated with certain conditions, healthcare providers can more accurately assess an individual's risk profile. This knowledge helps implement preventive measures, targeted screening, and lifestyle modifications to mitigate the potential impact of genetic predispositions.
- *Targeted Therapies:* Genetic testing can guide the selection of targeted therapies designed to act on specific molecular targets associated with a particular disease. For example, in oncology, specific genetic mutations may indicate eligibility for targeted cancer therapies that specifically inhibit the aberrant signaling pathways responsible for tumor growth.

- *Clinical Decision Support:* Genetic testing results can serve as vital clinical decision support tools. They provide additional information for healthcare providers, helping them make informed decisions about treatment options, medication selection, dosage adjustments, and monitoring strategies tailored to each individual's genetic profile.

It is important to note that genetic testing is just one component of personalized medicine, and treatment decisions are made by considering multiple factors such as patient history, clinical presentation, and other relevant information. Genetic testing results are typically interpreted and used with comprehensive clinical assessments to guide personalized treatment approaches.

*Tailored Safety Assessments:* Tailored safety assessments are an emerging approach in regulatory toxicology that can transform how safety assessments are conducted. Tailored safety assessments involve evaluating the safety and potential risks associated with treatments or exposure to chemicals or substances based on an individual's characteristics. This assessment considers factors like genetic variations, pre-existing medical conditions, age, sex, and other relevant factors. By understanding an individual's unique vulnerabilities and susceptibilities, tailored safety assessments can help determine appropriate exposure limits, precautions, and mitigation strategies to minimize potential harm.

Benefits of Personalized Medicine and Tailored Safety Assessments:

- *Improved Treatment Outcomes:* Personalized medicine allows for more precise and targeted treatments, increasing the likelihood of positive treatment outcomes.
- *Reduced Adverse Reactions:* Tailored safety assessments can identify individuals more susceptible to adverse reactions, enabling healthcare providers to take necessary precautions and minimize risks.
- *Optimization of Drug Dosages:* Personalized medicine enables the adjustment of medication dosages based on individual factors, improving efficacy while reducing potential side effects.
- *More Efficient Resource Allocation:* Resources can be utilized more efficiently by focusing interventions and safety measures on individuals at higher risk.
- *Enhanced Patient Satisfaction:* Personalized medicine acknowledges individual needs, preferences, and characteristics, leading to a more patient-centered approach and increased satisfaction.

It is important to note that while personalized medicine and tailored safety assessments hold significant promise, their implementation is still evolving. Continued research, advancements in technology, and collaboration between healthcare providers, researchers, and regulatory bodies are necessary to refine further and expand the application of these approaches.

#### *16.7.2.2 Integration of Real-World Evidence*

Real-world evidence (RWE) refers to data and findings obtained from the real-world setting outside of clinical trials. Integrating RWE in healthcare and drug development can provide valuable insights into interventions' effectiveness, safety, and utility in real-world populations.

**Examples of RWE studies that have successfully addressed data quality limitations and biases to generate meaningful and reliable findings:**

- *FDA Sentinel Initiative:* The FDA Sentinel Initiative is a prominent example of leveraging RWE to monitor the safety of medical products. The initiative overcomes data quality limitations by integrating electronic health records, administrative claims data, and other real-world data sources by employing rigorous data validation processes and standardized data elements. Advanced analytic techniques, such as sequential surveillance algorithms, enable the detection of safety signals and enhance the reliability of findings.
- *Comparative Effectiveness Research using RWE:* Many studies have utilized RWE to assess the comparative effectiveness of treatments in real-world settings. These studies often employ robust study designs and statistical methods, such as propensity score matching, instrumental variable analysis, or propensity score weighting, to address selection biases and confounding. Through careful adjustment for patient characteristics and treatment selection, these studies generate reliable estimates of treatment effects.
- *Disease Registries for Long-term Outcomes:* Disease registries collect longitudinal data on patients with specific conditions, allowing for long-term outcomes assessment. These registries address data quality limitations by implementing stringent data quality control measures, including data validation and auditing. For example, the National Cancer Database (NCDB) in the United States collects cancer-related data from various sources, ensuring comprehensive and reliable information for research and quality improvement initiatives.

- *Observational Studies with Rigorous Methodologies:* Various observational studies have successfully mitigated data quality limitations and biases through rigorous methodologies. These studies often employ large sample sizes, comprehensive data collection, and statistical analyses that account for confounders. For instance, the "ROSA" study (Registry of Spondyloarthritis) utilizes a nationwide registry coupled with standardized assessments to study patients' long-term outcomes, ensuring reliable findings.
- *Large-Scale Collaborative Efforts:* Collaborative initiatives that pool data from multiple sources increase data completeness and help overcome biases. Projects like the Observational Health Data Sciences and Informatics (OHDSI) and European Network of Centres for Pharmacoepidemiology and Pharmacovigilance (ENCePP) enable collaborations among researchers, facilitating the integration of diverse data sources and harmonizing methodologies. These collaborations enhance the reliability and generalizability of findings.

These examples highlight the potential of carefully designed RWE studies to overcome data quality limitations and biases. By implementing rigorous methodologies, ensuring data validation and standardization, and applying appropriate statistical techniques, meaningful and reliable findings can be generated, providing valuable insights into real-world outcomes and informing decision-making in healthcare.

**Future directions and advancements regarding the integration of RWE:**

- *RWE in Drug Development:* Using RWE to support drug development can provide insights into the effectiveness and safety of interventions in diverse populations, leading to more personalized therapies. By incorporating RWE from electronic health records, patient registries, claims data, and other sources, drug developers can refine patient selection criteria, optimize clinical trial design, and generate real-world data to support regulatory submissions.
- *RWE in Post-Market Surveillance:* The integration of RWE in post-market surveillance can enhance drug safety monitoring by providing insights into real-world effectiveness, safety, and adverse events. RWE can be used to identify potential safety signals, estimate incidence rates, and monitor drug utilization patterns to inform regulatory decisions and facilitate early intervention and risk mitigation.
- *RWE in Comparative Effectiveness Research:* RWE can play an essential role in comparative effectiveness research by enabling the comparison of different treatment options in diverse patient populations. Incorporating RWE from multiple data sources can provide insights into interventions' relative effectiveness and safety, identifying subgroups of patients who may benefit most from specific therapies.
- *RWE in Health Services Research:* The integration of RWE in health services research can inform health policy decisions by evaluating the impact of healthcare interventions on patient outcomes, quality of life, and healthcare utilization. RWE can also be used to identify disparities in healthcare access and outcomes, informing targeted interventions and policy changes to improve the delivery of care.
- *Advancements in Data Analytics:* The increasing availability of large data sets and advancements in data analytics techniques, such as machine learning, natural language processing, and data integration, can facilitate the integration of RWE and enhance its utility. These technologies enable more efficient data cleaning, integration, and analysis, enabling the generation of real-time insights into drug safety and effectiveness.
- *Regulatory Framework Adaptations:* The integration of RWE into regulatory decision-making will require adaptations to regulatory frameworks to accommodate the use of real-world data. Developing standardized approaches for data collection, analysis, and reporting, and establishing clear thresholds for evidence generation will be crucial for effectively integrating RWE into regulatory decision-making.

Overall, integrating RWE has significant potential to transform drug development, post-market surveillance, comparative effectiveness research, and health services research. By leveraging real-world data, healthcare stakeholders can generate insights into the effectiveness, safety, and utility of interventions, leading to more personalized, efficient, and effective healthcare delivery.

**Limitations and Challenged of Integration of RWE**

- *Data Quality and Standardization:* RWE is derived from various sources with varying data quality and standards. Incomplete, missing, or inaccurately recorded data can introduce biases and impact the reliability of findings. Addressing this challenge requires efforts in data quality assurance, including data cleaning, validation, and standardization processes. Adopting standardized data elements, coding systems, and interoperable health information systems can improve data quality and comparability.

- *Selection Bias and Confounding:* Non-random patient selection and confounding factors can influence the outcomes observed in RWE studies. Confounding factors, such as differences in patient characteristics, disease severity, or treatment selection, can lead to biased estimations. Employing rigorous study designs, such as propensity score matching, instrumental variable analysis, or stratification, can help address selection bias and confounding, allowing more accurate comparisons and estimations.
- *Data Privacy and Ethical Concerns:* Protecting patient privacy and ensuring data security are critical considerations when integrating RWE. Strict adherence to privacy regulations, obtaining informed consent, implementing robust data anonymization and de-identification techniques, and employing secure data storage and transfer protocols are essential to address ethical and privacy concerns.
- *Data Completeness and Representativeness:* RWE studies heavily rely on available data sources, which may suffer from data gaps and lack representativeness. Incomplete or biased data can limit the generalizability of findings. Strategies such as comprehensive data collection frameworks, long-term data collection initiatives, and the integration of multiple data sources (e.g., electronic health records, claims data, patient registries) can enhance data completeness and representativeness.
- *Temporal and Causal Inferences:* Establishing causal relationships and temporal sequences of events can be complex in RWE studies. Temporal biases, such as reverse causality and time-related confounding, can challenge the interpretation of findings. Advanced statistical methods, such as instrumental variable analysis, augmented inverse probability weighting, or fixed-effect models, can help mitigate these challenges and strengthen causal inferences.
- *Transparency and Reproducibility:* To ensure the reliability and validity of RWE findings, transparency and reproducibility are essential. Transparent reporting of study methodology, data sources, and analytical approaches promotes scientific rigor and allows for peer review and replication of results. Embracing open science practices, pre-registration of study protocols, and data-sharing initiatives can foster transparency and reproducibility in RWE studies.

Addressing these limitations and challenges requires collaboration among stakeholders, including researchers, policymakers, healthcare providers, and data custodians. Continuous improvement in data quality, study designs, analytical methods, and the development of standardized guidelines can enhance the reliability and validity of RWE, supporting evidence-based decision-making in healthcare.

## 16.8 CONCLUSION

The intricate landscape of biologic therapeutics necessitates a robust regulatory toxicology framework to ensure patient safety and efficacy. This review has underscored the unique challenges biologics poses compared to conventional small-molecule drugs, emphasizing the critical role of comprehensive preclinical and clinical evaluation. Ranging from immunogenicity and biosimilarity assessment to long-term safety monitoring, regulatory toxicology is indispensable in mitigating risks and optimizing therapeutic benefits.

As the field of biologics continues to evolve, so too must regulatory strategies. Collaborative efforts between regulatory authorities, industry, and academia are essential for developing innovative approaches to address emerging complexities, such as advanced therapies and personalized medicine. By fostering a culture of scientific rigor and transparency, the regulatory toxicology community can contribute significantly to the safe and effective translation of biologic research into life-saving treatments.

The field of regulatory toxicology for biologics is experiencing rapid evolution, driven by technological advancements, emerging therapeutic modalities, and a deeper understanding of biological systems. Advancements in omics technologies, such as genomics and proteomics, enable more sophisticated safety assessments. Additionally, the rise of cell and gene therapies, as well as complex biologic formulations, presents novel challenges and necessitates the development of tailored regulatory approaches. Therefore, to address these complexities, regulatory agencies, and the industry are collaborating to establish robust frameworks that ensure the safety and efficacy of these innovative therapeutics while fostering scientific progress.

## REFERENCES

1. Center for Biologics Evaluation and Research (CBER). *What Are "Biologics" Questions and Answers [Internet]*. Available from: www.fda.gov/about-fda/center-biologics-evaluation-and-research-cber/what-are-biologics-questions-and-answers
2. Sharma P, Joshi RV, Pritchard R, Xu K, Eicher MA. Therapeutic antibodies in medicine. *Molecules*. 2023 Sep 5;28(18):6438.

3. U.S. Food & Drug Administration. *Biological Product Definitions [Internet].* [cited 2024 Sep 16]. Available from: www.fda.gov/files/drugs/published/Biological-Product-Definitions.pdf

4. Nepovimova E, Kuca K. The history of poisoning: From ancient times until modern ERA. *Arch Toxicol.* 2019 Jan;93(1):11–24.

5. Pognan F, Beilmann M, Boonen HCM, Czich A, Dear G, Hewitt P, et al. The evolving role of investigative toxicology in the pharmaceutical industry. *Nat Rev Drug Discov.* 2023 Apr;22(4):317–35.

6. Kissner T, Blaich G, Baumann A, Kronenberg S, Hey A, Kiessling A, et al. Challenges of non-clinical safety testing for biologics: A report of the 9th BioSafe European Annual General Membership Meeting. *mAbs [Internet].* 2021 Jan 1 [cited 2024 Sep 16];13(1). Available from: www.tandfonline.com/doi/full/10.1080/19420862.2021.1938796

7. ICH guideline S6 (R1)—*Preclinical safety evaluation of biotechnology-derived pharmaceuticals.* In *Proceedings of the International Conference on Harmonisation of Technical Requirements for Registration of Pharmaceuticals for Human Use.* 2011.

8. Brennan FR, Morton LD, Spindeldreher S, Kiessling A, Allenspach R, Hey A, et al. Safety and immunotoxicity assessment of immunomodulatory monoclonal antibodies. *mAbs.* 2010 May;2(3):233–55.

9. Hourd P, Ginty P, Chandra A, Williams DJ. Manufacturing models permitting roll out/scale out of clinically led autologous cell therapies: Regulatory and scientific challenges for comparability. *Cytotherapy.* 2014 Aug;16(8):1033–47.

10. U.S. Food & Drug Administration. *Guidance for Industry E4 Dose-Response Information to Support Drug Registration [Internet].* 1996. Available from: www.fda.gov/regulatory-information/search-fda-guidance-documents/e4-dose-response-information-support-drug-registration

11. Erhirhie EO, Ihekwereme CP, Ilodigwe EE. Advances in acute toxicity testing: Strengths, weaknesses and regulatory acceptance. *Interdiscip Toxicol.* 2018 May 1;11(1):5–12.

12. Descotes J. Immunotoxicity of monoclonal antibodies. *mAbs.* 2009 Mar;1(2):104–11.

13. Descotes J. Safety immunopharmacology: Evaluation of the adverse potential of pharmaceuticals on the immune system. *J Pharmacol Toxicol Methods.* 2012 Sep;66(2):79–83.

14. Brinks V, Jiskoot W, Schellekens H. Immunogenicity of therapeutic proteins: The use of animal models. *Pharm Res.* 2011 Oct;28(10):2379–85.

15. Mukherjee P, Roy S, Ghosh D, Nandi SK. Role of animal models in biomedical research: A review. *Lab Anim Res [Internet].* 2022 Dec [cited 2024 Sep 16];38(1). Available from: https://labanimres.biomedcentral.com/articles/10.1186/s42826-022-00128-1

16. Prior H, Haworth R, Labram B, Roberts R, Wolfreys A, Sewell F. Justification for species selection for pharmaceutical toxicity studies. *Toxicol Res (Camb).* 2021 Jan 5;9(6):758–70.

17. *ICH Harmonised Tripartite Guideline Guidance on Nonclinical Safety Studies for the Conduct of Human Clinical Trials and Marketing Authorization for Pharmaceuticals M3(R2).* In *International Conference on Harmonisation of Technical Requirements for Registration of Pharmaceuticals for Human Use*, 2009.

18. Riss T, Niles A, Moravec R, Karassina N, Vidugiriene J. Cytotoxicity assays: In vitro methods to measure dead cells. In: Markossian S, Grossman A, Arkin M, Auld D, Austin C, Baell J, et al., editors. *Assay Guidance Manual [Internet].* Bethesda (MD): Eli Lilly & Company and the National Center for Advancing Translational Sciences; 2004 [cited 2024 Sep 16]. Available from: www.ncbi.nlm.nih.gov/books/NBK540958/

19. Larsson P, Engqvist H, Biermann J, Werner Rönnerman E, Forssell-Aronsson E, Kovács A, et al. Optimization of cell viability assays to improve replicability and reproducibility of cancer drug sensitivity screens. *Sci Rep*. 2020 Apr 2;10(1):5798.

20. U.S. Food and Drug Administration. Immunogenicity Assessment for Therapeutic Protein Products – Guidance for Industry. 2014. www.fda.gov/media/85017/download

21. Kuriakose A, Chirmule N, Nair P. Immunogenicity of biotherapeutics: Causes and association with posttranslational modifications. *J Immunol Res*. 2016;2016:1298473.

22. Suh K, Kyei I, Hage DS. Approaches for the detection and analysis of antidrug antibodies to biopharmaceuticals: A review. *J Sep Sci*. 2022 Jun;45(12):2077–92.

23. Mascarenhas-Melo F, Diaz M, Gonçalves MBS, Vieira P, Bell V, Viana S, et al. An overview of biosimilars—development, quality, regulatory issues, and management in healthcare. *Pharmaceuticals*. 2024 Feb 11;17(2):235.

24. Marks C, Hummer AM, Chin M, Deane CM. Humanization of antibodies using a machine learning approach on large-scale repertoire data. Martelli PL, editor. *Bioinformatics*. 2021 Nov 18;37(22):4041–7.

25. Ducancel F, Muller BH. Molecular engineering of antibodies for therapeutic and diagnostic purposes. *mAbs*. 2012 Jul;4(4):445–57.

26. Grabowski H. Follow-on biologics: Data exclusivity and the balance between innovation and competition. *Nat Rev Drug Discov*. 2008 Jun;7(6):479–88.

27. US FDA. *How Drugs Are Developed and Approved [webpage on the Internet]*. Silver Spring, MD: US FDA; 2014. [Internet]. [cited 2024 Aug 25]. Available from: www.fda.gov/Drugs/DevelopmentApprovalProcess/HowDrugsareDevelopedandApproved/ApprovalApplications/TherapeuticBiologicApplications/Biosimilars/default.htm

28. Teruyo A. *Recent Regulations of Biosimilars in Japan [Internet]*. DIA 2011, Chicago, Illinois; 2011. Available from: https://pbwt2.gjassets.com/content/uploads/2016/09/000163979.pdf

29. Horikawa H, Tsubouchi M, Kawakami K. Industry views of biosimilar development in Japan. *Health Policy*. 2009 Jul;91(2):189–94.

30. Suh SK, Park Y. Regulatory guideline for biosimilar products in Korea. *Biologicals*. 2011 Sep;39(5):336–8.

31. Health Canada. Guidance Document Information and Submission Requirements for Biosimilar Biologic Drugs, *Published by authority of the Minister of Health, Health Canada*. 2010.

32. Batran RA, Elmoshneb M, Hussein AS, Hussien OM, Adel F, Elgarhy R, et al. Biosimilars: Science, implications, and potential outlooks in the Middle East and Africa. *BTT*. 2022 Oct; 16:161–71.

33. Takebe T, Imai R, Ono S. The current status of drug discovery and development as originated in United States academia: The influence of industrial and academic collaboration on drug discovery and development. *Clin Transl Sci*. 2018 Nov;11(6):597–606.

34. Dingman R, Balu-Iyer SV. Immunogenicity of protein pharmaceuticals. *J Pharm Sci*. 2019 May;108(5):1637–54.

35. Joubert MK, Deshpande M, Yang J, Reynolds H, Bryson C, Fogg M, et al. Use of in vitro assays to assess immunogenicity risk of antibody-based biotherapeutics. Stoddart CA, editor. *PLoS ONE*. 2016 Aug 5;11(8):e0159328.

36. Schreitmüller T, Barton B, Zharkov A, Bakalos G. Comparative immunogenicity assessment of biosimilars. *Future Oncol.* 2019 Jan 31;15(3):319–29.

37. Brady SP, Monosson E, Matson CW, Bickham JW. Evolutionary toxicology: Toward a unified understanding of life's response to toxic chemicals. *Evol Appl.* 2017 Sep;10(8):745–51.

38. Najjar A, Kramer N, Gardner I, Hartung T, Steger-Hartmann T. Editorial: Advances in and applications of predictive toxicology: 2022. *Front Pharmacol [Internet].* 2023 Aug 2 [cited 2024 Sep 17];14. Available from: www.frontiersin.org/articles/10.3389/fphar.2023.1257423/full

39. Li S, Xia M. Review of high-content screening applications in toxicology. *Arch Toxicol.* 2019 Dec;93(12):3387–96.

40. Bedford R, Perkins E, Clements J, Hollings M. Recent advancements and application of in vitro models for predicting inhalation toxicity in humans. *Toxicol In Vitro.* 2022 Mar;79:105299.

41. Lynch S, Pridgeon CS, Duckworth CA, Sharma P, Park BK, Goldring CEP. Stem cell models as an *in vitro* model for predictive toxicology. *Biochem J.* 2019 Apr 15;476(7):1149–58.

42. Wang L, Hu D, Xu J, Hu J, Wang Y. Complex in vitro model: A transformative model in drug development and precision medicine. *Clin Transl Sci [Internet].* 2024 Feb [cited 2024 Sep 17];17(2):e13695. Available from: https://ascpt.onlinelibrary.wiley.com/doi/10.1111/cts.13695

43. Hemmerich J, Ecker GF. In silico toxicology: From structure–activity relationships towards deep learning and adverse outcome pathways. *WIREs Comput Mol Sci [Internet].* 2020 Jul [cited 2024 Sep 17];10(4):e1475. Available from: https://wires.onlinelibrary.wiley.com/doi/10.1002/wcms.1475

# Index

*Note*: Page numbers in *italics* indicate figures and in **bold** indicate tables on the corresponding pages.